MONOGRAPHS ON THE PHYSICS AND CHEMISTRY OF MATERIALS

General Editors

RICHARD J. BROOK   ANTHONY CHEETHAM

ARTHUR HEUER   SIR PETER HIRSCH

TOBIN J. MARKS   DAVID G. PETTIFOR

MANFRED RUHLE   JOHN SILCOX

ADRIAN P. SUTTON   MATTHEW V. TIRRELL

VACLAV VITEK

HIGH-ENERGY ELECTRON DIFFRACTION AND MICROSCOPY

L.-M. PENG

Department of Electronics
Peking University

S. L. DUDAREV

EURATOM/UKAEA Fusion Association
Culham Science Centre, Oxfordshire

M. J. WHELAN

Department of Materials
University of Oxford

OXFORD
UNIVERSITY PRESS

OXFORD
UNIVERSITY PRESS

Great Clarendon Street, Oxford OX2 6DP

Oxford University Press is a department of the University of Oxford.
It furthers the University's objective of excellence in research, scholarship,
and education by publishing worldwide in

Oxford   New York

Auckland   Cape Town   Dar es Salaam   Hong Kong   Karachi
Kuala Lumpur   Madrid   Melbourne   Mexico City   Nairobi
New Delhi   Shanghai   Taipei   Toronto

With offices in

Argentina   Austria   Brazil   Chile   Czech Republic   France   Greece
Guatemala   Hungary   Italy   Japan   Poland   Portugal   Singapore
South Korea   Switzerland   Thailand   Turkey   Ukraine   Vietnam

Oxford is a registered trade mark of Oxford University Press
in the UK and in certain other countries

Published in the United States
by Oxford University Press Inc., New York

© L.–M. Peng, S. L. Dudarev, and M. J. Whelan, 2004

The moral rights of the author have been asserted
Database right Oxford University Press (maker)

First published 2004

First published in paperback 2011

British Library Cataloguing in Publication Data

Data available

Library of Congress Cataloging in Publication Data

Data available

Typeset by the authors using LaTeX

ISBN 978–0–19–850074–2 (hbk.); 978–0–19–960224–7 (pbk.)

1 3 5 7 9 10 8 6 4 2

FOREWORD BY SIR PETER HIRSCH FRS

Over the last fifty years tremendous progress has been made in the development and application of electron microscopy, electron diffraction and electron spectroscopy to the study of materials. These techniques provide the materials scientist with a wide range of tools outstanding for their variety, versatility, and power. They can be applied to study the atomic and electronic structure of materials, some with atomic resolution, both inside the materials and at their surface.

This unique power of electron optical techniques is fundamentally due to the strong interaction of high energy electrons with solids. The corollary of this strong interaction is that multiple elastic scattering is prevalent. This in turn means that the interpretation of micrographs and diffraction patterns is invariably more complex than it would be under single scattering conditions, while at the same time multiple scattering leads to new and subtle features in the diffraction patterns which can yield additional information.

The strong electron-solid interaction also leads to inelastic scattering, which is utilized in spectroscopic techniques, such as energy loss spectroscopy, which can be applied to obtain chemical information about the solid. It also leads to diffuse scattering which in turn results in additional features in the diffraction patterns, such as Kikuchi patterns, and which can also be utilized as a distinct technique in high resolution electron microscopy. In a crystal the effective mean free path for a particular inelastic scattering mechanism depends on the form of the wavefield, and this is determined by multiple elastic scattering.

The variety of interactions, elastic and inelastic, between the high energy electrons and the solid, and multiple scattering effects lead inevitably to complexities in the experimentally observed images and diffraction patterns. A fundamental understanding of the elastic and inelastic scattering of electrons in solids is therefore required to put the interpretation of the experiments on a firm basis. This book provides a comprehensive and up-to-date treatment of the various theoretical tools available for the elucidation of the electron-solid interactions, and of the propagation of high-energy electrons in crystals, and for the interpretation of experimental results. Such theoretical tools are now becoming even more important for quantitative analysis since recent instrumental advances have made it possible to record intensities of energy filtered images and diffraction patterns. It covers uniquely theoretical treatments of both transmission and reflection diffraction by high-energy electrons within a common framework, and addresses the relation between them. Some of the theoretical tools have been developed during the last ten to twenty years, and have given new insight into the physical processes. The three authors are outstanding authorities in this field, and have been responsible for a number of these advances. Their comprehensive treatise fills an important gap in the literature of this field, and provides the materials scientist with a sound theoretical basis for the interpretation of experiments using electron microscopy, electron diffraction, and associated spectroscopic techniques, and for the further development of these techniques, which play such a crucial role in materials science.

Oxford, 29 July 2003

To Maria, Xiao Xiong, and Xiao Ying

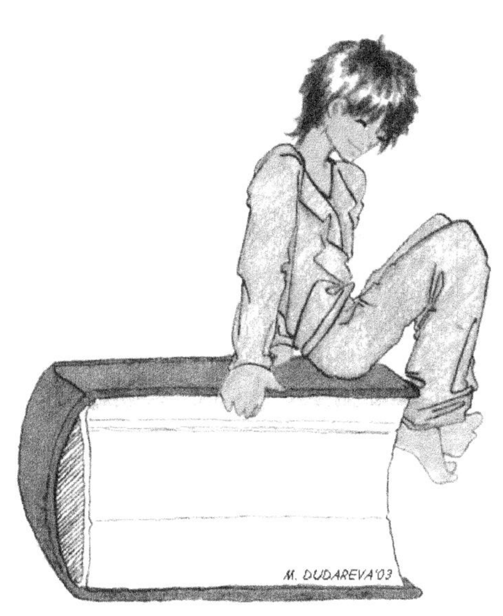

'This was the sort of stuff they liked:
short and obvious.'

J. R. R. Tolkien, *The Fellowship of the Ring*

PREFACE

Electron microscopy and the closely associated phenomenon of electron diffraction have long and prestigious histories. The development of the electron microscope as an instrument has its roots in experimental investigations in the 1920s of the cathode ray oscillograph, and the discovery of the focussing properties for an electron as a particle of a short axially symmetric magnetic field (for an authoritative account see the article by Mulvey (Mulvey, 1967)). This led to the demonstration by Ruska and Knoll in 1931 of two-stage electron optical magnification, which in turn led to the construction by Ruska in 1933 of a two-stage electron microscope with a resolution exceeding that of the optical microscope (for a review see Ruska (Ruska, 1980)). The idea that an electron is associated with a wave also goes back to the early 1920s, when in 1923 and 1924 de Broglie published his revolutionary concept that the electron can also be thought of as a wave disturbance with a characteristic wavelength determined by its momentum (de Broglie, 1923a, de Broglie, 1923b, de Broglie, 1924). In 1927 Davisson and Germer in the USA and independently Thomson and Reid in the UK demonstrated experimentally the wave nature of the electron by observing diffraction effects. Davisson and Germer used reflection diffraction geometry with low-energy electrons, while Thomson and Reid used transmission geometry with high-energy electrons (for a review see Thomson (Thomson, 1968)). Davisson and Thomson were jointly awarded the Nobel Prize in Physics in 1937 'for their experimental discovery of the diffraction of electrons by crystals'. Almost fifty years later, in 1986, Ruska was awarded the Nobel Prize in Physics 'for his fundamental work in electron optics, and for the design of the first electron microscope'.

Over the years the interpretation of images of materials obtained with the electron microscope has evolved from simple mass thickness ideas to sophisticated diffraction contrast and phase contrast concepts important for the study of defects in materials (Whelan, 1986). On the one hand, electron microscopy is a powerful tool for investigating the structure of a material, while on the other hand, various spectroscopies associated with the electron beam (X-ray and energy loss spectroscopies) provide information about chemical composition. Thus electron microscopy has become a technique of choice in materials science and related fields of study. Electron microscope laboratories have been established in many universities world-wide, and the trend towards increasing the imaging power of the electron microscope continues.

The field of electron microscopy has recently witnessed new developments, and many fairly exotic features have now become a routine part of the electron microscope examination of specimens. The introduction of field emission electron guns (FEGs), charge coupled device (CCD) cameras, energy-filtering facilities, and spherical aberration (C_s) correction combined with monochromatization has led to a level of energy resolution approaching 20 meV, while the use of diffuse scattering for high-resolution imaging of atomic structures has evolved into the Z-contrast imaging technique. These developments take electron microscopy further into the realm of electron holography, where electrostatic and magnetic fields can be imaged with nanometre resolution. Transmission electron diffraction has also proved to be remarkably successful when applied to surface science, and reflection high-energy electron diffraction (RHEED) has proved to be an invaluable tool for monitoring molecular beam epitaxial (MBE) growth of surfaces at the nanoscale.

The power of electron beam-based techniques relies on the fact that electrons strongly interact with matter. Moreover, the fact that electrons have a very short wavelength has made electron lithography combined with ion beams a widely used tool for fabricating small-scale quantum structures. The most recent developments in electron microscopy combine conventional electron imaging with scanning tunnelling microscopy (STM) and scanning probe microscopy (SPM), providing complementary information about the material at the quantum level.

The application of electron microscopy to the investigation of materials requires an understanding of how electrons interact with matter, and how elastic and inelastic scattering occurs in various parts of a specimen illuminated with a beam of high-energy electrons. Since the discovery of electron diffraction, electron microscopy techniques have mainly relied on the observation of the effects of elastic scattering of electrons by solids. Methods of acquisition of elastically scattered electrons, involving for example the use of a CCD camera, have become extremely efficient and now millions of electron diffraction data points can be collected in a matter of seconds. A serious challenge now facing electron microscopy is how to adequately retrieve the information contained in very high-quality images and diffraction patterns and how to quantify the results of observations. In many cases this means that many diffracted beams need to be included in the theoretical treatment, and that effects associated with diffuse scattering by the disorder present in any real material become significant. Inelastic scattering combined with more conventional diffraction approaches promises to provide valuable information about the dynamics of motion of atomic particles in the material. In this book we aim to bridge the gap between elementary texts focussed on applications, and classical treatises, to provide a sound basis for the development of electron diffraction and microscopy well into the new millennium.

The authors wish to acknowledge the inspiration and encouragement they have received over the years from colleagues too numerous to be mentioned individually. But nevertheless, they should like to single out Professors Sir Peter Hirsch, Archie Howie, Hatsujiro Hashimoto, Jiye Ximen, Kehsin Kuo, John Cowley, and Mikhail Ivanovich Ryazanov. We are also grateful to Professor Les Allen of Melbourne University who kindly read the manuscript and proposed some changes.

The preparation of the manuscript was greatly assisted by travel grants provided by the Royal Society, the Chinese Academy of Sciences, and the EU-

RATOM/UKAEA Fusion Association, all of which we gratefully acknowledge.

We thank Yuan and Galina for their patience and support during the time it has taken us to write this book.

LMP, SLD, and MJW
Beijing and Oxford, July 2003

1

BASIC CONCEPTS OF HIGH-ENERGY ELECTRON DIFFRACTION

1.1 Introduction

The phenomenon of electron diffraction was discovered by Davisson and Germer (Davisson and Germer, 1927) in the USA and independently by Thomson and Reid (Thomson and Reid, 1927) in Scotland, who observed that electrons scattered by a crystal form an interference pattern similar to the pattern formed by light waves scattered by a diffraction grating. These observations provided the most convincing and straightforward test of the basic principles of wave mechanics, and the technique of electron diffraction has since undergone extensive development and found many fields of application. Among the diffraction methods commonly used in the analysis of crystal structures, which include X-ray, neutron and electron diffraction, the latter exhibits the highest level of sensitivity. Electrons interact with a solid between 100 to 1000 times more strongly than do X-rays or neutrons. The high sensitivity of electron diffraction, combined with the availability of electron lenses, high-quality sources and efficient detectors, has made electron diffraction and microscopy an indispensable tool for studying the structure of materials (Hirsch et al., 1977, Cowley, 1990, Spence, 1988, Reimer, 1989, Cowley, 1993a, Williams and Carter, 1996).

The strength of the interaction of electrons with atoms means that solids are less transparent for electrons than they are for X-rays or neutrons. Even for the case of scattering by a thin crystal the probability is high that incident electrons will be scattered several times before they emerge back into the vacuum. It is therefore not appropriate in general to assume that the kinematical or, in other words, the single-scattering approximation is valid for electron diffraction. The situation calls for the development of a general theoretical framework suitable for the treatment of multiple scattering events and for the interpretation of experimental results. In this book we aim to provide an up-to-date account of the theoretical tools available for this purpose, and to describe some recent important applications in specific fields. Throughout this book we will use the conventional quantum-mechanical notation in which a plane wave takes the form $\exp[i(\mathbf{k} \cdot \mathbf{r} - \omega t)]$, where $\mathbf{k}$ is the wave vector of the electron $k = 2\pi/\lambda$ and $\omega = E/\hbar$, where λ is the wavelength and E is the energy of the electron (see for example Schiff (Schiff, 1968)). We shall use this notation instead of the crystallographic notation, where the same plane wave is described as $\exp[-i(\mathbf{k} \cdot \mathbf{r} - \omega t)]$ (see for example Cowley (Cowley, 1990)), and consider only high-energy electrons in the energy range of a few keV to a few MeV. If the incident beam energy is lower than

1

about 1 keV it becomes important to take into account the fact that the incident electrons are not distinguishable from electrons of the solid. The treatment of scattering then requires taking into account exchange effects (Ochkur, 1964), and also virtual inelastic scattering effects (Pendry, 1974). On the other hand, if the energy of electrons is very high, say greater than 10 MeV, Bremsstrahlung energy losses become significant, and in addition the specimen can be seriously damaged by electron-induced atomic displacements (Reimer, 1989).

1.2 The interaction between high-energy electrons and a solid

The incident high-energy electron interacts with a solid via the electrostatic potential of the Coulomb forces acting between the incident electron and the electrons and the positively charged nuclei of the solid. Using the International System of Units (SI units), we find the following expression for the electrostatic interaction:

$$V(\mathbf{r}; \mathbf{r}_1...\mathbf{r}_j...; \mathbf{R}_1...\mathbf{R}_n...) = \frac{1}{4\pi\epsilon_0} \left(\sum_n \frac{-Z_n e^2}{|\mathbf{r} - \mathbf{R}_n|} + \sum_j \frac{e^2}{|\mathbf{r} - \mathbf{r}_j|} \right), \qquad (1.1)$$

where ϵ_0 is the permittivity of the vacuum and $\mathbf{r}$ denotes the spatial coordinates of the incident high-energy electron. On the right-hand side of this equation the first term is the interaction of the incident electron with the nuclei ($\mathbf{R}_n$ and Z_n are the coordinates and the atomic number of the n-th nucleus) and the second term is the interaction with the electrons of the solid ($\mathbf{r}_j$ represents the coordinates of the j-th electron).

The Hamiltonian of the system consisting of the incident electron and the nuclei and electrons of the solid is given by

$$H = -\frac{\hbar^2}{2m}\nabla^2 + V(\mathbf{r}; \mathbf{r}_1......\mathbf{r}_j...; \mathbf{R}_1...\mathbf{R}_n...) + H_{cr}, \qquad (1.2)$$

where the first term represents the kinetic energy of the incident electron, the second term denotes the interaction potential between the incident electron and the solid as given by (1.1), and the third term is the Hamiltonian of the solid. This latter Hamiltonian is given by the sum of the kinetic energy operators of all the nuclei and electrons of the solid and their interaction potentials. The lowest energy state of the solid is its ground state. All other states, which by definition have energies which are higher than the energy of the ground state, are excited states. Writing the wave function of the l-th state of the solid in the form $\phi_l = \phi_l(\mathbf{r}_1, ..., \mathbf{R}_1, ...)$, we have

$$H_{cr}\phi_l = E_l\phi_l, \qquad (1.3)$$

where E_l is the corresponding energy eigenvalue of the l-th state of the solid.

1.3   Elastic and inelastic scattering, and the complex potential

Electrons may be scattered elastically or inelastically. In an elastic collision the solid remains in its original state so that the energy of the incident electron remains unchanged, i.e. $\phi_f = \phi_i$ (here the subscripts f and i refer to the final state and the initial state, respectively). On the other hand, in an inelastic collision the incident electron loses (or acquires) an amount of energy ΔE equal to $E_f - E_i$, and the solid undergoes a transition from the initial state ϕ_i to the final state ϕ_f. Taking into account the fact that both nuclei and electrons of the solid are moving particles, we separate the potential V into time-independent and time-dependent parts, i.e.

$$V = \langle V \rangle_t + \Delta V,$$

where $\langle ... \rangle_t$ denotes averaging over time,

$$\langle V(t) \rangle_t = \frac{1}{T} \int_0^T V(t)dt,$$

where $T \to \infty$, and ΔV is the fluctuation of the interaction potential about the time-averaged value $\langle V \rangle_t$. The time-independent part of the potential gives rise to elastic scattering, while the time-dependent part results in inelastic scattering. For high-energy electron diffraction the role played by the fluctuating part of the potential ΔV is often less significant than that of the time-independent part. In a number of cases the effect of inelastic scattering on elastic scattering may be taken into account using a first-order perturbation analysis that gives rise to the appearance of an imaginary part of the time-averaged potential. The effect of inelastic excitations on elastic scattering may then be described by treating the potential of interaction between the incident electron and the solid as a complex quantity. This complex potential is usually called the optical potential, by analogy with the complex refractive index characterizing optical properties of partially absorbing media (Yoshioka, 1957, Hodgson, 1963).

It is essential to recognize an important difference between the optical potential used in electron diffraction and the complex refractive index used in optics. In optics the imaginary part of the refractive index leads to the absorption of light. In electron diffraction problems the imaginary part of the potential serves as a measure of the strength of scattering by fluctuations of the potential and does not lead to any real absorption of electrons (the latter of course cannot occur since electrons are charged particles). The physical picture responsible for the appearance of the optical potential is as follows. After an inelastic collision the solid undergoes a transition from its initial state to a final state (different from the initial state), and this transition is also accompanied by a change in the state of the incident electron. In principle a subsequent scattering event may return the electron into its original state with the solid reverting to the state it occupied before the interaction with the incident electron took place. However, normally the probability of the solid reverting to the initial state is negligibly small (Rez, 1976) and, as far as elastic scattering is concerned, electrons can be considered to have been effectively removed or 'absorbed' from the elastic channels of scattering. It is this process that is responsible for the appearance of an imaginary addition to the time-independent part of the potential.

There are three main mechanisms of inelastic scattering of high-energy electrons. They are, respectively, collective excitations of valence electrons (plasmon excitations) which typically have energies of the order of 10–40 eV, single electron excitations with energies up to a few thousand eV, and lattice vibrations (phonon excitations) with energies of the order of 10^{-2} eV. A full treatment of multiple elastic and inelastic scattering is somewhat more complex than the treatment of pure elastic scattering (Høier, 1973, Dudarev and Ryazanov, 1988, Dudarev et al., 1993a) and we shall describe the necessary concepts later in this book. There exist many important situations, such as the case of high-resolution Z-contrast imaging of crystals (Pennycook and Jesson, 1990), where the treatment of inelastic scattering of electrons may be significantly simplified. In these cases the optical potential approach combined with the single inelastic scattering approximation provides a suitable basis for calculating the angular distribution of inelastically scattered electrons. We will briefly discuss this in section 1.7 and postpone the detailed treatment until Chapters 7 and 8.

1.4 The amplitude and the differential cross-section of scattering of electrons

In a typical electron diffraction experiment electrons are scattered by a specimen in all directions. At a large distance from the specimen (which is assumed to be situated at the origin) where the scattered electrons are counted by detectors, the wave function ψ acquires the asymptotic form

$$\psi = \psi_0 + \psi_s \longrightarrow \psi_0 + f\frac{\exp(ikr)}{r}, \tag{1.4}$$

where ψ_0 and ψ_s denote the incident wave and the scattered wave respectively, and f is called the amplitude of scattering (see Fig. 1.1). The wave function of incident electrons ψ_0 is usually assumed to have the form of a plane wave propagating in the direction of wave vector $\mathbf{k}_0$

$$\psi_0(\mathbf{r}) = \exp(i\mathbf{k}_0 \cdot \mathbf{r}). \tag{1.5}$$

For a given position of the source of electrons and the detector the asymptotic form of the solution of the problem of scattering for the incident plane wave is

$$\psi(\mathbf{k}_0, \mathbf{r}) = \exp(i\mathbf{k}_0 \cdot \mathbf{r}) + f(\mathbf{k}, \mathbf{k}_0)\frac{\exp(ikr)}{r}, \tag{1.6}$$

where the notation $f(\mathbf{k}, \mathbf{k}_0)$ emphasizes the fact that the process of scattering involves momentum transfer from $\mathbf{k}_0$ to $\mathbf{k}$, where $|\mathbf{k}| = |\mathbf{k}_0|$.

In a real experiment the incident electrons emitted by the source are collimated by an aperture as shown schematically in Fig. 1.1 into a fairly well-defined

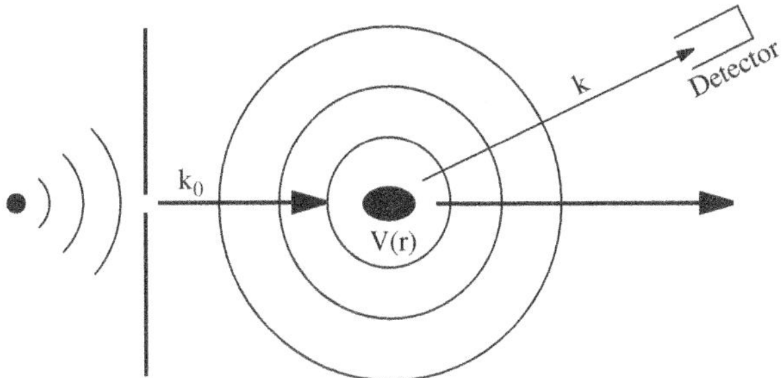

FIG. 1.1. Schematic diagram showing scattering of a collimated high-energy electron beam by a potential field $V(\mathbf{r})$.

beam. A collimated beam cannot of course be described by a plane wave of the form (1.5), but it can be represented as a linear combination (superposition) of plane waves with slightly different wave vectors. The total angular spread of the beam is of the order of the ratio of the wavelength of the electron to the diameter of the collimating aperture. In practice this ratio can be made extremely small. Since the amplitude $f(\mathbf{k}, \mathbf{k}_0)$ does not vary rapidly as a function of the angle of scattering, the small spread of directions of wave vectors in the incident wave does not affect the observed value of $f(\mathbf{k}, \mathbf{k}_0)$. This justifies the above assumption (1.5).

The differential cross-section is defined as the flux of electrons scattered in a solid angle do and normalized to the flux density of the incident beam (Newton, 1966). Assuming that the incident plane wave has the form (1.5), i.e. that the density of electrons is normalized to one electron per unit volume, the flux density of the incident beam is given by

$$\mathbf{j}_0 = \frac{i\hbar}{2m}[\psi_0 \nabla \psi_0^* - \psi_0^* \nabla \psi_0] = \frac{\hbar \mathbf{k}_0}{m}. \tag{1.7}$$

For a scattered spherical wave of the form

$$\psi_s(\mathbf{r}) = f(\theta, \phi)\frac{\exp(ikr)}{r}, \tag{1.8}$$

in the limit of large r we find

$$\mathbf{j}_s = \frac{i\hbar}{2m}[\psi_s \nabla \psi_s^* - \psi_s^* \nabla \psi_s] = \frac{\hbar k_0}{m} \cdot \frac{\mathbf{r}}{r^3}|f(\theta, \phi)|^2. \tag{1.9}$$

Here θ and ϕ are the polar and the azimuthal angles of the spherical system of coordinates where $\mathbf{r} = (x, y, z) = (r \sin\theta \cos\phi, r \sin\theta \sin\phi, r \cos\theta)$.

The number of electrons passing through a spherical surface element $dS = r^2 do$ of a sphere of a very large radius per unit time is given by

$$dI = \mathbf{j}_s \cdot d\mathbf{S} = \frac{\hbar k_0}{m}|f(\theta, \phi)|^2 do. \tag{1.10}$$

Dividing this by j_0, we find the differential cross-section of elastic scattering

$$\frac{d\sigma}{do} = |f(\theta, \phi)|^2. \tag{1.11}$$

What is observed experimentally in a diffraction pattern in the case of elastic scattering is just this term.

1.5   Elastic scattering by a time-independent potential – the one-body Schrödinger equation

A formal theory of electron diffraction inevitably deals with time-dependent processes (Newton, 1966). Electrons are emitted by an electron gun in an electron microscope, and are accelerated towards the specimen, where they interact via the electrostatic potential V. At a large distance from the specimen the scattered electrons propagate outward in all directions where they are counted by detectors (see Fig. 1.1). A formal treatment of electron diffraction must therefore be based on the time-dependent Schrödinger wave equation

$$i\hbar\frac{\partial \Phi}{\partial t} = H\Phi, \tag{1.12}$$

where $\Phi = \Phi(\mathbf{r}; \mathbf{r}_1...\mathbf{r}_i...; \mathbf{R}_1...\mathbf{R}_n...)$ is the wave function describing both the incident electron and the specimen. We first consider the case of purely elastic scattering of a high-energy electron by the specimen neglecting the excitations resulting from the interaction between the two. Assume that the incident electron and the solid form a closed system and the interaction with the environment is neglected. In this case the energy of the system is equal to the energy of the incident electron plus the energy of the solid. The total wave function is therefore an eigenfunction of the Hamiltonian of the entire system

$$\Phi(\mathbf{r}; \mathbf{r}_1...\mathbf{r}_i...; \mathbf{R}_1...\mathbf{R}_n..., t) = \Psi(\mathbf{r}; \mathbf{r}_1...\mathbf{r}_i...; \mathbf{R}_1...\mathbf{R}_n...)$$
$$\times \exp\left[-i\frac{(E + E_s)t}{\hbar}\right], \tag{1.13}$$

where E is the energy of the incident electron and E_s is that of the solid. Substituting this into equation (1.12), we find

$$H\Psi = (E + E_s)\Psi. \tag{1.14}$$

Although being now time independent, eqn (1.14) still describes a complex many-body problem involving the incident electron and the electrons and nuclei of the solid. However, in the case of purely elastic scattering, the above equations may be simplified and transformed into a one-body equation. Since the energy of the incident electron is very high, it can be effectively distinguished from the electrons of the solid (Ochkur, 1964, Rez, 1976). The exchange effects that are important in the case where the energies of the incident electron and electrons in the crystal are comparable can now be neglected. Furthermore, since elastic scattering does not affect the quantum state of the solid, the total wave function of the system may be represented by a product of two wave functions, $\psi(\mathbf{r})$ describing the incident electron and $\phi(\mathbf{r}_1...\mathbf{r}_i...; \mathbf{R}_1...\mathbf{R}_n...)$ describing the electrons and nuclei of the solid,

$$\Psi(\mathbf{r}; \mathbf{r}_1...\mathbf{r}_i...; \mathbf{R}_1...\mathbf{R}_n...) = \phi(\mathbf{r}_1...\mathbf{r}_i...; \mathbf{R}_1...\mathbf{R}_n...)\psi(\mathbf{r}). \tag{1.15}$$

In eqn (1.14) H is the Hamiltonian of the entire system given by

$$-\frac{\hbar^2}{2m}\nabla_r^2 + H_s(\mathbf{r}_i, ..., \mathbf{R}_n, ...) + H_{int}(\mathbf{r}, ..., \mathbf{r}_i, ..., \mathbf{R}_n, ...), \tag{1.16}$$

where the first term in eqn (1.16) is the Hamiltonian of the incident electron, H_s is the Hamiltonian of the solid, and H_{int} is the interaction between the incident electron and the solid. By substituting (1.16) and (1.15) into (1.14), multiplying the result from the left-hand side by ϕ^* and integrating it over the entire range of coordinates of the solid (taking ϕ to be normalized), we obtain a one-body Schrödinger equation for the incident electron

$$\left[-\frac{\hbar^2}{2m}\nabla^2 + V(\mathbf{r})\right]\psi(\mathbf{r}) = E\psi(\mathbf{r}), \tag{1.17}$$

where

$$V(\mathbf{r}) = \int \phi^*(\mathbf{r}_1, ..., \mathbf{R}_1, ...)H_{int}(\mathbf{r}, \mathbf{r}_1..., \mathbf{R}_1...)\phi(\mathbf{r}_1, ..., \mathbf{R}_1, ...)d\mathbf{r}_1...d\mathbf{R}_1.... \tag{1.18}$$

Assuming that the atomic nuclei are located at discrete points $\mathbf{R}_n$, the integration over the coordinates of electrons gives (Spence and Zuo, 1992)

$$V(\mathbf{r}) = \frac{1}{4\pi\epsilon_0}\left(\int \frac{e^2\rho(\mathbf{r}')}{|\mathbf{r}-\mathbf{r}'|}d\mathbf{r}' - \sum_n \frac{Z_n e^2}{|\mathbf{r}-\mathbf{R}_n|}\right), \tag{1.19}$$

where $\rho(\mathbf{r}')$ is the one-particle density of electrons (which is the probability density of finding an electron at $\mathbf{r}'$ multiplied by the number of electrons in the solid). This quantity in some cases (e.g. in the case where it is calculated using the Kohn–Sham equations of density functional theory) can be represented in the form of a sum of contributions of occupied effective one-electron states $\rho(\mathbf{r}') = \sum_j |\varphi_j(\mathbf{r}')|^2$. The Fourier components of this potential (1.19) may be retrieved from electron diffraction experiments, and this information may then be used to find out where the electrons and the atomic nuclei are in a solid. For example, using electron diffraction it is possible to test the accuracy of approximations used in density functional calculations (Dudarev et al., 2000).

To a good approximation the potential $V(\mathbf{r})$ may be written as

$$V(\mathbf{r}) = \sum_n \varphi_n(\mathbf{r} - \mathbf{R}_n) + \Delta V(\mathbf{r}) = V_{atomic}(\mathbf{r}) + \Delta V(\mathbf{r}), \qquad (1.20)$$

where the first term represents the sum of the potentials $\varphi_n(\mathbf{r})$ of individual non-interacting atoms, i.e. the atomic contribution, and the second term describes changes in the electron density due to chemical bonding between atoms. An interesting feature of the charge distribution in a solid is that a substantial part of the overlap between atoms is already taken into account by the first term based on the isolated atom superposition approximation. The amount of charge redistribution associated with the formation of chemical bonds described by the second term in (1.20) is normally small in comparison with the first term (typically 5% or less), and the superposition approximation often provides an excellent starting point for the interpretation of high-energy electron diffraction experiments. The potential $\varphi(\mathbf{r})$ of an isolated atom is given by the Fourier transform of the electron atomic scattering amplitude $f^{(e)}(s)$ (referred to as scattering factor henceforth in this book)

$$\varphi(\mathbf{r}) = -\frac{16\pi\hbar^2}{m_0} \int f^{(e)}(s) \exp(4\pi i s \cdot \mathbf{r}) d\mathbf{s}, \qquad (1.21)$$

where $s = \sin\theta/\lambda$, $\lambda = 2\pi/k_0$ is the wavelength of the incident electron and θ is half of the angle of scattering. Numerical values of $f^{(e)}(s)$ are given in the International Tables for X-ray Crystallography (Shmueli, 1993) and their parameterization can be found in (Peng et al., 1996e, Peng, 1998).

1.6   Selected area electron diffraction (SAED), convergent-beam electron diffraction (CBED), and Kikuchi patterns

There are two fundamentally distinct methods of forming an electron diffraction pattern. The ray diagram in Fig. 1.2 shows the formation of the so-called selected area electron diffraction (SAED) pattern in a transmission electron microscope. A nearly parallel electron beam is incident on the specimen and a regular array of diffraction spots is formed by transmission of the electron beam through a thin single crystal. To reduce the specimen area contributing to the diffraction pattern, an aperture in the image plane conjugate to the specimen plane is inserted. Ideally, if the diameter of the aperture is D and the magnification of the objective lens is M, the effective diameter of the area selected on the specimen should be equal to D/M. However, the spherical aberration of the objective lens as well as its defocus error make it impossible to reduce the size of the selected area beyond a certain limit (see e.g. Hirsch et al. (Hirsch et al., 1977)), so in practice the resolution of the technique is limited and typically is not less than several hundred nanometres.

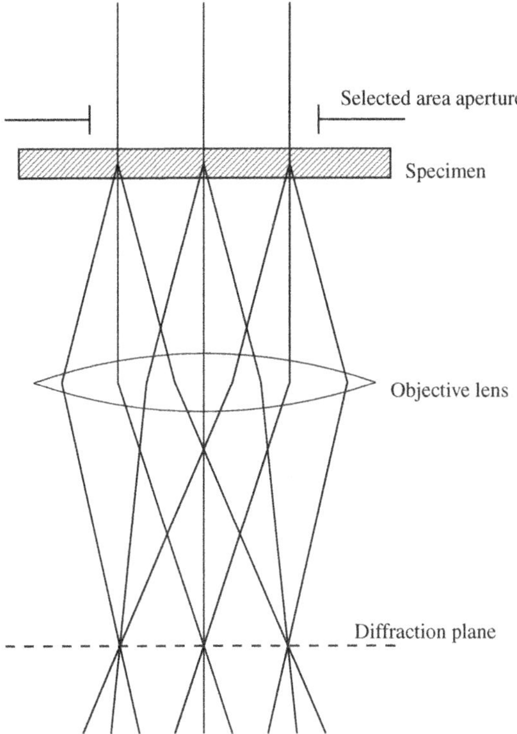

FIG. 1.2. Schematic diagram showing the principles of the formation of an SAED pattern. The selected area aperture shown in the figure is an effective or virtual one. The real selected area aperture is in a plane below which is not shown in the figure.

Figure 1.3 shows a many-beam bright-field (BF) transmission electron microscope (TEM) image taken from a silicon specimen near the [111] zone axis. The transmitted beam contributing to the image has been selected using an objective aperture. The three pairs of extinction contours of the 220 type seen in the image show that the sample is bent and the local angle of incidence changes appreciably across the sample. The overall intensity of the image is seen to decrease from right to left. This is because the thickness varies across the sample, with the thickness being greater in the left part of the image in comparison with the right part of the image. An ordinary SAED pattern obtained using this sample then needs to be corrected for the distortion in both the crystal orientation and the sample thickness, and doing this in practice poses a difficult problem.

An alternative method of forming an electron diffraction pattern involves using the convergent-beam electron diffraction (CBED) geometry. In this geometry a convergent electron beam defined by a circular condenser aperture is focussed onto the specimen (Fig. 1.4). Each diffraction spot is then spread out

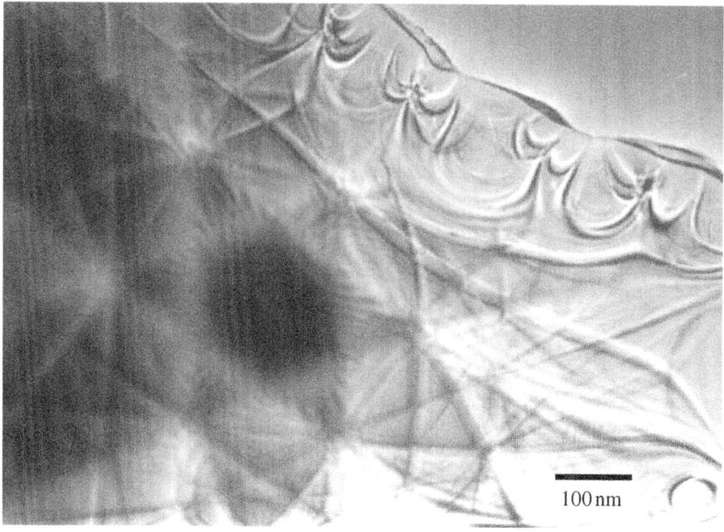

FIG. 1.3. Many-beam bright-field TEM image (near the $\langle 111 \rangle$ zone axis) of a bent silicon single crystal, showing three pairs of $\{220\}$ extinction contours. From (Peng, 1995).

into a circular disc, where each point A$'$ in a disc corresponds to a particular angle of incidence (point A). The variation of intensity across each disc reflects variation of the intensity of the relevant diffracted beam as a function of the angle of incidence. A graphical representation of this variation is called a rocking curve. CBED patterns are in fact two-dimensional rocking curves taken from a very small illuminated area, the size of which ranges from tens to hundreds of angstroms (10 Å$= 1$ nm).

Figure 1.5 shows a $\langle 100 \rangle$ zone axis transmission CBED pattern taken using an NiO specimen. The pattern shows the transmitted disc, four $\{200\}$ and four $\{220\}$ discs. The diameter of the discs in this case is large and the discs overlap. This CBED pattern has been taken using a small probe of the size of a few nanometres. This makes it reasonable to assume that the pattern is taken from a region where the crystal thickness and the orientation of the crystal are uniform. The CBED pattern obtained in this way is well suited for comparing with theoretical calculations.

In the case of elastic diffraction and a parallel incident beam the directions of diffracted beams satisfy the Bragg law (we will discuss this in more detail in Chapter 2). Diffraction spots seen in diffraction patterns of this type are sharp and well defined. However, as mentioned in the preceding section, electrons may be scattered inelastically as well as elastically. Elastic diffraction results from scattering by the periodic time-averaged potential, while inelastic diffuse scattering results from fluctuations of the potential. By definition the fluctuation

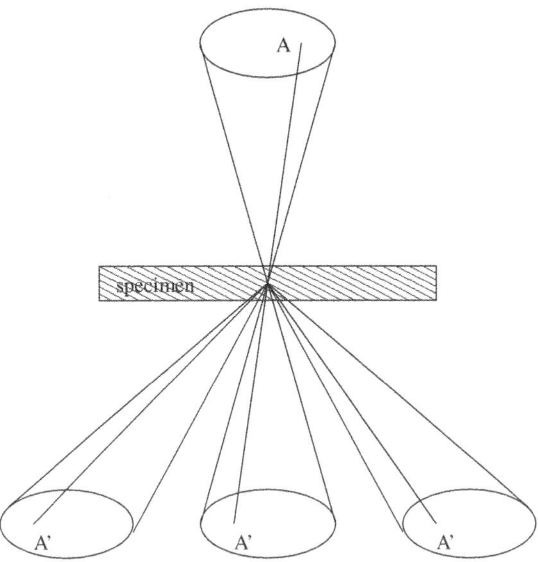

FIG. 1.4. Schematic diagram showing the formation of a CBED pattern in the transmission electron diffraction geometry. The circle around the point A represents the condenser lens aperture, which is considered to be filled with point sources such as at A. Each point source produces diffracted beams A' in the corresponding convergent beam discs.

part of the potential is aperiodic and therefore inelastic diffuse scattering does not contribute to the intensity of Bragg diffraction spots. Diffraction of diffusely or inelastically scattered electrons gives rise to the formation of a continuous pattern of lines, rings, and parabolas. This type of pattern is generated by Bragg diffraction of inelastically scattered electrons, and the corresponding patterns are called Kikuchi patterns after S. Kikuchi who discovered them in 1928. An example of Kikuchi patterns is given in Fig. 1.6.

1.7   Scattering by time-dependent fluctuations of the potential

As mentioned in the preceding sections, the potential of interaction between the incident electron and the solid is inevitably time dependent, and this dependence is associated with the motion of electrons and nuclei in the solid. It is always possible to separate this potential into time-independent and time-dependent parts

$$V(\mathbf{r}, t) = \langle V(\mathbf{r}, t) \rangle_t + \Delta V(\mathbf{r}, t), \tag{1.22}$$

in which the time-dependent component $\Delta V(\mathbf{r}, t)$ is the fluctuating part of the potential and the time-independent component $\langle V(\mathbf{r}, t) \rangle_t$ is simply the time average of the potential given by (1.19). The definition (1.22) shows that $\langle \Delta V(\mathbf{r}, t) \rangle_t = 0$.

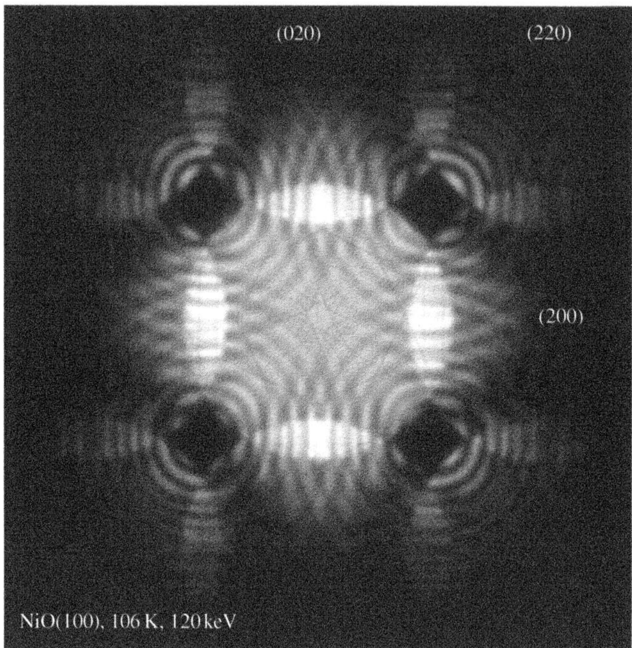

FIG. 1.5. A CBED pattern taken for an NiO single crystal along the [100] zone axis for a primary beam energy of 120 keV. From (Peng and Zuo, 1999) .

The time-dependent part of the potential

$$\Delta V(\mathbf{r}, t) = \Delta V\big(\mathbf{r}; \mathbf{r}_1(t)...\mathbf{r}_j(t)...; \mathbf{R}_1(t)...\mathbf{R}_n(t)...\big)$$

gives rise to the exchange of energy between the incident high-energy electrons and the solid. To evaluate the cross-section of electron energy losses we have to treat both electrons and nuclei in the solid quantum-mechanically. We describe this later in this book in Chapters 7 and 8. Here we give a simplified treatment of scattering by a time-dependent potential.

We assume for the moment that electrons and nuclei in the crystal are classical particles, the motions of which are not affected by the interaction with the incident high-energy electron. The time characterizing the thermal motion of nuclei or the motion of atomic electrons in the solid does not exceed $\sim 10^{-12}$ seconds. Since this time scale is smaller than the time required to record a diffraction pattern or an image, what is observed experimentally is the intensity averaged over a relatively long period of time. In the case of elastic diffraction, the process of scattering may be regarded as being due to the interaction of electrons with the time-averaged potential (1.19). The effect of time-dependent fluctuations on elastic scattering may be described by using the concept of the optical potential (i.e. a potential that has a real and an imaginary

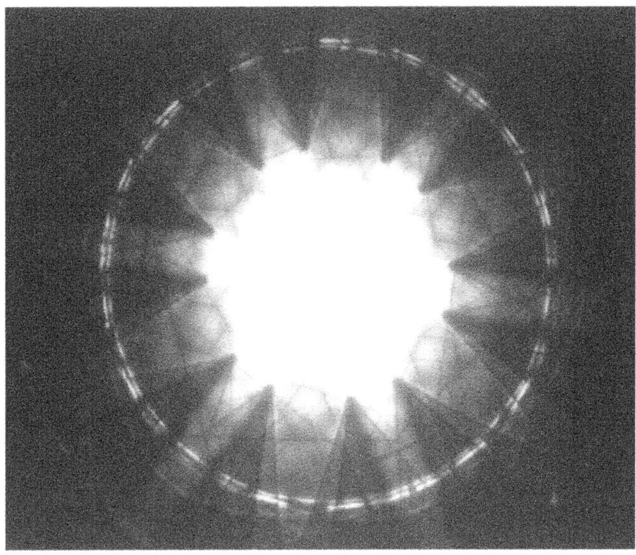

FIG. 1.6. A $\langle 111 \rangle$ zone axis electron diffraction pattern obtained from a GaAs single crystal for a primary beam energy of 120 keV, showing Kikuchi patterns consisting of lines, bands, and parabolas. From (Peng, 1995).

part). Extensive studies have revealed that among the three leading inelastic scattering mechanisms which contribute to the imaginary part of the optical potential, for elements of medium atomic number at moderate temperatures the contribution of phonon scattering is greater than the contribution of two other types of inelastic scattering (Yoshioka and Kainuma, 1962, Whelan, 1965a, Whelan, 1965b, Hall and Hirsch, 1965a, Humphreys and Hirsch, 1968, Radi, 1970, Allen and Rossouw, 1989).

The imaginary part of the optical potential associated with thermal diffuse scattering (TDS) can be calculated by using a computer routine published by Bird and King (Bird and King, 1990) or by using the parameterization by Peng et al. (Peng et al., 1996e, Peng, 1998). Contributions of plasmon and single electron excitations to the optical potential were evaluated by Radi (Radi, 1970) and, more recently, studied by Saldin and Spence (Saldin and Spence, 1994).

While the effect of fluctuations $\Delta V(\mathbf{r}, t)$ on the elastic scattering of electrons is described by the imaginary part of the optical potential, electrons that are inelastically or diffusely scattered by $\Delta V(\mathbf{r}, t)$ may be described perturbatively by using the distorted wave Born approximation (DWBA) (Mott and Massey, 1965, Landau and Lifshitz, 1977). In this case the time-dependent amplitude of scattering by fluctuations is given by (Dudarev et al., 1993a)

$$f(\mathbf{k}, \mathbf{k}_0, t) = f^{(el)}(\mathbf{k}, \mathbf{k}_0) - \frac{m}{2\pi\hbar^2} \int \psi_{-\mathbf{k}}(\mathbf{r}) \Delta V(\mathbf{r}, t) \psi_{\mathbf{k}_0}(\mathbf{r}) d\mathbf{r}, \qquad (1.23)$$

where the superscript (el) is used to denote the amplitude of elastic scattering which is independent of time. $\mathbf{k}_0$ and $\mathbf{k}$ are the wave vectors of the high-energy electron before and after the interaction with the specimen. The quantity $\psi_{\mathbf{k}_0}(\mathbf{r})$ is the wave function of the high-energy electron describing dynamical diffraction of the electron incident in the direction characterized by the wave vector $\mathbf{k}_0$.

The time dependence of the amplitude of scattering given by (1.23) follows the variation of the time-dependent part of the interaction potential. The time average of (1.23) equals the amplitude of elastic scattering $f^{(el)}(\mathbf{k}, \mathbf{k}_0)$.

The angular distribution of scattered electrons observed in a diffraction experiment is given by the time-averaged differential cross-section of scattering

$$\frac{d\sigma}{do} = \left(\frac{d\sigma}{do}\right)^{(el)} \tag{1.24}$$

$$+ \left(\frac{m}{2\pi\hbar^2}\right)^2 \int\int \psi_{-\mathbf{k}}(\mathbf{r})\psi^*_{-\mathbf{k}}(\mathbf{r}')\langle\Delta V^*(\mathbf{r}', t)\Delta V(\mathbf{r}, t)\rangle_t \psi_{\mathbf{k}_0}(\mathbf{r})\psi^*_{\mathbf{k}_0}(\mathbf{r}')d\mathbf{r}d\mathbf{r}'.$$

A detailed derivation of the DWBA that goes beyond the assumption that the motion of electrons and nuclei in a solid is not affected by the interaction with the incident electron, and includes the treatment of energy losses, will be given in Chapters 7 and 8. This approach can be used to simulate Kikuchi patterns similar to those shown in Fig. 1.6 in both transmission and reflection geometries of diffraction (Dudarev et al., 1993b).

1.8 Damping of coherence in inelastic scattering and the validity of the optical potential

The contrast of electron microscope images of crystal lattices and defects results from the interference of electron waves. In view of this, understanding mechanisms of evolution of coherence of electron beams propagating through a specimen represents an important point required for the correct interpretation of observations. Two distinct situations may be identified here, namely (1) small-angle inelastic scattering processes and (2) processes involving scattering through large angles. In the case of both plasmon and single electron excitations fluctuations of the Coulomb interaction potential $V(\mathbf{r}, \mathbf{r}_1, ...)$ may be regarded as being delocalized over many interatomic distances. Therefore inelastic scattering by these fluctuations is dominated by scattering through small angles. Howie (Howie, 1963) showed that in thin crystalline films electrons scattered inelastically through small angles behave almost like elastically scattered electrons, leading to the formation of almost the same features in diffraction contrast images of crystal defects. More recent investigations (Dudarev et al., 1992a, Peng et al., 1993, Dudarev et al., 1993a) show that inelastically scattered electrons undergoing scattering through small angles are not entirely coherent. The degree of coherence gradually decreases with increasing crystal thickness and electrons become almost incoherent after about two successive events of small-angle inelastic scattering or, in other words, in a crystal of thickness greater than twice the mean free path of inelastic scattering. In the case of silicon the mean free path of the electron–electron interactions ℓ_{e-e} is of the order of 100 nm, giving the upper limit of about 200 nm for the thickness of the crystal in which coherence between inelastically scattered electrons is retained. This estimate agrees with detailed calculations performed using the kinetic equation for the density matrix (Dudarev et al., 1992a, Dudarev et al., 1993d).

In the case of large-angle inelastic scattering, e.g. in the case of thermal diffuse scattering, the inelastically scattered electrons become incoherent long before the second scattering event by thermal fluctuations of the interaction potential. The large-angle inelastically scattered electrons may of course undergo further multiple elastic scattering. The diffracted beams resulting from this subsequent elastic scattering also interfere with each other, giving rise to fine features observed both in high-resolution electron microscopy (HREM) images (Cowley, 1988) and in conventional diffraction contrast (DC) images of defects (Peng and Cowley, 1989).

The optical potential can be used only for treating elastically scattered electrons where the effect of diffuse scattering on the elastic scattering is taken into account by adding an imaginary part to the Fourier components of the periodic potential of the crystal. The derivation of the optical potential is based on the assumption that scattering by fluctuations of the interaction potential is relatively weak and can be treated perturbatively. For many important applications this approximation is sufficiently accurate, and results of calculations performed using the optical potential method agree well with experimental observations. The actual problem associated with the analysis of experimental diffraction data consists in finding a way of separating contributions from the elastic and inelastic scattering processes.

In principle the inelastically scattered electrons may be separated from the elastically scattered electrons using energy filters that select electrons passing through the specimen undergoing no energy loss. However, the resolution of a typical energy filter installed in a transmission electron microscope does not exceed a few tenths of an eV, and this is much larger than the characteristic scale of energy losses associated with phonon excitations (several tens of meV). The elastically scattered electrons and electrons scattered by phonon excitations cannot therefore be separated in the final image plane even in the case where energy filtering is used.

In the case of diffraction contrast imaging of crystal defects the inelastically scattered electrons are separated from elastically scattered electrons using a very small aperture around the transmitted beam (bright field) or one of the diffracted beams (dark field) (see Fig. 1.7a). Since thermal diffuse scattering typically involves relatively large momentum transfer in the directions perpendicular to the incident and diffracted beams, the use of a small aperture is particularly suitable for filtering out electrons scattered by phonons. Energy-filtered diffraction contrast images of crystal defects may therefore be well described using the concept of the optical potential.

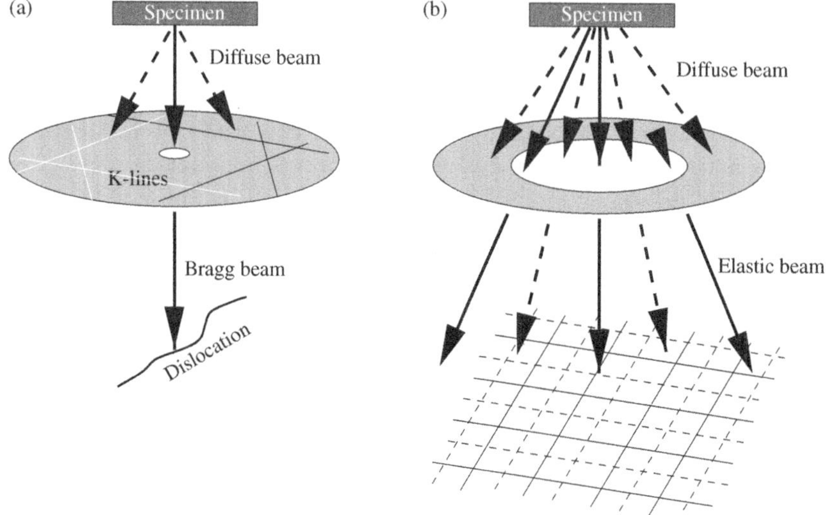

FIG. 1.7. Schematic diagram illustrating the principles of image formation in the case of (a) diffraction contrast imaging of crystal defects and (b) HREM imaging of crystal lattices. In (a) electrons are 'absorbed' from the image by scattering outside the objective lens aperture by both Bragg diffraction and diffuse scattering. In (b) the aperture is large enough to include Bragg-diffracted beams that interfere and form a lattice image. From (Peng, 1999).

On the other hand, a much larger size aperture is typically used in high-resolution imaging of crystal lattices. In this case some of the inelastically scattered electrons contribute to the formation of HREM images (see Fig. 1.7b). While the thermally scattered electrons do not interfere coherently with elastically scattered electrons, the inelastically scattered electrons undergo elastic dynamical diffraction and form high-resolution features which are superimposed on the HREM image formed by elastically scattered electrons (Cowley, 1988). For very thin samples the contribution of inelastically scattered electrons to the formation of HREM images may be neglected, and the effect of inelastic scattering on electrons passing through a small aperture may be taken into account using the optical potential. To treat the effects of multiple inelastic and elastic scattering in thick specimens one should use the density matrix approach described in Chapter 8.

In the CBED mode of electron diffraction (Spence and Zuo, 1992) a convergent electron beam, usually formed by a circular aperture, is focussed on the specimen (see Fig. 1.8). Each diffraction spot of the conventional electron diffraction pattern then becomes a disc, where each point in the disc corresponds to a particular angle of incidence (Fig. 1.4). Figure 1.8 shows that while for each direction of incidence the diffuse background is superimposed on a diffraction intensity peak, the total distribution of diffusely scattered electrons in a CBED

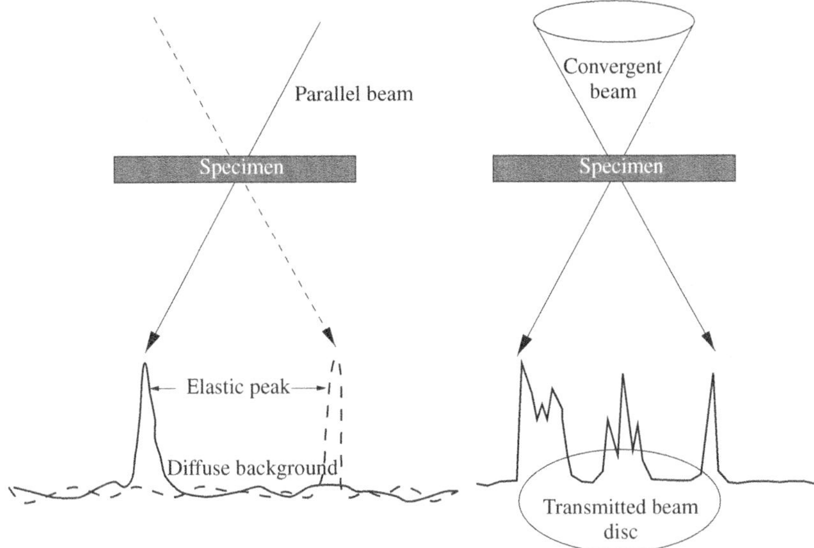

FIG. 1.8. Schematic diagram showing the effect of diffuse background in CBED. From (Peng, 1999).

pattern is a smooth function of the angle of scattering in comparison with the more rapidly varying rocking curves formed by elastically scattered electrons. This makes it possible to approximate the distribution of electrons diffusely scattered within each CBED disc by a constant background and to subtract this from the CBED pattern. Recent quantitative CBED results have confirmed the validity and high accuracy of this procedure (Zuo et al., 1997, Ren et al., 1997).

1.9 Relativistic corrections

The velocity of electrons used in high-energy electron diffraction studies of materials lies in the range between $0.5c$ and $0.99c$, where c is the speed of light, and where relativistic effects are not negligible. In principle the relativistic Dirac equation should therefore be used to describe the motion of high-energy electrons. Fortunately it was shown by Fujiwara (Fujiwara, 1961) and Howie (Howie, 1962) that in the case of high-energy electron diffraction we may neglect effects associated with the spin of electrons and instead of the Dirac equation we can use the relativistic Klein-Gordon equation

$$\left[-\frac{\hbar^2}{2m}\nabla^2 + V(\mathbf{r}) \right] \psi(\mathbf{r}) = E\left(\frac{m_0}{m}\right)\left(1 + \frac{E}{2m_0 c^2}\right)\psi(\mathbf{r}). \qquad (1.25)$$

This equation shows that the non-relativistic Schrödinger equation (1.17) can be corrected for relativistic effects by replacing the electron rest mass by its relativistic mass, and the beam energy E by

$$E \to E\left(\frac{m_0}{m}\right)\left(1 + \frac{E}{2m_0 c^2}\right) = \frac{\hbar^2 k_0^2}{2m}, \qquad (1.26)$$

where $\mathbf{k}_0$ is the relativistic wave vector. Fujiwara (Fujiwara, 1961) and Howie (Howie, 1962) showed that eqn (1.25) remains valid well into the MeV energy range.

1.10 Probability current density and conservation of probability

In electron diffraction the wave function is interpreted following the standard convention of quantum mechanics (Schiff, 1968) that the quantity

$$P(\mathbf{r}, t) = \psi^*(\mathbf{r}, t)\psi(\mathbf{r}, t) = |\psi(\mathbf{r}, t)|^2 \qquad (1.27)$$

gives the probability of finding the high-energy electron at point $\mathbf{r}$ at time t. To investigate how this probability varies with time, we consider its time derivative

$$\begin{aligned}
\frac{\partial P}{\partial t} &= \psi^* \frac{\partial \psi}{\partial t} + \frac{\partial \psi^*}{\partial t} \psi \\
&= \frac{1}{i\hbar}\left\{\psi^*\left[-\frac{\hbar^2}{2m}\nabla^2 \psi + V(\mathbf{r}, t)\psi\right] - \psi\left[-\frac{\hbar^2}{2m}\nabla^2 \psi^* + V^*(\mathbf{r}, t)\psi^*\right]\right\} \\
&= -\nabla \cdot \mathbf{j}(\mathbf{r}, t) + \frac{1}{i\hbar}[V(\mathbf{r}, t) - V^*(\mathbf{r}, t)]P(\mathbf{r}, t), \qquad (1.28)
\end{aligned}$$

where vector $\mathbf{j}$,

$$\mathbf{j}(\mathbf{r}, t) = \frac{i\hbar}{2m}\left[\psi\nabla\psi^* - \psi^*\nabla\psi\right], \qquad (1.29)$$

gives the probability flux density. If we write V in the form $V = V_R + iV_I$, where V_R and V_I are the real and imaginary parts of the potential, we find

$$\frac{\partial P(\mathbf{r}, t)}{\partial t} + \nabla \cdot \mathbf{j}(\mathbf{r}, t) = \frac{2V_I}{\hbar}P(\mathbf{r}, t). \qquad (1.30)$$

This equation is similar to the classical equation of continuity where the right-hand side of the equation represents a source or a sink of probability for a positive or negative value of V_I, respectively. Integration of the above equation over a volume Ω bounded by a surface S leads to

$$\frac{d}{dt}\int_\Omega P(\mathbf{r}, t)d\mathbf{r} = -\int_S \mathbf{j} \cdot d\mathbf{s} + \frac{2}{\hbar}\int_\Omega V_I(\mathbf{r}, t)P(\mathbf{r}, t)d\mathbf{r}. \qquad (1.31)$$

For a non-absorbing crystal ($V_I = 0$) eqn (1.31) shows that the total number of elastically scattered electrons is conserved. The probability of finding an electron in the volume Ω varies depending on the rate of current flow through the boundary S. If the wave function ψ vanishes asymptotically in the limit $\mathbf{r} \to \infty$ so that the surface integral is negligibly small, the rate of variation of probability of finding the electron in Ω depends on the sign of the imaginary part of the potential. Elastically scattered electrons are continuously scattered into inelastic channels and the rate of scattering back into the initial state is negligibly small. We may therefore treat such electrons as if they had effectively disappeared, i.e. it is as though they had been 'absorbed' by the specimen. This requires that we choose the sign of the imaginary part of the optical potential as $V_I \leq 0$. The total number of electrons absorbed from the elastic channel is measured by the total absorption cross-section given by

$$\sigma_{abs} = -\frac{2}{\hbar v} \int_{\Omega} V_I(\mathbf{r}, t) P(\mathbf{r}, t) d\mathbf{r}, \qquad (1.32)$$

where v is the velocity of the electrons.

1.11 Correlation between theory and experiment

Quantitative electron diffraction investigation of materials relies on the possibility of making detailed comparisons between theoretical predictions and experimental observations. A comparison may be made visually, as in the case of HREM, or quantitatively. The latter case requires calculating an image or a diffraction pattern formed by both elastically and inelastically scattered electrons. Recent studies (Dudarev et al., 1993a, Marthinsen et al., 1994) showed that reasonable agreement can be achieved between the calculated and experimentally observed diffraction patterns.

Alternatively an energy-filtering facility may be used in order to obtain a diffraction pattern. Figure 1.9 shows an energy-filtered CBED pattern taken from a silicon single crystal near a $\langle 110 \rangle$ zone axis using a 1.4 nm fine electron probe and a 10 eV energy window for filtering. To record the CBED pattern the crystal was tilted to satisfy conditions of the systematic diffraction case, i.e. the case where the diffraction process is dominated by reflections of the hhh type. The use of energy filtering greatly simplifies the interpretation of diffraction patterns, since the majority of inelastically scattered electrons are excluded by the filter. The effects of inelastic scattering on the intensity distribution seen in the diffraction pattern may be taken into account by using an appropriately chosen complex optical potential as discussed in the preceding section. Although the typical level of resolution of energy filters used in an electron microscope is not sufficient to remove the contribution of thermal diffuse scattering, in the case of thin crystals containing light elements with relatively high Debye temperatures (such as beryllium, diamond, and silicon) the effects of phonon scattering are relatively weak and the accuracy of calculations performed using the optical potential model is high.

Figure 1.10 shows the energy-filtered experimental (solid line) and the computed (dotted line) one-dimensional CBED rocking curves corresponding to the line AB shown in Fig. 1.9. Ten adjustable parameters were used in fitting the simulated diffraction data to the experimental curve. The fitting parameters included the real and imaginary parts of the 111 and 222 structure factors, the thickness of the crystal, and the spatial coordinates of points A and B in Fig. 1.9.

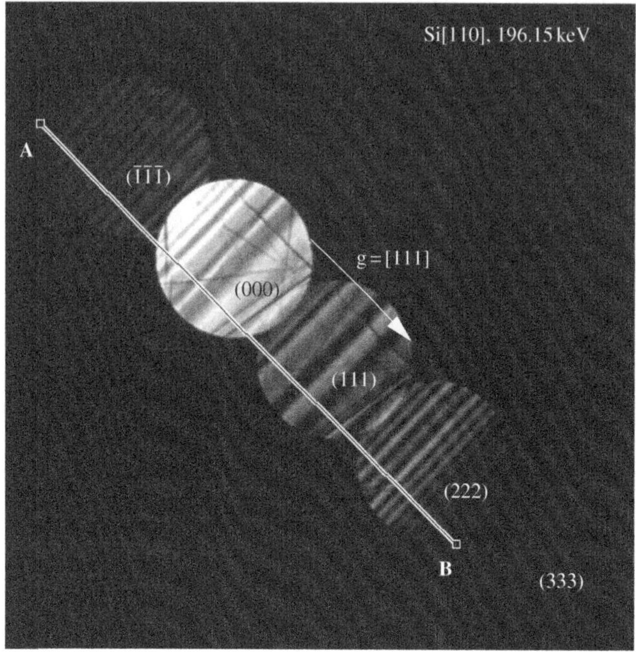

FIG. 1.9. Energy-filtered CBED pattern recorded from a silicon single crystal near the $\langle 110 \rangle$ zone axis. The primary beam energy is 196.15 keV, and the crystal thickness is 343 nm. From (Ren et al., 1997).

The plot below the figure shows the difference between the calculated and the observed intensity distributions, and the residual χ^2 value equals 2.87, indicating a very high level of agreement between the theory and experiment.

Figure 1.11 shows sets of energy-unfiltered experimental and simulated rocking curves taken across the 200 CBED disc observed using an Si single crystal at 80 keV. Figures 1.11a, 1.11b and 1.11c were recorded for three different crystal thicknesses: (a) $t = 3420$ Å , (b) $t = 2860$ Å and (c) $t = 3420$ Å. In the figure the dashed curves are calculated using the optical potential method. It is seen that the agreement between the theory and experimental results is relatively poor. This is because the optical potential method describes only the elastically scattered electrons. The contribution of inelastically scattered electrons to the diffraction pattern is completely neglected. The dotted curves are calculated using the quantum kinetic equation for the density matrix of high-energy electrons. The density matrix approach takes into account effects of multiple inelastic scattering and partial coherence (Dudarev et al., 1992a, Dudarev et al., 1993a). This figure shows that a fair agreement between the theoretical and the energy-unfiltered experimental rocking curves can be achieved, but the agreement is still poorer than that of Fig. 1.10. We note that the agreement between theoretical predictions

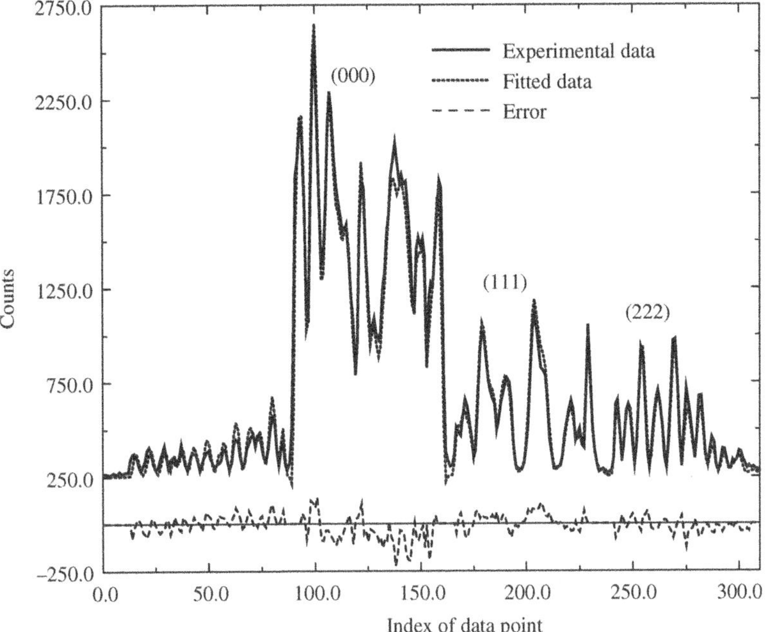

FIG. 1.10. Experimental (solid) and simulated (dotted) intensity distributions plotted along the line AB shown in Fig. 1.9. The dashed line shown below the intensity profiles shows the difference between the experimental and the simulated rocking curves. Point A of Fig. 1.9 corresponds to the first data point of this figure, and point B corresponds to the last data point of the figure. From (Ren et al., 1997).

and experimental observations shown in Fig. 1.11 is the best achieved so far for the energy-unfiltered rocking curves (Peng et al., 1993, Dudarev et al., 1993a). This shows that in order to perform a quantitative electron diffraction study it is highly desirable to use an energy-filtering facility.

1.12 Summary

Elastic and inelastic scattering of high-energy electrons can be used to probe the structure and dynamical properties of solids. Diffraction patterns formed by elastically scattered electrons contain information about the distribution of charge density and the equilibrium atomic structure. The cross-section of inelastically scattered electrons reflects the dynamics of motion of atoms and electrons in the solid.

In the case of elastic scattering the effective interaction between the incident electron and the solid does not depend on time and scattering is described by the time-independent Schrödinger equation in which the potential is a complex quantity. The imaginary part of this optical potential describes effective absorption of high-energy electrons associated with inelastic scattering by phonon and

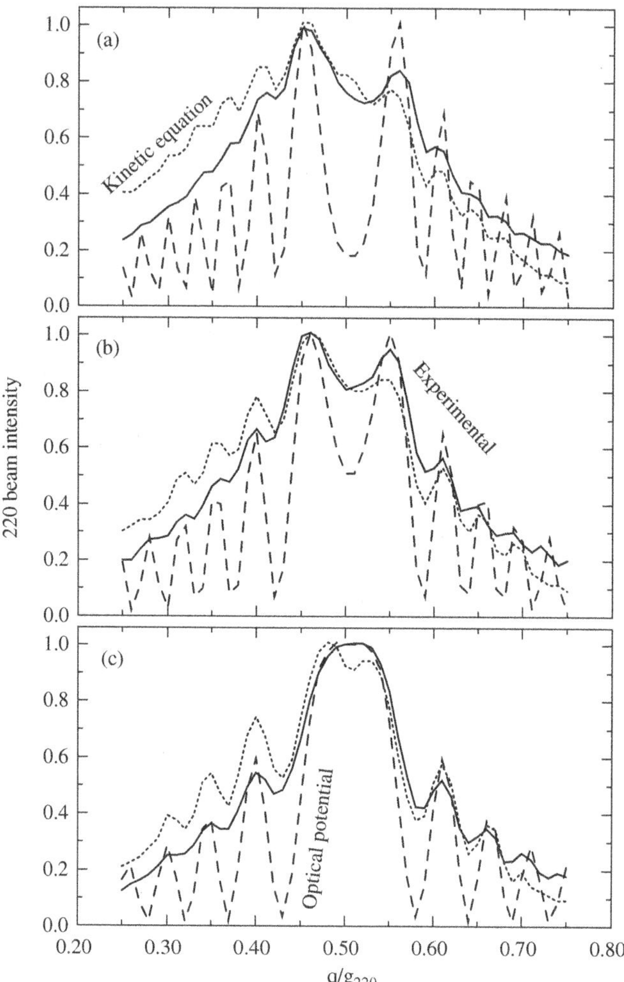

FIG. 1.11. Experimental (solid lines) and simulated (dotted lines) 220 diffracted beam intensity profiles plotted as a function of the transverse momentum q for a silicon crystal. The length of q is measured in the units of g_{220}, where g_{220} is the reciprocal lattice vector of the 220 reflection. The three sets of curves correspond to crystal thickness of (a) 342 nm; (b) 286 nm; (c) 234 nm. From (Dudarev et al., 1993a).

electron excitations. Relativistic corrections may be introduced into this equation by replacing the electron rest mass by its relativistic mass and by replacing the primary beam energy by $E(m_0/m)(1 + E/2m_0c^2)$. Diffraction of elastically scattered electrons can be studied by using an energy-filtering system. The observed diffraction patterns agree very well with intensity distributions predicted theoretically using the optical potential method.

The time dependence of the interaction potential gives rise to the exchange of energy between the incident electron and the solid, and inelastic scattering acts as a sink for elastically scattered electrons. To interpret experimental observations obtained using an ordinary electron microscope with no energy-filtering facility, it is necessary to use an approach that treats elastic and inelastic scattering on an equal footing. Such an approach is now available, and it will be described later in Chapters 7 and 8. Inelastically scattered electrons reveal more information about the structure of the specimen than do elastically scattered electrons. Inelastic scattering of electrons can be used to investigate not only the time-averaged properties but also to probe the dynamics of motion of atoms and electrons in the material.

2

KINEMATIC THEORY

2.1 Introduction

Since high-energy electrons interact strongly with solids it is likely that scattering will occur more than once as electrons propagate through the specimen. This also means that the single scattering, or kinematic, approximation cannot be applied to the interpretation of experimental observations. Nevertheless, there are some interesting situations where the single scattering approximation does provide a useful means for understanding experimental results and in particular for understanding the geometry of electron diffraction patterns. In general, to develop a useful single scattering approximation we need to represent the potential $V(\mathbf{r})$ by a sum of two terms

$$V = V_0 + \Delta V, \tag{2.1}$$

where ΔV is small enough so that the kinematic diffraction theory is applicable. Depending on the nature of V_0, we can distinguish between three different cases. In the first case we have $V_0 = 0$ and the kinematic scattering approximation is assumed to be applicable to the problem of scattering by the entire potential V. In section 2.3 we will show that unfortunately, even in the problem of scattering of high-energy electrons by an individual light atom, e.g. an oxygen atom, the conditions of validity of the single scattering approximation are not satisfied. This means that in the case where $V_0 = 0$ the kinematic theory should be used with extreme caution.

The second case corresponds to the situation where V_0 is constant. In this case the magnitude of V_0 may be large and an electron may be scattered by this constant potential many times. Fortunately, the processes involving multiple scattering by a constant potential V_0 can easily be taken into account. Since diffraction effects result from the spatial variation of the potential, multiple scattering by a constant potential is equivalent to the renormalization of the wavelength of the high-energy electron. In the case where V_0 is complex, the imaginary part of the potential gives rise to the attenuation of the beam of electrons. For example, a constant complex potential is often used to describe plasmon scattering. If a plane wave

$$\psi_0 = \exp(i\mathbf{k}_0 \cdot \mathbf{r}) \tag{2.2}$$

is incident on a solid characterized by a potential V of the form of (2.1) then the constant part of the potential V_0 transforms this wave into another plane wave with a new (in the general case, complex) wave vector $\mathbf{k}$. This 'transformed' incident wave may then undergo kinematic scattering by the potential ΔV, and the

24

treatment of this single scattering process is no more difficult than the treatment of the case where $V_0 = 0$. To distinguish the case $V_0 \neq 0$ from conventional kinematic diffraction corresponding to $V_0 = 0$, we shall call this case 'quasi-kinematic diffraction'.

The third case corresponds to a general form of the potential $V_0 = V_0(\mathbf{r})$. Assuming that dynamical diffraction of electrons by $V_0(\mathbf{r})$ is well understood, and that we have an explicit expression for the wave function $\psi(\mathbf{r})$ describing the relevant dynamical effects, the perturbation of this wave function by ΔV may then be treated kinematically. This third case is more complicated than either the kinematic or quasi-kinematic case. In the next section we shall consider only the first two cases and defer the discussion of the perturbation treatment to Chapters 7 and 10.

2.2 Kinematic and quasi-kinematic diffraction theory

2.2.1 Kinematic diffraction

To proceed further we will need to make use of the one-body Schrödinger wave equation introduced in Chapter 1

$$\left[-\frac{\hbar^2}{2m}\nabla^2 + V(\mathbf{r}) \right] \psi(\mathbf{r}) = E\psi(\mathbf{r}). \tag{2.3}$$

In free space where $V(\mathbf{r}) = 0$, solutions of this equation have the form of plane waves (2.2) with $k_0 = \sqrt{2mE/\hbar^2}$. Another possible solution of this equation in free space is a spherical wave of the form

$$G(\mathbf{r}, \mathbf{r}') = -\frac{m}{2\pi\hbar^2}\frac{\exp(ik_0|\mathbf{r} - \mathbf{r}'|)}{|\mathbf{r} - \mathbf{r}'|}. \tag{2.4}$$

This spherical wave is usually called the Green's function. This function satisfies the Schrödinger equation (2.3) everywhere in space except at point $\mathbf{r} = \mathbf{r}'$ and is generated by a source situated at $\mathbf{r}'$. Substituting (2.4) back into the Schrödinger equation (2.3), we arrive at

$$\left(E + \frac{\hbar^2}{2m}\nabla^2 \right) G(\mathbf{r}, \mathbf{r}') = \delta(\mathbf{r} - \mathbf{r}'), \tag{2.5}$$

where $\delta(\mathbf{r})$ is the Dirac delta function. Using this relation and the general properties of the Dirac delta function (see Appendix A), it can be proved that equation (2.3) is equivalent to an integral equation

$$\psi(\mathbf{r}) = \psi_0(\mathbf{r}) + \int G(\mathbf{r}, \mathbf{r}')V(\mathbf{r}')\psi(\mathbf{r}')d\mathbf{r}', \tag{2.6}$$

where $\psi_0(\mathbf{r})$ is the incident wave (2.2). The advantage of using this integral equation is that it leads to a clear physical interpretation of the process of scattering.

For an incident plane wave ψ_0 the total wave field consists of two components, namely the incident wave ψ_0 (the first term) and the scattered wave ψ_s (the second term). The scattered wave may be interpreted as a coherent superposition of spherical waves $G(\mathbf{r}, \mathbf{r}')$ generated by sources situated at points $\mathbf{r}'$. The effective strength of sources is given by $V(\mathbf{r}')\psi(\mathbf{r}')$. The spherical wave (2.4) is therefore the free space Green's function describing the free propagation (without further scattering) of a wave originating at $\mathbf{r}'$ to another point $\mathbf{r}$.

Equation (2.6) does not give a solution to the wave equation (2.3), since the unknown wave function $\psi(\mathbf{r}')$ enters the integral in the right-hand side of this equation. However, if the potential $V(\mathbf{r})$ is weak so that the amplitude of the scattered wave ψ_s is smaller than that of the incident wave ψ_0, we may approximate the wave function ψ entering the integral on the right-hand side of eqn (2.6) by ψ_0 and find

$$\psi(\mathbf{r}) \approx \psi_0(\mathbf{r}) + \int G(\mathbf{r}, \mathbf{r}')V(\mathbf{r}')\psi_0(\mathbf{r}')d\mathbf{r}'. \tag{2.7}$$

This equation is known as the first (order) Born approximation, or the kinematic approximation. For an incident plane wave of the form (2.2) at a large distance from the specimen ($r \gg r'$), where the potential field is zero, we can approximate $|\mathbf{r} - \mathbf{r}'|$ in the denominator of (2.4) by r. In the numerator we write $k_0|\mathbf{r} - \mathbf{r}'| \approx k(r - r'\cos\theta) = k_0 r - \mathbf{k} \cdot \mathbf{r}'$, where θ is the angle between $\mathbf{k}$ (along $\mathbf{r}$) and $\mathbf{r}'$. The Green's function (2.4) can now be approximated as

$$G(\mathbf{r}, \mathbf{r}') \approx -\frac{m}{2\pi\hbar^2}\frac{\exp(ik_0 r)}{r}\exp(-i\mathbf{k}\cdot\mathbf{r}').$$

Substituting this into eqn (2.7), we find

$$\psi(\mathbf{r}) \approx \psi_0(\mathbf{r}) + \frac{\exp(ik_0 r)}{r}f^{(B)}(\mathbf{k}, \mathbf{k}_0), \tag{2.8}$$

where $f^{(B)}(\mathbf{k}, \mathbf{k}_0)$ is the Born or the kinematic scattering amplitude given by

$$f^{(B)}(\mathbf{k}, \mathbf{k}_0) = f^{(B)}(\mathbf{k} - \mathbf{k}_0) = -\frac{m}{2\pi\hbar^2}\int V(\mathbf{r}')\exp[-i(\mathbf{k} - \mathbf{k}_0)\cdot\mathbf{r}']d\mathbf{r}'. \tag{2.9}$$

This shows that in the Born approximation the amplitude of scattering is equal to the Fourier transform of the potential $V(\mathbf{r})$. Given that the condition of validity of the kinematic approximation is satisfied, the Born scattering amplitudes may in principle be determined from scattering experiments (Kohl and Rose, 1985). An inverse Fourier transform of the Born scattering amplitude

$$\begin{aligned}
V(\mathbf{r}) &= -\frac{2\pi\hbar^2}{m}\int f^{(B)}(\mathbf{q})\exp(i\mathbf{q}\cdot\mathbf{r})\frac{d\mathbf{q}}{(2\pi)^3}\\
&= -\frac{16\pi\hbar^2}{m}\int f^{(B)}(4\pi\mathbf{s})\exp(4\pi i\mathbf{s}\cdot\mathbf{r})d\mathbf{s},
\end{aligned} \tag{2.10}$$

where $\mathbf{q} = \mathbf{k} - \mathbf{k}_0 = 4\pi\mathbf{s}$, gives the real-space distribution of the potential. In this way we may obtain information about atomic positions and the distribution of valence electrons around atomic nuclei.

2.2.2 *Quasi-kinematic diffraction*

We now consider the second case where V_0 is a constant potential. By writing the solution of eqn (2.3), where $V(\mathbf{r}) = 0$ and $k' = \sqrt{2m(E - V_0)/\hbar^2}$, as $\psi_0 = \exp(i\mathbf{k}' \cdot \mathbf{r})$, from eqn (2.6) we find the following expression for the total wave function:

$$\psi(\mathbf{r}) \approx \psi_0(\mathbf{r}) + \int G(\mathbf{r}, \mathbf{r}')\Delta V(\mathbf{r}')\psi_0(\mathbf{r}')d\mathbf{r}'. \tag{2.11}$$

Following a procedure similar to the one leading to eqn (2.8), we arrive at

$$\psi(\mathbf{r}) = \psi_0(\mathbf{r}) + \frac{\exp(ik_0 r)}{r} f(\mathbf{k}, \mathbf{k}_0), \tag{2.12}$$

where

$$f(\mathbf{k}, \mathbf{k}_0) = -\frac{m}{2\pi\hbar^2} \int \Delta V(\mathbf{r}') \exp[-i(\mathbf{k} - \mathbf{k}') \cdot \mathbf{r}']d\mathbf{r}'. \tag{2.13}$$

This expression is similar to eqn (2.9), but here $\mathbf{k}'$ differs from the wave vector $\mathbf{k}_0$ of the incident beam in both amplitude and direction. Furthermore, $\mathbf{k}'$ may be complex. In the transmission diffraction geometry, to a very good approximation we may take $k' \approx k_0$ and $\Delta V(\mathbf{r}) \approx \Delta V(\mathbf{X})$, where $\mathbf{X}$ is the two-dimensional spatial coordinate in the plane normal to the incident beam. By writing $\mathbf{k}' \approx \mathbf{k}_0 + i\mu\hat{\mathbf{z}}$, where μ is the mean absorption constant and $\hat{\mathbf{z}}$ is a unit vector along the z axis, we find an approximate expression for the amplitude of scattering

$$f(\mathbf{k}, \mathbf{k}_0) = -\frac{m}{2\pi\hbar^2} \int \Delta V(\mathbf{X}') \exp[-i(\mathbf{k} - \mathbf{k}_0) \cdot \mathbf{X}']d\mathbf{X}' \int \exp(-\mu z)dz. \tag{2.14}$$

This quasi-kinematic transmission diffraction case differs from the true kinematic case in that the intensity of the incident beam no longer remains constant. Instead, it is attenuated owing to the presence of inelastic scattering. Still, the amplitude of scattering is proportional to the Fourier transform of the potential $\Delta V(\mathbf{r})$ and the procedure for retrieving the potential from diffraction experiments remains the same as in the kinematic case.

2.3 Scattering by a single atom

In the preceding section we showed that at large distance from the specimen the scattered wave function ψ_s is proportional to the amplitude of scattering $f(\mathbf{k}, \mathbf{k}_0)$. If the kinematic approximation applies, this amplitude reduces to the Born scattering amplitude, which is simply equal to the Fourier transform of the potential $V(\mathbf{r})$. In the case of elastic diffraction by a single atom, the Born scattering amplitude may be related to the electron atomic scattering factor,

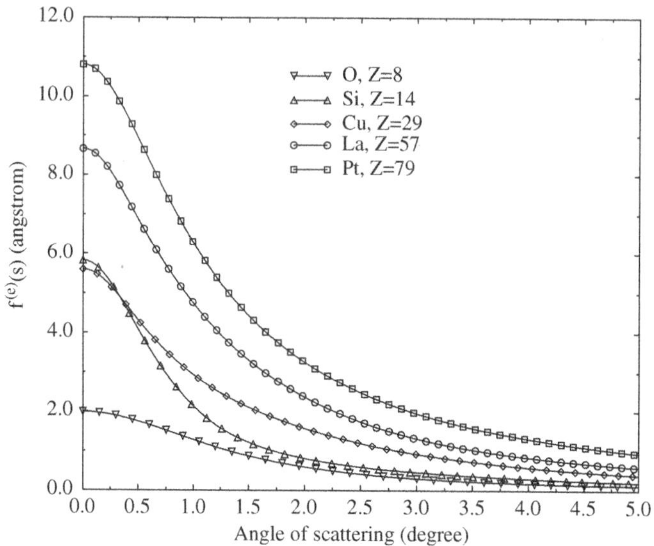

FIG. 2.1. Plots of the electron atomic scattering factor $f^{(e)}(s)$ for atoms of several elements of the periodic table calculated for the energy of the primary beam of 100 keV. The angle of scattering 2θ is related to s via $s = \sin\theta/\lambda$.

i.e. $f^{(B)} = (m/m_0)f^{(e)}$, tabulated in the International Tables for Crystallography (Shmueli, 1993). Alternatively, the atomic scattering factor can be calculated using an analytical fit (Smith and Burge, 1962, Doyle and Turner, 1968, Peng et al., 1996e, Peng, 1998) of the form

$$f^{(e)}(s) = \sum_{j=1}^{n} a_j \exp(-b_j s^2), \tag{2.15}$$

where $s = \sin\theta/\lambda$, 2θ is the angle of scattering, and a_j and b_j are fitting parameters. Smith and Burge (Smith and Burge, 1962) used six parameters ($n = 3$) while Doyle and Turner (Doyle and Turner, 1968) used eight parameters ($n = 4$). These authors found values of fitting parameters for only a limited number of atoms. More recently Peng et al. (Peng et al., 1996e, Peng, 1998) used 10 parameters, i.e. $n = 5$, and evaluated them for all the neutral atoms of the entire periodic table (Peng et al., 1996e) and also for 109 ions (Peng, 1998).

Figure 2.1 shows plots of atomic scattering factors calculated for atoms of several elements of the periodic table for the primary beam energy of 100 keV (λ=0.037 Å). The scattering factor is largest in the forward direction in the angular range $\theta < 5°$. Since the real-space resolution of a microscope is inversely proportional to the maximum momentum transferred in a diffraction process, the angle of scattering $\theta \sim 5°$ corresponds to the real-space resolution of approximately 0.4 Å in the plane perpendicular to the beam. The magnitude of the momentum transfer in the direction parallel to the beam is considerably smaller.

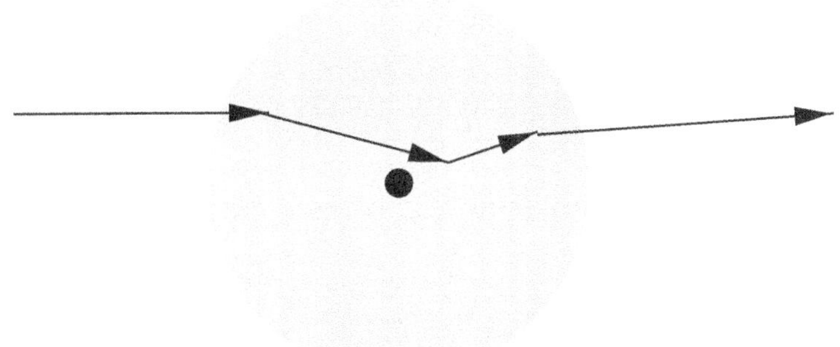

FIG. 2.2. Schematic diagram illustrating how a high-energy electron undergoes multiple scattering in the electrostatic field of a single atom.

For a scattering angle of the order of $5°$ the momentum transfer in the beam direction is approximately equal to 0.2 Å^{-1}. This gives spatial resolution of about 5 Å in the direction of the beam. In the case of high-energy electron diffraction by a thin film (where the thickness of the film is smaller than 20 Å), we may neglect the variation of the potential in the beam direction. This constitutes the projected potential approximation (PPA). This approximation is justified by the fact that the process of propagation of the incident high-energy electron is not very sensitive to the variation of the potential in the direction of the beam, or in other words by the fact that the amplitude of scattering is small for large angles corresponding to electron backscattering, as shown in Fig. 2.1.

Scattering by a single atom cannot in general be regarded as being kinematic, i.e. as a process that can be accurately described by the electron scattering factor $f^{(e)}(s)$. Indeed, this factor is nothing but the Fourier transform of the potential of a spherically symmetric neutral atom. Since the potential of an atom is delocalized in real space, the electron may undergo multiple scattering by this potential. The process of multiple scattering is illustrated schematically in Fig. 2.2. This process can be very well described by the phase object approximation (POA). In this approximation the electron wave function in a reference plane to the right of the atom in Fig. 2.2 is expressed as (Cowley, 1990)

$$\psi(\mathbf{r}) = \psi_0 \exp\{-i\sigma\phi_p(x, y)\}, \tag{2.16}$$

where ψ_0 is the amplitude of the incident wave, σ is the relativistic electron interaction constant given by $\sigma = m/(\hbar^2 k_0) = m\lambda/(2\pi\hbar^2)$, and $\phi_p(x, y) = \int \phi(x, y, z)dz$ is the atomic potential projected in direction z of the incident beam. Using eqns (2.10) and (2.15), we find the atomic potential

$$\phi(x,y,z) = -\frac{16\pi\hbar^2}{m_0} \int f^{(e)}(s) \exp(4\pi i s \cdot \mathbf{r}) ds$$

$$= -\frac{2\pi\hbar^2}{m_0} \sum_{j=1}^{n} a_j \left(\frac{4\pi}{b_j}\right)^{3/2} \exp\left[-\frac{4\pi^2}{b_j}(x^2+y^2+z^2)\right]. \quad (2.17)$$

The atomic potential projected in the z direction is given by

$$\phi_p(x,y) = \int_{-\infty}^{\infty} \phi(x,y,z)dz = -\frac{8\pi^2\hbar^2}{m_0} \sum_{j=1}^{n} \frac{a_j}{b_j} \exp\left[-\frac{4\pi^2}{b_j}(x^2+y^2)\right]. \quad (2.18)$$

Using eqn (2.16) and considering the case where the atom is illuminated by an incident plane wave ($\psi_0 = 1$), we find

$$\psi(x,y) = \exp\{i\varphi(x,y)\}, \quad (2.19)$$

where

$$\varphi(x,y) = -\sigma\phi_p(x,y) = 4\pi\lambda\left(\frac{m}{m_0}\right) \sum_{j=1}^{n} \left(\frac{a_j}{b_j}\right) \exp\left[-\frac{4\pi^2}{b_j}(x^2+y^2)\right]. \quad (2.20)$$

If the scattering potential is sufficiently weak so that $\varphi \ll 1$, eqn (2.19) may be expanded to the first order in φ, namely

$$\psi(x,y) \simeq 1 + i\varphi(x,y). \quad (2.21)$$

This corresponds to the weak phase object approximation (WPOA). To estimate the range of validity of the WPOA we examine the maximum phase shift φ_{max} corresponding to $x = y = 0$. Using eqn (2.20), we find

$$\varphi_{max} = 4\pi\lambda\left(\frac{m}{m_0}\right) \sum_{j=1}^{n} \left(\frac{a_j}{b_j}\right) = 4\pi\lambda\left(1 + \frac{E}{m_0 c^2}\right) \sum_{j=1}^{n} \left(\frac{a_j}{b_j}\right), \quad (2.22)$$

where E is the primary beam energy. We use data for an atom of copper from the tables provided by Peng et al. (Peng et al., 1996e) and find that $\varphi_{max}(\text{Cu}) = 1.35$ rad for $E = 100$ keV and $\varphi_{max}(\text{Cu}) = 0.858$ rad for $E = 500$ keV. This shows that the kinematic approximation is not valid for the case of scattering of high-energy electrons even by a single atom of Cu for either $E = 100$ keV or $E = 500$ keV.

Figure 2.3 shows several one-dimensional plots of the phase function $\varphi(x, y = 0)$ calculated for copper for several primary beam energies and temperatures. Calculations were carried out using eqn (2.20). In all cases shown in Fig. 2.3 the kinematic approximation breaks down near the centre of the atom ($x = y = 0$). It should be noted, however, that the function $\varphi(x,y)$ is sharply peaked at the centre of the atom and decreases rapidly with distance away from the centre. At

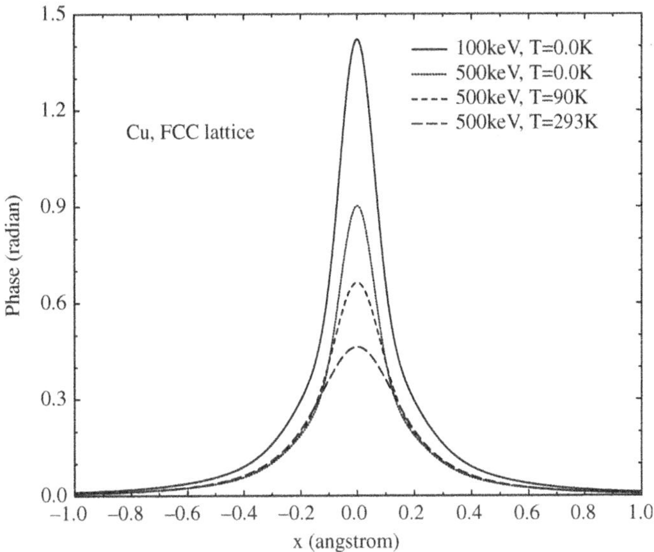

FIG. 2.3. One-dimensional phase plots calculated for a Cu atom for various primary beam energies and temperatures. The temperature dependence is taken into account using Debye-Waller factors, and values of these factors at various temperatures are taken to be the same as those calculated for crystalline copper.

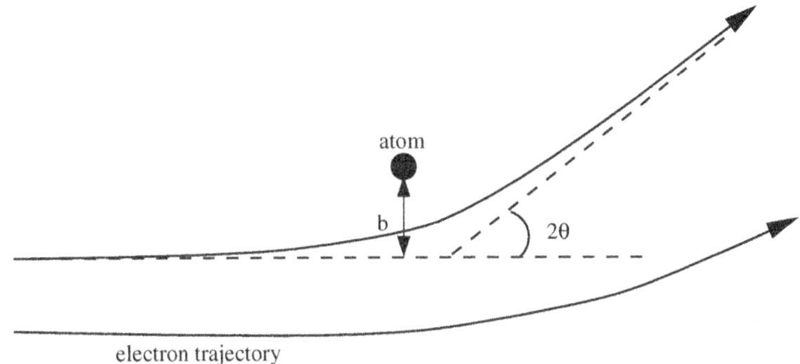

FIG. 2.4. Diagram illustrating the classical process of scattering of an electron by an atom. Here b is the impact parameter of scattering and 2θ is the angle of scattering.

a fraction of an angstrom away from the centre of the atom the magnitude of $\varphi(x, y)$ becomes smaller than one, and therefore for relatively large values of x and y the condition of validity of the WPOA is satisfied.

To understand how the kinematic approximation breaks down, we consider a classical collision between an electron and an atom shown in Fig. 2.4. A collision in quantum theory is of course different from a classical collision since a quantum-mechanical collision is a wave phenomenon. However, if the atomic potential is sufficiently weak, it is possible to establish a useful analogy between the quantum-mechanical description of a scattering process and the scattering of a beam of classical particles by a central potential (Messiah, 1972). In the classical theory of scattering an electron follows a hyperbolic trajectory due to the attractive Coulomb force acting between the electron and the nucleus. The collision process is characterized by an impact parameter b, which is the perpendicular distance from the scattering centre to the straight line representing the straight trajectory of the incident particle (see Fig. 2.4). The smaller the impact parameter b the larger is the angle of scattering 2θ. This shows that large-angle scattering is mainly associated with scattering events occurring close to the nucleus. It is a corollary of this argument that the kinematic diffraction approximation is more likely to break down for large-angle scattering events occurring close to the nucleus than for small-angle scattering events occurring far from the nucleus. In other words, small-angle scattered electrons are more likely to result from single scattering events, while large-angle scattered electrons are likely to have undergone multiple scattering by the potential of the atom.

Shown in Fig. 2.5 are kinematic and dynamical amplitudes of scattering for an oxygen atom at 100 keV. It is seen that while the magnitude of the dynamical amplitude of scattering for the oxygen atom is almost the same as the Born scattering amplitude (Fig. 2.5a), the phase associated with the dynamical scattered beam amplitude is substantial (see Fig. 2.5b and note that in the kinematic case the phase vanishes). The phase of the amplitude is given by the inverse tangent function, which is far more sensitive to small changes in the imaginary part of the amplitude of scattering while the absolute value of the amplitude is not (Fig. 2.5a). This shows that if the quantity observed in a diffraction experiment is not sensitive to the phase, as in conventional electron diffraction experiments where only the diffracted beam intensities are recorded and the specimen contains only light atoms, the results can be interpreted in terms of the Born amplitude. On the other hand, if the observed quantities are sensitive to the phase of the amplitude of scattering, as in electron holography experiments (Tonomura, 1987, Lichte, 1992, Cowley, 1993b), experimental results cannot be interpreted in terms of the Born amplitude. In this case a full dynamical treatment is required to predict the correct phase of diffracted beams.

An alternative treatment of atomic scattering uses the so-called partial wave (or phase-shift) analysis (Mott and Massey, 1965), where the scattering factor f is expressed in the form of an infinite sum of partial waves

$$f(2\theta) = \frac{1}{2ik} \sum_{l=0}^{\infty} (2l + 1) \exp(i\delta_l - 1) P_l(\cos 2\theta) = |f| \exp\{i\eta(2\theta)\}, \qquad (2.23)$$

where δ_l is the partial wave phase shift and P_l is a Legendre polynomial. Ibers and Hoerni (Ibers and Hoerni, 1954) used this approach to calculate $|f|$ and the phase η for atoms of various atomic numbers. They showed that the phase η of the atomic scattering can reach appreciable values.

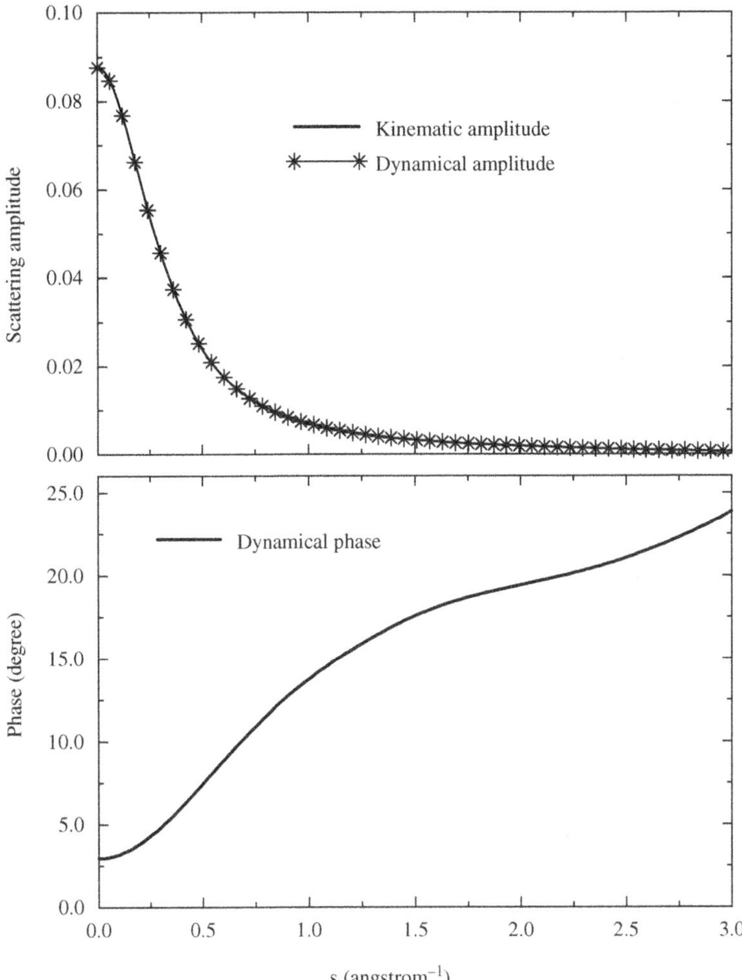

FIG. 2.5. (a) Absolute values of kinematic and dynamical amplitudes of scattering
for an oxygen atom at the primary beam energy of 100 keV and (b) phase of the
dynamical amplitude of scattering as a function of the angle of scattering.

2.4 Amplitude of scattering by an assemblage of atoms

In this section we consider scattering by an assemblage of atoms. We start by
considering how electrons are scattered by a string consisting of n atoms shown
schematically in Fig. 2.6. Assuming that the string of atoms is not very long (say
less than 20 Å), we use the phase object approximation discussed in the preceding section. Within the range of validity of the POA, the effective potential of
scattering associated with the string is given by $nV(\mathbf{r})$. To put things in perspective, in Fig. 2.7 we show the one-dimensional projected potentials calculated for

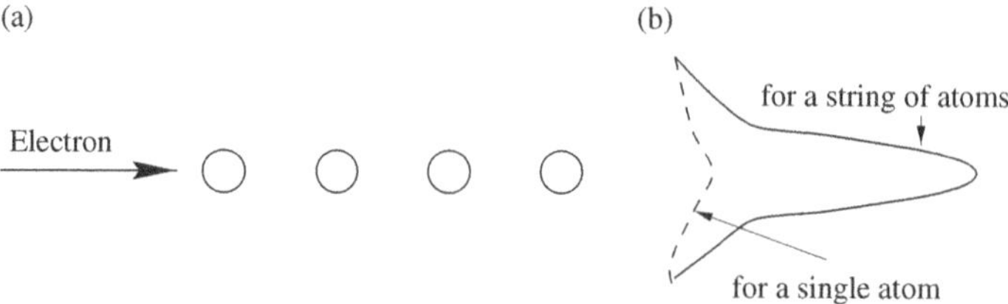

FIG. 2.6. Schematic diagrams showing (a) scattering of a high-energy electron by a string of atoms and (b) potentials of a single atom and of a string of atoms.

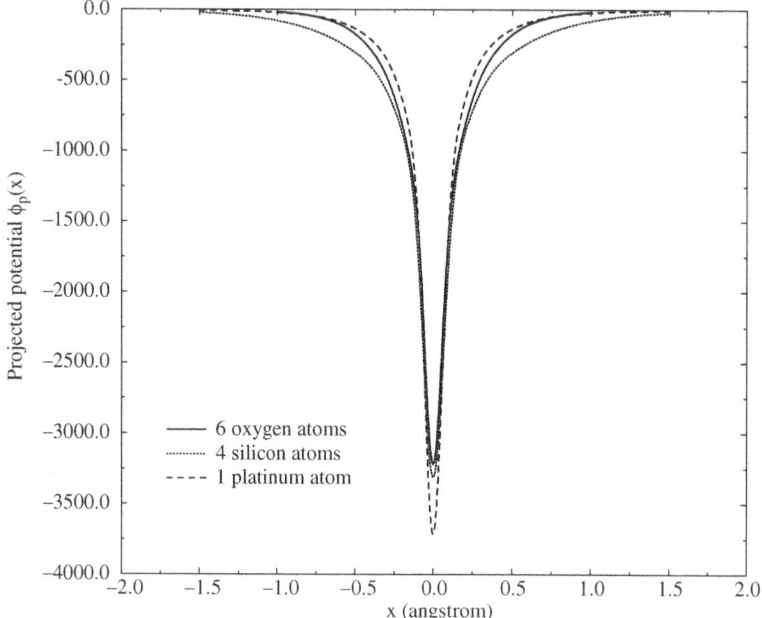

FIG. 2.7. One-dimensional plots of the projected potential $\phi_p(x)$ for a single atom of platinum, four silicon atoms and six oxygen atoms. In the figure $\phi_p(x)$ is given in eV·Å units.

a single platinum atom, for four silicon atoms, and for six oxygen atoms. The figure shows that the strength of the effective potential is approximately the same in all the three cases. Since the thickness of samples used in transmission electron diffraction (TED) experiments typically exceeds 50 Å or 20 atomic layers, this figure shows that even in the case of light atoms it is not appropriate to use the simple kinematic approach in order to evaluate the amplitude of scattering.

We now consider the case where the potentials of atoms overlap as shown in Fig. 2.8. While the potential of each atom may be strong, this figure shows that it is possible to treat the resulting projected potential as a superposition of the mean inner potential V_0 and the remaining spatially varying component $V - V_0$. If

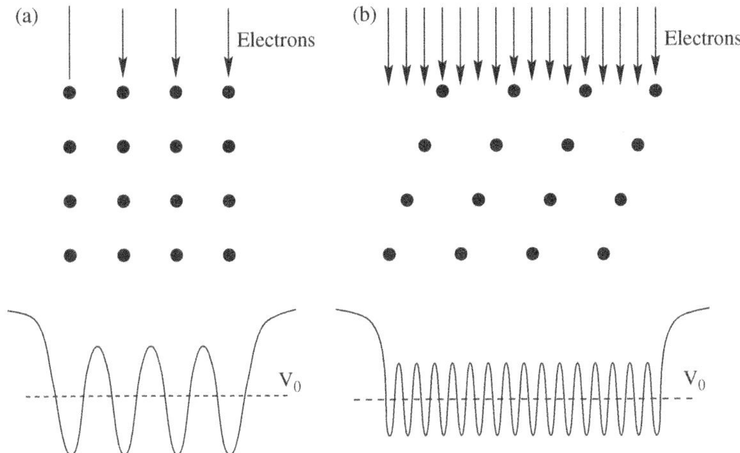

FIG. 2.8. Schematic diagrams showing the projected potentials for (a) a zone axis and (b) an inclined electron beam illumination.

this varying component is sufficiently small, we can apply the kinematic approach to this term only. The case then falls within the scope of the quasi-kinematic approximation discussed in section 2.2.2. Using the POA we find

$$\psi \approx \exp\left\{-i\sigma \int V(z)dz\right\}$$

$$= \exp(-i\sigma V_0 t)\left\{1 - i\sigma \int [V(z) - V_0]dz\right\}, \tag{2.24}$$

where V_0 may be very strong, but $V - V_0$ is still sufficiently weak to be described by the quasi-kinematic approximation.

Two important cases related to this situation are shown schematically in Figs 2.8b and 2.9. The first one (Fig. 2.8b) corresponds to the case of inclined incidence of electrons on a perfect crystal, and the second one corresponds to scattering of electrons by an amorphous sample (Fig. 2.9). In both cases the use of the quasi-kinematic approximation is justified, and the diffracted beam intensities may be related directly to the Fourier components of the potential $V(\mathbf{r})$ (note that V_0 only affects the phase of the transmitted beam).

It should be noted that while for a very thin film and the transmission geometry of diffraction there is no substantial difference in the diffraction patterns calculated using the kinematic and the quasi-kinematic approximations, the values of amplitudes of scattering calculated using the two approximations deviate appreciably in the case where the thickness of the film is substantial. For example, while the amplitude of scattering by an atom calculated using the kinematic approximation is independent of the coordinate of the atom in the direction of the incident beam, the amplitude of scattering calculated using the quasi-kinematic approximation (illustrated in Fig. 2.10) does vary as a function of this

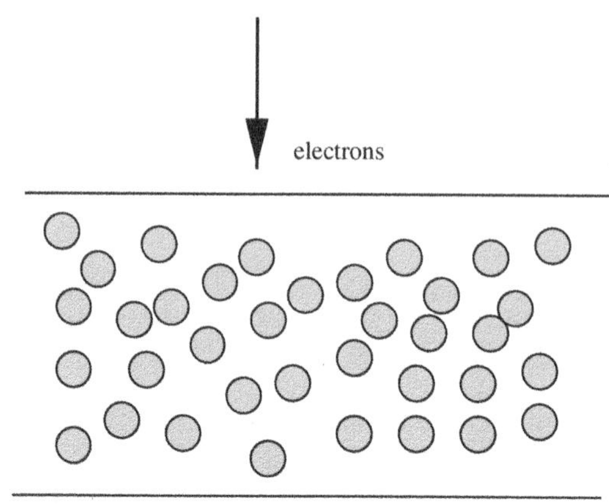

FIG. 2.9. Schematic diagrams of an amorphous sample and the corresponding projected potential.

coordinate. In principle this fact can be used for determining the position of an atom in the direction of the incident beam (subject to the somewhat limited resolution of this technique).

2.5 Diffraction by single crystals

Although the intensities of diffracted beams in the case of scattering of electrons by a crystalline solid may differ from those predicted by the kinematic theory, the geometry of diffraction patterns predicted by the kinematic theory is identical to that found using the dynamical theory. This is because the general principle of conservation of momentum and energy applies equally both to the kinematic and to the dynamical cases of electron diffraction. As a result the direction of propagation of electrons diffracted by a crystal turns out to be the same in both cases (we shall come back to this point again in Chapter 5). In this section we consider the geometry of diffraction patterns using the kinematic theory. Since results described in this section remain valid in the general case of dynamical electron diffraction, we introduce some concepts like the Ewald sphere and the reciprocal lattice that we shall later rely upon in the treatment of dynamical effects.

We first consider electron diffraction by a crystalline sample. A crystal is an object formed by repeated three-dimensional translations of a unit that may

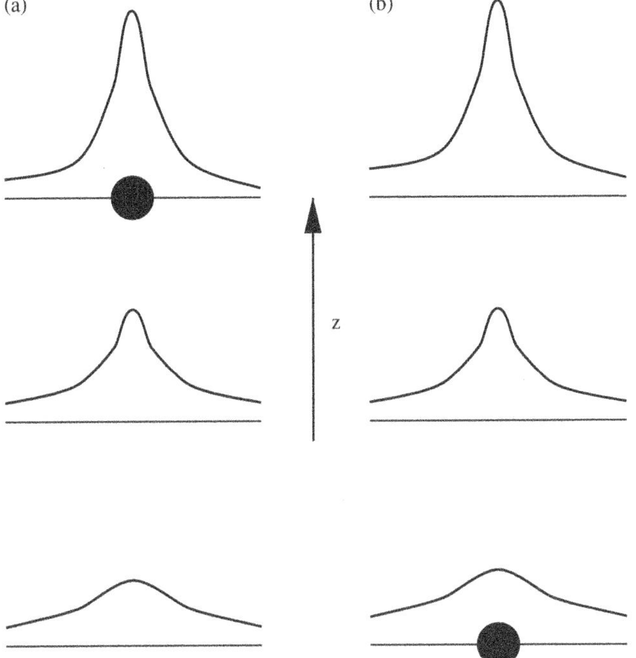

FIG. 2.10. Sketch illustrating the case where the incident high-energy electrons illuminate an atom situated near the (a) exit and (b) entrance faces of the sample.

consist of more than one atom. The operation of repetitive translation of this unit is defined by a lattice function

$$L(\mathbf{r}) = \sum_{n_1} \sum_{n_2} \sum_{n_3} \delta(\mathbf{r} - n_1\mathbf{a} - n_2\mathbf{b} - n_3\mathbf{c}), \qquad (2.25)$$

where $\mathbf{a}, \mathbf{b}$, and $\mathbf{c}$ are any three non-coplanar vectors called the primitive lattice vectors, and n_1, n_2, and n_3 are three independent integers. The three-dimensional space spanned by the vectors $\mathbf{a}$, $\mathbf{b}$, and $\mathbf{c}$ is normally referred to as the real space, and the parallelepiped defined by the three vectors is called the primitive unit cell.

Let us use $\phi(\mathbf{r})$ to describe the variation of the potential in the unit cell corresponding to $n_1 = n_2 = n_3 = 0$. The potential of the entire crystal is then given by the sum of contributions of all the cells translated in various directions

$$V(\mathbf{r}) = \sum_{n_1} \sum_{n_2} \sum_{n_3} \phi(\mathbf{r} - n_1\mathbf{a} - n_2\mathbf{b} - n_3\mathbf{c}) = \phi(\mathbf{r}) * L(\mathbf{r}), \qquad (2.26)$$

where the symbol $*$ represents the operation of convolution.

As discussed in section 2.2, in the kinematic approximation the amplitude of the scattered wave is proportional to the Fourier transform of the potential, namely

$$f(\mathbf{q}) = \mathcal{F}\{\phi(\mathbf{r}) * L(\mathbf{r})\} = \mathcal{F}\{\phi(\mathbf{r})\} \times \mathcal{F}\{L(\mathbf{r})\}$$

$$= \frac{1}{\Omega_0} F(\mathbf{q}) \sum_h \sum_k \sum_l \delta(\mathbf{q} - h\mathbf{a}^* - k\mathbf{b}^* - l\mathbf{c}^*), \qquad (2.27)$$

where Ω_0 is the volume of a unit cell, $F(\mathbf{q})$ is the amplitude of the wave scattered by a unit cell, $\mathcal{F}$ denotes the Fourier transform, and h, k, l are integers.

Vectors $\mathbf{a}^*$, $\mathbf{b}^*$ and $\mathbf{c}^*$ are related to the real space vectors $\mathbf{a}$, $\mathbf{b}$, and $\mathbf{c}$ by

$$\mathbf{a}^* = 2\pi \frac{\mathbf{b} \times \mathbf{c}}{\mathbf{a} \cdot (\mathbf{b} \times \mathbf{c})}, \quad \mathbf{b}^* = 2\pi \frac{\mathbf{c} \times \mathbf{a}}{\mathbf{a} \cdot (\mathbf{b} \times \mathbf{c})}, \quad \mathbf{c}^* = 2\pi \frac{\mathbf{a} \times \mathbf{b}}{\mathbf{a} \cdot (\mathbf{b} \times \mathbf{c})}. \qquad (2.28)$$

The space spanned by vectors $\mathbf{a}^*, \mathbf{b}^*$ and $\mathbf{c}^*$ is called reciprocal space, and the three vectors themselves are called primitive reciprocal lattice vectors.

The amplitude of scattering $F(\mathbf{q})$ by a unit cell is called the structure factor. By writing the potential of the j-th neutral atom as $\varphi_j(\mathbf{r})$ and its position vector as $\mathbf{r}_j$, and neglecting the effect of redistribution of valence electrons due to chemical bonding effects, we can write the potential of a unit cell in the form

$$\phi(\mathbf{r}) = \sum_j \varphi_j(\mathbf{r} - \mathbf{r}_j), \qquad (2.29)$$

where the summation is performed over atoms in the unit cell. The Fourier transform of this potential gives the structure factor

$$F(\mathbf{q}) = \sum_j f_j^{(e)}(s) \exp(-i\mathbf{q} \cdot \mathbf{r}_j), \qquad (2.30)$$

where $q = 4\pi s$. The crystal structure factors can be measured experimentally with high accuracy. By plotting a map showing the difference between a measured distribution of electrostatic potential and the distribution found using the superposition of potentials of isolated atoms, it is possible to find the real-space distribution of valence electron density and to identify the directions of chemical bonds (Smart and Humphreys, 1980, Spence, 1993).

In real space any three lattice points define a crystallographic or a Miller plane. Suppose that the plane intersects the three crystallographic axes at three points $\mathbf{r}_1 = \mathbf{a}/h$, $\mathbf{r}_2 = \mathbf{b}/k$, and $\mathbf{r}_3 = \mathbf{c}/l$. Let $\boldsymbol{\gamma}$ be the unit vector normal to the Miller plane (hkl) and $d = 2\pi/|\mathbf{g}|$ be the inter–planar distance. The equation defining the Miller plane is given by

$$\boldsymbol{\gamma} \cdot \mathbf{r} = d. \qquad (2.31)$$

Since $\mathbf{r}_1, \mathbf{r}_2$ and $\mathbf{r}_3$ are the three points lying in this plane, for these points we find

$$\gamma \cdot \frac{\mathbf{a}}{h} = \gamma \cdot \frac{\mathbf{b}}{k} = \gamma \cdot \frac{\mathbf{c}}{l} = d.$$

Using these and the following relations (which may be proved by successively forming the dot product of γ with $\mathbf{a}$, $\mathbf{b}$, and $\mathbf{c}$)

$$2\pi\gamma = (\gamma \cdot \mathbf{a})\mathbf{a}^* + (\gamma \cdot \mathbf{b})\mathbf{b}^* + (\gamma \cdot \mathbf{c})\mathbf{c}^*,$$

it can be readily shown that

$$\mathbf{g} = h\mathbf{a}^* + k\mathbf{b}^* + l\mathbf{c}^* = \frac{2\pi\gamma}{d}.$$

The equation of the Miller plane (hkl) then becomes

$$\frac{\mathbf{g}}{|\mathbf{g}|} \cdot (x\mathbf{a} + y\mathbf{b} + z\mathbf{c}) = d = \frac{2\pi}{|\mathbf{g}|},$$

or

$$hx + ky + lz = 1. \tag{2.32}$$

This equation applies for any h, k, l. Assuming that for a particular set of h', k', l' the largest common factor among h', k', and l' is m, i.e. $(h', k', l') = m(h, k, l)$, the equation of the crystallographic plane $(h'k'l')$ is given by

$$h'x + k'y + l'z = 1, \quad \text{or} \quad hx + ky + lz = m^{-1}. \tag{2.33}$$

By varying m from $-\infty$ to ∞ we construct a family of planes parallel to the plane defined by eqn (2.32). These planes are equally spaced, and each plane is identical to any other plane within the family through a lattice translation. The three indices h, k, and l define the family and are called the Miller indices, and the family of planes is represented by the three indices included within brackets (hkl). A simple interpretation of the indices is that the plane of the family, which is the closest to the origin (i.e. corresponding to $m = 1$), makes intercepts on the crystal axes of lengths a/h, b/k, and c/l. In reciprocal space the vector $\mathbf{g}$ is normal to the family of lattice planes (hkl), and the spacing between the set of $m(hkl)$ planes is equal to $d = 2\pi|\mathbf{g}|^{-1} \times m^{-1}$.

The expression for the diffracted beam amplitude (2.27) can be interpreted in the following way. The diffracted beam has a non-zero amplitude only when $\mathbf{q}$ coincides with one of the vectors

$$\mathbf{g} = h\mathbf{a}^* + k\mathbf{b}^* + l\mathbf{c}^*. \tag{2.34}$$

This is known as the Laue condition. Alternatively, the diffraction condition can be obtained following the original approach by W.L. Bragg (Bragg et al., 1975). Let θ be the angle between the primary beam and the family of lattice planes (hkl) (see Fig. 2.11a). The difference in the path length between waves diffracted by the two successive planes is equal to $2d\sin\theta$, where d is the spacing between

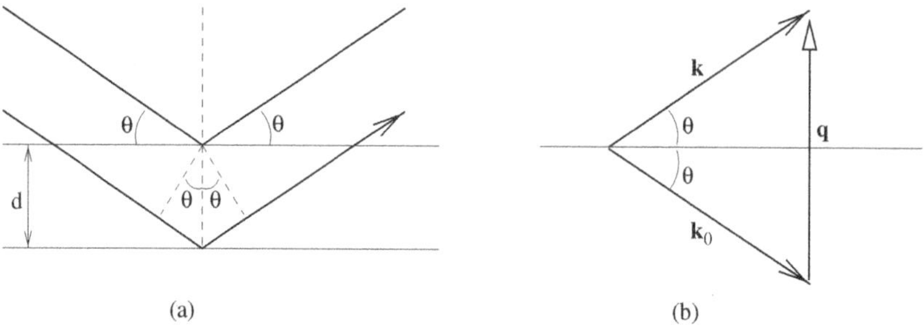

FIG. 2.11. Schematic diagrams showing (a) the reflection of an electron wave train by successive lattice planes and (b) momentum transfer involved in the reflection process.

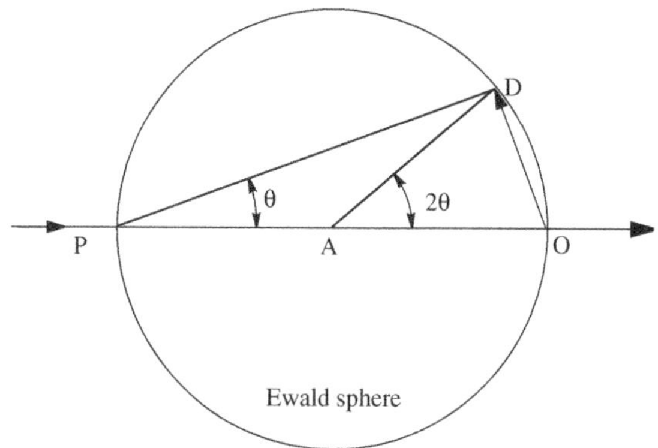

FIG. 2.12. Ewald sphere construction for diffraction of radiation characterized by a relatively long wavelength and by a relatively small radius of the Ewald sphere.

the lattice planes. Constructive interference occurs in the case where the difference in the path length is a multiple of the electron wavelength, i.e.

$$2d \sin \theta = n\lambda. \tag{2.35}$$

This is the well-known Bragg law, and the angle θ which satisfies this equation is called the Bragg angle. Figure 2.11b shows that the magnitude of the wave vector change $\mathbf{q}$ occurring in the diffraction process is given by $q = 2k \sin \theta$. By using the Bragg law (2.35) it is easy to show that

$$\mathbf{q} = n\mathbf{g}, \tag{2.36}$$

and this is the Laue condition satisfied for the reciprocal lattice point $n\mathbf{g}$. We see that the Bragg law (2.35) is equivalent to the Laue condition.

The geometry of electron diffraction patterns can be predicted using the Ewald sphere construction shown in Fig. 2.12. We draw a sphere of radius $2\pi/\lambda$

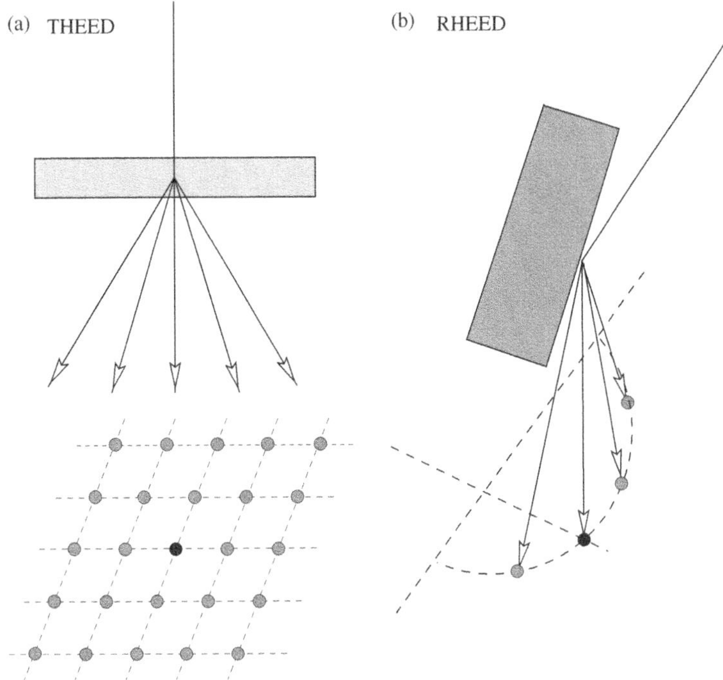

FIG. 2.13. Diffraction geometry for (a) TED and (b) RHEED. The dashed lines are only a visual aid.

centred at a point A in such a way that the incident beam direction is along the diameter PO. We place the origin of the reciprocal lattice at O and assume that the reciprocal lattice point $n\mathbf{g}$ lies on the sphere at D. We have that OD= $ng = 2k\sin\theta = 4\pi\sin\theta/\lambda$. Since $g = 2\pi/d$, we arrive at the Bragg law (2.35). The line PD in Fig. 2.12 may be interpreted as the trace of the reflecting plane (hkl), and the sphere itself is called the Ewald sphere. The necessary and sufficient criterion for the Bragg or Laue condition to be satisfied is that the relevant reciprocal lattice point lies on the Ewald sphere. Since the vector AD gives the direction of the diffracted beam, we may interpret the position of the centre of the sphere A as the point where the crystal is located. Figure 2.12 was drawn for the case of radiation of long wavelength, such as X-rays, so that the radius of the Ewald sphere is of the order of 1 Å^{-1}. For high-energy electrons the geometry of the Ewald sphere construction is very different. For example, for 100 keV primary beam energy, the radius of the Ewald sphere is about 170 Å^{-1}, and this is more than an order of magnitude greater than the length of the largest reciprocal lattice vector $\mathbf{g}$ ($|\mathbf{g}| \sim 12$ Å^{-1}) typically involved in a diffraction process.

In the vacuum the directions of diffracted beams are determined by the energy and momentum conservation laws. In the TED geometry, since the boundary conditions require that the components of the wave vectors parallel to the surface are conserved, the diffraction pattern has the form of an array of sharp spots as shown in Fig. 2.13a. In the RHEED geometry of diffraction the determination of the direction of diffracted beams also requires using the condition of energy conservation, or in other words, the fact that wave vectors of diffracted beams must lie on the Ewald sphere as shown in Fig. 2.13b.

2.6   Diffraction by a gas, an amorphous solid, and a liquid

In this section we consider kinematic diffraction by non-crystalline materials, including gases, liquids, and amorphous solids. A gas is a low-density compound that consists of almost non-interacting atoms or molecules. In a liquid the density is typically several orders of magnitude higher than the density of a gas and attraction between neighboring particles keeps them in close contact. The amplitude of the thermal motion of particles in a liquid is large, and thermal motion gives rise to rapid fluctuations of interatomic distances and rapid changes of the local environment of every particle.

A non-crystalline aperiodic atomic structure can also exist at zero absolute temperature where the thermal motion is absent. An amorphous solid can be thought of as a frozen liquid where the mobility of atoms is not high enough to transform it into a crystalline state. The reason why an amorphous solid retains its stability is associated with the fact that the process of crystallization requires a large number of atoms to move coherently to form a regular lattice, and such coherent movements do not occur at relatively low temperatures.

Using eqn (2.30) we find that the kinematic amplitude of scattering of electrons by a system consisting of N atoms is given by

$$F(\mathbf{q}) = \sum_{i=1}^{N} f_i \exp(-i\mathbf{q} \cdot \mathbf{r}_i), \qquad (2.37)$$

where $\mathbf{q} = \mathbf{k} - \mathbf{k}_0$ and for simplicity we have dropped the superscript (e) of the electron atomic scattering factor $f^{(e)}$. The diffracted beam intensity is equal to

$$I(\mathbf{q}) = F(\mathbf{q})F^*(\mathbf{q}) = \sum_{i=1}^{N}\sum_{j=1}^{N} f_i^* f_j \exp\{i\mathbf{q} \cdot (\mathbf{r}_i - \mathbf{r}_j)\}. \qquad (2.38)$$

Gases, liquids (excluding the so-called liquid crystals), and amorphous solids are statistically isotropic. They can be described by using concepts of interatomic distances r and probability distributions. We start by considering a pair of atoms as a rigid body, and allow it to take with equal probability all orientations in space. If the vector $\mathbf{r}_{ij} = \mathbf{r}_i - \mathbf{r}_j$ takes all orientations, its terminal point lies anywhere on the surface of a sphere of radius r_{ij}. The contribution to the amplitude of scattering from each pair of atoms is then given by

$$\langle \exp(i\mathbf{q} \cdot \mathbf{r}_{ij}) \rangle = \frac{1}{4\pi r_{ij}^2} \int_0^{2\pi} d\phi \int_0^{\pi} \exp(iq r_{ij} \cos\theta) r_{ij}^2 \sin\theta d\theta = \frac{\sin(q r_{ij})}{q r_{ij}}. \qquad (2.39)$$

Adding up all the terms in eqn (2.38), we arrive at

$$I(q) = \sum_{i=1}^{N} \sum_{j=1}^{N} f_i^* f_j \frac{\sin(qr_{ij})}{qr_{ij}}. \tag{2.40}$$

This is the Debye scattering equation, which forms the basis for the radial distribution function (RDF) method first introduced by Pauling and Brockway in 1935 (Pauling and Brockway, 1935). The fact that electrons can be multiply scattered by each atom may be partly taken into account for gases by using the dynamical scattering amplitude that includes a phase shift η, so that $f_i(s) = |f_i(s)|\exp\{i\eta_i(s)\}$ (see for example the end of section 2.3). In this case the product $f_i f_j$ is replaced in eqn (2.40) by $|f_i f_j|\cos\{\eta_i(s) - \eta_j(s)\}$. Neglecting the phase shift η leads to erroneous results for the structure of molecules containing heavy atoms. For example, the molecule of uranium hexafluoride UF_6 was initially thought of as being unsymmetrical with the U–F bonds differing in length and pointing in the opposite directions and therefore breaking the expected octahedral symmetry (Schomaker and Glauber, 1952, Glauber and Schomaker, 1953). This confusion was associated with the fact that in a pair of light and heavy atoms (as in the case of the U–F bond in UF_6) the difference between phases $\eta_i(s) - \eta_j(s)$ is approximately linear in s in the limit of relatively small s. This approximate linear dependence can at the same time be explained by using the first-order Born approximation for $f_{i,j}$ (i.e. by assuming that $\eta = 0$) and by taking the molecule to be a lower-symmetry object. For more detail the reader is referred to the paper by Glauber and Schomaker quoted above, and also to a paper by Hoerni and Ibers (Hoerni and Ibers, 1953).

In what follows we consider only a simple case where the specimen consists of atoms of only one type, where $f_i = f_j = f$. For a more general treatment, the reader is referred to the treatment by Warren (Warren, 1990). By separating in eqn (2.38) the terms with $i = j$, we find

$$I(q) = \sum_{i=1}^{N} f^2 + \sum_{i=1}^{N} f^2 \sum_{j \neq i} \exp(i\mathbf{q} \cdot \mathbf{r}_{ij}). \tag{2.41}$$

Here the first background term decreases monotonically as a function of q, while the second term oscillates about this background. Figure 2.14 shows schematically the distribution of scattered intensity calculated for a system of N atoms assuming that there is no correlation between the positions of atoms in the system (see Fig. 2.14a) and assuming the positions of atoms are correlated (see Fig. 2.14b).

We now introduce the radial distribution function $J(r) = 4\pi r^2 \rho(r)$ that gives the average number of atoms situated in the spherical layer at a distance between r and $r + dr$ from a given atom. Let ρ_a be the average density of atoms in the sample. By adding and subtracting a term involving ρ_a, we obtain two integral terms, where each term extends over the volume of the sample:

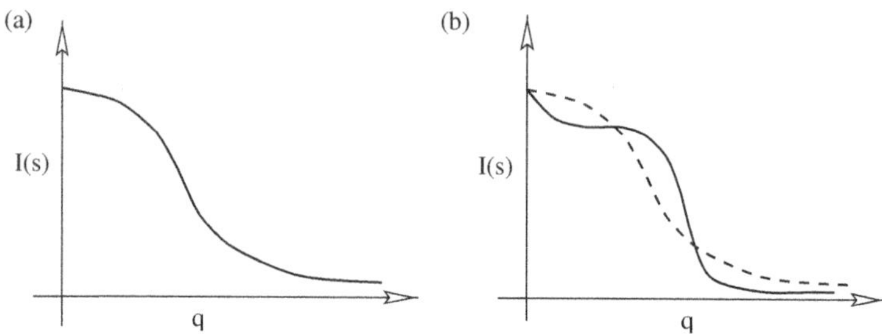

FIG. 2.14. Schematic diagrams showing distributions of intensity calculated for (a) uncorrelated and (b) correlated ensembles of atoms.

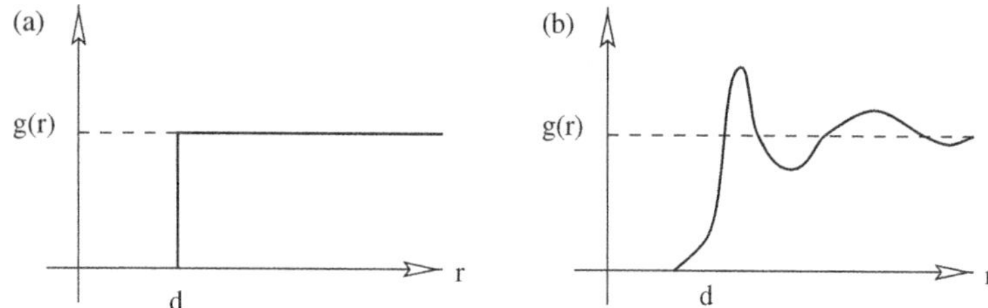

FIG. 2.15. The pair correlation function $g(r)$ for (a) a gas and (b) a non-crystalline solid. For $r \gg d$ the function $g(r)$ approaches unity.

$$I(q) = \sum_{i=1}^{N} f^2 + \sum_{i=1}^{N} f^2 \int_{V} [\rho(r_{ij}) - \rho_a] \exp(i\mathbf{q}\cdot\mathbf{r}_{ij}) dV_j + \sum_{i=1}^{N} f^2 \rho_a \int_{V} \exp(i\mathbf{q}\cdot\mathbf{r}_{ij}) dV_j.$$

$$(2.42)$$

For a fixed separation $\mathbf{r}_{ij} = \mathbf{r}$, we introduce the mean density of particles $\rho(\mathbf{r}) = \langle \rho(\mathbf{r}_{ij}) \rangle$, where the average is taken over all pairs of atoms $\mathbf{r}_{ij}$. If there is no preferred orientation in the sample, the function $\rho(\mathbf{r}) - \rho_a$ is independent of the direction of $\mathbf{r}$ and we can write $\rho(\mathbf{r}) - \rho_a = \rho(r) - \rho_a$. The term involving $[\rho(r_{ij}) - \rho_a]$ becomes

$$N f^2 \int_0^\infty \int_0^\pi [\rho(r) - \rho_a] \exp(iqr\cos\theta) 2\pi r^2 \sin\theta d\theta dr$$

$$= N f^2 \int_0^\infty 4\pi r^2 [\rho(r) - \rho_a] \frac{\sin qr}{qr} dr,$$

where the range of integration over r extends from zero to infinity. We have replaced the upper limit of integration by infinity because in the case of a non-crystalline material the function $g(r) = \rho(r)/\rho_a$ approaches unity for r greater than a few atomic distances (see Fig. 2.15), and this scale is many times smaller than the size of the sample.

The third term in eqn (2.42) depends on the shape of the sample and is usually negligible everywhere except for the case of scattering through very small angles. Assuming that the term depending on the shape of the sample is negligible, we find from eqn (2.42) that

$$I(q)/N = f^2 + f^2 \int_0^\infty 4\pi r^2 [\rho(r) - \rho_a] \frac{\sin qr}{qr} dr. \qquad (2.43)$$

The above equation can be readily inverted by using the Fourier transform:

$$\Phi(q) = 4\pi \int_0^\infty f(r) \sin(qr) dr, \qquad f(r) = \frac{1}{2\pi^2} \int_0^\infty \Phi(q) \sin(qr) dq.$$

By using these formulae we find the so-called reduced density function

$$G(r) = 4\pi r [\rho(r) - \rho_a] = \frac{2}{\pi} \int_0^\infty q \frac{I(q)/N - f^2}{f^2} \sin(qr) dq. \qquad (2.44)$$

Rearranging terms in this equation, we arrive at an important and widely used equation for the radial distribution function:

$$4\pi r^2 \rho(r) = 4\pi r^2 \rho_a + \frac{2r}{\pi} \int_0^\infty q \frac{I(q)/N - f^2}{f^2} \sin(qr) dq. \qquad (2.45)$$

Figure 2.16a shows the reduced density function $G(r)$ of tetrahedral amorphous carbon, Fig. 2.16b shows the corresponding radial distribution function $4\pi r^2 \rho(r)$, and Fig. 2.16c shows the pair-correlation function $\rho(r)/\rho_a$. By analysing these functions it is possible to find quantities such as the nearest neighbour distance and the coordination numbers. For further details see Cockayne and Mackenzie (Cockayne and Mackenzie, 1988).

2.7   Diffraction by polycrystals and textures

An ideal polycrystalline material is an assemblage of a very large number of randomly oriented crystallites as shown in Fig. 2.17a. Although the properties of individual crystallites are anisotropic, the polycrystalline material is macroscopically isotropic. The effect of randomness in the orientation of crystallites on the diffraction pattern is illustrated in Fig. 2.17b for two reciprocal lattice vectors $\mathbf{r}_1^*$ and $\mathbf{r}_2^*$. In the case of an ideal polycrystalline sample, we need to think of reciprocal lattice vectors as lying on spheres rather than forming a set of discrete points in reciprocal space. Figure 2.17 shows that diffraction directions are now given by the intersection of the Ewald sphere and these 'reciprocal lattice' spheres. The resulting diffraction pattern has the form of a series of concentric circles around the direction of the incident beam.

In principle, if the size of crystallites forming the polycrystalline material is very small, the radial distribution function method discussed in the preceding

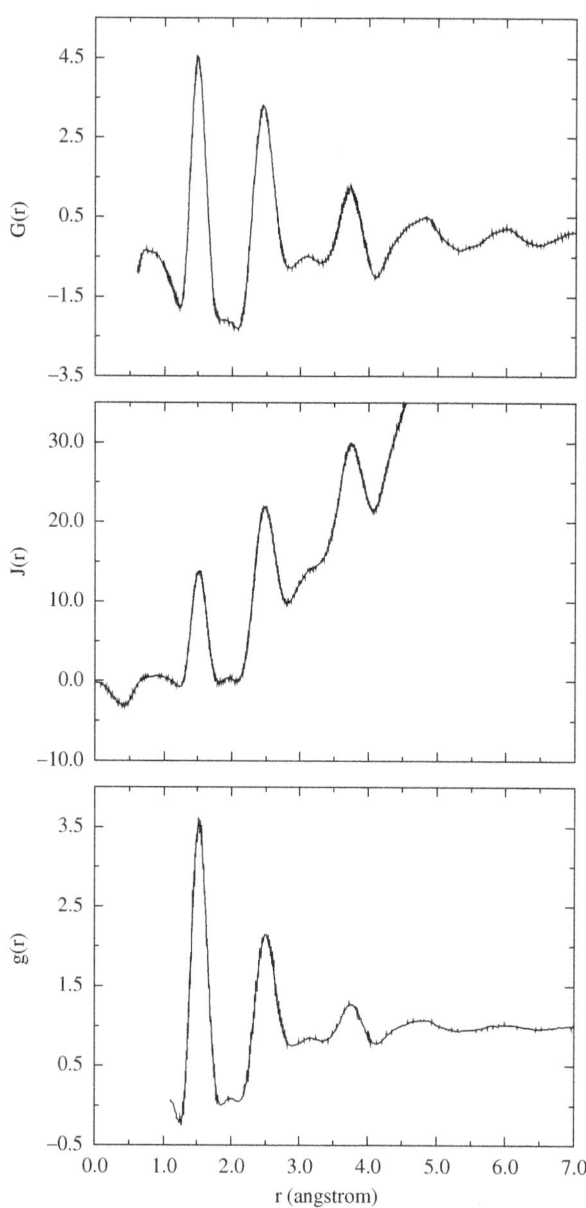

FIG. 2.16. (top) The reduced density function $G(r)$, (middle) the radial distribution function $J(r)$ and (bottom) the pair-correlation function $g(r)$ for tetrahedral amorphous carbon. Based on (Cockayne and Mackenzie, 1988).

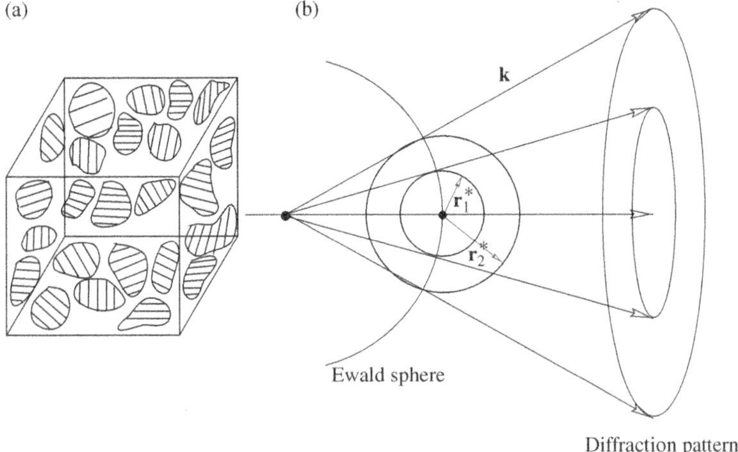

FIG. 2.17. Schematic diagram showing (a) a polycrystalline material and (b) the corresponding Ewald sphere construction and the diffraction pattern.

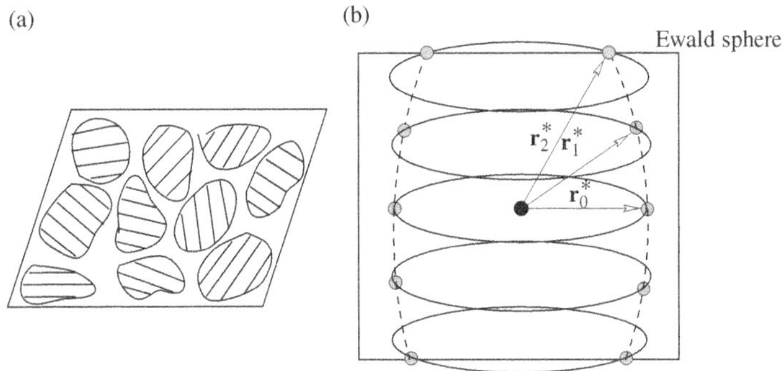

FIG. 2.18. Schematic diagrams showing (a) a texture sample and (b) the corresponding geometry of the diffraction pattern.

section may also be used for analysing the electron diffraction data. However, in practice the method works only for thin samples consisting of very small crystallites containing only light atoms. For heavy atoms and thick samples, coherent multiple scattering occurring within each individual crystallite and incoherent multiple scattering by successive crystallites strongly affect the observed results. An important case, where the material is characterized by an intermediate degree of randomness, corresponds to a textured material where crystallites have some preferential orientation shown schematically in Fig. 2.18a. In the case shown in this figure the preferential orientation is perpendicular to the plane of the image, i.e. all crystallites tend to share a common c-axis. This case is often realized when small crystals of a platelike nature are deposited on a flat surface. The reciprocal space representation of such a texture consists of a series of rings, and the intersections of the Ewald sphere with the rings form a pattern consisting of spots lying on a series of ellipses as shown in Fig. 2.18b. Note that in the case of high-energy electron diffraction the Ewald sphere is nearly flat, as opposed to the X-ray diffraction case. To an excellent approximation the Ewald sphere can be regarded as a plane parallel to the photographic film on which the diffraction pattern is recorded. The geometry of the diffraction pattern is therefore identical to the pattern formed by the intersection of the film with the three-dimensional reciprocal lattice of the sample. Dynamical diffraction effects may be taken into account to a certain extent using Blackman's method (Blackman, 1939). In some cases information about the structure of the specimen may be obtained from electron diffraction patterns by using the Fourier series method (see for example Vainshtein (Vainshtein, 1964)).

2.8 Fluctuation microscopy

Recently there has been a very interesting new development in the electron microscopy of amorphous materials. This development is associated with the work started by Treacy and Gibson (Treacy and Gibson, 1996) who realized that by reducing the size of the volume illuminated by high-energy electrons it is possible to eliminate 'self-averaging' effects and to design a robust method of determination of the medium-range order (MRO) in an amorphous material.

To make the idea of the method more transparent and to show how it is related to similar statistical approaches developed in other fields, we give a brief account of the history of fluctuation microscopy. In 1972 Rudee and Howie (Rudee and Howie, 1972) investigated the origin of lattice fringes observed in high-resolution electron microscope images of amorphous films of silicon and germanium. On the basis of largely qualitative arguments they concluded that high-resolution images show evidence of ordered regions present in the otherwise disordered structure. They called the ordered regions 'crystallites' and found that the average size of those crystallites was of the order of 14 Å. A more rigorous analysis of the problem given later by Howie, Krivanek, and Rudee (Howie et al., 1973) and by Cochran (Cochran, 1973) showed that (1) the description of the structure of an amorphous material using radial distribution functions (RDFs) is limited and different atomistic models give rise to virtually indistinguishable RDFs, (2) predictions based on the random network models do not agree with experimental electron microscope images and MRO is indeed present in the samples studied by high-resolution microscopy, and (3) application of hollow cone or microfocus illumination might prove useful for the identification of crystallites. However, putting these ideas on a quantitative basis proved difficult. Despite the fact that HREM studies of amorphous materials carried out later (Krivanek et al., 1976, Saito, 1984, Hamada and Fujita, 1986) agreed with the analysis given by Howie et al. (Howie et al., 1973), the concept of 'crystallites' remained an entity defined in qualitative rather than in quantitative terms. Developing a more robust approach to extract structural infor-

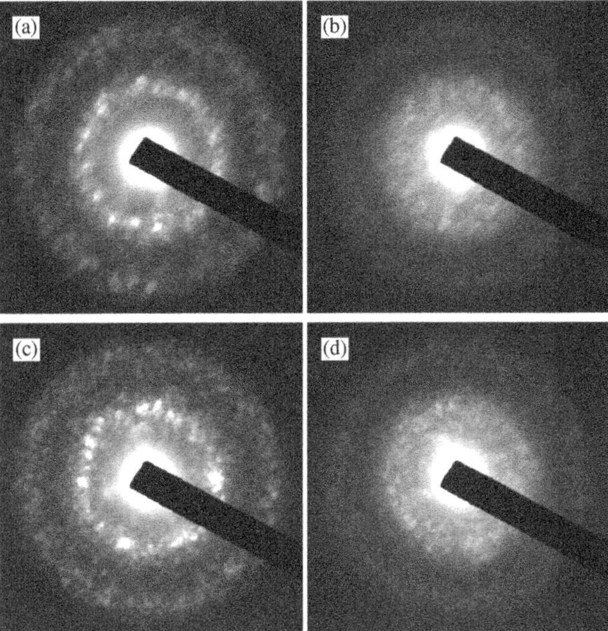

FIG. 2.19. Nanodiffraction patterns taken from different areas of an amorphous Si sample using different probes: (a) an ordered region with a 0.8 nm probe; (b) a disordered region with a 0.8 nm probe; (c) an ordered region with a 2.9 nm probe; (d) a disordered region with a 2.9 nm probe. Fluctuation microscopy uses a statistical measure of the magnitude of differences from place to place, such as those obtained by comparing (a) with (b), or (c) with (d). From (Voyles and Muller, 2002).

mation from amorphous materials remained an outstanding problem. Krivanek and Howie (Krivanek and Howie, 1975) noted somewhat pessimistically that 'it seems impossible at present to draw any reliable conclusions about the local atomic arrangement in amorphous materials by simple examination of electron micrographs'.

The novel 'fluctuation microscopy' approach developed by Treacy and Gibson and their colleagues (Treacy and Gibson, 1996, Gibson and Treacy, 1998, Voyles et al., 2000, Voyles and Muller, 2002) showed that a combination of k-space and real-space techniques makes it possible to extract more information from observations than does a simple examination of high-resolution images or conventional diffraction patterns. Fluctuation microscopy uses the fact that by illuminating a very small volume of an amorphous material (typically of the order of 10^3 atoms) it is possible to observe fluctuations of structure on the atomic scale. Indeed, it is well known that the relative significance of fluctuations diminishes as a function of the number of atoms N in the system as

$N^{-1/2}$ (Landau and Lifshitz, 1993). As a result a conventional diffraction pattern observed by illuminating a large area of an amorphous sample contains structural information averaged over a large number of atomic positions and orientations of interatomic bonds. By illuminating a very small volume of the sample containing a relatively small number of atoms it is possible to observe fluctuations of structure and investigate more subtle aspects of atomic ordering going beyond the concept of RDF.

It is worth noting at this point that an analogous experimental technique, but on a larger scale, had been developed many years before for studying the structure of deformed metals using microbeam X-ray diffraction (Kellar et al., 1950, Hirsch and Kellar, 1952), and that from the historical point of view that work led, via the use of selected area electron diffraction, to the widespread use of transmission electron microscopy for the study of materials (Whelan, 1986, Whelan, 2002).

Consider a coherent convergent beam of electrons described by a superposition of plane waves with wave vectors $\mathbf{k} = k\mathbf{n}$, namely

$$\Psi(\mathbf{r}) = \int_{\theta < \theta_0} \exp(ik\mathbf{n} \cdot \mathbf{r})d\mathbf{n}, \tag{2.46}$$

where $\mathbf{n} = (\sin\theta\cos\phi, \sin\theta\sin\phi, \cos\theta)$ is the unit vector in the direction of propagation of one of the plane waves forming the incident beam. θ and ϕ are the polar and azimuthal angles of incidence, and the point $\theta = 0$ (where $\mathbf{n}$ is parallel to the z axis) corresponds to the optical axis of the microscope. θ_0 denotes the angle of convergence of the incident beam. By taking into account that $d\mathbf{n} = \cos\theta d\theta d\phi$ and that $\int_0^{2\pi} \exp(i\xi\cos\phi)d\phi = 2\pi J_0(\xi)$, where $J_0(\xi)$ is the Bessel function of the zeroth order, we find after a simple integration that

$$\Psi(\mathbf{r}) \approx 2\pi \int_0^{\theta_0} \theta J_0(k\sqrt{x^2+y^2}\,\theta)d\theta = \frac{2\pi\theta_0}{k\sqrt{x^2+y^2}} J_1(k\theta_0\sqrt{x^2+y^2}), \tag{2.47}$$

where $J_1(x)$ is the Bessel function of the first order. Here we have assumed that $\theta \ll 1$, so that $\sin\theta \approx \theta$ and $\cos\theta \approx 1$.

Equation (2.47) shows that the diameter D of the region illuminated by the incident coherent convergent beam of electrons is inversely proportional to the angle θ_0 and is given by

$$D \approx (k\theta_0)^{-1}. \tag{2.48}$$

For $\theta_0 \sim 1$ mrad and the energy of the incident electrons of the order of 100 keV we find that $D \sim 1$ nm. When an amorphous specimen showing some degree of local crystalline order is illuminated by a convergent beam and the size of the probe D is sufficiently small, the diffraction pattern shows fluctuations that are related to the local order in the distribution of atoms in the material. Figure 2.19 gives an example of the application of fluctuation microcopy to the investigation of medium-range order in amorphous silicon.

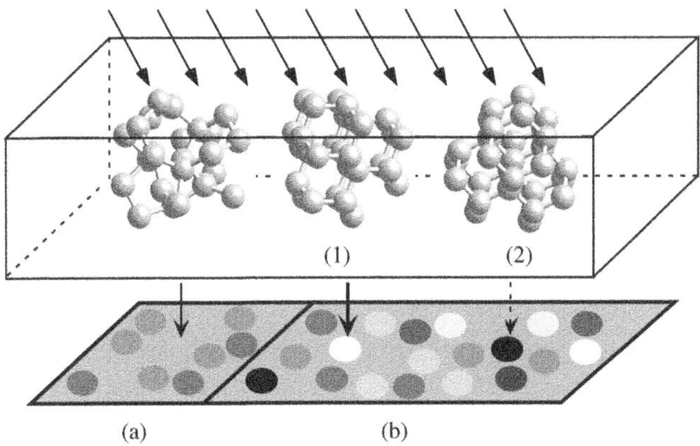

FIG. 2.20. An idealized view of fluctuation microscopy. (a) A dark-field image from a disordered specimen; (b) a dark-field image from a specimen with medium-range order, where (1) shows an ordered cluster near a Bragg diffraction condition and (2) shows an ordered cluster away from any strong Bragg condition. Typically, the ordered regions will not be as regularly crystalline as depicted here, but will be strongly deformed owing to the interaction of atoms forming the ordered regions with the surrounding amorphous structure. From (Voyles et al., 2000).

It is natural to introduce a quantitative measure V of fluctuations of intensity in the diffraction pattern by the relation (Voyles and Muller, 2002)

$$V = \frac{\langle I^2 \rangle}{(\langle I \rangle)^2} - 1, \tag{2.49}$$

where I is the observed scattered intensity and $\langle ... \rangle$ denotes averaging over positions of the electron probe on the surface of the specimen. A somewhat similar quantity has been investigated in recent years in connection with mesoscopic fluctuations of conductance observed in small samples (Washburn, 1991, Beenakker, 1997) and fluctuations of intensity of light reflected from a random medium (Rogozkin, 1997). By taking into account the fact that the intensity I is proportional to the square of the amplitude of scattering of electrons by the illuminated volume of the specimen, we see that the quantity V defined by eqn (2.49) contains terms proportional to the fourth power of the amplitude of scattering. Since the amplitude of scattering by each atom in the illuminated volume contains a phase factor that depends on the location of the atom in the material, we find that V contains information about the three- and four-body correlation functions (Gibson et al., 2000, Iwai et al., 1999).

In qualitative terms the origin of fluctuations can be understood by using the model shown in Fig. 2.20. According to this model, diffraction patterns taken from a fully disordered material exhibit relatively weak fluctuations of intensity, while the presence of a substantial degree of local order (i.e. the presence of 'crystallites' or 'regions of correlated structure' (Cowley, 2002)) in the material gives rise to strong intensity fluctuations.

Recent applications of fluctuation microscopy include the determination of local crystalline order in amorphous germanium (Gibson and Treacy, 1997), silicon dioxide (Miller and Gibson, 1998), and amorphous silicon (Voyles et al., 2001a, Voyles et al., 2001b).

2.9   Summary

In the kinematic theory of electron diffraction we assume that electrons undergo single scattering as they propagate through the specimen. This means that the transmitted electrons are scattered at different lattice sites, but once the electron is scattered it will propagate through the solid freely without any further scattering. The amplitudes of diffracted beams are proportional to the Fourier components of the electrostatic potential of the solid. In the case of scattering by a single atom the kinematic scattering amplitude equals the atomic scattering factor $f^{(e)}(s)$ of the atom, and the phase of the amplitude vanishes. In the case of a solid consisting of N atoms the amplitude is given by

$$F(\mathbf{q}) = \sum_{j=1}^{N} f_j^{(e)}(s) \exp(-i\mathbf{q} \cdot \mathbf{r}_j).$$

In the case of high-energy electron diffraction the conditions of validity of the kinematic approximation are almost never satisfied, even where we consider scattering of electrons by a single atom. The kinematic approximation almost invariably breaks down in the case of scattering by an assemblage of atoms. For light atoms such as an oxygen atom, the failure of the kinematic approximation manifests itself in the fact that the amplitude of scattering acquires a phase, which is entirely absent in the kinematic limit. In the ordinary diffraction experiments where only intensities are observed, the data may often be analyzed using the kinematic theory. In the case where observations are sensitive to the phase of the amplitude of scattering, e.g. in the case of electron holography, the kinematic approximation cannot even be applied to the treatment of scattering by a single light atom.

In the quasi-kinematic approximation the potential is represented by a sum of two terms, $V = V_0 + \Delta V$, where V_0 is a constant. In the transmission diffraction geometry the diffracted beam amplitudes are proportional to the Fourier transform of the potential (similar to the case of kinematic scattering), but the incident beam intensity no longer remains uniform everywhere inside the sample. The quasi-kinematic approximation can be used to analyse diffraction by non-crystalline solids and also diffraction by single crystals where the direction of the incident beam is not close to a crystallographic axis.

While the intensity of diffracted beams in most cases deviates significantly from that predicted by the kinematic theory, the geometry of diffraction patterns expected from a kinematic treatment of scattering is identical to that predicted using full dynamical diffraction theory. This is because both the kinematic and dynamical diffraction cases are subject to the same general principle of energy and momentum conservation, and the same boundary conditions have to be satisfied in both cases. In the quantitative analysis of diffraction data from amorphous or partially ordered materials, dynamical diffraction effects can be approximately taken into account by using the Blackman method.

3

DYNAMICAL THEORY I. GENERAL THEORY

3.1 Introduction

Elastic scattering of high-energy electrons is described by the one-particle Schrödinger equation (1.17). This equation was derived using the Born–Oppenheimer
approximation and the independent electron approximation. The Born–Oppenheimer approximation states that it is possible to separate the degrees of freedom
corresponding to the motion of electrons and ions in the material. Since the
mass of an electron is about four orders of magnitude smaller than that of an
ion, typical electron velocities are much higher than those of ions. Since ions
move much more slowly than electrons, we can assume that at any moment
of time the electron system remains in its ground state corresponding to the
instantaneous positions of ions (see for example (Ashcroft and Mermin, 1976)).
The coordinates of ions enter (1.17) as parameters determining the form of the
potential energy term $V(\mathbf{r})$ in this equation.

For a given configuration of ions the electrons do not move independently. The
motion of electrons is correlated since they repel each other electrostatically and
in addition the many-body wave function satisfies the Pauli principle. However, in
many cases the motion of electrons can be described using a self-consistent mean-
field approximation, where every electron is an effective independent particle
moving in the potential field created by all other electrons. A somewhat similar
idea forms the basis of the method developed by Kohn and Sham and based
on the Hohenberg and Kohn theorem (see for example (Parr and Yang, 1989,
Pettifor, 1995)). The latter states that the energy and the full many-body wave
function of the ground state of a system of interacting electrons are completely
determined by the electron charge density $\rho(\mathbf{r})$. Let $\psi_j(\mathbf{r})$ be the wave function
of the j-th electron. The charge density $\rho(\mathbf{r})$ is given by

$$\rho(\mathbf{r}) = \sum_{\text{occupied } j} \psi_j(\mathbf{r})\psi_j^*(\mathbf{r}). \tag{3.1}$$

In the Kohn–Sham method each electron in the solid satisfies an independent
one-particle Schrödinger equation (1.17), where the effective potential felt by an
electron is given by

$$V(\mathbf{r}) = V_H(\mathbf{r}) + V_N(\mathbf{r}) + V_{XC}(\mathbf{r}), \tag{3.2}$$

where $V_H(\mathbf{r})$ is called the Hartree potential. This potential describes the electrostatic interaction of an electron with all other electrons in the material, where the

54

electron charge density is equal to $\rho(\mathbf{r})$. $V_N(\mathbf{r})$ is the electrostatic potential acting between an electron and the nuclei. The third term in (3.2) $V_{XC}(\mathbf{r})$ is called the exchange-correlation potential. This quantity describes all the effects associated with the fact that electrons moving in the material are interacting quantum particles and their motion is correlated. Electrons with parallel spins keep away from each other in order to satisfy the Pauli exclusion principle, and electrons with anti-parallel spins keep away from each other too, in order to reduce the energy of their Coulomb repulsion. For interacting electrons the former effect leads to lowering of energy due to the exchange interaction, while the lowering of energy due to the latter effect is associated with electron-electron correlations (Pettifor, 1995). Calculating the exchange-correlation potential $V_{XC}(\mathbf{r})$ is difficult, and the exact form of this term is in fact unknown. Fortunately in the case of scattering of high-energy electrons by solids or liquids this term does not play a significant part. Because of this $V_{XC}(\mathbf{r})$ may be safely neglected in the treatment of high-energy electron diffraction as opposed to the case of low-energy electron diffraction (LEED) (Pendry, 1974).

For elastic scattering $V(\mathbf{r})$ can be calculated using the optical potential approximation (Yoshioka, 1957) and employing either the published numerical routines (Bird and King, 1990, Weickenmeier and Kohl, 1991) or parameterizations of the Doyle–Turner type (Doyle and Turner, 1968, Dudarev et al., 1995a, Peng et al., 1996b, Peng et al., 1996e). Details of how to carry out practical calculations are given in Chapter 13. Relativistic corrections are introduced by replacing the electron mass by its relativistic mass and the primary beam energy by $E(m_0/m)(1 + E/2m_0c^2)$ (Fujiwara, 1961, Howie, 1962). In this and the following two chapters we will be discussing solutions of eqn (1.17), as well as mathematical properties of these solutions, and also methods for calculating amplitudes of diffracted beams in both the transmission and reflection geometries of electron diffraction.

3.2 Role of symmetry in dynamical diffraction

In electron diffraction experiments we study the structure of specimens by measuring intensities of diffracted beams or by analysing images formed by the interference of diffracted beams. The amplitudes and intensities of diffracted beams depend on the thickness of the specimen and on the orientation of the incident beam, and the form of the function describing this dependence can be fairly complex. Nevertheless, it is possible to make broadly valid statements regarding the form of this function. These statements reflect the symmetry of the energy-filtered diffraction patterns, i.e. the patterns formed by elastically scattered electrons.

The symmetry of electron diffraction patterns and the way in which symmetry considerations affect the treatment of dynamical electron diffraction were investigated by Gjønnes and Moodie (Gjønnes and Moodie, 1965), Buxton et al. (Buxton et al., 1976), and by the group led by Tanaka (Tanaka et al., 1983a, Tanaka et al., 1983b). It has now been firmly established that by using convergent beam electron diffraction (CBED) and inspecting the symmetry of the distribution of intensity in CBED discs, and also by examining the dynamical extinction (or the Gjønnes and Moodie (G–M)) lines (Gjønnes and Moodie, 1965), it is possible to uniquely identify all the point groups characterizing the symmetry of the specimen. The practical way of identification of space groups is described in a review by Steeds (Steeds, 1983). The analysis of symmetry of dynamical diffraction patterns is largely based on the application of the reciprocity principle. In this section we describe this principle, but we defer the discussion of applications of this principle to point group and space group determination to Chapter 9.

We start from the integral form of the Schrödinger equation (2.6)

$$\psi(\mathbf{r}) = \psi_0(\mathbf{r}) + \int G(\mathbf{r}, \mathbf{r}')V(\mathbf{r}')\psi(\mathbf{r}')d\mathbf{r}', \tag{3.3}$$

where the first term represents the incident wave and the integral represents the scattered wave. The Green's function $G(\mathbf{r}, \mathbf{r}')$ represents the amplitude of the electron wave function at a point $\mathbf{r}$, due to a point source situated at $\mathbf{r}'$, and is given by eqn (2.4)

$$G(\mathbf{r}, \mathbf{r}') = -\frac{m}{2\pi\hbar^2}\frac{\exp(ik|\mathbf{r} - \mathbf{r}'|)}{|\mathbf{r} - \mathbf{r}'|}, \tag{3.4}$$

where $k = \sqrt{2mE/\hbar^2}$. For a point source of electrons situated at A the incident wave function can be written as

$$\psi_0(\mathbf{r}) = -\frac{m}{2\pi\hbar^2}\frac{\exp(ik|\mathbf{r} - \mathbf{r}_A|)}{|\mathbf{r} - \mathbf{r}_A|} = G(\mathbf{r}, \mathbf{r}_A). \tag{3.5}$$

Substituting for $\psi_0(\mathbf{r})$ from eqn (3.5) into eqn (3.3) we find

$$\psi(\mathbf{r}) = G(\mathbf{r}, \mathbf{r}_A) + \int G(\mathbf{r}, \mathbf{r}')V(\mathbf{r}')\psi(\mathbf{r}')d\mathbf{r}',$$

or in terms of the Born series

$$\psi(\mathbf{r}) = G(\mathbf{r}, \mathbf{r}_A) + \int G(\mathbf{r}, \mathbf{r}')V(\mathbf{r}')G(\mathbf{r}', \mathbf{r}_A)d\mathbf{r}'$$

$$+ \int\int G(\mathbf{r}, \mathbf{r}')V(\mathbf{r}')G(\mathbf{r}', \mathbf{r}'')V(\mathbf{r}'')G(\mathbf{r}'', \mathbf{r}_A)d\mathbf{r}'d\mathbf{r}''\dots\,.$$

We now exchange the positions of the source and the observer, i.e. transpose the position vectors of $\mathbf{r}$ and $\mathbf{r}_A$, namely

$$\psi(\mathbf{r}_A) = G(\mathbf{r}_A, \mathbf{r}) + \int G(\mathbf{r}_A, \mathbf{r}')V(\mathbf{r}')G(\mathbf{r}', \mathbf{r})d\mathbf{r}'$$

$$+ \int\int G(\mathbf{r}_A, \mathbf{r}')V(\mathbf{r}')G(\mathbf{r}', \mathbf{r}'')V(\mathbf{r}'')G(\mathbf{r}'', \mathbf{r})d\mathbf{r}'d\mathbf{r}''\dots.$$

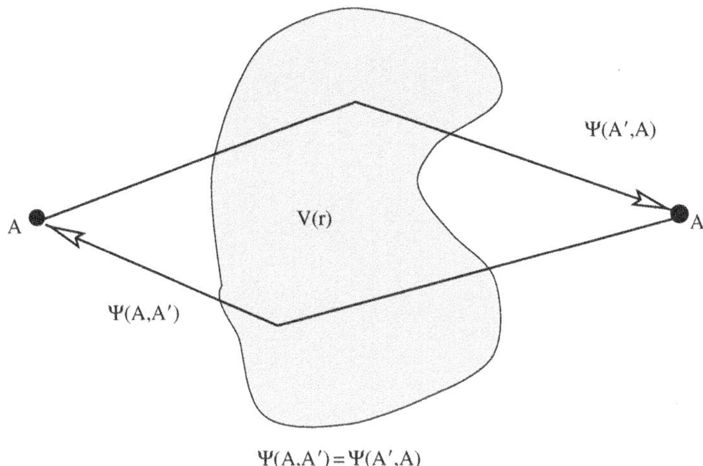

FIG. 3.1. Schematic diagram illustrating the reciprocity principle. The wave field at point A' due to a source at point A is equal to that at point A due to the same source placed at point A'.

Since the Green's function $G(\mathbf{r}, \mathbf{r}')$ given by eqn (2.4) is symmetric with respect to the transposition of $\mathbf{r}$ and $\mathbf{r}'$, we arrive at the following equality

$$\psi(\mathbf{r}_A) = G(\mathbf{r}, \mathbf{r}_A) + \int G(\mathbf{r}, \mathbf{r}')V(\mathbf{r}')G(\mathbf{r}', \mathbf{r}_A)d\mathbf{r}'$$

$$+ \int\int G(\mathbf{r}, \mathbf{r}')V(\mathbf{r}')G(\mathbf{r}', \mathbf{r}'')V(\mathbf{r}'')G(\mathbf{r}'', \mathbf{r}_A)d\mathbf{r}'d\mathbf{r}''... = \psi(\mathbf{r}),$$

i.e. the amplitude $\psi(\mathbf{r})$ of the wave field at point $\mathbf{r}$ generated by the source at point $\mathbf{r}_A$ is equal to the amplitude $\psi(\mathbf{r}_A)$ at $\mathbf{r}_A$ generated by the same source placed at point $\mathbf{r}$ (see Fig. 3.1). Our proof shows that this statement remains valid (Pogany and Turner, 1968) irrespective of the form of the potential $V(\mathbf{r})$.

The principle can be extended to include vector fields (Cowley, 1969), such as magnetic fields of electron lenses or deflection coils, provided that the reversal of the direction of the electron beam is accompanied by the reversal of the direction of magnetic fields (Gunning and Goodman, 1992).

The symmetries characterizing the distribution of intensity in a CBED disc can be derived from the reciprocity principle. To illustrate how this principle can be applied to the observed intensity distributions, we consider a crystal specimen characterized by a horizontal mirror plane shown in Fig. 3.2. Figure 3.2a shows an electron incident on the crystal from above the specimen. The point of observation is situated below the crystal. Applying the mirror symmetry operation to case (a), we obtain (b). Now the source is situated below the specimen. But according to the reciprocity principle the source and the point of observa-

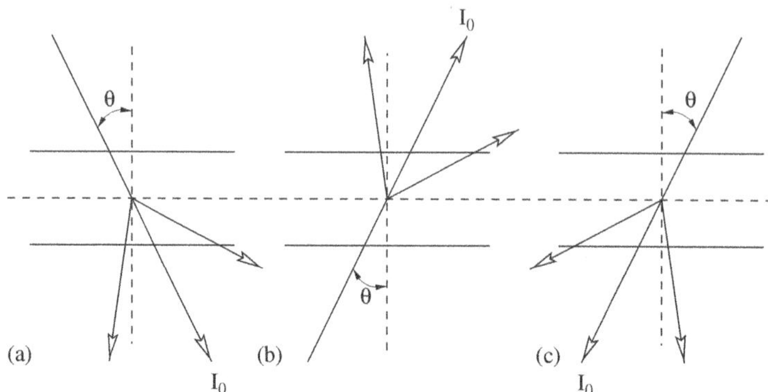

FIG. 3.2. Schematic diagram illustrating how the reciprocity principle can be applied
to reduce the symmetry of a diffraction pattern.

tion can be transposed to give (c). Comparing (a) and (c), we find the following
relationship:

$$I(\theta) = I(-\theta). \tag{3.6}$$

Since the argument leading to this relationship is independent of the beam azimuth, the presence of the horizontal mirror symmetry plane gives rise to the
two-fold symmetry of the distribution of intensity in the CBED disc corresponding to the transmitted beam. Buxton et al. (Buxton et al., 1976) and Tanaka
et al. (Tanaka et al., 1983a, Tanaka et al., 1983b) carried out a detailed analysis
of the effects of crystal symmetry on CBED patterns. By consulting the tables
published by these authors and by inspecting the symmetries of distributions of
intensity seen in CBED discs it is possible to determine the point group and the
space group of the crystal.

Information about the symmetry of the crystal that can be obtained by using
dynamical scattering of electrons is fundamentally different from the information
that can be retrieved from kinematic scattering. The symmetry of a transmission diffraction pattern reflects the symmetry of the wave function describing
electrons emerging from the exit face of the crystal. The symmetry of this wave
function reflects the entire history of propagation of the electron wave through
the crystal. This symmetry is therefore intimately related to the symmetry of
the illuminated part of the crystal, rather than to the symmetry of a single unit
cell, as it is in the case of kinematic scattering. For example, the symmetry of
diffraction patterns obtained from a wedge-shaped crystal is usually lower than
the symmetry of the same patterns obtained from a rectangular slab. This is because a wedge has, at best, a mirror plane symmetry, while additional symmetry
operations can normally be applied to a slab. Similarly a rectangular specimen
reveals lower symmetry if the surface of the crystal is not perpendicular to the
electron beam. The type of surface termination may also affect the overall crystal
symmetry seen in CBED patterns. For example, if the unit cell contains several

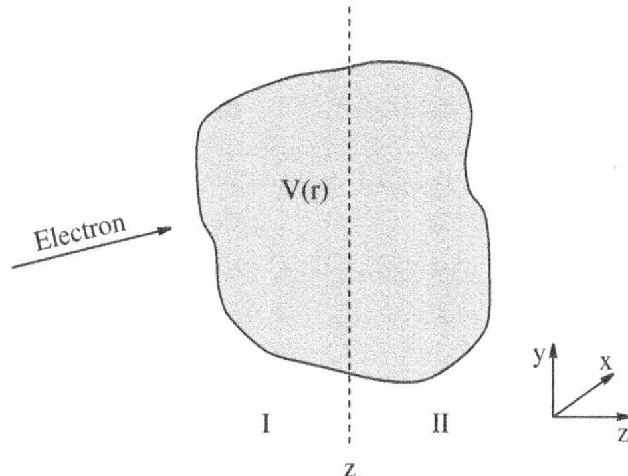

FIG. 3.3. Coordinate system used to define forward and backward scattering.

distinct layers of atoms, the termination of the crystal corresponding to a particular layer of atoms may correspond to an integral or a non-integral number of unit cells in the direction of the incident beam. The observed symmetry of the crystal therefore depends sensitively on the atomic structure of its surfaces.

In principle, inelastic scattering may also alter the symmetry of CBED patterns in comparison with what may be expected from image simulations performed using the elastic dynamical diffraction theory (Dudarev et al., 1993a). However, it is possible to show that the symmetry of diffraction patterns remains unaffected if the inelastic scattering processes involve only small energy losses (Pogany and Turner, 1968). Since the typical energy resolution of an energy filtering system for high-energy electron diffraction is of the order of 1 eV, energy-filtered CBED patterns always contain a contribution from thermal diffuse (phonon) scattering. However, since the energy losses associated with phonon scattering are small, the symmetry of CBED patterns containing both elastic and phonon contributions will be the same as in the case of purely elastic scattering.

3.3 Forward and backward scattering

In this section we use the Cartesian coordinate system shown in Fig. 3.3 and consider the wave field at a point $\mathbf{r} = (\mathbf{X}, z) = (x, y, z)$. We divide the entire space into two regions I and II by the (x, y) plane passing through point $\mathbf{r}$. The amplitude of the electron wave at point $\mathbf{r}$ is given by eqn (3.3), which we write in the form

$$\psi(\mathbf{X}, z) = \psi_0(\mathbf{X}, z) + \int_{-\infty}^{z} dz' \int d\mathbf{X}' G(\mathbf{X}, \mathbf{X}'; z, z') V(\mathbf{X}', z') \psi(\mathbf{X}', z')$$

$$+ \int_{z}^{\infty} dz' \int d\mathbf{X}' G(\mathbf{X}, \mathbf{X}'; z, z') V(\mathbf{X}', z') \psi(\mathbf{X}', z'). \tag{3.7}$$

In this equation the first term denotes the wave function of the incident electron, the second term represents the contribution of scattering at all points with $z' < z$, and the third term represents the contribution of scattering at points with $z' > z$. If the incident electrons are propagating in the positive direction of the z-axis as shown in Fig. 3.3, the contribution from region I is associated with forward scattering of electrons, while the contribution from region II where $z' > z$ results from backscattering of electrons. Because high-energy electron diffraction is dominated by small-angle scattering events, in most cases backscattering of electrons may be safely neglected. This approximation is called the forward scattering approximation, and it forms the basis for a simplified method describing the so-called transmission or Laue diffraction geometry. On the other hand, if electrons are incident on the surface of the crystal at a glancing angle, as is the case of reflection or Bragg diffraction geometry, both forward and backward scattering contribute to the wave function at point $\mathbf{r}$, and all three terms entering eqn (3.7) need to be included to describe the diffraction process.

3.4 The multislice method

In this section we derive the multislice method (Cowley and Moodie, 1957) and (Goodman and Moodie, 1974) for solving the Schrödinger wave equation (1.17). This method was originally developed in 1957 by Cowley and Moodie, who showed that the method makes it possible to find analytical expressions for the amplitudes of diffracted beams, to develop useful approximate solutions (Cowley and Moodie, 1962), and to discuss general principles of dynamical electron diffraction (Gjønnes and Moodie, 1965). Right from the very beginning this method does not assume that the specimen is crystalline or that it is characterized by any specific symmetry. The method can therefore be used to perform dynamical diffraction calculations for arbitrary aperiodic structures. The multislice approach makes extensive use of convolution operations, and requires that fast algorithms like the fast Fourier transform (FFT) are used for carrying out numerical calculations. The multislice method is very efficient numerically and is widely used for simulating HREM images. In this section we describe the principles of the method, and in the next chapter we will show how it relates to other algorithms developed for simulating electron microscope images.

Although the multislice method was originally derived following the analogy with classical optics, it was subsequently shown that in the limiting case of zero slice thickness the multislice formulation provides a way of solving the Schrödinger equation (Goodman and Moodie, 1974). For an incident plane wave $\psi_0(\mathbf{r}) = \exp(i\mathbf{k} \cdot \mathbf{r})$ equation (3.3) has the following form:

$$\psi(\mathbf{r}) = \exp(i\mathbf{k} \cdot \mathbf{r}) - \frac{m}{2\pi\hbar^2} \int \frac{\exp(ik|\mathbf{r} - \mathbf{r}'|)}{|\mathbf{r} - \mathbf{r}'|} V(\mathbf{r}')\psi(\mathbf{r}')d\mathbf{r}'. \qquad (3.8)$$

We choose the Cartesian system of coordinates in such a way that the z-axis is parallel to the direction of the incident beam so that $\mathbf{k} = (0, 0, k)$, and consider the amplitude of ψ at $\mathbf{r}$. Figure 3.3 shows that in this diffraction geometry only forward scattering contributes to the amplitude of the electron wave in the plane $z' = z$. We neglect the backscattering term in (3.7) and consider only the first two terms in that equation. Taking $\psi(\mathbf{r})$ in the form $\psi(\mathbf{r}) = \exp(i\mathbf{k} \cdot \mathbf{r})\phi(\mathbf{r})$ and using the small-angle approximation, we arrive at the following equation for the modulation function $\phi(\mathbf{r})$:

$$\phi(\mathbf{r}) = 1 - \frac{m}{2\pi\hbar^2} \int \frac{\exp[ik|\mathbf{r} - \mathbf{r}'| - i\mathbf{k} \cdot (\mathbf{r} - \mathbf{r}')]}{|\mathbf{r} - \mathbf{r}'|} V(\mathbf{r}')\phi(\mathbf{r}')d\mathbf{r}' \tag{3.9}$$

$$= 1 - i\frac{\pi}{E\lambda} \int \int_{z'=-\infty}^{z'=z} V(\mathbf{X}', z')\phi(\mathbf{X}', z')\frac{1}{i\lambda(z - z')} \exp\left(ik\frac{|\mathbf{X} - \mathbf{X}'|^2}{2(z - z')}\right) d\mathbf{X}'dz',$$

where $\lambda = 2\pi/k$ is the wavelength of the electrons, and $E = \hbar^2 k^2/(2m)$ is the primary electron beam energy. In the derivation we have used the fact that $\mathbf{k} \cdot (\mathbf{r} - \mathbf{r}') = k(z - z')$ and that $|\mathbf{r} - \mathbf{r}'| \approx (z - z') + (\mathbf{X} - \mathbf{X}')^2/2(z - z')$.

We now consider scattering by a slice of material between two planes $z = z_n$ and $z = z_{n+1}$, and find an expression relating the electron wave function at z_{n+1} and z_n. To do this we first define the interaction constant

$$\sigma = \pi/E\lambda \tag{3.10}$$

characterizing the strength of interaction of the high-energy electron with atoms in the material, and the Fresnel propagation function

$$p(\mathbf{X}, z) = \frac{1}{iz\lambda} \exp\left(ik\frac{\mathbf{X}^2}{2z}\right) \tag{3.11}$$

describing the spreading in $\mathbf{X}$ (or the x and y directions) of an electron wave originating from a point source situated at the origin and propagating in the positive direction of the z-axis.

Ishizuka (Ishizuka, 1982) showed that provided that the potential field $V(\mathbf{X}, z)$ does not vary appreciably over the thickness of the slice, the modulation function $\phi(\mathbf{X}, z)$ can be represented in the form

$$\phi(\mathbf{X}, z_{n+1}) = \int p(\mathbf{X} - \mathbf{X}', z_{n+1} - z_n)\phi(\mathbf{X}', z_n) \exp\left\{-i\sigma \int_{z_n}^{z_{n+1}} V(\mathbf{X}', z')dz'\right\}d\mathbf{X}'.$$

Denoting the operation of convolution by $*$, we see that

$$\phi_{n+1} = \phi(\mathbf{X}, z_{n+1}) = [q_n\phi_n] * p_n, \tag{3.12}$$

where $p_n = p(\mathbf{X}, z_{n+1} - z_n)$ and where $q_n(\mathbf{X})$ defines the transmission function of the slice

$$q_n(\mathbf{X}) = \exp\left\{-i\sigma \int_{z_n}^{z_{n+1}} V(\mathbf{X}, z')dz'\right\}. \tag{3.13}$$

In equation (3.12) for clarity the lateral coordinate $\mathbf{X}$ of the convolution operation is omitted from the left-hand and right-hand terms. The repetitive application of the above procedure to the entire stack of slices describing the specimen represents the multislice algorithm developed by Cowley and Moodie (Cowley and Moodie, 1957).

The transmission function can also be derived using a physical optics argument. Following Cowley (Cowley, 1990), we define the refractive index of the solid as

$$n(\mathbf{X}, z) = \sqrt{1 + \frac{V(\mathbf{X}, z)}{E}} \approx 1 + \frac{V(\mathbf{X}, z)}{2E}. \tag{3.14}$$

The additional phase acquired by the electron wave propagating through the slice compared with the phase of the wave propagating in vacuum is given by

$$\frac{2\pi}{\lambda} \int_{z_n}^{z_{n+1}} \left[n(\mathbf{X}, z') - 1 \right] dz' = \sigma \int_{z_n}^{z_{n+1}} V(\mathbf{X}, z') dz'. \tag{3.15}$$

The effect of the interaction of high-energy electrons with the atoms in the slice is represented by multiplying the wave function incident on the surface of the slice by (3.13).

We now divide the specimen into N thin slices parallel to the (x, y) plane. The effect of interaction of the electron with atoms in the n-th slice is described by the multiplication of the wave function incident on the slice at $z = z_n$ by $q_n(\mathbf{X})$. The wave function at the exit plane of the slice $z = z_{n+1}$ is obtained by applying the Huygen's principle, i.e. by convoluting the wave function at the preceding plane with the propagator given by (3.4). In the small-angle approximation this spherical wave propagator reduces to the expression given by (3.11), where the spherical wave front is approximated by a paraboloid. The generalization of eqn (3.12) to the case describing the propagation of the electron wave through an array of N slices has the form

$$\phi_{N+1} = [q_N [q_{N-1} [... [q_3 [q_2 [q_1 \phi_1] * p_1] * p_2] * p_3]...] * p_{N-1}] * p_N, \tag{3.16}$$

where the wave incident on the i-th slice is multiplied by the transmission function q_i, then convoluted with the Fresnel propagator p_i to give the wave incident on the $(i + 1)^{th}$ slice. The entire procedure is repeated until the wave reaches the exit face of the crystal (see Fig. 3.4). Goodman and Moodie developed a computational procedure (Goodman and Moodie, 1974) based on the multislice approach. This procedure has been described in several review articles, see e.g. a review by Self et al. (Self et al., 1983).

The complication associated with the general multislice formula (3.16) is mainly related to the calculation of Fresnel propagators $p_n(\mathbf{X})$. In a thin film where the spread of the electron wave by Fresnel diffraction can be ignored, the general multislice formula reduces to a simple expression

$$\phi_{N+1}(\mathbf{X}) = \prod_n q_n(\mathbf{X})\phi_0(\mathbf{X}) = \exp\left\{ -i\sigma \int_{z_0}^{z_{N+1}} V(\mathbf{X}, z') dz' \right\} \phi_0(\mathbf{X}), \tag{3.17}$$

which constitutes the phase object approximation (POA) similar to that given by eqn (2.16). The validity of this approximation can be assessed by applying

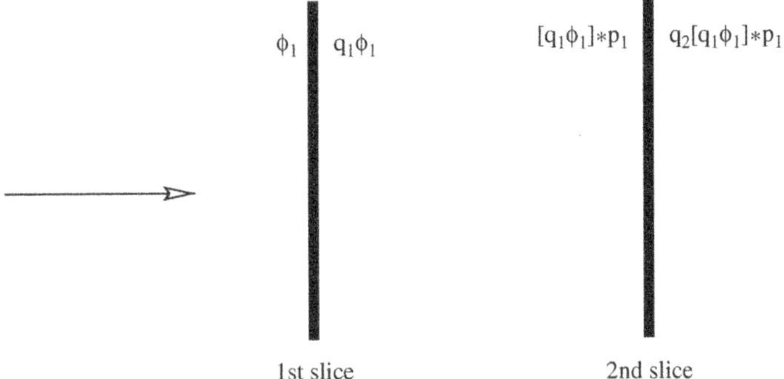

FIG. 3.4. Schematic diagram illustrating the propagation of a high-energy electron through an array of two slices.

the criterion that the thickness of the film should not exceed the Fresnel length D defined by

$$z_N \leq D = r_0^2/(2\lambda),   \tag{3.18}$$

where r_0 defines the scale of spatial resolution. Alternatively, the same condition can be derived by requiring that for a given thickness of the film the variation of the phase of the electron wave inside the region $R < r_0$ must not exceed π. It is significant that the validity of the POA depends on the required level of resolution. For example for 100 keV electrons, eqn (3.18) gives $z_N = 14$ Å for $r_0 = 1$ Å and $z_N < 4$ Å for $r_0 = 0.5$ Å.

In the multislice algorithm it is assumed that the projected potential approximation remains valid within each slice (i.e. it is assumed that the interaction potential does not vary in the direction normal to the slice). However, the potential can vary from one slice to another. In this way the variation of the potential in the direction normal to the surface influences the intensity of diffracted beams. The variation of the potential in the direction normal to the surface is described by the Fourier components corresponding to reciprocal lattice vectors with non-zero projections on the z-axis. Effects associated with the variation of the potential in the direction normal to the surface are referred to as higher order Laue zone (HOLZ) effects.

3.5   The general matrix method

The general matrix method has its origin in the Bloch wave theory of dynamical electron diffraction developed by Bethe (Bethe, 1928). Initially the method was developed to describe diffraction by three-dimensional ideal crystals. In this case solutions of the Schrödinger equation inside the crystal are Bloch waves. The method can also be applied to deformed or aperiodic materials. This can be achieved using either the periodic continuation approximation (Cowley, 1990) or the column approximation developed in Cambridge by Whelan, Hirsch and Howie (Whelan and Hirsch, 1957, Hirsch et al., 1960, Howie and Whelan, 1961).

3.5.1 *Fundamental equations*

We start from the one-particle Schrödinger equation (1.17), which we write in a slightly different form

$$[\nabla^2 + k_0^2 + U(\mathbf{r})]\psi(\mathbf{r}) = 0, \tag{3.19}$$

where

$$k_0^2 = 2mE/\hbar^2, \quad U(\mathbf{r}) = -2mV(\mathbf{r})/\hbar^2.$$

Following Bethe (Bethe, 1928) we begin by considering eigenstates of a high-energy electron in a periodic potential field. We represent the wave function in the form of a linear superposition of plane waves

$$b(\mathbf{k}, \mathbf{r}) = \sum_g C_g(\mathbf{k}) \exp[i(\mathbf{k} + \mathbf{g}) \cdot \mathbf{r}]. \tag{3.20}$$

We use a similar representation for the crystal potential

$$U(\mathbf{r}) = \sum_f U_f \exp(i\mathbf{f} \cdot \mathbf{r}), \tag{3.21}$$

where $\mathbf{g}$ and $\mathbf{f}$ are three-dimensional reciprocal lattice vectors. By substituting the expressions (3.20) and (3.21) into (3.19) and by using the relations

$$\nabla^2 b(\mathbf{k}, \mathbf{r}) = \nabla^2\{\sum_g C_g(\mathbf{k}) \exp[i(\mathbf{k}+\mathbf{g})\cdot\mathbf{r}]\} = -\sum_g (\mathbf{k}+\mathbf{g})^2 C_g(\mathbf{k}) \exp[i(\mathbf{k}+\mathbf{g})\cdot\mathbf{r}],$$

and

$$U(\mathbf{r})b(\mathbf{k}, \mathbf{r}) = \sum_g \sum_f U_f C_g(\mathbf{k}) \exp[i(\mathbf{k} + \mathbf{g} + \mathbf{f}) \cdot \mathbf{r}],$$

and by introducing the notation $\mathbf{h} = \mathbf{g} + \mathbf{f}$ and rearranging terms in the product as follows

$$U(\mathbf{r})b(\mathbf{k}, \mathbf{r}) = \sum_h \exp[i(\mathbf{k} + \mathbf{h}) \cdot \mathbf{r}] \sum_g U_{h-g} C_g(\mathbf{k})$$
$$= \sum_g \exp[i(\mathbf{k} + \mathbf{g}) \cdot \mathbf{r}][U_0 C_g(\mathbf{k}) + \sum_{h \neq g} U_{g-h} C_h(\mathbf{k})],$$

we arrive at

$$\sum_g \exp[i(\mathbf{k} + \mathbf{g}) \cdot \mathbf{r}]\{[k_0^2 + U_0 - (\mathbf{k} + \mathbf{g})^2]C_g(\mathbf{k}) + \sum_{h \neq g} U_{g-h} C_h(\mathbf{k})\} = 0.$$

Since the exponential terms associated with different reciprocal lattice vectors are independent, their coefficients can be individually equated to zero to give a set of equations (one for each reciprocal lattice vector $\mathbf{g}$)

$$[K^2 - (\mathbf{k} + \mathbf{g})^2]C_g(\mathbf{k}) + \sum_{h \neq g} U_{g-h}C_h(\mathbf{k}) = 0, \tag{3.22}$$

where

$$K^2 = k_0^2 + U_0. \tag{3.23}$$

Equations (3.22) are the basic equations of the dynamical theory of electron diffraction. For convenience we define the eigenvalue variable (sometimes referred to by the German word 'Anpassung') γ by

$$\mathbf{k} = \mathbf{K} + \gamma\mathbf{n},$$

where $\mathbf{n}$ is a unit vector pointing towards the crystal in the direction normal to the surface and the tangential component of vector $\mathbf{K}$ satisfies the condition $\mathbf{K}_t = \mathbf{k}_t$. Since

$$K^2 - (\mathbf{k} + \mathbf{g})^2 = -\gamma^2 - 2(\mathbf{K} + \mathbf{g}) \cdot \mathbf{n}\gamma + [K^2 - (\mathbf{K} + \mathbf{g})^2],$$

the set of equations (3.22) becomes

$$\{\gamma^2 + 2(\mathbf{K} + \mathbf{g}) \cdot \mathbf{n}\gamma - [K^2 - (\mathbf{K} + \mathbf{g})^2]\}C_g - \sum_{h \neq g} U_{g-h}C_h = 0. \tag{3.24}$$

Using matrix notation the same equations can be written in the form

$$\mathbf{AC} = 0, \tag{3.25}$$

where the matrix $\mathbf{A}$ has elements of the form

$$(A)_{gh} = \{\gamma^2 + 2(\mathbf{K} + \mathbf{g}) \cdot \mathbf{n}\gamma - [K^2 - (\mathbf{K} + \mathbf{g})^2]\}\delta_{gh} - U_{g-h}(1 - \delta_{gh}),$$

δ_{gh} is the Kronecker delta symbol, and $\mathbf{C}$ is a column vector of the form

$$(\mathbf{C})_h = C_h.$$

3.5.2 *The dispersion surface*

The fundamental equations of electron diffraction (3.22) relating the wave vector $\mathbf{k}$ to the total energy E (via k_0) are called the dispersion equations. These equations show that the wave vector of a Bloch wave lies on a certain surface called the dispersion surface. While all the possible values of wave vectors are determined by eqns (3.22), the choice of a particular set of wave vectors excited for a given diffraction case is determined by the boundary conditions. In general the set of wave vectors excited in a particular geometry of diffraction depends on the direction of the incident beam. Boundary conditions (see section 3.5.4) require that the components of the wave vector tangential (denoted by the subscript t) to the surface must be equal to those of the incident beam, i.e. $\mathbf{k}_t = \mathbf{k}_{0_t}$. The surface normal component of the wave vector k_z is a parameter that has to

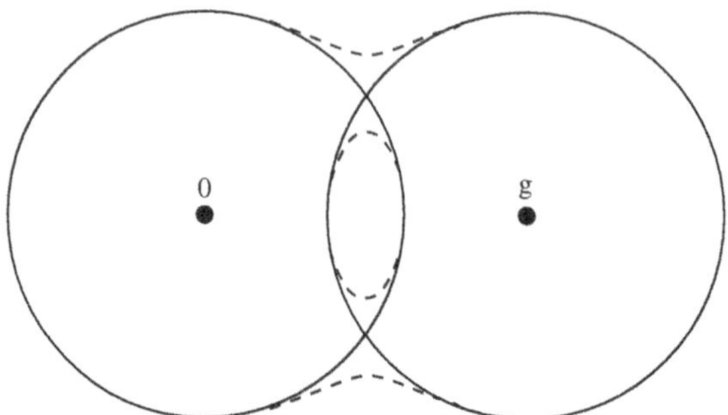

FIG. 3.5. The construction of the two-beam dispersion surface.

be chosen so that the basic eqn (3.22) (i.e. ultimately the Schrödinger equation (3.19)) is satisfied. The dispersion surface shows how the surface normal component of the wave vector varies as a function of k_{0_x} and k_{0_y} for a given energy E of the incident electrons. In other words, the dispersion surface is defined by equation $k_z = k_z(k_{0_x}, k_{0_y})$.

To understand the topology of the dispersion surface, we note that the energy of the incident high-energy electron is of the order of several kiloelectronvolts, while the energy associated with the potential of the crystal lattice is of the order of a few tens of volts. In the first approximation we may neglect the Fourier components of the potential U_g and consider only the case of the 'empty' lattice. Equations (3.22) then reduce to

$$K^2 = (\mathbf{k} + \mathbf{g})^2, \tag{3.26}$$

and the solution of these equations has the form of a set of spheres centered around reciprocal lattice points $\mathbf{g}$. In the two-beam approximation where only one diffracted beam associated with $\mathbf{g}$ is excited, the dispersion surface consists of two spheres with radii K centred around $\mathbf{0}$ and $\mathbf{g}$ (see Fig. 3.5).

The fact that the matrix elements U_{g-h} have finite values modifies this 'ideal' dispersion surface. The changes are appreciable only in the vicinity of points where spheres intersect, reflecting the fact that in any of these regions two or more beams are excited simultaneously. The modified dispersion surface has an approximately hyperbolic shape in the vicinity of the intersection between the spheres, while away from the intersection points the dispersion surface follows closely the empty lattice approximation.

3.5.3 *Translation properties of Bloch waves*

In principle the set of fundamental equations (3.22) describes an infinite number of diffracted beams. However, many of the beams satisfy the condition

$$|K^2 - (\mathbf{k} + \mathbf{h})^2| \gg \max\{...U_g...\}, \tag{3.27}$$

and for each of these beams eqn (3.22) reduces to

$$[K^2 - (\mathbf{k} + \mathbf{h})^2]C_h = 0. \tag{3.28}$$

The solution of the latter equation is trivial, namely $C_h = 0$. This means that beams satisfying condition (3.27) do not affect the solution of the remaining equations. Using this condition we divide the beams into two groups. The beams that satisfy condition (3.27) are excluded from consideration. In this way we truncate the infinite set of equations, transforming it into a finite set involving only the beams that do not satisfy condition (3.27). Coming back to the Ewald sphere construction introduced in Chapter 2, we note that the condition (3.27) means simply that a particular reciprocal lattice point $\mathbf{h}$ lies far away from the Ewald sphere. In practice only those reciprocal lattice points that lie on or near the Ewald sphere need to be included in the fundamental equations (3.22).

We now consider a diffraction problem where a total of N reciprocal lattice points lie on or close to the Ewald sphere. The set of infinite equations (3.25) is then truncated to a set of N equations. To obtain a non-trivial solution of these equations we set the determinant of the $N \times N$ matrix $\mathbf{A}$ to be equal to zero, i.e. $\det|\mathbf{A}| = 0$, resulting in $2N$ values $\gamma^{(j)}, (j = 1, ..., 2N)$ and $2N$ associated Bloch waves $b^{(j)}(\mathbf{k}^{(j)}, \mathbf{r})$.

Not all the Bloch waves generated by eqns (3.25) are distinct. The waves characterized by wave vectors differing only by a reciprocal lattice vector are physically equivalent. To prove this we consider the basic equation (3.22) for a wave vector $\mathbf{k}$

$$[K^2 - (\mathbf{k} + \mathbf{l})^2]C_l(\mathbf{k}) + \sum_{h \neq l} U_{l-h}C_h(\mathbf{k}) = 0.$$

Letting $\mathbf{l} = \mathbf{f} + \mathbf{g}$, we arrive at

$$[K^2 - (\mathbf{k} + \mathbf{f} + \mathbf{g})^2]C_{f+g}(\mathbf{k}) + \sum_{h \neq f+g} U_{f+g-h}C_h(\mathbf{k}) = 0.$$

Furthermore, letting $\mathbf{h}' = \mathbf{h} - \mathbf{g}$ and $\mathbf{k}' = \mathbf{k} + \mathbf{g}$, we find

$$[K^2 - (\mathbf{k}' + \mathbf{f})^2]C_{f+g}(\mathbf{k}) + \sum_{h' \neq f} U_{f-h'}C_{h'+g}(\mathbf{k}) = 0.$$

Alternatively, considering a wave vector $\mathbf{k}'$ and using the basic equation (3.22) we find that

$$[K^2 - (\mathbf{k}' + \mathbf{f})^2]C_f(\mathbf{k}') + \sum_{h' \neq f} U_{f-h'}C_{h'}(\mathbf{k}') = 0.$$

Since any two identical wave vectors correspond to identical eigenvectors $\mathbf{C}$, we find the condition of periodicity of amplitudes of Bloch waves $\{C_g\}$:

$$C_{f+g}(\mathbf{k}) = C_f(\mathbf{k} + \mathbf{g}). \tag{3.29}$$

We now consider a Bloch wave $b(\mathbf{k} + \mathbf{f}, \mathbf{r})$ associated with the wave vector $\mathbf{k} + \mathbf{f}$

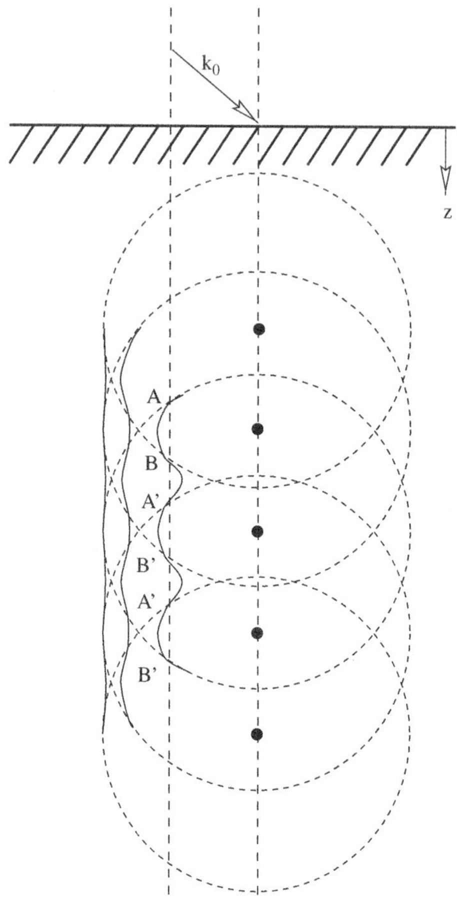

FIG. 3.6. Dispersion surface construction for the systematic reflection case.

$$b(\mathbf{k} + \mathbf{f}, \mathbf{r}) = \sum_g C_g(\mathbf{k} + \mathbf{f}) \exp[i(\mathbf{k} + \mathbf{f} + \mathbf{g}) \cdot \mathbf{r}].$$

Using (3.29) for $C_g(\mathbf{k} + \mathbf{g})$ and denoting $\mathbf{h} = \mathbf{f} + \mathbf{g}$ we arrive at

$$b(\mathbf{k} + \mathbf{f}, \mathbf{r}) = \sum_g C_{g+f}(\mathbf{k}) \exp\{i[\mathbf{k} + (\mathbf{f} + \mathbf{g})] \cdot \mathbf{r}\}$$

$$= \sum_h C_h(\mathbf{k}) \exp[i(\mathbf{k} + \mathbf{h}) \cdot \mathbf{r}] = b(\mathbf{k}, \mathbf{r}),$$

i.e. the Bloch wave $b(\mathbf{k}, \mathbf{r})$ is a function that is periodic in reciprocal space.

The wave vectors $\mathbf{k}$ of excited Bloch waves are determined by the boundary conditions. The component of the wave vector of an excited Bloch wave parallel

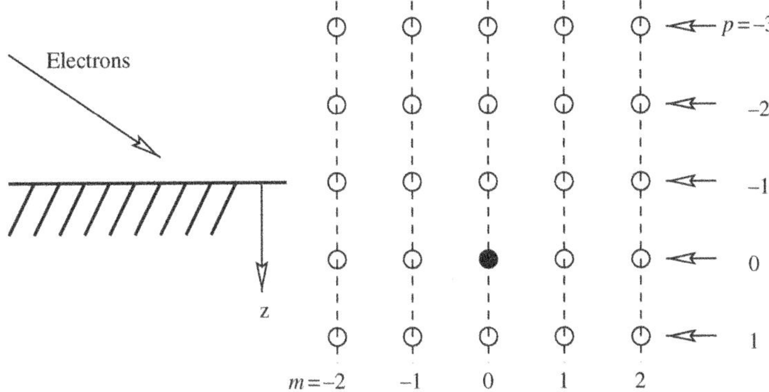

FIG. 3.7. Schematic diagram illustrating the notation used in indexing the reciprocal lattice rods and lattice points along the rods.

to the surface equals that of the wave vector of the incident beam. The surface normal component of the wave vector is determined by the intersection of a line parallel to the z axis with the dispersion surface shown in Fig. 3.6. This figure illustrates a special case where only one row of reciprocal lattice points perpendicular to the surface is involved in the diffraction process. This diffraction case is usually referred to as the systematic diffraction case. We see that the dispersion surface is periodic in the direction normal to the surface. We proved above that not all the intersection points between the vertical line and the dispersion surface correspond to distinct Bloch states. In the case shown in Fig. 3.6 there are only two distinct wave vectors that lie within the distance of $|\mathbf{g}|$ from each other. For example, points A' and B' are equivalent to points A and B. In principle any set of two distinct wave points can be used to describe the diffraction process. In electron diffraction it is common to choose points that are closest to the end of the wave vector $\mathbf{K}$ that is obtained from the incident wave vector $\mathbf{k}_0$ after correcting the latter for the mean inner potential U_0 of the crystal.

3.5.4 *Boundary conditions and formal solutions*

In the process of dynamical diffraction only the components of the wave vector parallel to the surface are conserved. Components of the wave vector normal to the surface $\mathbf{k}_z$ are chosen according to the form of the dispersion surface. It is convenient to decompose a reciprocal lattice vector $\mathbf{g}$ into the surface parallel $\mathbf{m}$ and surface normal p components, namely $\mathbf{g} = (\mathbf{m}, \mathrm{p})$. We also introduce a new notation using italic symbols (i.e. $g = (m, p)$ where g, m, and p are integers (see Fig. 3.7)) to index the reciprocal lattice vectors.

In the case where there are n distinct reciprocal lattice rods (lines of reciprocal lattice points normal to the surface of the crystal) involved in the process of diffraction, we have $2n$ independent Bloch waves. These Bloch waves form N excited diffracted beams within the crystal, where N denotes the total number of reciprocal lattice points contributing to the expansion of Bloch states in terms of plane waves. The wave function of the electron in the crystal has the form (Bethe, 1928, Fujimoto, 1959)

$$\psi(\mathbf{r}) = \sum_{j=1}^{2n} \alpha^{(j)} b^{(j)}(\mathbf{k}^{(j)}, \mathbf{r}) = \sum_{j=1}^{2n} \alpha^{(j)} \sum_{g=g_1}^{g_N} C_g^{(j)} \exp[i(\mathbf{K} + \gamma^{(j)}\mathbf{n} + \mathbf{g}) \cdot \mathbf{r}],$$

where $\alpha^{(j)}$ is the excitation amplitude of the j-th Bloch wave $b^{(j)}$. This amplitude is determined by the boundary conditions. The amplitude of the diffracted beam associated with the m-th reciprocal lattice rod is given by

$$\psi_m(z) = \sum_{j=1}^{2n} \alpha^{(j)} \sum_{p} C_{mp}^{(j)} \exp[i(K_z + \gamma^{(j)} + \mathrm{p})z].$$

The conditions of continuity of diffracted beam amplitudes and their surface normal derivatives have the form

$$\exp(-iK_z z)\psi_m(z) = \sum_{j=1}^{2n} \alpha^{(j)} \sum_{p} C_{mp}^{(j)} \exp[i(\gamma^{(j)} + \mathrm{p})z], \tag{3.30}$$

$$-i\exp(-iK_z z)\psi'_m(z) = \sum_{j=1}^{2n} \alpha^{(j)} \sum_{p} (K_z + \gamma^{(j)} + \mathrm{p})C_{mp}^{(j)} \exp[i(\gamma^{(j)} + \mathrm{p})z], \tag{3.31}$$

where $\psi'_m = d\psi_m/dz$. In what follows we shall use the notation ψ_m to refer to the left-hand side of (3.30) and use ψ'_m to denote that of (3.31). Equations (3.30) and (3.31) can then be written in matrix notation as

$$\boldsymbol{\Psi}(z) = \mathbf{SP}(z)\mathbf{C}\boldsymbol{\Upsilon}(z)\boldsymbol{\alpha}, \tag{3.32}$$

where $\boldsymbol{\Psi}$ is a $2n$-dimensional super vector

$$\boldsymbol{\Psi}(t) = \begin{pmatrix} \{\psi_m(z)\} \\ \{\psi'_m(z)\} \end{pmatrix},$$

$\mathbf{S}$ is a $(2n \times 2N)$ matrix the elements of which are

$$(S)_{mg} = \begin{cases} 1, & \text{if g belongs to the } m\text{-th rod} \\ 0, & \text{otherwise} \end{cases}$$

for $m \leq n, g \leq N$, and

$$(S)_{m+n,g+N} = (S)_{m,g}$$

$$(S)_{m+n,g} = (S)_{m,g+N} = 0,$$

$\mathbf{P}$ is a $(2N \times 2N)$ diagonal matrix

$$(P)_{gh} = \exp(ipz)\delta_{gh}$$

for $h, g \leq N, g = (m, p)$, and

$$(P)_{g+N,h+N} = (P)_{g,h},$$

$\mathbf{C}$ is a $(2N \times 2n)$ matrix

$$(C)_{hi} = C_h^{(i)}$$

for $h \leq N, h = (l, q)$, and

$$(C)_{h+N,i} = (K_z + \gamma^{(i)} + \mathrm{q})C_h^{(i)},$$

$\mathbf{\Upsilon}$ is a $(2n \times 2n)$ diagonal matrix

$$(\mathbf{\Upsilon})_{ij} = \exp(i\gamma^{(i)}z)\delta_{ij},$$

and $\boldsymbol{\alpha}$ is a $2n$-dimensional column vector

$$(\boldsymbol{\alpha})_j = \alpha^{(j)}.$$

If for a crystal slab of thickness t we choose that the plane $z = z_1$ represents the top surface and the plane $z = z_2$ represents the bottom surface of the crystal ($z_2 - z_1 = t$), we find from (3.32)

$$\mathbf{\Psi}(z_1) = \mathbf{SP}(z_1)\mathbf{C\Upsilon}(z_1)\boldsymbol{\alpha}, \tag{3.33}$$

$$\dot{\mathbf{\Psi}}(z_2) = \mathbf{SP}(z_2)\mathbf{C\Upsilon}(z_2)\boldsymbol{\alpha}. \tag{3.34}$$

In eqn (3.34) $\boldsymbol{\alpha}$ may be eliminated by using eqn (3.33), i.e.

$$\boldsymbol{\alpha} = [\mathbf{SP}(z_1)\mathbf{C\Upsilon}(z_1)]^{-1}\mathbf{\Psi}(z_1),$$

from which we find

$$\mathbf{\Psi}(z_2) = [\mathbf{SP}(z_2)\mathbf{C\Upsilon}(z_2)][\mathbf{SP}(z_1)\mathbf{C\Upsilon}(z_1)]^{-1}\mathbf{\Psi}(z_1)$$
$$= [\mathbf{SP}(z_2)\mathbf{C}]\mathbf{\Upsilon}(z_2)\mathbf{\Upsilon}^{-1}(z_1)[\mathbf{SP}(z_1)\mathbf{C}]^{-1}\mathbf{\Psi}(z_1).$$

Since

$$\mathbf{\Upsilon}(z_2)\mathbf{\Upsilon}^{-1}(z_1) = \mathbf{\Upsilon}(z_2 - z_1) = \mathbf{\Upsilon}(t),$$

we see that

$$\mathbf{\Psi}(z_2) = \mathbf{M}(t)\mathbf{\Psi}(z_1), \tag{3.35}$$

where

$$\mathbf{M}(t) = [\mathbf{SP}(z_2)\mathbf{C}]\mathbf{\Upsilon}(t)[\mathbf{SP}(z_1)\mathbf{C}]^{-1}. \tag{3.36}$$

This procedure can be generalized further to describe diffraction from an assembly of crystal slabs, where the thickness of each slab equals t_n, giving rise to

$$\mathbf{\Psi}(z) = \mathbf{M}(z)\mathbf{\Psi}(0), \tag{3.37}$$

where matrix $\mathbf{M}(z)$ is the scattering matrix given by

$$
\begin{aligned}
\mathbf{M}(z) = &...\{[\mathbf{S}_n\mathbf{P}_n(Z_n)\mathbf{C}_n]\boldsymbol{\Upsilon}_n(t_n)[\mathbf{S}_n\mathbf{P}_n(Z_{n-1})\mathbf{C}_n]^{-1}\} \\
&\times\{[\mathbf{S}_{n-1}\mathbf{P}_{n-1}(Z_{n-1})\mathbf{C}_{n-1}]\boldsymbol{\Upsilon}_{n-1}(t_{n-1})[\mathbf{S}_{n-1}\mathbf{P}_{n-1}(Z_{n-2})\mathbf{C}_{n-1}]^{-1}\}... \\
&...\{[\mathbf{S}_1\mathbf{P}_1(t_1)\mathbf{C}_1]\boldsymbol{\Upsilon}_1(t_1)[\mathbf{S}_1\mathbf{P}_1(0)\mathbf{C}_1]^{-1}\},
\end{aligned}
$$

and where

$$
Z_n = \sum_{k=1}^{n} t_k.
$$

If all the crystal slabs are identical, i.e. $\mathbf{S}_n = \mathbf{S}$, $\mathbf{P}_n = \mathbf{P}$, $\mathbf{C}_n = \mathbf{C}$, $\boldsymbol{\Upsilon}_n = \boldsymbol{\Upsilon}$, we find

$$
\mathbf{M}(z) = [\mathbf{SP}(z)\mathbf{C}]\boldsymbol{\Upsilon}(z)[\mathbf{SP}(0)\mathbf{C}]^{-1}.
$$

Taking into account that $z = z_2$, $z_1 = 0$, and $t = z_2 - z_1 = z$, we see that the latter equation is the same as (3.36).

We now consider a general case of diffraction from a crystal composed of a sequence of slabs. Since the region above the top surface contains only the incident beam and Bragg-reflected beams, the wave function in this region has the form

$$
\Psi^v_{upper}(\mathbf{r}) = \exp(i\mathbf{k}_0 \cdot \mathbf{r}) + \sum_{m=1}^{n} \mathcal{R}_m \exp[i(\mathbf{K}_{mt} - \mathbf{K}_{mz}) \cdot \mathbf{r}], \qquad (3.38)
$$

where the first term represents the incident beam, and $\mathcal{R}_m$ are the amplitudes of reflected beams ($m = 1$ denotes the reciprocal lattice rod passing through the origin in the reciprocal space). The wave vectors of reflected beams are

$$
\mathbf{K}_{mt} = (\mathbf{k}_0 + \mathbf{g})_t = \mathbf{k}_{0_t} + \mathbf{m}, \quad \mathbf{K}_{mz} = \sqrt{[\mathbf{k}_0^2 - (\mathbf{k}_0 + \mathbf{g})_t^2]}\mathbf{n},
$$

where the subscript t denotes the tangential component of the wave vector. For example,

$$
\mathbf{K}_{1t} = \mathbf{k}_{0_t}, \quad K_{1z} = k_{0_z}.
$$

Boundary conditions at the top surface have the form

$$
\begin{pmatrix} \{\psi_m(0)\} \\ \{\psi'_m(0)\} \end{pmatrix} = \begin{pmatrix} 1 + \mathcal{R}_1 \\ \mathcal{R}_2 \\ \vdots \\ \mathcal{R}_n \\ K_{1z}(1 - \mathcal{R}_1) \\ -K_{2z}\mathcal{R}_2 \\ \vdots \\ -K_{nz}\mathcal{R}_n \end{pmatrix}.
$$

Only the transmitted beams propagate in the region below the bottom surface. The wave function in this region is given by

$$\Psi_{lower}^{v}(\mathbf{r}) = \sum_{m=1}^{n} \mathcal{T}_m \exp\{i(\mathbf{K}_{mt} + \mathbf{K}_{mz}) \cdot \mathbf{r}\},\qquad(3.39)$$

where $\mathcal{T}_m$ is the transmitted beam amplitude associated with the m-th reciprocal lattice rod. The boundary conditions at the bottom surface are

$$\begin{pmatrix} \{\psi_m(t)\} \\ \{\psi_m'(t)\} \end{pmatrix} = \begin{pmatrix} \mathcal{T}_1 \exp(iK_{1z}t) \\ \mathcal{T}_2 \exp(iK_{2z}t) \\ \vdots \\ \mathcal{T}_n \exp(iK_{nz}t) \\ K_{1z}\mathcal{T}_1 \exp(iK_{1z}t) \\ K_{2z}\mathcal{T}_2 \exp(iK_{2z}t) \\ \vdots \\ K_{nz}\mathcal{T}_n \exp(iK_{nz}t) \end{pmatrix},$$

where t is the total thickness of the crystal slab system. Using matrix notation, eqn (3.37) can be rewritten as

$$\begin{pmatrix} \{\mathcal{T}_m \exp(iK_{mz}t)\} \\ \{K_{mz}\mathcal{T}_m \exp(iK_{mz}t)\} \end{pmatrix} = \begin{pmatrix} \{M_{11}\}_{mn} & \{M_{12}\}_{mn} \\ \{M_{21}\}_{mn} & \{M_{22}\}_{mn} \end{pmatrix} \begin{pmatrix} \{\delta_{1n} + \mathcal{R}_n\} \\ \{K_{nz}(\delta_{1n} - \mathcal{R}_n)\} \end{pmatrix},$$

where $\mathbf{M}_{11}, \mathbf{M}_{12}, \mathbf{M}_{21}$, and $\mathbf{M}_{22}$ are the $n \times n$ submatrices of $\mathbf{M}(z)$. By eliminating amplitudes of transmitted beams, we find

$$\{M_{11}\}\{\delta_{1n} + \mathcal{R}_n\} + \{M_{12}\}\{K_{nz}(\delta_{1n} - \mathcal{R}_n)\} =$$

$$\{M_{21}/K_{mz}\}_{mn}\{\delta_{1n} + \mathcal{R}_n\} + \{M_{22}/K_{mz}\}_{mn}\{K_{nz}(\delta_{1n} - \mathcal{R}_n)\}.$$

Rearranging terms in this equation, we find

$$\{[\{M_{11}\} - \{M_{12}K_{nz}\}] - [\{M_{21}/K_{mz}\} - \{M_{22}K_{nz}/K_{mz}\}]\}\{\mathcal{R}_n\}$$

$$+\{[\{M_{11}\} + \{M_{12}K_{nz}\}] - [\{M_{21}/K_{mz}\} + \{M_{22}K_{nz}/K_{mz}\}]\}\{\delta_{1n}\} = 0.$$

It is a simple matter now to derive a formal expression for the amplitudes of reflected beams $\{\mathcal{R}_m\}$

$$\{\mathcal{R}_m\} = -\frac{[\{M_{11}\} + \{M_{12}K_{nz}\}] - [\{M_{21}/K_{mz}\} + \{M_{22}K_{nz}/K_{mz}\}]}{[\{M_{11}\} - \{M_{12}K_{nz}\}] - [\{M_{21}/K_{mz}\} - \{M_{22}K_{nz}/K_{mz}\}]}\{\delta_{1n}\}.$$

$$(3.40)$$

It is essential here that the matrix in the denominator pre-multiplies the matrix of the numerator. Having found the amplitudes of the reflected beams, we can now find the amplitudes of the transmitted beams $\{\mathcal{T}_m\}$ as

$$\{\mathcal{T}_m\} = (\{M_{11}\exp(-iK_{mz}z)\}, \{M_{12}\exp(-iK_{mz}z)\}) \begin{pmatrix} \{\delta_{1n} + \mathcal{R}_n\} \\ \{K_{nz}(\delta_{1n} - \mathcal{R}_n)\} \end{pmatrix}$$

$$= [\{M_{11}\exp(-iK_{mz}z)\} + \{M_{12}K_{nz}\exp(-iK_{mz}z)\}]\{\delta_{1n}\} \qquad (3.41)$$

$$+ [\{M_{11}\exp(-iK_{mz}z)\} - \{M_{12}K_{nz}\exp(-iK_{mz}z)\}]\{\mathcal{R}_n\}.$$

This equation gives a formal solution of the problem of diffraction of high-energy electrons (Peng and Whelan, 1990b, Peng and Whelan, 1990c).

3.6 Summary

Dynamical diffraction of high-energy electrons is described by the one-particle Schrödinger equation. This equation can be solved by using the multislice approach or the general matrix method. These approaches can be simplified by making further approximations. Numerical methods for calculating the amplitudes of transmitted and reflected beams will be described in the following chapters.

4

DYNAMICAL THEORY II. TRANSMISSION HIGH-ENERGY ELECTRON DIFFRACTION

4.1 Introduction

The majority of practical applications of high-energy electron diffraction are associated with the use of transmission geometry of scattering. Transmission geometry was also the first considered by Laue (Friedrich et al., 1912, von Laue, 1912) in his work on X-ray diffraction, and later by Thomson and Reid in their studies of electron diffraction (Thomson and Reid, 1927). In the literature transmission high-energy electron diffraction (THEED) is often referred to as the Laue case of electron diffraction (see for example the book by Thomson and Cochrane (Thomson and Cochrane, 1939) where they describe the history of the development of the subject).

In THEED geometry the electrons are incident nearly normally at the top face of a thin crystal, the thickness of which typically does not exceed 1 μm (see Fig. 4.1). In almost all the theoretical treatments the specimen is assumed to have the form of a perfect crystal slab with parallel faces terminated by low-index atomic planes. In the Cartesian system of coordinates the (x, y) plane is normally chosen to be parallel to the surface with the z axis pointing into the crystal. In transmission geometry of diffraction the general dynamical theory discussed in Chapter 3 may be greatly simplified, since backscattering of electrons may be safely neglected.

4.2 Diffraction geometry

Using the coordinate system defined in Fig. 4.1 we may treat the reciprocal lattice as a stack of two-dimensional nets of reciprocal lattice points $\mathbf{G}$ lying in planes parallel to the (x, y) plane. In a general case the successive layers may be shifted with respect to each other and the shift may be conveniently described as $\Delta\mathbf{G} + g_z\hat{\mathbf{z}}$ (see Fig. 4.1), where $\Delta\mathbf{G}$ is a vector parallel to the (x, y) plane and $\hat{\mathbf{z}}$ is a unit vector pointing in the positive z direction. Let a general reciprocal lattice vector in the n-th layer be

$$\mathbf{g}^{(n)} = (\mathbf{G} + n\Delta\mathbf{G}, ng_z) = (\mathbf{G}^{(n)}, ng_z).$$

Denoting the period of translations along the z axis by c, and writing the position vector in the form $\mathbf{r} = (\mathbf{X}, z)$, where $\mathbf{X}$ is the vector component (x, y) parallel to the surface, we represent the scaled crystal potential $U(\mathbf{r}) = -2mV(\mathbf{r})/\hbar^2$ by the Fourier series

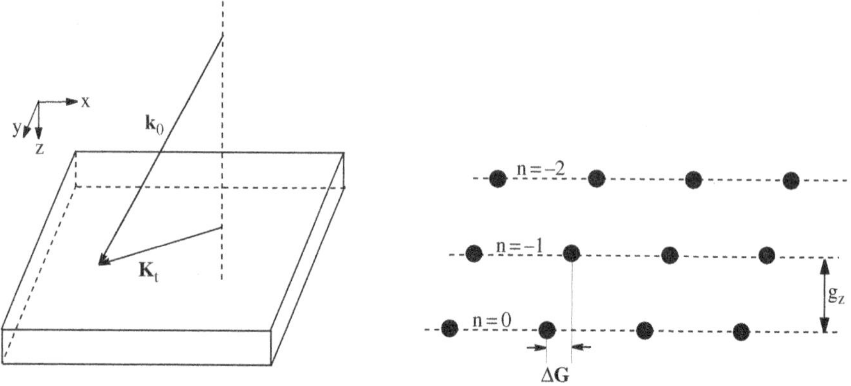

FIG. 4.1. Schematic diagram showing the real-space coordinate system and layers of the reciprocal lattice corresponding to the Laue case of electron diffraction.

$$U(\mathbf{r}) = \sum_g U_g \exp(i\mathbf{g} \cdot \mathbf{r})$$

$$= \sum_n \left\{ \sum_{G^{(n)}} U_{\mathbf{g}^{(n)}} \exp(i\mathbf{G}^{(n)} \cdot \mathbf{X}) \right\} \exp(ing_z z)$$

$$= \sum_n \exp(ing_z z) U^{(n)}(\mathbf{X}), \tag{4.1}$$

where

$$U^{(n)}(\mathbf{X}) = \sum_{G^{(n)}} U_{\mathbf{g}^{(n)}} \exp(i\mathbf{G}^{(n)} \cdot \mathbf{X}) \tag{4.2}$$

is called the conditional projected potential (Cochran and Dyer, 1952). By using the relation

$$\int_c \exp(ing_z z) dz = c\delta_{0,n}$$

and eqn (4.1), it may be readily shown that

$$U^{(n)}(\mathbf{X}) = \frac{1}{c} \int_0^c U(\mathbf{X}, z) \exp(-ing_z z) dz. \tag{4.3}$$

In particular we recover the familiar real-space definition of $U^{(0)}(\mathbf{X})$

$$U^{(0)}(\mathbf{X}) = \frac{1}{c} \int_0^c U(\mathbf{X}, z) dz,$$

where $U^{(0)}(\mathbf{X})$ is the potential projected along the z axis. It is convenient to classify reciprocal lattice vectors $\mathbf{g} = (\mathbf{G}^{(n)}, ng_z)$ into Laue zones, and to define reflections corresponding to $n = 0$ as belonging to the zero-order Laue zone

(ZOLZ), and reflections with $n \neq 0$ as belonging to higher order Laue zones (HOLZs) (see Fig. 4.2). The zone composed of vectors with $n = -1$ is called the first higher order Laue zone (FHOLZ), and that composed of vectors with $n = -2$ is called the second higher order Laue zone (SHOLZ).

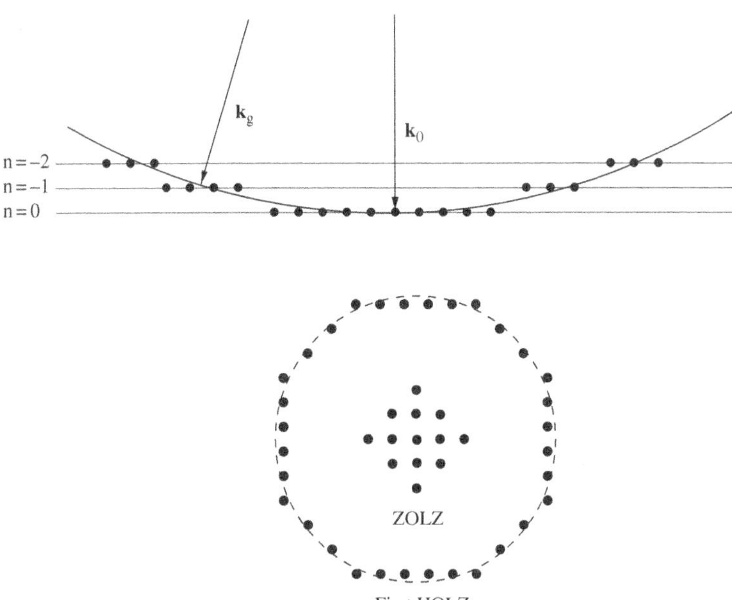

FIG. 4.2. Schematic diagram showing the Laue zones and the expected transmission electron diffraction pattern involving both ZOLZ and FHOLZ.

A diffraction pattern is produced by the intersection of diffracted beams emerging from the bottom face of a crystal slab with a recording medium such as a photographic film. Since a diffraction pattern is formed in the vacuum region far away from the sample, the geometry of the diffraction pattern is determined entirely by the principles of energy and momentum conservation. For a plane wave incident on the crystal slab the principle of energy conservation requires that all the diffracted beams satisfy the Ewald sphere construction introduced in Chapter 2, just as in the kinematic case (see also Fig. 4.2). On the other hand, the principle of momentum conservation requires that all reciprocal lattice points lying on or close to the Ewald sphere have surface parallel components of the form $\mathbf{G} + n\Delta\mathbf{G}$. By combining these two principles we see that the directions of diffracted beams are given by coordinates of the intersection of the Ewald sphere with the two-dimensional rods of the reciprocal lattice that have surface parallel components of the form $\mathbf{G} + n\Delta\mathbf{G}$ shown in Fig. 4.2. It should be noted that the reciprocal lattice points shown schematically as dots in Fig. 4.2 have in fact some finite width in the z direction. In the kinematic case the variation of intensity along the rod results from the fact that the thickness of the crystal slab is finite. To a good approximation this intensity variation may be modelled by a simple shape function (Cowley, 1990)

$$S(\kappa_z) = \sin^2\left(\frac{\kappa_z t}{2}\right) \Big/ \left(\frac{\kappa_z}{2}\right)^2,$$

where κ_z is the distance along a rod to the nearest reciprocal lattice point, and t is the thickness of the crystal slab. This function has a sharp peak at $\kappa_z = 0$ and is characterized by a half-width of the order of $\delta\kappa_z = 2\pi/t$. If the crystal slab consists of N layers of atoms, this estimate shows that the effective width of reciprocal lattice points is of the order of $1/N$ of the distance between the successive Laue zones. Dynamical diffraction may modify the kinematic form of the intensity variation along the rods. However, this does not alter the general rule that the larger the distance between a reciprocal lattice point and the Ewald sphere, the weaker is the intensity of the corresponding diffracted beam.

We showed in Chapter 2 that high-energy electron diffraction is dominated by forward scattering of electrons. An important consequence of this is that a transmission diffraction pattern is not very sensitive to variations of the potential in the direction of the incident beam. This fact explains the high accuracy of the projected potential approximation, which assumes that $U(\mathbf{r}) = U^{(0)}(\mathbf{X})$. The zone axis THEED patterns are dominated by zero-order Laue zone reflections. The variation of the potential in the direction of the incident beam gives rise to the appearance of HOLZ reflections in electron diffraction patterns (see Fig. 4.2).

Since all the HOLZ reflections corresponding to positive values of n lie outside the Ewald sphere, these reflections do not appear in transmission electron diffraction patterns observed experimentally. We can only see the HOLZ reflections corresponding to negative values of n. For the direction of incidence parallel to the zone axis the intersection of the Ewald sphere with a HOLZ forms a ring, and the radius K_{HOLZ} of this ring follows from a simple geometric consideration

$$K_{HOLZ}^2 = k^2 - (k + ng_z)^2 = -2k(ng_z) - (ng_z)^2,$$

where $k = |\mathbf{k}_0|$. In most cases it is possible to neglect the second term on the right-hand side of the above equation, i.e. to use the approximation

$$K^2 \approx -2kng_z. \tag{4.4}$$

This approximation replaces the Ewald sphere by a parabola and is usually called the high-energy approximation. The accuracy of this approximation is usually very high, and it can be used in most quantitative electron diffraction calculations. However, it leads to small errors in the limit of large angles of scattering, for example, in evaluating the radii of the HOLZ rings. In the latter case the error may be estimated to be of the order of ng_z/k, and typically this quantity does not exceed a few percent (Bird, 1989).

4.3 Basic concepts and the treatment of ZOLZ diffraction

There is a large body of literature describing various approaches to the problem of dynamical electron diffraction in THEED geometry, where only ZOLZ reflections are involved (Hirsch et al., 1977, Cowley, 1993a, Spence and Zuo, 1992). These approaches include the original Bethe theory (Bethe, 1928), the Howie-Whelan equations (Howie and Whelan, 1961), the Sturkey scattering matrix treatment (Sturkey, 1962), and the multislice approach (Cowley and Moodie, 1957). Although all these approaches are equivalent, see (Goodman and Moodie, 1974) and (Spence and Whelan, 1975), they were developed bearing in mind different applications, and each of them is more suitable for solving a particular type of problem than the others. For example, although the Sturkey scattering matrix approach provides the most compact means of expressing the effect of dynamical electron diffraction on the incident electron beam, numerically it is far less efficient in calculating diffracted beam amplitudes than, for example, Cowley and Moodie's multislice method. On the other hand, the Howie-Whelan equations are most suitable for simulating diffraction contrast images of crystal defects, particularly when they are combined with the column approximation introduced by Whelan and Hirsch (Whelan and Hirsch, 1957). In this section we will not enter into a detailed discussion of these approaches and instead we refer the interested reader to treatises by Hirsch et al. (Hirsch et al., 1977), Reimer (Reimer, 1989), and Cowley (Cowley, 1990). In what follows we will consider only the Bloch-wave-based methods, introducing the basic concepts and concentrating on their use in recent applications.

4.3.1 *Basic equations and Bloch waves*

The dynamical theory of diffraction was first developed for X-rays by Darwin (Darwin, 1914) and independently by Ewald (Ewald, 1916a, Ewald, 1916b, Ewald, 1916c). The first theory of dynamical diffraction of electrons was developed by Bethe (Bethe, 1928) shortly after the discovery of electron diffraction. The theoretical framework used by Bethe was similar to that used previously by Ewald (Ewald, 1916a). He addressed mainly the reflection case of electron diffraction used in the experiments carried out by Davisson and Germer (Davisson and Germer, 1927). Bethe's theory was later generalized to the case of THEED, where particularly important contributions were made by Blackman (Blackman, 1939), MacGillavry (MacGillavry, 1940), Fujimoto (Fujimoto, 1959) and Howie and Whelan (Howie and Whelan, 1961). At present Bethe's approach is usually referred to as the Bloch wave approach, since the potential of interaction of electrons with the specimen is assumed to vary periodically in space and the electron wave field obeys the Bloch theorem (Ashcroft and Mermin, 1976). The concept of a 'Bloch wave' was originally introduced by Ewald (Ewald, 1916a) in his formulation of the dynamical theory of X-ray diffraction, and in the literature the wave field in a crystal is sometimes referred to as the 'Ewald wave' (Authier et al., 1996).

In a periodic potential the translation symmetry of the crystal lattice requires that the electron wave function should have the form of a product of a phase factor $\exp(i\mathbf{k} \cdot \mathbf{r})$ and a periodic function $u(\mathbf{r})$. The wave function may be represented by a Fourier series (Ashcroft and Mermin, 1976)

$$b(\mathbf{k}, \mathbf{r}) = \exp(i\mathbf{k} \cdot \mathbf{r})u(\mathbf{r}) = \exp(i\mathbf{k} \cdot \mathbf{r}) \sum_g C_g \exp(i\mathbf{g} \cdot \mathbf{r}),$$

such that $b(\mathbf{k}, \mathbf{r} + \mathbf{R}) = \exp(i\mathbf{k} \cdot \mathbf{R})b(\mathbf{k}, \mathbf{r})$, where $\mathbf{R}$ denotes a Bravais lattice vector. A wave function satisfying the above condition is usually referred to as a Bloch wave. In a general case the three-dimensional wave vector $\mathbf{k}$ introduced above is not the most convenient quantum number characterizing the solution of the problem. For example, in the case of diffraction by a slab shown in Fig. 4.1, the surface parallel component of the wave vector $\mathbf{k}_t$ is indeed a convenient quantum number, since the periodicity of the crystal potential in the x and y directions is not affected by the presence of the surface, and the component of the wave vector parallel to the surface is conserved across the surface, i.e. $\mathbf{k}_t = \mathbf{K}_t$. However, the surface normal component of the wave vector $\mathbf{k}_z$ is not a conserved quantity and it has to be determined from the boundary condition of continuity of the wave function and its normal derivative across the surface and from the basic equations of electron diffraction (3.22)

$$[K^2 - (\mathbf{k} + \mathbf{g})^2]C_g(\mathbf{k}) + \sum_{h \neq g} U_{g-h}C_h(\mathbf{k}) = 0, \tag{4.5}$$

where $\mathbf{K}$ is the wave vector of the incident wave corrected for the refraction effect, i.e. $K^2 = k_0^2 + U_0$, and $\mathbf{k}_0$ is the wave vector of incident electrons. It is convenient to represent the (as yet unknown) wave vector $\mathbf{k}$ in the form

$$\mathbf{k} = \mathbf{K} + \gamma\hat{\mathbf{z}}, \tag{4.6}$$

where $\mathbf{K}_t = \mathbf{k}_t = \mathbf{k}_{0t}$. In the case of THEED this leads to a considerable simplification of the basic equations (4.5). Indeed, in the case where high-energy electrons are incident normally on the surface of the crystal the electron wave vector satisfies the inequality

$$K \approx K_z \gg \gamma + g_z.$$

This inequality is valid for all vectors $\mathbf{g}$ involved in the diffraction process and for all the Bloch waves. Noting that ZOLZ reflections correspond to $g_z = 0$, we can simplify eqns (4.5) further as

$$2K_z(\gamma - S_g)C_g - \sum_{h \neq g} U_{g-h}C_h = 0. \tag{4.7}$$

In this equation the quantity

$$S_g = [K^2 - (\mathbf{K} + \mathbf{g})^2]/2K_z = (K + |\mathbf{K} + \mathbf{g}|)(K - |\mathbf{K} + \mathbf{g}|)/(2K_z) \approx K - |\mathbf{K} + \mathbf{g}| \tag{4.8}$$

is equal to the shortest distance between the reciprocal lattice point $\mathbf{g}$ and the Ewald sphere shown in Fig. 4.2. This quantity is usually called the excitation error (the equivalent German term is 'Anregungsfehler'). Equation (4.7) has non-zero solutions only for a set of discrete values of $\gamma^{(j)}$, where the range of variation of index j depends on the size of the matrix U_{gh} ($=U_{g-h}$). For each eigenvalue $\gamma^{(j)}$ there is a corresponding eigenvector, the g-th component of which is given by $\{\mathbf{C}^{(j)}\}_g = C_g^{(j)}$. In terms of eigenvectors eqn (4.7) can be rewritten in the matrix form as

$$\mathbf{AC} = \gamma \mathbf{C} \tag{4.9}$$

where $\mathbf{A}$ is a square matrix given by

$$A_{gh} = S_g \delta_{gh} + \frac{U_{g-h}}{2K_z}(1 - \delta_{gh}). \tag{4.10}$$

If the total of N reciprocal lattice vectors is included in the expansion of the crystal potential $U(\mathbf{r})$ and the Bloch wave $b(\mathbf{k}, \mathbf{r})$, the size of matrix $\mathbf{A}$ is $N \times N$. Eigenvalues $\gamma^{(j)}$ can be found by solving the secular equation of the form

$$\det|\mathbf{A} - \gamma \mathbf{I}| = 0.$$

This determinant is nothing but a polynomial of the N-th degree in γ. This polynomial has N roots $\gamma^{(j)}$, $j = 1, ..., N$. Each eigenvalue $\gamma^{(j)}$ corresponds to the wave vector $\mathbf{k}^{(j)} = \mathbf{K} + \gamma^{(j)}\hat{\mathbf{z}}$ and a Bloch wave

$$b^{(j)}(\mathbf{k}^{(j)}, \mathbf{r}) = \sum_g C_g^{(j)}(\mathbf{K}_t) \exp[i(\mathbf{K} + \mathbf{g} + \gamma^{(j)}\hat{\mathbf{z}}) \cdot \mathbf{r}]. \tag{4.11}$$

The eigenvalue $\gamma^{(j)}$ is referred to in the German literature as the 'Anpassung' (Bethe, 1928).

4.3.2 Bound and free Bloch waves

The basic equations (4.5) of dynamical electron diffraction are similar to equations usually encountered in solid state physics in the problem of calculation of the electronic band structure of a crystalline material. These equations are in fact just a restatement of the original Schrödinger equation in momentum space, simplified by the fact that the Fourier component U_g of a periodic potential is only non-vanishing when $\mathbf{g}$ is one of the reciprocal lattice vectors. This observation illustrates a close link between the eigenvalues $\gamma^{(j)}$ introduced in the previous section and the band structure energies E_j studied in the electronic band structure calculations. In this section we explore this connection.

Consider a plane wave incident on the surface of the crystal. In this case above the crystal the wave function has the form

$$\psi_0(\mathbf{r}) = \exp(i\mathbf{k}_0 \cdot \mathbf{r}) = \exp(i\mathbf{K}_t \cdot \mathbf{X}) \exp(ik_{0z}z) = \Phi(\mathbf{X}) \exp(ik_{0z}z).$$

In this expression we have divided the wave function into two parts, i.e. a part $\Phi(\mathbf{X})$ that depends on the lateral coordinate $\mathbf{X}$, and a part $\exp(ik_{0_z}z)$ that depends on z. In a crystal both parts will be modified by the periodic potential. Since high-energy electron diffraction is dominated by forward scattering and is almost insensitive to the variation of the potential in the direction of the incident beam, we expect that in a crystal the z-dependent part of the wave function can be represented by an exponential factor $\exp(ik_z z)$ similar to that describing the propagation of electrons in the vacuum. It is therefore convenient to write the electron wave function in a crystal in the following form:

$$\psi(\mathbf{r}) = \Phi(\mathbf{X}, z)\exp(ik_0 z), \tag{4.12}$$

expecting that $\Phi(\mathbf{X}, z)$ is a slowly varying function of z.

In the above expression $\psi(\mathbf{r})$ is the exact solution of the Schrödinger equation, where the function $\Phi(\mathbf{X}, z)$ now also depends on z. Substituting this expression into the time-independent Schrödinger equation

$$\left[-\nabla^2 - U(\mathbf{r}) \right] \psi(\mathbf{r}) = k_0^2 \psi(\mathbf{r}), \tag{4.13}$$

we find

$$\left[-\left(\frac{\partial^2}{\partial x^2} + \frac{\partial^2}{\partial y^2} \right) - U(\mathbf{r}) \right] \Phi(\mathbf{X}, z) = 2ik_0 \frac{\partial \Phi(\mathbf{X}, z)}{\partial z}.$$

In deriving the above equation we have assumed that Φ is a slowly varying function of z and have neglected the second-order derivative $\partial^2 \Phi / \partial z^2$. For ZOLZ diffraction we may introduce a further approximation $U(\mathbf{r}) = U(\mathbf{X})$. We note that the Hamiltonian on the left-hand side of the above equation,

$$\hat{H}(\mathbf{X}) = -\left(\frac{\partial^2}{\partial x^2} + \frac{\partial^2}{\partial y^2} \right) - U(\mathbf{X}), \tag{4.14}$$

is independent of z. The wave function $\Phi(\mathbf{X}, z)$ may now be separated into the $\mathbf{X}$- and z-dependent parts, i.e. $\Phi(\mathbf{X}, z) = \tau(\mathbf{X})Z(z)$. This leads to (Berry, 1971, Bird, 1989)

$$\hat{H}(\mathbf{X})\tau_m(\mathbf{X}) = \mathcal{E}_m \tau_m(\mathbf{X})$$
$$2ik_0 \frac{\partial Z_m}{\partial z} = \mathcal{E}_m Z_m, \tag{4.15}$$

where $\mathcal{E}_m$ is an eigenvalue of the two-dimensional Hamiltonian $\hat{H}(\mathbf{X})$, i.e. the eigenvalue of an electron moving in two dimensions in the potential $U(\mathbf{X})$, and $\tau_m(\mathbf{X})$ is the corresponding two-dimensional eigenfunction. m is an index distinguishing solutions of the two-dimensional eigenvalue problem (4.15). For a given eigenvalue $\mathcal{E}_m$, the z-dependent part of the wave function Z_m equals

$$Z_m = A_m \exp\left(-i\frac{\mathcal{E}_m}{2k_0} z \right), \tag{4.16}$$

where A_m is a constant determined by the boundary condition. The m-th eigen-function $\psi_m(\mathbf{r})$ is therefore given by

$$\psi_m(\mathbf{r}) = \tau_m(\mathbf{X}) \exp\left[i\left(k_0 - \frac{\mathcal{E}_m}{2k_0}\right)z\right]. \tag{4.17}$$

In a crystal the two-dimensional Hamiltonian $\hat{H}$ is a periodic function of the coordinates x and y, and its eigenfunctions $\tau_m(\mathbf{X})$ and $\psi_m(\mathbf{r})$ naturally satisfy the Bloch theorem. These functions can therefore be represented by eqn (4.11), which may be rewritten as

$$b^{(j)}(\mathbf{X},z) = \left\{\sum_g C_g^{(j)}(\mathbf{K}_t)\exp[i(\mathbf{K}_t + \mathbf{g})\cdot\mathbf{X}]\right\}\exp[i(K_z + \gamma^{(j)})z].$$

The sum over g can be identified with $\tau_j(\mathbf{X})$ in eqn (4.17). By equating the argument of the remaining exponential term with that in eqn (4.17), we find

$$\mathcal{E}_j = 2k_0(k_0 - K_z - \gamma^{(j)}), \tag{4.18}$$

where k_0 and γ are given in units of Å^{-1} and the energy $\mathcal{E}_j$ is given in Å^{-2}. The usefulness of the energy concept stems from the fact that the value of $\mathcal{E}_j$ gives an indication as to whether the corresponding Bloch wave is similar to a bound state or to a free state. In the case where $\mathcal{E}_j < 0$ the corresponding Bloch wave resembles a bound state localized around atomic strings in each of the projected unit cells as illustrated in Figs 4.4 and 4.5.

We now consider an example that illustrates the concept of atomic strings and bound Bloch waves. Figure 4.3 shows a structural model of a 1T-VSe$_2$ unit cell and its projection in the [001] direction. This is a trigonal layered crystal (see e.g. (Bird, 1989)) and its lattice parameters are $a = b = 3.36$ Å, $c = 6.10$ Å. Each unit cell contains three atoms, namely one V atom (atom 1) situated in the plane $z = 0$, and two Se atoms situated at $z = 0.264c$ (atom 5) and $z = 0.736c$ (atom 6) respectively. There are three atomic strings in the [001] projected unit cell shown in Fig. 4.3b. The strings situated in the four corners of the unit cell (marked as 1,2,3,4) are equivalent and they consist of V atoms. The two strings situated inside the unit cell (marked as 5,6) consist of Se atoms.

For a 200 keV primary beam energy, there are three bound Bloch states at energies $\mathcal{E}_1 = -15.134$ Å^{-2}, $\mathcal{E}_2 = -14.704$ Å^{-2} and $\mathcal{E}_3 = -5.443$ Å^{-2}. Of these three states, the first two are associated with the Se atomic strings and the corresponding Bloch waves are localized around these strings (see Fig. 4.4 (top)). The third bound Bloch state is associated with the V atomic string (see Fig. 4.4 (bottom)). Figure 4.4 also shows the two-dimensional distributions of electron density of the three bound Bloch waves, i.e. the distribution of $|b^{(i)}(\mathbf{X})|^2$ (i=1,2,3) calculated for the exact zone axis incidence $\mathbf{K}_t = 0$. Figure 4.4 (top) gives the electron density map of Bloch wave number 2 (this map is almost identical to that of Bloch wave number 1, which is not shown in the figure)

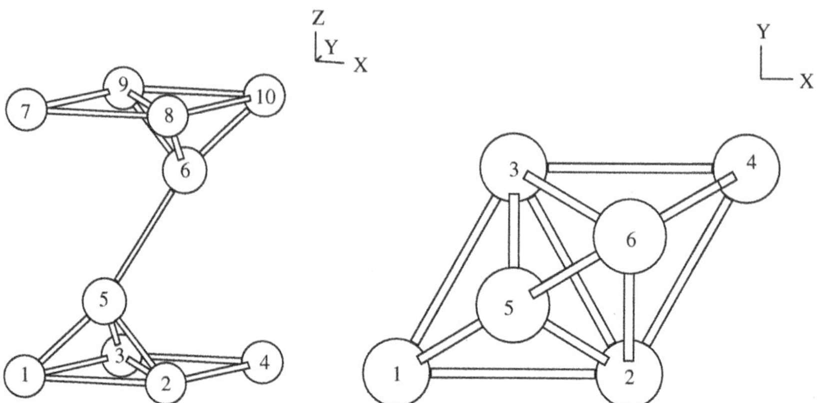

FIG. 4.3. (a) Atomic model for 1T-VSe$_2$ (left) and (b) its [001] projection (right).

showing two high-density regions (white spots) around the two Se atomic strings within the unit cell (marked in the figure by white lines). Figure 4.4 (bottom) is the electron density map of Bloch wave number 3. This density map shows peaks of electron density around the four corners of the unit cell that correspond to the atomic strings of V atoms.

The bound Bloch wave associated with the atomic string of V atoms has a higher energy (-5.443 Å^{-2}) than that associated with the atomic strings of Se atoms (-15.134 Å^{-2} and -14.704 Å^{-2}). This difference is associated with the fact that a Se atom is heavier (the atomic number of Se is 34) than a V atom (with atomic number 23). Figure 4.5 shows the real parts of the two bound Bloch waves associated with the atomic strings of Se atoms. It is seen that the two bound Bloch waves are in fact anti-bonding (Fig. 4.5a) and bonding (Fig. 4.5b) combinations of the two atomic 1s orbitals localized around atomic strings 5 and 6 shown in Fig. 4.3b. Since the symmetries of the two bound Bloch waves are different, we expect that their excitation probabilities will also be different.

We will see later that for the exact zone axis incidence the excitation probability of the anti-bonding Bloch state shown in Fig. 4.5a is equal to zero, while that of the Bloch state shown in Fig. 4.5b is finite and is equal to 0.608.

4.3.3 Dispersion surfaces and band structure

Both the eigenvalue $\gamma^{(j)}$ and energy $\mathcal{E}_j$ of the two-dimensional Hamiltonian $\hat{H}$ depend on the tangential component $\mathbf{K}_t$ of the incident electron wave vector, i.e. $\gamma^{(j)} = \gamma^{(j)}(\mathbf{K}_t)$, $\mathcal{E}_j = \mathcal{E}_j(\mathbf{K}_t)$. While the surface $\gamma^{(j)}(\mathbf{K}_t)$ is usually referred to as the dispersion surface, the multisheet surface $\mathcal{E}_j(\mathbf{K}_t)$ is the two-dimensional band structure. Figure 4.6 shows a one-dimensional cut of 10 branches of (a) the dispersion surface and (b) the band structure. We first note that there exists a one-to-one correspondence between branches of the band structure and branches of the dispersion surface. The correspondence between the two surfaces is, however, inverted, i.e. the energy bands with lower energy correspond

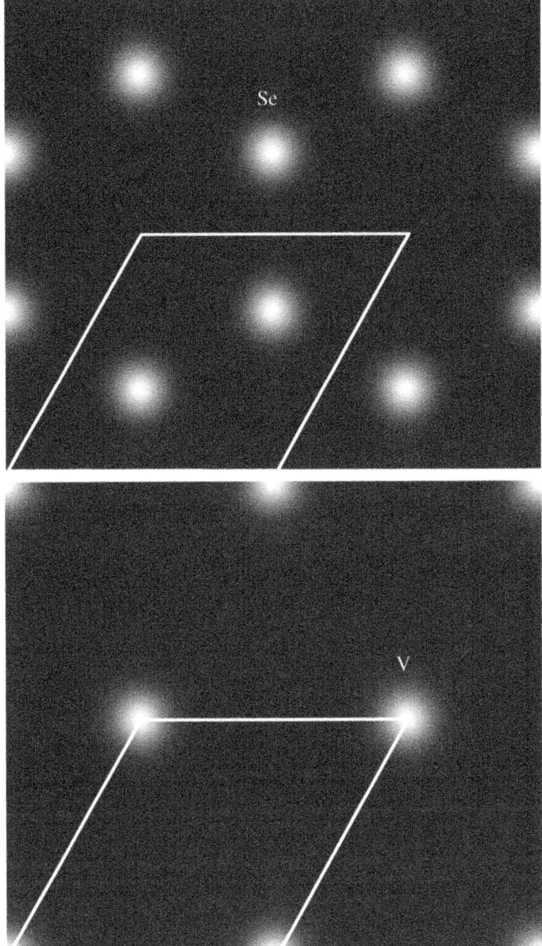

FIG. 4.4. Electron density maps of Bloch wave number 2 (top) and Bloch wave number 3 (bottom) calculated for a 1T-VSe$_2$ crystal for the exact zone axis incidence and primary beam energy of 200 keV.

to higher branches of the dispersion surface. These features reflect the definition (4.18) of the energy bands, i.e. $\mathcal{E}_j(\mathbf{K}_t) = 2k_0(k_0 - K_z - \gamma^{(j)}(\mathbf{K}_t))$.

A band structure similar to the one shown in Fig. 4.6b typically consists of one or a few low-lying bands with energies situated well below zero. These bands describe tightly bound Bloch states, and the corresponding Bloch waves are localized around atomic strings (see for example Fig. 4.4). Since these tightly bound waves are well localized in real space, we may expect that in reciprocal space they will be flat or dispersionless, i.e. these energy bands will show little dependence on $\mathbf{K}_t$. This is indeed the case for the lowest three energy bands of Fig.

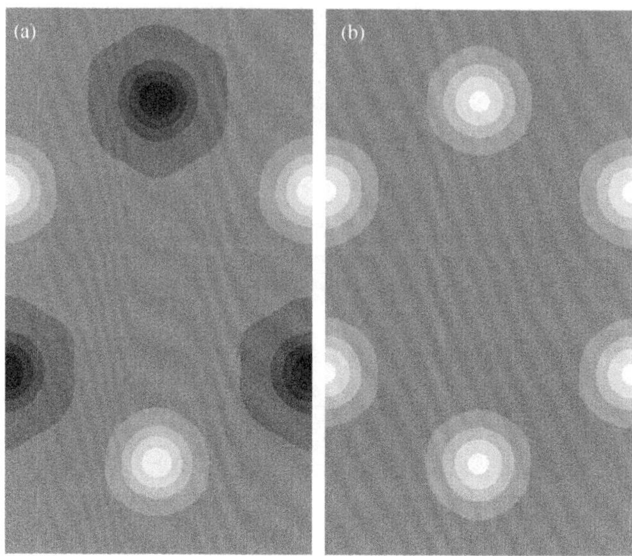

FIG. 4.5. Real parts of (a) Bloch wave number 1, $b^{(1)}(\mathbf{x})$, and (b) Bloch wave number 2, $b^{(2)}(\mathbf{X})$, calculated for a 1T-VSe$_2$ crystal and 200 keV energy of the incident electrons.

4.6b. We should also note that the tightly bound Bloch waves localized around a particular type of atomic string in a crystal are hardly affected by the presence of adjacent atomic strings. As a result the dispersion surface characterizing these Bloch states is nearly isotropic, i.e. the dispersion relation is nearly independent of the direction of $\mathbf{K}_t$. Although Fig. 4.6 is plotted along the [100] direction, the first three branches (j=1,2,3) remain almost the same when plotted along other directions, such as [110] and [010]. On the other hand, energy bands corresponding to higher energies, especially those with $\mathcal{E}_j(\mathbf{K}_t) > 0$, show strong dependence on the direction in which a curve is plotted. This is because Bloch states with $\mathcal{E}_j(\mathbf{K}_t) > 0$ describe less strongly bound Bloch waves that are delocalized in real space. In this case there exists a considerable overlap between wave fields associated with different atomic strings, and the dispersion depends strongly on the crystal environment. These branches are therefore anisotropic, and this may have a significant effect on diffraction phenomena (see (Peng and Zuo, 1999), and later discussion in section 4.3.5).

4.3.4 *Excitation of Bloch waves*

The spatial structure of Bloch waves is compatible with the translation symmetry of a crystal since they are eigenstates of the two-dimensional Hamiltonian $\check{H}$ given by (4.14). Although the number of possible Bloch waves is infinite, not all of them will be excited in a particular diffraction case and contribute to the total electron wave field in the crystal

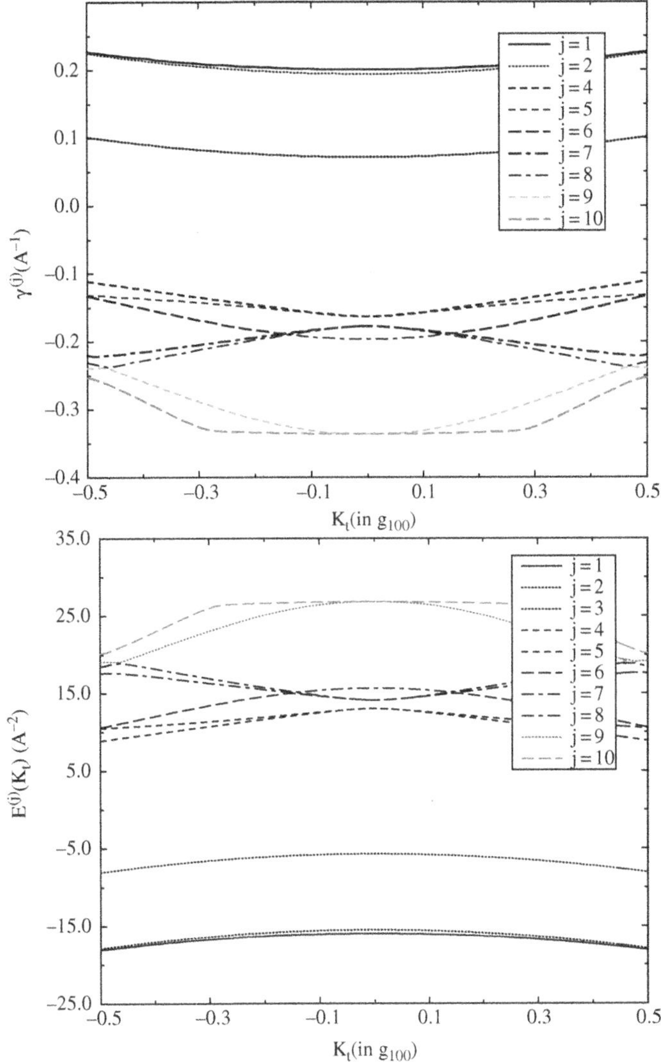

FIG. 4.6. (top) One-dimensional plot of the dispersion surface and (bottom) the band structure of 1T-VSe$_2$ along the [100] direction calculated for a primary beam energy of 200 keV.

$$\psi(\mathbf{r}) = \sum_j \alpha^{(j)}(\mathbf{K}_t) b^{(j)}(\mathbf{K}_t, \mathbf{r}). \qquad (4.19)$$

The contribution of a particular Bloch wave to the total wave field is described by the excitation amplitude $\alpha^{(j)}(\mathbf{K}_t)$ that is determined by the boundary conditions. Since in the THEED case backscattered electrons may be neglected, the boundary condition requires that the transmitted and diffracted wave amplitudes be continuous across the surface of the crystal. For a plane wave incident on the crystal surface we find that the amplitudes of the beams in the region above the crystal are given by $\psi_g = \delta_{g0}$, where $g = 0$ corresponds to the transmitted beam. In the crystal the diffracted beam amplitudes are

$$\psi_g(z) = \sum_j \alpha^{(j)} C_g^{(j)} \exp[i(K_z + \gamma^{(j)})z]. \tag{4.20}$$

At the top surface of the crystal where $z = 0$ the above equation gives the following expression for the g-th diffracted beam amplitude:

$$\psi_g(0) = \delta_{g0} = \sum_j \alpha^{(j)} C_g^{(j)}. \tag{4.21}$$

A particularly simple expression for $\alpha^{(j)}$ can be found in the case where we neglect the imaginary part of the potential associated with effects of inelastic scattering. In this case the crystal potential $V(\mathbf{r})$ and the corresponding re-scaled potential $U(\mathbf{r}) = -2mV(\mathbf{r})/\hbar^2$ are both real and $U(\mathbf{r}) = U^*(\mathbf{r})$, where the asterisk * denotes the complex conjugate. Using the Fourier expansions of both $U(\mathbf{r})$ and $U^*(\mathbf{r})$, we find

$$U(\mathbf{r}) = \sum_g U_g \exp(i\mathbf{g} \cdot \mathbf{r}) = U^*(\mathbf{r}) = \sum_g U_g^* \exp(-i\mathbf{g} \cdot \mathbf{r}) = \sum_g U_{-g}^* \exp(i\mathbf{g} \cdot \mathbf{r}).$$

Since these equalities are satisfied for all $\mathbf{r}$ we have $U_g = U_{-g}^*$. Using this result it can be readily shown that matrix $\mathbf{A}$ defined by eqn (4.9) is in fact a Hermitian matrix, i.e. $\mathbf{A} = \mathbf{A}^\dagger$, where $\mathbf{A}^\dagger$ denotes the transposed complex conjugate matrix. Since the matrix $\mathbf{A}$ is Hermitian all its eigenvalues are real. The eigenvectors are orthogonal to each other and form a complete set. They satisfy the orthogonality relations of the form

$$\sum_g C_g^{*(i)} C_g^{(j)} = \delta_{ij}, \quad \sum_j C_g^{*(j)} C_h^{(j)} = \delta_{gh}. \tag{4.22}$$

From these relations (4.22) it is easy to show that the matrix $\mathbf{C}$, defined by $(\mathbf{C})_{gj} = C_g^{(j)}$, is a unitary matrix, i.e. $\mathbf{C}^{-1} = \mathbf{C}^\dagger$. Furthermore, if the crystal possesses a centre of symmetry and the origin of coordinates is taken there then $\mathbf{C}$ becomes a real matrix, i.e. $\mathbf{C} = \mathbf{C}^*$. In this case $\mathbf{C}^{-1} = \tilde{\mathbf{C}}$ (where $\tilde{\mathbf{C}}$ denotes the transpose), or in other words $\mathbf{C}$ is an orthogonal matrix.

These relations may be used to find the excitation amplitudes $\alpha^{(j)}$. By comparing eqns (4.22) and (4.21) we immediately see that

$$\alpha^{(j)} = C_0^{*(j)}, \tag{4.23}$$

and obtain the following general solution for the diffracted beam amplitudes in the THEED geometry:

$$\psi_g(z) = \sum_j C_0^{*(j)} C_g^{(j)} \exp[i(K_z + \gamma^{(j)})z]. \qquad (4.24)$$

The excitation amplitude $\alpha^{(j)}$ of the Bloch wave j may be readily found numerically by following the procedure outlined above, once the basic equation (4.7) is solved and the corresponding eigenvectors $C_g^{(j)}$ are calculated. However, this procedure does not provide an explanation of why a particular Bloch wave may be strongly excited while the contribution of other Bloch waves to the electron wave field in the crystal may remain negligible. To understand this point we now examine the real-space approach that illustrates some important aspects of excitation of Bloch waves of high-energy electrons.

Equation (4.19) describes the real-space distribution of the wave field. At the top surface of the crystal (corresponding to the plane $z = 0$), this wave field matches the incident plane wave

$$\exp(i\mathbf{k}_0 \cdot \mathbf{X}) = \sum_j \alpha^{(j)} b^{(j)}(\mathbf{X}).$$

Multiplying the above equation by $b^{*(j)}(\mathbf{X})$ and noting that the two-dimensional Bloch waves are mutually orthogonal, i.e.

$$\int b^{*(j)}(\mathbf{X}) b^{(i)}(\mathbf{X}) d\mathbf{X} = A_c \delta_{i,j},$$

where the integration is performed over the area A_c of the two-dimensional unit cell, we find

$$\alpha^{(j)} = \frac{1}{A_c} \int \exp(i\mathbf{k}_0 \cdot \mathbf{X}) b^{*(j)}(\mathbf{X}) d\mathbf{X} = C_0^{*(j)},$$

which agrees with (4.23). In particular in the case of the zone axis incidence $\mathbf{k}_0 \cdot \mathbf{X} = 0$, the excitation amplitude $\alpha^{(j)}$ simply equals the integral of the Bloch wave over a two-dimensional unit cell. In the 1T-VSe$_2$ case discussed previously one of the most tightly bound Bloch waves corresponds to an anti-bonding combination of the two identical 1s atomic orbitals localized around the two Se atomic strings shown in Fig. 4.5a. The integration of this bound Bloch wave over a two-dimensional unit cell results in zero excitation amplitude ($\alpha^{(1)} = 0$) for this Bloch wave. On the other hand, for the bound Bloch wave corresponding to a bonding combination of the two 1s atomic orbitals (see Fig. 4.5b) the excitation amplitude is a maximum ($\alpha^{(2)} = 0.602$) for zone axis incidence. The bound Bloch wave number 3, which is localized around atomic strings of V atoms, corresponds to the 1s atomic state of a V atom. Since a V atom is lighter than an Se atom, we should expect that the amplitude of excitation of this Bloch wave, bound by the V atomic string, will be smaller than that of the Bloch wave number 2 bound by the Se atomic strings. In full agreement with our expectations, the integration of Bloch wave number 3 over the unit cell results in a smaller excitation amplitude ($\alpha^{(3)} = 0.488$) in comparison with the excitation amplitude of Bloch wave number 2 ($\alpha^{(2)} = 0.602$) localized around the heavier Se atomic strings.

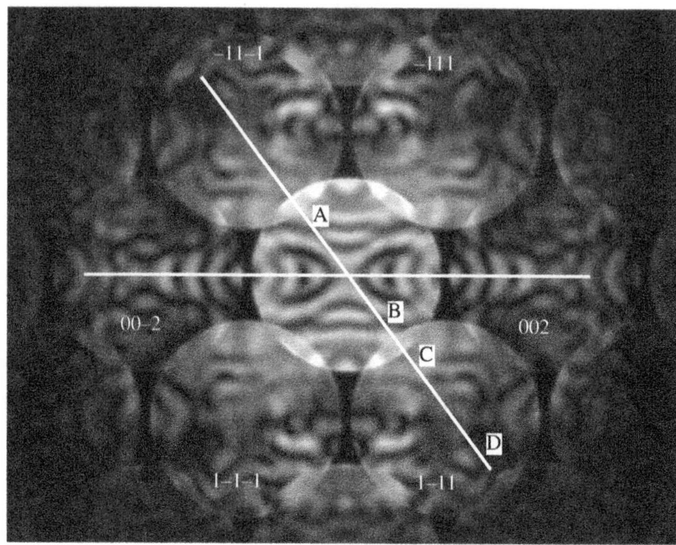

FIG. 4.7. Energy-filtered experimental Si[110] zone axis CBED pattern. The pattern was obtained for a primary beam energy of 200 keV, an energy window of 10 eV, and an electron probe size of 1.4 nm using a Philips CM200/FEG electron microscope.

In addition to the bound state of 1s type, Bloch states of 2p character may also be found around atomic strings formed by heavy atoms. These Bloch waves have positive amplitude on one side of the atomic string and negative amplitude on the other side of the string. These states are anti-symmetric with respect to either the y axis (for the $2p_x$ state) or the x axis (for the $2p_y$ state). The amplitude of excitation of a 2p Bloch state is given by an integral of an anti-symmetric function over the unit cell, which vanishes in the case where electrons are incident on the crystal in the direction parallel to the zone axis, i.e. $\alpha^{(2p)} = 0$ for $\mathbf{K}_t = 0$.

4.3.5 *Two and few Bloch wave approximations*

A calculation of diffracted beam amplitudes involves summation over Bloch waves, i.e. summation over j in eqn (4.24). In a dynamical diffraction calculation involving N beams, the number of Bloch waves resulting from solving the eigenvalue equation (4.7) equals the total number of beams N. As we mentioned earlier, only some Bloch waves are strongly excited in a crystal and contribute to the electron wave field. In the case of normal incidence, taking into account two or three Bloch waves is often sufficient to explain the main features observed in diffraction patterns. At the quantitative level many Bloch waves are required to account for all the observed features. Figure 4.7 shows an experimental CBED pattern taken from the Si[110] zone axis. For comparison in Fig. 4.8 we show calculated CBED rocking curves along the [1$\bar{1}$1] line scan in the CBED pattern. The first data point in Fig. 4.8 corresponds to point A of Fig. 4.7, the 98th data

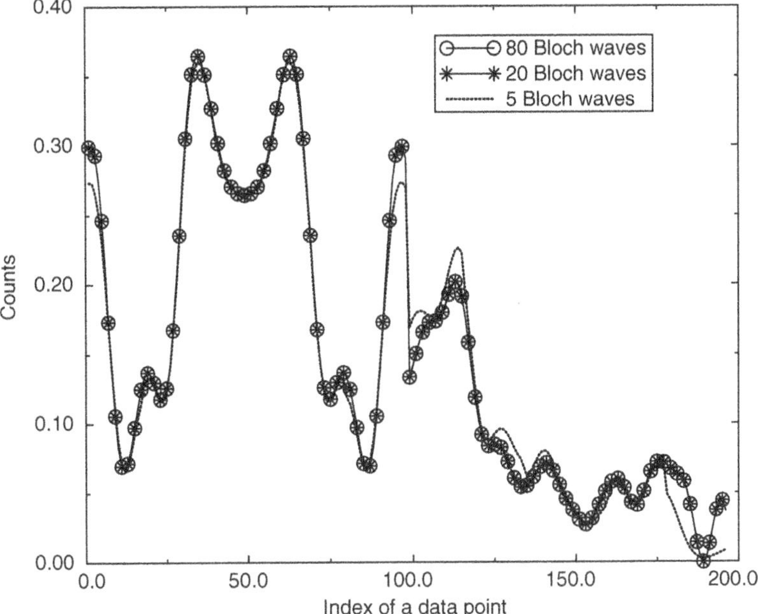

FIG. 4.8. CBED rocking curves for an Si single crystal. The primary beam energy is 200 keV and the thickness of the crystal is 369 nm. The three curves shown in the figure were calculated using 80 (circle plus solid line), 20 (star solid line), and 5 Bloch waves (dotted line) respectively. The curves exhibit variation of intensity along line AD in Fig. 4.7.

point corresponds to point B, the 99th data point corresponds to point C, and the 196th data point corresponds to point D. Figure 4.8 shows that although more than 80 beams are actually involved in the diffraction processes, only five Bloch waves are required in order to describe the main features of the rocking curve. By taking into account 20 Bloch waves it is possible to give a truly excellent description of all the features seen in the rocking curve.

The main features of a zone axis CBED pattern may be qualitatively explained by using only two or three Bloch waves. Figure 4.9a shows an experimental $[1\bar{1}1]$ zone axis CBED pattern obtained from an NiO sample. Figure 4.9b is a simulated zone-axis CBED pattern corresponding to the experimental conditions of Fig. 4.9a and crystal thickness $t = 2750$ Å. The Debye–Waller factors used in the simulation were $B_{Ni} = 0.1349$ Å^{-2} and $B_O = 0.2384$ Å^{-2}. These values correspond to the temperature of the specimen of 106.15 K (Gao et al., 1999). The computer simulation involved 193 beams, and it reproduced all the significant features seen in the pattern including the fine HOLZ lines within the bright-field (000) disc. We now show below that at the same time the pattern can be understood by using only a few Bloch waves.

Figure 4.10 shows plots of the band structure $\mathcal{E}_j(\mathbf{K}_t)$ and the corresponding

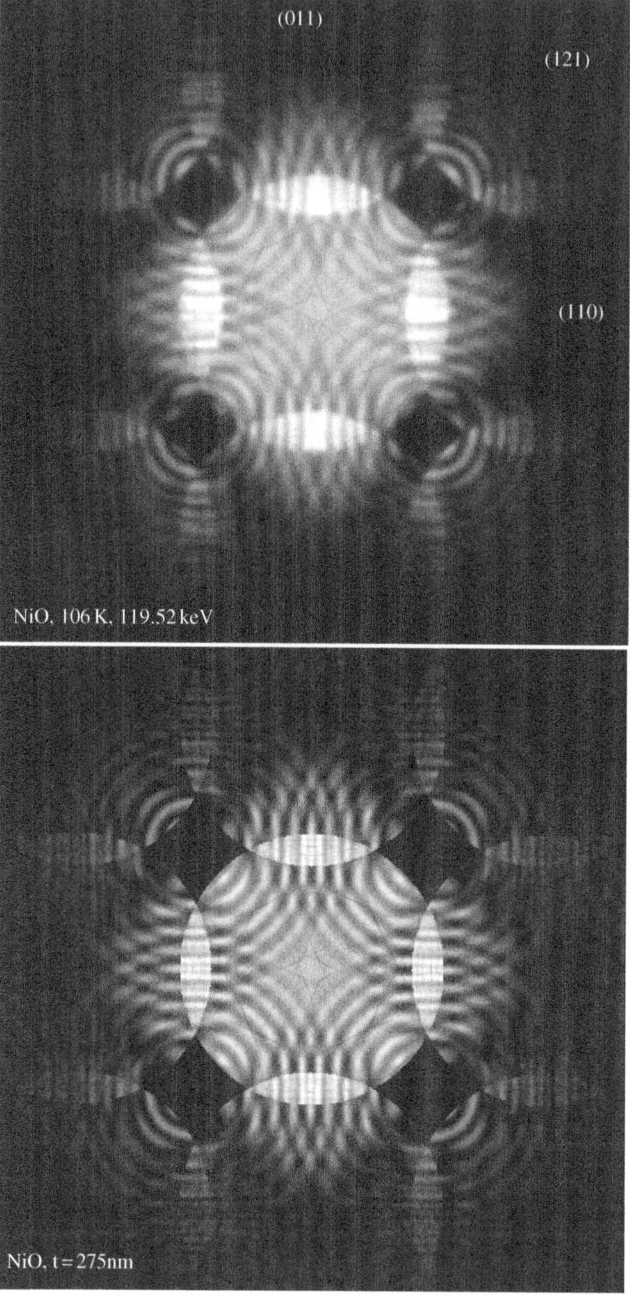

FIG. 4.9. Experimental (top) and simulated (bottom) $[1\bar{1}1]$ zone axis CBED patterns obtained from an NiO crystalline specimen. The pattern was recorded at a specimen temperature of 106 K and a primary beam energy of 119.15 keV, and the simulation was performed for a thickness of the crystal $t = 2750$ Å. From (Peng and Zuo, 1999).

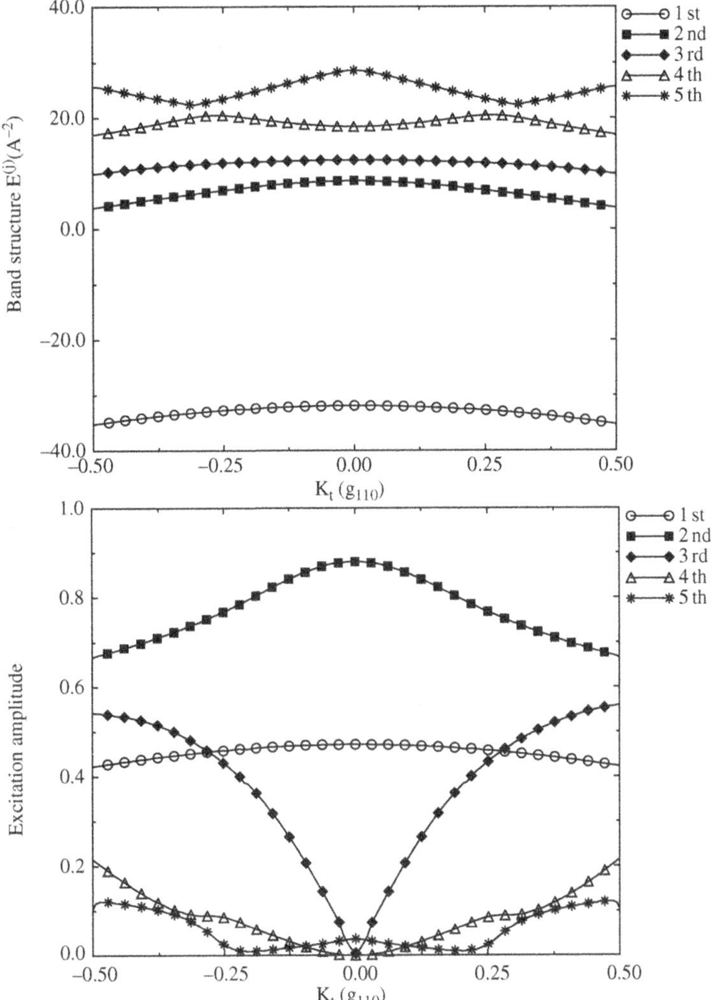

FIG. 4.10. Plots of the band structure $\mathcal{E}_j(\mathbf{K}_t)$ (top) and the corresponding excitation amplitudes $|\alpha^{(j)}(\mathbf{K}_t)|$ (bottom) of the first five most strongly excited Bloch waves corresponding to the [110] zone axis. From (Peng and Zuo, 1999).

excitation amplitudes $|\alpha^{(j)}(\mathbf{K}_t)|$ of the five most strongly excited Bloch waves calculated for an NiO crystal for $\mathbf{K}_t$ pointing in the $\langle 110 \rangle$ direction. The plots were calculated for conditions corresponding to the experimental CBED pattern of Fig. 4.9 (top). These plots show that among all the possible Bloch waves, one bound Bloch state with $\mathcal{E}_1 < 0$ and two free Bloch states corresponding to $\mathcal{E}_2 > 0$ and $\mathcal{E}_3 > 0$ dominate the pattern.

The most striking feature of Fig. 4.9 is the off-axis ring-pattern seen inside the

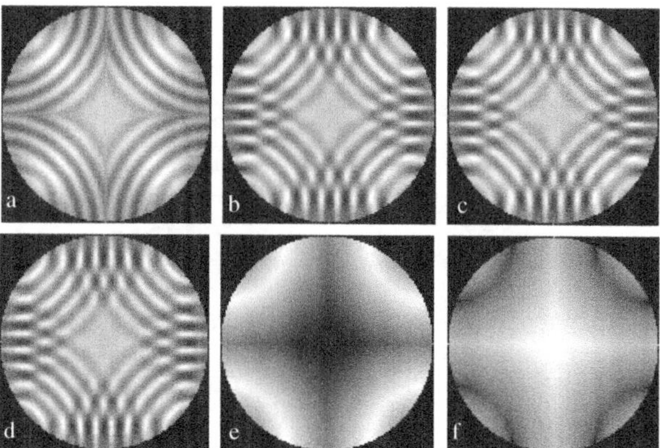

FIG. 4.11. CBED patterns of the transmitted disc formed by (a) two Bloch waves, numbers 2 and 3; (b) three Bloch waves, numbers 2, 3 and 4; (c) four Bloch waves, numbers 1, 2, 3 and 4; (d) all Bloch waves. (e) is the map of $\mathcal{E}_3(\mathbf{K}_t) - \mathcal{E}_2(\mathbf{K}_t)$ and (f) is the map of $\xi^{(3-2)}(\mathbf{K}_t)$. From (Peng and Zuo, 1999).

central CBED disc. This ring pattern looks like a simple fringe pattern formed as a result of interference between only two Bloch waves. This does not seem to agree with the plots of excitation amplitudes of Fig. 4.10 showing that at least three Bloch states contribute to the CBED pattern. Trial simulations using the three strongly excited Bloch waves reveal that among the three Bloch waves, only the two free Bloch waves with $\mathcal{E}_j > 0$ contribute significantly to the variation of intensity in the CBED pattern of NiO shown in Fig. 4.9 (bottom). These free Bloch states are the states numbered 2 and 3. Figure 4.11a shows a CBED pattern simulated using Bloch waves 2 and 3 alone. All the basic features of the off-axis ring pattern observed experimentally are reproduced here. By adding another free Bloch wave number 4 we obtain a CBED pattern (Fig. 4.11b) that is seen to be almost identical to that simulated using a large number of Bloch states (Fig. 4.11d). The inclusion of the bound Bloch state 1 in Fig. 4.11c does not have any noticeable effect on the intensity distributions in comparison with Fig. 4.11b, which was calculated using only 'free' Bloch waves. These results suggest that the basic features of the off-axis ring patterns may be reproduced by using a two Bloch wave approximation. In the two Bloch wave approximation the intensity of the transmitted beam has the form

$$
\begin{aligned}
I_0(\mathbf{K}_t) &= \left| \sum_{j=2}^{3} C_0^{(j)*} C_0^{(j)} \exp(i\gamma^{(j)} t) \right|^2 \\
&= |C_0^{(2)}(\mathbf{K}_t)|^4 + |C_0^{(3)}(\mathbf{K}_t)|^4 \\
&\quad + 2|C_0^{(2)}(\mathbf{K}_t)|^2 |C_0^{(3)}(\mathbf{K}_t)|^2 \cos\left\{ \left[\gamma^{(3)}(\mathbf{K}_t) - \gamma^{(2)}(\mathbf{K}_t) \right] t \right\}.
\end{aligned} \quad (4.25)
$$

This equation shows that fringes seen in the CBED pattern result mainly from interference between Bloch waves 2 and 3 and are described by the term $\cos[(\gamma^{(3)} - \gamma^{(2)})t]$.

The fringes observed in Fig. 4.9 and described by eqn (4.25) may be regarded as extinction distance contours, and the formation of the contours may be explained by using Fig. 4.11. This figure shows maps of the effective extinction distance associated with interference between Bloch waves 2 and 3, i.e. $\xi^{(3-2)}(\mathbf{K}_t) = 2\pi/\{[\gamma^{(3)}(\mathbf{K}_t) - \gamma^{(2)}(\mathbf{K}_t)]\}$. For a thin crystal where the thickness t of the crystal satisfies the condition $t \ll \xi^{(3-2)}(\mathbf{K}_t)$, the cosine function on the right-hand side of eqn (4.25) deviates very little from unity. In this case for all the points in the CBED disc we find

$$I_0(\mathbf{K}_t) \approx |C_0^{(2)}(\mathbf{K}_t)|^4 + |C_0^{(3)}(\mathbf{K}_t)|^4 + 2|C_0^{(2)}(\mathbf{K}_t)|^2|C_0^{(3)}(\mathbf{K}_t)|^2.$$

As the thickness of the crystal increases, the phase factor on the right-hand side of eqn (4.25) $[\gamma^{(3)} - \gamma^{(2)}]t$ first approaches π in the four corners of Fig. 4.11e at $t = 189$ Å. This value of t corresponds to a quarter of the extinction distance $\xi^{(3-2)}(\mathbf{K}_t)$, and the CBED pattern darkens first around these four corners. As the crystal thickness increases further, more fringes appear. This is because in a thicker crystal a small change in $\delta\gamma$ gives rise to a relatively large variation of the phase. Note that the CBED contrast never changes near the centre of the disc. This is because the symmetry of the Bloch wave 3 is of the $2p$ type, and this Bloch wave does not contribute to diffracted intensity at the center of the disc. The transmitted and diffracted beam intensities are in this case determined entirely by the Bloch wave number 2 and are therefore independent of the thickness of the crystal.

We now consider Bloch wave number 1. Figure 4.10 shows that this Bloch wave is strongly excited for all the points in the transmitted CBED disc. However, this Bloch wave does not contribute substantially to the observed contrast of CBED patterns. To understand this effect we note that Bloch state 1 is a tightly bound state characterized by a negative value of $\mathcal{E}_1$ and localized around atomic strings. This Bloch wave is therefore strongly absorbed as a result of large-angle thermal diffuse scattering. Quantitatively this is reflected in a large value of the imaginary part of the eigenvalue $Im\{\gamma^{(1)}\} = \mu^{(1)}$. The intensity of the Bloch wave is given by

$$I^{(1)}(\mathbf{r}) = |C_0^{(1)*} \sum_g C_g^{(1)} \exp\{i\mathbf{g} \cdot \mathbf{r}\} \exp(i\gamma^{(1)}z)|^2 = I^{(1)}(x, y, z = 0)\exp(-2\mu z).$$

The absorption length ℓ $(= (2\mu)^{-1})$ of this Bloch state is of the order of 400 Å. If the thickness of the crystal t equals ℓ, then 63% of the intensity of the Bloch wave will be absorbed through large-angle inelastic scattering. As a result, the tightly bound Bloch wave will hardly contribute to the contrast of the CBED pattern at $t = 2750$ Å corresponding to Fig. 4.9b. For zone axis incidence, the absorption length ℓ of the free Bloch waves exceeds 10000 Å. These free Bloch waves are sometimes called the channelled Bloch waves to emphasize the fact that they may channel through a crystal with no significant absorption.

Our analysis shows that although only a few *Bloch waves* are required to explain the main features of experimental zone axis CBED patterns, many *diffracted beams* are actually involved in the diffraction processes. This applies particularly to the case of tightly bound Bloch waves. These waves are localized around atomic strings and the Fourier expansion of these states includes a large number of plane waves. The widely used two-beam approximation (see e.g. (Hirsch et al., 1977)) may in fact be better understood as the two Bloch wave approximation, since it is rarely true that only two strongly excited beams form the diffraction pattern.

4.3.6 *Propagation of Bloch waves*

In this section we consider propagation of Bloch waves in the crystal. For a wave vector $\mathbf{k} = \mathbf{K} + \gamma\hat{\mathbf{z}}$ we have the following compact expression for the Bloch wave function:

$$b(\mathbf{r}) = \sum_g C_g \exp[i(\mathbf{k} + \mathbf{g}) \cdot \mathbf{r}].$$

The propagation of a Bloch wave is best described by the current density vector (here we use notation consistent with the Schrödinger equation (1.17))

$$\mathbf{j}(\mathbf{r}) = -\frac{i\hbar}{2m}[\psi^*(\mathbf{r})\nabla\psi(\mathbf{r}) - \psi(\mathbf{r})\nabla\psi^*(\mathbf{r})]$$

$$= \frac{\hbar}{2m}\sum_{g,h} C_g C_h^*(2\mathbf{k} + \mathbf{g} + \mathbf{h})\exp[i(\mathbf{g} - \mathbf{h}) \cdot \mathbf{r}]. \tag{4.26}$$

It can be readily shown that $\mathbf{j}(\mathbf{r} + \mathbf{R}) = \mathbf{j}(\mathbf{r})$, where $\mathbf{R}$ is a Bravais lattice vector, i.e. the current density is periodic in real space. By averaging the current density over a unit cell and noting that $\langle\exp[i(\mathbf{g} - \mathbf{h}) \cdot \mathbf{r}]\rangle = \delta_{gh}$, where $\langle...\rangle$ denote the volume average, we find the following simple expression for the average current density of a Bloch wave:

$$\bar{\mathbf{j}}(\mathbf{k}) = \frac{1}{\Omega_0}\int \mathbf{j}(\mathbf{r})d\mathbf{r} = \frac{\hbar}{m}\sum_g (\mathbf{k} + \mathbf{g})|C_g(\mathbf{k})|^2, \tag{4.27}$$

where Ω_0 is the volume of the unit cell and the integration is performed over the cell. In this expression we have explicitly stated that the average current density depends on the wave vector $\mathbf{k}$ of the Bloch state. The vector $\sum_g(\mathbf{k}+\mathbf{g})|C_g(\mathbf{k})|^2$ is normal to the branch of the dispersion surface for wave vector $\mathbf{k}$ (the dispersion surface was introduced in section (3.5.2)). The proof of this is as follows. The dispersion surface is a surface of constant energy $E(\mathbf{k})$, and therefore its normal is in the direction of $\nabla_{\mathbf{k}}E(\mathbf{k})$, where $\nabla_{\mathbf{k}}$ denotes the operator $(\partial/\partial k_x, \partial/\partial k_y, \partial/\partial k_z)$. From equations (3.22) and (3.23) the fundamental equation is

$$[k_0^2 - (\mathbf{k} + \mathbf{g})^2]C_g(\mathbf{k}) + \sum_h U_{g-h}C_h(\mathbf{k}) = 0. \tag{4.28}$$

Multiply this equation by C_g^* and sum over g. Since $\sum_g C_g C_g^* = 1$, we obtain

$$k_0^2 = \frac{2m}{\hbar^2} E(\mathbf{k}) = \sum_g (\mathbf{k} + \mathbf{g})^2 |C_g(\mathbf{k})|^2 - \sum_{g,h} U_{g-h} C_h(\mathbf{k}) C_g^*(\mathbf{k}).$$

Operating on this equation with $\nabla_\mathbf{k}$ gives

$$\nabla_\mathbf{k}(k_0^2) = \frac{2m}{\hbar^2} \nabla_\mathbf{k} E(\mathbf{k}) = \sum_g (\mathbf{k} + \mathbf{g})^2 \nabla_\mathbf{k} |C_g(\mathbf{k})|^2 - \sum_{g,h} U_{g-h} \nabla_\mathbf{k} [C_h(\mathbf{k}) C_g^*(\mathbf{k})]$$
$$+ 2 \sum_g (\mathbf{k} + \mathbf{g}) |C_g(\mathbf{k})|^2. \tag{4.29}$$

For a non-absorbing crystal $U_{g-h} = U_{h-g}^*$. Using this fact and the fundamental equation (4.28), manipulation of the first and second terms on the right-hand side of eqn (4.29) shows that their sum is $k_0^2 \nabla_\mathbf{k}[\sum_g |C_g(\mathbf{k})|^2]$, which is equal to zero since $\sum_g |C_g(\mathbf{k})|^2 = 1$. Therefore

$$\sum_g (\mathbf{k} + \mathbf{g}) |C_g(\mathbf{k})|^2 = \frac{m}{\hbar^2} \nabla_\mathbf{k} E(\mathbf{k}),$$

proving the result required, and equation (4.27) can then be written as

$$\bar{\mathbf{j}}(\mathbf{k}) = \hbar^{-1} \nabla_\mathbf{k} E(\mathbf{k}) = \frac{\hbar}{2m} \nabla_\mathbf{K} \left(\mathcal{E}(\mathbf{K}) \right), \tag{4.30}$$

where $\mathcal{E}(\mathbf{K}) = (2m/\hbar^2) E(\mathbf{k})$ is the energy expressed in units of $\hbar^2/2m$ as in equation (4.18). Equation (4.30) is a well-known result for the current density vector in the band theory of solids. As expected, the same result applies to the propagation of high-energy electrons in a solid. Similar considerations apply also to the case of X-ray diffraction by single crystals (Kato, 1958, Ewald, 1958). For tightly bound Bloch waves, where the dispersion surfaces are almost flat, electrons travel along crystal zone axes and are localized around atomic strings.

4.3.7 *Effects of absorption*

When the effect of inelastic scattering on the elastic wave field is no longer negligible, the potential of the crystal can no longer be regarded as a real quantity. As a result the matrix $\mathbf{A}$ is no longer Hermitian. In the early days of the development of electron diffraction, solving a general eigenvalue equation involving a general non-Hermitian complex matrix was considered to be a formidable mathematical problem. This has led to the development of a simplified approach for taking into account the effects of inelastic scattering (Hashimoto et al., 1962) that is based on a first-order perturbation treatment. If iU_g' is the contribution to the crystal potential associated with inelastic scattering and $i\Delta\gamma^{(j)}$ is the resulting perturbation of the j-th eigenvalue, from eqn (4.7) we find that

$$2K_z(S_g - \gamma^{(j)} - i\Delta\gamma^{(j)})C_g^{(j)} + iU_0'C_g^{(j)} + \sum_{h \neq g}(U_{g-h} + iU_{g-h}')C_h^{(j)} = 0.$$

Multiplying both sides of this equation by $C_g^{*(j)}$, summing over g, and using the orthogonality relations (4.22), we arrive at

$$\Delta\gamma^{(j)} = \frac{1}{2K_z}\left[U_0' + \sum_g \sum_{h \neq g} C_g^{*(j)}U_{g-h}'C_h^{(j)}\right]. \tag{4.31}$$

Substituting the above expression into eqn (4.24) we obtain the well-known formula for the diffracted beam amplitude

$$\psi_g(z) = \sum_j C_0^{*(j)}C_g^{(j)} \exp[i(K_z + \gamma^{(j)})z]\exp(-\Delta\gamma^{(i)}z). \tag{4.32}$$

In particular, eqn (4.31) shows that the mean absorption U_0' contributes equally to all the diffracted beams and this contribution is independent of the incident beam direction. Note that the eigenvectors $C_g^{(j)}$ entering the expression (4.31) are those found for a non-absorbing crystal, i.e. in this approximation the effect of absorption on the eigenvectors is neglected. This agrees fully with the conventional quantum-mechanical perturbation theory treatment of eigenstates and eigenvalues where the first-order correction applies to the eigenvalues only. Changes in the eigenvectors are second-order effects. For the majority of qualitative applications of THEED the treatment given above is sufficient.

4.4 The general treatment of THEED and HOLZ diffraction

4.4.1 *Kinematic geometry of HOLZ diffraction*

In the kinematic approximation the positions of HOLZ lines are given by the Bragg law. Graphically the geometry of HOLZ lines may be obtained from the intersection of spheres of radius k_0 drawn about the origin and the HOLZ reflection **g** shown in Fig. 4.12. Within the bright-field (BF) disc, a dark deficiency line occurs, which corresponds to the bright excess line within the HOLZ disc. When many HOLZ reflections are present, a network of deficiency lines will be observed in the BF disc, where each line corresponds to an excess line in a HOLZ disc. For moderately thick crystals each HOLZ disc shows several excess lines instead of one line as predicted by the kinematic theory. We will show in the next section that this effect of line splitting is associated with the dynamical nature of ZOLZ diffraction. Sometimes fluctuations of intensity of HOLZ reflections are observed along the lines. These fluctuations result from dynamical diffraction involving several HOLZ reflections.

4.4.2 *Formation of a HOLZ ring*

In the discussion of ZOLZ diffraction we have shown that for near zone axis incidence we can use the projected potential approximation. This approximation

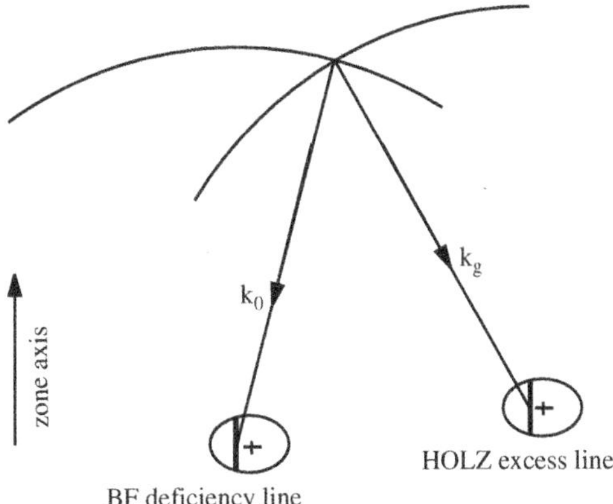

FIG. 4.12. Schematic diagram illustrating the kinematic geometry of HOLZ diffraction.

assumes that the three-dimensional potential of the crystal $U(\mathbf{r})$ may be approximated by a two-dimensional periodic potential $U^{(0)}(\mathbf{X})$. The electron wave function in the crystal may then be represented by a sum of two-dimensional Bloch waves that are eigenstates of the two-dimensional Hamiltonian $\hat{H}$, i.e. $\psi(\mathbf{X}) = \sum_j \alpha^{(j)} b^{(j)}(\mathbf{X})$. In this approximation the Bloch wave excitation amplitudes $\alpha^{(j)}$ are independent of the coordinate z of the electron, meaning that Bloch waves propagate through the crystal independently.

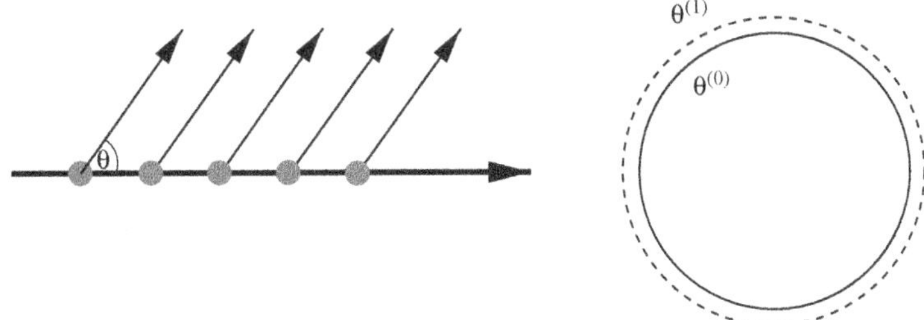

FIG. 4.13. Schematic diagram showing the diffraction of electrons localized in a tightly bound Bloch wave around an atomic string by atoms of the string. From (Peng, 1995).

Consider the process of propagation of a tightly bound Bloch state along the string of atoms illustrated schematically in Fig. 4.13. An electron wave travelling along an atomic string gives rise to scattered waves emerging from each atom. These waves interfere constructively in directions lying on a set of conical surfaces centred around the string of atoms as their axis. If we assume that λ_0 is the electron wavelength in the vacuum, λ is the wavelength in the crystal and λ' is the wavelength in the space between the atoms, then the path length difference between two successive wavelets is given by

$$\Delta L = c\mu' - c\cos\theta\mu,$$

where c is the interatomic spacing in the direction of the string, and $\mu = \lambda_0/\lambda$ and $\mu' = \lambda_0/\lambda'$ are the electron refraction indices in the crystal and along the atomic string respectively. Constructive interference will occur far away from the crystal if this path length difference is equal to a multiple of the electron wavelength λ_0, i.e. if $\Delta L = n\lambda_0$, where n is an integer. The semi-angles θ of the cones are therefore determined by the relation

$$\frac{c}{\lambda'} - \frac{c\cos\theta}{\lambda} = n. \qquad (4.33)$$

Neglecting relativistic effects and noting that $|V| \ll E$, where E is the energy of the incident electrons and V is the effective additional negative inner potential seen by the electron travelling along the atomic string, we find

$$\lambda = \lambda'\sqrt{1 - \frac{V}{E}} \approx \lambda'\left(1 - \frac{V}{2E}\right).$$

Equation (4.33) then reduces to

$$\cos\theta = 1 - \frac{n\lambda}{c} - \frac{V}{2E}. \qquad (4.34)$$

In a diffraction pattern this scattering will generate a set of concentric rings for each $n = 1, 2, 3\ldots$, and the radius of each ring will be slightly reduced owing to the presence of the positive third term in eqn (4.34). The rings result from the periodicity in the distribution of atoms in the direction of the electron beam and are in fact HOLZ rings.

While electrons occupying the ZOLZ Bloch states may be scattered out of these states and form HOLZ rings, the number of electrons scattered out of an individual Bloch state depends on the real-space distribution of electron density in that state. This leads to an imbalance between the numbers of electrons scattered out of and into various states and destroys the notion of an ideal ZOLZ approximation. Consequently electrons will be redistributed among the ZOLZ Bloch states, and the excitation amplitudes $\alpha^{(j)}$ will no longer remain constant as a function of the coordinate z in the direction of the zone axis. The redistribution of electrons among the ZOLZ Bloch waves will generate dark fine lines in CBED discs that can be observed experimentally. For example, these deficiency lines can be seen in Fig. 4.12. In the limit of a thin crystal these dark deficiency lines are generated by electrons that initially travelled in directions defined by these dark fine lines but have subsequently been scattered towards HOLZ rings. For crystals of small and moderate thickness the HOLZ rings are usually bright, and the fine lines seen in the ZOLZ discs are dark. This contrast sometimes changes for thicker crystals where fine lines in the ZOLZ discs may no longer appear as deficiency lines. This transition results from dynamical diffraction, i.e. electrons may be scattered not only out of the ZOLZ Bloch states towards HOLZ diffracted beams, but also back from the HOLZ beams into the ZOLZ Bloch states, and so form fairly complicated interference patterns.

While the independent Bloch wave picture is no longer valid when HOLZ interactions are present, the ZOLZ Bloch waves may be regarded as independent as far as the formation of the HOLZ rings is concerned. Since different Bloch waves have different real-space distributions, different Bloch waves effectively feel different potentials V when they travel along strings of atoms in a crystal. An estimate of the value of this potential is given by the matrix element (Peng and Gjønnes, 1989)

$$V^{(j)} = \langle b^{(j)}|V(\mathbf{r}) - V_0|b^{(j)}\rangle. \tag{4.35}$$

By substituting the Fourier expansions of the wave functions and the crystal potential into this equation, we find

$$V^{(j)} = \frac{1}{\Omega_0} \int_{\Omega_0} b^{*(j)}(\mathbf{r})[V(\mathbf{r}) - V_0]b^{(j)}(\mathbf{r})d\mathbf{r} \tag{4.36}$$

$$= \sum_g \sum_h \sum_{l \neq 0} C_g^{*(j)} V_l C_h^{(j)} \frac{1}{\Omega_0} \int_{\Omega_0} \exp[i(\mathbf{h} - \mathbf{g} + \mathbf{l}) \cdot \mathbf{r}]d\mathbf{r}$$

$$= \sum_g \sum_h \sum_{l \neq 0} C_g^{*(j)} V_l C_h^{(j)} \delta(\mathbf{h} - \mathbf{g} + \mathbf{l}) = \sum_g \sum_{h \neq g} C_g^{*(j)} V_{g-h} C_h^{(j)},$$

where Ω_0 is the volume of a unit cell. We can manipulate this expression further using the fundamental equation (4.7). Multiplying the equation by $C_g^{*(j)}$, summing over $\mathbf{g}$, and noting that $U_{g-h} = -2mV_{g-h}/\hbar^2$, we find

$$2K_z \sum_g C_g^{*(j)}(\gamma^{(j)} - S_g)C_g^{(j)} + \frac{2m}{\hbar^2} \sum_g \sum_{h \neq g} C_g^{*(j)} V_{g-h} C_h^{(j)} = 0,$$

and therefore

$$V^{(j)} = \sum_g \sum_{h \neq g} C_g^{*(j)} V_{g-h} C_h^{(j)} = -\frac{\hbar^2 K_z}{m}\left[\gamma^{(j)} - \sum_g C_g^{*(j)} S_g C_g^{(j)}\right]. \tag{4.37}$$

By substituting (4.37) into (4.34) we arrive at

$$\cos\theta^{(j)} = 1 - \frac{n\lambda}{c} - \frac{V^{(j)}}{2E} = 1 - \frac{n\lambda}{c} + \frac{\hbar^2 K_z}{2mE}\left[\gamma^{(j)} - \sum_g |C_g^{(j)}|^2 S_g\right], \tag{4.38}$$

where the last term involving $|C_g|^2 S_g$ may be neglected for all the strongly excited Bloch waves, since for these waves $|C_g|$ has appreciable amplitude only if

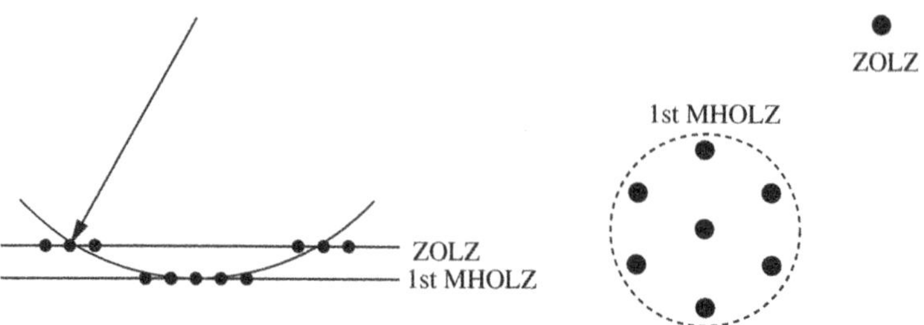

FIG. 4.14. Schematic diagram showing the inverse HOLZ diffraction geometry. From (Peng, 1995).

the corresponding excitation error S_g is small. This equation shows that each strongly excited Bloch wave will produce a HOLZ ring. Rings are formed by non-degenerate Bloch states, and states corresponding to different eigenvalues $\gamma^{(j)}$ are separated in a diffraction pattern. Summarizing we may say that HOLZ rings result from diffraction of electron waves by the potential variation along the incident beam direction, and that each HOLZ ring corresponds to a tightly bound Bloch wave, i.e. each HOLZ ring is generated by diffraction of a particular Bloch wave by the variation of the potential along the beam direction. Different Bloch waves with different eigenvalues $\gamma^{(j)}$ give rise to distinct HOLZ rings. We therefore observe the separation of different Bloch states in reciprocal space illustrated schematically in Fig. 4.13. A simple application of the reciprocity principle then suggests that if we send an electron beam down one of the directions $\theta^{(j)}$ (see Fig. 4.14), along which constructive interference between scattered electron waves by successive atoms along an atom string occurs, we can selectively excite different Bloch states and images of individual Bloch waves can be formed using this approach (Dudarev and Peng, 1993b, Rossouw et al., 1998). This case will be discussed in more detail later in Chapter 6 in connection with resonance scattering of electrons.

4.4.3 *Distribution of intensity in HOLZ patterns*

The arguments given in the previous section are based on an assumption that while the interaction among ZOLZ reflections is strong and ZOLZ diffraction is dynamical, interaction between HOLZ and ZOLZ reflections is much weaker and consequently HOLZ diffraction may be regarded as kinematic. This picture is sketched in Fig. 4.15. In the left-hand part of the figure coupling between HOLZ reflections is neglected and it is assumed that HOLZ diffraction occurs only once. Since the kinematic approximation is adopted for the HOLZ reflection, it is represented in this diagram by a sphere, exactly as in the kinematic case (see Fig. 4.12). On the other hand, the dynamical nature of ZOLZ diffraction is also taken into consideration. The kinematic sphere centred about the origin is replaced by a set of branches of the ZOLZ dispersion surface, and each intersection of the ZOLZ dispersion surface with the HOLZ sphere gives rise to a ring within the HOLZ disc. Whether or not these rings will be visible depends on several factors which we discuss in this section. HOLZ reflections may in principle interact with each other via ZOLZ reflections as shown in the right-hand part of Fig. 4.15. This results in intensity variations along the HOLZ rings. Experimentally this coupling is often observed around intersections of Kikuchi lines and bands with HOLZ rings. This effect may be avoided in experiments by slightly tilting the sample or the incident electron beam.

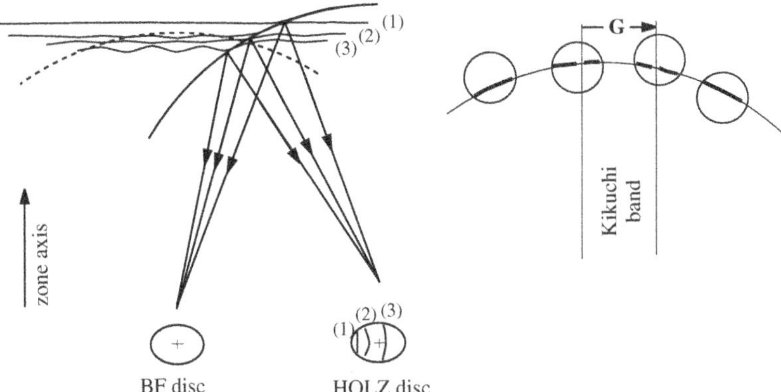

FIG. 4.15. Schematic diagram illustrating the formation of dynamical HOLZ rings and the effect of dynamical ZOLZ diffraction on HOLZ rings.

The fact that the interaction between HOLZ and ZOLZ reflections is much weaker than the interaction involving ZOLZ reflections alone suggests that a perturbation treatment may be developed to estimate the intensity of HOLZ effects and to identify the key parameters controlling HOLZ reflections. Starting from the Schrödinger equation (3.19), we write the total electron wave function as

$$\psi = \psi_0 + \sum_{g^{(n)}} \psi_{g^{(n)}}, \tag{4.39}$$

where the $g^{(n)}$ denote HOLZ reflections and $\psi_{g^{(n)}}$ are the corresponding diffracted beam amplitudes. The first-order Born approximation gives

$$[\nabla^2 + k_0^2]\psi = -U(\mathbf{r})\psi_0, \tag{4.40}$$

in which ψ_0 is the electron wave function associated with ZOLZ reflections $\mathbf{g}^{(0)} = \mathbf{G}$, given by

$$\psi_0 = \sum_j \alpha^{(j)} \sum_{g^{(0)}} C_{g^{(0)}}^{(j)} \exp(i\mathbf{k}_{g^{(0)}}^{(j)} \cdot \mathbf{r}), \tag{4.41}$$

and $\psi_{g^{(n)}}$ is that associated with the $g^{(n)}$ reflection, namely

$$\psi_{g^{(n)}} = \sum_j \phi_{g^{(n)}}^{(j)}(z) \exp(i\mathbf{k}_{g^{(n)}}^{(j)} \cdot \mathbf{r}). \tag{4.42}$$

Similarly we write the crystal potential as

$$U(\mathbf{r}) = U^{(0)}(\mathbf{X}) + \sum_{g^{(n)}} U_{g^{(n)}} \exp(i\mathbf{g}^{(n)} \cdot \mathbf{r}), \qquad (4.43)$$

where $U^{(0)}(\mathbf{X})$ is the projected potential. Substitution of (4.39) and (4.43) into (4.40) gives

$$(\nabla^2 + k_0^2) \left[\psi_0 + \sum_{g^{(n)}} \psi_{g^{(n)}} \right] = - \left[U^{(0)} + \sum_{g^{(n)}} U_{g^{(n)}} \exp(i\mathbf{g}^{(n)} \cdot \mathbf{r}) \right] \psi_0.$$

Noticing that $[\nabla^2 + k_0^2]\psi_0 = -U^{(0)}\psi_0$, for the term on the left-hand side of the above equation we obtain

$$\{\nabla^2 + k_0^2\} \sum_{g^{(n)}} \psi_{g^{(n)}} = \sum_{g^{(n)}} \sum_j \{\nabla^2 + k_0^2\} \phi_{g^{(n)}}^{(j)}(z) \exp(i\mathbf{k}_{g^{(n)}}^{(j)} \cdot \mathbf{r})$$

$$= \sum_{g^{(n)}} \sum_j \left\{ \frac{d^2}{dz^2} + 2i(k_{g^{(n)}}^{(j)})_z \frac{d}{dz} + [k_0^2 - (k_{g^{(n)}}^{(j)})^2] \right\} \phi_{g^{(n)}}^{(j)}(z) \exp(i\mathbf{k}_{g^{(n)}}^{(j)} \cdot \mathbf{r}).$$

The right-hand side of the equation has the form

$$\left[\sum_{g^{(n)}} U_{g^{(n)}} \exp(i\mathbf{g}^{(n)} \cdot \mathbf{r}) \right] \psi_0 = \sum_{g^{(n)}} U_{g^{(n)}} \exp(i\mathbf{g}^{(n)} \cdot \mathbf{r}) \sum_{j,g^{(0)}} \alpha^{(j)} C_{g^{(0)}}^{(j)} \exp(i\mathbf{k}_{g^{(0)}}^{(j)} \cdot \mathbf{r})$$

$$= \sum_{g^{(n)}} \sum_{g^{(0)}} \sum_j U_{g^{(n)}} \alpha^{(j)} C_{g^{(0)}}^{(j)} \exp\left\{ i[\mathbf{k}_{g^{(n)}}^{(j)} + \mathbf{g}^{(0)}] \cdot \mathbf{r} \right\}$$

$$= \sum_{g^{(n)}} \sum_{g^{(0)}} \sum_j U_{g^{(n)} - g^{(0)}} \alpha^{(j)} C_{g^{(0)}}^{(j)} \exp(i\mathbf{k}_{g^{(n)}}^{(j)} \cdot \mathbf{r}).$$

Taking into account that $\phi_{g^{(n)}}^{(j)}(z)$ are relatively slowly varying functions of z, and that $k_z^{(j)} \approx k_{0_z}$, we finally arrive at

$$\sum_{g^{(n)},j} \left[\left(2ik_{0_z} \frac{d}{dz} + k_0^2 - (k_{g^{(n)}}^{(j)})^2 \right) \phi_{g^{(n)}}^{(j)}(z) + \alpha^{(j)} \sum_{g^{(0)}} C_{g^{(0)}}^{(j)} U_{g^{(n)} - g^{(0)}} \right] e^{i\mathbf{k}_{g^{(n)}}^{(j)} \mathbf{r}} = 0.$$

Since the exponential terms associated with different reciprocal lattice vectors are independent, to satisfy the above equation at all points $\mathbf{r}$ their coefficients must be zero, i.e.

$$2ik_{0_z} \frac{d}{dz} \phi_{g^{(n)}}^{(j)}(z) + [k_0^2 - (k_{g^{(n)}}^{(j)})^2] \phi_{g^{(n)}}^{(j)}(z) + \alpha^{(j)} \sum_{g^{(0)}} C_{g^{(0)}}^{(j)} U_{g^{(n)} - g^{(0)}} = 0. \quad (4.44)$$

Integration of the above equation gives the intensity distribution of the jth HOLZ ring within the HOLZ disc associated with the $g^{(n)}$ reflection

$$I^{(j)} \propto \left| \alpha^{(j)} \sum_{g^{(0)}} C^{(j)}_{g^{(0)}} U_{g^{(n)} - g^{(0)}} \right|^2 \frac{\sin^2(s^{(j)}_{g^{(n)}} z/2)}{(s^{(j)}_{g^{(n)}})^2}, \tag{4.45}$$

with $s^{(j)}_{g^{(n)}}$ given by

$$2k_0 s^{j}_{g^{(n)}} = k_0^2 - (\mathbf{k}_0 + \mathbf{g}^{(n)} + \gamma^{(j)} \mathbf{n})^2.$$

Equation (4.45) shows that while the amplitude of the diffracted beam originating from the j-th Bloch wave and forming the $g^{(n)}$ reflection is determined by

$$\alpha^{(j)} \sum_{g^{(0)}} C^{(j)}_{g^{(0)}} U_{g^{(n)} - g^{(0)}} = \alpha^{(j)} \beta^{(j)}_{g^{(n)}},$$

with $\beta^{(j)}_{g^{(n)}} = \sum_{g^{(0)}} C^{(j)}_{g^{(0)}} U_{g^{(n)} - g^{(0)}}$, HOLZ diffraction geometry is determined by $s^{(j)}_{g^{(n)}}$, the effective excitation error associated with the j-th HOLZ line and $g^{(n)}$ reflection. The central maximum position of the HOLZ line obeys $s^{(j)}_{g^{(n)}} = 0$, which gives the radius of the HOLZ ring $R^{(j)}$ as

$$R^{(j)} = \sqrt{2k_{0_z} g_z^{(n)} + 2g_z^{(n)} \gamma^{(j)} - 2k_{0_z} \gamma^{(i)} - g_z^{(n)2} - \gamma^{(j)2}},$$

and also the kinematic HOLZ ring radius R_0 as

$$R_0 = \sqrt{2k_{0_z} g_z^{(n)} \left(1 - \frac{g_z^{(n)}}{2k_{0z}} \right)}.$$

An important conclusion which may be drawn from the above analysis is that there is no diffraction from a ZOLZ Bloch state into a HOLZ excess line unless the ZOLZ Bloch state is appropriately excited, i.e. unless $\alpha^{(j)} \neq 0$. When the relevant Bloch wave is excited, the amplitude of the diffracted beam is proportional to the product $\alpha^{(j)} \beta^{(j)}_{g^{(n)}}$. It is not difficult to show that $\beta^{(j)}_{g^{(n)}}$ is in fact the Fourier coefficient of the modified real-space potential $U^{(n)}(\mathbf{X}) b^{(j)}(\mathbf{X})$ (Vincent et al., 1984) given by

$$U^{(n)}(\mathbf{X}) b^{(j)}(\mathbf{X}) = \sum_{G^{(n)}} U_{g^{(n)}} \exp(i\mathbf{G}^{(n)} \cdot \mathbf{X}) \times \sum_{G^{(0)}} C^{(j)}_{g^{(0)}} \exp(i\mathbf{G}_0 \cdot \mathbf{X})$$

$$= \sum_{G^{(n)}} \left\{ \sum_{G^{(0)}} U_{g^{(n)} - g^{(0)}} C^{(j)}_{g^{(0)}} \right\} \exp(i\mathbf{G}^{(n)} \cdot \mathbf{X}) = \sum_{G^{(n)}} \beta^{(j)}_{g^{(n)}} \exp(i\mathbf{G}^{(n)} \cdot \mathbf{X}),$$

and the necessary conditions for the appearance of a HOLZ line associated with the j-th Bloch wave and the $g^{(n)}$ reflection are therefore that (1) the conditional projected potential $U^{(n)}$ is not equal to zero, (2) the j-th Bloch wave is appropriately excited, and (3) $U^{(n)}(\mathbf{X})$ overlaps with Bloch wave $b^{(j)}(\mathbf{X})$. Recalling the definition of the conditional projected potential (4.3)

$$U^{(n)}(\mathbf{X}) = \frac{1}{c}\int_0^c U(\mathbf{X}, z)\exp(-ing_z z)dz,$$

we see that the projected atomic potentials are averaged potentials weighted with the factor $\exp(-ing_z z)$. Consider a simple cubic structure containing one atom per unit cell. Calculating the projection in the direction parallel to a $\langle 100\rangle$ axis, we have $\exp(-ing_{100}z_n) = 1$, where $z_n = 0$ and c is the z-component of the nth atom and $g_{100} = 2\pi/c$. Using this we see that $U^{(0)}(\mathbf{X})$, $U^{(1)}(\mathbf{x})$, $U^{(2)}(\mathbf{X})\ldots$ are all identical. In general, however, $\exp(-ingz_n)$ is a complex number, and the conditional projected potential $U^{(n)}(\mathbf{X})$ is complex. Take the CsCl structure as an example. This structure has two non-equivalent atomic strings running along the $\langle 100\rangle$ zone axis. Atoms A are located at the corners of the unit cell and atoms B are situated at the centre of the cell (see Fig. 4.16). We assume that atom A is a stronger scatterer (Cs), and in the figure it is represented by an open circle. Atom B (Cl) is situated at the centre of the cell and is represented by a filled circle. The position of this atom corresponds to $z = c/2$ and it contributes to $U^{(1)}(\mathbf{X})$ with a negative weighting factor $\exp(-i\pi) = -1$ and to $U^{(2)}(\mathbf{X})$ with a positive factor $\exp(-i2\pi) = 1$ as indicated in Fig. 4.16 by positive and negative signs.

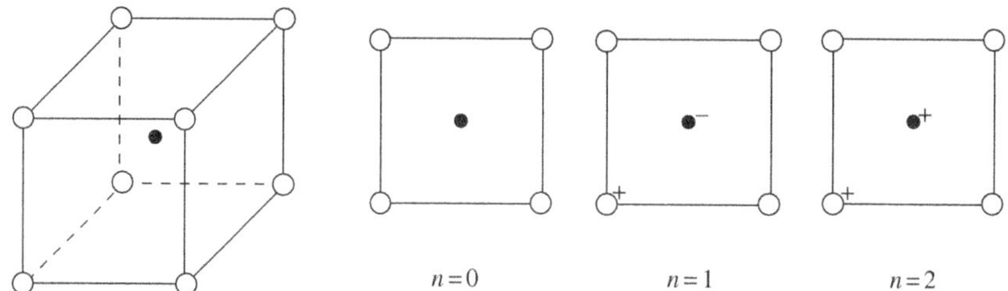

FIG. 4.16. Schematic diagram showing the structure of CsCl, and indicating the signs of the atomic contributions to the first three conditional projected potentials $U^{(n)}$ corresponding to $n = 0, 1, 2$.

To evaluate HOLZ amplitudes the conditional projected potential $U^{(n)}$ needs to be multiplied by the ZOLZ Bloch waves. For the sake of simplicity, we will consider only the most tightly bound 1s states. For the CsCl structure the strongest bound state corresponding to $j = 1$ is localized around the deeper potential well of the strings of atoms A situated at the corners of the cell, while the second Bloch state $j = 2$ is localized around the weaker potential well of the string of atoms B at the centre of the cell as shown in Figs 4.17a and b.

For the first-order HOLZ or FHOLZ reflection, $n = 1$, the two modified potential distributions $U^{(1)}b^{(1)}$ and $U^{(1)}b^{(2)}$ are sketched in Figs 4.17c and 4.17d.

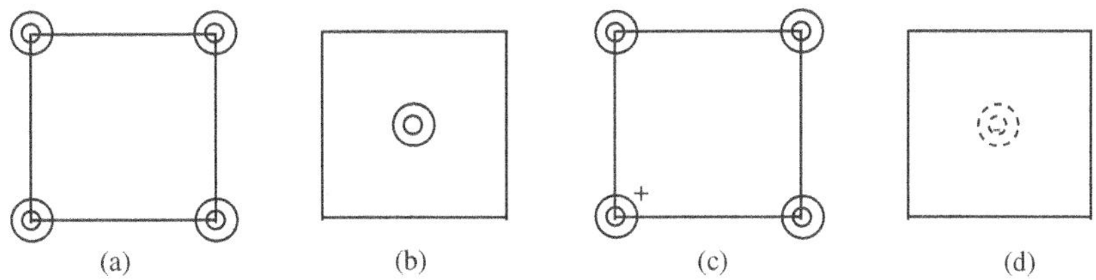

FIG. 4.17. Schematic diagram showing the two most tightly bound Bloch waves $b^{(1)}$ and $b^{(2)}$, and the two corresponding modified potential $U^{(1)}b^{(1)}$ and $U^{(1)}b^{(2)}$.

It is seen that the first modified potential shown in (c) is localized around the deeper well of atom A at the corners of the cell, and the corresponding HOLZ line results almost entirely from scattering by atoms A. On the other hand, the second modified potential shown in (d) is localized around the shallower well of strings formed by atoms B situated at the centre of the cell, giving rise to a distinct HOLZ line at a larger angle than the first HOLZ line. The HOLZ lines are therefore sensitive to the chemical composition of the crystal, and may be used in principle for locating substitutional atoms.

For relatively thin crystals the line width of HOLZ lines depends mainly on the crystal thickness through the shape transform function given by eqn (4.45). The resolution of the HOLZ fine line structure may therefore be improved by increasing the thickness of the specimen.

4.4.4 *General treatment of HOLZ diffraction*

In the previous section we discussed the main aspects of HOLZ diffraction, including the formation of the HOLZ rings and key parameters controlling HOLZ intensities. The treatment given above was approximate in the sense that the diffraction coupling between the ZOLZ and the HOLZ beams was treated kinematically and the dynamical coupling between the HOLZ reflections was neglected. While these approximations are useful for understanding the basic features of HOLZ diffraction, they are not appropriate for quantitative analysis of HOLZ intensities. In this section we will develop a general approach to HOLZ diffraction where we treat both the ZOLZ and the HOLZ reflections dynamically, making no formal distinction between them.

4.4.4.1 *Matrix representation* The treatment of HOLZ diffraction involving reflections with $g_z \neq 0$ is more complicated than that of ZOLZ diffraction, and the basic equation describing HOLZ diffraction is eqn (4.5). Substituting the definition $\mathbf{k} = \mathbf{K} + \gamma^{(j)}\hat{\mathbf{z}}$ into this equation and neglecting a second-order term in $\gamma^{(j)}$, we obtain a first-order eigenvalue equation which may be rewritten in the matrix form as

$$\mathbf{AC} = \mathbf{C\Upsilon}, \tag{4.46}$$

in which the matrix $\mathbf{C}$ is the usual eigenvector matrix with $\{\mathbf{C}\}_{gi} = C_g^{(i)}$, the matrix $\mathbf{\Upsilon}$ is the diagonal eigenvalue matrix with elements $\{\mathbf{\Upsilon}\}_{jj} = \gamma^{(j)}$, and the elements of matrix $\mathbf{A}$ are

$$\{\mathbf{A}\}_{gh} = \frac{S_g}{1 + g_z/K_z}\delta_{gh} + \frac{U_{gh}}{2K_z(1 + g_z/K_z)}(1 - \delta_{gh}).$$

When N diffracted beams (belonging to n rods of the reciprocal lattice) are excited in the crystal, the above eigenvalue equation gives rise to n distinct values of γ and thus to a total of n forward-propagating Bloch waves. To an excellent approximation in the Laue case, and for almost all systems of practical importance, we may assume that $n \approx N$, i.e. only one diffracted beam associated with each reciprocal lattice rod is excited.

The main difficulty in solving the eigenvalue equation (4.46) is associated with the fact that even for an ideal non-absorbing crystal the matrix $\mathbf{A}$ is no longer a Hermitian matrix. To overcome this difficulty Lewis et al. (Lewis et al., 1978) suggested making a transformation of the eigenvector $C_g^{(i)}$, namely

$$B_g^{(i)} = \sqrt{1 + g_z/K_z}\,C_g^{(i)}. \tag{4.47}$$

In terms of the new system of eigenvectors $\{B_g^{(i)}\}$ the basic equation (4.46) becomes

$$\mathbf{AB} = \mathbf{B}\boldsymbol{\Upsilon}^{(i)}, \tag{4.48}$$

with

$$\{A\}_{gh} = \frac{S_g\delta_{gh}}{1 + g_z/K_z} + \frac{U_{gh}}{2K_z\sqrt{1 + g_z/K_z}\sqrt{1 + h_z/K_z}}(1 - \delta_{gh}).$$

It may be readily verified that for an ideal non-absorbing crystal the newly defined matrix $\mathbf{A}$ is a Hermitian matrix and the associated eigenvalues are real. Since the eigenvectors form a complete and orthogonal set, they satisfy the relations

$$\sum_g B_g^{*(i)}B_g^{(j)} = \delta_{ij}, \quad \sum_i B_g^{*(i)}B_h^{(i)} = \delta_{gh}. \tag{4.49}$$

A general solution for the diffracted beam amplitudes is given by

$$\psi_g(z) = \sum_j \alpha^{(j)}B_g^{(j)}\exp[i(K_z + \gamma^{(i)} + g_z)z], \tag{4.50}$$

with $\alpha^{(j)} = B_0^{*(j)}$ for an incident plane wave. To a first-order approximation the same perturbation treatment that was previously discussed in connection with ZOLZ diffraction may again be applied to deal with the effect of inelastic scattering on the elastically scattered electron wave field.

For a general absorbing crystal the orthogonality relations (4.49) are no longer valid. We can nevertheless solve the eigenvalue equation numerically obtaining eigenvalues that are no longer real and eigenvectors that are no longer orthogonal to each other. Using the matrix form and neglecting a common phase factor $\exp(iK_z z)$ we may rewrite the solution (4.50) as

$$
\begin{pmatrix} \psi_1(z) \\ \psi_2(z) \\ \vdots \end{pmatrix} = \begin{pmatrix} \exp(ig_{1z}z) & & \\ & \exp(ig_{2z}z) & \\ & & \ddots \end{pmatrix} \begin{pmatrix} B_1^{(1)} & B_1^{(2)} & \dots & B_1^{(N)} \\ B_2^{(1)} & B_2^{(2)} & \dots & B_2^{(N)} \\ \dots & \dots & \dots & \dots \end{pmatrix}
$$

$$
\times \begin{pmatrix} \exp(i\gamma^{(1)}z) & & \\ & \exp(i\gamma^{(2)}z) & \\ & & \ddots \end{pmatrix} \begin{pmatrix} \alpha^{(1)} \\ \alpha^{(2)} \\ \vdots \end{pmatrix}
$$

or, in the more compact representation

$$\boldsymbol{\Psi}(z) = \mathbf{P}(z)\mathbf{B}\boldsymbol{\Upsilon}(z)\boldsymbol{\alpha}, \tag{4.51}$$

where $\boldsymbol{\Psi}(z)$ is a column vector of diffracted beam amplitudes with $\{\boldsymbol{\Psi}\}_g = \psi_g$, $\mathbf{B}$ is an $N \times N$ square matrix with $\{\mathbf{B}\}_{gj} = B_g^{(j)}$, $\boldsymbol{\Upsilon}(z) = \{\exp(i\gamma^{(j)}z)\}_d$, and $\mathbf{P}(z) = \{\exp(ig_z z)\}_d$ are two diagonal matrices. At the plane $z = z_1$ we have $\boldsymbol{\Psi}(z_1) = \mathbf{P}(z_1)\mathbf{B}\boldsymbol{\Upsilon}(z_1)\boldsymbol{\alpha}$. By inverting this equation we find

$$\boldsymbol{\alpha} = \boldsymbol{\Upsilon}(-z_1)\mathbf{B}^{-1}\mathbf{P}(-z_1)\boldsymbol{\Psi}(z_1).$$

Substitution of the above expression for $\boldsymbol{\alpha}$ into eqn (4.51) gives

$$\boldsymbol{\Psi}(z) = \mathbf{P}(z)\mathbf{B}\boldsymbol{\Upsilon}(z - z_1)\mathbf{B}^{-1}\mathbf{P}(-z_1)\boldsymbol{\Psi}(z_1). \tag{4.52}$$

Generalizing this to the case of dynamical electron diffraction by an assembly of crystal slabs where the thickness of each slab equals t_i, we obtain the following expression for the column vector of amplitudes of diffracted beams at the bottom face of the last slab:

$$\boldsymbol{\Psi}(z) = \mathbf{M}(z)\boldsymbol{\Psi}(0), \tag{4.53}$$

where

$$\mathbf{M}(z) = \prod_i [\mathbf{P}_i(z_i)\mathbf{B}_i\boldsymbol{\Upsilon}_i(t_i)\mathbf{B}_i^{-1}\mathbf{P}(-z_{i-1})], \tag{4.54}$$

and $z_i = \sum_{n=1}^{i} t_n$ is the z coordinate of the bottom face of the i-th crystal slab. Under the ZOLZ or projected potential approximation $g_z = 0$ for all reflections, and we have for all crystal slabs $\mathbf{P}_i(z) = \mathbf{I}$. The general expression (4.53) then reduces to that of Howie and Whelan for ZOLZ diffraction (Howie and Whelan, 1961). The HOLZ reflection affects the diffracted beam amplitudes in two ways. Firstly the presence of HOLZ reflections changes both the eigenvalues and eigenvectors of Bloch waves. It should be noted, however, that in most cases diffraction processes are dominated by only a few major Bloch waves, sometimes as few as two (Peng and Zuo, 1999) as discussed in section 4.3.5. HOLZ reflections are generally weak, and one would not expect that HOLZ reflections would have a profound effect on these dominant Bloch waves. Secondly,

HOLZ reflections introduce an additional phase factor $\exp(ig_z z)$ that plays a substantial part in the case of diffraction by a thick crystal. Although this phase factor does not alter the diffracted beam intensities, it affects the relative phase of diffracted beams and therefore the interference between these beams. Consequently we expect that phase factors should be taken into account when evaluating diffracted beam amplitudes from a multilayer structure, i.e. from a system composed of an assembly of crystal slabs.

In the case of an incident plane wave the boundary condition on the entrance surface has the form

$$\mathbf{\Psi}(0) = \begin{pmatrix} \psi_1(0) \\ \psi_2(0) \\ \vdots \end{pmatrix} = \begin{pmatrix} 1 \\ 0 \\ \vdots \end{pmatrix},$$

where the first element corresponds to the incident beam. We find that the diffracted beam amplitudes at the exit surface at depth z are given by

$$\mathbf{\Psi}(z) = \begin{pmatrix} \psi_1(z) \\ \psi_2(z) \\ \vdots \end{pmatrix} = \begin{pmatrix} M_{11} & M_{12} & \dots \\ M_{21} & M_{22} & \dots \\ \vdots & \vdots & \vdots \end{pmatrix} \begin{pmatrix} 1 \\ 0 \\ \vdots \end{pmatrix} = \begin{pmatrix} M_{11} \\ M_{21} \\ \vdots \end{pmatrix},$$

where the matrix $\mathbf{M}$ is given by eqn (4.54) for an assembly of crystal slabs. The amplitudes of diffracted beams ψ_g depend in general on the direction of the incident beam characterized by the projection of the incident wave vector on the ZOLZ plane, i.e. on $\mathbf{K}_t = (k_{0_x}, k_{0_y})$. For a CBED geometry the intensity distribution of the g-th disc is given by $I_g(\mathbf{K}_t) = |\psi_g(k_{0_x}, k_{0_y})|^2$.

To illustrate the treatment of HOLZ diffraction we now consider an application of eqn (4.54) to an Si/Ge_xSi_{1-x} strained layer superlattice (SLS). The SLS consists of alternating layers of Si and Si_xGe_{1-x}. The lattice constant of Si is 5.43 Å, while that of Ge is 5.66 Å. The mismatch between the two lattices of Si and Ge is about 4.2%, which is sufficiently large to introduce considerable strain in the superlattice. The Si_xGe_{1-x} alloy has a smaller lattice mismatch with Si and the strain can be reduced. In order to grow a defect-free Si/Ge_xSi_{1-x} superlattice, the lateral lattice constants of Si and Si_xGe_{1-x} layers must be equal, and consequently both lattices are subject to distortion. This distortion depends on several factors, including the composition value x and number of Si and Si_xGe_{1-x} layers within each period of the SLS. The challenge that we are facing is how to determine accurately the local lattice distortion and alloy composition. Detailed analysis shows that both the position and strength of HOLZ lines in the transmission CBED disc depend sensitively on the local strain and alloy composition. HOLZ diffraction therefore provides an ideal means for exploring these characteristics of SLS.

Figure 4.18 shows the [102] zone axis experimental (a) and simulated (b) large-angle CBED (LACBED) pattern for an Si/Ge_xSi_{1-x} SLS. The simulation was performed using 31 beams, the primary beam energy was equal to 99.4 keV

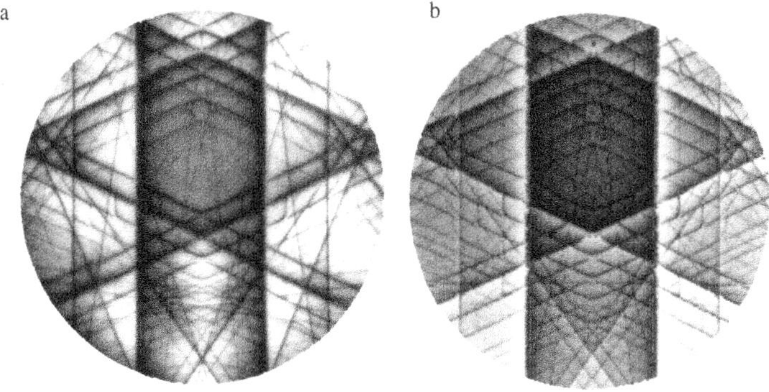

FIG. 4.18. (a) Experimental and (b) simulated [102] large-angle CBED patterns from an Si/Ge$_x$Si$_{1-x}$ SLS. From (Wang et al., 1992a).

and the crystal thickness was equal to 2450 Å. The surface normal components of the strained epilayer lattice constants are $a_n(\text{Si}) = 5.427$ Å and $a_n(\text{Ge}_x\text{Si}_{1-x}) = 5.586$ Å. By matching the simulated LACBED pattern (Fig. 4.18b) and the experimentally observed pattern (Fig. 4.18a), the alloying composition in the Ge$_x$Si$_{1-x}$ layer was determined to be $x = 0.37$, and in a single period of SLS the Si layer thickness was determined to be 200 Å, i.e. it contains 38 repeating unit layers of Si, and the Ge$_x$Si$_{1-x}$ part contains 7 layers of Ge$_x$Si$_{1-x}$ atoms (Wang et al., 1992b).

4.4.4.2 *The Howie–Whelan equations*   In the treatment of diffraction by an assembly of single crystal slabs we made the assumption that the entire specimen could be regarded as a perfect defect-free crystal that is infinite in extent and periodic in two dimensions in the (x, y) plane. Real crystals, however, contain defects such as dislocations and stacking faults, where the perfect crystal periodicity is destroyed. The most efficient way of treating this case consists in using the column approximation proposed by Whelan and Hirsch (Whelan and Hirsch, 1957). In this approximation a crystal slab is divided into many parallel columns with their long axis pointing in the z direction. Dynamical diffraction processes occurring in the individual columns are assumed to be independent, i.e. the intercolumn scattering is neglected. The initial justification for the column approximation was based on the two-beam treatment and the smallness of the scattering angle θ_B involved in the two-beam approximation. It was soon realized that the range of validity of the column approximation is wider than anticipated on the basis of the two-beam approximation. This may be understood from our previous discussion on the propagation of Bloch waves. Although an ideal two-beam condition can never be satisfied, and there always exist some other higher order reflections in addition to the Bragg reflection involved in the two-beam approximation, the total wave field can nevertheless be expressed in terms of two or a few strongly excited Bloch waves. In almost all THEED applications these strongly excited Bloch waves are nearly dispersionless and the direction of propagation of the corresponding Bloch waves is nearly parallel to the lattice planes. The inter column diffraction is therefore minimized in the THEED geometry.

Within the range of validity of the column approximation we may regard the diffracted beam amplitude at a given point in a column as if it had resulted from diffraction by an assembly of perfect crystal slabs. The effect of crystal defects on diffraction in the column is modelled by a relative shift $\mathbf{R}(z)$ of the otherwise perfect crystal slabs with respect to each other, where the magnitude of this shift depends on the position of a slab along the column. In other words, we may further divide a column into slices, where each slice is shifted by $\mathbf{R}(z)$ with respect to the entrance surface. Since every slice is assumed to be perfect, their Bloch states must remain the same except for a phase shift in their eigenvectors due to the shift of the origin. Using the local system of coordinates a Bloch wave may be written as $b(\mathbf{r}') = \sum_g C'_g \exp[i(\mathbf{k} + \mathbf{g}) \cdot \mathbf{r}]$, where the prime is used to denote that corresponding quantities are defined in the local coordinate system. The same Bloch wave may also be expressed using the global coordinate system associated with the entrance surface of the crystal. By writing $\mathbf{r}' = \mathbf{r} - \mathbf{R}(z)$ we find

$$b(\mathbf{r}') = \sum_g C_g \exp[i(\mathbf{k} + \mathbf{g}) \cdot (\mathbf{r} - \mathbf{R})] = b'(\mathbf{r}) = \sum_g C'_g \exp[i(\mathbf{k} + \mathbf{g}) \cdot \mathbf{r}].$$

This leads to

$$C'_g = \exp[-i\mathbf{k} \cdot \mathbf{R}]C_g \exp[-i\mathbf{g} \cdot \mathbf{R}], \tag{4.55}$$

in which the common phase factor $\exp[-i\mathbf{k}\cdot\mathbf{R}]$ neither changes the relative phase shift between the diffracted beams nor affects their intensities, and may therefore be neglected. Defining a diagonal matrix $\mathbf{Q}$ with elements $\{\mathbf{Q}\}_g = \exp[-i\mathbf{g} \cdot \mathbf{R}]$, we see that eqn (4.52), relating diffracted beam amplitude vectors $\boldsymbol{\Psi}(\mathbf{z})$ at the exit and entrance surfaces of a very thin slice of crystal, acquires the form

$$\boldsymbol{\Psi}(z + \Delta z) = \mathbf{P}(z + \Delta z)\mathbf{Q}(z)\mathbf{B}\boldsymbol{\Upsilon}(\Delta z)\mathbf{B}^{-1}\mathbf{Q}^{-1}(z)\mathbf{P}^{-1}(z)\boldsymbol{\Psi}(z).$$

By expanding this expression to first order in Δz, such that $\{\boldsymbol{\Upsilon}(\Delta z)\}_j \approx 1 + i\Delta z\gamma^{(j)}$ and $\{\mathbf{P}(\Delta z)\}_g \approx 1+ig_z z$, and noticing from (4.46) that $\mathbf{AB} = \mathbf{B}\{\gamma^{(i)}\}_d$, we arrive at

$$\frac{d\boldsymbol{\Psi}(z)}{dz} = i\{g_z\}_d\boldsymbol{\Psi}(z) + i\mathbf{P}(z)\mathbf{Q}(z)\mathbf{A}\mathbf{Q}^{-1}(z)\mathbf{P}^{-1}(z)\boldsymbol{\Psi}(z). \tag{4.56}$$

This is the generalized Howie–Whelan equations for an absorbing crystal where HOLZ effects are explicitly included. Alternatively we may derive a differential equation for the Bloch wave excitation vector α starting from eqn (4.46)

$$\boldsymbol{\Psi}(z) = \mathbf{P}(z)\mathbf{Q}(z)\mathbf{B}\boldsymbol{\Upsilon}\alpha.$$

Differentiating the above equation, we find

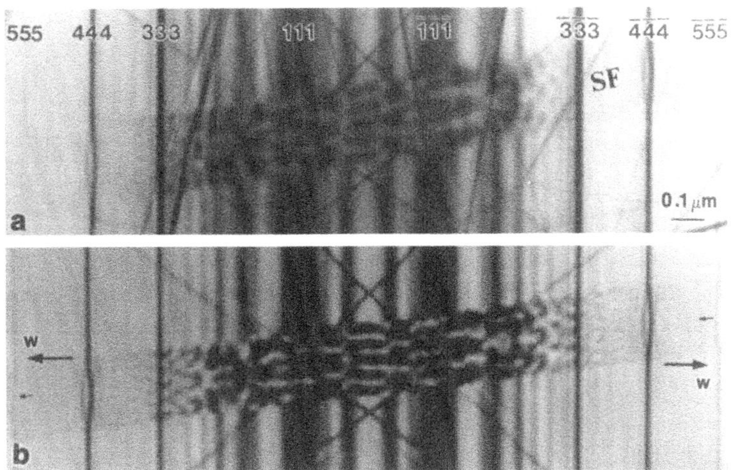

FIG. 4.19. (a) Experimental and (b) simulated bright-field $[0\bar{1}1]$ large-angle CBED pattern for a silicon crystal containing a stacking fault. The simulation was made for a primary beam energy of 100 keV and a crystal thickness of 1650 Å. From (Wang et al., 1992b).

$$\frac{d\mathbf{\Psi}(z)}{dz} = \{ig_z\}_d\mathbf{P}(z)\mathbf{Q}(z)\mathbf{B}\mathbf{\Upsilon}\alpha + \left\{ -i\frac{d(\mathbf{g}\cdot\mathbf{R})}{dz} \right\}_d \mathbf{P}(z)\mathbf{Q}(z)\mathbf{B}\mathbf{\Upsilon}\alpha$$
$$+ \mathbf{P}(z)\mathbf{Q}(z)\mathbf{B}\{i\gamma^{(i)}\}_d\mathbf{\Upsilon}\alpha + \mathbf{P}(z)\mathbf{Q}(z)\mathbf{B}\mathbf{\Upsilon}\frac{d\alpha}{dz}.$$

Comparing the above equation with (4.56) we arrive at

$$\frac{d\alpha}{dz} = i\mathbf{\Upsilon}^{-1}\mathbf{B}^{-1}\left\{ \frac{d(\mathbf{g}\cdot\mathbf{R})}{dz} \right\}_d \mathbf{B}\mathbf{\Upsilon}\alpha, \tag{4.57}$$

and the form of this equation is identical to that involving only the ZOLZ reflections. The dynamical HOLZ effects are fully included in this equation through matrices $\mathbf{\Upsilon}$ and $\mathbf{B}$.

We now consider an application of the generalized Howie–Whelan equation (4.56) to the simplest case of a planar defect, namely a stacking fault. Figure 4.19a shows an experimental BF $[0\bar{1}1]$ LACBED pattern of a silicon crystal containing a stacking fault (denoted as SF in the figure). Figure 4.19b is the corresponding pattern simulated using eqn (4.56). The simulation was made using 15 beams, assuming the displacement vector $\mathbf{R} = \frac{1}{3}[1\bar{1}1]$, the crystal thickness of $t = 1650$ Å, and the defocus value of $\Delta f = 30$ μm. Four HOLZ reflections were included in the simulation (Wang et al., 1992a). The agreement between the experimental image (a) and the simulation (b) is very good.

4.4.4.3 *Scattering matrix method*

In the matrix notation the generalized Howie–Whelan equation (4.56) may be written as

$$\frac{d\mathbf{\Psi}(z)}{dz} = i\mathbf{A}'(z)\mathbf{\Psi}(z), \tag{4.58}$$

where

$$\mathbf{A}' = \{g_z\}_d + \mathbf{P}(z)\mathbf{Q}(z)\mathbf{A}\mathbf{Q}^{-1}(z)\mathbf{P}^{-1}(z). \tag{4.59}$$

A formal solution of this matrix equation has the form

$$\mathbf{\Psi}(z) = \exp\left[i\int_0^z \mathbf{A}'(z)dz\right]\mathbf{\Psi}(0), \tag{4.60}$$

which is the generalized Sturkey scattering matrix formula (Sturkey, 1962) that describes an absorbing crystal and that includes HOLZ effects. In the above solution the operator expression $\exp[i\int_0^z \mathbf{A}'(z)dz]$ is understood as follows:

$$\exp\left[i\int_0^z \mathbf{A}'(z)dz\right] = \sum_{i=1}^{\infty}\frac{1}{n!}\left[i\int_0^z \mathbf{A}'(z)dz\right]^n.$$

Under the ZOLZ or projection potential approximation, matrix $\mathbf{A}'$ reduces to $\mathbf{A}$ and is independent of z. The general scattering matrix formula (4.60) then reduces to

$$\mathbf{\Psi}(z) = \exp(i\mathbf{A}z)\mathbf{\Psi}(0). \tag{4.61}$$

Mathematically we can always regard a crystal as an assembly of many thin slices. We can therefore approximate the integral in eqn (4.60) as follows:

$$\int_0^{\infty} \mathbf{A}'(z)dz \approx \sum_i \mathbf{A}'(z_i)\Delta z. \tag{4.62}$$

This approximation amounts to replacing a continuous function by a step function, and the accuracy of the approximation may be improved indefinitely by reducing the slice thickness Δz. For each thin slice the matrix $\mathbf{A}'(z_i)$ is assumed to be constant, i.e. HOLZ effects are assumed to be absent within each slice. For the i-th slice the matrix $\mathbf{A}'(z_i)$ reduces to the usual matrix $\mathbf{A}(z_i)$ and the general scattering matrix formula then takes the form

$$\mathbf{\Psi}(z) = \prod_k \exp\{i\mathbf{A}(z_k)\Delta z\}\mathbf{\Psi}(0). \tag{4.63}$$

Since within the k-th thin slice the matrix $\mathbf{A}$ satisfies the basic equation of dynamical electron diffraction (4.7) stating that $\mathbf{AC} = \mathbf{B}\{\gamma^{(i)}\}_d$, we find

$$\exp(i\mathbf{A}z) = \sum_{i=1}^{\infty}\frac{1}{n!}(iz)^n(\mathbf{A})^n = \sum_{i=1}^{\infty}\frac{1}{n!}(iz)^n[\mathbf{B}\{\gamma^{(i)}\}_D\mathbf{B}^{-1}]^n$$

$$= \sum_{i=1}^{\infty}\frac{1}{n!}(iz)^n[\mathbf{B}\{\gamma^{(i)}\}_D\mathbf{B}^{-1}][\mathbf{B}\{\gamma^{(i)}\}_D\mathbf{B}^{-1}]...[\mathbf{B}\{\gamma^{(i)}\}_D\mathbf{B}^{-1}]$$

$$= \sum_{i=1}^{\infty}\frac{(iz)^n}{n!}\mathbf{B}\{[\gamma^{(i)}]^n\}_D\mathbf{B}^{-1} = \mathbf{B}\{\exp(i\gamma^{(i)}z)\}_D\mathbf{B}^{-1},$$

and therefore,

$$\mathbf{\Psi}(z) = \prod_k \exp(i\mathbf{A}_k t_k)\mathbf{\Psi}(0) = \prod_k [\mathbf{B}_k \mathbf{\Upsilon}_k(t_k)\mathbf{B}_k^{-1}]\mathbf{\Psi}(0). \qquad (4.64)$$

This equation is equivalent to the Howie–Whelan equations. The difference between it and the original Howie–Whelan approach (Howie and Whelan, 1961) is that now HOLZ effects are included provided that the slice thickness is smaller than the lattice constant along the z axis.

4.5 Summary

In the THEED geometry the fast electrons are incident almost normally on a thin crystal. The (x, y) plane is chosen to be parallel to the surface with the z axis pointing into the crystal. In this case the scaled crystal potential $U(\mathbf{r}) = -2mV(\mathbf{r})/\hbar^2$ can be decomposed as

$$U(\mathbf{X}, z) = \sum_n \exp(ing_z z)U^{(n)}(\mathbf{X}),$$

where $U^{(n)}(\mathbf{X}) = \sum_{G^{(n)}} U_{g^{(n)}} \exp(i\mathbf{G}^{(n)} \cdot \mathbf{X})$ is the conditional projected potential and $\mathbf{G}^{(n)}$ is a two-dimensional reciprocal lattice vector of the n-th HOLZ.

In THEED fast electrons are diffracted by the two-dimensional projected potential $U^{(0)}(\mathbf{X})$

$$U^{(0)}(\mathbf{X}) = \frac{1}{c} \int_0^c U(\mathbf{X}, z)dz.$$

Only ZOLZ reflections, i.e. reciprocal lattice vectors $\mathbf{G}^{(0)}$, are involved in this diffraction process. Although for near zone axis incidence many hundreds of electron beams may need to be included in the Fourier expansion of the potential and the wave function, only a few two-dimensional (ZOLZ) Bloch waves describing the two-dimensional motion of electrons in the potential field $U^{(0)}(\mathbf{X})$ are appreciably excited. Depending on whether the corresponding energy eigenvalues $\mathcal{E}_m$ are negative or positive, we may describe the ZOLZ Bloch waves as bound with their wave fields localized around atom strings or as free with relatively delocalized wave fields. The averaged current flow of a ZOLZ Bloch wave is perpendicular to the corresponding dispersion surface. For tightly bound Bloch states the dispersion surface is almost flat, showing little dependence on $\mathbf{K}_t$, and electrons associated with these Bloch waves travel through the crystal down the zone axis and remain localized around atomic strings. In general the rate of absorption of these tightly bound Bloch waves is much higher than the rate of absorption of the delocalized Bloch states.

The variation of the crystal potential along the z axis gives rise to HOLZ diffraction effects. The n-th HOLZ beam corresponds to the reciprocal lattice vector $\mathbf{g}^{(n)} = (\mathbf{G}^{(n)}, ng_z)$, and the amplitude of this beam depends on the value of the matrix element of the conditional projected potential $U^{(n)}(\mathbf{X})$. To a good approximation HOLZ diffraction may be described as resulting from scattering of tightly bound ZOLZ Bloch waves localized around atom strings by atoms situated along the string, where the rate of scattering depends on the overlap between the ZOLZ Bloch wave $b(\mathbf{X})$ and $U^{(n)}(\mathbf{X})$. Since $U^{(n)}(\mathbf{X})$ is only appreciable near the axis of atomic strings, only tightly bound Bloch waves contribute to the HOLZ diffracted beam amplitudes. In general HOLZ diffraction is much weaker than ZOLZ diffraction, and in most cases a kinematic treatment of HOLZ diffraction suffices. HOLZ diffraction does not alter the form of the ZOLZ Bloch waves, but it modifies the excitation amplitudes of the ZOLZ Bloch waves. If necessary HOLZ diffraction effects can be treated dynamically, and some of the well-established ZOLZ diffraction approaches may be readily generalized to include HOLZ diffraction.

5

DYNAMICAL THEORY III. REFLECTION HIGH-ENERGY ELECTRON DIFFRACTION

5.1 Introduction

In their pioneering experiments leading to the independent and simultaneous discovery with Thomson and Reid (Thomson and Reid, 1927) of the phenomenon of electron diffraction, Davisson and Germer (Davisson and Germer, 1927) used the reflection geometry of diffraction and investigated how the intensities of reflected beams vary as a function of the primary beam energy. The modern technique that evolved from Davisson and Germer's early experiments is usually referred to as low-energy electron diffraction (LEED). The main subject of this chapter is reflection high energy electron diffraction (RHEED). In RHEED high-energy electrons (the energy of the electrons can be $\sim$10 keV and higher) are incident on the surface of a crystal at a glancing angle of a few degrees. Although the geometry of RHEED is almost the same as that used by Davisson and Germer, the energy of the electrons used in LEED is much lower, typically of the order of several tens of electronvolts. The angles of reflection employed in LEED are much larger too, and typically they are of the order of $45°$. The first dynamical theory of electron diffraction was developed by Bethe (Bethe, 1928) in order to give a quantitative account of the experimental observations by Davisson and Germer (Davisson and Germer, 1927). A matter of particular importance was Bethe's introduction of the concept of an inner potential. This point proved to be very significant for the correct interpretation of the experiments of Davisson and Germer. In Bethe's original treatment the crystal surface is considered as a plane separating forward- and backward-scattered electrons. Since in RHEED the electrons are incident on the surface at a glancing angle, both forward and backward scattering are equally important and must be included in the theoretical treatment. A difficult point for the theoretical treatment of RHEED is associated with the need to include the backscattered electrons. This leads to either a set of first-order non-linear differential equations, or to a second-order eigenvalue problem. The majority of approaches to dynamical RHEED calculations use Bethe's coordinate system and differ from each other mainly by the methods of linearization of the basic equations of dynamical theory of electron diffraction. Alternatively, a plane separating forward- and backward-scattered electrons may be taken to be perpendicular to the surface of the crystal. Using the fact that scattering of electrons by atoms is dominated by forward scattering, all the numerical procedures developed for THEED calculations may then be employed in the RHEED geometry. The two settings for dynamical RHEED

117

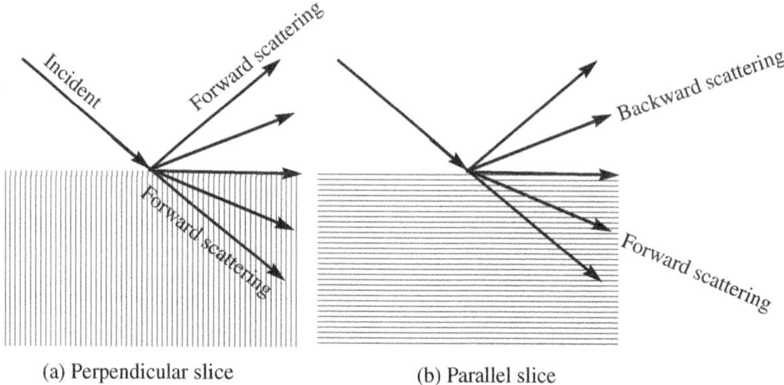

FIG. 5.1. Schematic diagram showing the two alternative approaches to dynamical RHEED calculations. In (a) diffraction is dominated by forward scattering from the slices and in (b) both forward and back-scattering from the slices are present.

calculations are illustrated in Fig. 5.1.

This chapter is organized as follows. In section 5.2 we briefly discuss the general notation used in surface crystallography and the geometry of a RHEED pattern. In section 5.3 we outline various approaches to dynamical RHEED calculations and describe in depth one particular approach based on the concept of Bloch waves. Section 5.4 gives worked examples that use the FORTRAN routines listed in Appendix C. The interested reader may use these routines to follow the examples given in section 5.4 and to reproduce the relevant figures.

5.2 Surface structure notation and RHEED geometry

5.2.1 *The nature of the surface*

In this chapter we shall be considering electron diffraction from surfaces. It is therefore appropriate to start with an introduction to the terminological conventions used in surface crystallography. In the mathematical sense a surface cannot have structure. In this book, by surface structure we mean the structure of the solid in the vicinity of the surface. A surface may therefore be defined as a substrate, having the proper three-dimensional periodicity of the bulk, plus a region of the solid in the vicinity of the mathematical surface which is usually called the *selvedge* (Wood, 1964)) by analogy with the selvedge of a piece of cloth (a narrow band woven in such a way as to prevent fraying of the edge).

Figure 5.2 is a schematic diagram of an epitaxial film illustrating the terminology of surface crystallography. A surface may be clean, consisting of only those atoms that constitute the substrate material, or it may have other atoms deposited on it. In either case it is possible that the few atomic layers of the selvedge may adopt atomic configurations different from those of the bulk. The substrate is periodic in the directions both parallel and normal to the surface.

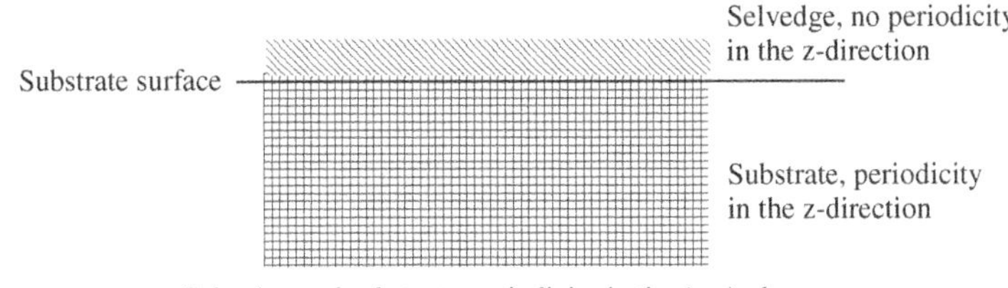

FIG. 5.2. Diagram showing an epitaxial film on a substrate and illustrating the relevant terminology.

The substrate surface is marked in Fig. 5.2 as the mathematical plane at which the periodicity in the direction normal to the surface vanishes. The atomic layers of the selvedge may differ from the bulk in the layer spacing normal to the surface, which is normally called *surface relaxation*, and in the displacement of atoms in the direction parallel to the surface. The latter effect is called *surface reconstruction*. The selvedge is characterized by a two-dimensional periodicity in the plane parallel to the surface. This periodicity may be the same as that of the bulk material (this is the case for a relaxed surface), or it may differ from the periodicity of the substrate as in the case of a reconstructed surface. The periodicity characterizing a reconstructed surface is usually coherent with the periodicity of the substrate, meaning that both the selvedge and the substrate share common periods of translation which may be greater than those of the substrate or the selvedge.

5.2.2 *The five surface nets*

Since the surface is periodic in two dimensions, its equivalent points form a two-dimensional net in which the area units are unit meshes. Corresponding to the 14 Bravais lattices of a three dimensionally periodic structure, five two-dimensional nets are sufficient to describe any surface structure, and these five nets are shown in Fig. 5.3. These are the oblique structure that has no symmetry; the primitive (denoted by the letter p) and the centred rectangular (denoted by the letter c) structures that are the two symmetrical non-equivalent lattices characterized by mirror symmetry; the square structure characterized by a four-fold rotation axis; and the hexagonal structure that has a six-fold rotation axis. Among the five nets the centred rectangular net is the only non-primitive net. Centring of any other net leads only to a net that can be equally well classified by a primitive net of the same symmetry.

5.2.3 *The relation between the surface mesh and the substrate mesh*

If the structure of the surface atomic layers of a solid is different from that of the substrate, then the surface may in principle be either coherent or incoherent with the substrate. Nevertheless, because the structure of the substrate is usually known, it is common to take it as a reference with its net parallel to the surface. A very convenient notation for surface mesh structures was proposed in 1964 by

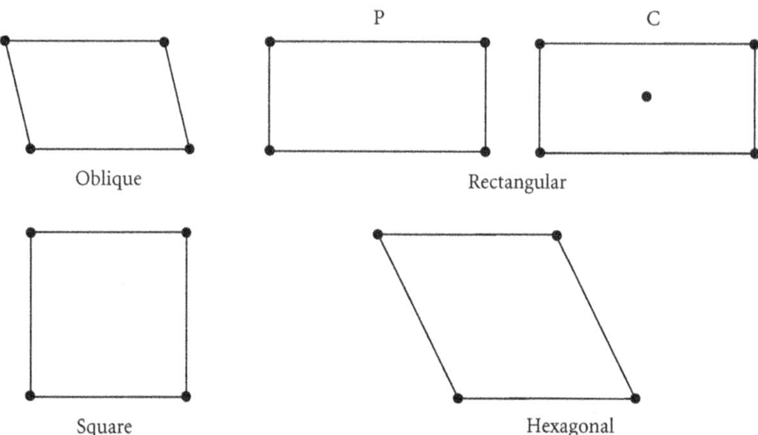

FIG. 5.3. Diagram showing the five two-dimensional Bravais nets.

Wood (Wood, 1964). In his notation the vectors defining the mesh of the surface structure are expressed in terms of the mesh of the underlying structure. The notation defines the ratio of the lengths of the surface and substrate meshes, i.e.

$$p = |\mathbf{a}_s|/|\mathbf{a}|, \quad q = |\mathbf{b}_s|/|\mathbf{b}|, \tag{5.1}$$

in which the primitive translation vectors of the substrate net are $\mathbf{a}$ and $\mathbf{b}$, and those of the surface are $\mathbf{a}_s$ and $\mathbf{b}_s$. A shorthand term for a surface mesh with

$$\mathbf{a}_s = 2\mathbf{a}, \quad \mathbf{b}_s = 2\mathbf{b},$$

is therefore '2 × 2'. If a centred mesh is chosen, this shorthand term should be preceded by c. For a full identification, the shorthand term should be preceded by the symbol of the substrate, e.g. Ni, and the orientation of the substrate surface, i.e. {100}, followed by the symbol of atoms deposited on the surface. For a {100} surface of nickel with oxygen on it, the notation is then Ni{100}(1 × 1)–O. It often occurs that the surface mesh vectors $\mathbf{a}_s$ and $\mathbf{b}_s$ are not parallel to the substrate vectors $\mathbf{a}$ and $\mathbf{b}$. The notation then defines the angle ϕ through which one mesh must be rotated in order to be aligned with the other. If the deposit A on the {hkl} surface of material X gives rise to the formation of a structure characterized by primitive translation vectors of length $|\mathbf{a}_s| = p|\mathbf{a}|$ and $|\mathbf{b}_s| = q|\mathbf{b}|$ and the angle of rotation ϕ, the general Wood notation is

$$\mathrm{X}\{hkl\}(p \times q)\, R\phi - \mathrm{A}. \tag{5.2}$$

It should be noted that the Wood shorthand notation may be used only when the symmetries of surface and substrate meshes are the same. If the surface and the substrate are characterized by different Bravais nets, a matrix notation must be used. Assuming that the primitive mesh vectors of the selvedge are $\mathbf{a}'$ and $\mathbf{b}'$, we can relate them to those of the substrate by

$$\mathbf{a}' = G_{11}\mathbf{a} + G_{12}\mathbf{b}$$
$$\mathbf{b}' = G_{21}\mathbf{a} + G_{22}\mathbf{b} \tag{5.3}$$

where G_{ij} are the four components of a matrix G that relates the selvedge and the substrate meshes. The area of the substrate unit mesh is given by $|\mathbf{a} \times \mathbf{b}|$, and the determinant of the matrix G is simply equal to the ratio of the areas of the two meshes $|\mathbf{a}' \times \mathbf{b}'|/|\mathbf{a} \times \mathbf{b}|$. The determinant $\det\{G\}$ therefore provides a convenient measure for the classification of surface structures.

In the first case both $\det\{G\}$ and G_{ij} are integers. The selvedge mesh has the same translational symmetry as the substrate. In the second case $\det\{G\}$ is a rational fraction (i.e. some of the elements G_{ij} are rational fractions). The selvedge and substrate meshes are then rationally related. The structures are commensurate, and the true surface mesh is larger than either the substrate or the selvedge meshes. This type of surface structure is usually called a coincidence net since the larger surface mesh has a dimension over which the selvedge mesh comes into coincidence with the substrate mesh. The true surface mesh now has primitive translation vectors $\mathbf{a}''$ and $\mathbf{b}''$ that are related to the substrate and selvedge meshes by matrices $\mathbf{P}$ and $\mathbf{Q}$

$$\begin{pmatrix} \mathbf{a}'' \\ \mathbf{b}'' \end{pmatrix} = \mathbf{P} \begin{pmatrix} \mathbf{a} \\ \mathbf{b} \end{pmatrix} = \mathbf{Q} \begin{pmatrix} \mathbf{a}' \\ \mathbf{b}' \end{pmatrix}. \tag{5.4}$$

The determinants of $\mathbf{P}$ and $\mathbf{Q}$ are chosen in such a way so that $\det\{P\}$ and $\det\{Q\}$ have the smallest possible integral values and are related by $\det\{G\} = \det\{P\}/\det\{Q\}$. In the third case when $\det\{G\}$ is an irrational number, the selvedge and substrate meshes are incommensurate and no true surface mesh exists.

5.2.4 *Surface reciprocal lattice rods*

The reciprocal lattice of a three-dimensional crystal has been discussed in earlier chapters. In the kinematic approximation a diffracted beam is generated when a reciprocal lattice point lies on the surface of the Ewald sphere. In the two-dimensional case the lack of periodicity in the direction normal to the surface relaxes one of the Laue conditions, and diffraction occurs continuously with changes of the direction of the incident beam or rotation of the surface. The two-dimensional surface reciprocal lattice may therefore be conveniently represented by an array of lines or one-dimensional 'rods' of infinite extent in the direction normal to the surface. A diffracted beam is generated when the Ewald sphere intersects a 'rod'. Figure 5.4 shows schematically an array of surface reciprocal lattice rods, the Ewald sphere, and the resulting geometry of the RHEED pattern. While this figure depicts diffracted beams as discrete spots, in practice the spots are often streaked in the rod direction owing to effects such as incident beam divergence and thermal diffuse scattering. This may be particularly marked for the 0-layer of diffraction spots.

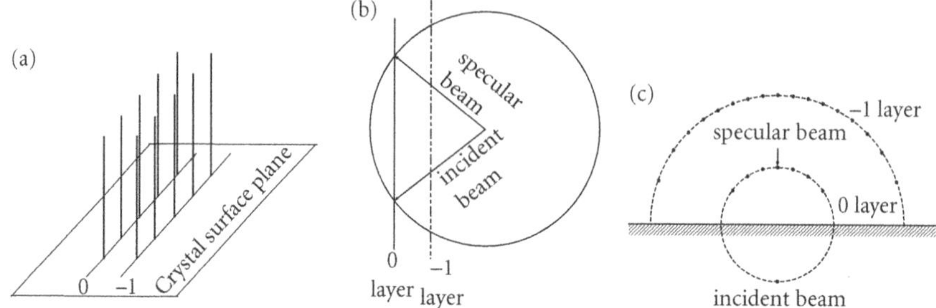

FIG. 5.4. (a) Perspective diagram showing reciprocal space rods normal to the crystal surface. The rods are arranged in layers labelled 0, -1,.... . (b) Ewald sphere construction for the incident beam in the plane perpendicular to the 0th layer of rods of (a), showing the directions of the incident beam and the specular reflected beam. The traces of the 0, -1 layers of rods are indicated by a solid and a dotted line respectively. (c) Schematic diagram of the appearance of diffraction spots for the 0 and -1 layers of reciprocal lattice rods.

While the geometry of a transmission dynamical electron diffraction pattern may differ from that predicted by the kinematic theory, e.g. in the dynamical approximation it is possible to generate a diffracted beam associated with a reciprocal lattice point situated at some distance away from the Ewald sphere, the geometry of a RHEED pattern predicted in the kinematic approximation remains correct even in the presence of strong dynamical diffraction effects. The reason why the geometry of a RHEED pattern remains the same in both kinematic and dynamical theories is that it is entirely determined by two conservation laws, namely those of the conservation of energy and the conservation of the momentum. The principle of energy conservation requires that for elastic scattering, the diffracted electron beam energy must be equal to that of the incident beam, i.e. for all diffracted beams we have

$$E = \frac{\hbar^2 \mathbf{k}_0^2}{2m} = \frac{\hbar^2 \mathbf{k}_g^2}{2m}. \tag{5.5}$$

This condition requires that for all the diffracted beams vector $\mathbf{k}_g$ must lie on the Ewald sphere defined by the incident electron wave vector $\mathbf{k}_0$. On the other hand, the principle of momentum conservation requires that a diffracted beam must have the same tangential component of the wave vector as that of the incident electron beam, or differ from it by a two-dimensional reciprocal lattice vector $\mathbf{G}$, namely

$$\mathbf{k}_{g_t} = \mathbf{k}_{0_t} + \mathbf{G}. \tag{5.6}$$

This equation further constrains a diffracted beam to be on one of the surface reciprocal lattice rods. These two conditions therefore require the diffracted beam directions to be determined by the intersections of the Ewald sphere and the surface reciprocal lattice rods. This is the same condition as that predicted by the kinematic theory discussed above.

5.3 RHEED theory

The existing dynamical RHEED theories may be classified into four distinct types, namely, (1) multislice theory, (2) semi-reciprocal theory, (3) Green's function theory and (4) Bloch wave theory. Formally it has been shown that some of these formulations are entirely equivalent, see e.g. (Goodman and Moodie, 1974) and (Peng and Whelan, 1990b). Numerical comparisons were also made for some specific cases, such as for specular reflected beam intensities calculated using the semi-reciprocal methods by Ichimiya (Ichimiya, 1983) and by Maksym and Beeby (Maksym and Beeby, 1981) for RHEED from the Si(100) surface. Qualitative agreement was found between the RHEED rocking curves calculated by the two methods (Kawamura et al., 1988). More recently very good agreement was achieved between calculated one-rod RHEED rocking curves using the multislice method (Peng and Cowley, 1986) and the semi-reciprocal approach of Maksym and Beeby (Gotsis and Maksym, 1997). A comprehensive 'Round Robin of RHEED Calculations' was organized by A. Ichimiya in 1995. RHEED rocking curves for Au(001), Au(111), Si(001), and Si(111) surfaces were calculated for the [1$\bar{1}$0] and [11$\bar{2}$] azimuths by several groups around the world, and the results were quantitatively compared by A. Ichimiya and presented at the 'Winter Workshop on Electron Diffraction and Imaging at Surfaces', held in Arizona in January 1996. Contributors to this study included G. Anstis, S. L. Dudarev, W. Hanada, A. Ichimiya, U. Korte, S. Lordi, P. Maksym, Z. Mitura, L.-M. Peng, A. Smith, and N. L. Yakovlev. Initial assessment of the results revealed that the calculated data sets contributed from Oxford (by Dudarev), Osnabrück (by Korte), Leicester (by Maksym), and Beijing (by Peng) were nearly exactly the same. While results contributed by others differed from each other in terms of absolute intensities, they nevertheless exhibited many common features.

In this section we will give a brief account of all four approaches, and in particular we will investigate the Bloch wave method. We will give a detailed account of the formulation and relevant numerical procedures, and we will also describe applications of the method to dynamical RHEED calculations. All four types of approaches are presented as self-contained sections, and can be read independently. Those who are mainly interested in investigating the computational aspect of dynamical RHEED are invited to go directly to Appendix C and follow the FORTRAN routines listed there. These routines were used to produce figures illustrating various theoretical concepts described in this chapter.

5.3.1 *The THEED approach to RHEED*

In principle the problem of RHEED from a surface of a crystal may be addressed using the same general numerical procedure as the one used in the transmission electron diffraction (TED) geometry. The basic idea behind this type of approach is as follows. If the crystal surface is positioned in such a way that incident electrons are incident on the surface in the direction nearly parallel to the z axis, all the diffracted beams will also be contained within a small segment of solid angle around the z axis (see Fig. 5.1a). The diffraction is then just forward scattering of electrons, and backscattering effects may safely be neglected.

The main assumption made in practically all the theoretical treatments of THEED is that the illuminated sample is two-dimensionally periodic in the (x, y) plane, which is nearly perpendicular to the direction of the incident beam. In order to treat crystals containing a defect such as the crystal surface, a periodic continuation approximation is usually made (Cowley, 1992). In this approximation a periodic array of large simulation unit cells is artificially constructed, where each simulation cell contains a crystal block the surface of which is nearly parallel to the direction of the incident beam. This treatment requires, however, that the size of the simulation cell in the direction perpendicular to the incident beam must be sufficiently large and many beams are needed to achieve a convergent result. In principle any numerical procedure based on the THEED dynamical theory may be adopted for RHEED calculation using this scheme. However, since at least two orders of magnitude more beams are required for a RHEED calculation than for a THEED calculation, the numerical algorithm on which a practical calculation of the kind described above could be based must be extremely fast and robust. Up to now only the Cowley and Moodie multislice method has been employed to address cases of practical interest (Peng and Cowley, 1986, Peng and Cowley, 1988c, Ma, 1991, Anstis and Gan, 1994, Gotsis and Maksym, 1997).

Figure 5.5 illustrates the construction of an artificial large periodic array of supercells used in the very first dynamical RHEED calculation by Peng and Cowley (Peng and Cowley, 1986) who used the transmission multislice method (Cowley and Moodie, 1957). One of the difficulties that Peng and Cowley encountered at an early stage of the work was the determination of the initial state of the electron wave function. If we approximate the initial state of the electron by a plane wave, the wave vector of which makes a small angle with the surface of the crystal, the simulation will describe the case of illumination of the corner of a crystal, with a large part of the beam entering the crystal through the surface nearly perpendicular to the incident beam. This part of the beam will then dominate the wave field in the crystal over a very large distance in the direction of incidence. In order to avoid this unwanted complication, a mask F is applied, which lets through only the part of the incident beam that illuminates the appropriate crystal surface at a glancing angle. We also note that if the mask contains a sharp discontinuity corresponding to a sudden increase in its transparency from 0 to 100% at a certain point then this will introduce severe undesirable Fresnel diffraction effects. The transition region over which the transparency of the mask varied was therefore made more gradual by rounding off the discontinuity by a Gaussian function. The model therefore simulates the effect of a beam of finite width striking the crystal face at a glancing angle. Initially, the beam propagates entirely in the vacuum. The wave field in the crystal

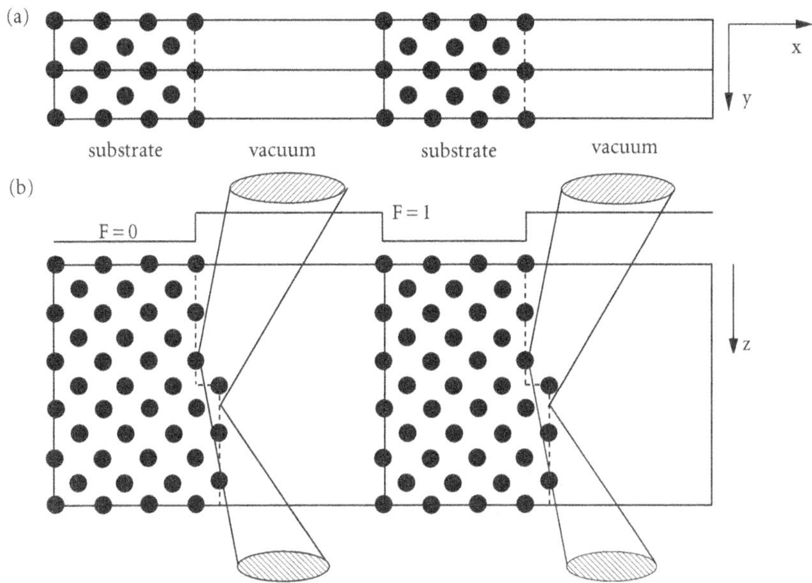

FIG. 5.5. (a) Top and (b) side view of the simulation supercell for dynamical RHEED calculations based on the periodic continuation method. The fast electrons are incident on the stepped surface at glancing angle θ. The mask function F on the entrance face $z = 0$ prevents the irradiation of the top face of the crystal. The crystal surface of interest in RHEED is the y, z surface.

and in the vacuum is then gradually generated as the electron wave propagates in the forward direction from slice to slice. After a certain distance a steady state is reached and the result of the simulation becomes representative of the case of an infinite plane wave incident at a glancing angle on an infinite crystal surface. Effects of scattering by defects of the crystal structure or by imperfections on the crystal surface may now be explored by introducing changes in the nature of the slices in the model, as illustrated in Fig. 5.5b by a surface step. Using this model, Peng and Cowley (Peng and Cowley, 1986) calculated electron wave functions both inside and outside the crystal. These wave functions were subsequently used to describe the standing-wave pattern formed in the top few atomic layers of the crystal, to calculate the intensities of diffraction spots in a RHEED pattern and to simulate the contrast of a reflection electron microscopy (REM) image of the surface.

While the method of calculation of the electron wave field described above gives only an approximate representation of the ideal case corresponding to an incident plane wave, this method is still well suited for solving the case where the incident wave is not a plane wave. An important example is given by the case of coherent convergent-beam illumination widely used in a scanning transmission electron microscope (STEM). In this case the condenser aperture of the STEM is illuminated by a highly coherent source of electrons (e.g. a field-emission electron gun) and the wave field in the aperture plane may be regarded as coherent at all points within the aperture. A coherent small electron probe may then be formed by focussing the beam on the surface, and a coherent convergent-beam electron diffraction (CBED) pattern may be obtained. Since the cross-section of the focussed coherent electron beam is small, the wave field can easily avoid the areas covered by the mask F positioned in front of the crystal and reach the surface of the crystal as shown in Fig. 5.5. The multislice method is therefore ideal for simulating a coherent CBED RHEED pattern from a surface containing defects, which may not be readily treated by the other methods that we discuss in the following sections of this chapter.

The early work on RHEED carried out using the multislice method was focussed mainly on the computation of the electron wave field in the crystal and on the corresponding RHEED patterns. Attempts to calculate dynamical RHEED rocking curves were, however, not successful. This is because, among other factors, the total size of the supercell required to obtain a convergent RHEED solution varies as a function of the direction of the incident beam. The extent of the supercell in the direction parallel to the direction of incidence turns out to be particularly large in the case where the so-called 'surface resonance' condition is satisfied. In this case electrons are trapped near the surface of the crystal and they propagate over a large distance before being rescattered out of the crystal and contributing to the intensity of reflected beams (a detailed consideration of this case is given later in Chapter 6). In a special case called the one-rod or systematic reflection case, electrons are reflected only by atomic planes parallel to the surface of the crystal and no surface resonance processes are involved. In this case Gotsis and Maksym (Gotsis and Maksym, 1997) were able to show that the specular RHEED rocking curve calculated using the multislice method agrees very well with that calculated using the semi-reciprocal approach discussed in the following section.

5.3.2 *The semi-reciprocal formulation*

Historically the first quantitative RHEED dynamical calculation made for a real surface structure with a surface selvedge was based on a semi-reciprocal approach (Maksym and Beeby, 1981). The semi-reciprocal approach to RHEED calculations was first proposed by Tournarie (Tournarie, 1962) and later developed by Kambe (Kambe, 1967), Lynch and Moodie (Lynch and Moodie, 1972), Maksym and Beeby (Maksym and Beeby, 1981), Ichimiya (Ichimiya, 1983), and Zhao et al. (Zhao et al., 1988). This approach may be derived starting from the Schrödinger equation

$$[\nabla^2 + k_0^2 + U(\mathbf{r})]\psi(\mathbf{r}) = 0, \tag{5.7}$$

where $U(\mathbf{r}) = -2mV(\mathbf{r})/\hbar^2$ is the scaled interaction potential ($V(\mathbf{r})$ is negative), and $\mathbf{k}_0$ is the wave vector of the incident electron. The term 'semi-reciprocal' is derived from the fact that a two-dimensional rather than three-dimensional Fourier expansion is used for both the potential and the wave function

$$U(\mathbf{r}) = \sum_G U_G(z) \exp(i\mathbf{G} \cdot \mathbf{x}), \quad \psi(\mathbf{r}) = \sum_G \psi_G(z) \exp[i(\mathbf{G} + \mathbf{k}_{0_t}) \cdot \mathbf{x}],$$

where $\mathbf{k}_0 = (\mathbf{k}_{0_t}, k_{0_z})$ and $\mathbf{r} = (\mathbf{x}, z)$. Substituting the above expansions into the basic equation (5.7) we obtain a set of linear coupled differential equations

$$\frac{d^2}{dz^2}\psi_G(z) + \Gamma_G^2 \psi_G(z) + \sum_H U_{GH}(z)\psi_H(z) = 0, \tag{5.8}$$

where

$$\Gamma_G^2 = k_0^2 - (\mathbf{k}_{0_t} + \mathbf{G})^2.$$

In matrix notation the above equation may be written in a more compact form as

$$\frac{d^2}{dz^2}\boldsymbol{\psi}(z) = -\mathbf{M}(z)\boldsymbol{\psi}(z), \tag{5.9}$$

where $\boldsymbol{\psi}(z)$ is a column vector the G-th element of which is $\{\boldsymbol{\psi}(z)\}_G = \psi_G(z)$, and the matrix $\mathbf{M}(z)$ is a square matrix with its GH-th element given by

$$M_{GH}(z) = \Gamma_G^2 \delta_{GH} + U_{GH}(z). \tag{5.10}$$

5.3.2.1 *The Tournarie transformation* The second-order differential equation (5.9) may be transformed trivially into a set of two first-order differential equations as follows:

$$\frac{d\boldsymbol{\psi}(z)}{dz} = \boldsymbol{\psi}'(z); \quad \frac{d\boldsymbol{\psi}'(z)}{dz} = -\mathbf{M}(z)\boldsymbol{\psi}(z).$$

This transformation was first used by Tournarie (Tournarie, 1962), and was later extensively applied (Lynch and Smith, 1983, Smith and Lynch, 1988) to dynamical RHEED calculations. The approach developed by the latter authors starts with a definition of a supervector

$$\boldsymbol{\phi}(z) = \begin{pmatrix} \boldsymbol{\psi}(z) \\ \boldsymbol{\psi}'(z) \end{pmatrix}. \tag{5.11}$$

In terms of this supervector we have

$$\frac{d}{dz}\boldsymbol{\phi}(z) = \frac{d}{dz}\begin{pmatrix} \boldsymbol{\psi}(z) \\ \boldsymbol{\psi}'(z) \end{pmatrix} = \begin{pmatrix} d\boldsymbol{\psi}(z)/dz \\ -\mathbf{M}(z)\boldsymbol{\psi}(z) \end{pmatrix} = \mathbf{S}(z)\boldsymbol{\phi}(z),$$

where the matrix $\mathbf{S}(z)$ may be expressed in terms of the $\mathbf{M}(z)$ matrix defined by eqn (5.10) as

$$\mathbf{S}(z) = \begin{pmatrix} \mathbf{0} & \mathbf{I} \\ -\mathbf{M}(z) & \mathbf{0} \end{pmatrix}. \tag{5.12}$$

Formally the above equation may be solved to give

$$\boldsymbol{\phi}(z) = \exp\left[\int_0^\infty \mathbf{S}(z)dz\right]\boldsymbol{\phi}(0). \tag{5.13}$$

In general the matrix $\mathbf{S}(z)$ depends on z. This z dependence, however, may be circumvented by treating the crystal as an assembly of many very thin slices, each having a z-independent potential field. This procedure is equivalent to replacing an integral by a sum

$$\int_0^z \mathbf{S}(z)dz = \sum_i \mathbf{S}(z_i)\Delta z_i,$$

and the accuracy of this approximation may be improved to any desired degree by reducing the slice thickness Δz. The solution (5.13) now takes the form

$$\phi(z) = \prod_i \exp\{\mathbf{S}(z_i)\Delta z_i\}\phi(0). \tag{5.14}$$

For each slice its properties are entirely characterized by its matrix $\mathbf{S}(z_i)$. For each matrix $\mathbf{S}(z_i)$ we may solve the eigenvalue equation

$$\mathbf{S}(z)\mathbf{c}(z) = \mathbf{c}(z)\lambda. \tag{5.15}$$

An eigenvector matrix $\mathbf{C}$ may then be constructed by equating the j-th column of $\mathbf{C}$ to the j-th eigenvector $\mathbf{c}^{(j)}$, i.e. $\mathbf{C} = (\mathbf{c}^{(1)}, \mathbf{c}^{(2)}, ...)$. Using this definition we obtain

$$\mathbf{S}(z) = \mathbf{C}(z)\mathbf{\Lambda}(z)\mathbf{C}^{-1}(z),$$

where $\mathbf{\Lambda}$ is the diagonal eigenvalue matrix associated with the slice. The j-th element of the matrix of eigenvalues is given by $\{\mathbf{\Lambda}\}_j = \lambda^{(j)}$. In terms of the eigenvector and eigenvalue matrices, we obtain from eqn (5.14)

$$\phi(z) = \prod_i \exp\{\mathbf{C}(z_i)\mathbf{\Lambda}(z_i)\Delta z_i \mathbf{C}^{-1}(z_i)\}\phi(0) \tag{5.16}$$

$$= \prod_i \mathbf{C}(z_i)\{\exp(\mathbf{\Lambda}(z_i)\Delta z_i)\}_d \mathbf{C}^{-1}(z_i)\phi(0),$$

where the subscript d denotes a diagonal matrix, and where we have used the following sequence of transformations:

$$\exp(C\Lambda C^{-1}) = 1 + C\Lambda C^{-1} + (C\Lambda C^{-1})(C\Lambda C^{-1})/2! + ...$$
$$= 1 + C\Lambda C^{-1} + (C\Lambda^2 C^{-1})/2! + ... = C(1 + \Lambda + \Lambda^2 + ...)C^{-1}$$
$$= C\exp(\Lambda)C^{-1}.$$

The RHEED problem is now solved by first finding the eigenvalues of $\mathbf{S}$ for each thin slice, by multiplying the respective transfer matrices to obtain the transfer matrix for the entire crystal, and then by matching the solution to the appropriate boundary conditions.

It should be noted that the accuracy with which an eigenvalue equation may be solved depends sensitively on the structure of the matrix $\mathbf{S}$. The more symmetric the matrix is, the better is the accuracy with which the eigenvalue equation may be solved. Since the matrix $\mathbf{S}$ has no diagonal elements, solving the eigenvalue problem (5.15) numerically is inefficient. Matrix $\mathbf{M}$, on the other hand, is a much better matrix from the numerical point of view since it is highly symmetric. Indeed, in the case where only elastic scattering is present and there is no absorption, this matrix is Hermitian. Lynch and Moodie (Lynch and Moodie, 1972) have shown that the transfer matrix for a slice

$$\mathbf{\Omega}_i = \exp[\mathbf{S}(z_i)\Delta z_i] \tag{5.17}$$

may be expressed in terms of the $\mathbf{M}(z_i)$ matrix and the corresponding eigenvalue and eigenvector matrices $\mathbf{\Lambda}_i$ and $\mathbf{T}(z_i)$ defined by

$$\mathbf{M}(z_i)\mathbf{T}(z_i) = \mathbf{T}(z_i)\mathbf{\Lambda}_i.$$

They showed that

$$\mathbf{\Omega} = \frac{1}{2}\begin{bmatrix} (\mathbf{P}+\mathbf{P}^{-1}) & -i\mathbf{M}^{-1/2}(\mathbf{P}-\mathbf{P}^{-1}) \\ i\mathbf{M}^{-1/2}(\mathbf{P}-\mathbf{P}^{-1}) & (\mathbf{P}+\mathbf{P}^{-1}) \end{bmatrix}, \tag{5.18}$$

where $\mathbf{P} = \exp(i\mathbf{M}^{1/2}\Delta z)$, and the notation $\mathbf{M}^{1/2}$ means $\mathbf{M}^{1/2} = \mathbf{T}\mathbf{\Lambda}^{1/2}\mathbf{T}^\dagger$, where $\mathbf{\Lambda}^{1/2}$ is the diagonal matrix of positive square roots of the eigenvalues of $\mathbf{M}$, and $\mathbf{T}$ is the unitary matrix of eigenvectors of $\mathbf{M}$. Computer algorithms based on the above transformation have been developed by Lynch and Smith (Lynch and Smith, 1983). These algorithms have been applied extensively to the calculation of intensity distributions in convergent-beam RHEED patterns (Smith and Lynch, 1988).

5.3.2.2 *The Ichimiya transformation* Alternatively the second-order differential equation (5.8) may be linearized using the procedure proposed by Ichimiya (Ichimiya, 1983). Noticing the fact that we may write the basic equation (5.8) as

$$\left(\frac{d}{dz}+i\Gamma_G\right)\left(\frac{d}{dz}-i\Gamma_G\right)\psi_G(z) + \sum_H U_{GH}(z)\psi_H(z) = 0,$$

a transformation

$$\left(\frac{d}{dz}+i\Gamma_G\right)\psi_G(z) = i\tau_G(z), \quad \left(\frac{d}{dz}-i\Gamma_G\right)\psi_G(z) = -i\rho_G(z), \tag{5.19}$$

then leads to

$$\psi_H(z) = \frac{1}{2\Gamma_H}[\tau_H(z)+\rho_H(z)]$$

and

$$i\left(\frac{d}{dz}-i\Gamma_G\right)\tau_G(z) + \sum_H U_{GH}(z)\psi_H(z) = 0,$$

$$-i\left(\frac{d}{dz}+i\Gamma_G\right)\rho_G(z) + \sum_H U_{GH}(z)\psi_H(z) = 0.$$

A combination of these equations gives

$$\frac{d}{dz}\begin{pmatrix} \{\tau_G\} \\ \{\rho_G\} \end{pmatrix} = i\begin{pmatrix} \Gamma_G\delta_{GH} + \dfrac{U_{GH}}{2\Gamma_H} & \dfrac{U_{GH}}{2\Gamma_H} \\ -\dfrac{U_{GH}}{2\Gamma_H} & -\Gamma_G\delta_{GH} - \dfrac{U_{GH}}{2\Gamma_H}U_{GH} \end{pmatrix}\begin{pmatrix} \{\tau_H\} \\ \{\rho_H\} \end{pmatrix}.$$

In matrix form this equation can be written as

$$\frac{d}{dz}\phi(z) = \mathbf{A}(z)\phi(z), \tag{5.20}$$

where

$$\phi(z) = \begin{pmatrix} \tau \\ \rho \end{pmatrix}, \quad \mathbf{A}(z) = i\begin{pmatrix} \Gamma_G\delta_{GH} + \dfrac{U_{GH}}{2\Gamma_H} & \dfrac{U_{GH}}{2\Gamma_H} \\ -\dfrac{U_{GH}}{2\Gamma_H} & -\Gamma_G\delta_{GH} - \dfrac{U_{GH}}{2\Gamma_H}U_{GH} \end{pmatrix}.$$

Again a formal solution of this matrix equation has the form of (5.13). However, unlike the matrix $\mathbf{S}$ introduced in the previous section, the $\mathbf{A}$ matrix entering Ichimiya's equations is highly symmetric and may therefore be used directly for the evaluation of the transfer matrix Ω of each individual slice. This is the method adopted by Ichimiya and his colleagues (Ichimiya, 1983), and the consideration given above shows that his method is formally equivalent to the approach discussed in the previous section.

5.3.2.3 *The Maksym and Beeby transformation*

A somewhat more complicated numerical procedure was developed by Maksym and Beeby who used the following transformation (Maksym and Beeby, 1981):

$$\psi_G(z) = Q_G^+(z)\exp(i\Gamma_G z) + Q_G^-(z)\exp(-i\Gamma_G z)$$

and obtained a set of first-order coupled differential equation for Q_G^+ and Q_G^-

$$\frac{d}{dz}\begin{pmatrix} \{Q_G^+\} \\ \{Q_G^-\} \end{pmatrix} = \begin{pmatrix} \dfrac{U_{GH}}{2i\Gamma_G}e^{-i(\Gamma_G-\Gamma_H)z} & \dfrac{U_{GH}}{2i\Gamma_G}e^{-i(\Gamma_G+\Gamma_H)z} \\ -\dfrac{U_{GH}}{2i\Gamma_G}e^{i(\Gamma_G+\Gamma_H)z} & \dfrac{U_{GH}}{2i\Gamma_G}e^{i(\Gamma_G-\Gamma_H)z} \end{pmatrix}\begin{pmatrix} \{Q_G^+(z)\} \\ \{Q_G^-(z)\} \end{pmatrix}.$$

Solutions of the above system of equations may be found by direct numerical integration (Maksym and Beeby, 1981) rather than by using the transfer matrix approach.

5.3.3 *The Green's function approach*

In this section we consider the Green's function approach. Although this approach has not yet been fully implemented in many-beam dynamical RHEED calculations, it provides some important insights into the nature of dynamical diffraction of electrons in RHEED. Extensive treatments of the Green's function approach were given by Tournarie (Tournarie, 1962) and by Kambe (Kambe, 1967).

We define the Green's function $\mathsf{G}(\mathbf{r}, \mathbf{r}')$ of the wave equation as a function satisfying the following equation:

$$(\nabla^2 + k_0^2)\mathsf{G}(\mathbf{r}, \mathbf{r}') = \delta(\mathbf{r} - \mathbf{r}'). \tag{5.21}$$

It describes the free space propagation of a wave from a point source at $\mathbf{r}'$ to the observation point at $\mathbf{r}$, and has the form

$$\mathsf{G}(\mathbf{r}, \mathbf{r}') = -\frac{1}{4\pi}\frac{\exp(ik_0|\mathbf{r} - \mathbf{r}'|)}{|\mathbf{r} - \mathbf{r}'|}. \tag{5.22}$$

This function differs from Green's function (2.4) only by a constant factor, i.e. $\mathsf{G}(\mathbf{r}, \mathbf{r}') = \hbar^2 G(\mathbf{r}, \mathbf{r}')/2m$.

In the case of an inhomogeneous wave equation

$$[\nabla^2 + k_0^2]\psi(\mathbf{r}) = -U(\mathbf{r})\psi(\mathbf{r}), \tag{5.23}$$

Green's function (5.22) can be used to transform it into an integral equation

$$\psi(\mathbf{r}) = \psi^{(0)}(\mathbf{r}) - \int \mathsf{G}(\mathbf{r}, \mathbf{r}')U(\mathbf{r}')\psi(\mathbf{r}')d\mathbf{r}', \tag{5.24}$$

where $\psi^{(0)}(\mathbf{r})$ is a solution of the wave equation corresponding to $U(\mathbf{r}) = 0$, e.g. a plane wave or a spherical wave.

For the convenience of later discussion we quote here also the $\mathbf{k}$-space or the spectral representation of the Green's function

$$\mathsf{G}(\mathbf{r}, \mathbf{r}') = \frac{1}{(2\pi)^3} \int \frac{\exp[i\mathbf{k} \cdot (\mathbf{r} - \mathbf{r}')]}{k_0^2 - k^2 + i0} d\mathbf{k}. \tag{5.25}$$

This Green's function describes propagation of the electron in three-dimensional space from $\mathbf{r}'$ to $\mathbf{r}$. In electron diffraction by crystals, however, a description in terms of scattering between planes z' and z is more relevant. In particular in RHEED, scattering in both directions (i.e. $z > z'$ and $z < z'$) must be included. In order to achieve this let us decompose vectors $\mathbf{k}$ and $\mathbf{r}$ in the above equations into components parallel and normal to the surface, i.e. in real space $\mathbf{r} = (\boldsymbol{\rho}, z)$, in reciprocal space $\mathbf{k}_0 = (\mathbf{k}_{0_\rho}, k_{0z})$, and $\mathbf{k} = (\mathbf{k}_{0_\rho} + \mathbf{s}, k_{0_z} + \zeta)$, as shown in Fig. 5.6a. The Green's function (5.25) can then be written as

$$
\begin{aligned}
\mathsf{G}(\mathbf{r}, \mathbf{r}') &= \frac{1}{(2\pi)^3} \int\int \frac{\exp[i(\mathbf{k}_{0_\rho} + \mathbf{s}) \cdot (\boldsymbol{\rho} - \boldsymbol{\rho}') + i(k_{0_z} + \zeta)(z - z')]}{k_0^2 - (\mathbf{k}_{0_\rho} - \mathbf{s})^2 + (k_{0_z} + \zeta)^2 + i0} ds\, d\zeta \\
&= \frac{1}{(2\pi)^3} \int \exp[i(\mathbf{k}_{0_\rho} + \mathbf{s}) \cdot (\boldsymbol{\rho} - \boldsymbol{\rho}')]ds \int_{-\infty}^{\infty} \frac{\exp[i(k_{0_z} + \zeta)(z - z')]d\zeta}{k_0^2 - (\mathbf{k}_{0_\rho} + \mathbf{s})^2 - (k_{0_z} + \zeta)^2 + i0} \\
&= -\frac{1}{(2\pi)^3} \int \exp[i(\mathbf{k}_{0_\rho} + \mathbf{s}) \cdot (\boldsymbol{\rho} - \boldsymbol{\rho}')]ds \int_{-\infty}^{\infty} \frac{\exp[i(k_{0_z} + \zeta)(z - z')]}{(\zeta - \zeta^s)(\zeta - \zeta_s)}d\zeta,
\end{aligned}
$$

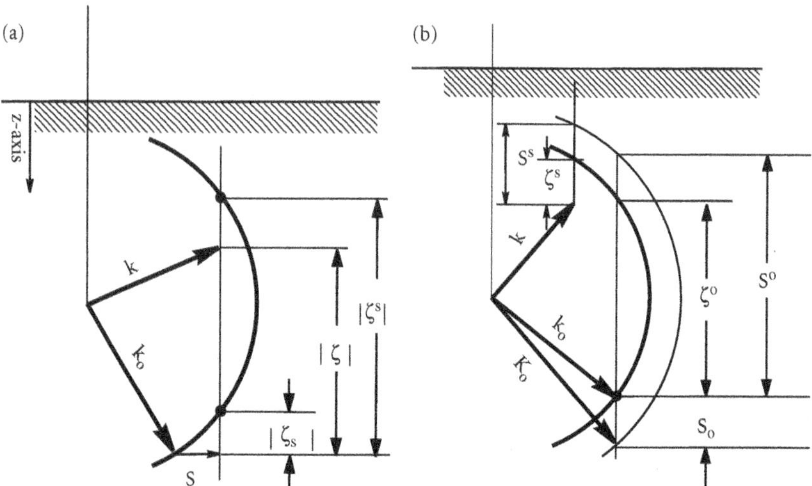

FIG. 5.6. Schematic diagram showing the coordinate systems and magnitudes of the excitation errors (a) without and (b) with the crystal inner potential energy effect being taken into account. In (b) $\mathbf{K}_0$ is the incident electron wave vector corrected for the crystal inner potential energy, i.e. $K_0 = \sqrt{k_0^2 + U_0}$.

where the two poles are given by

$$\zeta_s = -k_{0_z} + \sqrt{k_0^2 - (\mathbf{k}_{0_\rho} + \mathbf{s})^2 + i0}, \quad \zeta^s = -k_{0_z} - \sqrt{k_0^2 - (\mathbf{k}_{0_\rho} + \mathbf{s})^2 + i0}.$$

The integration over the z component of vector $\mathbf{k}$ (i.e. ζ) in the above equation should be carried out in the complex ζ-plane following a contour which is different for forward $(z > z')$ and backward $(z < z')$ scattering, as discussed for the transmission case by Fujiwara (Fujiwara, 1959) and by Gjønnes (Gjønnes, 1972) (see Fig. 5.7). This procedure gives

$$G(\mathbf{r}, \mathbf{r}') = -\frac{1}{(2\pi)^3} \int \exp[i(\mathbf{k}_{0_\rho} + \mathbf{s}) \cdot (\boldsymbol{\rho} - \boldsymbol{\rho}') + ik_{0_z}(z - z')]d\mathbf{s}$$

$$\times \left\{ 2\pi i \frac{\exp[i\zeta_s(z - z')]}{(\zeta_s - \zeta^s)} \Theta(z - z') - 2\pi i \frac{\exp[i\zeta^s(z - z')]}{(\zeta^s - \zeta_s)} \Theta(z' - z) \right\},$$

in which $\Theta(z - z')$ is the step function defined by $\Theta(z - z') = 1$ for $z - z' > 0$ and $\Theta(z - z') = 0$ for $z - z' < 0$. We now have the following explicit expressions for the forward $(z > z')$ and backward- $(z < z')$ scattering Green's functions:

$$G^F(\boldsymbol{\rho}, \boldsymbol{\rho}'; z, z') = \frac{i}{(2\pi)^2} \int \frac{\exp[i(k_{0_z} + \zeta_s)(z - z')]}{\zeta^s - \zeta_s} \exp[i(\mathbf{k}_{0_\rho} + \mathbf{s}) \cdot (\boldsymbol{\rho} - \boldsymbol{\rho}')]d\mathbf{s},$$

$$G^B(\boldsymbol{\rho}, \boldsymbol{\rho}'; z, z') = \frac{i}{(2\pi)^2} \int \frac{\exp[i(k_{0_z} + \zeta^s)(z - z')]}{\zeta^s - \zeta_s} \exp[i(\mathbf{k}_{0_\rho} + \mathbf{s}) \cdot (\boldsymbol{\rho} - \boldsymbol{\rho}')]d\mathbf{s}.$$

$$(5.26)$$

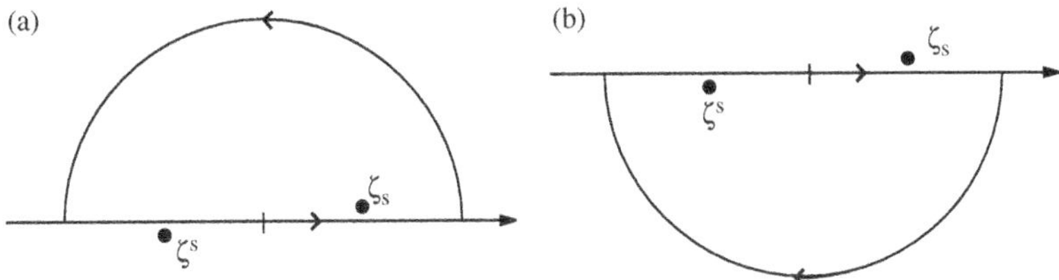

FIG. 5.7. Integration contours for (a) forward scattering with $z - z' > 0$ and (b) backward scattering with $z - z' < 0$.

For a periodic potential $U(\mathbf{r}) = \sum_h U_h \exp(i\mathbf{h} \cdot \mathbf{r})$, we may expand the wave function inside the crystal in the Fourier series

$$\psi(\mathbf{r}) = \sum_g \psi_g(z) \exp[i(\mathbf{k}_0 + \mathbf{g}) \cdot \mathbf{r}]. \tag{5.27}$$

Assuming a plane wave incidence and a crystal plate of thickness D, substitution of the above expansions for $U(\mathbf{r})$ and $\psi(\mathbf{r})$ into the integral equation of scattering (5.24) then gives

$$\psi(\mathbf{r}) = \exp(i\mathbf{k}_0 \cdot \mathbf{r}) \ - \sum_{g,h} U_{gh} \int d\boldsymbol{\rho}' \int_0^z dz' \psi_h(z') \exp[i(\mathbf{k}_0 + \mathbf{g}) \cdot \mathbf{r}']\mathsf{G}^F(\mathbf{r}, \mathbf{r}')$$

$$- \sum_{g,h} U_{gh} \int d\boldsymbol{\rho}' \int_z^D dz' \psi_h(z') \exp[i(\mathbf{k}_0 + \mathbf{g}) \cdot \mathbf{r}']\mathsf{G}^B(\mathbf{r}, \mathbf{r}'),$$

where $U_{gh} = U_{\mathbf{g}-\mathbf{h}}$. Substitution of the Green's function (5.26) into the above equations then gives a set of coupled integral equations for the diffracted beam amplitudes $\psi_g(z)$:

$$\psi_g(z) = \delta_{0g} + \frac{i}{\Delta_g} \sum_h U_{gh} \int_0^z \psi_h(z') \exp[i\zeta_g(z - z')]dz'$$

$$+ \frac{i}{\Delta_g} \sum_h U_{gh} \int_z^D \psi_h(z') \exp[i\zeta^g(z - z')]dz', \tag{5.28}$$

where each beam appears with two excitation errors, for forward and backward scattering respectively

$$\zeta_g = -k_{0_z} - g_z + \sqrt{k_0^2 - (\mathbf{k}_0 + \mathbf{g})_\rho^2}, \quad \zeta^g = -k_{0_z} - g_z - \sqrt{k_0^2 - (\mathbf{k}_0 + \mathbf{g})_\rho^2}, \tag{5.29}$$

and

$$\Delta_g = \zeta_g - \zeta^g = 2\sqrt{k_0^2 - (\mathbf{k}_0 + \mathbf{g})_\rho^2}. \tag{5.30}$$

We now look for a solution of the form

$$\psi_g(z) = C_g \exp(i\gamma z),$$

and obtain a set of integral equations for the eigenvalue γ,

$$C_g \exp(i\gamma z) = \delta_{0g} + \frac{i}{\Delta_g} \sum_h U_{gh} C_h \exp(i\zeta_g z) \int_0^z \exp[-i(\zeta_g - \gamma)z']dz'$$

$$+ \frac{i}{\Delta_g} \sum_h U_{gh} C_h \exp(i\zeta^g z) \int_z^D \exp[-i(\zeta^g - \gamma)z']dz'. \qquad (5.31)$$

Let

$$A = \frac{i}{\Delta_g} \sum_h U_{gh} C_h \exp(i\zeta_g z) \int_0^z \exp[-i(\zeta_g - \gamma)z']dz'$$

$$B = \frac{i}{\Delta_g} \sum_h U_{gh} C_h \exp(i\zeta^g z) \int_z^D \exp[-i(\zeta^g - \gamma)z']dz',$$

and differentiate eqn (5.31) with respect to z. We arrive at the following two equations:

$$C_g \exp(i\gamma z) = \delta_{0g} + A + B \qquad (5.32)$$

$$\gamma C_g \exp(i\gamma z) = \zeta_g A + \zeta^g B. \qquad (5.33)$$

From these equations, noticing that $\delta_{0g}\zeta_g = 0$ and $\delta_{0g}\zeta^g = -2\delta_{0g}k_{0_z}$, we can obtain explicit expressions for the two integrals A and B

$$A = -\frac{1}{\Delta_g}(\zeta^g - \gamma)C_g \exp(i\gamma z) + \delta_{0g}, \quad B = \frac{1}{\Delta_g}(\zeta_g - \gamma)C_g \exp(i\gamma z). \qquad (5.34)$$

We now consider how to obtain an eigenvalue equation for the eigenvalue γ and eigenvector C_g. On differentiating eqn (5.33) once more we have

$$(i\gamma)^2 C_g \exp(i\gamma z) = \frac{i}{\Delta_g} \sum_h U_{gh} C_h \Big\{ (i\zeta_g)^2 \int_0^z \exp[i\zeta_g(z - z') + i\gamma z']dz'$$

$$+ (i\zeta^g)^2 \int_z^D \exp[i\zeta^g(z - z') + i\gamma z']dz' + i\Delta_g \exp(i\gamma z) \Big\}.$$

Substituting expressions (5.34) into the above equation we obtain

$$(i\gamma)^2 C_g = \frac{1}{\Delta_g}\Big\{ (\zeta_g)^2(\zeta^g - \gamma) - (\zeta^g)^2(\zeta_g - \gamma) \Big\}C_g - \sum_h U_{gh} C_h,$$

which after some rearrangement leads to the following eigenvalue system:

$$(\gamma - \zeta_g)(\gamma - \zeta^g)C_g - \sum_h U_{gh} C_h = 0. \qquad (5.35)$$

Alternatively this eigenvalue system may be rewritten as

$$C_g = \frac{1}{\Delta_g} \sum_h \frac{U_{gh}C_h}{\gamma - \zeta_g} - \frac{1}{\Delta_g} \sum_h \frac{U_{gh}C_h}{\gamma - \zeta^g}, \qquad (5.36)$$

where the two sums represent the contributions from forward and backward scattering respectively. When the crystal inner potential energy U_0 is taken into consideration, we may define excitation errors inside the crystal (see Fig. 5.6b)

$$S_g = -k_{0_z} - g_z + \sqrt{k_0^2 + U_0 - (\mathbf{k}_0 + \mathbf{g})_\rho^2},$$

$$S^g = -k_{0_z} - g_z - \sqrt{k_0^2 + U_0 - (\mathbf{k}_0 + \mathbf{g})_\rho^2}.$$

The eigenvalue equation (5.35) becomes

$$(\gamma - S_g)(\gamma - S^g)C_g - \sum_{h \neq g} U_{gh}C_h = 0, \qquad (5.37)$$

and the Bragg condition for the g-th diffracted beam corresponds to $S^g = 0$, i.e. $K^2 = \mathbf{k}_0^2 + U_0 = (\mathbf{k}_0 + \mathbf{g})^2$. Noticing that

$$S_g S^g = -[k_0^2 + U_0 - (\mathbf{k}_0 + \mathbf{g})^2], \quad S_g + S^g = -2(\mathbf{k}_0 + \mathbf{g})_z,$$

the above eigenvalue equation then reduces to the usual second-order eigenvalue equation for three-dimensional Bloch wave coefficients C_g,

$$\left\{\gamma^2 + 2(\mathbf{k}_0 + \mathbf{g})_z \gamma - [k_0^2 + U_0 - (\mathbf{k}_0 + \mathbf{g})^2]\right\} C_g - \sum_{g \neq h} U_{gh}C_h = 0. \qquad (5.38)$$

In general the eigenvalue equation (5.38) will give rise to $2N$ eigenvalues γ and $2N$ Bloch waves when N beams are included. In many cases, however, not all the $2N$ Bloch waves are independent and physically allowed, and many of them may indeed be excluded. As an illustration we consider a symmetric two-beam case with one direct beam and one reflected beam and with the atom planes parallel to the crystal surface, i.e. $\mathbf{g}_\rho = 0$. Using the definition (5.30) we have for the two beams

$$\zeta_0 = 0, \quad \zeta^0 = -2k_{0_z},$$

$$\zeta_g = -g, \quad \zeta^g = -2k_{0_z} - g.$$

The corresponding eigenvalue equation (5.35) gives

$$\begin{pmatrix} \gamma(\gamma + 2k_{0_z}) - U_0 & -U_g \\ -U_g & (\gamma + g)(\gamma + 2k_{0_z} + g) - U_0 \end{pmatrix} \begin{pmatrix} C_0 \\ C_g \end{pmatrix} = 0. \qquad (5.39)$$

Let $\gamma' = \gamma + k_{0_z} + g/2$. Then the two diagonal matrix elements in the above equation become

$$\gamma(\gamma + 2k_{0_z}) - U_0 = \left[(\gamma' - g/2) - k_{0_z}\right]\left[(\gamma' - g/2) + 2k_{0_z}\right] - U_0$$
$$(\gamma + g)(\gamma + 2k_{0_z} + g) - U_0 = \left[(\gamma' + g/2) - k_{0_z}\right]\left[(\gamma' + g/2) + 2k_{0_z}\right] - U_0,$$

and the four corresponding eigenvalues are given by

$$(\gamma')_{1,4} = \pm\sqrt{U_0 + (\tfrac{1}{2}g)^2 + (k_{0_z})^2 + \sqrt{U_g^2 + g^2 k_{0_z}^2 + g^2 U_0}}$$

$$(\gamma')_{2,3} = \pm\sqrt{U_o + (\tfrac{1}{2}g)^2 + (k_{0_z})^2 - \sqrt{U_g^2 + g^2 k_{0_z}^2 + g^2 U_0}}.$$

Of the four Bloch waves, 2 and 3 have complex eigenvalues around the Bragg condition ($S^g = 0$, i.e. $k_{0_z} + g = \sqrt{k_{0_z}^2 + U_0}$), where $(\gamma')_{2,3} \sim \pm i\sqrt{U_g}$. The other Bloch waves 1 and 4 have eigenvalues with only a small imaginary part, which is zero when absorption is neglected, and with a large real part. Eigenvalues $\gamma^{1,4}$ are seen to be close to g and $-g$ respectively (see Figs 5.8a,d). Of the corresponding Bloch waves, 2 and 3 are found to include the beams 0 and g, whereas Bloch waves 1 and 4 include plane wave components close to the beams $-g$ and $2g$ respectively. The latter two may therefore be considered as unwanted solutions if only two beams are assumed in the description. In the present framework they are associated with the forward excitation error for the beam $g(S_g)$ and the backscattering excitation error for beam $0(S^0)$. It is convenient to distinguish between the two contributions from forward and backward scattering by writing

$$C_g = c_g - c^g. \tag{5.40}$$

For forward scattered beams, the backscattering excitation errors S^g are large and the corresponding amplitudes c^g are small and may be neglected. For these beams we have $c_g \gg c^g$, and from eqn (5.35)

$$(S_g - \gamma)c_g + (1/\Delta_g)\sum_{h \neq g} U_{gh}(c_h - c^h) = 0. \tag{5.41}$$

For backscattered beams, $c^g \gg c_g$, and the forward-scattering excitation errors S_g are large so that the corresponding amplitudes c_g may be neglected. For these beams we have

$$(S^g - \gamma)c^g + (1/\Delta_g)\sum_{h \neq g} U_{gh}(c_h - c^h) = 0. \tag{5.42}$$

In the above equations each reflection **g** is now represented by two parts, representing contributions from forward and backward scattering respectively, and with the corresponding excitation errors S_g and S^g. It should be noted that the separation (5.40) of c_g and c^g is based on physical criteria in terms of forward- and backward-scattering contributions, rather than on application of a standard mathematical transformation. The advantage of this procedure may be seen in

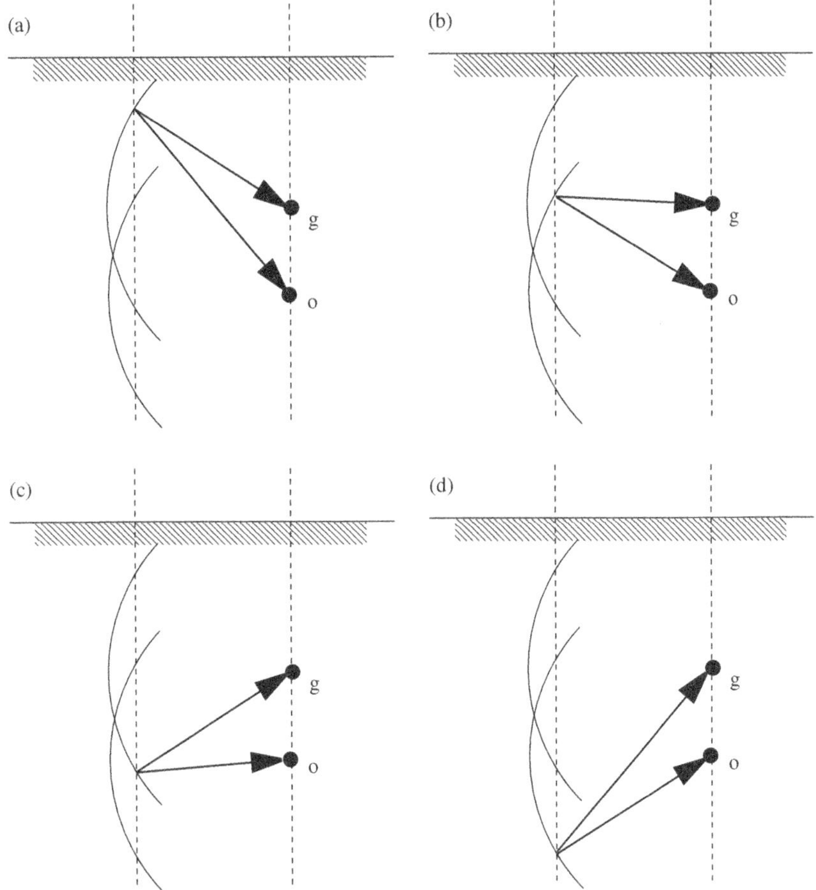

F IG. 5.8. A symmetric two-beam case illustrating plane wave components for (a) Bloch wave 1, (b) Bloch wave 2, (c) Bloch wave 3, and (d) Bloch wave 4.

the facility of entering beams as either backscattered or forward scattered in the calculations, according to the magnitude of their excitation errors S_g and S^g. In the symmetric two-beam case discussed above, the backscattering contribution to the forward zero beam and the forward-scattering contribution to the **g** beam may be excluded. An approximate two-beam form is obtained

$$\begin{pmatrix} (S_0 - \gamma) & -U_g/2k_{0_z} \\ U_g/2k_{0_z} & (S^g - \gamma) \end{pmatrix} \begin{pmatrix} c_0 \\ c^g \end{pmatrix} = 0,$$

which yields two eigenvalues

$$\gamma_{1,2} = \frac{1}{2}[(S^g + S_0) \pm \sqrt{(S^g - S_0)^2 - (U_g/k_{0_z})^2}],$$

where only the forward-scattering excitation error S_0 and the backward-scattering excitation error S^g appear, while the forward-scattering contribution c_g to C_g and the backward-scattering contribution c^0 to C_0 are neglected. The extension of the two-beam case to more beams along the systematic row is straightforward, as is the similar scheme for calculations near the Bragg condition for other reflections in the row (Peng et al., 1992).

5.3.4 The Bloch wave method

Although the methods introduced in the preceding sections can be used for fast numerical calculations of dynamical RHEED beam intensities, these methods do not make explicit use of the three-dimensional periodicity of the substrate and are applicable only to a crystal slab of finite thickness rather than to a semi-infinite crystal. In this section we discuss Bloch wave methods suitable for the treatment of RHEED from a surface of either a semi-infinite crystal or a crystal slab. We will start our development from the original Bethe three-dimensional Bloch wave approach, which makes explicit use of the crystal periodicity in the direction normal to the surface, and then introduce additional complexity associated with the presence of the surface selvedge.

5.3.4.1 *Three-dimensional Bloch waves* In his original theoretical formulation (Bethe, 1928) Bethe assumed that the crystal is three-dimensionally periodic and that the potential field is terminated abruptly at a certain mathematical plane (see the curve corresponding to the truncated potential model in Fig. 5.9). This assumption of course contradicts the basic principles of electrostatics, as one cannot possibly consider this plane as charged without coming into conflict with the atomic foundation of the whole theory (von Laue, 1931). The Bethe theory nevertheless may serve as a good starting point for the description of diffraction of high-energy electrons in the crystal bulk. The treatment of dynamical diffraction processes occurring in the surface region of the crystal by Bethe's method is, however, not entirely satisfactory. In this section we will first discuss Bethe's original formulation of the theory, and then consider the means for incorporating effects arising from the smooth transition of the internal electrostatic field of the crystal to the zero field in the vacuum region.

The Bethe theory starts from the fundamental equation of the dynamical theory of electron diffraction for a three-dimensional periodic potential field $U(\mathbf{r})$

$$[K^2 - (\mathbf{k} + \mathbf{g})^2]C_g + \sum_{h \neq g} U_{gh}C_h = 0, \qquad (5.43)$$

where $\mathbf{K}$ is the electron wave vector corrected for the mean inner potential of the crystal, i.e. $K^2 = k_0^2 + U_0$. The C_g are the Fourier coefficients appearing in the expansion of $\psi(\mathbf{r})$ as a Bloch function, and the U_g are the Fourier coefficients of the crystal potential $U(\mathbf{r})$ defined in section 5.3.2,

$$\psi(\mathbf{r}) = \sum_g C_g \exp\{i(\mathbf{k} + \mathbf{g}) \cdot \mathbf{r}\}, \quad U(\mathbf{r}) = \sum_g U_g \exp(i\mathbf{g} \cdot \mathbf{r}).$$

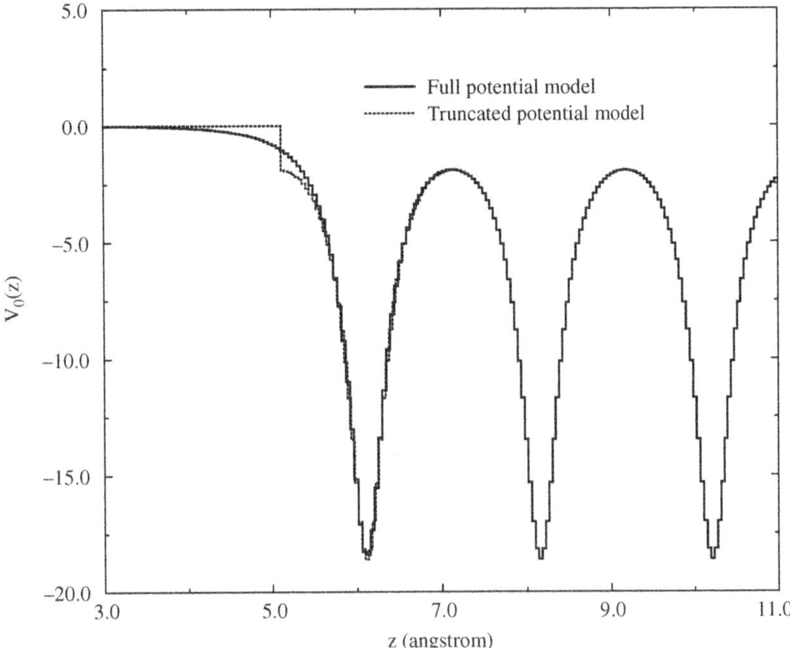

F IG. 5.9. Plots of the average potential $V_0(z)$ (in eV units) at the near surface region for the full and for the truncated potential models. The potential distribution is calculated for an Ag(001) surface, with the z axis parallel to the $[00\bar{1}]$ direction.

The set of equations (5.43) is in fact a second-order equation for eigenvalues and eigenvectors. Since the surface parallel component of the electron wave vector $\mathbf{k}$ is conserved (subject to addition of the surface component of a reciprocal lattice vector) in reflection dynamical diffraction, the only as yet undetermined component of the wave vector is its surface normal component k_z. Using this fact we can reduce the general three-dimensional eigensystem (5.43) to a one-dimensional eigensystem by introducing a variable γ such that

$$\mathbf{k} = \mathbf{K} + \gamma\hat{\mathbf{z}}, \tag{5.44}$$

where $\hat{\mathbf{z}}$ is a unit vector pointing in the positive z direction. The set of equations (5.43) then becomes

$$[\gamma^2 + 2(\mathbf{K} + \mathbf{g})_z\gamma - 2KS_g]C_g - \sum_{h \neq g} U_{gh}C_h = 0, \tag{5.45}$$

where

$$2KS_g \approx K^2 - (\mathbf{K} + \mathbf{g})^2, \tag{5.46}$$

and S_g is the usual excitation error for reflection $\mathbf{g}$ equal to the distance in reciprocal space between the point $\mathbf{K} + \mathbf{g}$ and the Ewald sphere of radius K.

Eqn (5.46) is approximate because a term involving S_g^2 has been neglected. In matrix notation eqn (5.45) can be rewritten as

$$(\gamma^2 \mathbf{I} + \gamma \mathbf{D} + \mathbf{Q})\mathbf{C} = 0, \tag{5.47}$$

in which $\mathbf{I}$ is an identity matrix, $\mathbf{D}$ is a diagonal matrix of the form

$$\{\mathbf{D}\}_g = 2(\mathbf{K} + \mathbf{g})_z, \tag{5.48}$$

and $\mathbf{Q}$ is a general matrix whose elements are

$$\{\mathbf{Q}\}_{gh} = -2k_0 S_g \delta_{gh} - U_{gh}(1 - \delta_{gh}), \tag{5.49}$$

where δ_{gh} is the Kronecker δ symbol, and $\mathbf{C}$ is a column vector

$$\{\mathbf{C}\}_h = C_h. \tag{5.50}$$

The matrix equation (5.47) is nothing but a special case of a general high-order eigenvalue problem

$$(\gamma^m \mathbf{C}_0 + \gamma^{m-1}\mathbf{C}_1 + ... + \gamma \mathbf{C}_{m-1} + \mathbf{C}_m)\mathbf{X} = 0. \tag{5.51}$$

It can be solved by forming two supermatrices $\mathbf{A}$ and $\mathbf{B}$,

$$\mathbf{A} = \begin{pmatrix} -\mathbf{C}_1 & -\mathbf{C}_2 & -\mathbf{C}_3 & ... \\ \mathbf{I} & 0 & 0 & ... \\ 0 & \mathbf{I} & 0 & ... \\ ... & ... & ... & ... \end{pmatrix} \quad \text{and} \quad \mathbf{B} = \begin{pmatrix} \mathbf{C}_0 & 0 & 0 & 0 \\ 0 & \mathbf{I} & 0 & 0 \\ 0 & 0 & \mathbf{I} & 0 \\ ... & ... & ... & ... \end{pmatrix},$$

in such a way that the original high-order problem (5.51) is transformed into a first-order problem

$$\mathbf{A}\mathbf{Z} = \gamma \mathbf{B}\mathbf{Z}, \tag{5.52}$$

where

$$\mathbf{Z}^T = (\gamma^{m-1}\mathbf{X}, \gamma^{m-2}\mathbf{X}, ..., \mathbf{X}),$$

and the superscript T denotes the transpose of the vector. The first-order linear system (5.52) can now be solved using standard numerical computer routines, such as NAG (NAG93, 1993) or EISPACK (Smith et al., 1976).

For electron diffraction, $m = 2$, and eqn (5.47) may be transformed into

$$\begin{pmatrix} -\mathbf{D} & -\mathbf{Q} \\ \mathbf{I} & 0 \end{pmatrix} \begin{pmatrix} \gamma \mathbf{C} \\ \mathbf{C} \end{pmatrix} = \gamma \begin{pmatrix} \gamma \mathbf{C} \\ \mathbf{C} \end{pmatrix}. \tag{5.53}$$

This equation was first introduced by Colella (Colella, 1972) in his treatment of the RHEED problem. In the case where there is only a finite number N of reciprocal lattice points lying on or close to the Ewald sphere, only N diffracted beams will be appreciably excited in the crystal. The set of infinite equations

(5.53) may then be truncated down to a set of $2N$ equations, giving in general $2N$ values of $\gamma^{(j)}, (j = 1, ..., 2N)$ and $2N$ associated Bloch waves $b^{(j)}(\mathbf{k}^{(j)}, \mathbf{r})$ $(j = 1, ..., 2N)$. The wave function of the electron in the crystal can be written as a linear combination of $2N$ Bloch waves

$$\psi(\mathbf{r}) = \sum_{j=1}^{2N} \alpha^{(j)} b^{(j)}(\mathbf{k}^{(j)}, \mathbf{r}) = \sum_{j=1}^{2N} \alpha^{(j)} \sum_{g} C_g^{(j)} \exp[i(\mathbf{K} + \gamma^{(j)}\hat{\mathbf{z}} + \mathbf{g}) \cdot \mathbf{r}], \quad (5.54)$$

where $\alpha^{(j)}$ is the excitation amplitude of the j-th Bloch wave. We note, however, that to the accuracy of a phase factor, the Bloch wave solutions of eqn (5.45) are periodic in reciprocal space, i.e.

$$b^{(j)}(\mathbf{k}, \mathbf{r}) = b^{(j)}(\mathbf{k} + \mathbf{g}, \mathbf{r}). \quad (5.55)$$

It is a corollary of this property that not all the Bloch waves resulting from eqn (5.45) are distinct. Those whose wave vectors $\mathbf{k}$ differ by only a reciprocal lattice vector $\mathbf{g}$ are physically equivalent (for more detail, see Chapter 3 and a book by Ashcroft and Mermin (Ashcroft and Mermin, 1976)).

When n reciprocal lattice rods (containing in total N reciprocal lattice points) are involved in the diffraction process, there are $2n$ independent Bloch waves, and a total of N diffracted beams propagating in the crystal. The total electron wave field within the crystal can be represented as

$$\psi(\mathbf{r}) = \sum_{j=1}^{2n} \alpha^{(j)} \sum_{g=g_1}^{g_N} C_g^{(j)} \exp[i(\mathbf{K} + \gamma^{(j)}\hat{\mathbf{z}} + \mathbf{g}) \cdot \mathbf{r}], \quad (5.56)$$

and the diffracted beam amplitude associated with the G-th reciprocal lattice rod is given by

$$\psi_G(z) = \sum_{j=1}^{2n} \alpha^{(j)} \sum_{g_G} C_{g_G}^{(j)} \exp[i(K_z + \gamma^{(j)} + g_{G_z})z]. \quad (5.57)$$

Here we have used the notation $\mathbf{g}_G = (\mathbf{G}, g_{G_z})$, where $\mathbf{G}$ denotes the surface parallel component of a reciprocal lattice rod and g_{G_z} denotes the component of a particular reciprocal lattice point along the rod.

The boundary conditions are the conditions of continuity imposed on the diffracted beam amplitudes and their surface normal derivatives

$$\exp(-iK_z z)\psi_G(z) = \sum_{j=1}^{2n} \alpha^{(j)} \sum_{g_G} C_{g_G}^{(j)} \exp[i(\gamma^{(j)} + g_{G_z})z], \quad (5.58)$$

$$-i\exp(-iK_z z)\psi_G'(z) = \sum_{j=1}^{2n} \alpha^{(j)} \sum_{g_G} (K_z + \gamma^{(j)} + g_{G_z})C_{g_G}^{(j)} \exp[i(\gamma^{(j)} + g_{G_z})z],$$

$$(5.59)$$

where ψ'_G denotes the derivative of ψ_G with respect to z. Neglecting a common phase factor identical for all the beams, eqns (5.58) and (5.59) can be rewritten in matrix notation as (Peng and Whelan, 1990b)

$$\boldsymbol{\Psi}(z) = \mathbf{S}\mathbf{P}(z)\mathbf{C}\boldsymbol{\Upsilon}(z)\boldsymbol{\alpha}, \tag{5.60}$$

where $\boldsymbol{\Psi}$ is a $2n$-dimensional supervector

$$\boldsymbol{\Psi}(z) = \begin{pmatrix} \{\psi_G(z)\} \\ \{\psi'_G(z)\} \end{pmatrix}, \tag{5.61}$$

$\mathbf{S}$ is a $(2n \times 2N)$ matrix, the elements of which are

$$\{\mathbf{S}\}_{Gg} = \begin{cases} 1, & \text{if } g \text{ belongs to the } G\text{-th rod} \\ 0, & \text{otherwise} \end{cases} \tag{5.62}$$

for $G \leq n, g \leq N$, and

$$\{\mathbf{S}\}_{G+n,g+N} = \{\mathbf{S}\}_{G,g}; \quad \{\mathbf{S}\}_{G+n,g} = \{\mathbf{S}\}_{G,g+N} = 0,$$

$\mathbf{P}$ is a $(2N \times 2N)$ diagonal matrix

$$\{\mathbf{P}\}_{gh} = \exp(ig_z z)\delta_{gh} \tag{5.63}$$

for $h, g \leq N, g = (G, p)$, and

$$\{\mathbf{P}\}_{g+N,h+N} = \{\mathbf{P}\}_{g,h}, \tag{5.64}$$

$\mathbf{C}$ is a $(2N \times 2n)$ matrix

$$\{\mathbf{C}\}_{hi} = C_h^{(i)} \tag{5.65}$$

for $h \leq N$, and

$$\{\mathbf{C}\}_{h+N,i} = (K_z + \gamma^{(i)} + h_z)C_h^{(i)},$$

$\boldsymbol{\Upsilon}$ is a $(2n \times 2n)$ diagonal matrix

$$\{\boldsymbol{\Upsilon}\}_{ij} = \exp(i\gamma^{(i)}z)\delta_{ij}, \tag{5.66}$$

and $\boldsymbol{\alpha}$ is a $2n$-dimensional column vector $(\boldsymbol{\alpha})_j = \alpha^{(j)}$.

If z_t and z_b are the coordinates of the top and bottom surfaces of a crystal slab, then from eqn (5.60) we have

$$\boldsymbol{\Psi}(z_t) = \mathbf{S}\mathbf{P}(z_t)\mathbf{C}\boldsymbol{\Upsilon}(z_t)\boldsymbol{\alpha},$$

which gives

$$\boldsymbol{\alpha} = \boldsymbol{\Upsilon}(-z_t)[\mathbf{S}\mathbf{P}(z_t)\mathbf{C}]^{-1}\boldsymbol{\Psi}(z_t).$$

Substitution of the above equation into eqn (5.60) gives an expression relating the supervectors $\boldsymbol{\Psi}$ at the top and the bottom surfaces of the crystal slab

$$\boldsymbol{\Psi}(z_b) = \mathbf{M}(z_b, z_t)\boldsymbol{\Psi}(z_t), \tag{5.67}$$

in which the matrix $\mathbf{M}$ is called the scattering matrix of the slab and is given by

$$\mathbf{M}(z_b, z_t) = [\mathbf{S}\mathbf{P}(z_b)\mathbf{C}]\boldsymbol{\Upsilon}(\Delta z)[\mathbf{S}\mathbf{P}(z_t)\mathbf{C}]^{-1},$$

and where $\Delta z = z_b - z_t$ is the thickness of the crystal slab.

We now consider RHEED from the surface of a semi-infinite bulk crystal. For the n-rod case we have a total of $2n$ non-equivalent eigenvalues and corresponding Bloch waves. However, only n out of $2n$ eigenvalues are physically allowed, since only half of all the Bloch waves excited in the crystal correspond to Bloch waves attenuating into the crystal bulk (Lamla, 1938a, Lamla, 1938b, Miyake et al., 1954, Marks and Ma, 1988). The boundary condition at the interface between the substrate and the selvedge gives an equation for the component of the solution corresponding to the G-th reciprocal lattice rod:

$$\psi_G(z_s) = \sum_{j=1}^{n} \alpha^{(j)} \sum_{g_G} C_{g_G}^{(j)} \exp[i(\gamma^{(j)} + g_{G_z})z_s] \tag{5.68}$$

$$\psi'_G(z_s) = \sum_{j=1}^{n} \alpha^{(j)} \sum_{g_G} (K_z + \gamma^{(j)} + g_{G_z}) C_{g_G}^{(j)} \exp[i(\gamma^{(j)} + g_{G_z})z_s], \tag{5.69}$$

or in matrix form

$$\boldsymbol{\Psi}(z_s) = \mathbf{S}^b \mathbf{P}^b(z_s) \mathbf{C}^b \boldsymbol{\Upsilon}^b(z_s) \boldsymbol{\alpha}, \tag{5.70}$$

where z_s is the thickness of the selvedge, and the superscript b refers to the crystal bulk. Matrices $\mathbf{S}^b$ and $\mathbf{P}^b$ are given by eqns (5.62)–(5.66), but $\mathbf{C}^b$ and $\boldsymbol{\Upsilon}^b$ now reduce to $2N \times n$ and $n \times n$ matrices respectively, and the vector $\boldsymbol{\alpha}$ becomes n dimensional. Equation (5.70) can now be written as

$$\boldsymbol{\Psi}(z_s) = \begin{pmatrix} \{\mathbf{B}_1\boldsymbol{\alpha}\} \\ \{\mathbf{B}_2\boldsymbol{\alpha}\} \end{pmatrix}, \quad \text{with} \quad \begin{pmatrix} \mathbf{B}_1 \\ \mathbf{B}_2 \end{pmatrix} = \mathbf{S}^b \mathbf{P}^b(z_s) \mathbf{C}^b \boldsymbol{\Upsilon}^b(z_s). \tag{5.71}$$

In the vacuum region above the crystal only the incident beam (indexed by $G = 0$) and reflected beams are present. By neglecting scattering in the selvedge, i.e. by assuming $z_s = 0$, we obtain for the supervector $\boldsymbol{\Psi}$

$$\boldsymbol{\Psi} = \begin{pmatrix} \{\delta_{G,0} + \mathcal{R}_G\} \\ \{k_{G_z}(\delta_{G,0} - \mathcal{R}_G)\} \end{pmatrix}. \tag{5.72}$$

The boundary conditions at the entrance surface ($z = 0$) require that

$$\begin{pmatrix} \{\mathbf{B}_1\} \\ \{\mathbf{B}_2\} \end{pmatrix} \boldsymbol{\alpha} = \begin{pmatrix} \{\delta_{G,0} + \mathcal{R}_G\} \\ \{k_{G_z}(\delta_{G,0} - \mathcal{R}_G)\} \end{pmatrix}, \tag{5.73}$$

and this leads to an explicit expression for the amplitudes of beams reflected from a truncated surface

$$\{\mathcal{R}_G\} = -\frac{\{(\mathbf{B}_1)_{GH}^{-1} - (\mathbf{B}_2)_{GH}^{-1} k_{H_z}\}}{\{(\mathbf{B}_1)_{GH}^{-1} + (\mathbf{B}_2)_{GH}^{-1} k_{H_z}\}} \{\delta_{H,0}\}. \tag{5.74}$$

Here we have used the notation defined in Chapter 3, namely that the reciprocal of the matrix in the denominator pre-multiplies the matrix in the numerator, i.e. $\mathbf{A}/\mathbf{B} = \mathbf{B}^{-1}\mathbf{A}$.

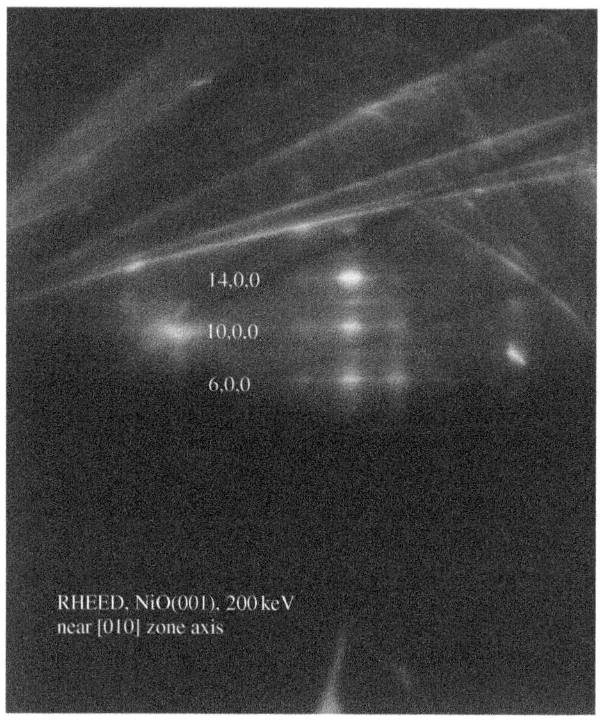

FIG. 5.10. Experimental RHEED pattern from the NiO(001) surface. The primary beam energy is 200 keV, and the electrons are incident on the surface near the [010] zone axis. From (Peng et al., 1997).

5.3.4.2 *The systematic diffraction case and the effect of surface refraction*   The simplest diffraction case describing reflection of an electron from a crystal surface is the so-called 'systematic' case (Peng and Whelan, 1991a) or the 'one-rod' case (Ichimiya, 1987). In this diffraction geometry the incident electrons are diffracted only by the atomic planes parallel to the surface of the crystal, and the diffracted beam intensities are sensitive only to the projection of coordinates of reflecting atomic planes on the direction normal to the surface. Experimentally the systematic diffraction case may be realized by tilting the incident electron beam away from a principal zone axis by about 20–30 mrad so that the resulting RHEED pattern is dominated by the horizontal Kikuchi lines. An example of a RHEED pattern corresponding to the systematic diffraction case is shown in Fig. 5.10. This pattern was obtained using the NiO(001) surface and the direction of incidence was close to the [010] zone axis. The horizontal Kikuchi lines are an indication of the Bragg condition being satisfied by atomic planes parallel to the surface. Bragg reflections may be readily indexed as shown in the figure.

The angles corresponding to the horizontal Kikuchi lines shown in this figure differ from those predicted by the standard Bragg law

$$2d \sin \theta = m\lambda, \qquad\qquad (5.75)$$

where d is the spacing of the atomic planes, θ is half the angle of scattering, and m is an integer denoting the order of the interference maximum, as indicated in Fig. 5.10. There are two effects that are responsible for the discrepancy between the experimentally observed positions of Bragg reflections and those predicted by the kinematic Bragg law (5.75).

The first one is the so-called refraction effect resulting from the difference between the effective refractive indices of the vacuum and the crystal. For high-energy electrons the Snell refraction law is given by (Hirsch et al., 1977)

$$n = \frac{\lambda_0}{\lambda} = \frac{\cos \theta_0}{\cos \theta}, \qquad\qquad (5.76)$$

where n denotes the refractive index, λ_0 and λ are the electron wavelengths in the vacuum and in the crystal, and $2\theta_0$ and 2θ refer to the scattering angles measured in the vacuum and in the crystal respectively. For a given kinetic energy E_{kin} the wavelength is given by

$$\lambda = \frac{0.3878314}{\sqrt{E_{kin}(1 + 0.97846707 \times 10^{-3} E_{kin})}},$$

where E_{kin} is measured in keV and the electron wavelength is given in Å. For a given primary beam energy E_0 and crystal inner potential V_0 (negative for electrons), $E_{kin} = E_0 = \hbar^2 k_0^2/2m$ in the vacuum and $E_{kin} = \hbar^2 k_0^2/2m - V_0$ in the crystal. A typical value of the crystal inner potential is of the order of (minus) 10 or 20 eV. The glancing scattering angle θ of high-energy electrons in the crystal is larger than the glancing angle θ_0 in the vacuum as shown in Fig. 5.11. All the observed Bragg peaks therefore appear at angles smaller than those predicted by the kinematic expression (5.75). In particular for zero glancing angle, i.e. for $\theta_0 = 0$, we obtain a finite value of θ. This means that a certain range of glancing angles is inaccessible in RHEED experiments. For the NiO(001) surface, for example, the crystal inner potential is equal to -21.2 eV (Peng et al., 1997). Since even for zero glancing angle of incidence, the effective glancing angle in the crystal is larger than the 200 Bragg angle, the first Bragg peak that can be observed in RHEED is therefore 400 as shown in Fig. 5.10 below the 600 line. It is in fact a general rule of electron diffraction that the effective glancing angle inside the crystal is larger than that in the vacuum, and this rule results from the fact that electrons are negatively charged and therefore feel an attractive inner potential of the crystal. In other diffraction cases, e.g. in X-ray diffraction where the refractive index is less than unity, this rule does not apply.

For the second effect we now consider the origin of Bragg peaks in a RHEED rocking curve. In the systematic reflection case where only one rod of the reciprocal lattice is involved in the diffraction process, the wave function of the electron in the crystal may be expressed as a sum of two Bloch waves

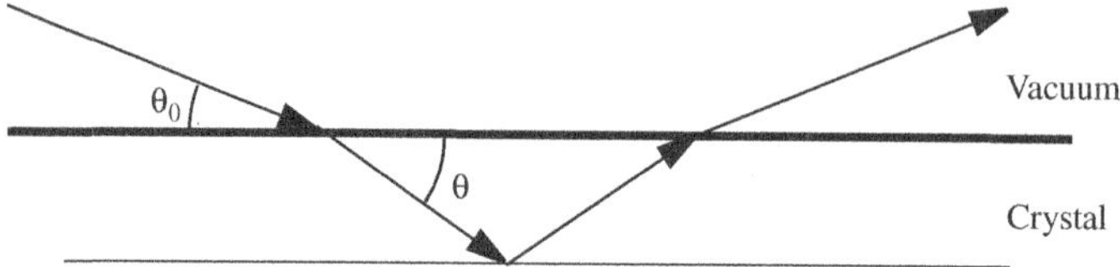

FIG. 5.11. Schematic diagram showing the refraction effect occurring when a beam of high-energy electrons is incident on the surface of a crystal.

$$\psi(z) = \sum_{j=1}^{2} \alpha^{(j)} \sum_{g} C_g^{(j)} \exp[i(K_z + \gamma^{(j)} + g)z],$$

where $\gamma^{(j)}$ are the corresponding eigenvalues shown in Fig. 5.12 by the intersection of the line normal to the surface with the two branches of the dispersion curve. We note that for an infinite number of reciprocal lattice points lying on a rod normal to the surface there is an infinite number of eigenvalues. The Bloch waves associated with these eigenvalues are, however, almost all degenerate and indistinguishable. In fact for a systematic reflection case there are only two distinct eigenvalues lying within the first Brillouin zone boundary and contributing to the Bloch sum. In the vicinity of the Bragg condition shown in Fig. 5.12 for the $2g$ reflection these eigenvalues become imaginary. For a symmetric crystal slab these two imaginary eigenvalues form a pair, where one of the eigenvalues is positive and the other one is negative. The Bloch wave associated with the eigenvalue that has a negative imaginary part is localized near the bottom face of the slab and decays exponentially towards the entrance surface. In the case of a semi-infinite crystal this Bloch wave does not contribute to the solution of the diffraction problem. In this case the total electron wave function is determined entirely by the evanescent Bloch wave which is localized near the entrance surface of the crystal. Since electrons do not accumulate in the crystal bulk, they are all reflected back into the vacuum giving rise to a Bragg peak of finite width in the systematic RHEED rocking curve. This effect of total reflection was first studied by Darwin (Darwin, 1914) for the case of reflection of X-rays from the surface of a crystal. Dynamical electron diffraction modifies the dispersion surface shown in Fig. 5.12 and therefore the value of the allowed eigenvalue or wave vector. Consequently the position of the Bragg peak is also altered.

5.3.4.3 *Two-dimensional Bloch waves* For an N-beam case the computer time required for finding a numerical solution of eqn (5.45) scales as $\mathcal{O}[(n{\times}n_z)^3]$, where n is the number of distinct reciprocal lattice rods and n_z is the average number of reciprocal lattice points per rod in the direction of the surface normal. For a convergent result n_z needs to be taken to be of the order of 10, giving rise to computations lasting 1000 times longer than for computations involving only a two-dimensional potential field averaged in the z direction such as is encountered in the transmission diffraction case. A more efficient and stable algorithm for the

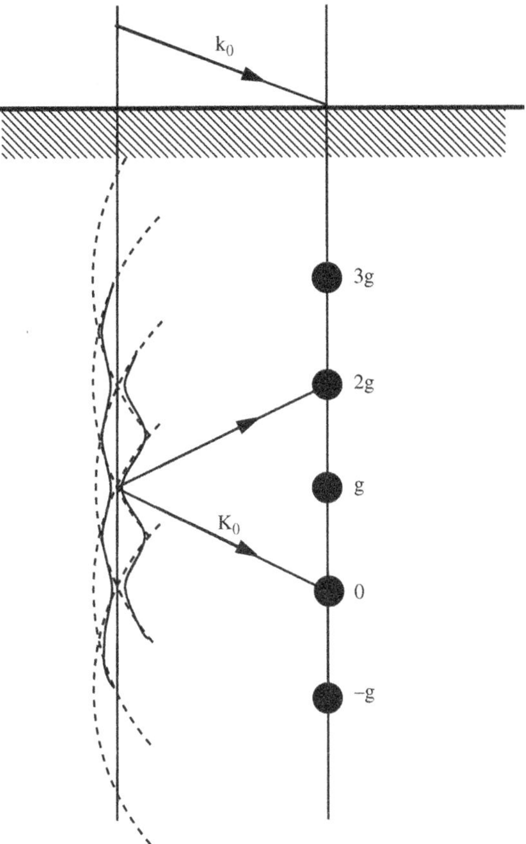

FIG. 5.12. Schematic diagram showing the dispersion curve and the physically allowed eigenvalues as intersections of branches of the dispersion curve with a vertical solid line passing through the end point of the incident electron wave vector $\mathbf{K}_0$.

calculation of diffracted beam intensities from the surface of a crystal is based on the use of two-dimensional Bloch waves. In this scheme the variation of the potential in the direction normal to the surface through the selvedge as well as through a bulk repeat unit slab is simulated by dividing the selvedge and the bulk unit slab into many slices. In this case we assume that for each slice the two-dimensional potential field varying in the plane parallel to the surface is constant in the direction normal to the surface (see Fig. 5.9). For a typical metal such as a single crystal of silver, a very accurate result may be obtained by cutting a 0.4 nm thick unit cell into approximately 100 slices. This scheme proves to be approximately 10 times faster than the three-dimensional Bloch wave method. Other advantages associated with the use of two-dimensional Bloch waves include the reduction of large asymmetric matrices of size about 1000×1000 elements into smaller and nearly symmetric matrices of size about 100×100 elements. As a rule, numerical results obtained using symmetric eigensystems of smaller size are more reliable and accurate. The most important gain associated with the use of two-dimensional Bloch waves is the ability to deal with a variation of the potential in the direction normal to the surface of the crystal. Therefore the two-dimensional Bloch wave approach solves the difficulty encountered in the original three-dimensional diffraction theory developed by Bethe.

The concept of two-dimensional Bloch waves may be introduced by considering the basic equation (5.45) for each slice of the assembly. Since the potential field of each slice is constant in the direction normal to the surface, for all the reflections entering eqn (5.45) we have $\mathbf{g} = (\mathbf{G}, g_z) = \mathbf{G}$ and

$$\{(\gamma + K_z)^2 - [K^2 - (\mathbf{K} + \mathbf{G})_t^2]\}C_G - \sum_{H \neq G} U_{GH} C_H = 0,$$

or

$$(\mathcal{K}^2 - \Gamma_G^2)C_G - \sum_{H \neq G} U_{GH} C_H = 0, \tag{5.77}$$

where $\mathcal{K} = \gamma + K_z$, $\Gamma_G^2 = K^2 - (\mathbf{K} + \mathbf{G})_t^2$, and the subscript t denotes the tangential component of the vector parallel to the surface. In the limiting case where $U_{GH} = 0$, i.e. in the empty lattice approximation, we obtain the following solution of the above equation:

$$\mathcal{K} = \pm\Gamma_G, \quad C_G = 1, \quad C_H = 0 \ (H \neq G).$$

The amplitude of the diffracted beam associated with the G-th reciprocal lattice rod (5.57) in this case is reduced to

$$\psi_G(z) \propto \exp(\pm i\Gamma_G z).$$

Since Γ_G measures the length of the intersection between the Ewald sphere and the Gth reciprocal lattice rod, it is convenient to classify all the waves associated with reciprocal lattice rods into propagating and evanescent waves according to whether the corresponding values of Γ_G are real or imaginary. Rods of the reciprocal lattice lying within the Ewald sphere intersect with the sphere giving rise to real values of Γ_G. On the other hand, rods situated outside the Ewald sphere and corresponding to $|\mathbf{K}| < |\mathbf{k}_{0_t} + \mathbf{G}|$ cannot intersect the sphere, and the relevant Γ_G take imaginary values giving rise to evanescent waves localized at either the top or the bottom surfaces of the thin slice. A special diffraction phenomenon occurs in the case where the G-th rod is tangential to the Ewald sphere. This case corresponds to $\Gamma_G \approx 0$ and the generation of a wave propagating nearly parallel to the surface of the crystal, and the relevant diffraction condition is called a surface resonance condition. Many interesting and highly unusual phenomena occur in the vicinity of surface resonance conditions, and we will discuss these in Chapter 6.

Although $\mathcal{K}$ enters eqn (5.77) in the form of a second-order term, this eigensystem is in fact a first order eigensystem in $\mathcal{K}^2$ giving rise to $2n$ distinct eigenvalues

$$\mathcal{K}^{(i)} = \mathcal{K}^{(1)}, -\mathcal{K}^{(1)}, \mathcal{K}^{(2)}, -\mathcal{K}^{(2)}, ..., \mathcal{K}^{(n)}, -\mathcal{K}^{(n)} \quad (i = 1, ..., 2n),$$

and leading to $2n$ independent Bloch waves and n independent eigenvectors:

$$\mathbf{\Phi}^{(i)} = (C_1^{(i)}, C_2^{(i)}, ... C_n^{(i)}), \quad (i = 1, ..., n).$$

In general, for an n-beam RHEED case the eigensystem (5.77) generates n Bloch waves propagating in the forward direction and n propagating in the backward direction. The wave vectors $\mathcal{K}^{(i)}$ of these solutions are symmetric with respect to a mirror reflection of each crystal slice. This is a natural consequence of the assumption that within each crystal slice the potential field is constant in the direction normal to the surface.

Noting that all the rods of the reciprocal lattice satisfy $g_{G_z} = 0$, we obtain the following expressions for the diffracted beam amplitudes and their surface normal derivatives (5.58) and (5.59):

$$\psi_G(z) = \sum_{j=1}^{n} C_G^{(j)} \exp(i\mathcal{K}^{(j)}z)a^{(j)} + \sum_{j=1}^{n} C_G^{(j)} \exp(-i\mathcal{K}^{(j)}z)a^{(j)} \tag{5.78}$$

$$-i\psi_G'(z) = \sum_{j=1}^{n} (\mathcal{K}^{(j)}C_G^{(j)}) \exp(i\mathcal{K}^{(j)}z)a^{(j)} + \sum_{j=1}^{n} (-\mathcal{K}^{(j)}C_G^{(j)}) \exp(-i\mathcal{K}^{(j)}z)a^{(j)}.$$

$$\tag{5.79}$$

By defining a $2n \times 2n$ diagonal matrix $\mathbf{\Upsilon}(z)$

$$\{\mathbf{\Upsilon}(z)\}_{j,j} = \exp(i\mathcal{K}^{(j)}z), \quad \{\mathbf{\Upsilon}(z)\}_{j+n,j+n} = \exp(-i\mathcal{K}^{(j)}z) \quad (j = 1, ..., n),$$

a supermatrix $\mathbf{C}$

$$\mathbf{C} = \begin{pmatrix} \{C_G^{(j)}\} & \{C_G^{(j)}\} \\ \{\mathcal{K}^{(j)}C_G^{(i)}\} & \{-\mathcal{K}^{(j)}C_G^{(i)}\} \end{pmatrix},$$

and a vector

$$\{\boldsymbol{\alpha}\}_j = \{\boldsymbol{\alpha}\}_{j+n} = a^{(j)}(j = 1, 2, ..., n),$$

we obtain a compact supervector $\mathbf{\Psi}(z)$

$$\mathbf{\Psi}(z) = \begin{pmatrix} \{\psi_G(z)\} \\ \{-i\psi_G'(z)\} \end{pmatrix} = \mathbf{C}\mathbf{\Upsilon}(z)\boldsymbol{\alpha}. \tag{5.80}$$

For a thin slice of thickness t we have

$$\mathbf{\Psi}(z + t) = \mathbf{C}\mathbf{\Upsilon}(z + t)\boldsymbol{\alpha} = \mathbf{C}\mathbf{\Upsilon}(z + t)[\mathbf{C}\mathbf{\Upsilon}(z)]^{-1}\mathbf{\Psi}(z) = \mathbf{M}(t)\mathbf{\Psi}(z), \tag{5.81}$$

where $\mathbf{M}(z) = \mathbf{C}\mathbf{\Upsilon}(t)\mathbf{C}^{-1}$.

By writing the inverse of matrix $\{C_G^{(j)}\}$ as $\{\overline{C}_G^{(j)}\}$ we have

$$\sum_j C_G^{(j)}\overline{C}_H^{(j)} = \delta_{GH}, \quad \sum_G \overline{C}_G^{(i)} C_G^{(j)} = \delta_{ij}.$$

The inverse matrix of $\mathbf{C}$ is given by

$$\mathbf{C}^{-1} = \frac{1}{2}\begin{pmatrix} \{\overline{C}_G^{(j)}\} & \{\overline{C}_G^{(j)}/\mathcal{K}^{(j)}\} \\ \{\overline{C}_G^{(j)}\} & \{-\overline{C}_G^{(j)}/\mathcal{K}^{(j)}\} \end{pmatrix},$$

and the transfer matrix $\mathbf{M}$ for a slice of thickness t is

$$\mathbf{M}(t) = \mathbf{C}\mathbf{\Upsilon}(t)\mathbf{C}^{-1} = \begin{pmatrix} \{\sum_i C_G^{(j)}\cos(\mathcal{K}^{(j)}t)\overline{C}_H^{(j)}\} & \{i\sum_i C_G^{(j)}\sin(\mathcal{K}^{(j)}t)\overline{C}_H^{(j)}\} \\ \{i\sum_i C_G^{(j)}\sin(\mathcal{K}^{(j)}t)\overline{C}_H^{(j)}\} & \{\sum_i C_G^{(j)}\cos(\mathcal{K}^{(j)}t)\overline{C}_H^{(j)}\} \end{pmatrix}.$$
$$(5.82)$$

The inverse of the transfer matrix $\mathbf{M}(t)$ may be obtained simply by inverting the direction of propagation, i.e. $\mathbf{M}^{-1}(t) = \mathbf{M}(-t)$. In the case of scattering by a crystal slab, such as the surface selvedge or a repeat unit slab of the bulk crystal, we divide the slab into an assembly of thin slices, each characterized by a potential independent of the coordinate z in the direction normal to the surface. The total transfer matrix $\mathbf{M}$ of the entire slab is then represented by a product of submatrices $\mathbf{M}_n$ for all slices, i.e.

$$\mathbf{M}(t) = \prod_n \mathbf{M}_n(\Delta z_n), \tag{5.83}$$

where Δz_n is the thickness of the n-th slice, and $t = \sum_n \Delta z_n$ is thickness of the slab.

5.3.4.4 *RHEED from the surface of a crystal slab*

We now consider the calculation of amplitudes of diffracted beams reflected from the surface of a crystal slab (Maksym and Beeby, 1981, Ichimiya, 1983, Smith and Lynch, 1988). If the thickness of the slab is much larger than the distance characterizing the absorption of electrons, the treatment of scattering of electrons from a slab gives rise to the same reflection coefficients as those corresponding to the case of scattering from the surface of a semi-infinite crystal. In principle we could construct a transfer matrix for the bulk crystal slab or the selvedge using eqn (5.83). In practice, however, the transfer matrix defined by (5.83) contains exponentially divergent terms of the form $\exp(i\mathcal{K}^{(j)}z)$. Evanescent waves with negative imaginary $\mathcal{K}^{(j)}$ give rise to a rapid divergence of the elements of the transfer matrix $\mathbf{M}$ considered as a function of the thickness of the crystal slab. A numerically stable method of calculation of amplitudes of waves reflected from an assembly of slices is based on the use of the so-called $\mathbf{R}$-matrix, which is defined as the matrix relating the vector of surface normal derivatives of the wave function and the vector of components of the wave function itself (Ichimiya, 1983, Zhao et al., 1988). Assuming that our system consists of m repeating bulk unit slabs and the selvedge, we start from the bottom surface of the last bulk slab, i.e. the m-th slab, and obtain

$$
\begin{pmatrix} \{\psi_G\} \\ \{\psi'_G\} \end{pmatrix}_{m-1} = \begin{pmatrix} \mathbf{M}_{11} \ \mathbf{M}_{12} \\ \mathbf{M}_{21} \ \mathbf{M}_{22} \end{pmatrix}_m^{-1} \begin{pmatrix} \{\mathcal{T}_H \exp(ik_{H_z}t)\} \\ \{k_{H_z}\mathcal{T}_H \exp(ik_{H_z}t)\} \end{pmatrix}
$$
$$
= \begin{pmatrix} \{\overline{M}_{11} + \overline{M}_{12}k_{H_z}\}_{GH} \\ \{\overline{M}_{21} + \overline{M}_{22}k_{H_z}\}_{GH} \end{pmatrix}_m \{\mathcal{T}_H \exp(ik_{H_z}t)\}, \qquad (5.84)
$$

where $\mathcal{T}_H$ is the transmitted beam amplitude of the H-th beam, t is the thickness of a slab, and $\overline{\mathbf{M}}(t) = \mathbf{M}^{-1}(t) = \mathbf{M}(-t)$.

By introducing the $\mathbf{R}$-matrix

$$
\{\mathbf{R}\} = \{\overline{M}_{21} + \overline{M}_{22}k_{H_z}\}\{\overline{M}_{11} + \overline{M}_{12}k_{H_z}\}^{-1}, \qquad (5.85)
$$

we obtain

$$
\{\boldsymbol{\Psi'}\}_{m-1} = \{\mathbf{R}\}_m\{\boldsymbol{\Psi}\}_{m-1}, \qquad (5.86)
$$

where $\mathbf{R}_m$ is the $\mathbf{R}$-matrix of the m-th bulk unit slab.

The $\mathbf{R}$-matrix can now be propagated upward through the crystal composed of repeating unit slabs with indices $k = 1, ..., m-1$. Generalizing the definition of the $\mathbf{R}$-matrix, we obtain for the k-th slice

$$
\begin{pmatrix} \{\psi_G\} \\ \{\psi'_G\} \end{pmatrix}_{k-1} = \begin{pmatrix} \overline{\mathbf{M}}_{11} \ \overline{\mathbf{M}}_{12} \\ \overline{\mathbf{M}}_{21} \ \overline{\mathbf{M}}_{22} \end{pmatrix}_k \begin{pmatrix} \{\psi_H\} \\ \{\psi'_H\} \end{pmatrix}_k = \begin{pmatrix} \overline{\mathbf{M}}_{11} \ \overline{\mathbf{M}}_{12} \\ \overline{\mathbf{M}}_{21} \ \overline{\mathbf{M}}_{22} \end{pmatrix}_k \begin{pmatrix} \{\mathbf{I}\} \\ \{\mathbf{R}\} \end{pmatrix}_{k+1} \{\psi_H\}_k,
$$
$$
(5.87)
$$

giving rise to

$$
\{\boldsymbol{\Psi'}\}_{k-1} = \{\mathbf{R}\}_k\{\boldsymbol{\Psi}\}_{k-1}, \qquad (5.88)
$$

where

$$
\{\mathbf{R}\}_k = \{\overline{\mathbf{M}}_{21} + \overline{\mathbf{M}}_{22}\mathbf{R}_{k+1}\}\{\overline{\mathbf{M}}_{11} + \overline{\mathbf{M}}_{12}\mathbf{R}_{k+1}\}^{-1}. \qquad (5.89)
$$

At the surface of the crystal we have

$$
\begin{pmatrix} \{\delta_{G,0} + \mathcal{R}_G,\} \\ \{k_{G_z}(\delta_{G,0} - \mathcal{R}_G)\} \end{pmatrix} = \begin{pmatrix} \{\psi_G\} \\ \{\psi'_G\} \end{pmatrix}_0 = \begin{pmatrix} \{\delta_{G,H}\} \\ \{R_{GH}\}_1 \end{pmatrix} \{\psi_H\}_0, \qquad (5.90)
$$

where $\mathcal{R}_G$ is the amplitude of the G-th reflected beam. Using the above equation we obtain

$$
\{\mathcal{R}\}_G = -\frac{\{\delta_{G,H} - R_{GH}^{-1}k_{H_z}\}}{\{\delta_{G,H} + R_{GH}^{-1}k_{H_z}\}}\delta_{H,0}, \qquad (5.91)
$$

where here, as before, the division between two matrices $\mathbf{A}$ and $\mathbf{B}$ is defined as $\mathbf{A}/\mathbf{B} = \mathbf{B}^{-1}\mathbf{A}$. In the case where evanescent waves characterized by a large imaginary part of $\gamma^{(i)}$ are present, the transfer matrix may diverge even for a single bulk unit slab. In this case both the selvedge and the bulk unit slab should be divided into subslabs. The solution of the diffraction problem and the amplitudes of waves reflected from the surface of the crystal may be found using the same procedure based on the propagation of the $\mathbf{R}$-matrix from the bottom to the top of the entire crystal following the procedure described above.

5.3.4.5 *RHEED from the surface of a semi-infinite crystal* RHEED beam intensities from the surface of a semi-infinite crystal may be calculated using the two-dimensional Bloch wave method (Peng et al., 1996c). In a periodic bulk crystal we have

$$\boldsymbol{\Psi}(z + c) = \mathbf{M}(c)\boldsymbol{\Psi}(z), \tag{5.92}$$

where the matrix $\mathbf{M}(c)$ appearing in (5.92) is the transfer matrix associated with a bulk repeat unit slab, and the constant c is the lattice constant along the surface normal direction.

For a Bloch wave $b(\mathbf{r})$ in a crystal, it can be readily shown that both the G-th Fourier component $b_G(z)$ and its surface normal derivative $b'_G(z) = d[b_G(z)]/dz$ satisfy the Bloch theorem

$$b_G^{(j)}(z + c) = \exp(i\gamma^{(j)}c)b_G^{(j)}(z), \quad b_G^{'(j)}(z + c) = \exp(i\gamma^{(j)}c)b_G^{'(j)}(z), \tag{5.93}$$

i.e. they are both Bloch waves. We may therefore define a supervector $\mathbf{b}^{(j)}$ by

$$\mathbf{b}^{(j)} = \begin{pmatrix} \{b_G^{(j)}\} \\ \{-ib_G^{'(j)}\} \end{pmatrix}. \tag{5.94}$$

Since eqn (5.92) holds for any function satisfying the Bloch theorem, it also holds for $\mathbf{b}^{(j)}$ defined by (5.94), giving rise to

$$\mathbf{b}^{(j)}(z + c) = \exp(i\gamma^{(j)}c)\mathbf{b}^{(j)}(z) = \mathbf{M}(c)\mathbf{b}^{(j)}(z) \tag{5.95}$$

in which $\gamma^{(j)}$ is the j-th eigenvalue associated with the j-th supervector $b^{(j)}(z)$. For a dynamical RHEED calculation involving n reciprocal lattice rods, the transfer matrix $\mathbf{M}$ is a $2n \times 2n$ matrix. This $2n \times 2n$ matrix corresponds to a total of $2n$ Bloch waves, of which n propagate into the bulk of the crystal slab, and in the presence of absorption attenuate exponentially into the crystal. The other n waves propagate and attenuate in the opposite direction. By introducing a $2n \times 2n$ matrix $\mathbf{C} = \{\mathbf{b}^{(1)}, ..., \mathbf{b}^{(2n)}\}$ and a diagonal matrix $\boldsymbol{\Upsilon}(z) = \{\exp(i\gamma^{(j)}z)\}$, we obtain

$$\mathbf{M}(z) = \mathbf{C}\boldsymbol{\Upsilon}(z)\mathbf{C}^{-1}. \tag{5.96}$$

For a crystal slab consisting of m repeating unit slabs, substitution of eqn (5.96) into (5.92) results in

$$\boldsymbol{\Psi}(z + mc) = \mathbf{C}\boldsymbol{\Upsilon}(mc)\mathbf{C}^{-1}\boldsymbol{\Psi}(z). \tag{5.97}$$

For convenience we assume that among the $2n$ Bloch waves the first n solutions are the evanescent waves, and the remaining solutions are the anti-evanescent waves. For a semi-infinite crystal the anti-evanescent Bloch waves cannot contribute to the total wave field and must be discarded. Assuming that the interface between the selvedge and the bulk crystal is situated at $z = z_s$

and that the transfer matrix associated with the selvedge is $\mathbf{M}_s$, we obtain for a crystal slab consisting of m repeating unit slabs

$$\boldsymbol{\Psi}(z + mc) = \mathbf{C}\boldsymbol{\Upsilon}(mc)\mathbf{C}^{-1}\mathbf{M}_s\boldsymbol{\Psi}(0) \tag{5.98}$$

$$= \mathbf{C}\begin{pmatrix} \exp[i\gamma^{(j)}mc] & \\ & \exp[i\gamma^{(j+n)}mc] \end{pmatrix}\mathbf{C}^{-1}\mathbf{M}_s\begin{pmatrix} \delta_{G,0} + \mathcal{R}_G \\ k_{G_z}(\delta_{G,0} - \mathcal{R}_G) \end{pmatrix},$$

where $\mathcal{R}_G$ is the amplitude of the reflected beam associated with the Gth reciprocal lattice rod. For a semi-infinite crystal, since all the physically allowed quantities must have a finite amplitude, the lower half of the column vector $\mathbf{C}^{-1}\mathbf{M}_s\boldsymbol{\Psi}(0)$ on the right-hand side of eqn (5.98) must vanish. In terms of the two lower submatrices $\mathbf{M}'_{21}$ and $\mathbf{M}'_{22}$ of the matrix $\mathbf{M}' = \mathbf{C}^{-1}\mathbf{M}_s$ and the surface reflected beam amplitude vector $\{\mathcal{R}_G\}$, we can express this condition as

$$M'_{21}\{\delta_{G,0} + \mathcal{R}_G\} + M'_{22}\{k_{G_z}(\delta_{G,0} - \mathcal{R}_G)\} = 0, \tag{5.99}$$

which gives

$$\{\mathcal{R}_G\} = -\left(\{M'_{21} - M'_{22}k_{H_z}\}_{GH}\right)^{-1}\left(\{M'_{21} + M'_{22}k_{H_z}\}_{H,0}\right). \tag{5.100}$$

For a selvedge with a simple structure, the transfer matrix $\mathbf{M}_s$ associated with the selvedge is simple and convergent. In the case of a complex surface structure, however, the thickness of the selvedge may be larger than that of a unit bulk repeat slab and the selvedge transfer matrix $\mathbf{M}_s$ may diverge. In this case we again need to use the **R**-matrix method and propagate this matrix, rather than the transfer matrix, through the selvedge. Assuming that the crystal consists of m bulk slabs and that the interface between the selvedge and the bulk crystal is at $z = z_s$, we obtain the following expression relating the solution at the bottom of the crystal to the solution at the selvedge interface:

$$\boldsymbol{\Psi}(z_s + mc) = \mathbf{C}\boldsymbol{\Upsilon}(mc)\mathbf{C}^{-1}\boldsymbol{\Psi}(z_s). \tag{5.101}$$

By writing

$$\mathbf{C}^{-1} = \overline{\mathbf{C}} = \begin{pmatrix} \overline{\mathbf{C}}_{11} & \overline{\mathbf{C}}_{12} \\ \overline{\mathbf{C}}_{21} & \overline{\mathbf{C}}_{22} \end{pmatrix}, \tag{5.102}$$

and following the same procedure that leads to eqn (5.99), we obtain

$$\overline{\mathbf{C}}_{21}\{\Psi_G(z_s)\} + \overline{\mathbf{C}}_{22}\{-i\Psi'_G(z_s)\} = 0. \tag{5.103}$$

This gives an explicit expression for the **R**-matrix at the substrate/selvedge interface

$$\mathbf{R} = -\left(\overline{\mathbf{C}}_{22}\right)^{-1}\left(\overline{\mathbf{C}}_{21}\right). \tag{5.104}$$

This **R**-matrix can now be propagated upwards to the surface of the crystal to obtain the amplitudes of beams reflected from the surface.

Alternatively the set of Bloch waves propagating in the crystal bulk may be obtained by taking advantage of the fact that in the crystal only half of the $2n$ Bloch waves are physically allowed and the remaining half must have zero excitation amplitudes $\alpha^{(i)} = 0$. In the bulk crystal we therefore obtain

$$\{\psi_G(z)\} = \sum_{i=1}^{n} \alpha^{(i)} C_G^{(i)} \exp(i\mathcal{K}z), \qquad (5.105)$$

or in matrix notation

$$\{\psi_G(z)\} = \mathbf{c}\gamma\alpha,$$

where $\mathbf{c} = \{C_G^{(j)}\}$ is an $n \times n$ matrix, γ is a diagonal matrix with elements $\{\gamma\}_j = \exp(i\mathcal{K}^{(j)}z)$, and α is an n-dimensional vector with $\{\alpha\}_j = \alpha^{(j)}$. Considering the relationship between solutions of the RHEED problem at $z + c$ and z, where c is the lattice constant in the direction parallel to the surface normal, we obtain

$$\{\psi_G(z + c)\} = \mathbf{c}\gamma(d)\mathbf{c}^{-1}\{\psi_G(z)\}. \qquad (5.106)$$

On the other hand, using eqn (5.92) we find

$$\{\psi_G(z + c)\} = \mathcal{M}\{\psi_G(z)\}, \qquad (5.107)$$

where

$$\mathcal{M} = \prod_k \Big(\mathbf{M}_{11}(t_k) + \mathbf{M}_{12}(t_k)\mathbf{R}\Big)_{k-1}, \qquad (5.108)$$

and where $\mathbf{R}_k$ is the $\mathbf{R}$-matrix of the k-th slice. k is the index of slices composing the repeat bulk unit slab and $c = \sum_k t_k$. Comparing eqns (5.106) and (5.107) we see that the eigenvectors $\{C_G^{(j)}\}$ and eigenvalues $\{\exp(i\mathcal{K}^{(i)}d)\}$ of Bloch waves in the crystal bulk can be obtained by the direct diagonalization of matrix $\mathcal{M}$ (Dudarev, 1997).

5.3.4.6 *Computation of the electron wave function*   Having found the reflected beam amplitudes $\mathcal{R}_G$, we can now calculate the wave function. In the vacuum above the surface the wave function is given by

$$\psi_G(z) = \delta_{G,0} \exp(ik_{G_z} z) + \mathcal{R}_G \exp(-ik_{G_z} z). \qquad (5.109)$$

In the selvedge layer and in the substrate the wave function may be obtained by propagating the vector $\{\psi_G\}$ downwards through all the slices composing the crystal. Considering the first slice we obtain

$$\{\psi_G\}_1 = \{\mathbf{M}_{11}\}_1\{\psi_G\}_0 + \{\mathbf{M}_{12}\}_1\{\psi_G'\}_0 = \Big(\mathbf{M}_{11} + \mathbf{M}_{12}\mathbf{R}\Big)_1 \{\psi_G\}_0, \qquad (5.110)$$

where $\mathbf{R}_1$ is the $\mathbf{R}$ matrix of the first slice below the vacuum/selvedge interface given by

$$\{R_G\}_1 = k_{G_z} \frac{\delta_{G,0} - \mathcal{R}_G}{\delta_{G,0} + \mathcal{R}_G}.$$

Generalizing this relation we obtain for the k-th slice

$$\{\psi_G\}_{k+1} = \left(\mathbf{M}_{11} + \mathbf{M}_{12}\mathbf{R}\right)_{k+1} \{\psi_G\}_k. \tag{5.111}$$

Note that all the relevant matrices $\mathbf{M}_{11}$ and $\mathbf{M}_{12}$ for all the slices have already been calculated in the process of evaluation of $\mathcal{R}_G$. Therefore in this scheme the calculation of the electron wave function does not require any extra effort beyond some simple multiplication of $n \times n$ matrices.

Alternatively the wave function of electrons in the crystal bulk may be obtained using the diagonalization of the transfer matrix $\mathcal{M}$ corresponding to a repeat bulk unit slab. This procedure gives n eigenvalues of the form $\exp(i\mathcal{K}^{(i)}d)$ and n eigenvectors $\{C_G^{(j)}\}$. The wave function in the bulk is then given by

$$\psi_G(z) = \sum_j \alpha^{(j)} C_G^{(j)} \exp(i\mathcal{K}^{(j)}z), \tag{5.112}$$

where $\alpha^{(j)}$ are the usual excitation amplitudes of the bulk Bloch waves. In terms of the electron wave functions $\psi_G(z_0)$ corresponding to a certain coordinate z_0 in the crystal, this vector $\boldsymbol{\alpha}$ is given by

$$\boldsymbol{\alpha} = \mathbf{c}^{-1}\{\exp(-i\mathcal{K}^{(j)}d)\}_d\{\psi_G(z_0)\}, \tag{5.113}$$

and $\{\psi_G(z_0)\}$ may be found by simply propagating $\{\psi_G\}$ from the vacuum/crystal interface using eqn (5.111).

5.3.4.7 *The effects of weak and evanescent beams: the Bethe potentials* To obtain a convergent dynamical RHEED solution, a large number of HOLZ rods need to be used. These include evanescent beams associated with the minus HOLZ reflections lying outside the Ewald sphere. The inclusion of evanescent beams in dynamical RHEED calculations presents a difficult problem that becomes particularly acute in cases corresponding to large negative values of Γ_G^2. The contribution of evanescent waves of this type to the solution of the RHEED problem is small, and formally the presence of these evanescent waves leads to the excitation of Bloch waves characterized by values of $\mathcal{K}$ close to $\pm i\sqrt{|\Gamma_G^2|}$. We see that since the expression for $\psi_G(z)$ now includes exponential terms of the form $\exp(\sqrt{|\Gamma_G^2|}t)$, any numerical error in Bloch wave excitation amplitudes $\alpha^{(j)}$ and eigenvectors $C_G^{(j)}$ is exponentially amplified. This leads to the problem of exponential loss of accuracy of RHEED theory that we have already encountered in the context of the **R**-matrix treatment of diffraction by a crystal slab (Zhao et al., 1988). On one hand we know that the contribution from the evanescent waves corresponding to large $|\Gamma_G|$ must be small. On the other hand we also know that the collective contribution resulting from all the evanescent waves is not negligible. The inclusion of evanescent beams corresponding to large values of $|\Gamma_G|$ in numerical calculations is difficult, since it requires the thickness of slices to be very small so that the relevant calculations take significantly more computer time, in addition to the increase in the size of matrices associated with the inclusion of weak beams in the computational scheme.

In this section we show how the Bethe potential method can be applied to the problem of RHEED providing a very convenient and powerful way for dealing with evanescent beams and HOLZ effects. The Bethe potentials were first introduced by Bethe (Bethe, 1928) in the two-beam diffraction case in order to take into account the effect of weak beams on the two strong beams. In this section we will rederive the Bethe potentials and incorporate them into the two-dimensional Bloch wave formalism. We will show that the Bethe potentials method does indeed provide an answer to the question about how to deal with evanescent waves characterized by large imaginary parts of the wave vector and with the HOLZ effect (Peng et al., 1996b, Peng et al., 1996a). We start our derivation of the Bethe potentials from the fundamental equation (5.77)

$$(\mathcal{K}^2 - \Gamma_G^2)C_G - \sum_{L \neq G} U_{GL}C_L = 0. \tag{5.114}$$

Suppose that we separate the diffracted beams into two groups, where one group represents a set of n strong beams $\{\mathbf{G}\}$ and the other group represents a set of m weak beams $\{\mathbf{H}\}$. Using this notation we rewrite the fundamental equation (5.114) as

$$\begin{pmatrix} \mathcal{K}^2 - \Gamma_G^2 & -U_{GG'} & \cdots & -U_{GH} & -U_{GH'} & \cdots \\ -U_{G'G} & \mathcal{K}^2 - \Gamma_{G'}^2 & \cdots & -U_{G'H} & -U_{G'H'} & \cdots \\ \cdots & \cdots & \cdots & \cdots & \cdots & \cdots \\ -U_{HG} & -U_{HG'} & \cdots & \mathcal{K}^2 - \Gamma_H^2 & -U_{HH'} & \cdots \\ -U_{H'G} & -U_{H'G'} & \cdots & -U_{H'H} & \mathcal{K}^2 - \Gamma_{H'}^2 & \cdots \\ \cdots & \cdots & \cdots & \cdots & \cdots & \cdots \end{pmatrix} \begin{pmatrix} C_G \\ C_{G'} \\ \cdots \\ C_H \\ C_{H'} \\ \cdots \end{pmatrix} = 0. \tag{5.115}$$

By expanding the lower half of the equation, we find

$$\begin{pmatrix} C_H \\ C_{H'} \\ \cdots \end{pmatrix} = \begin{pmatrix} \mathcal{K}^2 - \Gamma_H^2 & -U_{HH'} & \cdots \\ -U_{H'H} & \mathcal{K}^2 - \Gamma_{H'}^2 & \cdots \\ \cdots & \cdots & \cdots \end{pmatrix}^{-1}$$

$$\times \begin{pmatrix} U_{HG} & U_{HG'} & \cdots \\ U_{H'G} & U_{H'G'} & \cdots \\ \cdots & \cdots & \cdots \end{pmatrix} \begin{pmatrix} C_G \\ C_{G'} \\ \cdots \end{pmatrix}. \tag{5.116}$$

Suppose that diffraction processes in a slice are determined by the Bloch waves associated with the set of strong beams. In this case we can make the following approximation in the treatment of weak beams:

$$\mathcal{K}^2 - \Gamma_H^2 \approx K_z^2 - \Gamma_H^2. \tag{5.117}$$

This approximation works well either for large angles of incidence where $\mathcal{K}_z \approx K_z$ or for beams corresponding to large values of Γ_H^2. Substituting (5.116) into (5.115), we obtain a matrix equation for the set of strong beams $\{\mathbf{G}\}$

$$\begin{pmatrix} \mathcal{K}^2 - (\Gamma_G^2 + \Delta\Gamma_G^2) & -(U_{GG'} + \Delta U_{GG'}) & \cdots \\ -(U_{G'G} + \Delta U_{G'G}) & \mathcal{K}^2 - (\Gamma_{G'}^2 + \Delta\Gamma_{G'}^2) & \cdots \\ & \cdots & \end{pmatrix} \begin{pmatrix} C_G \\ C_{G'} \\ \cdots \end{pmatrix} = 0, \qquad (5.118)$$

where

$$\begin{pmatrix} \Delta\Gamma_G^2 & \Delta U_{GG'} & \cdots \\ \Delta U_{G'G} & \Delta\Gamma_{G'}^2 & \cdots \\ \cdots & \cdots & \cdots \end{pmatrix} = \begin{pmatrix} U_{GH} & U_{GH'} & \cdots \\ U_{G'H} & U_{G'H'} & \cdots \\ \cdots & \cdots & \cdots \end{pmatrix}$$

$$\times \begin{pmatrix} K_z^2 - \Gamma_H^2 & -U_{HH'} & \cdots \\ -U_{H'H} & K_z^2 - \Gamma_{H'}^2 & \cdots \\ \cdots & \cdots & \cdots \end{pmatrix}^{-1} \begin{pmatrix} U_{HG} & U_{HG'} & \cdots \\ U_{H'G} & U_{H'G'} & \cdots \\ \cdots & \cdots & \cdots \end{pmatrix}. \qquad (5.119)$$

The correction to the crystal potential associated with the $\Delta\Gamma^2$ and ΔU terms may be evaluated numerically using the above equation. For a dynamical RHEED calculation involving n strong beams and m weak beams the original $(m+n) \times (m+n)$ matrix equation (5.115) is reduced to an $n \times n$ matrix equation (5.118). The calculations of $\Delta\Gamma_G^2$ and $\Delta U_{GG'}$ involves the inversion of an $n \times n$ matrix in eqn (5.119). Nevertheless, this proves to be considerably more efficient than solving the original $(m+n) \times (m+n)$ eigenvalue equation (5.115) directly.

One further simplifying step associated with the Bethe approximation (5.119) is based on the assumption

$$\begin{pmatrix} K_z^2 - \Gamma_H^2 & -U_{HH'} & \cdots \\ -U_{H'H} & K_z^2 - \Gamma_{H'}^2 & \cdots \\ \cdots & \cdots & \cdots \end{pmatrix}^{-1} \approx \begin{pmatrix} (K_z^2 - \Gamma_H^2)^{-1} & 0 & \cdots \\ 0 & (K_z^2 - \Gamma_{H'}^2)^{-1} & \cdots \\ \cdots & \cdots & \cdots \end{pmatrix},$$

which gives the well-known expression for the Bethe potentials (Bethe, 1928)

$$\Delta\Gamma_G^2 = \sum_{H \neq G} \frac{|U_{GH}|^2}{K_z^2 - \Gamma_H^2}, \qquad \Delta U_{GG'} = \sum_{H \neq G, G'} \frac{U_{GH} U_{HG'}}{K_z^2 - \Gamma_H^2}. \qquad (5.120)$$

The range of validity of the generalized Bethe approximation (5.120) for $\Delta\Gamma_G$ and $\Delta U_{GG'}$ is determined by the requirement

$$\mathcal{K} = \gamma + K_z \approx K_z, \qquad (5.121)$$

which is always satisfied for large angles of incidence, e.g in the case of THEED, where the angle of incidence $\theta \approx 90°$; and

$$|K_z^2 - \Gamma_H^2| \gg |U_{HH'}|, \qquad (5.122)$$

for all the weak beams $\{\mathbf{H}\}$. Since K_z^2 is always positive, the inequality (5.122) shows clearly that the Bethe approximation is more likely to be valid for an

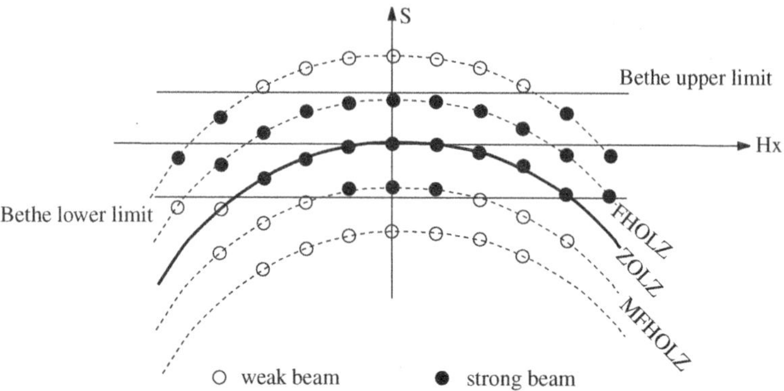

FIG. 5.13. Sketch showing the domain of validity of the Bethe approximation. All the reflections lying above the line 'Bethe upper limit' and below the line 'Bethe lower limit' satisfy the Bethe approximation.

evanescent beam characterized by $\Gamma_H^2 < 0$ than for a propagating beam with $\Gamma_H^2 > 0$.

The domain of validity of the Bethe approximation (5.122) is shown schematically in Fig. 5.13. The quantity $S(H_x)$ plotted in this figure is defined as

$$S(H_x) = \Gamma_H^2 - K_z^2 = -H^2 - 2\mathbf{K} \cdot \mathbf{H} = -(2K_y H_y + H_y^2 + H_x^2), \qquad (5.123)$$

where we have assumed that the direction of incidence is nearly parallel to a principal zone axis of the crystal. Figure 5.13 illustrates the case where there is no shift $\Delta\mathbf{G}$ between planes of reciprocal lattice perpendicular to the zone axis. For the ZOLZ beams we have $\mathbf{K} \cdot \mathbf{H} = 0$, while for the positive HOLZ beams $\mathbf{K} \cdot \mathbf{H} < 0$ and for the negative HOLZ beams $\mathbf{K} \cdot \mathbf{H} > 0$. The $S(H_x)$ curves associated with the negative HOLZ beams appear above the curve associated with the ZOLZ, and those associated with the positive HOLZ beams appear below the ZOLZ curve. The two solid lines marked in the figure as the Bethe upper and the Bethe lower limits are boundaries of the domain of validity of the Bethe approximation (5.122). The domain boundaries divide the reciprocal space into three regions, and of these the regions above and below the boundaries approximately satisfy the Bethe approximation (5.122).

Figure 5.13 also shows that for the ZOLZ and the negative HOLZ beams, only a small number of low-index reflections need to be treated using full dynamical theory. The higher order beams lying below the lower solid line can be treated approximately using the Bethe potentials. For the positive HOLZ beams the situation is somewhat different. While some of the low-index positive HOLZ beams characterized by small values of H_x satisfy the Bethe approximation, some intermediate-index positive HOLZ beams still have to be treated as strong beams. The figure also shows that the number of beams requiring full dynamical treatment is smaller in the negative HOLZ case than in the positive HOLZ case.

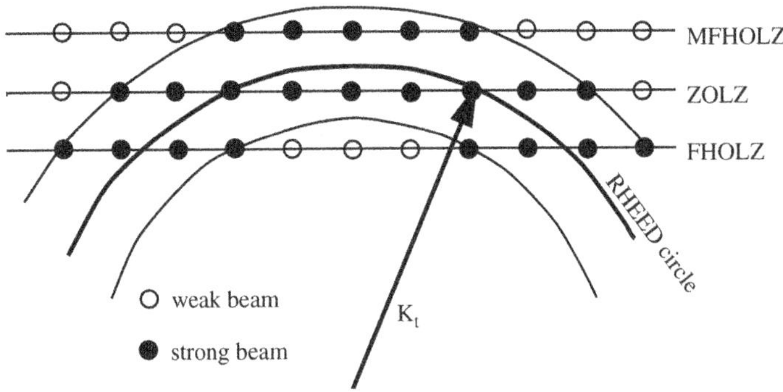

FIG. 5.14. An alternative view of the domain of validity of the Bethe approximation. The Bethe approximation is valid in the region outside the two concentric circles drawn on both sides of the RHEED circle.

An alternative view of the domain of validity of the Bethe approximation is illustrated by Fig. 5.14. This figure shows the RHEED circle (the intersection of the Ewald sphere and the (x, y) plane) and the two-dimensional reciprocal lattice. We note that quantity S can be represented in the following way:

$$S = \Gamma_H^2 - K_z^2 \approx 2K_t(K_t - |\mathbf{K}_t + \mathbf{H}|). \tag{5.124}$$

In this expression the radius of the circle is given by K_t, and the distance between a reciprocal lattice rod $\mathbf{H}$ and the circle is equal to $K_t - |\mathbf{K}_t + \mathbf{H}|$. The Bethe approximation is valid in the region outside the two concentric circles above and below the RHEED circle.

The most time-consuming part of a convergent calculation involving many reciprocal lattice rods is the diagonalization of the matrix equation (5.77). For a general $n \times n$ complex matrix the time required to diagonalize this matrix scales as n^3. On the other hand, calculations involving the Bethe potentials scale linearly with the number of rods n. Assuming that among the n rods m require full dynamical treatment, the computer time needed to obtain a convergent numerical solution scales as $m^3 + n$. We then see that the saving of computer time associated with the use of the Bethe potentials is approximately proportional to $n^3 - m^3 - n$. This is only a rough estimate. The actual time required for obtaining a fully convergent solution of a RHEED problem depends on the actual structure of the dynamical matrix and on the symmetry of the diffraction pattern.

5.4 Worked examples

In this section we give several examples illustrating various aspects of practical RHEED calculations. These examples cover RHEED both from a surface of a crystal composed of neutral atoms and from a surface of an ionic crystal. In our discussion we outline procedures for evaluating scattering potentials of individual atoms, for constructing the potential for a given surface structural model, and for calculating dynamical RHEED rocking curves. We also provide computer routines that solve numerical problems given in the examples. The interested reader may use these routines to obtain the data shown in the figures, thereby illustrating the methods and approaches which have been described.

5.4.1 *RHEED from the surface of a metal: the Ag(001) surface*

The first example we discuss refers to the (001) surface of crystalline silver. Since silver is a metal, to a very good approximation we may assume that the crystal is composed of neutral atoms. Indeed, the majority of dynamical RHEED calculations performed so far have been carried out using this neutral atom approximation. The ionicity of interatomic bonds has a profound effect on the dynamical diffraction process and the resulting RHEED rocking curves. We shall discuss this point later in connection with RHEED from surfaces of oxide materials.

5.4.1.1 *The structural model and the scattering potential* In this section we will consider the scattering potential for an ideal bulk terminated Ag(001) surface. Crystalline silver has the face-centred cubic (fcc) structure with lattice constant $a_0 = 4.09$ Å and with four atoms in the unit cell. Figure 5.15a shows the top [001] view of the Ag(001) surface. The square marked by the dotted line is the projection of the unit cell, while the solid line indicates the reduced unit cell with smaller lattice constants $a' = b' = 4.09/\sqrt{2} = 2.892$ Å. Both unit cells have the same atom layer spacing $c/2 = 2.045$ Å in the direction normal to the surface. To simulate the variation of the potential in the surface layer of the crystal we construct a large unit cell characterized by the following lattice parameters: $a' = a/\sqrt{2} = 2.892$ Å in the $[\bar{1}10]$ direction, $b' = b/\sqrt{2} = 2.892$ Å in the [110] direction, and $c' = 4c = 16.36$ Å in the [001] direction. The z axis of the Cartesian system of coordinates is chosen to be parallel to the [001] direction. Atoms of the surface layer (i.e. atoms shown as open circles in Fig. 5.15a) are located at $z = 0.357c'$, and the layer immediately below the surface layer is situated at $z = 0.5c'$. The third subsurface layer is located at $z = 0.625c'$, the fourth layer is at $0.750c'$, and the fifth atomic layer is situated at $0.875c'$. Taking into account five layers of atoms proves to be sufficient in order to simulate the variation of the potential along the surface normal direction from the vacuum region corresponding to $z < 0$ into the selvedge region corresponding to $0 < z < 0.4375c' = 7.1575$ Å (the latter is divided into 150 slices) and then into one bulk unit slab corresponding to the region $0.4375c' < z < 0.6875c' = 11.2475$ Å (the bulk slab is divided into 100 slices).

Figure 5.16 shows the Laue zone of reciprocal lattice points. These points are indexed on the basis of the conventional unit cell marked by the dotted line in Fig. 5.15a. In terms of the surface unit cell defined by the solid line in Fig. 5.15, these lattice points may be extended into two-dimensional rods of the reciprocal lattice and reindexed using the two indices HK, with 10 equivalent to the $1\bar{1}0$ and 01 equivalent to the 110. The indexing system that we use calls all reflections HK, with non-zero negative K, the higher-order Laue zone (HOLZ) reflections and all reflections with non-zero positive K, minus HOLZ (MHOLZ)

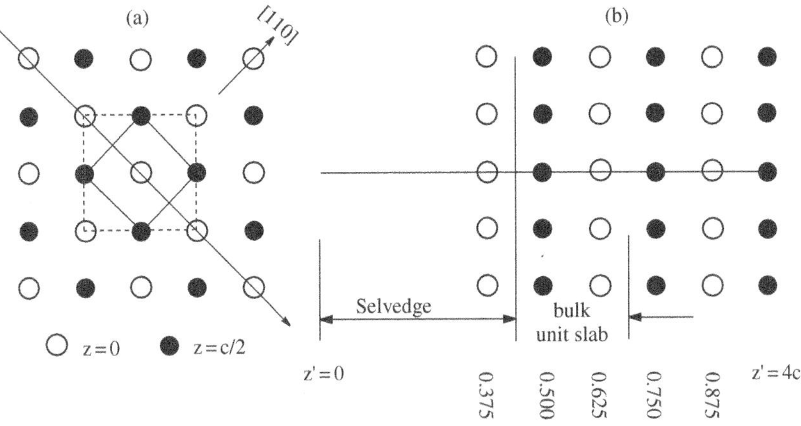

FIG. 5.15. (a) The top [001] view and (b) the side [100] view of the structural model of the Ag(001) surface.

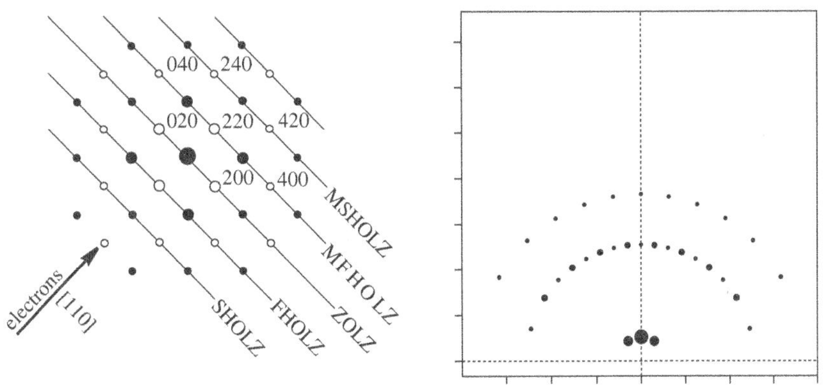

FIG. 5.16. Left: reflections of the type hk0 (shown as filled circles) and hk1 (shown as open circles). Right: kinematic RHEED pattern from the Ag(001) surface. The three Laue circles correspond, respectively, to surface rods of the type $H0$, $H\bar{1}$, and $H\bar{2}$. These reflections belong to ZOLZ, first-order HOLZ, and second-order HOLZ shown in the left part of the figure.

reflections. All reflections identified as $H0$ reflections belong to the zero-order Laue zone (ZOLZ). In Fig. 5.16 the term FHOLZ refers to the first-order HOLZ, SHOLZ to the second-order HOLZ, MFHOLZ to the minus first-order HOLZ, and MSHOLZ to the minus second-order HOLZ.

The interaction potential entering the eigensystems (5.45) and (5.77) can be calculated by representing the atomic scattering factors in the form of a sum of several Gaussian functions (Doyle and Turner, 1968)

$$f^{(e)}(s) = \sum_j a_j \exp(-b_j s^2), \tag{5.125}$$

where $s = \sin\theta/\lambda$, 2θ is the angle of scattering, and a_j and b_j are fitting parameters. In the case of neutral atoms these parameters are now known for all the elements of the periodic table for the representation involving five Gaussian functions (Peng et al., 1996e) and four Gaussian functions (Peng, 1999) (the latter representation is somewhat less accurate than the former). Assuming that the potential of interaction between the incident electron and the crystal can be approximated by the sum of potentials of individual atoms, and using the mixed real and reciprocal space representation (Tournarie, 1962, Kambe, 1967), we obtain (for more detail see section 13.6.1)

$$U(\mathbf{x}, z) = \sum_G U_G(z) \exp(i\mathbf{G} \cdot \mathbf{x}), \tag{5.126}$$

where the Fourier components $U_G(z)$ are given by

$$U_G(z) = \frac{8\pi}{A_0} \left(\frac{m}{m_0}\right) \sum_{n,j} \exp(-i\mathbf{G} \cdot \mathbf{x}_n) \tag{5.127}$$

$$\times \left\{ a_{j,n}^{(Re)} \sqrt{\frac{\pi}{b_{j,n}^{(Re)} + B_n}} \exp\left[-\frac{(b_{j,n}^{(Re)} + B_n)G^2}{(4\pi)^2} - \frac{4\pi^2(z - z_n)^2}{b_{j,n}^{(Re)} + B_n} \right] \right.$$

$$\left. + i a_{j,n}^{(Im)} \sqrt{\frac{\pi}{b_{j,n}^{(Im)} + B_n/2}} \exp\left[-\frac{(b_{j,n}^{(Im)} + B_n/2)G^2}{(4\pi)^2} - \frac{4\pi^2(z - z_n)^2}{b_{j,n}^{(Im)} + B_n/2} \right] \right\},$$

where $a_j^{(Re)}, b_j^{(Re)}$ and $a_j^{(Im)}$, $b_j^{(Im)}$ are fitting parameters for the real and imaginary parts of the electron atomic scattering factors, respectively, B_n is the Debye–Waller factor of the n-th atom, and the summation over n is carried out over all the atoms belonging to a surface unit cell. m and m_0 are the electron mass at the beam energy used and its rest mass respectively, while A_0 denotes the area of the surface unit cell. In the case of the Ag(001) surface shown in Fig. 5.15, $A_0 = a' \times b' = (2.892)^2$ Å^2. The parameters given by Dudarev et al. (Dudarev et al., 1995a) and Peng et al. (Peng et al., 1996e) make it possible to obtain values of $U_G(z)$ expressed in Å^{-2} units. Note that the fitting parameters $a_j^{(Im)}$ and $b_j^{(Im)}$ given by Dudarev et al. and Peng et al. (Dudarev et al., 1995a, Peng et al., 1996e, Peng et al., 1996d) were calculated for a primary beam energy of 100 keV. The values of these parameters may be readily scaled to take into account relativistic effects. Assuming that the energy E of electrons is expressed in keV, we obtain

$$a_j^{(Im)}(E) = \frac{\beta(100 \text{ keV})}{\beta(E)} a_j^{(Im)}(100 \text{ keV}), \tag{5.128}$$

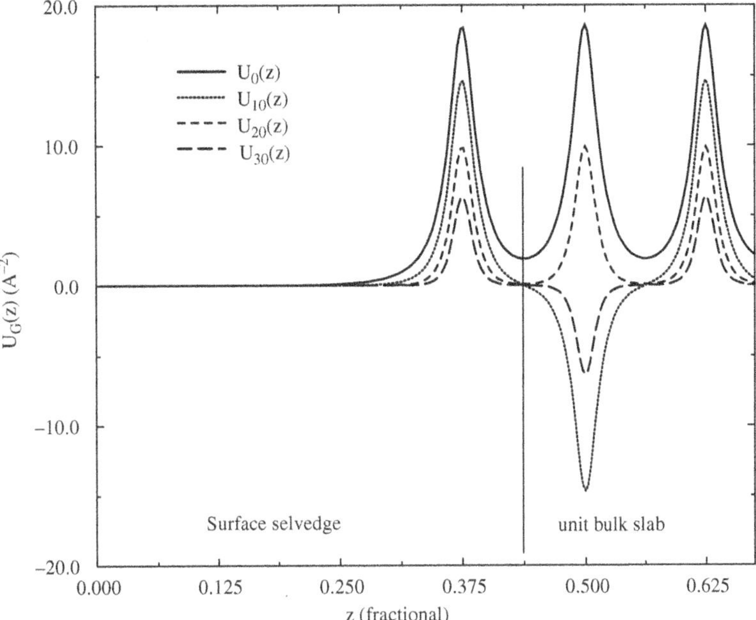

FIG. 5.17. Dependence of the four first Fourier components $U_G(z)$ on the coordinate z in the direction normal to the surface. The potential was evaluated for the case of 20 keV electrons incident on the (001) surface of crystalline silver. The direction of incidence is close to the [01] direction (which is equivalent to the [110] direction).

where $\beta = (1-\gamma^{-2})^{1/2}$, $\gamma = (m/m_0) = 1.0 + 1.9569314 \times 10^{-3}E$, and the primary beam energy E is expressed in keV units. The parameters b_j do not depend on the energy of electrons and no relativistic correction is required.

The zero-order Fourier component of the potential $U_0(z) = -2mV_0(z)/\hbar^2$, where V is the conventional potential entering the Schrödinger equation, must always be greater than zero, reflecting the fact that incident high-energy electrons are on average attracted to atoms in the crystal. The other Fourier components of the potential may vary and be either positive or negative depending on the phase factor $\exp(-i\mathbf{G} \cdot \mathbf{R}_n)$ entering eqn (5.127). This point is illustrated in Fig. 5.17 for the $U_{10}(z)$ and the $U_{30}(z)$ components of the potential. This figure shows that the Fourier components of the potential $U_G(z)$ decrease rapidly away from the centre of atomic layers, and in general it is a very good approximation when only the nearest neighbour contributions to $U_G(z)$ are included in the procedure of numerical evaluation of this quantity. Curves shown in Fig. 5.17 were computed by including only five atomic layers in the summation in eqn (5.127). This proves to be more than sufficient for simulating the variation of the potential across the surface selvedge layer and across the repeat bulk unit slab. In Appendix C we give an input data file for the case of the Ag(001) surface, and a FORTRAN routine for the computation of the Fourier components of the potential $U_G(z)$ shown in Fig. 5.17.

5.4.1.2 *RHEED calculations* Figure 5.18 shows calculated RHEED rocking curves representing the specular (00) and the two side (10) and (20) beams. Calculations were performed using 37 reciprocal lattice rods that include 11 ZOLZ rods: $(0,0)$, $(\pm1,0)$, $(\pm2,0)$, $(\pm3,0)$, $(\pm4,0)$, and $(\pm5,0)$; 15 FHOLZ rods: $(0,1)$, $(\pm1,\bar{1})$, $(\pm2,\bar{1})$, $(\pm3,\bar{1})$, $(\pm4,\bar{1})$, $(\pm5,\bar{1})$, $(\pm6,\bar{1})$, and $(\pm7,\bar{1})$; and 11 SHOLZ rods: $(0,\bar{2})$, $(\pm2,\bar{2})$, $(\pm4,\bar{2})$, $(\pm6,\bar{2})$, $(\pm8,\bar{2})$, and $(\pm10,\bar{2})$. The three curves shown in the figure were calculated for a crystal slab consisting of a surface selvedge layer shown in Fig. 5.17 and several bulk unit layers. For comparison the results obtained using 3 bulk unit layers, 12 bulk unit layers, and 24 bulk unit layers are shown together in the same figure. The plots show clearly that all the major features present in RHEED rocking curves are associated with processes occurring very near to the surface of the crystal. In fact it is possible to say that the inclusion of more than three bulk layers is required only to obtain a highly convergent result. The inclusion of 20 bulk unit slabs in the computational scheme makes the calculated amplitudes of reflected beams indistinguishable from those corresponding to the case of a semi-infinite crystal. Almost all the RHEED rocking curves published in the literature were in fact calculated in this way, i.e. by using a sufficiently thick crystal slab instead of solving the RHEED problem for a semi-infinite crystal.

Figure 5.19 shows rocking curves representing amplitudes of the specularly reflected $(0,0)$ beam and a side $(1,0)$ beam. Eleven rods belonging to the ZOLZ were actually included in the calculation, but within the range of variation of the glancing angle of incidence shown in the figure only the specular beam and two side $(\pm1,0)$ beams have appreciable intensities. This figure shows that although the amplitudes of some other beams may not be appreciable, the effect of these beams on the (0,0) and $(\pm1,0)$ beams may be very strong and this must be taken into account in a calculation in order to obtain a convergent result. In the case where only 18 rods were included in the calculation, the resulting RHEED rocking curves did not differ substantially from the result obtained using the full set of 37 rods. The small difference between the 18-rod and the 37-rod cases can be explained very accurately using the Bethe potential method discussed in the previous section.

5.4.2 *RHEED from a surface of an ionic crystal: the NiO(001) and UO$_2$(111) surfaces*

5.4.2.1 *Atomic scattering factors of ions* The amplitude of scattering of a high-energy electron by an ion differs very significantly from the amplitude of scattering by a neutral atom (see also section 13.6.1). In the case of X-ray diffraction the atomic scattering factors of both neutral atoms and ions satisfy the condition that

$$\lim_{s \to 0} f^{(X)}(s) = Z_0, \tag{5.129}$$

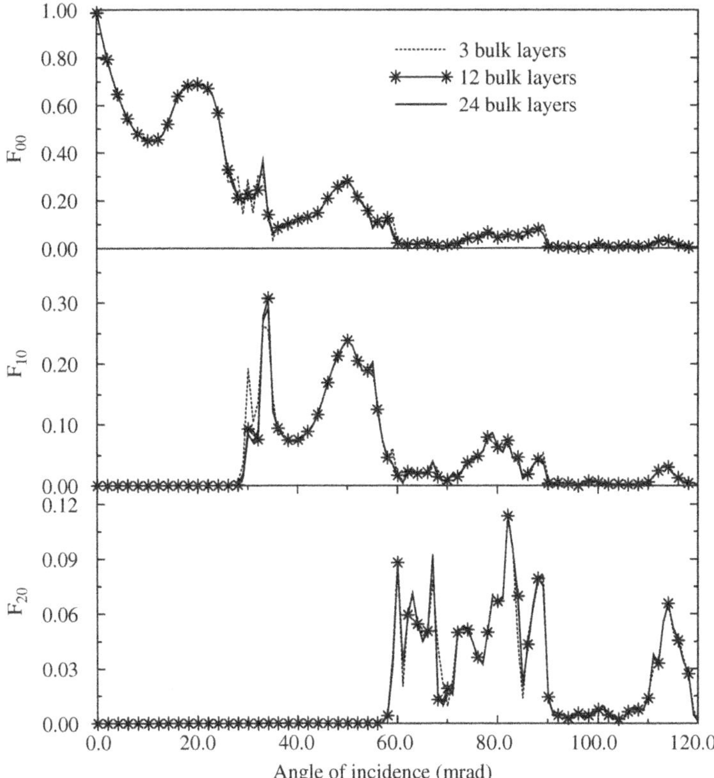

FIG. 5.18. Calculated amplitudes of the (00), (10), and (20) beams reflected from the Ag(001) surface. The curves show amplitudes of reflected beams plotted as a function of the glancing angle of incidence and calculated using atomistic models that include a surface selvedge layer and three unit bulk layers (dotted line), a selvedge layer and 12 unit bulk layers (starred-solid line), and a selvedge layer and 24 unit bulk layers (solid line).

where Z_0 is the number of electrons per atom. The atom can be either in a neutral or in a charged ionic state. For a neutral atom $Z_0 = Z$, where Z is the atomic number of the corresponding element. For an ion $Z_0 \neq Z$, and the difference between the two quantities represents the excess or the deficit of charge on the nucleus resulting from charge transfer associated with the formation of chemical bonds in the crystal. The atomic scattering factor for electron diffraction is related to that for X-ray diffraction by the Mott formula (Mott and Massey, 1965)

$$f^{(e)}(s) = \frac{m_0 e^2}{8\pi^2 \hbar^2} \frac{1}{4\pi\epsilon_0} \frac{Z - f^{(X)}(s)}{s^2}. \tag{5.130}$$

For an ion the number of electrons is not equal to the charge of the nucleus, i.e. $Z \neq Z_0$, and so it follows from eqn (5.130) that as s approaches zero the

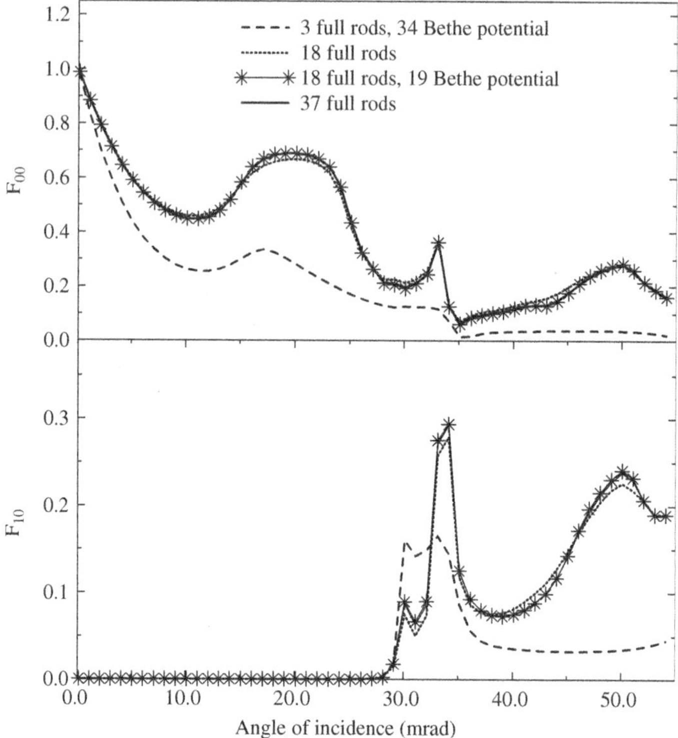

FIG. 5.19. Calculated amplitudes of the (00) and (10) beams reflected from the Ag(001) surface. Calculations were performed using the Bethe potential approximation for several different sets of strong/weak beams for the range of glancing angles of incidence where only the specular and the $(\pm 1, 0)$ beams have appreciable amplitudes.

scattering factor diverges, i.e. $f^{(e)}(s) \sim (Z - Z_0)/s^2$.

Electron scattering factors of ions have been calculated numerically and tabulated by several authors including Doyle and Turner (Doyle and Turner, 1968), Cowley (Cowley, 1992), and Rez et al. (Rez et al., 1994). In principle *ab initio* numerical values obtained by the above authors may be approximated by the same analytical formula (5.125) as in the case of neutral atoms. However, since the expression (5.125) *does not* diverge at zero angle of scattering as it should do for an ion, the conventional fitting procedure proves to be unsuitable for the accurate representation of scattering factors of ions. It may be argued that since a crystal is always composed of positive and negative ions, contributions associated with the excess and deficit of charge on the positive and negative ions must always exactly cancel each other at zero angle of scattering, and therefore the precise value of $\lim_{s \to 0} f^{(e)}(s)$ for ions is unimportant. Following this logic one might suppose that the values of $f^{(e)}(s)$ may still be fitted using (5.125)

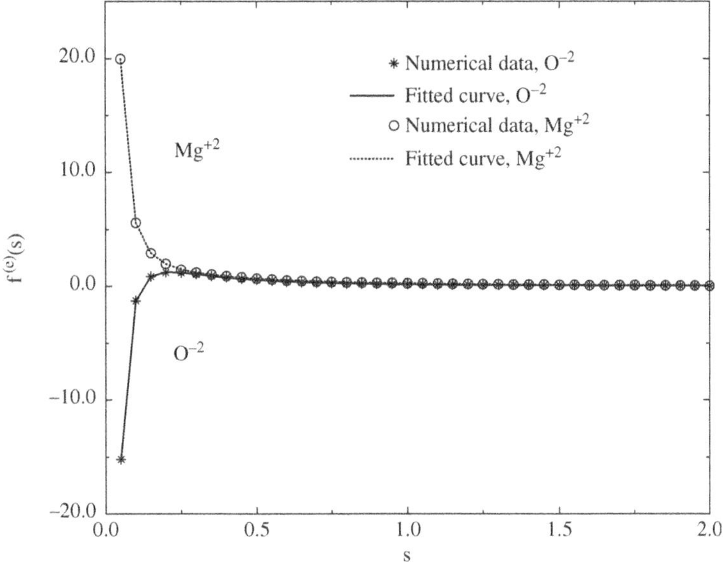

FIG. 5.20. Numerical and fitted electron scattering factors for O^{2-} and Mg^{2+}. From (Peng et al., 1998).

without introducing any significant error in the dynamical electron diffraction calculations. However, it is easy to demonstrate that this is not so. Figure 5.20 shows an example of Gaussian fitting performed for oxygen O^{2-} and magnesium Mg^{2+} ions. It is seen that fitting appears to give good results for $s > 0.05$ Å^{-1}. For smaller values of s the fitting becomes inaccurate. We show below that it is this interval of values of s which plays an essential part in dynamical RHEED calculations. This leads to the conclusion that although a straightforward fitting of electron scattering factors by a sum of several Gaussian terms may serve as a good approximation for the majority of applications in transmission electron diffraction (in the transmission case scattering through zero angle influences all the diffracted beam amplitudes in the same way giving rise to the same phase factor for all the beams), this procedure is certainly not sufficiently accurate for the case of RHEED where scattering at zero angle manifests itself as refraction of electrons at the surface.

An examination of the origin of the divergent behaviour of the electron scattering factor of an ion (see for example (Doyle and Turner, 1968)) shows that the divergent part arises from the contribution of the unscreened long-range Coulomb potential. This may be readily demonstrated by rearranging eqn (5.130) as

$$f^{(e)}(s) = \frac{m_0 e^2}{8\pi^2 \hbar^2} \frac{1}{4\pi\epsilon_0} \left\{ \frac{Z_0 - f^{(X)}(s)}{s^2} + \frac{\Delta Z}{s^2} \right\} = f_0^{(e)}(s) + \frac{m_0 e^2}{8\pi^2 \hbar^2} \frac{1}{4\pi\epsilon_0} \frac{\Delta Z}{s^2},$$

$$(5.131)$$

where $\Delta Z = Z - Z_0$ represents the ionic charge and the second term on the

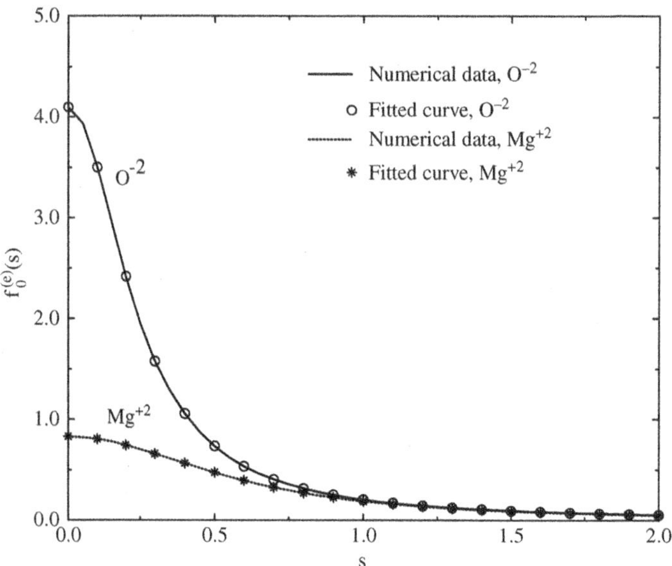

FIG. 5.21. Comparison of exact and Gaussian fitted values of $f_0^{(e)}(s)$ with the divergent contribution of ionic charge subtracted for O^{2-} and Mg^{2+} ions. From (Peng et al., 1998).

right-hand side represents the Coulomb part of the scattering factor. The first term on the right-hand side (i.e. $f_0^{(e)}(s)$) results from scattering of electrons by the screened atomic field. Condition (5.129) ensures that $f_0^{(e)}(s)$ remains finite in the limit $s \to 0$. Figure 5.21 shows numerical values of $f_0^{(e)}(s)$ for ions of Mg^{2+} and O^{2-} and the fitted curves of $f_0^{(e)}(s)$ obtained using a linear combination of Gaussians (5.125). It is seen that for all the angles of scattering the Gaussian fit exhibits a high degree of numerical accuracy.

5.4.2.2 *The construction of the scattering potential for an ionic crystal* The potential of an ionic crystal can be calculated using a purely numerical algorithm by constructing a three-dimensional super unit cell. The size of the super unit cell in the plane parallel to the surface equals the size of the surface unit cell. The super unit cell may be arbitrarily large in the direction normal to the surface. In the super unit cell approach the potential is calculated using a conventional Fourier expansion both in the plane of the surface and in the direction normal to it. The Fourier components of a three-dimensional potential are given by (Hirsch et al., 1977)

$$V_g = -\frac{\hbar^2}{2m_0} \cdot \frac{4\pi}{\Omega} \sum_n f_n^{(e)}(s) \exp(-i\mathbf{g} \cdot \mathbf{r}_n - B_n s^2), \qquad (5.132)$$

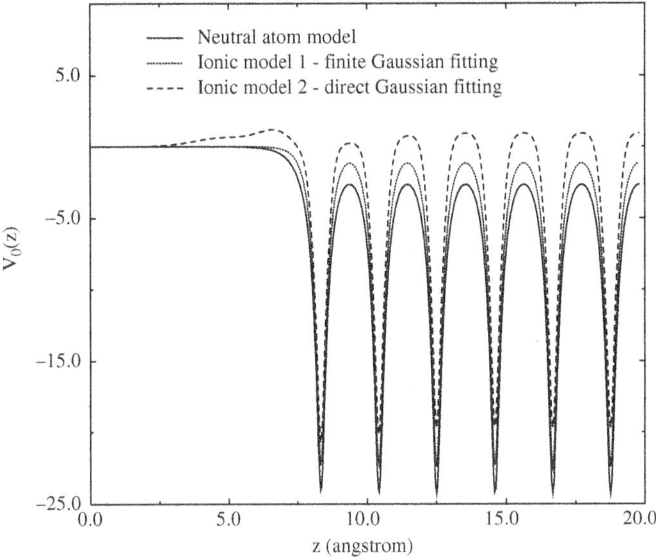

FIG. 5.22. Averaged potential distribution in an NiO single crystal plotted along the [010] direction. The potential $V_0(z)$ is in electron volts. From (Peng et al., 1998).

where Ω is the volume of the super unit cell, $g = 4\pi s$, and $\mathbf{g} = (\mathbf{G}, \ell)$ is a three-dimensional reciprocal lattice vector, $\mathbf{r}_n$ is the coordinate of the n-th atom, and the summation is carried out over all atoms within the super unit cell. After all the components V_g have been calculated, the distribution of the potential in real space can be obtained by using the inverse Fourier transform. For $V_G(z)$ given by eqn (5.126) we obtain

$$V_G(z) = \sum_\ell V_{G,\ell} \exp(i\ell z).$$
(5.133)

As before we separate $V_G(z)$ into two parts, namely the contribution of ionic charge $\Delta V_G(z)$ and the part associated with the screened atomic field $V_G^0(z)$. While the contribution resulting from $f_0^{(e)}(s)$ always remains finite, the part of the potential associated with the presence of ionic charges gives rise to divergent terms. In the case where the termination of the crystal lattice satisfies the condition of charge neutrality and where it has vanishing dipole moment, the averaged potential is given by

$$\Delta V_0(z) = \sum_{\ell \neq 0} \Delta V_\ell^0 \exp(i\ell z) + \frac{e^2}{4\pi\Omega} \sum_n \Delta Z_n (8\pi^2 z_n^2 + B_n).$$
(5.134)

Figure 5.22 shows the distribution of the averaged potential calculated for the (100) surface of nickel monoxide for the primary beam energy of 200 keV. The super unit cell was chosen to be four times larger than the projection of the bulk unit cell of NiO in the $\langle 100 \rangle$ direction, and within this super unit cell only the central part was assumed to be occupied by Ni^{2+} and O^{2-} ions. In the many-beam dynamical RHEED calculations described below, the region between $z = 0$ and $z = 7.294$ Å was treated as the surface region, while the remaining part of the super unit cell situated between $z = 7.294$ Å and 11.462 Å approximated the behaviour of the potential in the crystal bulk.

5.4.2.3 *The effect of ionic charge on the electrostatic potential* The scattering potential may be constructed using the three-dimensional super unit cell discussed in the previous section. However, this purely numerical approach does not provide much physical insight, and it is desirable to develop a more transparent analytical treatment of the problem.

Although an analytical expression describing the Fourier components of the potential $V_G(z)$ at a finite temperature is not available at present, a simple analytical representation of $V_G(z)$ can still be obtained in the case of zero thermal vibration amplitude, i.e. in the case where the Debye–Waller factor $B = 0$. Equation (5.130) represents the electron scattering factor of an ion by a sum of two terms, where the second term describes the contribution due to the ionic charge and the first term $f_0^{(e)}(s)$ is the contribution from the screened atomic field. As was shown above, the term $f_0^{(e)}(s)$ remains finite for all angles of scattering and it can therefore be fitted accurately by five Gaussian terms as in eqn (5.125). The contribution to $V_G(z)$ resulting from this term therefore takes a form that is identical to expression (5.126) and that vanishes in the vacuum region outside the crystal. The contribution to the potential resulting from the second term of eqn (5.130) describes effects associated with the long-range Coulomb field giving rise to a real-space term of the form

$$\Delta\phi(\mathbf{r}) = \frac{e\Delta Z}{4\pi\epsilon_0 r}. \tag{5.135}$$

The corresponding contribution to $V_G(z)$ is

$$\Delta V_G(z) = -\frac{e^2}{2\epsilon_0 S_0} \cdot \frac{1}{G} \sum_n \Delta Z_n \exp\left[-i\mathbf{G} \cdot \mathbf{R}_n - G|z - z_n| \right], \tag{5.136}$$

for $G \neq 0$ and

$$\Delta V_0(z) = -\frac{e^2}{2\epsilon_0 S_0} \sum_n \Delta Z_n |z - z_n|.. \tag{5.137}$$

It should be pointed out that in deriving the above expression for $\Delta V_0(z)$ we have used the condition of charge neutrality, i.e. $\sum_n \Delta Z_n = 0$. If we define the positive direction of the z axis as pointing outwards then at a certain distance above the crystal surface for all z_n we have $z > z_n$ and

$$\Delta V_0(z) = -\frac{e^2}{2\epsilon_0 S_0} \sum_n \Delta Z_n z_n. \tag{5.138}$$

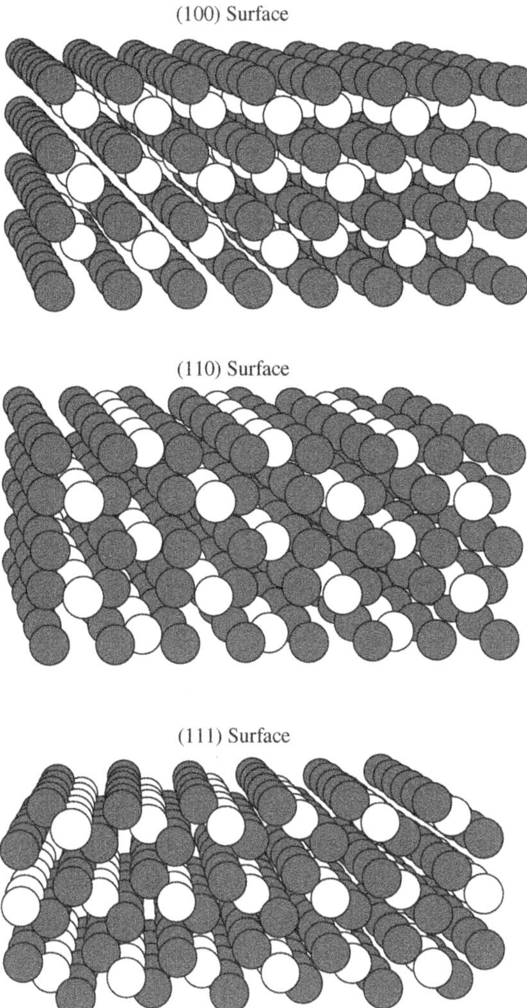

FIG. 5.23. Ionic models for the (100), (110), and (111) surfaces of the fluorite structure. In this drawing the light spheres correspond to the anions (U^{4+}) in uranium dioxide, and the dark spheres correspond to the cations (O^{2-}). From (Peng et al., 1998).

In other words the value of the potential in the vacuum is proportional to the z projection of the *total* dipole moment of the crystal $\sum_n \Delta Z_n z_n$. Depending on the geometry of termination of the crystal lattice we can identify three distinct classes of ionic surfaces. The first case corresponds to the situation where both positively and negatively charged ions are located in the same atomic plane. An example of this kind is given by the (001) and (110) terminations of the sodium chloride structure or the (110) plane of the fluorite structure (see Fig. 5.23). For each atomic plane z_n is constant and the z projection of the dipole moment vanishes, i.e.

$$\mu = \sum_n \Delta Z_n z_n = \text{const} \sum_n \Delta Z_n = 0.$$

In the vacuum region the potential rapidly approaches zero, and this case presents no particular difficulty in dynamical RHEED calculations.

In the second case the positively and negatively charged ions are located on different atomic planes, so that each atomic plane parallel to the surface is charged, but the total dipole moment of the repeat unit is still equal to zero. Examples of surface terminations of this type include the (111) surface of the fluorite structure, provided that the surface is terminated on the anion plane (Tasker, 1979) (see Fig. 5.23). The fluorite structure consists of units of anion–cation–anion atomic planes situated periodically in the direction perpendicular to the (111) plane. These units consist of sets of $O^{2-}-U^{4+}-O^{2-}$ atomic planes in uranium dioxide or sets of $F^{1-}-Ca^{2+}-F^{1-}$ atomic planes in calcium fluoride. For each set of planes the total dipole moment vanishes and the scattering potential is equal to zero in the vacuum region. Dynamical RHEED calculation can again be performed using one of the conventional numerical techniques, although care needs to be taken in the correct representation of the relevant Coulomb terms.

In the third case positively and negatively charged ions do not lie in the same plane, neither is the total dipole moment of a bulk repeat unit of atomic planes equal to zero. Examples of this case include the (100) surface of the fluorite structure (see Fig. 5.23) and the (111) surface of the sodium chloride structure. In uranium dioxide the repeat unit set of atomic planes in the $\langle 100 \rangle$ direction consists of two ionic planes $2O^{2-}-U^{4+}$, and the total dipole moment of this set of planes does not vanish. As a result the potential of the crystal diverges as the thickness of the crystal slab increases. Other related physical quantities also show divergent behaviour (e.g. the surface energy of the (100) surface of UO_2 is infinite (Tasker, 1979)) and this makes a simple termination of the bulk structure physically impossible. The problem can be formally eliminated by choosing a suitable surface reconstruction. For example, in the case of the (100) surface of UO_2 the problem associated with the bulk dipole moment can be eliminated by transferring half of the top oxygen layer from one surface of the crystal slab to the other surface. In a real crystal the tendency towards lowering the surface energy indeed leads to surface reconstruction and/or to the generation of surface defects. One of the possible scenarios involves the formation of surface steps characterized by different types of termination so that oppositely charged surface planes would be exposed (Tasker, 1979). Therefore in any realistic situation the total dipole moment of a repeat unit set of atomic planes associated with a particular termination of the crystal lattice must vanish in order to eliminate the divergent behaviour of the potential in the limit of large thickness of the crystal slab. In should be noted that surface reconstruction may lead to the appearance of a *surface* dipole moment and may therefore influence the apparent value of the

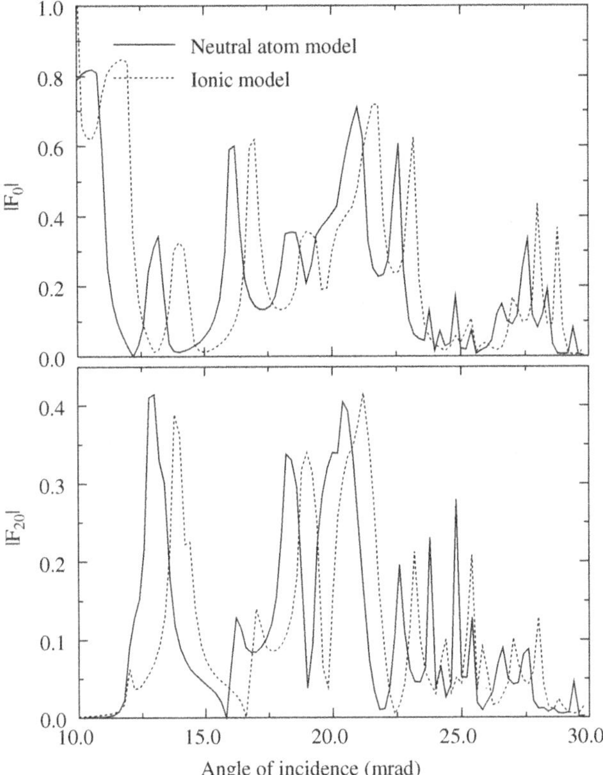

F IG. 5.24. Calculated RHEED rocking curves representing (a) the specularly reflected
beam and (b) the (20) side beam reflected from the (100) surface of NiO. Calculations were performed using the neutral atom model (solid line) and the ionic model
(dotted line) for the 200 keV incident beam energy and the [001] beam azimuth.
The curves shown in the figure represent amplitudes of the (a) specularly reflected
beam $|F_0|$ and (b) the (20) side beam $|F_{20}|$. From (Peng et al., 1998).

inner potential of the crystal observed in experiments on refraction of electrons
by a crystal surface.

5.4.2.4 *The effect of charge transfer on RHEED rocking curves*   Once all the
Fourier components of the potential have been evaluated, dynamical RHEED
calculations may be readily performed following the approach described in the
preceding section. Figure 5.24 shows two RHEED rocking curves calculated for
the primary beam energy of 200 keV for the (001) surface of NiO. Electrons
are incident on the surface in the direction parallel to the [010] zone axis. The
two curves shown in the figure were calculated using the neutral atom and the
ionic models for (a) the specularly reflected beam and (b) the side (20) beam. An
examination of this figure shows that the transfer of charge from nickel to oxygen

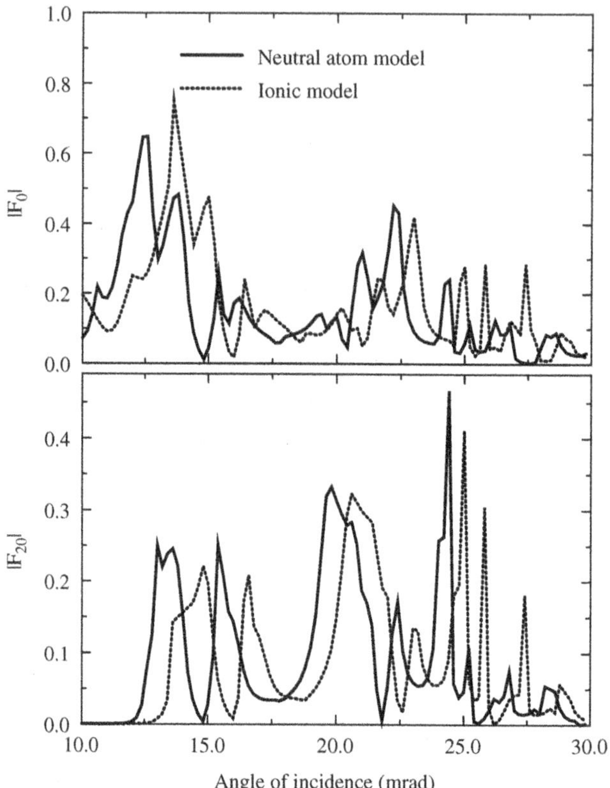

FIG. 5.25. Calculated RHEED rocking curves for (top) the specularly reflected beam and (bottom) the (20) side beam. Calculations were performed using the neutral atom model (solid line) and the ionic model (dotted line) for a single crystal of UO_2. The calculations were made for 200 keV primary beam energy, and for the beam incident along the $[\overline{1}12]$ azimuth. From (Peng et al., 1998).

ions results in a shift of the rocking curves towards higher angles, and this finding is consistent with what can be expected from the analysis of the distribution of potential shown in Fig. 5.22. This figure shows that the transfer of charge between ions leads mainly to a reduction in the value of the inner potential. It has a less dramatic effect on the shape of the potential. Since a smaller value of the inner potential is equivalent to a smaller value of the refractive index, the main effect of the charge transfer on RHEED rocking curves is therefore associated with the shift of all the diffraction features towards the region of higher glancing angles as illustrated in Fig. 5.24.

Figure 5.25 shows two RHEED rocking curves calculated for the (111) surface of uranium dioxide UO_2. Electrons are incident on the surface in the direction of the $[\overline{1}12]$ zone axis, and Fig. 5.25 shows the calculated absolute amplitudes of the specular and the side (20) beams. Although for the (111) surface the negatively charged O^{2-} and the positively charged U^{4+} planes are shifted with respect to each other in the direction normal to the surface, the z projection of the total dipole moment of each repeat unit set of planes of the form $O^{2-}-U^{4+}-O^{2-}$ is equal to zero. The figure shows that there exists a clear correlation between the rocking curves calculated using the neutral atom and the ionic models.

The effect of charge transfer in an ionic crystal does not only result in a homogeneous shift of peaks in rocking curves. It also changes the relative peak heights and the relative positions of peaks. Although in many cases there still exists a one-to-one correspondence between the peaks in the RHEED rocking curves calculated using the neutral atom and the ionic models, charge transfer often leads to the appearance of new features and sometimes any correspondence between the two curves appears to be lacking. This shows the importance of using an adequate representation of electron scattering factors, which takes into account the effects of charge transfer occurring in a real crystal. In a recently studied case of reflection diffraction of electrons from an NiO surface (Peng et al., 1997) it was found that the use of a mixed neutral atom/ionic model provides a reasonably good description of the one-rod RHEED case. It is expected that in the case of many-beam RHEED, this scheme will also provide a good starting point for the subsequent refinement of the atomic structure of the surface and the distribution of electronic charge density near the surface.

5.5   RHEED from growing surfaces: intensity oscillations

One of the most interesting recent applications of RHEED to structural analysis of crystal surfaces is associated with the discovery by Wood (Wood, 1981) and by Harris, Joyce, and Dobson (Harris et al., 1981), of the effect of oscillations of the intensity of electron beams reflected from a growing surface at grazing incidence. Although Wood was the first to observe this effect during molecular beam epitaxial (MBE) growth of a semiconductor crystal, it was Harris et al. (Harris et al., 1981) who realized that the period of oscillations is equal to the time required to deposit a new monolayer of atoms on the surface. RHEED oscillations have provided a very convenient way of monitoring the growth of semiconductor crystals, and in some cases many thousands of periods of oscillations were observed (Sakamoto et al., 1985).

The significance of RHEED oscillations as a method of structure determination is associated with the fact that temporal variations of the intensity of diffracted beams reflect the dynamics of evolution of the crystal structure, as opposed to conventional diffraction techniques aimed at characterizing the structure of a crystalline material at equilibrium. The development of a quantitative theory of RHEED oscillations has proved to be a challenging problem, which still remains largely unsolved. A number of approaches described in the literature in the last 20 years address various aspects of RHEED oscillations and provide a reasonably complete qualitative picture of this phenomenon.

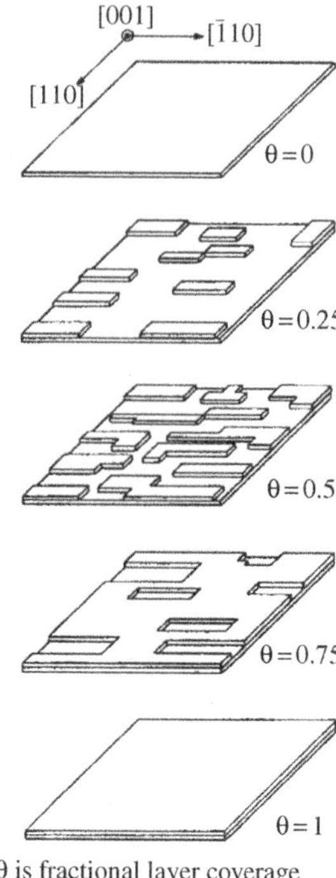

FIG. 5.26. Real-space view of the process of formation of a new atomic layer on a growing singular surface. From (Neave et al., 1983).

Neave et al. (Neave et al., 1983) proposed that oscillations of RHEED reflections are associated with periodic changes in the roughness of the growing surface, as atoms forming a new monolayer gradually fill vacant surface lattice sites (see e.g. Fig. 5.26 illustrating the process of formation of a new layer of atoms on a growing surface). Van Hove et al. (Van Hove et al., 1983) realized that in the case of a vicinal surface there are two competing modes of growth, namely (1) the mode associated with the advancement of steps due to the attachment of atoms to step edges, and (2) the mode associated with the nucleation of new islands on flat terraces. In the case (1) RHEED oscillations do not occur since the 'macroscopic' morphology of the growing surface does not change. The case (2) is similar to that describing growth of a singular surface, and therefore RHEED intensities in this case do exhibit oscillatory behaviour. Neave et al. (Neave et al., 1985) showed by monitoring RHEED oscillations how it is possible to investigate the transition between these two modes of growth. Figure 5.27 illustrates how the transition between the step-flow mode of growth and the two-dimensional nucleation of islands on terraces is manifested in the form of temporal variation of the intensity of the specular beam reflected from the growing surface.

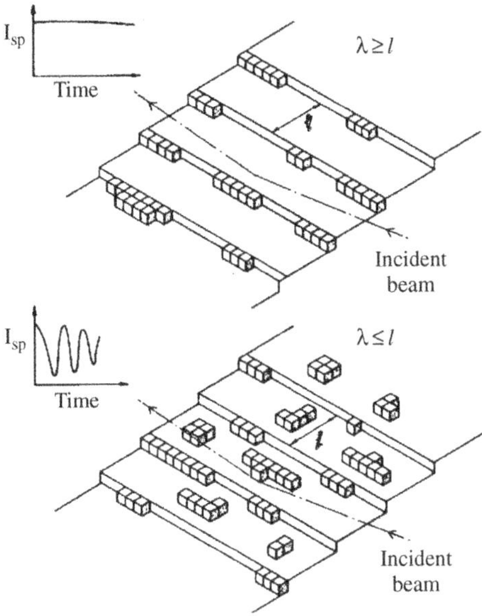

FIG. 5.27. Schematic illustration of the two possible modes of growth of a vicinal surface and the corresponding changes in the intensity of the specular RHEED reflection. From (Neave et al., 1985).

The need to establish a link between the observed variation of RHEED intensity and the structure and morphology of growing surfaces has led Dobson et al. (Dobson et al., 1988) to the formulation of the 'step density' model of RHEED oscillations. In this model it is assumed that the origin of oscillations is associated with the periodic increase of diffuse intensity as the occupancy of every new atomic layer passes through half a monolayer coverage (where the roughness of the surface on the atomic scale reaches its peak value), and that the increase in diffuse intensity is matched by the decrease of the intensity of Bragg peaks. The model has been successfully applied to the interpretation of a number of fairly complex experimental observations (Shitara et al., 1992, Sudijono et al., 1992, Vvedensky et al., 1993). The validity of the model was confirmed (Holmes et al., 1997) by a direct comparison of profiles of temporal variation of the intensity of RHEED reflections with changes in the density of steps on the growing surface observed using a scanning tunneling microscope (STM).

The step density model does not, however, give a fully comprehensive description of RHEED oscillations. Although the contribution of diffuse scattering to the observed RHEED patterns is very strong (Larsen and Meyer-Ehmsen, 1990), there is a significant point that the step density model does not address. This point concerns the observed dependence of the *phase* of oscillations on the polar and azimuthal angles of incidence of electrons on the surface. If, say, the intensity of a RHEED reflection considered as a function of time t varies as

$$I(t) = \frac{1}{2}I_0 + I_1 \cos\left(\frac{2\pi}{T}t - \Phi_1\right), \qquad (5.139)$$

where T is the period of oscillations, then the quantity Φ_1 can be conveniently defined as the phase of oscillations (Mitura et al., 1998, Mitura et al., 2002). Adopting the step density model and assuming that growth starts at the moment when the surface is atomically flat, we find that $\Phi_1 = 0$. Indeed, since the deposition of atoms on the surface inevitably leads to the increase of surface roughness and the intensity of diffuse scattering, the brightness of Bragg peaks should start decreasing at $t = 0$. The latter condition is equivalent to saying that the phase of oscillations Φ_1 must vanish irrespective of the orientation of the incident beam of electrons. Measurements of the phase of RHEED oscillations performed independently by several groups world-wide (Joyce et al., 1987, Zhang et al., 1987, Resh et al., 1989, Crook et al., 1989) showed that this was actually not the case, and for many surfaces the observed phase of the oscillations exhibited a strong variation as a function of both the polar and the azimuthal angles of incidence (Horio and Ichimiya, 1993, Horio and Ichimiya, 1994a, Braun et al., 1998). On the one hand this shows that the range of validity of the step density model is somewhat limited (this conclusion was also derived from numerical studies of RHEED from a rough surface (Korte and Maksym, 1997, Maksym et al., 1998)) and the model should only be applied to the treatment of RHEED oscillations observed at certain orientations of the incident beam. On the other hand this shows that RHEED oscillations could potentially be used as a fairly unique method of structure determination. Indeed, by investigating how the phase of oscillations varies as a function of angles of incidence, it may be possible to determine the pathways of evolution of the structure of growing surfaces and understand how interatomic bonds are formed during the growth of new atomic layers.

The fact that the phase of RHEED oscillations varies as a function of azimuthal and polar angles of incidence points to the significance of interference between electron waves reflected from atomic planes exposed simultaneously to the incident beam during the growth (e.g. Fig. 5.26 shows that at least two atomic layers are exposed to the incident beam at any point in time for $\Theta < 1$). Lent and Cohen and colleagues (Lent and Cohen, 1984, Pukite et al., 1985, Cohen et al., 1989), and Ichimiya (Ichimiya, 1987) and Ishibashi (Ishibashi, 1991) developed a model where the interference between electrons scattered by several layers of atoms exposed during the growth was taken into account. This model, which is based on the Kirchhoff approximation (Voronovich, 1994, Dudarev, 1997), neglects an important contribution to the diffuse RHEED intensity associated with scattering of electrons by step edges (Lehmpfuhl et al., 1991). In qualitative terms, the presence of scattering by step edges (Lehmpfuhl et al., 1991) points to the possible existence of a fundamental link between the model of scattering by a stepped surface by Lent and Cohen (Lent and Cohen, 1984)

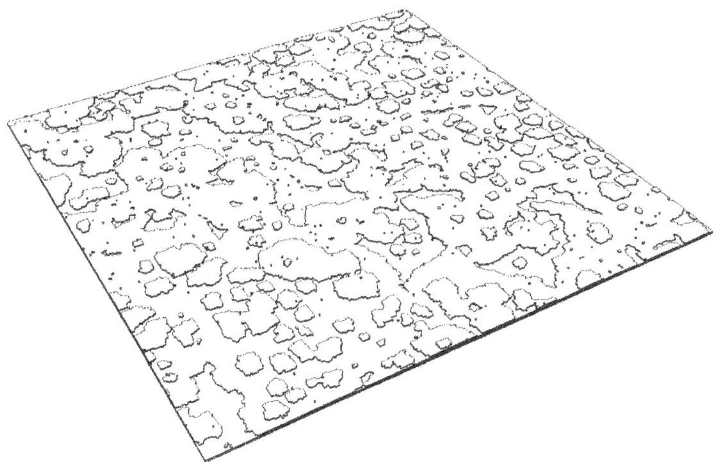

FIG. 5.28. A snapshot showing an MBE growing surface simulated using the kinetic Monte-Carlo approach. From (Vvedensky et al., 1993).

and Ichimiya (Ichimiya, 1987) and the step density model by Dobson et al. (Dobson et al., 1988). Unfortunately a full-scale treatment of the interaction of high-energy electrons with step edges still lies beyond the reach of the modern theory of scattering by rough surfaces (Chitanvis and Lax, 1984, Rodriguez, 1989, Voronovich, 1994, Dudarev, 1997).

An alternative theoretical approach to RHEED from a growing surface is based on the statistical treatment of multiple scattering of electrons described in (Dudarev et al., 1992b, Dudarev et al., 1994). Formally the treatment of the diffuse component of RHEED is similar to the treatment of diffuse scattering by randomly distributed defects (in the case of scattering by a growing surface the disorder is associated with the random occupation of lattice sites by atoms arriving on the surface). To find the intensity of RHEED reflections, as well as the intensity of diffuse scattering, we need to solve the kinetic equation for the density matrix described in Chapter 8. Solving the kinetic equation for the density matrix in the case of RHEED represents a formidable mathematical problem. In recent years significant progress has been achieved in understanding one of the main aspects of this problem, namely, dynamical diffraction of high-energy electrons from a growing surface and the origin of the phase of RHEED oscillations. These results are discussed below.

To appreciate the difficulty associated with the treatment of diffraction from a growing surface we examine Fig. 5.28 showing the morphology of the surface at some point during the growth. This figure gives an example of a configuration of a surface simulated using the kinetic Monte-Carlo approach.

Clearly, in a general case it is nearly impossible to follow all the processes of scattering and rescattering of electrons by islands of growth and the edges of atomic steps present on the surface. However, approximate treatments can be developed for two limits, namely, (1) for the case where the size of flat regions on the surface is large and the density of steps in small and (2) where the size of terraces is very small. In the case (1) the Kirchhoff approximation (Dudarev, 1997) or its simplified versions (Lent and Cohen, 1984, Pukite et al., 1985, Ichimiya, 1987, Ishibashi, 1991) can be used in order to link the morphology of the surface with the observed intensity of RHEED reflections (note that in this case scattering by step edges is completely neglected (Lehmpfuhl et al., 1991)). In the case (2) intensities of RHEED reflections can be evaluated by using a full-scale multiple scattering treatment of diffraction by the statistically *averaged* surface, i.e. by the surface where the strength of the potential of an atom occupying a given lattice site α depends not only on the type of the atom but also on the occupation probability Θ_α of this site. In this case the potential entering the Schrödinger equation describing RHEED from a growing surface is chosen in the form

$$V(\mathbf{r}) = \sum_\alpha \Theta_\alpha \left[V_\alpha^{(Re)}(\mathbf{r} - \mathbf{r}_\alpha) + i V_\alpha^{(Im)}(\mathbf{r} - \mathbf{r}_\alpha) + i(1 - \Theta_\alpha) V_\alpha^{(rand)}(\mathbf{r} - \mathbf{r}_\alpha) \right],$$

$$(5.140)$$

where $V_\alpha^{(Re)}(\mathbf{r} - \mathbf{r}_\alpha)$ denotes the usual electrostatic potential of an atom occupying site α, $V_\alpha^{(Im)}(\mathbf{r} - \mathbf{r}_\alpha)$ represents the imaginary part of the potential of atom α associated with thermal diffuse scattering, and $\Theta_\alpha(1 - \Theta_\alpha) V_\alpha^{(rand)}(\mathbf{r} - \mathbf{r}_\alpha)$ is an additional imaginary part of the potential resulting from random statistical fluctuations of occupancies of lattice sites (Dudarev et al., 1992b, Mitura et al., 1998). Note that the third term on the right-hand side of eqn (5.140) depends on the occupancies of atomic layers via $\Theta_\alpha(1 - \Theta_\alpha)$ and the imaginary part of the potential resulting from the statistical disorder reaches its maximum at $\Theta_\alpha = 1/2$ in qualitative agreement with the step density model (Dobson et al., 1988).

The potential (5.140) describes the *effective* interaction between the incident high-energy electron and the *statistically averaged* distribution of atoms in the surface layers of the growing crystal. The approximation (5.140) is valid in the limit where the size of islands of growth on the surface (or, in other words, the spatial scale describing correlations between occupancies of various lattice sites) is small (Dudarev et al., 1992b). An important advantage associated with the treatment of RHEED using the statistically averaged effective potential (5.140) is that once we know the functional form of the potential we can explicitly find RHEED intensities by solving the relevant Schrödinger equation numerically. Peng and Whelan (Peng and Whelan, 1990a, Peng and Whelan, 1991a, Peng and Whelan, 1991b, Peng and Whelan, 1991c) carried out a comprehensive study of the RHEED problem for a growing surface by investigating how the reflectivity of the surface varies as a function of occupancies of atomic layers, and how the RHEED intensity oscillates as a result of variation of these occupancies during the growth. Mitura et al. (Mitura et al., 1991) developed a similar approach and were able to achieve good qualitative agreement between the theoretically predicted and the experimentally observed temporal variations of RHEED

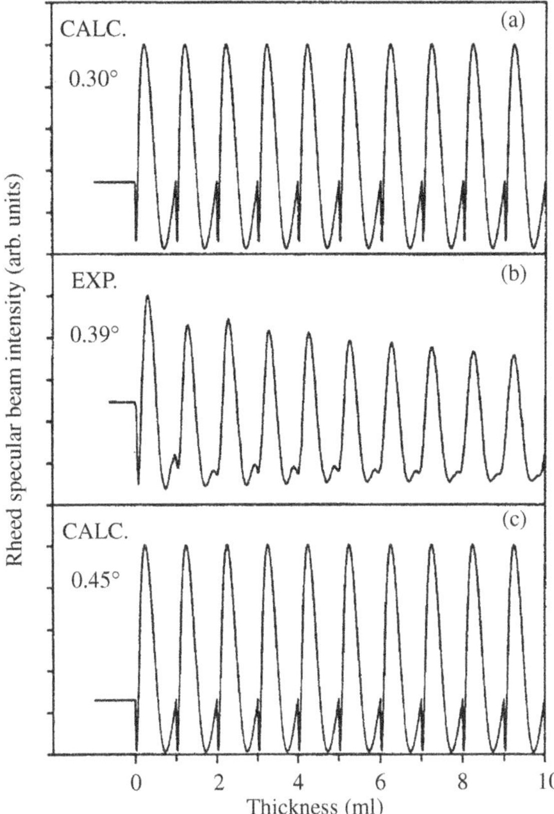

FIG. 5.29. Oscillations of the intensity of the specular RHEED reflection observed and calculated for the case of molecular beam epitaxial growth of the Pb-35%In(111) surface. From (Mitura et al., 1992).

intensities in the case of growth of a lead–indium alloy (Mitura et al., 1992), where the shape of oscillations is characterized by the presence of extra maxima shown in Fig. 5.29. Horio and Ichimiya (Horio and Ichimiya, 1994b) later found extra maxima in RHEED intensity oscillation curves observed in the case of growth of Ge on the Ge(111) surface. Calculations using the statistically averaged potential were able to reproduce the experimental observations very well. Mitsuishi et al. (Mitsuishi et al., 1995) arrived at a similar conclusion by studying a more difficult high-symmetry diffraction case where interactions between several RHEED reflections needed to be taken into account in order to explain the shape of the oscillation curve.

A somewhat indirect but very interesting confirmation of the validity of the statistical concepts underlying Peng and Whelan's treatment of RHEED oscillations comes from the work of Reginski et al. (Reginski et al., 1995) who observed oscillations for the orientation of the incident beam corresponding to the excitation of a surface resonance. We know that surface resonance scattering (described in Chapter 6) is a collective phenomenon associated with the formation of a localized quantum state in the potential of layers of atoms close to the surface. The fact that RHEED oscillations were best seen at the resonance orientation of the incident beam (see Fig. 5 of (Reginski et al., 1995)) illustrates the fundamental significance of the concept of the statistically averaged potential in the treatment of dynamical electron diffraction from a growing surface.

Probably the most direct confirmation of the validity of the dynamical diffraction model of RHEED oscillations comes from the recent work by Mitura et al. (Mitura et al., 1998, Mitura et al., 1999, Mitura et al., 2002), where the phase of RHEED intensity oscillations calculated for the GaAs(001) surface was directly compared with experimental data (Crook et al., 1989). The results of this comparison are shown in Fig. 5.30, where the phase of oscillations is plotted as a function of the polar angle of incidence of electrons on the surface. The very good agreement between the predicted and experimentally observed curves shows that the oscillatory behaviour of the intensity of RHEED reflections occurring during crystal growth is primarily associated with interference effects and dynamical diffraction of high-energy electrons.

The remaining problem that has not yet been fully resolved concerns the origin of diffuse 'Kikuchi' features often observed in RHEED patterns during MBE growth of crystals, see (Joyce et al., 1987, Larsen et al., 1987, Zhang et al., 1987) and (Dobson et al., 1988, Crook et al., 1989, Toyoshima et al., 1992). The term 'Kikuchi features' in the context of RHEED does not mean conventional Kikuchi lines and bands, as are often observed in electron diffraction experiments. In the case where RHEED oscillations are used to monitor crystal growth, scattering by the highly disordered surface layer of the crystal gives rise to strong diffuse features that are only to a certain extent related to conventional Kikuchi patterns. Here the use of the term 'Kikuchi features' reflects the fact that certain peaks observed in RHEED patterns cannot be clearly attributed to the 'ordinary' dynamical diffraction of the incident electrons.

In the case where the size of islands of growth on the surface is not very large, diffuse scattering by a disordered layer of atoms can be described by using the distorted wave Born approximation in a similar manner to that by which thermal diffuse scattering is described in the treatment of 'conventional' Kikuchi patterns. A schematic illustration of an application of the DWBA to the treatment of diffuse scattering by islands of growth on the surface is given in Fig. 5.31.

Figure 5.31 shows that if we take the solution of a dynamical diffraction problem in the statistically averaged potential (5.140) as the zero-order approximation then diffuse scattering results from fluctuations of the potential in the partially filled surface atomic layers of the crystal. A more extensive discussion of applications of DWBA to the treatment of diffuse scattering by a disordered surface is given in (Dudarev et al., 1994, Dudarev, 1997). The reader may also find it interesting to compare diffuse scattering occurring in the case of RHEED

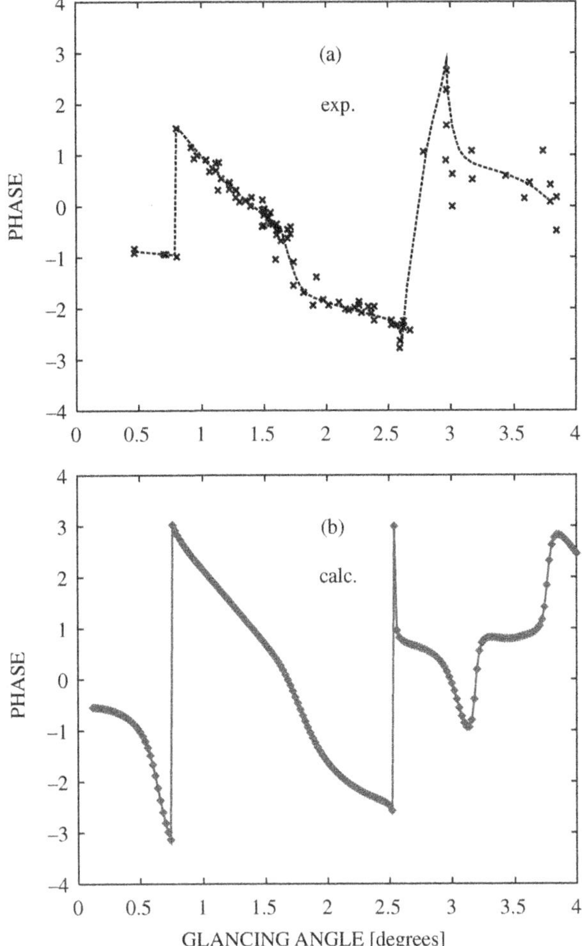

FIG. 5.30. Dependence of the phase of oscillations on the glancing angle of incidence of electrons on the surface. Experimental points are from (Crook et al., 1989). The theoretical curve was calculated using the statistically averaged potential (5.140). From (Mitura et al., 1998).

with that observed in the case of X-ray reflection from rough surfaces. In the case of X-ray diffuse scattering the observed cross-section exhibits interesting anomalies that can be readily interpreted using the DWBA (Gorodnichev et al., 1988). The advantage of using the DWBA is associated with the fact that this approximation takes into account the finite depth of penetration of electrons into the crystal. In other words, the DWBA naturally takes into account the fact that diffuse scattering is affected by dynamical diffraction of electrons in the average

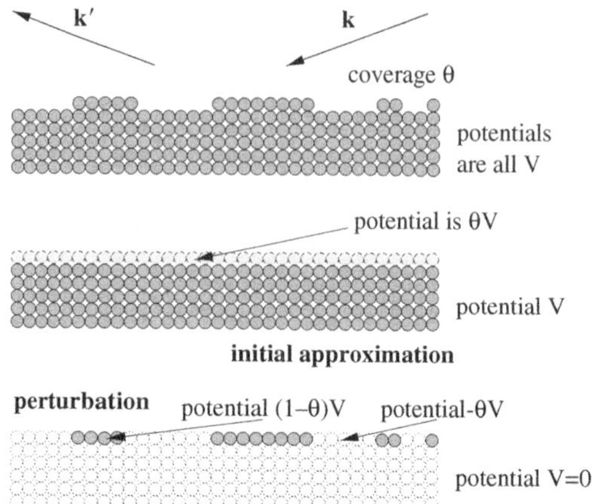

FIG. 5.31. A sketch illustrating how the distorted wave Born approximation is applied to the treatment of diffuse scattering by a statistically disordered growing surface. The perturbation part of the potential is associated with the difference between the instantaneous distribution of islands of growth on the surface and the statistically averaged potential given by equation (5.140).

potential (5.140). If we neglect dynamical diffraction effects and adopt the Kirchhoff approximation, in which the amplitude of scattering is given by a 'kinematic' sum of waves reflected by flat terraces on the surface, we will find that the distribution of diffuse intensity reflects the distribution of sizes of islands of growth on the surface (Lent and Cohen, 1986, Cohen et al., 1987). A similar approach to diffraction spot profile analysis has recently been investigated by Wollschläger et al. (Wollschläger et al., 1998).

As an alternative to a calculation based on the perturbation treatment of diffuse scattering, the cross-section can be evaluated numerically using a supercell method. In this case, a statistically disordered surface is modelled by a periodically repeated large two-dimensional supercell that in principle can be as large as the one shown in Fig. 5.28. By using a large supercell it is possible to reproduce statistical features characterizing the morphology of a real growing surface. Carrying out RHEED calculations for a large simulation cell will require taking into account a large number of diffracted beams and this should stimulate the development of new efficient numerical approaches to RHEED calculations (Korte et al., 1996, Meyer-Ehmsen, 1998, Maksym et al., 1998).

In conclusion we note an interesting possibility of combining RHEED oscillations with electron energy loss spectroscopy (Atwater et al., 1993). Electron energy loss spectra recorded and analysed in parallel with RHEED oscillations should make it possible not only to count the number of atomic layers grown on the surface but also to understand the microscopic mechanisms of growth; for example, to determine the order in which various atoms are incorporated in the crystal lattice. Modelling of electron energy loss spectra can be performed using a combination of a numerical treatment of dynamical diffraction of electrons and the DWBA. Recent publications (Horio, 1996, Horio and Hashimoto, 1997, Staib et al., 1999) show that applications of energy-filtered RHEED are becoming more common and this will undoubtedly increase the capabilities of this method for analysis of surface structures.

5.6 Summary

In this chapter we have reviewed several approaches to dynamical RHEED calculations from perfect and growing surfaces. The Bloch wave theory was discussed in particular detail, and the connection between this method and other approaches to dynamical RHEED has been illustrated using several worked examples. We have also reviewed recent applications of RHEED to the study of growing surfaces, and have provided listings of several FORTRAN routines and example input files in Appendix C. These routines were used for generating the data shown in the figures in this chapter. They may be obtained by contacting one of the authors (LMP) at plm@ele.pku.edu.cn.

6

RESONANCE EFFECTS IN TRANSMISSION AND REFLECTION HIGH-ENERGY ELECTRON DIFFRACTION

6.1 The origin of resonances

Resonance scattering of electrons is probably one of the most remarkable and fundamental phenomena known in electron diffraction. In its simplest manifestation resonance diffraction gives rise to a dramatic increase of the reflectivity of a flat crystal surface that occurs in a certain relatively narrow range of angles of incidence. A typical reflection electron diffraction pattern (of which Fig. 6.1 is an example), in addition to Bragg diffraction spots, also shows a system of intersecting *parabolic* and *semi-circular* Kikuchi lines. These lines are associated with resonance scattering of electrons. Both the shape of resonance Kikuchi lines and their brightness make them very distinct and unusual, and the reason why they appear in a diffraction pattern requires a careful and thorough examination. Understanding the nature of resonance effects also helps in interpreting a number of closely related phenomena: for example, the rapid variation of the cross-section of inelastic scattering of electrons observed in the vicinity of resonance conditions (Horio and Ichimiya, 1983, Marten and Meyer-Ehmsen, 1988, Spence and Kim, 1988, Horio, 1998).

Effects associated with resonance scattering are frequently observed in atomic, molecular, and nuclear collisions. This has stimulated the development of a broad range of theoretical approaches describing the large variety of the available experimental information. All these approaches are unified in that they treat the rapid variation of the cross-section of scattering as a phenomenon that in one way or another is associated with the formation of a *quasi-bound state* of the system in the vicinity of the resonance condition (Taylor, 1972). In the case of resonance diffraction of high-energy electrons by a crystal surface electrons are temporarily 'trapped' by a metastable localized Bloch state. This process proves to be ultimately responsible for the appearance of the unusual Kikuchi patterns seen in Fig 6.1, as well as for other anomalies in the behaviour of the cross-section of electron scattering observed in the vicinity of resonance conditions.

The term 'resonance scattering' is normally used (Feshbach, 1958) to describe the case where the amplitude of scattering contains the characteristic Breit–Wigner term

$$f(E) \sim A + B \frac{\Gamma_{el}/2}{E - E_R + i\Gamma_{tot}/2}. \tag{6.1}$$

Here A and B are constant factors and E is a variable parameter that has the dimensions of energy but in electron diffraction is often related to the angle of

186

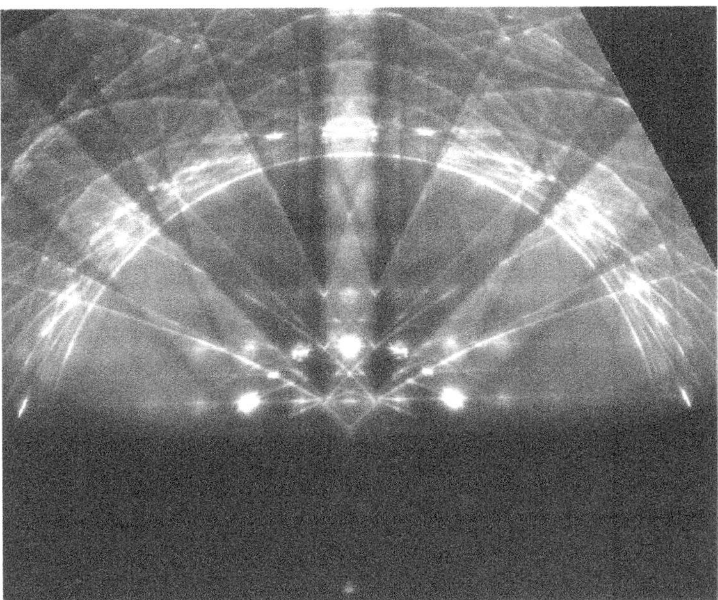

FIG. 6.1. Reflection high-energy electron diffraction pattern observed for the (110) surface of a crystal of GaAs. The direction of the incident beam is close to the [001] axis. The energy of electrons $E_0 = 120$ keV. Resonance diffraction is responsible for the formation of Kikuchi rings and parabolas seen in this pattern.

incidence. Γ_{tot} determines the 'width' of the resonance (i.e. the range of variation of the parameter E where the second term in (6.1) is dominant). We shall see that under certain conditions the reflectivity of a crystal surface follows a behaviour very similar to that described by formula (6.1).

Resonance effects also occur in transmission electron diffraction. Propagation of high-energy electrons through thin crystals may involve the trapping of electrons by metastable localized states. In the transmission diffraction case, instead of following the Breit–Wigner law (6.1), the amplitude of scattering oscillates as a function of the thickness of the crystal. Nevertheless, the physical picture of the process of resonance scattering occurring in the transmission case proves to be very similar to the reflection diffraction case. The comparison of the two cases given below shows that it is appropriate to treat resonance effects observed in transmission and reflection geometries as manifestations of essentially the same type of phenomenon.

6.2 Transmission resonance diffraction of high-energy electrons

In this section we investigate resonance effects occurring in the transmission geometry of diffraction. Reflection resonance diffraction of electrons will be studied later in this chapter. We begin with the transmission case because the treatment of this case does not require going beyond the conventional formulation of the dynamical diffraction theory. The treatment of resonance scattering of high-energy electrons from a crystal surface requires using methods that may be somewhat less familiar to the reader. Details of mathematical techniques needed for understanding RHEED resonances will be explained later in parallel with the analysis of the physical contents of the process of scattering responsible for the appearance of unusual resonance features in RHEED patterns.

6.2.1 *The geometry of transmission resonance diffraction*

Consider the propagation of a beam of high-energy electrons through a thin crystal. We assume that the thickness of the crystal does not exceed the inelastic mean free path of electrons with respect to thermal diffuse scattering and energy losses. Transmission resonance diffraction of high-energy electrons is a particular type of elastic scattering, which manifests itself only in a very narrow range of angles of incidence $\delta\theta \sim 10^{-3}$ mrad, where θ is the angle between the wave vector of incident electrons and a low-index crystallographic direction. Until recently the fact that the direction of incidence needed to be controlled with a very high precision impeded the explicit experimental observation of the effect (this difficulty has now been resolved by Rossouw et al. (Rossouw et al., 1998)).

There have been many 'indirect' observations of transmission resonance effects reported in the literature. For example, very bright Kikuchi rings and parabolas were observed in electron diffraction patterns by Emslie (Emslie, 1934) and more recently by Peng et al. (Peng et al., 1988, Peng and Gjønnes, 1989). Kikuchi patterns are known to be associated with dynamical diffraction of inelastically (or diffusely) scattered electrons. Therefore, the presence of unusually shaped (i.e. not the conventional line- or band-type) Kikuchi patterns in electron micrographs confirms the existence of some unconventional diffraction process, the nature of which we now investigate.

In the treatment of transmission high-energy electron diffraction it is normally assumed that the periodic potential influences the motion of the electron only in the case where the angle between the wave vector $\mathbf{k}_0$ of the electron and a crystallographic axis does not exceed the so-called *Lindhard angle* $\theta \leq \theta_L$ (the concept of a critical angle was introduced by Lindhard in the theory of channelling of ions in single crystals). This angle is defined by the condition that the kinetic energy of the *transverse* motion of the electron $E_\perp^{(\text{kin})} = E_0 \sin^2\theta$ is equal to the average height of barriers $|\overline{V}|$ which separate potential wells associated with the rows of atoms in the crystal

$$\theta \leq \theta_L \approx (|\overline{V}|/E_0)^{1/2}. \tag{6.2}$$

In (6.2) E_0 is the kinetic energy of the incident high-energy electron. In the quantum-mechanical treatment of the channelling problem inequality (6.2) is equivalent to the condition that in the crystal the incident plane wave rapidly undergoes a transition into a superposition of a large number of strongly excited diffracted beams (Kagan and Kononets, 1970). In the case where condition (6.2) is satisfied, the kinematic solution of the diffraction problem involves a large set of nearly divergent terms that are proportional to the Fourier components of the potential of the crystal. Each of these divergent terms corresponds to a reciprocal lattice vector $\mathbf{G}$ belonging to a row or a plane of vectors $\mathbf{G}$ that are perpendicular to the wave vector of the incident electron (Kagan and Kononets, 1970). In terms of the geometry of diffraction scattering, condition (6.2) corresponds to the configuration (a) shown in Fig. 6.2, where the Ewald sphere of radius k_0 ($\mathbf{k}_0^2 = 2mE/\hbar^2$) is tangential to a plane or to a row of reciprocal lattice vectors.

It is a simple matter to see that case (a) shown in Fig. 6.2 represents only *one of the many* possible configurations where the incident plane wave is radically transformed by dynamical diffraction effects. For example, configuration (b) shown in the same figure also gives rise to the simultaneous excitation of a large number of Bragg reflections. The number of diffracted beams excited in case (b) is comparable with the number of beams excited in case (a). By denoting the distance between successive planes of reciprocal lattice points by g, it is easy to find the angle θ_r between the direction of the wave vector of incident electrons $\mathbf{k}_0$ and the crystallographic axis

$$\theta_r = 2\sin^{-1}[(g/2k_0)^{1/2}]. \tag{6.3}$$

Does the case shown in Fig. 6.2b correspond to any appreciable probability of exciting the diffracted beams? This question can be answered in the following way. The momentum transfer corresponding to scattering through angle θ_r is given by $\hbar\delta k \sim \hbar k_0\theta_r \approx \hbar(2k_0g)^{1/2}$. Since any matrix element of the periodic potential of the crystal contains a Debye–Waller factor of the form $\exp\left[-\tfrac{1}{2}\langle(\mathbf{G}\cdot\mathbf{u})^2\rangle\right]$, where $\mathbf{G}$ is the reciprocal lattice vector and $\mathbf{u}$ is the vector of thermal displacement, all the processes corresponding to $\delta k \geq \hbar/(\langle u^2\rangle)^{1/2}$ are suppressed by the thermal motion of atomic nuclei (a detailed discussion of this point is given in (Kagan and Kononets, 1970)). A comparison of the latter inequality with eqn (6.3) leads to a condition that determines the upper limit of the interval of energies of incident electrons for which the amplitude of diffraction scattering corresponding to case (b) shown in Fig. 6.2b is appreciable

$$E_0 \leq \frac{\hbar^2}{8mg^2(\langle u^2\rangle)^2}. \tag{6.4}$$

At room temperature the right-hand side of (6.4) is of the order of 100 keV. This shows that in the non-relativistic range of energies of electrons ($E_0 \leq 100$ keV) the occurrence of an unusual type of diffraction scattering, corresponding to case (b) shown in Fig. 6.2b, is indeed possible.

We now prove that diffraction scattering corresponding to case (b) in Fig. 6.2 is characterized by the many-beam resonance coupling between the quasi-free state of motion of the electron in the direction $\mathbf{k}_0$ and a group of Bloch states tightly bound by the potential of atomic rows or atomic planes. In other words we will show that transmission resonance diffraction is capable of achieving what

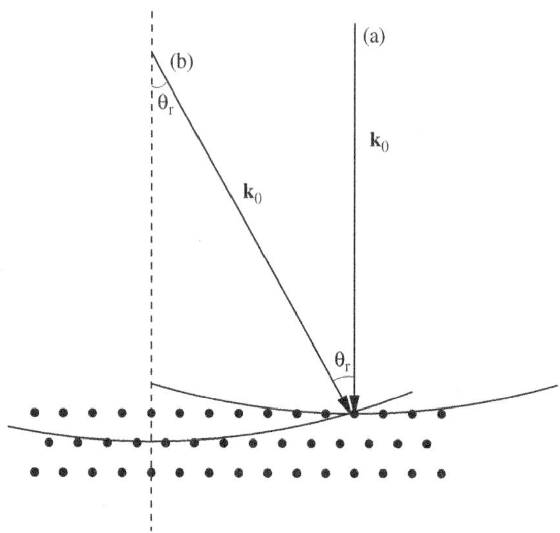

FIG. 6.2. Relative positions of the Ewald sphere and the reciprocal lattice. Case (a) corresponds to the normal incidence of electrons on the surface of the crystal. Case (b) corresponds to the orientation of the incident beam where the solution of the transmission diffraction problem exhibits resonance behaviour. From (Rossouw et al., 1998).

is beyond the capacity of conventional dynamical diffraction, namely that in the resonance mode the incident plane wave can be *fully* converted into a tightly bound Bloch state and vice versa. (Note that in the case of normal incidence of electrons on the surface of the crystal the probability of such 'conversion' is of the order of 50%.)

The theoretical investigation of the transmission resonance diffraction problem (Dudarev and Peng, 1993b, Dudarev and Peng, 1993a, Rossouw et al., 1998) was stimulated by the thorough experimental analysis of differential cross-sections of diffuse scattering of electrons through relatively large angles (Peng et al., 1988, Peng and Gjønnes, 1989, Yao and Cowley, 1989) Experimentally it was found that in addition to the conventional pattern of intersecting Kikuchi lines and bands, electron diffraction patterns also showed a number of almost perfect Kikuchi rings and parabolas. Kikuchi et al. (Kikuchi and Nakagawa, 1933) and Emslie (Emslie, 1934), who also saw the 'anomalous' patterns, tried to explain the observations by suggesting that ring and parabolic patterns result from *one-dimensional diffraction of electrons* by the rows of atoms running parallel to the direction of motion of the electron. Emslie (Emslie, 1934) also noted that inelastic scattering might play a part in the process of formation of Kikuchi rings and parabolas. However, no further argument has been developed since the earlier work by Kikuchi and Emslie to support their ideas (see also the discussion of Fig. 4.13 in Chapter 4). It still remained unclear why electron diffraction patterns showed simultaneously the usual Kikuchi lines and bands originating from the 'ordinary' three-dimensional diffraction of electrons and at the same time the same patterns showed the 'anomalous' Kikuchi patterns associated with a hypothetical 'one-dimensional' diffraction of electrons. Understanding the connection between the conventional notion of transmission diffraction of electrons and the 'one-dimensional' picture of scattering proposed by Kikuchi and Emslie requires analysing a dynamical diffraction problem, which takes into account both the many-beam diffraction of electrons and the periodicity of the lattice in a direction nearly parallel to the direction of incidence.

6.2.2 *Transmission resonance: a formal solution*

To illustrate the difference between the conventional many-beam transmission diffraction of electrons and *transmission resonance diffraction* we begin by considering the case where electrons are incident on the crystal surface at a small angle to a certain system of atomic planes. The wave function $\psi(\mathbf{r})$ describing the propagation of electrons through the crystal satisfies the Schrödinger equation

$$-\frac{\hbar^2}{2m}\nabla^2\psi(\mathbf{r}) + V(\mathbf{r})\psi(\mathbf{r}) = \frac{\hbar^2\mathbf{k}_0^2}{2m}\psi(\mathbf{r}), \qquad (6.5)$$

supplemented by the boundary condition $\psi(\mathbf{r}) = \exp(i\mathbf{k}_0 \cdot \mathbf{r})$ for $z < 0$. The crystal is assumed to have the form of a plate where one of the surfaces coincides with the plane $z = 0$ and the other corresponds to $z = t$. The potential $V(\mathbf{r})$ is assumed to be real, although this assumption is not essential for the treatment of the problem given below. The vector $\mathbf{k}_0$ makes a small angle θ with the z axis. If condition (6.2) is satisfied and the energy of electrons is sufficiently high so that

$$\hbar^2|\mathbf{k}_0|g/m \gg |\overline{V}|, \qquad (6.6)$$

we may neglect all the Fourier components of the potential except those corresponding to reciprocal lattice vectors $\mathbf{G}$ perpendicular to the direction of incidence (see Fig. 6.2a). Condition (6.6) means that the excitation of a diffracted beam corresponding to a reciprocal lattice vector, which is *not* orthogonal to $\mathbf{k}_0$, would require a very large change in the kinetic energy of the electron, and the probability of such a process taking place is very low. The solution of eqn (6.5) has the form of a superposition of transmitted and diffracted waves with slowly varying amplitudes

$$\psi(\mathbf{r}) = \sum_{\mathbf{G}} \phi_{\mathbf{G}}(z) \exp[i(\mathbf{k}_0 + \mathbf{G}) \cdot \mathbf{r}]. \qquad (6.7)$$

Substituting (6.7) into (6.5) and neglecting a small term in $d^2\phi_{\mathbf{G}}(z)/dz^2$, we obtain a system of coupled differential equations for amplitudes $\phi_{\mathbf{G}}(z)$

$$\hbar v \cos\theta \frac{d}{dz}\phi_{\mathbf{G}}(z) + i(\mathcal{E}_{\mathbf{G}} - \mathcal{E}_0)\phi_{\mathbf{G}}(z) + i\sum_{s} V_{\mathbf{GH}}\phi_{\mathbf{H}}(z) = 0, \qquad (6.8)$$

where $v = \hbar k_0/m$ is the velocity of the electron, $\mathcal{E}_{\mathbf{G}} = \hbar^2(\mathbf{k}_0 + \mathbf{G})^2/2m$, and $V_{\mathbf{GH}}$ are the Fourier components of the periodic potential of the crystal

$$V_{\mathbf{GH}} = V(\mathbf{G} - \mathbf{H}) = \frac{1}{\Omega_0} \int_{\Omega_0} d^3 r V(\mathbf{r}) \exp[-i(\mathbf{G} - \mathbf{H}) \cdot \mathbf{r}], \qquad (6.9)$$

where Ω_0 is the volume of a unit cell. The condition $E_0 \gg |\overline{V}|$ makes it possible to neglect the waves reflected from the surface of the crystal. The boundary conditions for eqn (6.8) have the form

$$\phi_{\mathbf{G}}(z = 0) = \delta_{\mathbf{G}0}. \qquad (6.10)$$

Equations (6.8) may be solved by making a linear transformation of amplitudes $\{\phi_{\mathbf{G}}(z)\}$

$$\phi_j(z) = \sum_{\mathbf{G}} \left[C^\dagger(\mathbf{q}) \right]_{j\mathbf{G}} \phi_{\mathbf{G}}(z), \qquad (6.11)$$

where the unitary matrix $C_{j\mathbf{G}}(\mathbf{q})$ satisfies the condition

$$\sum_{\mathbf{G},\mathbf{H}} \left[C^\dagger(\mathbf{q}) \right]_{j\mathbf{G}} \left[\frac{\hbar^2(\mathbf{q} + \mathbf{G})^2}{2m} \delta_{\mathbf{GH}} + V_{\mathbf{GH}} \right] C_{\mathbf{H}j'}(\mathbf{q}) = E_j(\mathbf{q})\delta_{jj'}. \qquad (6.12)$$

This matrix $C_{\mathbf{G}j}(\mathbf{q})$ depends only on the projection $\mathbf{q}$ of the wave vector of the incident electron $\mathbf{k}_0$ on the (x, y) plane. This matrix defines the set of Bloch functions of transverse motion of the electron in the two-dimensional (averaged along z) potential $V(x, y)$

$$b_j(\mathbf{q}, \mathbf{R}) = \sum_{\mathbf{G}} C_{\mathbf{G}j}(\mathbf{q}) \exp[i(\mathbf{q} + \mathbf{G}) \cdot \mathbf{R}]. \qquad (6.13)$$

Here $\mathbf{R}$ is a two-dimensional vector lying in the (x, y) plane. By using eqns (6.12) and (6.13), the solution of eqn (6.8) may be represented in the form

$$\phi_j(z) = C_{0j}^*(\mathbf{q}) \exp \left[\frac{iz}{\hbar v \cos \theta} \left(\frac{\hbar^2 \mathbf{q}^2}{2m} - E_j(\mathbf{q}) \right) \right]. \qquad (6.14)$$

Note that the value of the function $\phi_j(z)$ at $z = 0$ coincides with the well-known result of the theory of instant perturbations (Kagan and Kononets, 1970)

$$\phi_j(0) = C_{0j}^*(\mathbf{q}) = \frac{1}{A} \int_A b_j^*(\mathbf{q}, \mathbf{R}) \exp(i\mathbf{q} \cdot \mathbf{R})d^2 R, \qquad (6.15)$$

where A is the area of the crystal unit cell projected onto the (x, y) plane. This equation has a simple meaning, namely that at the moment when the electron enters the crystal the incident plane wave transforms into a linear combination of Bloch states. The amplitudes of excitation of various Bloch states are given by (6.15). By using eqns (6.7), (6.14) and (6.15), we obtain the wave function (6.5) describing the transmission dynamical diffraction of electrons

$$\psi(\mathbf{r}) = \exp[i(\mathbf{k}_0)_z z] \sum_j b_j(\mathbf{q}, \mathbf{R}) C_{0j}^*(\mathbf{q}) \exp\left[\frac{iz}{\hbar v \cos\theta}\left(\frac{\hbar^2 \mathbf{q}^2}{2m} - E_j(\mathbf{q})\right)\right],$$
(6.16)

where $(\mathbf{k}_0)_z = k_0 \cos\theta$. In particular, the probability density $|\psi(\mathbf{r})|^2$ associated with (6.16) equals (cf. formula (4.5) of (Kagan and Kononets, 1970))

$$|\psi(\mathbf{r})|^2 = \sum_{j,j'} C_{0j}^*(\mathbf{q}) C_{0j'}(\mathbf{q}) b_j(\mathbf{q}, \mathbf{R}) b_{j'}^*(\mathbf{q}, \mathbf{R}) \exp\left[-\frac{iz}{\hbar v \cos\theta}(E_j(\mathbf{q}) - E_{j'}(\mathbf{q}))\right].$$
(6.17)

Equations (6.5)–(6.17) give the solution of the problem of elastic scattering of high-energy electrons in a thin crystal for the case where the angle between the wave vector $\mathbf{k}_0$ of incident electrons and the crystallographic axis does not exceed the Lindhard angle θ_L given by (6.2).

6.2.3 *Transmission resonance: diffraction via tightly bound states*

Consider now the geometry of scattering shown in Fig. 6.2b. In the intermediate range of angles θ (where θ is the angle between the wave vector $\mathbf{k}_0$ of the incident electron and the z axis)

$$\theta_L \ll \theta \ll \theta_r,$$
(6.18)

the wave function of the electron remains almost unaffected by the interaction with the periodic potential of the crystal. In this range of angles of incidence the wave function $\psi(\mathbf{r})$ may be represented in the form of a linear combination of the incident plane wave and a large number of weakly excited diffracted beams.

A new radical change in the behaviour of the incident high-energy electron occurs when the angle of incidence θ reaches the vicinity of the resonance angle θ_r. For $\theta \approx \theta_r$ the wave function of the electron undergoes a transition associated with the excitation of diffracted beams $\mathbf{G}$, where the corresponding reciprocal lattice vectors are characterized by a non-zero projection on the z axis

$$\mathbf{G} \cdot \hat{\mathbf{z}} = g,$$
(6.19)

where $\hat{\mathbf{z}}$ is a unit vector along the z-axis. Now the motion of electrons in the periodic potential of the crystal cannot be separated into the independent longitudinal and transverse components. In eqn (6.5) it is now necessary to take into account the dependence of $V(\mathbf{r})$ on z. Fortunately this still does mean that we have to study a fully three-dimensional diffraction problem. Instead the inequality (6.6) makes it possible to introduce a simplifying approximation and in the vicinity of $\theta \approx \theta_r$ retain only three terms in the Fourier expansion of the periodic potential of the crystal

$$V(\mathbf{r}) = V^{(-1)}(\mathbf{R}) \exp(-igz) + V^{(0)}(\mathbf{R}) + V^{(1)}(\mathbf{R}) \exp(igz).$$
(6.20)

Only two terms need to be taken into account in the expansion of the wave function

$$\psi(\mathbf{r}) = \exp(ikz)\psi^{(0)}(\mathbf{R}, z) + \exp[i(k+g)z]\psi^{(1)}(\mathbf{R}, z), \qquad (6.21)$$

where $k = (\mathbf{k}_0)_z$.

The meaning of the approximation introduced by eqns (6.20) and (6.21) can be easily understood if we refer back to the kinematic treatment of electron diffraction described in Chapter 2. In the kinematic case the intensity of a diffracted beam $\mathbf{g}$ depends on how significantly the modulus of the wave vector $\mathbf{k} + \mathbf{g}$ differs from the modulus of the wave vector of the incident electrons. The closer this wave vector is to the Ewald sphere, the greater is the amplitude of that beam. In the case shown in Fig. 6.2b we take into account a large number of diffracted beams represented by reciprocal lattice points close to the Ewald sphere. Since projections of wave vectors of all these beams on the z axis is approximately equal to $k+g$, it is convenient to express their total contribution to the wave function in the form given by the second term in eqn (6.21). Coupling the first and the second terms in the expansion of the wave function (6.21) requires taking into account the first and the third terms in the expansion of the potential (6.20).

By substituting (6.21) and (6.20) into (6.5) we obtain a system of equations for functions $\psi^{(0)}(\mathbf{R}, z)$ and $\psi^{(1)}(\mathbf{R}, z)$

$$\frac{\hbar^2 k^2}{2m}\psi^{(0)}(\mathbf{R}, z) - i\frac{\hbar^2 k}{m}\frac{\partial}{\partial z}\psi^{(0)}(\mathbf{R}, z) - \frac{\hbar^2}{2m}\nabla^2_{\mathbf{R}}\psi^{(0)}(\mathbf{R}, z)$$
$$+ V^{(0)}(\mathbf{R})\psi^{(0)}(\mathbf{R}, z) + V^{(-1)}(\mathbf{R})\psi^{(1)}(\mathbf{R}, z) = E\psi^{(0)}(\mathbf{R}, z)$$
$$\frac{\hbar^2(k+g)^2}{2m}\psi^{(1)}(\mathbf{R}, z) - i\frac{\hbar^2(k+g)}{m}\frac{\partial}{\partial z}\psi^{(1)}(\mathbf{R}, z) - \frac{\hbar^2}{2m}\nabla^2_{\mathbf{R}}\psi^{(1)}(\mathbf{R}, z) \qquad (6.22)$$
$$+ V^{(0)}(\mathbf{R})\psi^{(1)}(\mathbf{R}, z) + V^{(1)}(\mathbf{R})\psi^{(0)}(\mathbf{R}, z) = E\psi^{(1)}(\mathbf{R}, z).$$

Here the function $\psi^{(0)}(\mathbf{R}, z)$ is the amplitude of the transmitted wave, which in the crystal is coupled with the set of diffracted beams described by function $\psi^{(1)}(\mathbf{R}, z)$. At the entrance surface $z = 0$ the second component of the wave function is absent and the boundary condition on (6.22) has the form

$$\psi^{(0)}(\mathbf{R}, 0) = \exp(i\mathbf{q}\cdot\mathbf{R}), \qquad \psi^{(1)}(\mathbf{R}, 0) = 0. \qquad (6.23)$$

In eqns (6.22) $V^{(0)}(\mathbf{R})$ is the component of the potential averaged in the direction parallel to the z axis. The terms that are proportional to $V^{(\pm 1)}(\mathbf{R})$ give rise to the rapidly oscillating (as a function of z) components of the wave function. Equations (6.22) contain only the terms where the rapid oscillations of the wave function and the potential cancel each other through the $V(\mathbf{r})\psi(\mathbf{r})$ term in the Schrödinger equation. It is also important to note that for angles of incidence that are significantly larger than the Lindhard angle, we may neglect the effect of the average potential $V^{(0)}(\mathbf{R})$ on the first of the two components of the wave function (6.21). Taking also into account that $k^2 + q^2 = 2mE/\hbar^2$ and $g/k \ll 1$ and using (6.22), we arrive at

$$
\begin{aligned}
i\frac{k}{m}\frac{\partial}{\partial z}\psi^{(0)}(\mathbf{R}, z) ={}& -\frac{\hbar^2 \mathbf{q}^2}{2m}\psi^{(0)}(\mathbf{R}, z) - \frac{\hbar^2}{2m}\nabla_{\mathbf{R}}^2\psi^{(0)}(\mathbf{R}, z) \\
& + V^{(-1)}(\mathbf{R})\psi^{(1)}(\mathbf{R}, z), \\
i\frac{k}{m}\frac{\partial}{\partial z}\psi^{(1)}(\mathbf{R}, z) ={}& \left[\frac{\hbar^2(k+g)^2}{2m} - \frac{\hbar^2 k^2}{2m} - \frac{\hbar^2 \mathbf{q}^2}{2m}\right]\psi^{(1)}(\mathbf{R}, z) \\
& - \frac{\hbar^2}{2m}\nabla_{\mathbf{R}}^2\psi^{(1)}(\mathbf{R}, z) + V^{(0)}(\mathbf{R})\psi^{(1)}(\mathbf{R}, z) + V^{(1)}(\mathbf{R})\psi^{(0)}(\mathbf{R}, z).
\end{aligned}
\tag{6.24}
$$

This system of coupled equations describes the transmission resonance diffraction of high-energy electrons. In what follows we will discuss solutions of this system of equations. However, before proceeding further with the formal analysis we would like to draw the attention of the reader to the simple geometrical interpretation of eqns (6.24). In eqns (6.24) it is only the term in the square brackets on the right-hand side of the second equation that depends on the orientation of the incident beam of electrons with respect to the crystal lattice. In the case where this term is large, the coupling between the two equations is weak and there is no significant reconstruction of the wave function of the electron. However, if this term is close to zero, or in other words, if the projection $\mathbf{q}_0$ of the wave vector $\mathbf{k}_0$ of the incident electron on the (x, y) plane satisfies the condition

$$
\frac{\hbar^2(k+g)^2}{2m} - \frac{\hbar^2 k^2}{2m} - \frac{\hbar^2 \mathbf{q}_0^2}{2m} = 0,
\tag{6.25}
$$

the second component of wave function (6.21), $\psi^{(1)}(\mathbf{R}, z)$, becomes comparable with its first component, $\psi^{(0)}(\mathbf{R}, z)$, and the motion of the electron is characterized by the strong dynamical coupling between the incident plane wave and the set of diffracted beams defined by condition (6.19). Equation (6.25) has a simple geometrical meaning: it is equivalent to the condition that the Ewald sphere is tangential to the second of the planes of reciprocal lattice vectors shown in Fig. 6.2 (which in conventional notation illustrated in Fig. 4.2 is the plane of the minus first higher order Laue zone, MFHOLZ).

To some extent eqns (6.24) resemble the pair of Howie–Whelan equations of dynamical diffraction theory, where the amplitudes of the transmitted and diffracted waves are coupled through one of the Fourier components of the periodic potential of the crystal. However, there is an important difference between the two systems of equations. In the first place this difference manifests itself by the presence of the second derivative of the wave function, which is retained on the right-hand side of both equations (6.24). This second-derivative term is not particularly significant as far as the first of the two equations is concerned (since the first component of the wave function, $\psi^{(0)}(\mathbf{R}, z)$, describes the propagating plane wave with slowly varying amplitude, the corresponding term equals simply the kinetic energy of motion of the electron in the (x, y) plane). However, this term plays a very important part in the second of the two equations (6.24). The combination of the two terms

$$-\frac{\hbar^2}{2m}\nabla_{\mathbf{R}}^2 + V^{(0)}(\mathbf{R})$$

is the Hamiltonian describing the two-dimensional motion of the electron in the crystal potential averaged in the direction parallel to the direction of the zone axis. The eigenfunctions of this Hamiltonian define the set of two-dimensional Bloch functions $b_j(\mathbf{q}, \mathbf{R})$ that represent a natural choice of a complete system of functions to be used in searching for the solution of (6.24). Taking this into account, we assume that the solution of (6.24) has the form

$$\psi^{(0)}(\mathbf{R}, z) = \alpha(z)\exp(i\mathbf{q}\cdot\mathbf{R})$$
$$\psi^{(1)}(\mathbf{R}, z) = \sum_{\mathbf{Q},j}\beta_{\mathbf{Q},j}(z)b_j(\mathbf{Q}, \mathbf{R}), \qquad (6.26)$$

where the summation over $\mathbf{Q}$ and j is performed over the quasi-momentum and the band index of the spectrum of Bloch states of the transverse motion of electrons. Substituting (6.26) into (6.24), we obtain a system of equations for the slowly varying amplitudes $\alpha(z)$ and $\beta_{\mathbf{Q},j}(z)$

$$i\frac{\hbar^2 k}{m}\frac{d}{dz}\alpha(z) = \sum_{\mathbf{Q},j}\beta_{\mathbf{Q},j}(z)\left[\int d\mathbf{R}\exp(-i\mathbf{q}\cdot\mathbf{R})V^{(-1)}(\mathbf{R})b_j(\mathbf{Q}, \mathbf{R})\right],$$

$$i\frac{\hbar^2 k}{m}\frac{d}{dz}\beta_{\mathbf{Q},j}(z) = [E_j(\mathbf{Q}) - D(\mathbf{q})]\beta_{\mathbf{Q},j}(z)$$
$$+ \alpha(z)\int d\mathbf{R}b_j^*(\mathbf{Q}, \mathbf{R})V^{(1)}(\mathbf{R})\exp(i\mathbf{q}\cdot\mathbf{R}), \qquad (6.27)$$

where parameter $D(\mathbf{q})$ is equal to

$$D(\mathbf{q}) = \frac{\hbar^2\mathbf{q}^2}{2m} + \frac{\hbar^2 k^2}{2m} - \frac{\hbar^2(k+g)^2}{2m}.$$

Equations (6.27) for the amplitudes $\alpha(z)$ and $\beta_{\mathbf{Q},j}(z)$ are coupled through the matrix elements of the crystal potential of the form

$$\int d\mathbf{R}b_j^*(\mathbf{Q}, \mathbf{R})V^{(1)}(\mathbf{R})\exp(i\mathbf{q}\cdot\mathbf{R}).$$

The matrix element of the potential of the crystal $V^{(1)}(\mathbf{R})$ entering this expression is a periodic function of the coordinate $\mathbf{R}$. The Bloch function $b_j^*(\mathbf{Q}, \mathbf{R})$ is a periodic function of $\mathbf{R}$ multiplied by a phase factor of the form $\exp(i\mathbf{Q}\cdot\mathbf{R})$. Taking this into account we see that the matrix element of the potential taken between a plane wave $\exp(i\mathbf{q}\cdot\mathbf{R})$ and a Bloch function $b_j^*(\mathbf{Q}, \mathbf{R})$ is nonzero only if

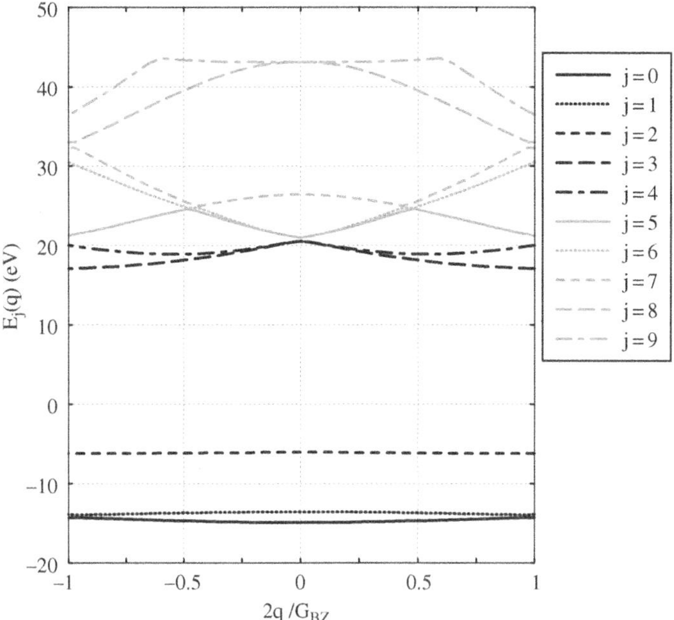

FIG. 6.3. A cross-section of the band structure of the transverse motion of high-energy electrons in the potential of atomic strings parallel to the $\langle 0001 \rangle$ axis of 1T-VSe$_2$ crystal. From (Dudarev and Peng, 1993b).

$\mathbf{q} = \mathbf{Q}$, or if the difference between the two vectors is proportional to a reciprocal lattice vector. Assuming for simplicity that it is the first of the two conditions that is satisfied for the given crystal structure and orientation of the incident beam, we obtain from (6.27)

$$i\frac{\hbar^2 k}{m}\frac{d}{dz}\alpha(z) = \sum_j \left[V^{(1)}(j) \right]^* \beta_j(z)$$

$$i\frac{\hbar^2 k}{m}\frac{d}{dz}\beta_j(z) = [E_j(\mathbf{q}) - D(\mathbf{q})]\beta_j(z) + \alpha(z)V^{(1)}(j), \qquad (6.28)$$

where $\beta_j(z) = \beta_{\mathbf{q},j}(z)$. The boundary conditions for eqns (6.28) are $\alpha(0) = 1$ and $\beta_j(0) = 0$.

Coupled equations of the form (6.28) are well known in the theory of resonance scattering (Goldberger and Watson, 1964). Solutions of these equations depend on the so-called resonance detuning parameters that in the present case are defined as

$$\delta_j(\mathbf{q}) = E_j(\mathbf{q}) - D(\mathbf{q}). \qquad (6.29)$$

These parameters depend on the angle between the wave vector of the electron and the crystallographic axis through the dependence of $D(\mathbf{q})$ on the projection $\mathbf{q}$ of the wave vector $\mathbf{k}_0$ of the incident electrons on the (x, y) plane and through the dispersion of the energies $E_j(\mathbf{q})$ of Bloch states.

Figure 6.3 gives an example illustrating how the first 10 functions $E_j(\mathbf{q})$ depend on $\mathbf{q}$ in the two-dimensional Brillouin zone of a 1T-VSe$_2$ crystal. Calculations were performed assuming that the incident high-energy electrons move nearly parallel to strings of atoms running in the $\langle 0001 \rangle$ direction. This figure shows that the spectrum of Bloch states contains three narrow bands of states (two bands at ~ -15 eV and one at around -7 eV). These three bands correspond to Bloch states of the tightly bound type, i.e. the states where the probability density is strongly peaked in the vicinity of the centre of atomic strings. The remaining curves shown in Fig. 6.3 correspond to Bloch states that describe the nearly free motion of electrons in the potential of atomic strings.

The band structure of the transverse motion of electrons shown in Fig. 6.3 is typical for problems of high-energy electron diffraction. Differences between curves $E_j(\mathbf{q})$ corresponding to different crystal structures are mainly associated with the character of dispersion of the lowest band of states $E_0(\mathbf{q})$. As a rule the higher the average charge of nuclei in the material, the weaker is the dispersion of the low-lying Bloch states. For example the lowest Bloch state corresponding to the potential of strings of atoms running parallel to the $\langle 100 \rangle$ axis in molybdenum is characterized by a dispersion that does not exceed 4% over the entire Brillouin zone. In this case the energy of the Bloch state $E_0(\mathbf{q})$ is confined within the interval -85 eV $\leq E_0(\mathbf{q}) \leq -82$ eV, and is separated from the rest of the spectrum by a gap of the order of 100 eV.

The dispersion of energies of Bloch states shown in Fig. 6.3 determines the form of solutions of eqns (6.28). To understand the connection between the functions shown in Fig. 6.3 and the way electrons move in the crystal let us consider how the parameter $D(\mathbf{q})$ varies as a function of the angle θ between the wave vector of the incident electrons $\mathbf{k}_0$ and the z axis (see Fig. 6.2). At $\mathbf{q} = 0$ (i.e. at $\theta = 0$) $D(\mathbf{q})$ entering equation (6.28) is large and negative. For example, for $E_0 = 100$ keV we find $D(0) \approx -\hbar^2 k_0 g/m \sim -2.5 \times 10^3$ eV. However, as the length of vector $\mathbf{q}$ increases, the magnitude of $D(\mathbf{q})$ rapidly increases too. As a result of this the parameters $\delta_0(\mathbf{q}), \delta_1(\mathbf{q})$, go through zero one after another in rapid succession. Analysing the solutions of (6.28) we find that the most interesting case is encountered in the vicinity of the point where $|\delta_0(\mathbf{q})| = 0$, or more precisely, in the region where $|\delta_0(\mathbf{q})| \leq |V^{(1)}(j = 0)|$. Figure 6.3 shows that in this region

$$|\delta_0(\mathbf{q})| \ll |\delta_1(\mathbf{q})|, \quad |\delta_0(\mathbf{q})| \ll |\delta_2(\mathbf{q})|, \; \dots \; . \tag{6.30}$$

These inequalities make it possible to solve (6.28) using the *isolated resonance approximation*, i.e. the approximation where we take into account only one Bloch state in the expansion (6.26). In this case we obtain two equations

$$i\frac{\hbar^2 k}{m}\frac{d}{dz}\alpha(z) = \left[V^{(1)}(0)\right]^* \beta_0(z)$$

$$i\frac{\hbar^2 k}{m}\frac{d}{dz}\beta_0(z) = [E_0(\mathbf{q}+\overline{\mathbf{g}}) - D(\mathbf{q})]\beta_0(z) + \alpha(z)V^{(1)}(0), \qquad (6.31)$$

complemented with the boundary condition $\alpha(0) = 1$ and $\beta_0(0) = 0$. The solution of (6.31) corresponding to this boundary condition is (note the similarity with the two-beam case of the conventional dynamical diffraction theory (Hirsch et al., 1977))

$$\alpha(z) = \exp\left(-i\frac{z\mathcal{E}}{2\hbar v\cos\theta}\right)\left\{\cos\left[\frac{z}{2\hbar v\cos\theta}(\mathcal{E}^2 + 4|V|^2)^{1/2}\right]\right.$$

$$\left. + \frac{i\mathcal{E}}{(\mathcal{E}^2 + 4|V|^2)^{1/2}}\sin\left[\frac{z}{2\hbar v\cos\theta}(\mathcal{E}^2 + 4|V|^2)^{1/2}\right]\right\}, \qquad (6.32)$$

$$\beta_0(z) = -\frac{2iV}{(\mathcal{E}^2 + 4|V|^2)^{1/2}}\exp\left(-i\frac{z\mathcal{E}}{2\hbar v\cos\theta}\right)\sin\left[\frac{z}{2\hbar v\cos\theta}(\mathcal{E}^2 + 4|V|^2)^{1/2}\right].$$

In these formulae $\mathcal{E} = E_0(\mathbf{q}) - D(\mathbf{q})$ and $V = V^{(1)}(0)$.

Despite the apparent similarity between this solution and the solution of the two-beam dynamical diffraction case (Hirsch et al., 1977), eqns (6.32) describe a process that is entirely different from oscillations of intensity of transmitted and diffracted beams observed in the transmission two-beam geometry of diffraction. In the transmission resonance diffraction case (6.32) the oscillating solution describes periodic variations of the probability of finding the incident electron either in the transmitted beam *or* in the tightly bound Bloch state, where the latter is localized in the potential of atomic strings running nearly parallel to the direction of the vertical axis shown in Fig. 6.2. To make this distinction clear, we consider an example where the wave function of the Bloch state is approximated by the linear combination of localized orbitals. In this case the wave function of the transmission resonance diffraction problem is given by

$$\psi(\mathbf{r}) = \alpha(z)\exp(i\mathbf{k}_0\mathbf{r}) + \beta_0(z)\exp[i(k+g)z]\sqrt{\tilde{A}}\sum_{A}\exp[i(\mathbf{q}+\overline{\mathbf{g}})\cdot\mathbf{R}_A]\Phi(\mathbf{R}-\mathbf{R}_A),$$

$$(6.33)$$

where the function $\Phi(\mathbf{R} - \mathbf{R}_A)$ describes a localized two-dimensional orbital bound in the potential of the A-th atomic string running parallel to the z axis. $\tilde{A}$ is the area corresponding to the projection of an atomic string onto the surface of the crystal. Each of the localized orbitals $\Phi(\mathbf{R})$ is normalized by condition $\int_{\tilde{A}} d^2 R|\Phi(\mathbf{R})|^2 = 1$.

Formula (6.33) describes a periodic transformation of the incident plane wave $\exp(i\mathbf{k}_0 \cdot \mathbf{r})$ into a tightly bound Bloch state characterized by the negative energy of transverse motion. Note that the phenomenon of transmission resonance diffraction is essentially a *many-beam* effect, since the expansion of the tightly bound Bloch state entering eqn (6.33) as a Fourier series in reciprocal lattice vectors in the (x, y) plane contains a large number of terms.

It is also possible to interpret the bulk resonance diffraction phenomenon as a process where the incident electron accelerates in the direction parallel to the direction of atomic strings, and where the increase of the kinetic energy of longitudinal motion is exactly matched by the decrease in the energy of transverse motion of the electron in the (x, y) plane. At the exact resonance condition (i.e. at $\mathcal{E} = 0$) the incident plane wave propagates through the crystal gradually losing its intensity until at $z_0 = \pi v \cos\theta / 2|V|$ it is completely transformed into a Bloch state localized around the atomic strings. At twice this distance, $2z_0$, the reverse transformation takes place and the plane wave re-emerges again. This is not surprising since resonance scattering is an *elastic* and, therefore, a reversible process.

Transmission resonance diffraction of electrons gives rise to the transformation of the incident plane wave into a Bloch state localized near centres of atomic strings. This leads to an increase in the cross-sections of all types of inelastic interactions that are characterized by a low-impact collision parameter of the electrons with the atoms in the crystal. For example the cross-section of the characteristic X-ray emission, thermal diffuse scattering, and characteristic energy losses all increase in the vicinity of the resonance condition. This effect has indeed been observed experimentally and we will discuss the relevant experimental findings and their interpretation later in this book.

6.3 Resonance diffraction from a crystal surface

In the previous sections of this chapter we considered resonance scattering of electrons in the transmission geometry of diffraction. We learnt that to find the wave function of the electron propagating through a crystal it was necessary to solve a system of coupled first-order differential equations. In the simplest case where only one Bloch state was retained in the expansion of the resonance wave function, finding the wave function requires solving a system of two differential equations, and an analytical treatment of the problem was possible.

We now consider a more difficult case where resonance scattering of electrons occurs in the reflection geometry of diffraction. It is a curious fact that it is in the reflection geometry where resonance scattering was first observed experimentally in 1933 and has been most often studied since then. The theoretical treatment of resonance scattering in the reflection geometry of diffraction proves to be far more difficult than the treatment of the transmission diffraction case. However, in recent years substantial progress has been made in the understanding of the phenomenon, and it is now possible to give a reasonably complete analysis of both the origin of the effect and its experimental manifestations.

Resonance effects were first observed in reflection high-energy electron diffraction by Kikuchi and Nakagawa (Kikuchi and Nakagawa, 1933) in Japan in 1933 during their pioneering experimental studies of reflection diffraction of fast electrons from surfaces at grazing incidence (see Fig. 6.4).

Following the work by Kikuchi and Nakagawa (Kikuchi and Nakagawa, 1933), Miyake et al. (Miyake et al., 1954), Kohra et al. (Kohra et al., 1962), Miyake and

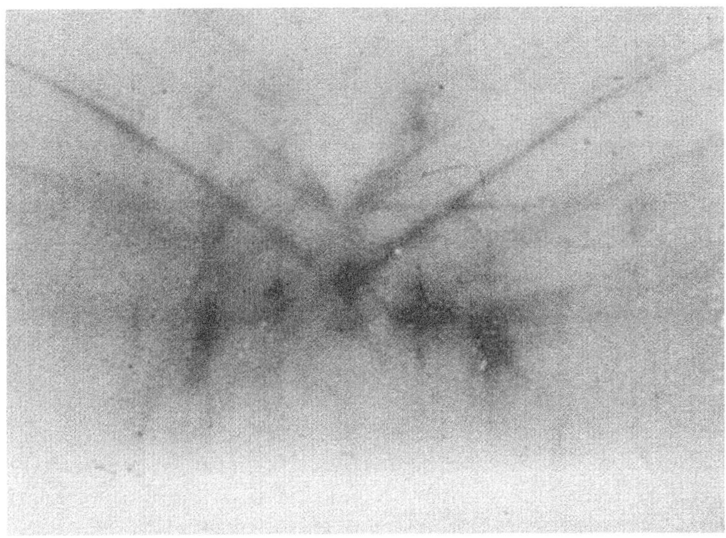

FIG. 6.4. One of the earliest experimental RHEED patterns taken by Kikuchi and Nakagawa. From (Kikuchi and Nakagawa, 1933).

Hayakawa (Miyake and Hayakawa, 1970, Hayakawa and Miyake, 1974) studied various aspects of the phenomenon and showed that resonance anomalies are associated with the geometry of scattering, where one of the diffracted beams propagates nearly parallel to the surface of the crystal. More recently, Marten and Meyer-Ehmsen (Marten and Meyer-Ehmsen, 1985) carried out a number of thorough investigations of resonance scattering of high-energy electrons using precise experimental techniques (a well written account of this work is given by Meyer-Ehmsen (Meyer-Ehmsen, 1988)). In the 1980s it was found that the resonance orientation of the incident beam is particularly suitable for reflection electron microscope (REM) observations of surface defects (Peng and Cowley, 1986, Hsu and Peng, 1986, Hsu and Peng, 1987). The work carried out by the group led by John Cowley at Arizona State University (see e.g. (Wang et al., 1987, Peng and Cowley, 1988b, Peng and Cowley, 1988d, Lu et al., 1991) and the analysis of the geometry of resonance diffraction patterns given by Maria Gajdardziska-Josifovska and John Cowley (Gajdardziska-Josifovska and Cowley, 1991)), and also contributions made by Ichimiya et al. (Ichimiya et al., 1980), Lehmpfuhl and Dowell (Lehmpfuhl and Dowell, 1986), Bleloch et al. (Bleloch et al., 1989), James et al. (James et al., 1989), Zuo and Liu (Zuo and Liu, 1992), and Horio (Horio, 1998) have provided the bulk of presently available experimental information about reflection diffraction of electrons from surfaces under resonance conditions. The interpretation of experimental data for a long time remained a debatable subject. It was unclear whether RHEED resonances were associated with scattering via surface states (McRae, 1979, Peng, 1994) or whether the nature of the phenomenon was more complex and scattering via bulk Bloch states also played an important part. Progress in the theoretical understanding of resonance scattering of high-energy electrons achieved in recent years makes it possible now to answer many of these questions with a reasonable degree of certainty.

Before proceeding towards the analysis of equations describing the resonance scattering of electrons from surfaces, let us briefly examine the qualitative similarity between the transmission and reflection resonance diffraction of electrons. Transmission resonance diffraction involves acceleration of the incident electron in the direction parallel to the axis of atomic strings. This acceleration is matched by the corresponding reduction of the energy of transverse motion of electrons in the plane perpendicular to the axis of atomic strings. Similarly, surface resonance scattering of electrons also involves their acceleration in a certain direction, but in the case of RHEED this direction lies *in the plane* of the surface. The difference in the boundary conditions between the transmission and reflection diffraction cases introduces a number of new aspects into the treatment of surface resonance scattering of electrons. For example instead of oscillating solutions, surface resonance diffraction gives rise to the appearance of the 'classical' Breit–Wigner peak in the dependence of the reflectivity of the crystal surface on the azimuthal and polar angles of incidence.

6.3.1 *The geometry of surface resonance scattering*

We start our treatment of resonance scattering of electrons from crystal surfaces from a consideration of the geometry of scattering. Assume that a high-energy electron is incident on the surface of the crystal (the plane of the surface is parallel to the plane $z = 0$ as shown in Fig. 6.5, which illustrates the geometry of resonance scattering of electrons from the (111) surface of an fcc crystal). The component of the wave vector $\mathbf{k}_0$ of the incident electron parallel to the surface is denoted by $\mathbf{k}_\parallel$ (i.e. $\mathbf{k}_0 = (\mathbf{k}_\parallel, k_z)$). Using the same notation we write the wave vector of electrons specularly reflected from the surface in the form $\mathbf{k} = (\mathbf{k}_\parallel, -k_z)$. In the absence of periodicity of the crystal potential in the (x, y) plane (i.e. assuming that $V(\mathbf{r}) = V(z)$) the (x, y) component of the wave vector is conserved, i.e. $\mathbf{k}_\parallel$ is constant. The motion of the electron in this case represents a superposition of the free motion in the direction parallel to the surface and propagation in the direction normal to the surface, which is influenced by the one-dimensional potential $V(z)$. The energy $E_\parallel^{(0)}$ associated with the motion parallel to the surface equals $\hbar^2 \mathbf{k}_\parallel^2 / 2m$ and the energy of motion towards and away from the surface equals $E_z^{(0)} = \hbar^2 k_z^2 / 2m$. Note that since for any grazing angle of incidence ζ, $k_z = k \sin \zeta$, the latter quantity $E_z^{(0)}$ always remains positive.

Now we consider what happens if we take the periodicity of the potential in the x and y directions into account. This periodicity gives rise to momentum transfers in the plane parallel to the surface. Consider, for example, elastic scattering associated with the surface reciprocal lattice vector $\mathbf{g}$ shown in Fig. 6.5. The projection of the wave vector of the *scattered* electron on the (x, y) plane is

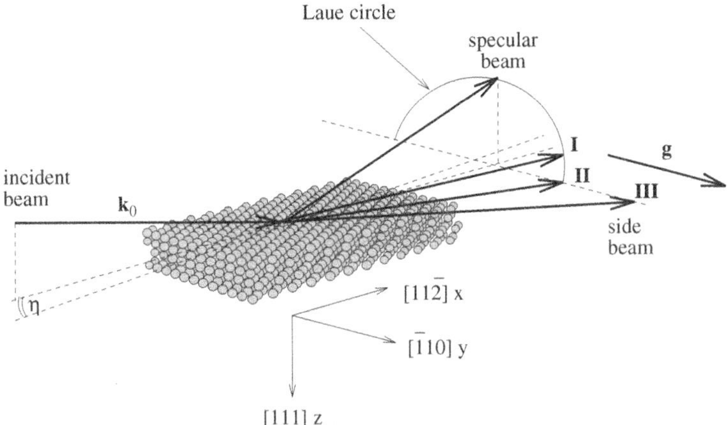

FIG. 6.5. Diagram illustrating the geometry of resonance scattering of high-energy electrons from the (111) surface of an fcc crystal. From (Dudarev and Whelan, 1996).

equal to $\mathbf{k}_\parallel + \mathbf{g}$ and the energy of motion parallel to the surface $E_\parallel^{(\mathbf{g})}$ is given by $\hbar^2(\mathbf{k}_\parallel + \mathbf{g})^2/2m$. Bearing in mind that we are considering the process of *elastic* scattering, where the total energy is conserved, we find that the energy of motion in the direction *normal* to the surface $E_z^{(\mathbf{g})}$ equals $E_\parallel^{(0)} + E_z^{(0)} - E_\parallel^{(\mathbf{g})}$. This can be written formally as $\hbar^2 K_\mathbf{g}^2/2m$, where $K_\mathbf{g}^2 = \mathbf{k}_0^2 - (\mathbf{k}_\parallel + \mathbf{g})^2$. Taking into account that there is no condition determining the sign of $E_z^{(\mathbf{g})}$, which therefore can be either positive or negative, we distinguish between three possible cases: case (I), where $K_\mathbf{g}^2 > 0$, case (II), where $K_\mathbf{g}^2 = 0$, and case (III), where $K_\mathbf{g}^2 < 0$.

Figure 6.5 shows how these three cases may be understood in terms of the geometry of scattering. Which case is realized in a particular experiment depends on the mutual orientation of the wave vector $\mathbf{k}_0$ of electrons incident on the surface and on the orientation of the reciprocal lattice vector $\mathbf{g}$. In case (I) the side beam is reflected back into the vacuum, case (II) corresponds to the side beam being parallel to the surface, and in case (III) (where $E_z^{(\mathbf{g})}$ is negative) the side beam becomes evanescent, i.e. the relevant component of the wave function attenuates exponentially *into the vacuum*. In terms of electron diffraction experimental observations, case (III) corresponds to the situation where the diffraction spot disappears below the shadow edge. It is in this case where the reflectivity of the crystal surface shows anomalous behaviour. In what follows we shall be using the term 'surface resonance scattering' to refer to peaks of intensity of the *specular* beam occurring as a function of the angle of incidence in the geometry of scattering corresponding to case (III), where $E_z^{(\mathbf{g})} < 0$.

6.3.2 *The two-rod approximation*

It is often the case that what is required for solving a particular physical problem is the development of an adequate mathematical approach to it. In the case of resonance scattering it has been known since the late 1950s (Feshbach, 1958) that the treatment of resonance scattering requires using the *integral* form of the Schrödinger equation. How could this idea be applied to the treatment of RHEED from a crystal surface?

We begin by introducing some basic notation. We assume that the surface of the crystal is parallel to the (x, y) plane and that the potential $V(\mathbf{r})$ is a periodic function of x and y. We make no assumption regarding the functional form of $V(\mathbf{r})$ considered as a function of z, and take into account the fact that in all the cases encountered in practice $V(\mathbf{r})$ becomes a periodic function of x, y, *and z* in the limit $z \to \infty$. Since any function which is periodic in the (x, y) plane may be represented by a two-dimensional Fourier series, we write

$$V(\mathbf{r}) = \sum_{\mathbf{g}} V_{\mathbf{g}}(z) \exp(i g \cdot \mathbf{R}), \qquad (6.34)$$

where $\mathbf{R} = (x, y)$. The Fourier components of the potential $V_{\mathbf{g}}(z)$ entering (6.34) vanish in the vacuum region, which corresponds to the limit $z \to -\infty$.

For a plane wave incident on the surface of the crystal $\psi_{inc}(\mathbf{r}) = \exp(i\mathbf{k}_0 \cdot \mathbf{r})$, the solution of the Schrödinger equation (6.5) also has the form of a two-dimensional Fourier series

$$\psi(\mathbf{r}) = \sum_{\mathbf{g}} \Phi_{\mathbf{g}}(z) \exp[i(\mathbf{k}_{\|} + \mathbf{g}) \cdot \mathbf{R}], \qquad (6.35)$$

where $\Phi_{\mathbf{g}}(z)$ are some as yet undetermined functions. These functions satisfy boundary conditions of the form

$$[\Phi_{\mathbf{g}}(z)]_{inc} = \begin{cases} \exp(ik_z z), & \text{when } \mathbf{g} = 0 \\ 0, & \text{when } \mathbf{g} \neq 0. \end{cases} \qquad (6.36)$$

Substituting (6.35) into the Schrödinger equation, we find that the functions $\Phi_{\mathbf{g}}(z)$ satisfy a system of coupled second-order differential equations

$$-\frac{\hbar^2}{2m}\frac{d^2}{dz^2}\Phi_{\mathbf{g}}(z) + \sum_{\mathbf{g}'} V_{\mathbf{g}-\mathbf{g}'}(z)\Phi_{\mathbf{g}'}(z) = \frac{\hbar^2 K_{\mathbf{g}}^2}{2m}\Phi_{\mathbf{g}}(z). \qquad (6.37)$$

The quantities $K_{\mathbf{g}}^2$ entering the right-hand side of this equation are given by $K_{\mathbf{g}}^2 = \mathbf{k}_0^2 - (\mathbf{k}_{\|} + \mathbf{g})^2$. The analysis of the geometry of resonance scattering given in the previous section suggests that solutions of (6.37) are likely to exhibit resonance behaviour in the region of parameters where one or several $K_{\mathbf{g}}^2$ are negative.

To find the amplitudes $R_{\mathbf{g}}$ of waves reflected from the surface of the crystal we need to investigate the asymptotic behaviour of functions $\Phi_{\mathbf{g}}(z)$ in the limit $z \to -\infty$

$$\Phi_{\mathbf{g}}(z) = \begin{cases} \exp(iK_0 z) + R_0 \exp(-iK_0 z), & \text{when } \mathbf{g} = 0 \\ R_{\mathbf{g}} \exp(-iK_{\mathbf{g}} z), & \text{when } \mathbf{g} \neq 0. \end{cases} \tag{6.38}$$

In principle it is possible to solve eqns (6.37) numerically and to find the coefficients $R_{\mathbf{g}}$. However, our aim here is not to study the numerical solutions of (6.37), but to find out whether these solutions contain the Breit–Wigner terms of the form (6.1). To achieve this we follow the treatment described by Feshbach (Feshbach, 1958) and transform (6.37) into the integral form. How do we carry out this transformation? To answer this question we need to introduce some new formalism. This formalism, which involves manipulations with wave functions and Green's functions, is in fact not hard to understand and to master. It proves to be extremely powerful and helpful in the treatment of electron diffraction and scattering.

Suppose that the problem of scattering leads to an operator equation

$$\hat{H}\psi = E\psi, \tag{6.39}$$

where Hamiltonian $\hat{H}$ is given by a sum of two terms $\hat{H} = \hat{H}_0 + \hat{V}$. Suppose that we know the solution of an equation which is similar to (6.39) and which also satisfies the appropriate boundary condition, e.g. asymptotic to an incident plane wave at infinity, but which corresponds to a different Hamiltonian $\hat{H}_0$. In other words, suppose that we know the wave function ψ_0 that satisfies the equation

$$\hat{H}_0\psi_0 = E\psi_0, \tag{6.40}$$

and is subject to the same boundary condition at infinity. We will now prove that in this case ψ satisfies the following equation:

$$\psi = \psi_0 + \frac{1}{E - \hat{H}_0 + i\delta}\hat{V}\psi. \tag{6.41}$$

The infinitely small imaginary term $i\delta$ in the denominator in (6.41) shows that we are considering an 'ordinary' retarded process of scattering, where the incident plane wave gives rise to a divergent scattered wave. The proof of the equivalence of (6.41) and (6.39) is straightforward

$$(E - \hat{H}_0)\psi = (E - \hat{H}_0)\left(\psi_0 + \frac{1}{E - \hat{H}_0 + i\delta}\hat{V}\psi\right) = 0 + \hat{V}\psi = (\hat{H} - \hat{H}_0)\psi. \tag{6.42}$$

The theory of resonance scattering is based on the observation that in some cases it is easier to solve the integral equation (6.41) than the somewhat more familiar differential Schrödinger equation (6.39). The operator $\hat{G} = (E - \hat{H}_0 + i\delta)^{-1}$ entering (6.41) is called the 'Green's function' for the Hamiltonian operator $\hat{H}_0$.

The answer to the question of whether solving eqn (6.41) is easier than solving (6.39) depends on whether or not we can find an exact expression or a reasonably good approximation for this Green's function. In the case where the operator $\hat{H}_0$ is Hermitian, and a complete and orthonormal set of eigenfunctions $\{\psi_n\}$ of this operator is known, $(E - \hat{H}_0 + i\delta)^{-1}$ may be represented in the form

$$\hat{G} = \frac{1}{E - \hat{H}_0 + i\delta} \equiv \sum_n \psi_n \frac{1}{E - E_n + i\delta} \psi_n^*, \qquad (6.43)$$

where E_n are the (real) eigenvalues of $\hat{H}_0$.

Equation (6.43) gives a straightforward recipe for calculating the Green's function. But one must remember that in electron diffraction the interaction potential is *not* Hermitian, making formula (6.43) inapplicable. The best known case, where (6.43) works well, is the case of the *free* motion of the electron where $\hat{H}_0 = -(\hbar^2/2m)\nabla^2$ *is* Hermitian. In this case $G(\mathbf{r}, \mathbf{r}', E)$ can be evaluated using formula (6.43). The normalised eigenfunctions of $\hat{H}_0 = -(\hbar^2/2m)\nabla^2$ are plane waves $\Omega^{-1/2}\exp(i\mathbf{q}\cdot\mathbf{r})$ and the eigenvalues of $-(\hbar^2/2m)\nabla^2$ are equal to $\hbar^2\mathbf{q}^2/2m$. Carrying out summation over eigenstates in (6.43), we obtain

$$
\begin{aligned}
G(\mathbf{r}, \mathbf{r}', E) &= \frac{1}{\Omega} \sum_{\mathbf{q}} \exp(i\mathbf{q} \cdot \mathbf{r}) \frac{1}{E - \hbar^2\mathbf{q}^2/2m + i\delta} \exp(-i\mathbf{q} \cdot \mathbf{r}') \\
&= \int \frac{d^3q}{(2\pi)^3} \exp(i\mathbf{q} \cdot \mathbf{r}) \frac{1}{E - \hbar^2\mathbf{q}^2/2m + i\delta} \exp(-i\mathbf{q} \cdot \mathbf{r}') \\
&= -\left(\frac{m}{2\pi\hbar^2}\right) \frac{\exp(iK|\mathbf{r} - \mathbf{r}'|)}{|\mathbf{r} - \mathbf{r}'|},
\end{aligned}
\qquad (6.44)
$$

where Ω is the normalization volume and $K = \sqrt{2mE/\hbar^2}$. In the one-dimensional case a similar calculation gives

$$
\begin{aligned}
G(z, z', E) &= \frac{1}{\Omega} \sum_q \exp(iqz) \frac{1}{E - \hbar^2 q^2/2m + i\delta} \exp(-iqz') \\
&= -i\left(\frac{m}{\hbar^2 K}\right) \exp(iK|z - z'|).
\end{aligned}
\qquad (6.45)
$$

We now apply the transformation (6.41) to the system of equations (6.37). We find

$$\Phi_0(z) = \Phi_{K_0}^{(+)}(z) + \sum_{\mathbf{g}' \neq 0} \int dz'\, G\left(z, z', \frac{\hbar^2 K_0^2}{2m}\right) V_{-\mathbf{g}'}(z')\Phi_{\mathbf{g}'}(z')$$

$$\Phi_{\mathbf{g}}(z) = \sum_{\mathbf{g}' \neq \mathbf{g}} \int dz'\, G\left(z, z', \frac{\hbar^2 K_{\mathbf{g}}^2}{2m}\right) V_{\mathbf{g}-\mathbf{g}'}(z')\Phi_{\mathbf{g}'}(z'), \quad \mathbf{g} \neq 0, \quad (6.46)$$

where the zero-order Hamiltonian $\hat{H}_0$ is chosen to have the form

$$\hat{H}_0(z) = -\frac{\hbar^2}{2m}\frac{d^2}{dz^2} + V(z). \qquad (6.47)$$

In eqn (6.46) $\Phi_{K_0}^{(+)}(z)$ is the solution of the one-dimensional problem of scattering

$$-\frac{\hbar^2}{2m}\frac{d^2}{dz^2}\Phi_{K_0}^{(+)}(z) + V(z)\Phi_{K_0}^{(+)}(z) = \frac{\hbar^2 K_0^2}{2m}\Phi_{K_0}^{(+)}(z), \qquad (6.48)$$

which corresponds to the plane wave incident from $z \to -\infty$

$$[\Phi_{K_0}^{(+)}(z)]_{inc} = \exp(iK_0 z). \qquad (6.49)$$

$G(z, z', E)$ is the one-dimensional Green's function describing the propagation of an electron between points z' and z in the potential $V(z)$

$$\left[E - \left(-\frac{\hbar^2}{2m}\frac{\partial^2}{\partial z^2} + V(z)\right)\right] G(z, z', E) = \delta(z - z'). \qquad (6.50)$$

Note that in addition to the kinetic energy, the Hamiltonian (6.47) includes the zero-order Fourier component of the potential of the crystal (i.e. potential $V(\mathbf{r})$ averaged in the plane (x, y) parallel to the surface of the crystal).

To solve eqns (6.46) we need to know the form of the kernel $G(z, z', E)$ of these equations. Although the general form of the solution of eqn (6.50) valid for an arbitrarily chosen potential $V(z)$ remains unknown, it proves to be possible to make a good guess about the behaviour of this function as a function of its third argument E. Indeed, in the limiting case $E \to \pm\infty$ the effect of the interaction potential becomes negligible, and the Green's function approaches its 'free' limit (6.45). In this limit $G(z, z', E)$ diverges as E^{-1} as energy E approaches zero. Further analysis shows that in the case of non-vanishing interaction potential $V(z)$ the Green's function exhibits similar behaviour and is maximum in the region of small values of its energy variable. Applying this to eqns (6.46) we see that on the right-hand side of these equations we should only retain terms corresponding to the smallest values of parameter $\hbar^2 K_{\mathbf{g}}^2/2m$. In the simplest case only one term needs to be retained, and the resulting system of equations represents the so-called *two-rod approximation* (Ohtsuki, 1970)

$$\Phi_0(z) = \Phi_{K_0}^{(+)}(z) + \int dz'\, G\left(z, z', \frac{\hbar^2 K_0^2}{2m}\right) V_{-\mathbf{g}}(z')\Phi_{\mathbf{g}}(z')$$

$$\Phi_{\mathbf{g}}(z) = \int dz'\, G\left(z, z', \frac{\hbar^2 K_{\mathbf{g}}^2}{2m}\right) V_{\mathbf{g}}(z')\Phi_0(z'), \qquad (6.51)$$

This approximation describes the case where only a single $\mathbf{g}$ vector parallel to the surface of the crystal is taken into account in the Fourier expansion of the wave function (6.35), and in this respect the process of scattering described by the system of two equations (6.51) corresponds *exactly* to the geometry of scattering shown in Fig. 6.5.

Equations (6.51) provide the basis for the analytical treatment of the phenomenon of resonance scattering in RHEED. In some sense these two equations are analogous to the two-beam approximation of the dynamical theory of diffraction of electrons in the transmission geometry of diffraction (these two approximations are not equivalent though, since eqns (6.51) describe *many-beam* effects). Both approximations, being rather qualitative than quantitative in their nature, still prove to be very valuable in terms of understanding important details of the process of dynamical diffraction.

In what follows we investigate solutions of eqns (6.51). First we consider resonance diffraction of electrons via a surface state. Subsequently we analyse a more general case, where diffraction involves virtual trapping and release of the electron by localized bulk states. We show that in a typical electron diffraction experiment both types of resonance scattering contribute to the observed variation of the intensity of diffraction beams considered as a function of the azimuthal and polar angles of incidence.

6.3.3 *Resonance scattering via a surface state*

Here we investigate a model of resonance scattering that describes virtual trapping of the incident high-energy electron by a surface state, see (McRae, 1979) and (Meyer-Ehmsen, 1988). Although this does not make it possible to describe diffraction intensities on the *quantitative* level, the simplicity of the model makes it a good introduction to the basic phenomena associated with resonance scattering.

Our aim here is to simplify eqns (6.51) as much as possible in order to isolate and retain terms describing resonance effects. Having this in mind we consider the case where the grazing angle of incidence (see Fig. 6.5) is sufficiently large and the energy of motion of the incident electron in the direction *normal to the surface* $\hbar^2 K_0^2/2m$ is many times the potential $V(z)$. In this case we may neglect the effects of one-dimensional reflection of electrons by $V(z)$ in eqn (6.48) and choose the wave function $\Phi_{K_0}^{(+)}(z)$ in the form of a plane wave $\Phi_{K_0}^{(+)}(z) = \exp(iK_0 z)$. The expression for the Green's function corresponding to the same approximation follows from (6.45), namely $G(z, z', \hbar^2 K_0^2/2m) = -i\left(m/\hbar^2 K_0\right)\exp(iK_0|z - z'|)$. We also assume that the spectrum of states associated with the potential $V(z)$ contains a surface state, the wave function of which is localized near the surface of the crystal, and that the energy of this surface state ϵ_s is well separated from the energy of all other states of the spectrum.

We assume that the argument $\hbar^2 K_{\mathbf{g}}^2/2m$ of the Green's function in the second of the two equations (6.51) is close to ϵ_s. In this case $G\left(z, z', \hbar^2 K_{\mathbf{g}}^2/2m\right)$ may be represented in the form

$$G\left(z, z', \hbar^2 K_{\mathbf{g}}^2/2m\right) = \frac{\psi_s(z)\psi_s^*(z')}{\hbar^2 K_{\mathbf{g}}^2/2m - \epsilon_s + i\Gamma_{in}/2} + \overline{G}\left(z, z', \hbar^2 K_{\mathbf{g}}^2/2m\right), \quad (6.52)$$

where Γ_{in} denotes the inelastic 'width' of the surface state and $\psi_s(z)$ is the wave function of this state. The wave function of the surface state is assumed to be

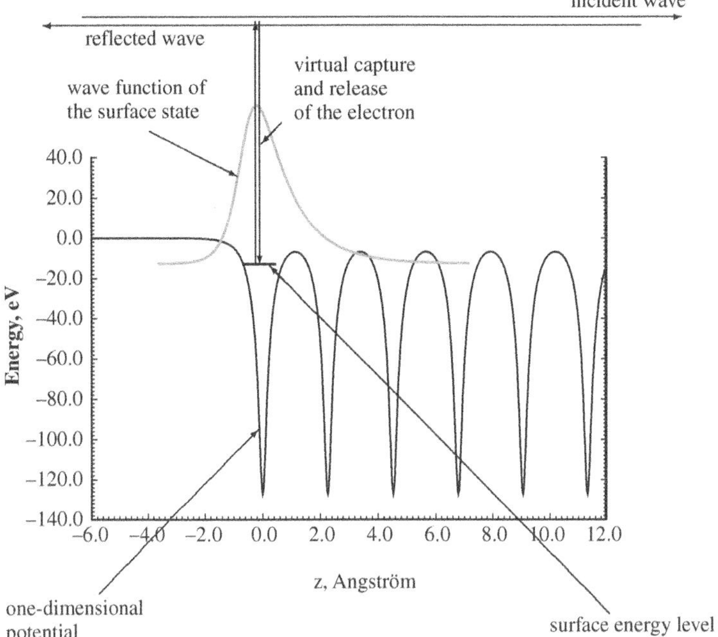

FIG. 6.6. A schematic diagram illustrating the process of resonance scattering via a surface state. The vertical axis shows the energy of motion normal to the surface, and vertical arrows illustrate coupling between the two equations in (6.51).

localized near the surface, i.e. $\lim_{\to\pm\infty}\psi_s(z) = 0$ and $\int dz|\psi_s(z)|^2 = 1$.

The second term $\overline{G}$ on the right-hand side of (6.52) is analytic in the vicinity of the pole $E = \epsilon_s - i\Gamma_{in}/2$ and the contribution of this term to the total Green's function in the vicinity of $E = \epsilon_s - i\Gamma_{in}/2$ is negligible. Using (6.52) we transform eqns (6.51) and obtain

$$\Phi_0(z) = \exp(iK_0 z) - \frac{im}{\hbar^2 K_0}\int dz'\,\exp(iK_0|z - z'|)V_{-\mathbf{g}}(z')\Phi_{\mathbf{g}}(z')$$

$$\Phi_{\mathbf{g}}(z) = \frac{\psi_s(z)}{\hbar^2 K_{\mathbf{g}}^2/2m - \epsilon_s + i\Gamma_{in}/2}\int dz'\,\psi_s^*(z')V_{\mathbf{g}}(z')\Phi_0(z'). \qquad (6.53)$$

Substituting the first equation into the second one, we arrive at a closed equation for $\Phi_{\mathbf{g}}(z)$

$$\Phi_{\mathbf{g}}(z) = \frac{\psi_s(z)}{\hbar^2 K_{\mathbf{g}}^2/2m - \epsilon_s + i\Gamma_{in}/2}\int dz'\,\psi_s^*(z')V_{\mathbf{g}}(z') \qquad (6.54)$$

$$\times\left[\exp(iK_0 z') - \frac{im}{\hbar^2 K_0}\int dz''\,\exp(iK_0|z' - z''|)V_{-\mathbf{g}}(z'')\Phi_{\mathbf{g}}(z'')\right].$$

This equation shows that any function $\Phi_{\mathbf{g}}(z)$ satisfying (6.53) *must be proportional to the wave function of the surface state* $\psi_s(z)$, i.e. $\Phi_{\mathbf{g}}(z) = \alpha\psi_s(z)$, where α is a (still unknown) constant factor. This is where we see exactly what is occurring when the incident electron is reflected from the surface. $\Phi_{\mathbf{g}}(z)$ is the component of the wave function which describes the side beam of electrons shown in Fig. 6.5. What we see here is that the electron which now moves parallel to the surface in the direction $\mathbf{k}_{\parallel} + \mathbf{g}$ is trapped by the surface state $\psi_s(z)$, which restricts the range of motion of electron in the direction *normal* to the surface. In other words a 'classical' picture of resonance scattering is emerging where an electron interacting with the crystal undergoes virtual trapping and release by a state, the wave function of which is localized near the surface.

Substituting $\Phi_{\mathbf{g}}(z) = \alpha\psi_s(z)$ into (6.54) we find a linear equation for parameter α

$$\alpha = \frac{1}{\hbar^2 K_{\mathbf{g}}^2/2m - \epsilon_s + i\Gamma_{in}/2} \int dz'\ \psi_s^*(z')V_{\mathbf{g}}(z') \tag{6.55}$$

$$\times \left[\exp(iK_0 z') - \frac{im}{\hbar^2 K_0}\alpha \int dz''\ \exp(iK_0|z' - z''|)V_{-\mathbf{g}}(z'')\psi_s(z'')\right],$$

the solution of which is

$$\alpha = \frac{\int dz\ \psi_s^*(z)V_{\mathbf{g}}(z)\exp(iK_0 z)}{\dfrac{\hbar^2 K_{\mathbf{g}}^2}{2m} - \epsilon_s + \dfrac{i}{2}(\Gamma_{in} + \Gamma_{el})}, \tag{6.56}$$

where $\Gamma_{el} = (2m/\hbar^2 K_0)\int\int dz dz'\ \psi_s^*(z)V_{\mathbf{g}}(z)\exp(iK_0|z - z'|)V_{-\mathbf{g}}(z')\psi_s(z')$ is the elastic width of the resonance. Substituting (6.56) into (6.53) we can calculate $\Phi_0(z)$ at any point z, namely

$$\Phi_0(z) = \exp(iK_0 z) - \frac{im}{\hbar^2 K_0}\alpha \int dz'\ \exp(iK_0|z - z'|)V_{-\mathbf{g}}(z')\psi_s(z'). \tag{6.57}$$

Analysing the asymptotic behaviour of this function in the limit $z \to -\infty$ we find the coefficient of resonance reflection of electrons from the surface

$$R_0^{(res)} = -\left(\frac{im}{\hbar^2 K_0}\right)\frac{\displaystyle\int dz\ \psi_s^*(z)V_{\mathbf{g}}(z)\exp(iK_0 z)\int dz'\ \exp(iK_0 z')V_{-\mathbf{g}}(z')\psi_s(z')}{\left[\dfrac{\hbar^2 K_{\mathbf{g}}^2}{2m} - \epsilon_s + \dfrac{i}{2}(\Gamma_{in} + \Gamma_{el})\right]}. \tag{6.58}$$

This formula can be simplified even further if we assume that the potential of atomic layers is short ranged, i.e. that $V_{\pm\mathbf{g}}(z) = V_{\pm\mathbf{g}}\delta(z)$. In this case from (6.58) we find

$$R_0^{(res)} = -\left(\frac{im}{\hbar^2 K_0}\right)\frac{V_{\mathbf{g}}V_{-\mathbf{g}}|\psi_s(0)|^2}{\left[\dfrac{\hbar^2 K_{\mathbf{g}}^2}{2m} - \epsilon_s + i\dfrac{\Gamma_{in}}{2} + \dfrac{im}{\hbar^2 K_0}V_{\mathbf{g}}V_{-\mathbf{g}}|\psi_s(0)|^2\right]}. \tag{6.59}$$

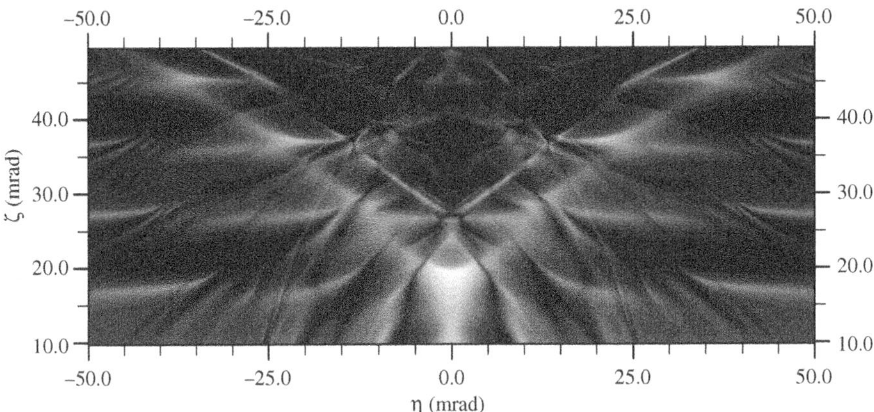

FIG. 6.7. A calculated RHEED pattern showing the intensity of specular beam of electrons reflected from the Pt(111) surface plotted as a function of the polar ζ and azimuthal η angles of incidence. The direction of incidence is close to the $\langle 11\bar{2}\rangle$ axis. Note two double resonance parabolas emerging from the centre of the pattern and ending in the upper right and upper left corners, and also the similarity between this pattern and the RHEED intensity distribution observed experimentally in (Smith et al., 1992) and reproduced in Fig. 6.8. The pattern shown in this figure was calculated numerically using five functions $\Phi_{\mathbf{g}}(z)$ in eqn (6.37). The energy of the incident electrons $E_0{=}100$ keV. From (Dudarev and Whelan, 1997).

Neglecting the imaginary parts of the Fourier components of the potential $V_{\mathbf{g}}$ and $V_{-\mathbf{g}}$, we arrive at

$$R_0^{(res)} = -\frac{i}{2}\frac{\Gamma_{el}}{\dfrac{\hbar^2 K_{\mathbf{g}}^2}{2m} - \epsilon_s + i\dfrac{\Gamma_{tot}}{2}}, \tag{6.60}$$

where $\Gamma_{tot} = \Gamma_{el} + \Gamma_{in}$. The intensity of the specular diffraction beam is given by

$$\left|R_0^{(res)}\right|^2 = \frac{\Gamma_{el}^2/4}{\left(\dfrac{\hbar^2 K_{\mathbf{g}}^2}{2m} - \epsilon_s\right)^2 + \Gamma_{tot}^2/4}. \tag{6.61}$$

Formula (6.61) shows that the intensity of the specular beam of electrons elastically scattered from the surface follows the Breit–Wigner resonance law, and in this respect resonance scattering of high-energy electrons from a crystal surface is similar to resonance phenomena observed in nuclear and atomic physics (Goldberger and Watson, 1964).

FIG. 6.8. Experimental large-angle convergent beam RHEED pattern from the Pt(111) surface. The direction of incidence is close to the $\langle 11\bar{2}\rangle$ axis. The energy of the incident electrons E_0=100 keV. From (Smith et al., 1992).

The 'elastic' width Γ_{el} of the resonance curve (6.61) depends on the polar grazing angle of incidence ζ via $\Gamma_{el} \sim K_0^{-1}$, where $K_0 = k\sin\zeta$, and is independent of the azimuthal angle of incidence η. At the same time, formula (6.61) shows that the intensity of the specular beam depends on the angles of incidence via the parameter $\hbar^2 K_{\mathbf{g}}^2/2m$, which is a function of both ζ and η. In the Cartesian system of coordinates, where the z axis is parallel to the inner normal to the surface, the wave vector of incident electrons has components $\mathbf{k}_0 = (k\cos\zeta\cos\eta, k\cos\zeta\sin\eta, k\sin\zeta)$ and the parameter $K_{\mathbf{g}}^2$ is given by

$$K_{\mathbf{g}}^2 = k^2[\sin^2\zeta - 2(g/k)\cos\zeta\sin\eta - (g/k)^2]. \qquad (6.62)$$

This equation shows that by varying η we can vary the difference $\hbar^2 K_{\mathbf{g}}^2/2m - \epsilon_s$ in the denominator of (6.61) without influencing the magnitude of the elastic width Γ_{el} of the resonance. Formula (6.61) shows that the intensity of the specular beam reaches its maximum value

$$\left| R_0^{(res)} \right|^2_{max} = \frac{\Gamma_{el}^2}{(\Gamma_{el} + \Gamma_{in})^2} \qquad (6.63)$$

for the direction of incidence where the energy $\hbar^2 K_{\mathbf{g}}^2/2m$ of the side beam (see Fig. 6.5) equals ϵ_s. Comparing this with (6.62) we see that in the (ζ, η) plane the maximum of the surface reflectivity follows a curve, the shape of which is close to the parabola $\zeta^2 - 2(g/k)\eta - (g/k)^2 = 2m\epsilon_s/\hbar^2$. The resonance 'width' of the parabola increases with decreasing ζ following $\Gamma_{el} \sim (\sin\zeta)^{-1}$. Both the parabolic shape of the resonance curve and the dependence of its width on the polar angle ζ agree well with the result of direct numerical simulation of RHEED patterns shown in Fig. 6.7.

Note also that since ϵ_s equals the energy of a *localized* state, the surface reflectivity is maximum at a point where $K_{\mathbf{g}}^2$ is negative. Equations (6.61) and (6.63) also show that the intensity of the specular beam always remains finite and that the total width of the resonance peak is equal to the sum of elastic and inelastic widths Γ_{el} and Γ_{in}.

6.3.4 *Resonance diffraction via localized bulk states*

In the previous section we considered a model where the electron incident on the surface of the crystal underwent virtual trapping and release by a bound state, the wave function of which was localized near the surface of the crystal. The model also assumed that the energy of this surface state ϵ_s was well separated from the energies of all other states.

This model is realistic (McRae, 1979, Meyer-Ehmsen, 1988) and has the important advantage of being exactly solvable. At the same time the restrictions imposed by the assumptions on which the model is based look somewhat too severe. To clarify this point we compare the 'surface state' model of resonance scattering with the treatment of *transmission* resonance diffraction (6.21)–(6.33). The only really vital assumption associated with the treatment of the transmission resonance case is associated with the notion of states tightly bound by the potential of atomic strings. Experimental results reported by James et al. (James et al., 1989) show that states tightly bound by the potential of atomic strings (or atomic planes) exist in both transmission *and* reflection geometry of diffraction. This suggests that, in reflection diffraction geometry too, electrons may undergo resonance scattering associated with their *collective* trapping by tightly bound states localized in the potential of atomic strings or atomic planes.

The boundary conditions that we encounter in the problem of RHEED are very different from, and more difficult to deal with than, the boundary conditions associated with the transmission diffraction case. Nevertheless, the problem proves to be solvable (Dudarev and Whelan, 1994b, Dudarev and Whelan, 1994a, Dudarev and Whelan, 1995, Derlet and Smith, 1997, Derlet and Smith, 1999). Below we briefly outline the approach by which its solution can be obtained.

Figure 6.9 illustrates how the incident electron interacts with states localized in the potential wells corresponding to atomic planes parallel to the surface of the crystal. A comparison of Figs 6.9 and 6.6 shows the difference between the *collective* mode of resonance scattering involving many tightly bound bulk states and resonance scattering via a surface state.

As before, the treatment of the problem is based on the two-rod approximation (6.51). However, the Green's function now has the form of a *sum* of

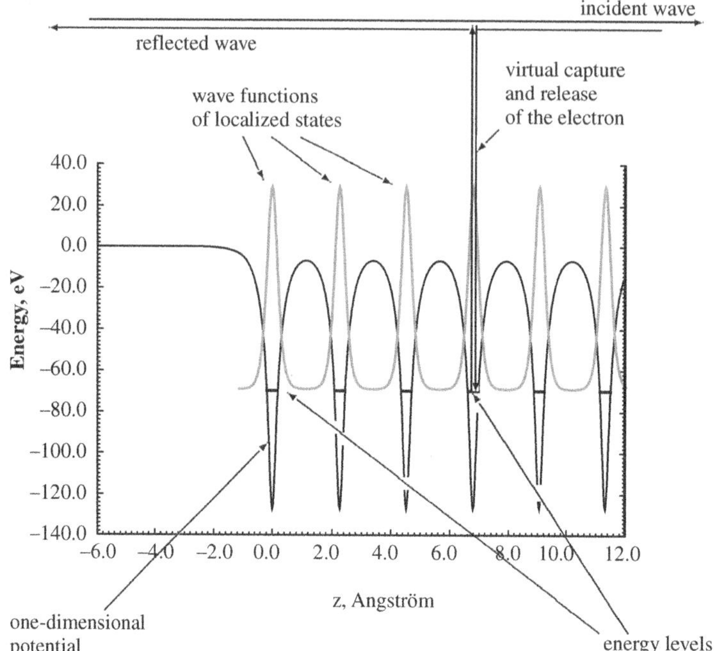

FIG. 6.9. A schematic diagram illustrating the process of resonance scattering involving *collective* trapping of the incident electron by bulk electronic states localized in the one-dimensional potential wells corresponding to atomic planes. The vertical axis corresponds to the energy of motion of the electron in the direction normal to the surface, and the arrows indicate the coupling between the two equations in (6.51).

contributions of states localized in the semi-infinite set of potential wells shown in Fig. 6.6

$$G\left(z, z', \frac{\hbar^2 K_{\mathbf{g}}^2}{2m}\right) = \sum_n \frac{\phi_n(z)\phi_n(z')}{\hbar^2 K_{\mathbf{g}}^2/2m - \epsilon_0 + i\Gamma_{in}/2} + \overline{G}\left(z, z', \frac{\hbar^2 K_{\mathbf{g}}^2}{2m}\right), \quad (6.64)$$

where $\overline{G}(z, z', E)$ is analytic in the vicinity of the point $E = \epsilon_0 - i\Gamma_{in}/2$. The summation in (6.64) is performed over potential wells centred at $z_n = nd$. Each $\phi_n(z)$ represents the wave function of the ground state for the relevant well, and is assumed to be real.

Substituting (6.64) into (6.51) and using the same approximations as those used in (6.53), we find

$$\Phi_0(z) = \exp(iK_0 z) - \frac{im}{\hbar^2 K_0} \int dz' \, \exp(iK_0|z - z'|)V_{-\mathbf{g}}(z')\Phi_{\mathbf{g}}(z'), \quad (6.65)$$

$$\Phi_{\mathbf{g}}(z) = \frac{1}{\hbar^2 K_{\mathbf{g}}^2/2m - \epsilon_0 + i\Gamma_{in}/2} \sum_n \phi_n(z) \int dz' \, \phi_n(z')V_{\mathbf{g}}(z')\Phi_0(z').$$

The second of these equations shows that $\Phi_{\mathbf{g}}(z)$ has the form of a linear combination of localized orbitals

$$\Phi_{\mathbf{g}}(z) = \sum_n \alpha_n \phi_n(z), \tag{6.66}$$

with some still unknown coefficients α_n. Comparing this with (6.54) and (6.55) we see that instead of a linear equation (6.55) we now arrive at a system of linear equations for the coefficients α_n

$$\alpha_n = \Lambda \left(\frac{\hbar^2 K_{\mathbf{g}}^2}{2m} - \epsilon_0 - i\frac{m}{\hbar^2 K_0}\Lambda^2 + i\frac{\Gamma_{in} + \Gamma_{el}}{2} \right)^{-1}$$

$$\times \left[\exp(iK_0 nd) - i\frac{m}{\hbar^2 K_0}\Lambda \sum_{l=0}^{\infty} \exp(iK_0 d|n - l|)\alpha_l \right], \tag{6.67}$$

where Λ and Γ_{el} are given by

$$\Lambda = \int_{-\infty}^{\infty} dz' \phi_n(z')V_{\mathbf{g}}(z') \exp[iK_0(z' - z_n)] \tag{6.68}$$

and

$$\Gamma_{el} = \frac{2m}{\hbar^2 K_0} \int_{-\infty}^{\infty} \int_{-\infty}^{\infty} dz'dz'' \phi_n(z')V_{\mathbf{g}}(z') \exp(iK_0|z' - z''|)V_{-\mathbf{g}}(z'')\phi_n(z''). \tag{6.69}$$

The crucial step in obtaining the solution of the problem consists in finding a way of solving the system of algebraic equations (6.67). It turns out that a simple technique based on mapping (6.67) onto a one-dimensional tight-binding Hamiltonian makes it possible to find the entire series of coefficients α_n and to calculate the resonance contribution to the reflectivity of the surface.

To achieve this we consider an auxiliary Schrödinger equation of the form

$$-\frac{\hbar^2}{2m}\frac{d^2}{dz^2}\psi(z) + U\sum_{l=0}^{\infty} \delta(z - ld)\psi(z) = \frac{\hbar^2 K_0^2}{2m}\psi(z). \tag{6.70}$$

At first glance it may seem that eqns (6.70) and (6.67) have very little in common. But it is a simple matter to show that in fact these two equations are closely related. The *integral* form of (6.70) corresponding to the incident plane wave of the form $[\psi(z)]_{inc} = J\exp(iK_0 z)$ is

$$\psi(z) = J\exp(iK_0 z) - i\frac{m}{\hbar^2 K_0}U \int dz' \exp(iK_0|z - z'|)\sum_{l=0}^{\infty}\delta(z' - ld)\psi(z'), \quad (6.71)$$

and this clearly resembles (6.67). Now suppose that we are interested in finding $\psi(z)$ at one of the points $z = nd$. From (6.71) we obtain

$$\psi(nd) = J\exp(iK_0 nd) - i\frac{m}{\hbar^2 K_0}U\sum_{l=0}^{\infty}\exp(iK_0 d|n - l|)\psi(ld). \quad (6.72)$$

A simple comparison of (6.72) with (6.67) shows that the appropriate choice of parameters J and U, i.e.

$$J = \Lambda\left[(\hbar^2 K_{\mathbf{g}}^2/2m) - \epsilon_0 - i(m/\hbar^2 K_0)\Lambda^2 + i(\Gamma_{in} + \Gamma_{el})/2\right]^{-1}$$

and $U = J\Lambda$, makes these two equations *identical*. The recognition of this fact solves the problem. Since eqn (6.70) is solvable, see (Ignatovich, 1986) and (Dudarev and Whelan, 1994a, Dudarev and Whelan, 1995), so is eqn (6.67). We can now find the amplitude of the coefficient of resonance reflection of electrons from the surface

$$R_0^{(res)} = \exp(-iK_0 d)\frac{\left[\dfrac{(r + \exp(-iK_0 d))^2 - t^2}{(r - \exp(-iK_0 d))^2 - t^2}\right]^{1/2} - 1}{\left[\dfrac{(r + \exp(-iK_0 d))^2 - t^2}{(r - \exp(-iK_0 d))^2 - t^2}\right]^{1/2} + 1}, \quad (6.73)$$

where

$$r = -\frac{i(mU/\hbar^2 K_0)}{1 + i(mU/\hbar^2 K_0)} \quad \text{and} \quad t = \frac{1}{1 + i(mU/\hbar^2 K_0)}. \quad (6.74)$$

Formula (6.73) gives the coefficient of resonance reflection of electrons from a crystal surface for the case where the process of scattering involves *collective* trapping and release of the incident electron by an ensemble of states localized in the potentials of atomic planes parallel to the surface of the crystal. It is not surprising therefore that the expression for $R_0^{(res)}$ does not follow the Breit–Wigner law (6.60). The Breit–Wigner formula (6.1) is valid in the case where only a single bound state (e.g. a surface state) is involved in the process of resonance scattering. Interference between waves reflected by atomic planes parallel to the crystal surface affects the shape of the resonance peak and gives rise to deviations from the Breit–Wigner law (6.1).

Figure 6.10 shows the dependence of the resonance part $R_0^{(res)}$ of surface reflectivity on the azimuthal angle η calculated using the analytical formula (6.73) for the (111) surface of platinum. The value of the grazing angle of incidence ζ is

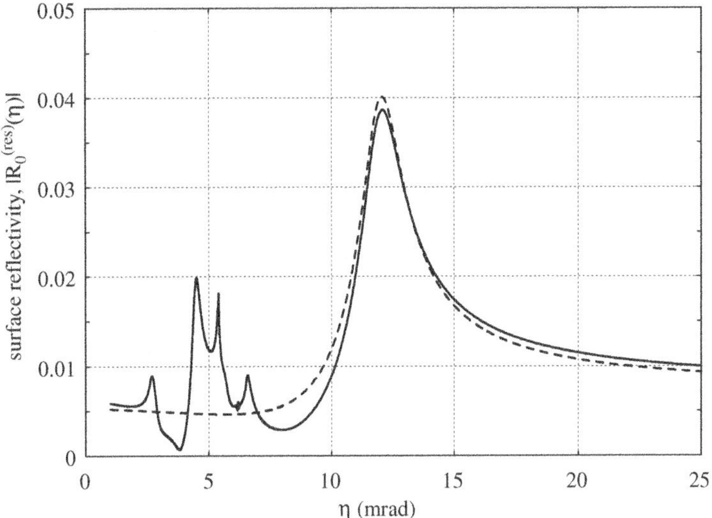

FIG. 6.10. The reflectivity of the Pt(111) surface plotted as a function of the azimuthal angle η for the glancing angle $\zeta = 58.6$ mrad and the energy of incident electrons $E_0 = 100$ keV. The solid curve was obtained by the numerical integration of eqns (6.37). The dashed curve was calculated using analytical expression (6.73). $\eta = 0$ corresponds to the $[11\bar{2}]$ azimuth in the (111) plane.

chosen in such a way as to minimize the contribution of non-resonance reflection of electrons from the surface (i.e. the contribution of residual reflectivity that does not vanish in the limit $V_{\mathbf{g}}(z) \to 0$). The curve calculated analytically is compared with the result of direct numerical integration of eqns (6.37). The two curves agree well, and this confirms the validity of our approximations and also the correctness of our understanding of the nature of resonance scattering.

6.3.5 *Interference between resonance and potential scattering*

The analysis given in the previous two sections shows that the term 'surface resonance' is directly applicable only to one particular model of diffraction involving trapping and release of the incident electron by a surface state. It is only in this case where the dependence of surface reflectivity (6.60) on the angle of incidence follows the Breit–Wigner resonance law (6.1). If more than one localized state is involved in the virtual trapping of electrons (as in Fig. 6.9 and in formula (6.73)), the form of the 'resonance' feature becomes more complex and deviations from the Breit–Wigner law become more pronounced.

In addition to effects of 'pure' resonance diffraction that we have analysed using the models described above, in practice there is one more contribution to the amplitude of the wave reflected from the surface, and it is often important to take this contribution into account. In the original Breit–Wigner formula (6.1)

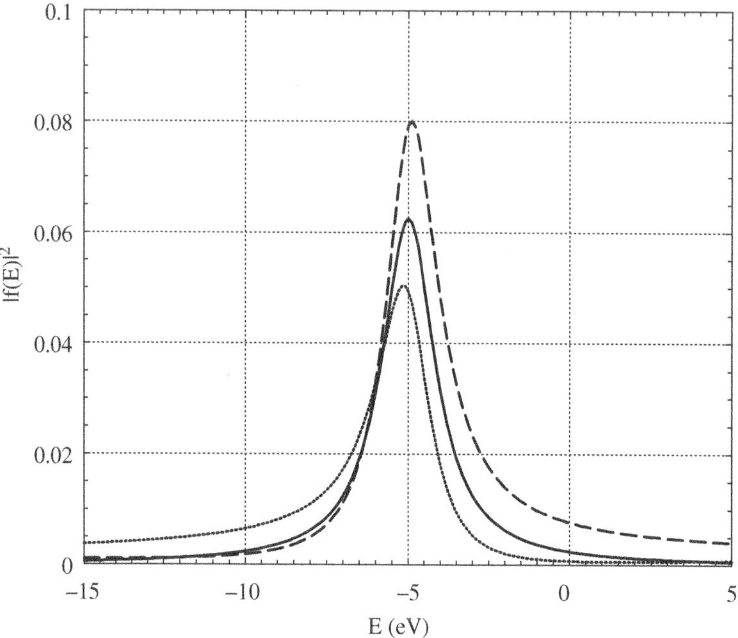

FIG. 6.11. A schematic illustration of the effect of potential scattering on the shape of the resonance peak. The curves were calculated using eqn (6.1) and assuming $\Gamma_{el} = 0.5$ eV, $\Gamma_{in} = 2$ eV, $E_s = -5$ eV, and $B = -1$. The solid curve corresponds to $A = 0$, the dotted curve corresponds to $A = 0.03 + i0.03$, and the dashed curve corresponds to $A = -0.03 - i0.03$. Note the asymmetry of the peak resulting from the interference between the resonance and potential channels of scattering.

this contribution is represented by the first constant term. Despite being a mere constant, this term has a noticeable effect on the shape of the resonance peak. Figure 6.11 illustrates how the presence of the first term in the Breit–Wigner law affects the shape of the resonance peak.

It is common to refer to the first term in the Breit–Wigner formula (6.1) as the 'potential' part of the amplitude of scattering, as opposed to the 'resonance' part of the amplitude of scattering given by the second term in (6.1). This terminology is widely accepted in the literature (Goldberger and Watson, 1964), although admittedly it sometimes has a confusing effect on a reader since the resonance scattering itself originates from the interaction of the incident electron with atomic *potentials*. In the context of our discussion the term *'potential scattering'* means *'non-resonance scattering'*, and this terminology is used to distinguish between the first and the second terms in the Breit–Wigner formula (6.1).

Resonance scattering in RHEED is associated with momentum transfers $\pm\hbar\mathbf{g}$ in the plane parallel to the surface of the crystal. Both the surface state model

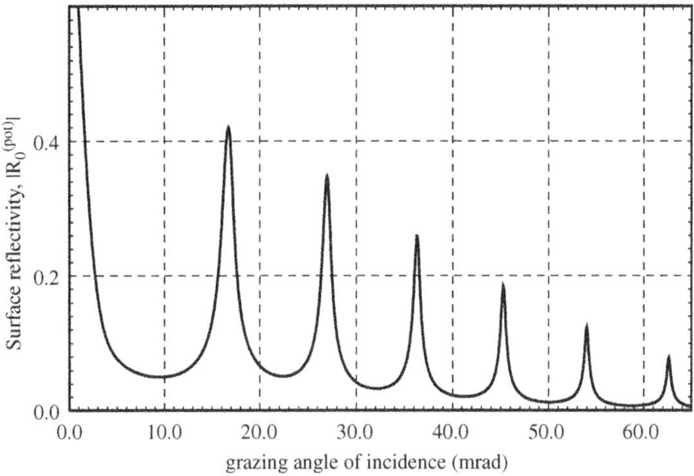

FIG. 6.12. The 'potential' part of the surface reflectivity calculated for the (111) surface of platinum and plotted as a function of the glancing angle ζ. Energy of electrons $E_0 = 100$ keV and the absolute temperature $T = 293$ K.

of resonance diffraction (6.58) and the 'collective resonance' model (6.73) predict that the resonance part of the surface reflectivity vanishes in the limit $V_{\pm \mathbf{g}}(z) \to 0$. The only contribution to the reflectivity of the surface that survives in this limit is associated with the potential $V(z) = A^{-1} \int_A dx dy V(\mathbf{r})$ averaged in the plane parallel to the surface of the crystal. So far all the effects associated with this part of the crystal potential have been neglected, both in the Green's function (6.50) and also in the wave function of the incident beam (6.48). In practice, however, the effects associated with the interference between the potential and the resonance modes of diffraction turn out to be quite significant. We will now develop a simple model that should enable us to analyse these effects and also to discuss their experimental manifestation.

We begin by defining the meaning of the term 'potential' as applied to the reflectivity of a crystal surface. In order to do this we re-examine eqns (6.51). If we approximate $\Phi_{K_0}^{(+)}(z)$ by the plane wave $\exp(iK_0 z)$ and assume that the propagation of the electron between z' and z in the first of the two equations (6.51) is not affected by the potential $V(z)$, we arrive at eqns (6.53) and (6.65) describing the surface state and 'collective' resonance models respectively. It is evident that solutions of eqns (6.53) and (6.65) are not valid for the original equations (6.51). The difference between the approximate and the 'exact' solutions of (6.51) is associated with the interaction of the electron with the potential $V(z)$. The presence of this potential modifies eqns (6.65) in two ways. Firstly, the interaction of electrons with potential $V(z)$ gives rise to the appearance of a new term in the asymptotic expansion of $\Phi_{K_0}^{(+)}(z)$ in the limit $z \to -\infty$. In addition to the incident plane wave, $\Phi_{K_0}^{(+)}(z)$ now contains a wave reflected from the surface

$$\Phi_{K_0}^{(+)}(z) = \exp(iK_0 z) + R_0^{(pot)} \exp(-iK_0 z), \qquad z \to -\infty, \qquad (6.75)$$

where coefficient $R_0^{(pot)}$ depends on the form of the potential $V(z)$. The presence of $V(z)$ also modifies the behaviour of the wave function of incident electrons $\Phi_{K_0}^{(+)}(z)$ in the crystal bulk (i.e. in the limit $z \to +\infty$). The second modification of the system of the two equations of resonance scattering is associated with the change in the form of the Green's function $G\left(z, z', \hbar^2 K_0^2/2m\right)$ describing how the high-energy electron propagates between z' and z in the one-dimensional potential $V(z)$.

Figure 6.12 shows how the potential part of the surface reflectivity depends on the grazing angle of incidence ζ. Figure 6.12 shows that although for some values of ζ the amplitude of $R_0^{(pot)}$ is very small indeed (e.g. for $\zeta = 50$ mrad we have $|R_0^{(pot)}| = 0.012$), in some other cases the magnitude of $R_0^{(pot)}$ is substantial (e.g. for $\zeta = 45.25$ mrad we obtain $|R_0^{(pot)}| = 0.185$).

What is the origin of peaks of reflectivity shown in Fig. 6.12? The answer to this question is simple: peaks of reflectivity shown in Fig. 6.12 correspond to forbidden gaps in the spectrum of states of one-dimensional motion of electrons in potential $V(z)$ in the crystal bulk. The energy $\hbar^2 K_0^2/2m$ of motion of the electron normal to the surface is related to the grazing angle of incidence ζ via $\hbar^2 K_0^2/2m = \hbar^2 k^2 \sin^2 \zeta/2m$. In this way the sequence of intervals of energy corresponding to forbidden gaps in the spectrum of states describing the motion of electrons in the potential $V(z)$, which becomes periodic in the limit $z \to \infty$, is mapped onto a sequence of intervals of grazing angles of incidence. Each forbidden gap in the spectrum of Bloch states corresponds to a peak in the surface reflectivity curve. The point that we want to make here is that when the energy $\hbar^2 K_0^2/2m$ of motion of the incident electron in the direction normal to the surface falls into one of the forbidden gaps in the spectrum of Bloch states, the Green's function in the first of the two equations (6.51) can no longer be approximated by the free space expression (6.45). A more general expression for the Green's function $G\left(z, z', \hbar^2 K_0^2/2m\right)$ is now required, which describes how an electron propagates from point z' to z under the influence of the external potential $V(z)$.

Here we encounter an important problem that needs to be considered in more detail. It turns out that in the case of one-dimensional motion of the electron, its Green's function can be found for an *arbitrary* potential $V(z)$. All that we need to know in order to calculate the Green's function is a set of two linearly independent solutions of eqn (6.48). We have already discussed one of these solutions, namely $\Phi_{K_0}^{(+)}(z)$. Another solution, which is called $\Phi_{K_0}^{(-)}(z)$, satisfies the same equation (6.48), but is subject to a different boundary condition at $z \to -\infty$,

$$\Phi_{K_0}^{(-)}(z) = \exp(-iK_0 z). \qquad z \to -\infty. \qquad (6.76)$$

It can be shown (Dudarev and Whelan, 1995) that the Green's function can be expressed in terms of these two functions $\Phi_{K_0}^{(+)}(z)$ and $\Phi_{K_0}^{(-)}(z)$ as

$$G\left(z, z', \frac{\hbar^2 K_0^2}{2m}\right) = -\frac{im}{\hbar^2 K_0} \begin{cases} \Phi_{K_0}^{(+)}(z)\Phi_{K_0}^{(-)}(z'), \text{ when } z > z' \\ \Phi_{K_0}^{(-)}(z)\Phi_{K_0}^{(+)}(z'), \text{ when } z < z'. \end{cases} \quad (6.77)$$

This expression satisfies the condition on the derivative of $G\left(z, z', \hbar^2 K_0^2/2m\right)$

$$\frac{\hbar^2}{2m}\left[\frac{\partial}{\partial z'}G\left(z, z', \frac{\hbar^2 K_0^2}{2m}\right)\bigg|_{z=z'+0} - \frac{\partial}{\partial z'}G\left(z, z', \frac{\hbar^2 K_0^2}{2m}\right)\bigg|_{z=z'-0}\right] = 1, \quad (6.78)$$

which follows from the integration of eqn (6.50) over a small interval of z containing the point $z = z'$. To prove that the coefficient on the right-hand side of eqn (6.77) is indeed equal to $-im/\hbar^2 K_0$ we note that the following combination of two functions

$$W\left[\Phi_{K_0}^{(-)}(z); \Phi_{K_0}^{(+)}(z)\right] = \Phi_{K_0}^{(-)}(z)\frac{d}{dz}\Phi_{K_0}^{(+)}(z) - \Phi_{K_0}^{(+)}(z)\frac{d}{dz}\Phi_{K_0}^{(-)}(z)$$

is their Wronskian which, since the Schrödinger equation does not contain a first order derivative, is independent of z (Jeffreys and Jeffreys, 1950). It can therefore be evaluated at any point z, e.g. at $z = -\infty$. Using for $\Phi_{K_0}^{(-)}(z)$ and $\Phi_{K_0}^{(+)}(z)$ their asymptotic expansions corresponding to $z \to -\infty$, we obtain that $W = 2iK_0$.

Using (6.77) and looking for a solution of eqns (6.51) in the form (6.66), we arrive at the following system of algebraic equations for coefficients α_n:

$$\alpha_n = \frac{\Lambda\tau}{(\hbar^2 K_{\mathbf{g}}^2/2m) - \epsilon_0 - M_{00} - (im/\hbar^2 K_0)\tau v\Lambda^2 + i\Gamma_{in}/2} \quad (6.79)$$

$$\times \left\{\exp(i\kappa nd) - \frac{im}{\hbar^2 K_0}\Lambda\sum_{l=0}^{\infty}\left[v\exp(i\kappa d|n-l|) + \rho\exp(i\kappa d(n+l))\right]\alpha_l\right\},$$

which differs from (6.67) and which contains several new parameters κ, τ, v, and ρ. These parameters are defined via the asymptotic expansion of functions $\Phi_{K_0}^{(+)}(z)$ and $\Phi_{K_0}^{(-)}(z)$ in the limit $z \to +\infty$, see e.g. (Dudarev and Whelan, 1995).

Bearing in mind the similarity between the system of algebraic equations for α_n and the one-dimensional Schrödinger equation with δ-function potentials (see eqns (6.65)–(6.72) that we discussed above), we *guess* that the solution of eqn (6.80) can be related to an appropriate solution of the Schrödinger equation of the form

$$-\frac{\hbar^2}{2m}\frac{d^2}{dz^2}\psi(z) + W\Theta(-z)\psi(z) + U\sum_{l=0}^{\infty}\delta(z - ld)\psi(z) = \frac{\hbar^2\kappa^2}{2m}\psi(z), \quad (6.80)$$

where $\Theta(z)$ is the step function defined by $\Theta(z) = 1$ for $z \geq 1$ and $\Theta(z) = 0$ for $z < 0$, and W and U are arbitrary parameters. Indeed, the integral form of eqn (6.80) for $z \geq 0$ is

$$\psi(z) = J\exp(i\kappa z) - \frac{imU}{\hbar^2\kappa}\sum_{l=0}^{\infty}\left\{\exp[i\kappa|z-ld|] + \left(\frac{\kappa-q}{\kappa+q}\right)\exp[i\kappa(z+ld)]\right\}\psi(ld),$$

$$(6.81)$$

where J is an arbitrary constant and $q^2 = \kappa^2 - 2mW/\hbar^2$. Assuming $z = nd$ we arrive at an equation identical to (6.80). As before, it turns out that eqn (6.80) is simple enough to be solved exactly. Details of calculations are given in (Dudarev and Whelan, 1995). The expression for the surface reflectivity proves to be remarkably simple

$$R_0^{(res)} = \frac{\tau R}{\upsilon - \rho R}, \tag{6.82}$$

where R is given by formula (6.73), where

$$r = t - 1 = -i\frac{m\Lambda^2}{\hbar^2 K_0}\upsilon\tau\left(\frac{\hbar^2 K_{\mathbf{g}}^2}{2m} - \epsilon_0 - M_{00} + i\frac{\Gamma_{in}}{2}\right)^{-1}. \tag{6.83}$$

Equation (6.82) does not assume that the potential contribution to the surface reflectivity is small. This makes the range of validity of formula (6.82) much wider than that of eqn (6.73), which corresponds to the limit $\tau = \upsilon \to 1$ and $\rho \to 0$. In the limiting case where $\hbar^2 K_0^2/2m \gg |V(z)|$ the Green's function $G(z, z', \hbar^2 K_0^2/2m)$ can be approximated by (6.45) and formula (6.82) reduces to eqn (6.73).

Equation (6.82) describes a case where electrons scattered by the surface via resonance and potential channels of scattering form an interference pattern. It turns out that intensity distributions observed experimentally in RHEED convergent beam diffraction patterns (CBEDs) (see for example (Spence and Zuo, 1992, Lehmpfuhl and Dowell, 1986, Smith et al., 1992)), exhibit the formation of interference patterns that are very similar to those predicted by formula (6.82).

To illustrate how well the treatment developed above describes experimental electron diffraction observations, we calculate the dependence of the intensity of the specular beam on the polar and azimuthal angles of incidence in the vicinity of the point where a *horizontal Kikuchi line* and a *resonance parabola* intersect each other.

We begin by explaining the meaning of the terms 'horizontal Kikuchi line' and 'resonance parabola'. An examination of Fig. 6.12 shows that the potential part of the coefficient of reflection of electrons from the surface $|R_0^{(pot)}(\zeta)|$ considered as a function of the grazing angle of incidence ζ has the form of a superposition of strong peaks separated by intervals where the reflectivity of the surface is relatively low. The position of the peaks is approximately given by the Bragg condition $K_0 d = (N + \frac{1}{2})\pi$, where N is an integer and d is the distance between atomic planes parallel to the surface. The peaks of $|R_0^{(pot)}(\zeta)|$ correspond to bright horizontal lines often seen in RHEED CBED patterns, and these lines are the horizontal Kikuchi lines. Segments of these horizontal lines are clearly seen in Fig. 6.7.

The origin of the term 'resonance parabola' is associated with the form of the condition determining the orientation of the incident beam corresponding to the excitation of resonance. This condition has the form (Ichimiya et al., 1980, Meyer-Ehmsen, 1988)

$$\frac{\hbar^2 K_{\mathbf{g}}^2}{2m} = \epsilon_0. \tag{6.84}$$

Taking into account that $K_{\mathbf{g}}^2 = \mathbf{k}_0^2 - (\mathbf{k}_\| + \mathbf{g})^2$, we rewrite (6.84) as

$$k_z^2 - 2(\mathbf{k}_\| \cdot \mathbf{g}) - \mathbf{g}^2 = \frac{2m\epsilon_0}{\hbar^2}. \tag{6.85}$$

For a given vector $\mathbf{g}$ this equation generates a family of parabolas (k_z versus $(\mathbf{k}_\| \cdot \mathbf{g})/|\mathbf{g}|$).

Interference effects in the vicinity of the point of intersection between a horizontal Kikuchi line and a resonance parabola were studied by Zuo and Liu (Zuo and Liu, 1992) who showed that these effects give rise to a particularly strong enhancement of the intensity of the specular beam (see Fig. 6.13). The intensity distribution observed in RHEED CBED patterns is similar to the one observed in the transmission geometry of diffraction where interference effects give rise to the splitting of intersecting Kikuchi lines (Gjønnes et al., 1988, James et al., 1994).

The difference between the two cases (transmission and reflection) lies in the fact that usually in transmission electron diffraction the effect of Kikuchi line splitting can be described using either a three- or a four-beam approximation. In the present case we are dealing with a process of scattering where at least one of the states is a tightly bound state, the Fourier expansion of which involves several tens of reciprocal lattice vectors. The distribution of intensity calculated numerically in the vicinity of the point of intersection of the 666 horizontal Kikuchi line and the resonance parabola associated with the $\bar{2}20$ side reflection is shown in Fig. 6.14 for the Pt(111) surface. Figure 6.15 shows the same part of the RHEED CBED pattern but now the calculations were carried out using the analytical expression (6.82). The distributions of intensity shown in Figs 6.14 and 6.15 are very similar, and this confirms the validity of the tight-binding model of resonance scattering of electrons from a crystal surface.

6.3.6 *The time delay of the incident electron in the resonance state*

An interesting question arising from the examination of intensity distributions shown in Figs 6.14 and 6.15 concerns the origin of the enhancement of the intensity of the specular beam seen in the vicinity of the resonance condition (6.85). One possible interpretation may be based on the supposition that the peaks of

FIG. 6.13. Experimental convergent-beam RHEED pattern near the 880 and 620 Bragg condition. The pattern was recorded under the condition of 100 kV electron accelerating voltage and room temperature. From (Zuo and Liu, 1992).

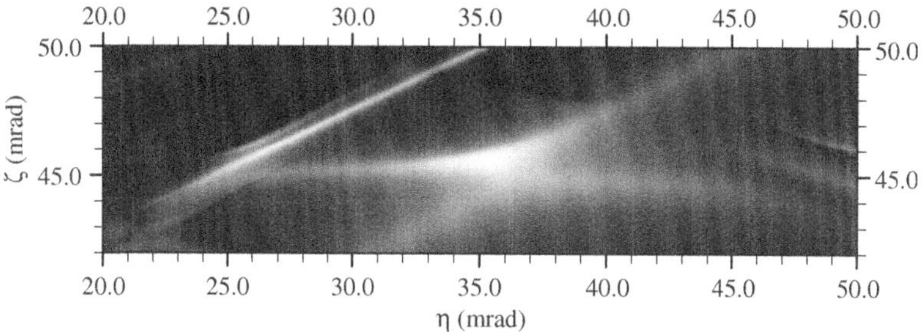

FIG. 6.14. Distribution of intensity in the RHEED CBED pattern calculated numerically for the Pt (111) surface (Dudarev and Whelan, 1996). Calculations were performed using eqns (6.37) in the two-rod approximation (6.51). Azimuth $\eta = 0$ corresponds to the $\langle 11\overline{2} \rangle$ direction, and the energy of incident electrons is 100 keV.

intensity shown in Figs 6.14 and 6.15 are associated with the minima of the effective depth of penetration l_{eff} of the wave function of the incident electrons in the crystal. One may speculate that in the case where the penetration depth is low, the rate of inelastic absorption of electrons is substantially reduced also, leading to the enhancement of the intensity of the beam reflected from the surface. The effective penetration depth of electrons in the crystal can be evaluated using the solution of eqn (6.80) (see also (Dudarev and Whelan, 1995, Rez, 1995))

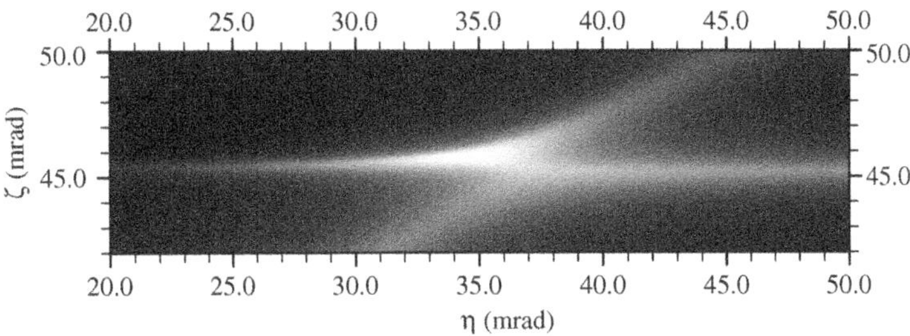

FIG. 6.15. Distribution of intensity in the RHEED CBED pattern calculated analytically using formula (6.82) for the same range of angles of incidence as that shown in Fig. 6.14.

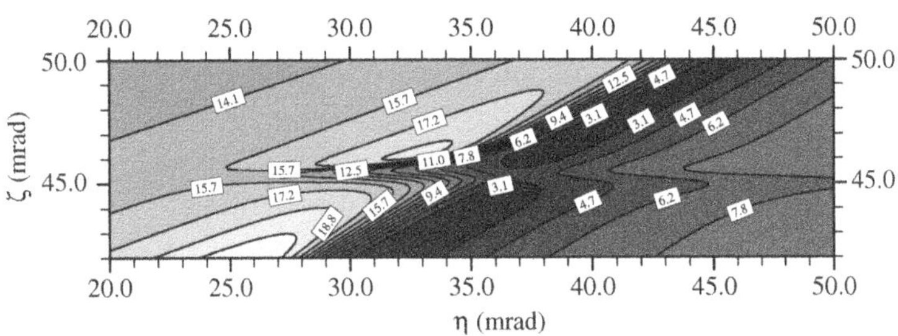

FIG. 6.16. Dependence of the effective depth of penetration of electrons in the crystal bulk l_{eff} on the azimuthal and polar angles of incidence calculated using formula (6.86) for the Pt(111) surface. Values of l_{eff} (Å) corresponding to the contour lines are given in boxes.

$$l_{eff} = -d\left[\Re\ln(\alpha_{n+1}/\alpha_n)\right]^{-1} = \left\{\Im\left[\kappa - i\frac{1}{d}\ln\left(1 - \frac{r}{R}\right) + i\frac{1}{d}\ln t\right]\right\}^{-1}. \quad (6.86)$$

The dependence of the effective penetration depth l_{eff} on the two angles of incidence ζ and η is shown in Fig. 6.16. This figure shows that l_{eff} is almost insensitive to the potential channel of scattering and that the magnitude of l_{eff} is determined mainly by the distance from the resonance parabola. The penetration depth remains large ($l_{eff} \sim 6$ Å, i.e. four times the minimum value $l_{eff} \sim 1.5$ Å) in the vicinity of the point $\zeta_0 = 45.8$ mrad, $\eta_0 = 34.4$ mrad, which corresponds to the maximum value of surface reflectivity $|R_0(\zeta_0, \eta_0)|^2 = 0.73$. This shows that although the dependence of the effective depth of penetration of electrons into the crystal does exhibit some degree of correlation with the angular dependence of the reflectivity of the crystal surface, the position of the maximum intensity of the specular beam *does not* correspond to the minimum depth of penetration of electrons into the crystal bulk.

Another parameter characterizing the process of resonance diffraction is the effective displacement Δ of the electron beam in the direction *parallel* to the surface. This parameter is important for electron microscope imaging of surfaces since it determines the magnitude of splitting of electron microscope images (Kambe, 1988) obtained at grazing incidence. Artmann (Artmann, 1948) showed that by considering an incident *wave packet* it is possible to demonstrate that a beam of electrons reflected from the surface is displaced in the direction parallel to the surface, and the magnitude of the displacement is given by

$$\delta \mathbf{R} = -\frac{\partial}{\partial \mathbf{k}_{\|}} \phi_0(\mathbf{k}_{\|}), \tag{6.87}$$

where $\phi_0(\mathbf{k}_{\|}) = \Im \ln R_0(\mathbf{k}_{\|})$ is the *phase* of the coefficient of reflection of electrons from the surface. By re-expressing this derivative in terms of two angles of incidence and considering the projection of $\delta \mathbf{R}$ on the direction of $\mathbf{k}_{\|}$, we obtain (Dudarev and Whelan, 1995)

$$\Delta = \frac{(\mathbf{k}_{\|} \cdot \delta \mathbf{R})}{|\mathbf{k}_{\|}|} = \frac{1}{k \sin \zeta} \frac{\partial}{\partial \zeta} \left\{ \Im \left[\ln R_0(\zeta, \eta) \right] \right\}. \tag{6.88}$$

This formula shows that all that we need to do to evaluate the displacement of the incident beam of electrons in the direction parallel to the surface is to differentiate the calculated *complex-valued* rocking curve.

It should be emphasized that expression (6.88) was derived by considering the evolution of the centre of a wave packet interacting with the potential of the crystal. Because of that the meaning of the parameter Δ cannot be interpreted literally in terms of classical mechanics. For example, our analysis shows that in some cases the value of Δ calculated according to eqn (6.88) is negative, and this somewhat anomalous behaviour follows from the quantum-mechanical treatment of the problem. Similar fundamental difficulties follow from an attempt to give a classical interpretation to the time delay associated with the process of quantum-mechanical tunnelling of a wave packet through a potential barrier (Landauer and Martin, 1994).

The idea that the resonance enhancement of the intensity of the specular beam is associated with the anomalously large distance of propagation in the direction parallel to the surface was originally proposed by Cowley (Cowley, 1982). Figure 6.17 shows a contour map which displays the dependence of Δ on the angles of incidence in the vicinity of the 666 Bragg/$\overline{2}$20 resonance condition. This contour map was calculated analytically using eqn (6.88).

The comparison of contour maps shown in Figs 6.14, 6.15, and 6.17 fully agrees with the hypothesis of Cowley (Cowley, 1982). The position of the maximum of Δ considered as a function of the polar and azimuthal angles of incidence nearly coincides with the position of the maximum of the surface reflectivity $|R_0(\zeta_0, \eta_0)|^2$. This confirms that the origin of the peak of reflectivity is indeed associated with trapping of the incident electron by the resonance state. The electron trapped by the localized resonance orbital continues its motion in the direction parallel to the surface and this gives rise to the anomalously high value

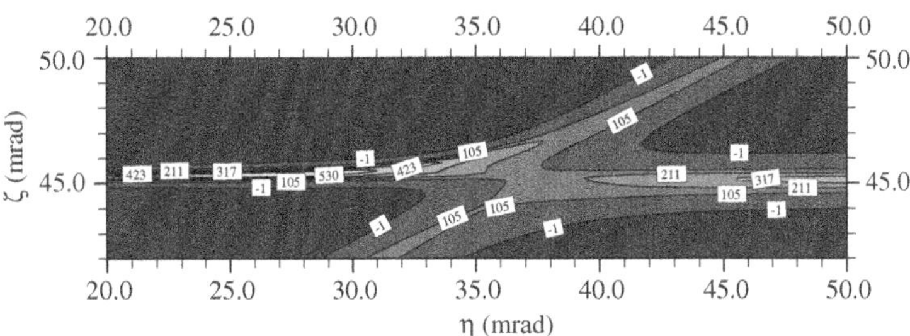

FIG. 6.17. Dependence of the distance Δ of propagation of electrons in the direction parallel to the surface of the crystal on the azimuthal and polar angles of incidence calculated using formula (6.88) for the (111) surface of platinum. Values of Δ (Å) corresponding to the contour lines are given in boxes.

of the displacement of the electron beam in the direction parallel to the surface of the crystal.

6.4 Summary

In this chapter we have given a comprehensive introduction to the treatment of resonance scattering of high-energy electrons. We have shown that resonance scattering occurs both in transmission and reflection geometries of diffraction, where it is responsible for the occurrence of characteristic rings and parabolae in electron diffration patterns. The treatment of resonance scattering given in this chapter shows that this type of scattering leads to 'trapping' of incident electrons by a resonating state, giving rise to an anomalously large displacement of the electron beam in the direction parallel to the surface of the crystal in the reflection geometry of diffraction.

7

DIFFUSE AND INELASTIC SCATTERING – ELEMENTARY PROCESSES

7.1 Diffuse and inelastic scattering

It is hard to give a precise definition of the term 'diffuse scattering' in electron diffraction. What is observed experimentally in a diffraction pattern is the distribution of electrons over the angle of scattering. An energy spectrometer makes it possible to observe the energy spectrum of scattered electrons for a given direction of their motion. It is only approximately that we may say that a part of the observed cross-section of scattering (called 'the cross-section of diffuse scattering') should be treated theoretically in a radically different way from how we treat the remaining 'diffraction' part. For crystalline samples that are small in size it is not possible to distinguish between electron 'diffraction' and 'diffuse scattering' of electrons. For a small sample the diffraction pattern has the form of a superposition of a series of relatively broad peaks where the positions of peak maxima are given by the Bragg law. It is only in the limit of a large number of atoms in the crystal that the peaks associated with elastic scattering by the periodic potential become infinitely sharp and it becomes possible to distinguish between diffraction and diffuse scattering of high-energy electrons.

It is a more straightforward matter to define the term 'inelastic scattering' applicable to the cases of both very small and very large crystals. Inelastic scattering is normally understood as the type of scattering associated with a change in the quantum state of the solid (this change may be associated with either the excitation or the de-excitation of internal degrees of freedom of the solid). It is indeed possible to say, as we did in Chapter 1, that inelastic scattering results from temporal variations of the interaction potential. At the same time a rigorous treatment of inelastic scattering requires going beyond the simplified consideration of scattering by time-dependent fluctuations developed in Chapter 1. The time dependence of the interaction potential acting on the high-energy electron results from the motion of electrons and nuclei in the solid. The electrostatic potential associated with the incident high-energy electron perturbs the trajectories of electrons and nuclei in the solid and creates electron-hole or collective excitations. It is therefore essential to treat the problem of scattering by a time-dependent potential as a problem of *interaction* between the incident electron and the electrons and nuclei in the solid. In other words it is necessary to take into account the fact that a high-energy electron propagating through the solid influences the very same time-dependent potential that is acting on it.

Consider a crystal that is infinite in the (x, y) plane as shown in Fig. 7.1.

228

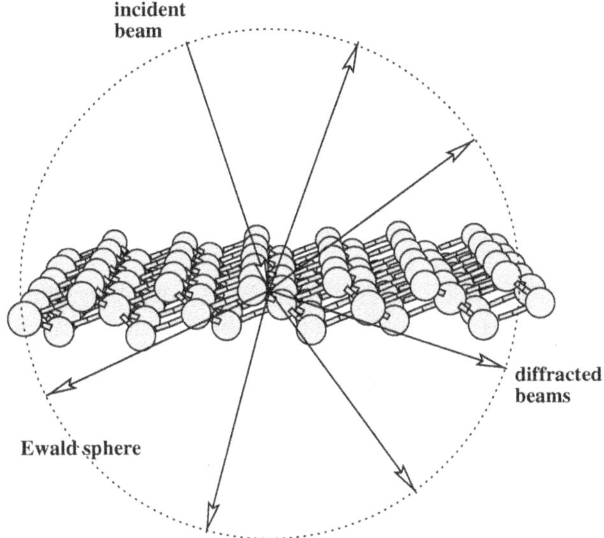

FIG. 7.1. Schematic illustration of the process of scattering of electrons by a crystal which is infinite in the x and y directions. Elastic scattering by a potential which is periodic in the (x, y) plane results in the formation of a series of sharp diffraction spots, the positions of which, on a micrograph, are determined by the condition of conservation of the projection of the momentum $\mathbf{k}_{0t}$ on the (x, y) plane (plus a reciprocal lattice vector $\mathbf{G}$, see text) supplemented by the condition of conservation of energy.

Assume that the potential of interaction between the incident electron and the crystal is a periodic function of x and y, i.e.

$$V(\mathbf{r}) = V(\mathbf{R}, z) = V(\mathbf{R} + n_1 \mathbf{a}_1 + n_2 \mathbf{a}_2, z)$$

where n_1 and n_2 are integer numbers and $\mathbf{a}_i = (a_{ix}, a_{iy})$ are the two-dimensional translation vectors. The condition of periodicity of the potential in the (x, y) plane is equivalent to the statement that at any given z the potential may be expanded as a two-dimensional Fourier series

$$V(\mathbf{R}, z) = \sum_{\mathbf{G}} V_{\mathbf{G}}(z) \exp(i \mathbf{g} \cdot \mathbf{R}), \tag{7.1}$$

where $\mathbf{G}$ is a vector belonging to the two-dimensional reciprocal lattice $\mathbf{G} = l_1 \mathbf{G}_1 + l_2 \mathbf{G}_2$ and l_1 and l_2 are arbitrary integers. For the incident plane wave $\exp(i \mathbf{k}_0 \cdot \mathbf{r}) = \exp[i(\mathbf{k}_0)_t \cdot \mathbf{R} + i k_{0z} z]$ diffraction by the potential (7.1) leads to the formation of scattered waves characterized by wave vectors, the (x, y) components of which equal either the projection $\mathbf{k}_{0t}$ of the wave vector of the incident wave on the (x, y) plane, or $\mathbf{k}_{0t} + \mathbf{G}$, where $\mathbf{G}$ is one of the reciprocal lattice vectors involved in the Fourier expansion (7.1).

The z component of the wave vector of a diffracted beam in the vacuum is determined by the condition that the total length of this vector in the vacuum must be equal to the length of the wave vector of incident electrons. This condition follows from the conservation of energy and in terms of geometry of scattering it means simply that the ends of all the wave vectors lie on the Ewald sphere (see Fig. 7.1). The z component of the wave vector of the diffracted beam for which $\mathbf{k}_t = \mathbf{k}_{0t} + \mathbf{G}$ therefore equals $\mathbf{k}_z = \pm\sqrt{\mathbf{k}_0^2 - (\mathbf{k}_{0t} + \mathbf{G})^2}$, where the plus and minus signs correspond to scattering in the forward and backward directions respectively.

This consideration shows that scattering by a potential, which is periodic in two dimensions, gives rise to a differential cross-section that has the form of a series of sharp diffraction spots (each corresponding to a particular reciprocal vector $\mathbf{G}$) with zero intensity between the spots. This statement evidently does not depend on the exact functional form of the dependence of the potential on the coordinate z in the direction perpendicular to the surface of the crystal. It is therefore the violation of the periodicity of the potential in the (x, y) plane that is responsible for the observed non-zero scattered intensity present in regions between diffraction spots in the diffraction pattern.

The non-periodic fluctuations of the interaction potential are associated with the presence of defects in the crystal, and also with phonon and electronic excitations. Close to the surface of the crystal the two-dimensional periodicity of the potential may be violated because of surface roughness or because of the presence of a disordered layer of impurities segregated at the surface. Broadly speaking, deviations of the potential from periodicity may be static (immobile defects or surface disorder) or dynamic (phonons or electronic excitations). Any potential which is not representable by equation (7.1) gives rise to the appearance of diffuse intensity between diffraction peaks, and it is possible to identify several types of diffuse scattering, each of them related to a particular type of deviation of $V(\mathbf{r})$ from the form given by equation (7.1).

In this chapter we give a coherent account of the effects of diffuse scattering which we will treat using the distorted wave Born approximation. We go a step further in the next chapter where we discuss the effects of multiple diffuse and inelastic scattering of high-energy electrons.

7.2 The distorted wave Born approximation

The treatment of diffuse and inelastic scattering by crystals is often considered to be a difficult topic in electron diffraction. Indeed, it is relatively hard, although possible (Allen and Josefsson, 1995), to work out a recipe for calculating the cross-section of diffuse scattering for every particular case if one tries to solve the problem from scratch starting from the Schrödinger equation. At the same time there is a general approach that makes the treatment of diffuse scattering fairly transparent and in fact no more difficult than the treatment of elastic dynamical diffraction of high-energy electrons by the periodic part of the potential of the crystal. Using this approach it is possible to derive in a simple and straightforward way the great majority of results on diffuse scattering that are presently available in the literature.

This method is based on the *distorted* wave Born approximation (DWBA) (Landau and Lifshitz, 1977). This approximation is significantly different from (and is significantly more powerful than) the usual Born approximation, which is often used in quantum mechanics for calculating the cross-section of scattering by a weak interaction potential. The DWBA *does not* require the interaction potential to be weak. Instead the idea of the DWBA is based on the possibility of representing the potential $V(\mathbf{r})$ in the form of a sum of two terms, $V(\mathbf{r}) = V_0(\mathbf{r}) + \delta V(\mathbf{r})$. The first term $V_0(\mathbf{r})$ may be arbitrarily large with no possibility of treating it as a perturbation. On the other hand $\delta V(\mathbf{r})$ is assumed to be sufficiently small so that only terms linear in $\delta V(\mathbf{r})$ have to be retained in the expression for the amplitude of scattering. We shall now derive the DWBA in a form suitable for applications in electron diffraction and microscopy.

We start from the integral form of the Schrödinger equation (2.6)

$$\psi(\mathbf{r}) = \psi_0(\mathbf{r}) + \int G(\mathbf{r} - \mathbf{r}')V(\mathbf{r}')\psi(\mathbf{r}')d\mathbf{r}', \qquad (7.2)$$

where $\psi_0(\mathbf{r})$ denotes the incident wave and $G(\mathbf{r} - \mathbf{r}')$ is the divergent spherical wave (2.4).

The integral equation (7.2) has a simple meaning. Its solution $\psi(\mathbf{r})$ satisfies the self-consistent condition that $\psi(\mathbf{r})$ equals the sum of the incident wave $\psi_0(\mathbf{r})$ and waves arriving at $\mathbf{r}$ from all the points $\mathbf{r}'$ where they underwent scattering by the potential $V(\mathbf{r}')$. In fact (7.2) can be derived simply by applying the Huygens principle and representing $\psi(\mathbf{r})$ by the sum of the incident wave $\psi_0(\mathbf{r})$ plus the wave that was scattered by potential $V(\mathbf{r})$ once (this contribution is given by $\int G(\mathbf{r}-\mathbf{r}')V(\mathbf{r}')\psi_0(\mathbf{r}')d\mathbf{r}'$) plus the wave scattered by this potential twice (this is described by the term $\int G(\mathbf{r} - \mathbf{r}')V(\mathbf{r}') \int G(\mathbf{r}' - \mathbf{r}'')V(\mathbf{r}'')\psi_0(\mathbf{r}'')d\mathbf{r}'d\mathbf{r}''$) plus the wave that was scattered by $V(\mathbf{r})$ three times, etc. It is clear that calculating $\psi(\mathbf{r})$ in this way is fairly impractical (there are too many three-dimensional integrals to compute) but the resulting series nevertheless formally solves eqn (7.2). Indeed every term of the above series can be obtained by iterating eqn (7.2) as discussed in Chapter 3.

To make eqn (7.2) look more symmetric we rewrite it assuming that the incident wave is generated by a point source situated at $\mathbf{r}_s$,

$$\mathcal{G}(\mathbf{r}, \mathbf{r}_s) = G(\mathbf{r} - \mathbf{r}_s) + \int G(\mathbf{r} - \mathbf{r}')V(\mathbf{r}')\mathcal{G}(\mathbf{r}', \mathbf{r}_s)d\mathbf{r}'. \qquad (7.3)$$

We see that $\mathcal{G}(\mathbf{r}, \mathbf{r}_s)$ is a solution of eqn (7.2) corresponding to the incident *spherical* wave (2.4).

We now write the potential in (7.3) as $V(\mathbf{r}) = V_0(\mathbf{r}) + \delta V(\mathbf{r})$ and solve the resulting equation iteratively:

$$\begin{aligned}
\mathcal{G} = G &+ G * (V_0 + \delta V)G + G * (V_0 + \delta V)G * (V_0 + \delta V)G \\
&+ G * (V_0 + \delta V)G * (V_0 + \delta V) * (V_0 + \delta V)G + \ldots \qquad (7.4)
\end{aligned}$$

Here the convolution sign $*$ denotes integration over the intermediate coordinates $\mathbf{r}'$, $\mathbf{r}''$, etc.

Now we assume that δV is small and that only terms of the zero and first order in δV need to be retained in (7.4). The sum of the zero-order terms equals

$$
\begin{aligned}
\mathcal{G}^{(0)} &= G + G * V_0 G + G * V_0 G * V_0 G + ... \\
&= G + G * V_0 [G + G * V_0 G + ...] \\
&= G + G * V_0 \mathcal{G}^{(0)}.
\end{aligned}
\tag{7.5}
$$

The real-space form of this equation is

$$
\mathcal{G}^{(0)}(\mathbf{r}, \mathbf{r}_s) = G(\mathbf{r} - \mathbf{r}_s) + \int G(\mathbf{r} - \mathbf{r}')V_0(\mathbf{r}')\mathcal{G}^{(0)}(\mathbf{r}', \mathbf{r}_s)d\mathbf{r}'.
\tag{7.6}
$$

The solution $\mathcal{G}^{(0)}(\mathbf{r}, \mathbf{r}_s)$ of this equation describes a spherical electron wave originating at $\mathbf{r}_s$ and propagating through the potential field $V_0(\mathbf{r})$. Given that now we understand the meaning of $\mathcal{G}^{(0)}(\mathbf{r}, \mathbf{r}_s)$ it is possible to write down a differential equation that is equivalent to (7.6)

$$
\left[\frac{\hbar^2 k_0^2}{2m} + \frac{\hbar^2}{2m}\nabla^2 - V_0(\mathbf{r}) \right] \mathcal{G}^{(0)}(\mathbf{r}, \mathbf{r}_s) = \delta(\mathbf{r} - \mathbf{r}_s).
\tag{7.7}
$$

This equation can also be derived formally from (7.6). To perform the derivation we act on both sides of eqn (7.6) by the operator $(\hbar^2/2m)[k_0^2 + \nabla^2]$ and note that

$$
\frac{\hbar^2}{2m}[k_0^2 + \nabla^2]G(\mathbf{r} - \mathbf{r}_s) = \delta(\mathbf{r} - \mathbf{r}_s).
$$

Examining the series (7.4) it is easy to see that the sum of all terms that are *linear* in δV is given by

$$
\begin{aligned}
\mathcal{G}^{(1)} &= [G + G * V_0 G + ...] * \delta V [G + G * V_0 G + ...] \\
&= \mathcal{G}^{(0)} * \delta V \mathcal{G}^{(0)}.
\end{aligned}
\tag{7.8}
$$

The first-order correction to (7.6) is therefore equal to

$$
\mathcal{G}^{(1)}(\mathbf{r}, \mathbf{r}_s) = \int \mathcal{G}^{(0)}(\mathbf{r}, \mathbf{r}')\delta V(\mathbf{r}')\mathcal{G}^{(0)}(\mathbf{r}', \mathbf{r}_s)d\mathbf{r}',
\tag{7.9}
$$

where $\mathcal{G}^{(0)}(\mathbf{r}, \mathbf{r}_s)$ is the solution of eqn (7.6), or equivalently eqn (7.7).

Equation (7.9) provides us with an important computational tool. Imagine that we know how a spherical wave propagates through some (arbitrarily chosen) potential $V_0(\mathbf{r})$. Now by using (7.9) we can find out what will happen to this wave if it encounters a fluctuation of this potential. A notable feature of function $\mathcal{G}^{(1)}(\mathbf{r}, \mathbf{r}_s)$ given by (7.9) is that it is symmetric with respect to the transposition of coordinates $\mathbf{r}$ and $\mathbf{r}_s$. This fully agrees with the principle of reciprocity, which holds for scattering by both $V_0(\mathbf{r})$ and $V_0(\mathbf{r}) + \delta V_0(\mathbf{r})$.

Formula (7.9) shows how a small change in the potential affects the amplitude of a wave originating from a point source. Although this finding is interesting on its own merit, in what follows we focus our attention on understanding how a small perturbation of the potential affects the amplitude of scattering. This question is directly related to the problem of interpretation of electron diffraction patterns observed in transmission or reflection geometries of scattering.

The definition of the amplitude of scattering (Landau and Lifshitz, 1977) states that it is equal to the probability amplitude of the process where the incident electron changes its momentum from $\mathbf{k}_0$ to $\mathbf{k}$. In practice this means that if we imagine how the process of scattering takes place in real space, then the calculation of the amplitude of scattering requires finding the relative weights of the incident and scattered plane waves in the spherical wave $\mathcal{G}(\mathbf{r}, \mathbf{r}_s)$.

To find these weights we need to evaluate the projection of the spherical wave $\mathcal{G}(\mathbf{r}, \mathbf{r}_s)$ given by (7.3) onto the incident $\exp(i\mathbf{k}_0 \cdot \mathbf{r}_s)$ and the scattered $\exp(i\mathbf{k} \cdot \mathbf{r})$ plane waves, respectively, i.e.

$$f(\mathbf{k}, \mathbf{k}_0) \sim \langle \exp(i\mathbf{k}\cdot\mathbf{r}) | \mathcal{G}(\mathbf{r}, \mathbf{r}_s) | \exp(i\mathbf{k}_0 \cdot \mathbf{r}_s) \rangle$$
$$= \int d\mathbf{r} \int d\mathbf{r}_s \exp(-i\mathbf{k}\cdot\mathbf{r}) \mathcal{G}(\mathbf{r}, \mathbf{r}_s) \exp(i\mathbf{k}_0 \cdot \mathbf{r}_s). \qquad (7.10)$$

The application of (7.10) to $\mathcal{G}^{(0)}(\mathbf{r}, \mathbf{r}')$ (see eqns (7.6) and (7.7)) gives rise to $f^{(0)}(\mathbf{k}, \mathbf{k}_0)$, which is the amplitude of scattering by the potential $V_0(\mathbf{r})$.

Let us now evaluate the projection of (7.9) onto the incident and scattered plane waves. Projecting the second of the two Green's functions entering (7.9) on the incident wave $| \exp(i\mathbf{k}_0 \cdot \mathbf{r}_s) \rangle$ does not present any difficulty since this operation is identical to the one which is required to transform (7.3) back to (7.2). In other words,

$$\mathcal{G}(\mathbf{r}, \mathbf{r}_s) | \exp(i\mathbf{k}_0 \cdot \mathbf{r}_s) \rangle = \psi_{\mathbf{k}_0}^{(0)}(\mathbf{r}), \qquad (7.11)$$

where $\psi_{\mathbf{k}_0}^{(0)}(\mathbf{r})$ is the solution of (7.2) for the incident wave $\exp(i\mathbf{k}_0 \cdot \mathbf{r}_s)$.

To evaluate $\langle \exp(i\mathbf{k}\cdot\mathbf{r}) | \mathcal{G}(\mathbf{r}, \mathbf{r}')$ we take into account the reciprocity condition $\mathcal{G}(\mathbf{r}, \mathbf{r}') = \mathcal{G}(\mathbf{r}', \mathbf{r})$ and the known relation between the bra and ket vectors (Dirac, 1981) $\langle \exp(i\mathbf{k}\cdot\mathbf{r}) | \longleftrightarrow | \exp(-i\mathbf{k}\cdot\mathbf{r}) \rangle$. We find the equation

$$\langle \exp(i\mathbf{k}\cdot\mathbf{r}) | \mathcal{G}(\mathbf{r}, \mathbf{r}') = \mathcal{G}(\mathbf{r}', \mathbf{r}) | \exp(-i\mathbf{k}\cdot\mathbf{r}) \rangle = \psi_{-\mathbf{k}}^{(0)}(\mathbf{r}'), \qquad (7.12)$$

which leads to the final result:

$$f^{(1)}(\mathbf{k}, \mathbf{k}_0) = -\frac{m}{2\pi\hbar^2} \int \psi_{-\mathbf{k}}^{(0)}(\mathbf{r}') \delta V(\mathbf{r}') \psi_{\mathbf{k}_0}^{(0)}(\mathbf{r}') d\mathbf{r}'. \qquad (7.13)$$

Here the numerical coefficient is chosen in such a way that in the limiting case $V_0(\mathbf{r}) = 0$ formula (7.13) reproduces the conventional Born approximation for the amplitude of scattering (Landau and Lifshitz, 1977).

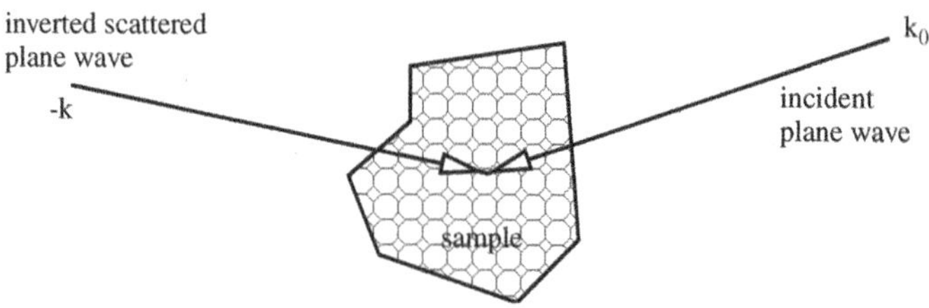

FIG. 7.2. This sketch illustrates how to use the distorted wave Born approximation in order to find the *first-order* correction to the amplitude of scattering due to a small perturbation $\delta V(\mathbf{r})$ of the potential. To evaluate the relevant matrix element we need to know two solutions of the zero-order Schrödinger equation, one corresponding to the 'ordinary' incident plane wave $\exp(i\mathbf{k}_0 \cdot \mathbf{r})$ and the other corresponding to the inverted scattered plane wave $\exp(-i\mathbf{k} \cdot \mathbf{r})$. These two solutions are very similar, and the only difference between them is associated with different directions of incidence (note that both $\mathbf{k}_0$ and $-\mathbf{k}$ point towards the sample).

We have now found a general rule for calculating the amplitude of scattering by a potential $V(\mathbf{r})$ representable in the form of a sum of two terms $V(\mathbf{r}) = V_0(\mathbf{r}) + \delta V(\mathbf{r})$. This rule is

$$f(\mathbf{k}, \mathbf{k}_0) = f^{(0)}(\mathbf{k}, \mathbf{k}_0) - \frac{m}{2\pi\hbar^2} \int \psi_{-\mathbf{k}}^{(0)}(\mathbf{r})\delta V(\mathbf{r})\psi_{\mathbf{k}_0}^{(0)}(\mathbf{r})d\mathbf{r} + ..., \qquad (7.14)$$

where $f^{(0)}(\mathbf{k}, \mathbf{k}_0)$ is the amplitude of scattering by the 'unmodified' potential $V_0(\mathbf{r})$ and $\psi_{-\mathbf{k}}^{(0)}(\mathbf{r})$ and $\psi_{\mathbf{k}_0}^{(0)}(\mathbf{r})$ are the solutions of the Schrödinger equation (7.2) for this potential $V_0(\mathbf{r})$ and boundary conditions taken in the form of incident plane waves $\exp(-i\mathbf{k} \cdot \mathbf{r})$ and $\exp(i\mathbf{k}_0 \cdot \mathbf{r})$ respectively. Formula (7.14) is called the DWBA.

The DWBA turns out to be particularly convenient for calculating the cross-sections of diffuse scattering of low- and high-energy electrons and X-rays by crystalline samples (Dmitrienko and Kaganer, 1990). Indeed, for a crystalline sample it is quite natural to separate the periodic part of the potential from the non-periodic part, where the latter may be associated with defects or other types of deviations from the long-range crystalline order. Assuming that in the above formulae $V_0(\mathbf{r})$ represents the periodic and $\delta V(\mathbf{r})$ the non-periodic part of the potential, from (7.14) we obtain a simple recipe for calculating the cross-section of diffuse scattering. Note that $\psi_{-\mathbf{k}}^{(0)}(\mathbf{r})$ and $\psi_{\mathbf{k}_0}^{(0)}(\mathbf{r})$ take into account *multiple* diffraction scattering of the electron before and after its interaction with the fluctuating part $\delta V(\mathbf{r})$ of the potential of the crystal.

The important point that we have not yet addressed in our discussion of the DWBA concerns the range of validity of this approximation. As we already stated above, to be able to retain only terms linear in $\delta V(\mathbf{r})$ in eqn (7.14) it is necessary to impose certain conditions on the magnitude of the fluctuating part of the potential. Since in the limiting case $V_0(\mathbf{r}) = 0$ the DWBA coincides with the conventional Born approximation, at least the standard condition of validity of the Born approximation (Landau and Lifshitz, 1977) has to be satisfied in the case of the DWBA too, namely

$$\max|\delta V(\mathbf{r})| \leq (ka)\frac{\hbar^2}{ma^2},\tag{7.15}$$

where a is the characteristic size of the region where $\delta V(\mathbf{r})$ is substantial. It is hard to derive a more rigorous condition of validity of (7.14). The most important corollary following from (7.15) is the fact that in order for the DWBA to remain valid the perturbation $\delta V(\mathbf{r})$ must remain *localized* in real space. A delocalized (i.e. characterized by a large value of a) or a periodic perturbation cannot be treated using (7.14) unless there are some special circumstances that justify the use of the DWBA in that particular case of interest. One has also to be aware of the fact that effects of *multiple* scattering by fluctuations often make a very significant contribution to the distribution of intensity in electron diffraction patterns, and the treatment of these effects requires going beyond the DWBA.

In what follows we consider three examples illustrating the application of the DWBA to the treatment of diffuse scattering in electron diffraction. Firstly we consider scattering by static deviations of the potential from periodicity associated with point defects. Secondly we illustrate how the DWBA can be used to treat the thermal diffuse (i.e. the phonon) scattering of high-energy electrons. Thirdly we show how the DWBA can be applied to the treatment of scattering by electronic excitations. As a concluding remark for this section we direct the reader who wishes to see a more thorough derivation of the DWBA to Appendix C of (Dudarev et al., 1993a).

7.3   Diffuse scattering by point defects

The best starting point for illustrating how the DWBA applies to real problems of scattering and diffraction is the problem of scattering of high-energy electrons by point defects. Impurities, interstitial atoms, and vacancies give rise to local deviations from periodicity in the distribution of atoms in the crystal lattice. Figure 7.3 shows in a schematic form several types of defect (Hall et al., 1966) that result in the appearance of diffuse intensity in electron diffraction patterns.

There are two reasons that make the DWBA particularly suitable for evaluating the cross-section of diffuse scattering by point defects. Firstly deviations of the interaction potential from periodicity due to the defects shown in Fig 7.3 are *localized*, and therefore the conditions of validity of the DWBA (7.15) are satisfied. Secondly since it is only the *deviation* of the potential from periodicity that needs to be treated as a perturbation, we can use the recipe provided by the DWBA and treat scattering by the periodic potential of the crystal non-perturbatively using the dynamical diffraction theory approach.

The presence of a defect results in a drastic change of the crystal potential $V(\mathbf{r})$ in the vicinity of a defect. This change is associated with the displacement

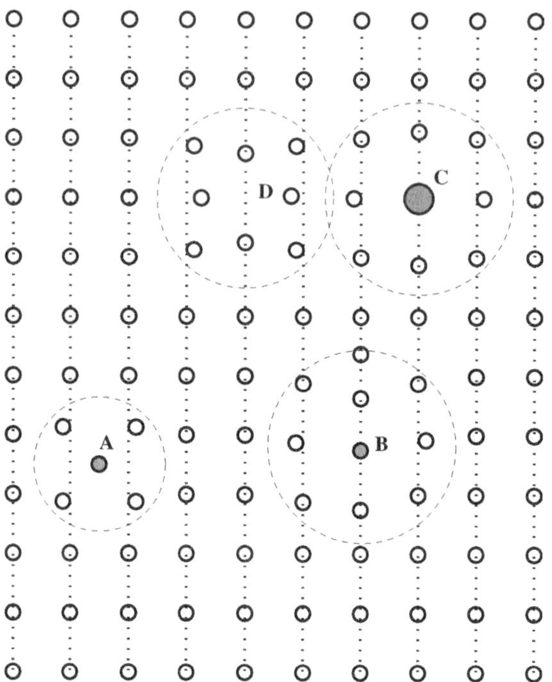

FIG. 7.3. Diagram showing how atoms in a crystal lattice are displaced near an interstitial (at A and B) or a substitutional impurity (at C), or near a vacancy (at D).

of atomic nuclei and electron cores near the defect and also with the contribution from the potential of the interstitial or the missing atom (in the case of a vacancy). Additional contributions to the local deviation of the potential from periodicity are due to the redistribution of valence electron density around defect sites and to the formation of localized electronic states. However, in most cases the part played by the latter effects is less significant than that resulting from the displacement of atomic cores and the addition or removal of atoms from the crystal lattice. We therefore neglect the second term on the right-hand side of eqn (1.20) and assume that the potential of the crystal is representable in the form of a sum of potentials of individual atoms

$$V(\mathbf{r}) = \sum_{n,j} \varphi_j(\mathbf{r} - \mathbf{r}_n - \mathbf{R}_j), \qquad (7.16)$$

where in a perfect crystal (i.e. in a crystal containing no defects) $\mathbf{r}_n$ is the position of the origin of the n-th unit cell and $\mathbf{R}_j$ is the coordinate of the j-th atom in that cell. The presence of an individual interstitial or a substitutional atom or a vacancy in a crystal does not alter the periodic part of the potential (7.16) at large distances from the defect (although the presence of high volume concentration of defects actually does give rise to a noticeable average expansion or contraction of the lattice (Hall et al., 1966)). Subtracting $V(\mathbf{r})$ given by (7.16) from the potential of the crystal containing an interstitial atom (cases A and B in Fig. 7.3) we arrive at

$$\delta V^{(int)}(\mathbf{r}) = \varphi^{(int)}(\mathbf{r} - \mathbf{r}_n - \mathbf{R}_{int}) + \sum_{j,m \text{ near } int} [\varphi_j(\mathbf{r} - \mathbf{r}_m - \mathbf{R}'_j) - \varphi_j(\mathbf{r} - \mathbf{r}_m - \mathbf{R}_j)],$$

$$(7.17)$$

where $\varphi_{int}(\mathbf{r} - \mathbf{r}_n - \mathbf{R}_{int})$ is the potential of the atom embedded in the crystal lattice. The remaining terms in (7.17) are due to the displacement of atoms both in the cell n containing the interstitial site and also in the neighbouring unit cells. The deviation of the interaction potential $V(\mathbf{r})$ from periodicity due to the displacement of atoms around defect sites give rise to a special type of diffuse scattering investigated by Huang (Huang, 1947). The nature of this type of scattering is associated with long-range elastic forces giving rise to long-range displacement fields around impurities, vacancies, and interstitial atoms.

For the case of a substitutional atom (case C shown in Fig. 7.3) we find

$$\delta V^{(sub)}(\mathbf{r}) = \varphi^{(sub)}(\mathbf{r} - \mathbf{r}_n - \mathbf{R}_j) - \varphi_j(\mathbf{r} - \mathbf{r}_n - \mathbf{R}_j)$$

$$+ \sum_{j'(\neq j), m(\neq n) \text{ near } sub} [\varphi_{j'}(\mathbf{r} - \mathbf{r}_m - \mathbf{R}'_{j'}) - \varphi_{j'}(\mathbf{r} - \mathbf{r}_m - \mathbf{R}_{j'})], \quad (7.18)$$

where sub refers to the substitutional atom and j, j' refer to atoms of the matrix.

In the case D shown in Fig. 7.3 where one lattice site is vacant (this is a vacancy defect), the first term in $\delta V(\mathbf{r})$ has minus sign in front of it

$$\delta V^{(vac)}(\mathbf{r}) = -\varphi_j(\mathbf{r} - \mathbf{r}_n - \mathbf{R}_j) + \sum_{j'(\neq j), m(\neq n) \text{ near } vac} [\varphi_{j'}(\mathbf{r} - \mathbf{r}_m - \mathbf{R}'_{j'}) - \varphi_{j'}(\mathbf{r} - \mathbf{r}_m - \mathbf{R}_{j'})].$$

$$(7.19)$$

Here $\varphi_j(\mathbf{r} - \mathbf{r}_n - \mathbf{R}_j)$ is the potential of the atom that occupied the respective lattice site before the vacancy was created. Note that in the case of the vacancy defect neighbouring atoms are normally displaced inwards towards the centre of the defect, while embedding of an interstitial atom in a crystal always leads to the expansion of the lattice.

Equations (7.17)–(7.19) show that the presence of defects in a crystal gives rise to two radically different types of diffuse scattering. One type of scattering is associated with strong but very localized fluctuations of the potential near the centre of the defect. The other type of diffuse scattering results from random displacements of adjacent atoms in the matrix due to the random distribution of defects in the crystal lattice. We now consider these two types of scattering separately beginning with the case where displacements of atoms surrounding the defect may be neglected. Note that in practice it is the second type of diffuse scattering which gives the dominant contribution to the effective 'absorption' of electrons propagating through a crystalline specimen (Hall et al., 1966).

We shall return to the problem of scattering by randomly displaced atoms of the matrix later in this chapter, where we consider diffuse scattering of electrons by thermal atomic vibrations. At present we focus our attention on diffuse scattering by the *centres* of point defects. The treatment of this type of scattering is relatively simple, and it is instructive to begin the investigation of the effects of diffuse scattering with an example showing in an explicit form how diffraction effects influence the cross-section of scattering by localized perturbations.

In the case of scattering of electrons by centres of point defects the form of the fluctuation part of the potential entering the expression for the differential cross-section is nearly identical in all the three cases (7.17)–(7.19) considered above. Namely, $\delta V(\mathbf{r}) = \varphi(\mathbf{r} - \mathbf{r}_n - \mathbf{R}_d)$, where $\mathbf{R}_d$ refers to the position of the defect in the crystal unit cell and $\varphi(\mathbf{r})$ is either the potential of the interstitial atom itself, or the difference between the potential of the substitutional atom and the potential of the atom of the matrix, or just the negative of the potential of the atom removed from a lattice site.

From the point of view of practical applications to problems of electron microscopy and diffraction, it makes little sense to try to evaluate the cross-section of diffuse scattering for wave vectors $\mathbf{k}$ and $\mathbf{k}_0$ both satisfying Bragg conditions. In this case the intensity of diffuse scattering (the second term in (7.14)) is too small in comparison with the intensity of Bragg diffraction spots (represented by the first term in (7.14)). Therefore in what follows we assume that the pair $(\mathbf{k}, \mathbf{k}_0)$ *does not* satisfy the Bragg condition and that the first term of (7.14) vanishes. Our calculations therefore refer to the distribution of intensity in the diffraction pattern in the region *between* diffraction spots where the 'ordinary' dynamical diffraction of electrons makes no contribution at all. In this case we disregard the zero-order term in (7.14) and find

$$\left(\frac{d\sigma}{do}\right)_{\text{diffuse}} = \left(\frac{m}{2\pi\hbar^2}\right)^2 \left| \int \psi_{-\mathbf{k}}^{(0)}(\mathbf{r})\varphi(\mathbf{r} - \mathbf{r}_n - \mathbf{R}_d)\psi_{\mathbf{k}_0}^{(0)}(\mathbf{r})d\mathbf{r} \right|^2, \qquad (7.20)$$

where $\psi_{\mathbf{k}_0}^{(0)}(\mathbf{r})$ is the wave function of electrons incident in the direction $\mathbf{k}_0$ and $\psi_{-\mathbf{k}}^{(0)}(\mathbf{r})$ is the wave function of electrons incident in the direction $-\mathbf{k}$ (see Fig. 7.3).

If the periodic component of the crystal potential is negligible, wave functions of the incident and scattered electrons are described by plane waves $\psi_{\mathbf{k}_0}^{(0)}(\mathbf{r}) = \exp(i\mathbf{k}_0 \cdot \mathbf{r})$ and $\psi_{-\mathbf{k}}^{(0)}(\mathbf{r}) = \exp(-i\mathbf{k} \cdot \mathbf{r})$. The cross-section of diffuse scattering in this case is proportional to the square of the Fourier component of the perturbation

$$\left(\frac{d\sigma}{do}\right)_{\text{diffuse}} = \left(\frac{m}{2\pi\hbar^2}\right)^2 |\varphi(\mathbf{k} - \mathbf{k}_0)\exp[-i(\mathbf{k} - \mathbf{k}_0) \cdot (\mathbf{r}_n + \mathbf{R}_d)]|^2$$

$$= \left(\frac{m}{2\pi\hbar^2}\right)^2 |\varphi(\mathbf{k} - \mathbf{k}_0)|^2, \qquad (7.21)$$

where $\varphi(\mathbf{k})$ denotes the Fourier transform of $\varphi(\mathbf{r})$, and contains no information about the position of the centre of the defect in the crystal lattice. By analysing the dependence of (7.21) on the angle of scattering it is not possible to say whether a particular defect is an interstitial atom or a vacancy. Nevertheless, even in the absence of information about the position of defects in the lattice the investigation of diffuse scattering by point defects by electron diffraction proves useful.

An interesting example illustrating the possibility of using diffuse scattering for imaging doped regions in semiconductor devices was given by Hull and colleagues (Hull et al., 1995a, Hull et al., 1995b). In this work Hull et al. investigated how the cross-section of large-angle diffuse scattering ($\theta \sim 50$ mrad, which at $E = 200$ keV corresponds to $s \sim 1\,\text{Å}^{-1}$) depends on the concentration of dopant atoms in crystalline indium phosphate. A substantial difference in the intensity of diffuse scattering was observed between cases where the beam of electrons was focussed on regions containing no impurities and on regions characterized by high concentration of dopant atoms. Hull et al. attributed the origin of the observed effects to scattering by fluctuations of the potential associated with centres of point defects. At the same time they noted that a substantial contribution to the cross-section comes from scattering by atoms displaced from their equilibrium positions by the elastic field associated with the defects.

Now we consider a case where either before or after the event of diffuse scattering the electron undergoes dynamical diffraction by the periodic part $V_0(\mathbf{r})$ of the potential of the crystal. In this case the cross-section of diffuse scattering is sensitive to the *location* of defects in the crystal lattice. The origin of this sensitivity is associated with the fact that wave functions describing the incident and scattered electrons feel the periodicity of the lattice. As a result the amplitude of scattering of the incident high-energy electron by a point defect turns out to be dependent on the position of the centre of the defect in the unit cell.

To illustrate the origin of the dependence we calculate the wave functions $\psi_{\mathbf{k}_0}^{(0)}(\mathbf{r})$ and $\psi_{-\mathbf{k}}^{(0)}(\mathbf{r})$ entering (7.20) using the two-beam approximation. In this approximation (Hashimoto et al., 1962, Hirsch et al., 1977) the solution of the dynamical diffraction problem is given by a sum of two Bloch functions (this applies to both to $\psi_{\mathbf{k}_0}^{(0)}(\mathbf{r})$ and to $\psi_{-\mathbf{k}}^{(0)}(\mathbf{r})$)

$$\psi_{\mathbf{k}_0}^{(0)}(\mathbf{r}) = \alpha^{(1)} b^{(1)}(\mathbf{k}_0^{(1)}, \mathbf{r}) + \alpha^{(2)} b^{(2)}(\mathbf{k}_0^{(2)}, \mathbf{r}) \tag{7.22}$$

where

$$b^{(j)}(\mathbf{k}_0^{(j)}, \mathbf{r}) = \frac{1}{\sqrt{2}} \left\{ \left[1 - \frac{(-1)^{j+1} w}{\sqrt{1+w^2}} \right]^{1/2} \exp[i\mathbf{k}_0^{(j)} \cdot \mathbf{r}] \right.$$
$$\left. + (-1)^{j+1} \left[1 + \frac{(-1)^{j+1} w}{\sqrt{1+w^2}} \right]^{1/2} \exp[i(\mathbf{k}_0^{(j)} + \mathbf{g}) \cdot \mathbf{r}] \right\}. \tag{7.23}$$

Here the two-beam parameter w characterizing the deviation of $\mathbf{k}_0$ from the Bragg condition is given by $w = (\hbar^2/4m|V_{\mathbf{g}}|)[\mathbf{k}_{0_t}^2 - (\mathbf{k}_{0_t} + \mathbf{g})^2]$, where we have assumed that the surface of the crystal is parallel to the (x, y) plane, $V_{\mathbf{g}}$ is a Fourier component of the potential of the crystal, and $\mathbf{k}_0^{(1,2)} = (\mathbf{k}_{0_t}, k_{0_z}^{(1,2)})$, where

$$\left[k_{0_z}^{(j)}\right]^2 = k_{0_z}^2 + \frac{1}{2}\Big[(\mathbf{k}_{0_t})^2 - (\mathbf{k}_{0_t} + \mathbf{g})^2$$
$$+ (-1)^{j+1}\sqrt{[(\mathbf{k}_{0_t})^2 - (\mathbf{k}_{0_t} + \mathbf{g})^2]^2 + 16m^2|V_{\mathbf{g}}|^2/\hbar^4}\Big]. \tag{7.24}$$

The excitation amplitudes $\alpha^{(j)}$ for the incident electrons are determined by the condition that at the surface of the crystal $z = 0$ the solution matches the incident plane wave,

$$\psi_{\mathbf{k}_0}^{(0)}(\mathbf{r})\Big|_{surf} = \exp(i\mathbf{k}_0 \cdot \mathbf{r}).$$

This gives

$$\alpha^{(j)} = \frac{1}{\sqrt{2}}\left[1 - \frac{(-1)^{j+1}w}{\sqrt{1+w^2}}\right]^{1/2}. \tag{7.25}$$

The excitation amplitudes of the two Bloch states forming the wave function of the incident electron (7.22) depend on the sign and on the magnitude of w. For $w > 0$ it is the Bloch state (2) which gives the dominant contribution to (7.22). On the other hand the Bloch state (1) has a higher excitation probability in the region $w < 0$. Depending on the type of the Bloch state occupied for a given direction of incidence, the magnitude of the cross-section of diffuse scattering varies quite substantially as a function of the position of the centre of the defect in a crystal unit cell.

As an illustration let us evaluate the cross-section of diffuse scattering for the somewhat artificial case where the incident electron occupies Bloch state (2). Substituting the expression for $b^{(2)}(\mathbf{k}_0^{(2)}, \mathbf{r})$ from (7.23) into (7.20) and approximating the wave function of the final state by a plane wave (i.e. assuming that $\psi_{-\mathbf{k}}^{(0)}(\mathbf{r}) = \exp(-i\mathbf{k} \cdot \mathbf{r}))$, we find

$$\left(\frac{d\sigma}{do}\right)_{\text{diffuse}}^{(2)} = \left(\frac{m}{2\pi\hbar^2}\right)^2 \left| \frac{1}{\sqrt{2}}\left[1 + \frac{w}{\sqrt{1+w^2}}\right]^{1/2} \varphi(\mathbf{k} - \mathbf{k}_0^{(2)}) \right. \tag{7.26}$$
$$\left. - \frac{1}{\sqrt{2}}\left[1 - \frac{w}{\sqrt{1+w^2}}\right]^{1/2} \varphi(\mathbf{k} - \mathbf{k}_0^{(2)} - \mathbf{g})\exp[-i\mathbf{g} \cdot (\mathbf{r}_n + \mathbf{R}_d)] \right|^2.$$

The second term on the right-hand side of this equation contains a phase factor of the form $\exp[-i\mathbf{g} \cdot (\mathbf{r}_n + \mathbf{R}_d)]$, which depends on the coordinate of the position of the centre of the defect in the unit cell. If $w = 0$ (i.e. if the incident beam satisfies the Bragg condition exactly) and if the defect is situated right at the origin of the unit cell where $\exp[-i\mathbf{g} \cdot (\mathbf{r}_n + \mathbf{R}_d)] = 1$, the two terms in (7.27) almost cancel each other (for the relatively large angles of scattering the difference between $\varphi(\mathbf{k} - \mathbf{k}_0^{(2)})$ and $\varphi(\mathbf{k} - \mathbf{k}_0^{(2)} - \mathbf{g})$ is small) and almost no diffuse scattering is observed. On the other hand, if the centre of the defect is displaced by a half of the lattice period away from the origin of the unit cell, where $\exp[-i\mathbf{g}\cdot(\mathbf{r}_n+\mathbf{R}_d)] = -1$, the cross-section of scattering is maximum and is approximately four times the value given by (7.20). The fact that dynamical diffraction of electrons in a crystal has a very strong effect on diffuse scattering was first noted by Hall et al. (Hall et al., 1966). Experimental investigations of diffuse scattering by defects under dynamical diffraction conditions have been recently carried out by Perovic et al. (Perovic et al., 1991) and by Rossouw et al. (Rossouw et al., 1994). A comprehensive summary of the current status of electron microscope observations of radiation-induced defects is given in a book by Jenkins and Kirk (Jenkins and Kirk, 2001).

7.4 The Van Hove dynamic form factor

Our next goal is to understand how high-energy electrons are scattered by thermal vibrations of atoms and how they dissipate energy through the creation of electronic excitations. In principle it is possible to address these two issues separately beginning each time with the Schrödinger equation for the entire system consisting of the incident high-energy electron and particles of the solid. This route was followed more than 40 years ago by Yoshioka (Yoshioka, 1957).

However, since then some more convenient and more powerful mathematical tools have been developed for the treatment of the same problem. The main idea of the modern approach to the treatment of scattering by electronic excitations and by thermal vibrations of atoms consists in separating the effects associated with scattering by the average potential $V_0(\mathbf{r})$ from the treatment of scattering resulting from inelastic transitions between many-particle eigenstates of the solid.

The idea of factorizing the cross-section of inelastic scattering and separating out a quantity called the *dynamic form factor* was proposed by Van Hove (Van Hove, 1954) several years before Yoshioka developed his approach to inelastic scattering of electrons in a crystalline environment. But since Van Hove's treatment was purely kinematical (i.e. it did not address the problem of how to treat elastic diffraction before and after an inelastic collision) and because it was more focussed on problems of neutron and X-ray scattering, it is not surprising that for a relatively long time Van Hove's ideas had little impact on the development of electron microscopy and electron diffraction.

To introduce the Van Hove approach we re-examine formula (7.14) by looking at it from a slightly different angle. We recall that the Coulomb potential of interaction between the incident high-energy electron and the crystal is a function of the coordinate of the incident electron $\mathbf{r}$ and the coordinates of electrons and atomic nuclei forming the solid, $V(\mathbf{r}) = V(\mathbf{r},\mathbf{r}_1,...,\mathbf{r}_i,....;\mathbf{R}_1,...,\mathbf{R}_n,...)$. Electrons and nuclei in the solid are constantly moving and it is the time-averaged periodic part of this interaction potential that gives rise to electron diffraction.

How do we evaluate the periodic part of the interaction potential from first principles? To answer this question we must remember the fundamental principle of Gibbs's statistical mechanics that states that averaging over a long period of time is equivalent to averaging over the *statistical ensemble*. Statistical averaging does not require tracing the trajectory of every electron and every atomic nucleus in a solid and solving the respective (Newton or Heisenberg) equations of motion. Instead statistical averaging assumes that all the configurations $\{\mathbf{r}_1, ..., \mathbf{r}_i, ...; \mathbf{R}_1, ..., \mathbf{R}_n, ...\}$ are possible and can be sampled by the moving particles given a sufficiently long interval of time. The contribution of each of these configurations to the average value is proportional to the statistical weight of the configuration. This statistical weight is given by $\exp(-\mathcal{H}/k_B T)$, where $\mathcal{H} = \mathcal{H}(\mathbf{r}_1, ..., \mathbf{r}_i, ...; \mathbf{R}_1, ..., \mathbf{R}_n, ...)$ is the Hamiltonian function of the solid (note that this function does not depend on the coordinate of the incident electron) and T is the absolute temperature. For example, the average interaction potential is given by

$$\langle V(\mathbf{r}, \mathbf{r}_1, ..., \mathbf{r}_i, ...; \mathbf{R}_1, ..., \mathbf{R}_n, ...)\rangle = V_0(\mathbf{r}) \qquad (7.27)$$

$$= \frac{1}{\mathcal{Z}} \int V(\mathbf{r}, \mathbf{r}_1, ..., \mathbf{r}_i, ...; \mathbf{R}_1, ..., \mathbf{R}_n, ...)$$

$$\times \exp\left[-\frac{\mathcal{H}(\mathbf{r}_1, ..., \mathbf{r}_i, ...; \mathbf{R}_1, ..., \mathbf{R}_n, ...)}{k_B T}\right] d\mathbf{r}_1...d\mathbf{r}_i...d\mathbf{R}_1...d\mathbf{R}_n...,$$

where $\mathcal{Z}$ is the partition function. The partition function is just a numerical factor that is required to satisfy the normalization condition $\langle 1 \rangle = 1$, namely

$$\mathcal{Z} = \int \exp\left[-\frac{\mathcal{H}(\mathbf{r}_1, ..., \mathbf{r}_i, ...; \mathbf{R}_1, ..., \mathbf{R}_n, ...)}{k_B T}\right] d\mathbf{r}_1...d\mathbf{r}_i...d\mathbf{R}_1...d\mathbf{R}_n... .$$

Equation (7.27) is valid for the case where electrons and nuclei move according to the laws of classical mechanics. For the quantum-mechanical treatment we introduce the normalized wave functions $\Psi_l(\mathbf{r}_1, ..., \mathbf{r}_i, ...; \mathbf{R}_1, ..., \mathbf{R}_n, ...)$ of the ground ($l = 0$) and excited ($l > 0$) states of the solid. The calculation of the average potential involves carrying out summation over all the many-particle eigenstates of the solid

$$\langle V(\mathbf{r}, \mathbf{r}_1, ..., \mathbf{r}_i, ...; \mathbf{R}_1, ..., \mathbf{R}_n, ...)\rangle = V_0(\mathbf{r}) \qquad (7.28)$$

$$= \frac{1}{\mathcal{Z}} \sum_l \int V(\mathbf{r}, \mathbf{r}_1, ..., \mathbf{r}_i, ...; \mathbf{R}_1, ..., \mathbf{R}_n, ...)$$

$$\times \exp\left(-\frac{\epsilon_l}{k_B T}\right) |\Psi_l(\mathbf{r}_1, ..., \mathbf{r}_i, ...; \mathbf{R}_1, ..., \mathbf{R}_n, ...)|^2 d\mathbf{r}_1...d\mathbf{r}_i...d\mathbf{R}_1...d\mathbf{R}_n...,$$

where $\mathcal{Z} = \sum_l \exp(-\epsilon_l/k_B T)$. Despite the seemingly simple appearance, expression (7.28) is more difficult to evaluate than (7.27). While (7.27) gives an explicit recipe of how to calculate the average value (all that one has to do is to compute a multidimensional integral), eqn (7.28) requires finding many-particle wave functions $\Psi_l(\mathbf{r}_1, ..., \mathbf{r}_i, ...; \mathbf{R}_1, ..., \mathbf{R}_n, ...)$ before carrying out summation over the quantum states of the solid. However, since electrons and vibrational modes of atomic nuclei (phonons) are quantum objects, (7.27) is rarely used for practical purposes. It is often possible to separate contributions to (7.28) from phonon and electronic excitations. Then the 'electronic part' of averaging over the quantum states is performed using density functional theory discussed in Chapter 1.

The fluctuating part of the potential is given by the difference between the instantaneous and the average values of (7.28). Let us evaluate the matrix element of (7.28) corresponding to a transition between two quantum states of the solid. Since the average potential $V_0(\mathbf{r})$ does not depend on the coordinates of particles of the solid, the matrix elements of the type

$$\int d\mathbf{r}_1...d\mathbf{r}_i d\mathbf{R}_1...d\mathbf{R}_n... \; \Psi_{l'}^*(\mathbf{r}_1, ..., \mathbf{r}_i, ...; \mathbf{R}_1, ..., \mathbf{R}_n, ...) V_0(\mathbf{r})$$
$$\times \; \Psi_l(\mathbf{r}_1, ..., \mathbf{r}_i, ...; \mathbf{R}_1, ..., \mathbf{R}_n, ...)$$

vanish unless we have $l = l'$. In other words,

$$\langle l'|V_0(\mathbf{r})|l\rangle = \delta_{l'l}V_0(\mathbf{r}). \tag{7.29}$$

Therefore, the matrix element of the fluctuating part of the potential is equal to

$$\langle l'|\delta V(\mathbf{r}, \mathbf{r}_1, ..., \mathbf{r}_i, ...; \mathbf{R}_1, ..., \mathbf{R}_n, ...)|l\rangle = \langle l'|V(\mathbf{r}, \mathbf{r}_1, ..., \mathbf{r}_i, ...; \mathbf{R}_1, ..., \mathbf{R}_n, ...)|l\rangle$$
$$- \delta_{l'l}V_0(\mathbf{r}). \tag{7.30}$$

It follows from this equation and from the definition of $V_0(\mathbf{r})$ that

$$\sum_l \exp(-\epsilon_l/k_B T)\langle l|\delta V(\mathbf{r}, \mathbf{r}_1, ..., \mathbf{r}_i, ...; \mathbf{R}_1, ..., \mathbf{R}_n, ...)|l\rangle = 0. \tag{7.31}$$

This condition is entirely equivalent to the statement that the fluctuating part of the potential vanishes after averaging over an infinitely long interval of time.

We can now generalize (7.14) and write down an expression for the amplitude of *inelastic* scattering of high-energy electrons by a crystal

$$f_{l'l}(\mathbf{k}, \mathbf{k}_0) = \delta_{l'l}f^{(0)}(\mathbf{k}, \mathbf{k}_0) \tag{7.32}$$
$$- \frac{m}{2\pi\hbar^2}\int \psi_{-\mathbf{k}}^{(0)}(\mathbf{r})\langle l'|\delta V(\mathbf{r}, \mathbf{r}_1, ..., \mathbf{r}_i, ...; \mathbf{R}_1, ..., \mathbf{R}_n, ...)|l\rangle \psi_{\mathbf{k}_0}^{(0)}(\mathbf{r})d\mathbf{r}.$$

To evaluate the cross-section of inelastic scattering we need to take into account the fact that when a high-energy electron enters the crystal, the quantum state of the latter is not defined uniquely. The probability of finding the crystal in the quantum state $|l\rangle$ before the interaction with the incident electron equals $\mathcal{Z}^{-1}\exp(-\epsilon_l/k_B T)$. Furthermore, since we do not record the final quantum state of the solid after the inelastic interaction with the incident high-energy electron (we only study the direction of scattering and the energy of electrons), the cross-section of inelastic scattering should be *averaged* over the initial and *summed* over the final states of the solid,

$$\left(\frac{d^2\sigma}{dodE}\right)_{\text{inel}} = \frac{k}{k_0}\left(\frac{m}{2\pi\hbar^2}\right)^2 \frac{1}{\mathcal{Z}}\sum_{l',l}\exp\left(-\frac{\epsilon_l}{k_BT}\right) \tag{7.33}$$

$$\times\left|\int \psi^{(0)}_{-\mathbf{k}}(\mathbf{r})\langle l'|\delta V(\mathbf{r},\mathbf{r}_1,...,\mathbf{r}_i,...;\mathbf{R}_1,...,\mathbf{R}_n,...)|l\rangle\psi^{(0)}_{\mathbf{k}_0}(\mathbf{r})d\mathbf{r}\right|^2$$

$$\times\,\delta\left(\frac{\hbar^2\mathbf{k}^2}{2m} + \epsilon_{l'} - \frac{\hbar^2\mathbf{k}_0^2}{2m} - \epsilon_l\right).$$

Here *do* denotes an element of the solid angle and

$$E = \frac{\hbar^2}{2m}\left(\mathbf{k}_0^2 - \mathbf{k}^2\right)$$

is the energy lost by the incident electron in the inelastic collision.

We can now introduce the mixed dynamic form factor $\bar{s}(\mathbf{r}',\mathbf{r},\omega)$ representing the main entity of the Van Hove theory

$$\bar{s}(\mathbf{r}',\mathbf{r},\omega) = \frac{1}{\mathcal{Z}}\sum_{l',l}\exp\left(-\frac{\epsilon_l}{k_BT}\right)\langle l|\delta V(\mathbf{r}',\mathbf{r}_1,...,\mathbf{r}_i,...;\mathbf{R}_1,...,\mathbf{R}_n,...)|l'\rangle$$

$$\times\langle l'|\delta V(\mathbf{r},\mathbf{r}_1,...,\mathbf{r}_i,...;\mathbf{R}_1,...,\mathbf{R}_n,...)|l\rangle\delta\left(\frac{\epsilon_{l'}-\epsilon_l}{\hbar}-\omega\right). \tag{7.34}$$

Using this definition and taking into account the property of the delta function $\delta(cf(x)) = (1/|c|)\delta(f(x))$, we find

$$\left(\frac{d^2\sigma}{dodE}\right)_{\text{inel}} = \frac{k}{k_0}\left(\frac{m}{2\pi\hbar^2}\right)^2 \frac{1}{\hbar} \tag{7.35}$$

$$\times\int\int \psi^{(0)}_{-\mathbf{k}}(\mathbf{r})\left[\psi^{(0)}_{-\mathbf{k}}(\mathbf{r}')\right]^*\psi^{(0)}_{\mathbf{k}_0}(\mathbf{r})\left[\psi^{(0)}_{\mathbf{k}_0}(\mathbf{r}')\right]^*\bar{s}(\mathbf{r}',\mathbf{r},E/\hbar)d\mathbf{r}d\mathbf{r}'.$$

This is the first of the two fundamental statements underpinning Van Hove's treatment of inelastic scattering. Formula (7.35) shows that the cross-section of inelastic scattering of any type can be represented by an integral of the product of four wave functions describing dynamical diffraction of electrons by the average potential $V_0(\mathbf{r})$, and a quantity (7.34) which contains information about the spectrum of excitations of the solid and about the spatial localization of these excitations. Hence, the problem of calculation of the cross-section of inelastic scattering requires addressing two entirely separate issues: (1) solving the diffraction problem for the average potential $V_0(\mathbf{r})$ and finding the wave functions of the initial and final states $\psi^{(0)}_{\mathbf{k}_0}(\mathbf{r})$ and $\psi^{(0)}_{-\mathbf{k}}(\mathbf{r})$, and (2) evaluating the mixed dynamic form factor $\bar{s}(\mathbf{r}',\mathbf{r},\omega)$. It is this factorization of the cross-section of inelastic scattering that makes the Van Hove approach considerably more convenient and easier to use than the treatment developed by Yoshioka.

The second, and perhaps the most beautiful, statement of Van Hove's theory clarifies the link between time-dependent fluctuations of the potential and inelastic scattering of electrons. To understand this connection we Fourier transform (7.34) and introduce the *time-dependent* mixed dynamic form factor

$$\bar{s}_t(\mathbf{r}', \mathbf{r}, t) = \int\limits_{-\infty}^{\infty} \exp(-i\omega t)\bar{s}(\mathbf{r}', \mathbf{r}, \omega)d\omega. \tag{7.36}$$

Using (7.34) we arrive at

$$\bar{s}_t(\mathbf{r}', \mathbf{r}, t) = \frac{1}{\mathcal{Z}} \sum_{l',l} \exp\left(-\frac{\epsilon_l}{k_B T}\right) \left\langle l \left| \exp\left(i\frac{\epsilon_l}{\hbar}t\right) \delta V(\mathbf{r}', \mathbf{r}_1, ...\mathbf{r}_i...; \mathbf{R}_1, ...\mathbf{R}_n...) \right. \right.$$

$$\times \left. \left. \exp\left(-i\frac{\epsilon_{l'}}{\hbar}t\right) |l'\rangle\langle l'| \delta V(\mathbf{r}, \mathbf{r}_1, ..., \mathbf{r}_i, ...; \mathbf{R}_1, ..., \mathbf{R}_n, ...) \right| l \right\rangle. \tag{7.37}$$

Since ϵ_l and $\epsilon_{l'}$ are energies of eigenstates of the Hamiltonian of the crystal $\hat{H}_{cr}$, we write

$$\langle l| \exp(i\epsilon_l t/\hbar) = \langle l| \exp(i\hat{H}_{cr}t/\hbar)$$

and

$$\exp(-i\epsilon_{l'}t/\hbar)|l'\rangle = \exp(-i\hat{H}_{cr}t/\hbar)|l'\rangle.$$

Now we recall the definition of the Heisenberg form of a quantum-mechanical operator

$$\hat{V}(t) = \exp(i\hat{H}_{cr}t/\hbar)\hat{V}\exp(-i\hat{H}_{cr}t/\hbar).$$

Using this definition we arrive at

$$\bar{s}_t(\mathbf{r}', \mathbf{r}, t) = \langle \delta\hat{V}(\mathbf{r}', \mathbf{r}_1(t), ..., \mathbf{r}_i(t), ...; \mathbf{R}_1(t), ..., \mathbf{R}_n(t), ...)$$

$$\times \delta\hat{V}(\mathbf{r}, \mathbf{r}_1(0), ..., \mathbf{r}_i(0), ...; \mathbf{R}_1(0), ..., \mathbf{R}_n(0), ...)\rangle, \tag{7.38}$$

where the angle brackets denote averaging over the statistical ensemble

$$\langle...\rangle = \mathcal{Z}^{-1} \sum_l \exp(-\epsilon_l/k_B T)\langle l|...|l\rangle.$$

Using (7.38) we find the final expression for the double differential cross-section of inelastic scattering

$$\left(\frac{d^2\sigma}{dodE}\right)_{\text{inel}} = \frac{k}{k_0} \left(\frac{m}{2\pi\hbar^2}\right)^2 \int\limits_{-\infty}^{\infty} \frac{dt}{2\pi\hbar} \exp\left(i\frac{E}{\hbar}t\right) \tag{7.39}$$

$$\times \int\int \psi_{-\mathbf{k}}^{(0)}(\mathbf{r}) \left[\psi_{-\mathbf{k}}^{(0)}(\mathbf{r}')\right]^* \psi_{\mathbf{k}_0}^{(0)}(\mathbf{r}) \left[\psi_{\mathbf{k}_0}^{(0)}(\mathbf{r}')\right]^* \bar{s}_t(\mathbf{r}', \mathbf{r}, t)d\mathbf{r}d\mathbf{r}'.$$

This equation, which is central to Van Hove's treatment of inelastic scattering, shows that the cross-section of inelastic scattering is given by the Fourier transform of the temporal correlation function of the fluctuating part of the interaction potential. The time dependence of the fluctuating part of the potential is due to the motion of electrons and atomic nuclei in the solid. A generalisation of eqn (7.39) to the case of a partially coherent incident beam of electrons is given in Chapter 8.

7.5 Thermal diffuse scattering

In this section we show how to evaluate the Van Hove dynamic form factor for thermal vibrations of atomic nuclei. Diffuse scattering due to fluctuations of the interaction potential associated with the thermal motion of atomic nuclei is called thermal diffuse scattering (TDS). It is important to realize that TDS is *not* the phonon scattering in the same sense as it is known in the theory of electron phonon interactions in solids. TDS is not necessarily accompanied by creation or annihilation of single phonons. Instead the Van Hove formulae (7.35) and (7.39) describe interactions followed by creation and annihilation of several phonons in each inelastic collision, i.e. multiphonon processes are included.

TDS is sometimes also called quasielastic scattering. The term 'quasielastic' is used to emphasize the fact that although interaction with atomic vibrations does lead to the excitation of phonons, the energy lost or acquired in a collision ($\sim k_B T \approx 0.025$ eV) is many times smaller than the typical resolution of electron spectrometers and therefore in practice electrons scattered by atomic vibrations can hardly be distinguished from truly elastically scattered (diffracted) electrons by means of energy filtering.

In the treatment of scattering by atomic vibrations we assume that TDS is not accompanied by the excitation of electronic states in the solid. The electronic subsystem is assumed to remain in its ground state where the electron density follows the instantaneous positions of atomic nuclei (the Born–Oppenheimer approximation). In this approximation the fluctuating part of the potential $\delta \hat{V}(t)$ is given by

$$\delta \hat{V}(\mathbf{r}, \mathbf{R}_1(t), ..., \mathbf{R}_n(t), ...)$$
$$= \sum_{n,j} \varphi_j(\mathbf{r} - \mathbf{r}_n - \mathbf{r}_{nj} - \hat{\mathbf{u}}_{nj}(t)) - \langle \varphi_j(\mathbf{r} - \mathbf{r}_n - \mathbf{r}_{nj} - \hat{\mathbf{u}}_{nj}(t)) \rangle, \quad (7.40)$$

where $\hat{\mathbf{u}}_{nj}$ is the operator of displacement of an atom from its equilibrium position $\mathbf{r}_n - \mathbf{r}_{nj}$ and $\langle \hat{\mathbf{u}}_{nj} \rangle = 0$. The Fourier integral representation of (7.40) is

$$\delta \hat{V}(\mathbf{r}, \mathbf{R}_1(t), ..., \mathbf{R}_n(t), ...) = \sum_{n,j} \int \frac{d\mathbf{q}}{(2\pi)^3} \varphi_j(\mathbf{q}) \exp[i\mathbf{q} \cdot (\mathbf{r} - \mathbf{r}_n - \mathbf{r}_{nj})]$$
$$\times \{ \exp[-i\mathbf{q} \cdot \mathbf{u}_{nj}(t)] - \langle \exp[-i\mathbf{q} \cdot \mathbf{u}_{nj}(t)] \rangle \}, (7.41)$$

where $\varphi_j(\mathbf{q})$ is the Fourier transform of $\varphi_j(\mathbf{r})$. The Van Hove dynamic form factor for thermal diffuse scattering can now be found by substituting (7.41) into (7.37)

$$\bar{s}_t(\mathbf{r}, \mathbf{r}', t) = \sum_{n,j} \sum_{n',j'} \int \int \frac{d\mathbf{q}}{(2\pi)^3} \frac{d\mathbf{q}'}{(2\pi)^3} \varphi_j(\mathbf{q}) \varphi_{j'}(\mathbf{q}')$$

$$\times \exp[i\mathbf{q} \cdot (\mathbf{r} - \mathbf{r}_n - \mathbf{r}_{nj}) + i\mathbf{q}' \cdot (\mathbf{r}' - \mathbf{r}_{n'} - \mathbf{r}_{n'j'})]$$

$$\times \{\langle \exp[-i\mathbf{q} \cdot \hat{\mathbf{u}}_{nj}(t) - i\mathbf{q}' \cdot \hat{\mathbf{u}}_{n'j'}(0)]\rangle - \langle \exp[-i\mathbf{q} \cdot \hat{\mathbf{u}}_{nj}(t)]\rangle$$

$$\times \langle \exp[-i\mathbf{q}' \cdot \hat{\mathbf{u}}_{n'j'}(0)]\rangle\}. \tag{7.42}$$

The practical question now consists in finding a way of evaluating the average exponential functions of operators of atomic thermal displacements. Afanas'ev and colleagues (Afanas'ev and Kagan, 1968, Afanas'ev et al., 1968) showed how this can be performed using the normal (phonon) mode representation of the displacement operators $\hat{\mathbf{u}}_{nj}(t)$. In the harmonic approximation each atomic displacement $\mathbf{u}_{nj}(t)$ can be represented in the form of a sum of contributions of normal modes (Marshall and Lovesey, 1971)

$$\hat{\mathbf{u}}_{n,j}(t) = \sum_{\alpha, \mathbf{f}} \left(\frac{\hbar}{2N \mathcal{M}_j \omega_\alpha(\mathbf{f})} \right)^{1/2} \left(\mathbf{e}_j(\mathbf{f}, \alpha) \hat{a}_{\mathbf{f}, \alpha} \exp\left[i\mathbf{f} \cdot \mathbf{r}_{nj} - i\omega_\alpha(\mathbf{f})t \right] \right.$$

$$\left. + \mathbf{e}_j^*(\mathbf{f}, \alpha) \hat{a}_{\mathbf{f}, \alpha}^\dagger \exp\left[-i\mathbf{f} \cdot \mathbf{r}_{nj} + i\omega_\alpha(\mathbf{f})t \right] \right). \tag{7.43}$$

Here N is the number of unit cells in the crystal (which is of the order of 10^{23} cm^{-3}) and the summation is over branches of the phonon spectrum $\omega_\alpha(\mathbf{f})$. In eqn (7.43) $\mathbf{e}_j(\mathbf{f}, \alpha)$ is the vector of polarization of the α-th branch and $\mathcal{M}_j$ is the mass of the j-th atom. $\hat{a}_{\mathbf{f}, \alpha}$ and $\hat{a}_{\mathbf{f}, \alpha}^\dagger$ are the annihilation and creation operators of the phonon mode $(\mathbf{f}, \alpha)$.

Let us first consider the term $\langle \exp[-i\mathbf{q} \cdot \hat{\mathbf{u}}_{nj}(t)]\rangle$. Since, according to (7.43),

$$\hat{\mathbf{u}}_{nj}(t) = \frac{1}{\sqrt{N}} \sum_{\mathbf{f}, \alpha} \hat{\mathbf{u}}_{nj}(\mathbf{f}, \alpha, t),$$

we find

$$\langle \exp[-i\mathbf{q} \cdot \hat{\mathbf{u}}_{nj}(t)]\rangle = \prod_{\mathbf{f}, \alpha} \left\langle \exp\left[-\frac{i}{\sqrt{N}} \mathbf{q} \cdot \hat{\mathbf{u}}_{nj}(\mathbf{f}, \alpha, t) \right] \right\rangle \tag{7.44}$$

$$= \prod_{\mathbf{f}, \alpha} \left[1 - \frac{i}{\sqrt{N}} \langle \mathbf{q} \cdot \hat{\mathbf{u}}_{nj}(\mathbf{f}, \alpha, t)\rangle - \frac{1}{2N} \langle (\mathbf{q} \cdot \hat{\mathbf{u}}_{nj}(\mathbf{f}, \alpha, t))(\mathbf{q} \cdot \hat{\mathbf{u}}_{nj}(\mathbf{f}, \alpha, t))\rangle + ... \right].$$

Transforming the sum over phonon wave vectors into the integral over the Brillouin zone

$$\sum_{\mathbf{f}, \alpha} \to N\Omega_0 \int_{\mathrm{BZ}} \frac{d\mathbf{f}}{(2\pi)^3},$$

where Ω_0 is the volume of the unit cell, and also taking into account the fact that all the cubic, quartic, etc., terms in expansion (7.45) contain a macroscopically large number N in the denominator and may therefore be neglected, we obtain

$$\langle \exp[-i\mathbf{q} \cdot \hat{\mathbf{u}}_{nj}(t)] \rangle = \prod_{\mathbf{f},\alpha} \left[1 - \frac{1}{2N} \langle (\mathbf{q} \cdot \hat{\mathbf{u}}_{nj}(\mathbf{f},\alpha,t))(\mathbf{q} \cdot \hat{\mathbf{u}}_{nj}(\mathbf{f},\alpha,t)) \rangle \right]$$

$$= \exp \left[-\frac{1}{2N} \sum_{\mathbf{f},\alpha} \langle (\mathbf{q} \cdot \hat{\mathbf{u}}_{nj}(\mathbf{f},\alpha,t))^2 \rangle \right] \qquad (7.45)$$

$$= \exp \left[-\frac{1}{2} \langle (\mathbf{q} \cdot \hat{\mathbf{u}}_{nj}(t))^2 \rangle \right] = \exp \left[-\frac{1}{2} \langle (\mathbf{q} \cdot \hat{\mathbf{u}}_{nj}(0))^2 \rangle \right].$$

Taking into account the fact that in the thermal equilibrium

$$\langle \hat{a}^\dagger_{\mathbf{f},\alpha} \hat{a}_{\mathbf{f},\alpha} \rangle = \frac{1}{\exp[\hbar\omega_\alpha(\mathbf{f})/k_B T] - 1},$$

$$\langle \hat{a}_{\mathbf{f},\alpha} \hat{a}^\dagger_{\mathbf{f},\alpha} \rangle = \frac{1}{\exp[\hbar\omega_\alpha(\mathbf{f})/k_B T] - 1} + 1,$$

$$\langle \hat{a}_{\mathbf{f},\alpha} \hat{a}_{\mathbf{f},\alpha} \rangle = 0,$$

$$\langle \hat{a}^\dagger_{\mathbf{f},\alpha} \hat{a}^\dagger_{\mathbf{f},\alpha} \rangle = 0, \qquad (7.46)$$

we arrive at

$$\langle \exp[-i\mathbf{q} \cdot \hat{\mathbf{u}}_{nj}(t)] \rangle \qquad (7.47)$$

$$= \exp \left[-\Omega_0 \sum_\alpha \int_{\mathrm{BZ}} \frac{d\mathbf{f}}{(2\pi)^3} \left(\frac{\hbar}{2\mathcal{M}_j \omega_\alpha(\mathbf{f})} \right) |\mathbf{q} \cdot \mathbf{e}_j(\mathbf{f},\alpha)|^2 \coth\left(\frac{\hbar\omega_\alpha(\mathbf{f})}{2k_B T} \right) \right],$$

where $\coth(x) = (e^x + e^{-x})/(e^x - e^{-x})$. A similar calculation leads to an explicit expression for the dynamic form factor (Dudarev et al., 1993c)

$$\bar{s}_t(\mathbf{r}, \mathbf{r}', t) = \sum_{n,j} \sum_{n',j'} \int \int \frac{d\mathbf{q}d\mathbf{q}'}{(2\pi)^6} \varphi_j(\mathbf{q}) \varphi_{j'}(\mathbf{q}') \exp[i\mathbf{q} \cdot (\mathbf{r} - \mathbf{r}_n - \mathbf{r}_{nj})$$

$$+ i\mathbf{q}' \cdot (\mathbf{r}' - \mathbf{r}_{n'} - \mathbf{r}_{n'j'})] \exp\left[-\frac{1}{2} \langle (\mathbf{q} \cdot \mathbf{u}_{nj}) \rangle - \frac{1}{2} \langle (\mathbf{q}' \cdot \mathbf{u}_{n'j'}) \rangle \right]$$

$$\times [\exp(-\langle \{\mathbf{q} \cdot \mathbf{u}_{nj}(t)\}\{\mathbf{q}' \cdot \mathbf{u}_{n'j'}(0)\} \rangle) - 1]. \qquad (7.48)$$

Note that this form factor not only depends on the mean square amplitude of thermal displacements but also contains the *time-dependent* correlation function $Y_{nj,n'j'}(\mathbf{q}, \mathbf{q}', t) = \langle (\mathbf{q} \cdot \mathbf{u}_{nj}(t))(\mathbf{q}' \cdot \mathbf{u}_{n'j'}(0)) \rangle$ of thermal displacements of atoms occupying different sites in the crystal lattice.

Fortunately for practical purposes the time dependence of this correlation function is not important. Indeed, as mentioned earlier, in most cases electron spectrometers used in conjunction with electron microscopes are not able to distinguish between elastically and quasielastically scattered electrons. This means that the diffuse intensity recorded by detectors is always integrated over a relatively large (in comparison with the characteristic phonon energy loss of ~ 0.025

eV) energy window of the order of 1 eV. Hence to calculate the *observed* cross-section of thermal diffuse scattering we need to integrate (7.39) over a relatively wide range of energies E (or, equivalently, over a range of frequencies $\omega = E/\hbar$) centred around $\omega = 0$. Noting that

$$\int_{-\infty}^{\infty} \exp(i\omega t)d\omega = 2\pi\delta(t),$$

from (7.39) we find

$$\left(\frac{d\sigma}{do}\right)_{\text{TDS}} = \left(\frac{m}{2\pi\hbar^2}\right)^2 \int \psi_{-\mathbf{k}}^{(0)}(\mathbf{r}) \left[\psi_{-\mathbf{k}}^{(0)}(\mathbf{r}')\right]^* \psi_{\mathbf{k}_0}^{(0)}(\mathbf{r}) \left[\psi_{\mathbf{k}_0}^{(0)}(\mathbf{r}')\right]^* \overline{s}_t(\mathbf{r}',\mathbf{r},0)d\mathbf{r}d\mathbf{r}'.$$

$$(7.49)$$

Here, $\overline{s}_t(\mathbf{r}',\mathbf{r},0)$ depends only on how displacements of different atoms are correlated *at the same moment of time.*

In principle, by following a similar procedure, it is possible to derive a general formula for the correlation function of thermal displacements (see for example eqn (20) of (Dudarev et al., 1993c)). However, to avoid writing lengthy equations, in what follows we focus attention on the two important limiting cases that illustrate the part played by thermal diffuse scattering in electron microscopy and diffraction.

The first case corresponds to the non-correlated thermal motion of atoms (or in other words to the Einstein model of thermal vibrations). This limiting case was first considered by Hall and Hirsch in 1965 (Hall and Hirsch, 1965a). In this case

$$\langle[\mathbf{q}\cdot\mathbf{u}_{nj}(0)][\mathbf{q}'\cdot\mathbf{u}_{n'j'}(0)]\rangle = \delta_{nn'}\delta_{jj'}\langle[\mathbf{q}\cdot\mathbf{u}_{nj}(0)][(\mathbf{q}'\cdot\mathbf{u}_{nj}(0)]\rangle$$

and the dynamic form factor is given by a relatively simple expression of the form

$$\overline{s}_t(\mathbf{r},\mathbf{r}',0) = \sum_{n,j} \int\int \frac{d\mathbf{q}d\mathbf{q}'}{(2\pi)^6}\varphi_j(\mathbf{q})\varphi_j(\mathbf{q}')\exp[i\mathbf{q}\cdot(\mathbf{r}-\mathbf{r}_n-\mathbf{r}_{nj}) \qquad (7.50)$$

$$+ i\mathbf{q}'\cdot(\mathbf{r}'-\mathbf{r}_n-\mathbf{r}_{nj})]\{\exp[-M_{nj}(\mathbf{q}+\mathbf{q}')] - \exp[-M_{nj}(\mathbf{q})-M_{nj}(\mathbf{q}')]\},$$

where $M_{nj}(\mathbf{q}) = \langle(\mathbf{q}\cdot\mathbf{u}_{nj})^2\rangle/2$. Note that in the Einstein approximation the dynamic form factor depends entirely on the average square amplitude of atomic displacements, and contains no correlation terms. In the absence of dynamical diffraction effects (i.e. if the wave functions of the incident and scattered electrons have the form of plane waves) formula (7.50) leads to a very curious result

$$\left(\frac{d\sigma}{do}\right)_{\text{TDS}} = \left(\frac{m}{2\pi\hbar^2}\right)^2 \sum_{n,j} |\varphi_j(\mathbf{k}-\mathbf{k}_0)|^2 \{1 - \exp[-2M_{nj}(\mathbf{k}-\mathbf{k}_0)]\}, \quad (7.51)$$

which shows that even if atoms vibrate independently and if dynamical diffraction effects are absent, the cross-section of scattering is still *not* equal to the sum of cross-sections of scattering by individual atomic potentials. Instead the cross-section of TDS contains an additional negative term, which vanishes in the high-temperature limit, and which serves as a reminder to us that it is only the *fluctuating* part of the potential that is responsible for the diffuse scattering. The average potential $V_0(\mathbf{r})$ remains excluded from the dynamic form factor, and the trace of this exclusion is seen in (7.51) even if no diffraction effects are present (Dudarev and Ryazanov, 1985, Dudarev and Whelan, 1994c). In other words, a crystal is more transparent for high-energy electrons moving through it than is an amorphous solid characterized by the same volume density of atoms.

The second limiting case corresponds to thermal displacements correlated over large distances. Long-range correlations result from virtual excitation and de-excitation of acoustic phonons (sound waves). In any crystalline solid there are three branches of acoustic phonons and, if the solid is isotropic, the dispersion relations of these branches read $\omega_1(\mathbf{f}) = c_l f$ and $\omega_{2,3}(\mathbf{f}) = c_t f$. Subscripts '$l$' and '$t$' refer to longitudinal (displacements in the direction of $\mathbf{f}$) and transverse (displacements in the plane orthogonal to $\mathbf{f}$) phonons. There are one longitudinal and two transverse phonon modes, and c_l and c_t are their velocities. Assuming that only acoustic phonons contribute to the correlation function of atomic displacements, we arrive at (Dudarev et al., 1993c)

$$\langle (\mathbf{q} \cdot \mathbf{u}_{nj}(0))(\mathbf{q}' \cdot \mathbf{u}_{n'j'}(0)) \rangle = \frac{k_B T}{8\pi\rho} \left[\frac{(\mathbf{q} \cdot \mathbf{q}')}{|\mathbf{r}_n - \mathbf{r}_{n'}|} \left(\frac{1}{c_l^2} + \frac{1}{c_t^2} \right) \right. \tag{7.52}$$
$$\left. + \frac{(\mathbf{q} \cdot [\mathbf{r}_n - \mathbf{r}_{n'}])(\mathbf{q}' \cdot [\mathbf{r}_n - \mathbf{r}_{n'}])}{|\mathbf{r_n} - \mathbf{r}_{n'}|^3} \left(\frac{1}{c_t^2} - \frac{1}{c_l^2} \right) \right],$$

where $\rho = \mathcal{M}/\Omega_0$ is the mass density of the solid and $\mathcal{M} = \sum_j \mathcal{M}_j$ is the total mass of atoms in a crystal unit cell. By substituting this into (7.48) and (7.49) and assuming that no dynamical diffraction effects occur in either the initial $\psi_{\mathbf{k}_0}^{(0)}$ or the final $\psi_{-\mathbf{k}}^{(0)}$ states of scattered electrons, we find

$$\left(\frac{d\sigma}{do} \right)_{\text{TDS}} = \left(\frac{m}{2\pi\hbar^2} \right)^2 \sum_{nj,n'j'} \varphi_j(\mathbf{k}_0 - \mathbf{k})\varphi_{j'}(\mathbf{k} - \mathbf{k}_0)$$
$$\times \exp\left[-M_{nj}(\mathbf{k} - \mathbf{k}_0) - M_{n'j'}(\mathbf{k} - \mathbf{k}_0) \right]$$
$$\times \exp[i(\mathbf{k} - \mathbf{k}_0) \cdot (\mathbf{r}_n + \mathbf{r}_{nj} - \mathbf{r}_{n'} - \mathbf{r}_{n'j'})]$$
$$\times \langle (\{\mathbf{k} - \mathbf{k}_0\} \cdot \mathbf{u}_{nj})(\{\mathbf{k} - \mathbf{k}_0\} \cdot \mathbf{u}_{n'j'}) \rangle. \tag{7.53}$$

The long-range correlations between displacements of atoms belonging to different unit cells ($n \neq n'$) seen in (7.52) give rise to a characteristic divergence of the TDS cross-section of scattering (7.53) in the vicinity of points in $\mathbf{k}$ space where $\mathbf{k} \approx \mathbf{k}_0 + \mathbf{g}$ and $\mathbf{g}$ is a reciprocal lattice vector. Indeed, since for $|\mathbf{k} - \mathbf{k}_0 - \mathbf{g}| \ll |\mathbf{g}|$

$$\sum_{n,n'} \frac{\exp[i(\mathbf{k} - \mathbf{k}_0 - \mathbf{g}) \cdot (\mathbf{r}_n - \mathbf{r}_{n'})]}{|\mathbf{r}_n - \mathbf{r}_{n'}|} \approx \frac{N}{\Omega_0} \frac{4\pi}{|\mathbf{k} - \mathbf{k}_0 - \mathbf{g}|^2},$$

(the calculation was performed using the relationship $\int |\mathbf{r}|^{-1} \exp(i\mathbf{q} \cdot \mathbf{r})d\mathbf{r} = 4\pi/|\mathbf{q}|^2$), where N is the number of unit cells in the crystal, we find the cross-section of TDS

$$\left(\frac{d\sigma}{do}\right)_{\text{TDS}} \approx \frac{N}{\Omega_0}\left(\frac{m}{2\pi\hbar^2}\right)^2 \frac{k_B T}{2\rho}\left(\frac{1}{c_l^2} + \frac{1}{c_t^2}\right)\sum_{j,j'}\varphi_j(-\mathbf{g})\varphi_{j'}(\mathbf{g}) \tag{7.54}$$

$$\times \exp\left[-M_{nj}(\mathbf{g}) - M_{n'j'}(\mathbf{g})\right]\exp[i\mathbf{g}\cdot(\mathbf{r}_{nj} - \mathbf{r}_{\mathbf{n}'j'})]\frac{g^2}{|\mathbf{k} - \mathbf{k}_0 - \mathbf{g}|^2},$$

that diverges at $\mathbf{k} = \mathbf{k}_0 + \mathbf{g}$. Dynamical diffraction effects prevent the cross-section from being divergent even in the most 'unfavourable' situation where the incident beam satisfies the exact Bragg condition $\mathbf{k}^2 = (\mathbf{k}_0 + \mathbf{g})^2$. A solution of the problem of why and how this occurs was given by Afanas'ev and Kagan (Afanas'ev and Kagan, 1968) and by Rez et al. (Rez et al., 1977). The main point of the treatment consists in taking into account dynamical diffraction effects in the initial and final states $\psi_{\mathbf{k}_0}^{(0)}(\mathbf{r})$ and $\psi_{-\mathbf{k}}^{(0)}(\mathbf{r})$ of diffuse scattering. Dynamical diffraction effects effectively eliminate the contribution from very long-range correlations of atomic displacements resulting in the appearance of the divergent term in (7.54).

TDS is responsible for a number of features seen in transmission and reflection high-energy electron diffraction patterns, and it also accounts for the dominant part of the cross-section of electron backscattering. However, to investigate all these points in sufficient detail and to make the discussion *quantitative* it is necessary to take into account dynamical diffraction effects and also effects of *multiple* diffuse scattering. We will consider this in Chapter 8. In the following section we apply the Van Hove approach to the calculation of electron energy loss spectra.

7.6 Electron energy losses

The Van Hove treatment of inelastic scattering of high-energy electrons described above shows that this type of scattering is associated with the motion of atomic nuclei and electrons in the crystal. The motion of atomic nuclei leads to TDS, which is responsible for a significant part of the diffuse intensity seen in electron diffraction patterns. However, atomic nuclei are heavy particles and they move much more slowly than the valence or inner-shell electrons. The spectrum of energy losses of high-energy electrons is related to the correlation function of fluctuations of the interaction potential via the Fourier transform (7.39). This means that the shorter the time scales characterizing the motion of particles in the solid the higher are the corresponding energy losses. In other words it is the interaction of the incident high-energy electron with *electrons* in the solid that gives the dominant contribution to the cross-section of energy losses.

This theoretical argument agrees with experiment, and the fact that electrons move much faster than the nuclei indeed results in the characteristic scale of energy losses associated with the interaction of the incident high-energy electron with atomic and valence electrons in the solid being between two and three orders of magnitude greater than the scale of energy losses associated with diffuse scattering of high-energy electrons by vibrating atomic nuclei.

We can therefore assume that the motion of nuclei may be neglected in the treatment of inelastic collisions of the incident high-energy electron with electrons in the solid. In other words, electronic excitations see nuclei as if the latter were frozen at their instantaneous positions $\mathbf{r}_n + \mathbf{r}_{nj} + \mathbf{u}_{nj}$. Here it is important to note the difference between the *instantaneous* and *equilibrium* positions of the nuclei. Despite the fact that we have neglected the motion of nuclei, the cross-section of inelastic scattering by electronic excitations remains a function of the coordinates of atoms. In the calculation of the cross-section of energy losses, these coordinates play the part of a set of external parameters. The dependence of the cross-section of energy losses on the coordinates of atomic nuclei needs to be taken into account only at the final stage of the calculation, where this cross-section is averaged over thermal displacements of atoms.

Assuming that atomic nuclei are frozen at their instantaneous positions, in (7.38) we retain only the terms that describe the Coulomb interaction between the incident electron and electrons of the solid. The operator of the density of electrons has the form

$$P(\mathbf{r}, t) = \sum_j \delta(\mathbf{r} - \mathbf{r}_j(t)) = \int \frac{d\mathbf{q}}{(2\pi)^3} \sum_j \exp[i\mathbf{q} \cdot (\mathbf{r} - \mathbf{r}_j(t))]$$

$$= \int \frac{d\mathbf{q}}{(2\pi)^3} \exp[i\mathbf{q} \cdot \mathbf{r}] P(\mathbf{q}, t), \tag{7.55}$$

where $P(\mathbf{q}, t) = \sum_j \exp[-i\mathbf{q} \cdot \mathbf{r}_j(t)]$ is the Fourier component of the density operator. In the treatment of inelastic scattering by electronic excitations it is common to use the Fourier representation of the potential and to talk about transitions accompanied by the transfer of momentum $\hbar\mathbf{q}$ and energy $E = \hbar\omega$. We therefore represent the Coulomb interaction between the incident electron and electrons of the solid in the form of a Fourier series and express it in terms of the density of electrons

$$V(\mathbf{r}, \mathbf{r}_1(t), ..., \mathbf{r}_i(t), ...) = \frac{1}{4\pi\epsilon_0} \sum_i \frac{e^2}{|\mathbf{r} - \mathbf{r}_i(t)|} = \frac{e^2}{4\pi\epsilon_0} \int \frac{d\mathbf{q}}{(2\pi)^3} \frac{4\pi}{q^2} \exp(i\mathbf{q}\cdot\mathbf{r}) P(\mathbf{q}, t).$$

$$\tag{7.56}$$

We can now define the *k-space* mixed dynamic form factor (Kohl and Rose, 1985)

$$S(\mathbf{q}, \mathbf{q}', \omega) = \frac{1}{\mathcal{Z}} \sum_{l,l'} \exp(-\epsilon_l/k_B T) \int \frac{dt}{2\pi} \exp(i\omega t) \langle l|\delta P(\mathbf{q}, t)|l'\rangle \langle l'|\delta P(-\mathbf{q}', 0)|l\rangle.$$

$$\tag{7.57}$$

This quantity is proportional to the Fourier transform of the real-space mixed dynamic form factor (7.34),

$$S(\mathbf{q}, \mathbf{q}', \omega) = \left(\frac{4\pi\epsilon_0}{e^2}\right)^2 \frac{q^2 (q')^2}{(4\pi)^2} \int d\mathbf{r} d\mathbf{r}' \exp(-i\mathbf{q} \cdot \mathbf{r} + i\mathbf{q}' \cdot \mathbf{r}') \bar{s}(\mathbf{r}, \mathbf{r}', \omega), \tag{7.58}$$

and, according to (7.57), it is related to the correlation function of temporal fluctuations of electron density. There is an important connection between $S(\mathbf{q}, \mathbf{q}', \omega)$ and the matrix of dielectric function $\varepsilon(\mathbf{q}, \mathbf{q}', \omega)$ (Kohl and Rose, 1985)

$$S(\mathbf{q}, \mathbf{q}', \omega) = \frac{i(\epsilon_0 \hbar / 2\pi e^2)}{[1 - \exp(-\hbar\omega/k_B T)]} \{q^2 \varepsilon^{-1}(\mathbf{q}, \mathbf{q}', \omega) - (q')^2 [\varepsilon^{-1}(\mathbf{q}', \mathbf{q}, \omega)]^*\}.$$

(7.59)

This equation is often used in quantitative studies of inelastic scattering of electrons by crystals. The fact that it is the *matrix* of the dielectric function that enters the theoretical treatment stems from the periodicity of the motion of valence electrons in the potential field of the lattice (Platzman and Wolf, 1973, Josefsson, 1993). For systems where electronic excitations are insensitive to the periodicity of the lattice, all the off-diagonal terms of the mixed dynamic form factor vanish, so that $S(\mathbf{q}, \mathbf{q}', \omega) = (2\pi)^3 \delta(\mathbf{q} - \mathbf{q}') S(\mathbf{q}, \omega)$, and the only quantity that needs to be known in this case is the dielectric function $\varepsilon(\mathbf{q}, \omega)$. The dynamic form factor $S(\mathbf{q}, \omega)$ is proportional to the imaginary part of the inverse of this function (Pines and Nozieres, 1966, Kohl and Rose, 1985)

$$S(\mathbf{q}, \omega) = -\frac{\epsilon_0 \hbar q^2}{\pi e^2} [1 - \exp(-\hbar\omega/k_B T)]^{-1} \Im \left[\frac{1}{\varepsilon(\mathbf{q}, \omega)} \right].$$

(7.60)

By substituting (7.58) and (7.36) into (7.39) and assuming that the initial and the final states of the electron are not affected by dynamical diffraction effects, we find

$$\left(\frac{d^2\sigma}{do\,dE} \right)_{\text{inel}} = \frac{k}{k_0} \left(\frac{4\gamma^2\Omega}{\hbar a_0^2} \right) \frac{1}{(\mathbf{k} - \mathbf{k}_0)^4} S\left(\mathbf{k}_0 - \mathbf{k}, \frac{E}{\hbar} \right),$$

(7.61)

where Ω is the volume of the sample, $\gamma = m/m_0$ is the relativistic factor, and a_0 is the Bohr radius $a_0 = 4\pi\epsilon_0 \hbar^2 / m_0 e^2 = 0.529$ Å.

There is a very substantial body of literature on methods of experimental investigation and approximations used in theoretical calculations of the dynamic form factor (7.57). The crucial point here consists in that both the initial $|l\rangle$ and the intermediate $|l'\rangle$ states of the electronic subsystem of the solid are *many-body* states. It is only to a first approximation that electrons in the solid can be considered as weakly interacting particles occupying single-particle states, and there are numerous examples where the approximation of independent (or weakly interacting) electrons fails. Hence the problem of the theoretical interpretation of electron energy loss spectra presents a formidable task even within the range of validity of the DWBA (7.39) (i.e. even where the probability of double, triple, etc., successive inelastic interactions between the incident high-energy electron and electrons in the solid is small). We shall now briefly review several theoretical models that approximate the observed energy loss spectra and explain the origin of characteristic features seen in these spectra. We begin by considering inelastic scattering by plasmons.

7.6.1 *Plasmons*

Plasmons are collective excitations of the electronic subsystem of the solid, i.e. they involve many valence electrons simultaneously (Pines and Nozieres, 1966). The motion of electrons with respect to the positively charged ions of the crystal lattice creates a time-dependent electric field which in turn influences the motion of electrons. The type of oscillations of the electron gas associated with the periodic exchange of energy between the moving electrons and the electric field depends on the chemical composition of the solid (e.g. on the density of electrons) and on the shape of the sample. The boundary conditions at the surface of the sample determine the energy $\hbar\omega_{pl}$ of plasmon excitations and the law of dispersion $\omega = \omega_{pl}(\mathbf{q})$ of these excitations. The dependence of the plasmon frequency on the wave vector for bulk plasmon excitations is given by the solution of the equation (Pines and Nozieres, 1966)

$$\varepsilon(\mathbf{q}, \omega_{pl}(\mathbf{q})) = 0. \tag{7.62}$$

Assuming that the imaginary part $\varepsilon_2(\mathbf{q}, \omega)$ of $\varepsilon(\mathbf{q}, \omega) = \varepsilon_1(\mathbf{q}, \omega) + i\varepsilon_2(\mathbf{q}, \omega)$ is a small quantity, we expand the denominator in (7.60) in the vicinity of $\omega_{pl}(\mathbf{q})$ in the Taylor series as $\varepsilon(\mathbf{q}, \omega) = [\partial\varepsilon_1(\mathbf{q}, \omega_{pl}(\mathbf{q}))/\partial\omega][\omega - \omega_{pl}(\mathbf{q})] + i\varepsilon_2(\mathbf{q}, \omega_{pl}(\mathbf{q}))$ (note that the zero-order term in the Taylor series vanishes due to (7.62)), and find

$$S(\mathbf{q}, \omega) = \frac{(\epsilon_0 \hbar q^2/\pi e^2)}{[1 - \exp(-\hbar\omega/k_B T)]} \frac{\varepsilon_2(\mathbf{q}, \omega_{pl}(\mathbf{q}))}{\left(\dfrac{\partial\varepsilon_1(\mathbf{q}, \omega_{pl}(\mathbf{q}))}{\partial\omega}\right)^2 (\omega - \omega_{pl}(\mathbf{q}))^2 + \varepsilon_2^2(\mathbf{q}, \omega_{pl}(\mathbf{q}))},$$

$$\tag{7.63}$$

In the limit $\varepsilon_2(\mathbf{q}, \omega) \to 0$ this equation can be simplified using the formula $\lim_{\eta\to 0}(x + i\eta)^{-1} = \wp(1/x) - i\pi\delta(x)$. We now see that

$$S(\mathbf{q}, \omega) = \frac{\epsilon_0 \hbar q^2}{e^2} \frac{1}{1 - \exp(-\hbar\omega/k_B T)} \frac{1}{\left|\dfrac{\partial\varepsilon_1(\mathbf{q}, \omega)}{\partial\omega}\right|_{\omega=\omega_{pl}(\mathbf{q})}} \delta(\omega - \omega_{pl}(\mathbf{q})). \tag{7.64}$$

The cross-section of scattering (7.61) corresponding to inelastic scattering by plasmon excitations is therefore given by

$$\left(\frac{d^2\sigma}{d o\, dE}\right)_{\text{plasmon}} = \frac{k}{k_0} \left(\frac{4\gamma^2}{a_0^2}\right) \Omega \left(\frac{\epsilon_0 \hbar}{e^2 |\partial\epsilon_1/\partial\omega|_{\omega_{pl}}}\right) \frac{1}{(\mathbf{k}_0 - \mathbf{k})^2} \delta[E - \hbar\omega_{pl}(\mathbf{k}_0 - \mathbf{k})]. \tag{7.65}$$

Here we have omitted the temperature factor $\exp(-\hbar\omega/k_B T)$, which in most cases does not affect the observed energy loss spectra because $\hbar\omega/k_B T \gg 1$.

There are two important terms in (7.65). The first one is the energy δ-function, which shows that the spectrum of electrons scattered in the direction characterized by wave vector $\mathbf{k}$ has the form of a sharp peak at $E = \hbar\omega_{pl}(\mathbf{k}_0 - \mathbf{k})$ (note that here E is the energy *lost* by the electron and therefore $\hbar\omega_{pl}(\mathbf{k}_0 - \mathbf{k})$

is the distance between the elastic peak and the plasmon peak in the energy loss spectrum). The second important factor is the dependence of the cross-section on the momentum transferred in the inelastic collision $d\sigma \sim (\mathbf{k}_0 - \mathbf{k})^{-2}$. At first glance it may seem that this term diverges at $\mathbf{k} = \mathbf{k}_0$. However, owing to the fact that the two vectors have different lengths ($\mathbf{k}$ is shorter than $\mathbf{k}_0$) this term instead gives rise to a special type of behaviour of the cross-section of scattering in the region of small angles of scattering.

Let ϑ be the angle between $\mathbf{k}$ and $\mathbf{k}_0$, so that $(\mathbf{k} - \mathbf{k}_0)^2 = k^2 + k_0^2 - 2kk_0 \cos\vartheta$. Taking into account that $k^2 = k_0^2 - 2m\omega_{pl}(\mathbf{k}_0 - \mathbf{k})/\hbar$ and considering the case of small angles of scattering $\vartheta \ll 1$, where $\cos\vartheta \approx 1 - \vartheta^2/2$, we find

$$(\mathbf{k} - \mathbf{k}_0)^2 \approx k_0^2\vartheta^2 + \frac{m^2\omega_{pl}^2(\mathbf{k}_0 - \mathbf{k})}{\hbar^2 k_0^2}.$$

Substituting this into (7.65) we arrive at

$$\left(\frac{d^2\sigma}{dod E}\right)_{\text{plasmon}} = \frac{\Omega}{2\pi a_0} \frac{1}{\vartheta^2 + \vartheta_E^2} \delta(E - \hbar\omega_{pl}), \qquad (7.66)$$

where $\vartheta_E = m\omega_{pl}/\hbar k_0^2$ and a_0 is the Bohr radius. The appearance of a characteristic cut-off angle ϑ_E in the denominator of eqn (7.66) is not associated with screening of the Coulomb interaction between the incident electron and electrons in the solid. The appearance of ϑ_E is due entirely to the fact that the incident electron loses energy $\hbar\omega_{pl}$ in the inelastic interaction, and the larger the energy lost in the collision the larger is the value of ϑ_E in eqn (7.66).

It is instructive to analyse the degree of *real-space localization* of plasmon excitations. In electron microscopy it is popular to say that plasmon excitations are delocalized and this is often quoted as the reason why plasmons have a relatively minor effect on the contrast of energy-unfiltered electron microscope images. But what is the quantitative measure of delocalization of plasmon excitations? To answer this question we evaluate the real-space dynamic form factor $\bar{s}(\mathbf{r}, \mathbf{r}', \omega)$ for the case where $S(\mathbf{q}, \omega)$ is given by (7.64). A simple calculation leads to

$$\bar{s}(\mathbf{r}, \mathbf{r}', \omega) = \left(\frac{e^2}{\epsilon_0}\right) \int \frac{d\mathbf{q}}{(2\pi)^3} \frac{1}{q^4} \exp[i\mathbf{q} \cdot (\mathbf{r} - \mathbf{r}')] S(\mathbf{q}, \omega)$$
$$= \frac{e^2}{4\pi^2\epsilon_0} \delta(\omega - \omega_{pl}) \left|\frac{\partial\varepsilon_1}{\partial\omega}\right|^{-1} \frac{1}{|\mathbf{r} - \mathbf{r}'|}, \qquad (7.67)$$

where we have neglected the dependence of the dielectric function and the plasmon frequency on the wave vector. This dependence modifies the behaviour of (7.67) in the region of small $|\mathbf{r} - \mathbf{r}'|$ where it eliminates the divergent term. However, the long-range asymptotic behaviour of (7.67) ($\bar{s}(\mathbf{r}, \mathbf{r}', \omega) \sim |\mathbf{r} - \mathbf{r}'|^{-1}$) remains unchanged even if the dependence of the dielectric function on $\mathbf{q}$ and the presence of the non-vanishing imaginary part of this function are taken into account. It is the absence of any characteristic spatial scale in the dependence of the dynamic form factor (7.67) on $|\mathbf{r} - \mathbf{r}'|$ that constitutes the real meaning of the concept of delocalization of plasmon excitations.

7.6.2 *Ionization of inner electronic shells*

Another type of characteristic features seen in electron energy loss spectra is associated with the excitation of relatively tightly bound inner-shell electrons into the continuum of unoccupied states above the Fermi energy. Many-body effects resulting from the interaction between electrons in the solid may also play an important part here, but it turns out that the effective one-electron approximation provides a reasonably good starting point for the interpretation of the spectra (Fuggle and Inglesfield, 1992).

To understand the details of the process of inner-shell ionization we neglect the temperature factor $\exp(-\hbar\omega/k_B T)$ in (7.57) (this factor is not particularly important in the case where the excitation energy $\hbar\omega$ is many times greater than $k_B T$ (~ 0.025 eV at $T = 300$ K)), and write $S(\mathbf{q}, \mathbf{q}', \omega)$ in the form

$$
\begin{aligned}
S(\mathbf{q}, \mathbf{q}', \omega) &= \sum_{n \neq 0} \langle 0| \sum_j \exp(-i\mathbf{q} \cdot \mathbf{r}_j)|n\rangle \langle n| \sum_{j'} \exp(i\mathbf{q}'\mathbf{r}_{j'})|0\rangle \delta\left(\frac{\epsilon_n - \epsilon_0}{\hbar} - \omega\right) \\
&= \sum_n \langle 0| \sum_j \exp(-i\mathbf{q} \cdot \mathbf{r}_j)|n\rangle \langle n| \sum_{j'} \exp(i\mathbf{q}'\mathbf{r}_{j'})|0\rangle \delta\left(\frac{\epsilon_n - \epsilon_0}{\hbar} - \omega\right) \\
&\quad - \langle 0| \sum_j \exp(-i\mathbf{q} \cdot \mathbf{r}_j)|0\rangle \langle 0| \sum_{j'} \exp(i\mathbf{q}'\mathbf{r}_{j'})|0\rangle \delta(\omega).
\end{aligned} \tag{7.68}
$$

Index '0' here refers to the initial ground state of the electronic subsystem of the solid and summation over n is performed over the excited electronic states. The last term in (7.68), which corresponds to $\epsilon_n = \epsilon_0$, does not give rise to energy losses. In what follows we consider only the range of frequencies $\omega > 0$, where the term proportional to $\delta(\omega)$ may be neglected.

Since (7.68) involves transitions between many-electron states, it is necessary to use the appropriate notation (see §64 of (Landau and Lifshitz, 1977) for more detail)

$$
\sum_j \exp(i\mathbf{q} \cdot \mathbf{r}_j) = \int \hat{\Psi}^\dagger(\mathbf{r}) \exp(i\mathbf{q} \cdot \mathbf{r})\hat{\Psi}(\mathbf{r})d\mathbf{r}, \tag{7.69}
$$

where operators $\hat{\Psi}(\mathbf{r})$ and $\hat{\Psi}^\dagger(\mathbf{r})$ describe annihilation and creation of an electron at point $\mathbf{r}$. By using (7.69), we represent (7.68) in the form

$$S(\mathbf{q}, \mathbf{q}', \omega) = \int d\mathbf{r} d\mathbf{r}' \exp(-i\mathbf{q} \cdot \mathbf{r} + i\mathbf{q}' \cdot \mathbf{r}')$$

$$\times \sum_n \langle 0|\hat{\Psi}^\dagger(\mathbf{r})\hat{\Psi}(\mathbf{r})|n\rangle \langle n|\hat{\Psi}^\dagger(\mathbf{r}')\hat{\Psi}(\mathbf{r}')|0\rangle \delta \left(\frac{\epsilon_n - \epsilon_0}{\hbar} - \omega \right)$$

$$= \int d\mathbf{r} d\mathbf{r}' \exp(-i\mathbf{q} \cdot \mathbf{r} + i\mathbf{q}' \cdot \mathbf{r}')\langle 0|\hat{\Psi}^\dagger(\mathbf{r})\hat{\Psi}(\mathbf{r})$$

$$\times \delta \left(\frac{H_{cr} - \epsilon_0}{\hbar} - \omega \right) \hat{\Psi}^\dagger(\mathbf{r}')\hat{\Psi}(\mathbf{r}')|0\rangle, \tag{7.70}$$

where we have taken into account the fact that quantum states $|n\rangle$ form a complete set, i.e. that $\sum_n |n\rangle\langle n| = 1$. Since $|n\rangle$ is an eigenstate of the Hamiltonian of the crystal we have $H_{cr}|n\rangle = \epsilon_n|n\rangle$.

The term in angular brackets in (7.70) has a simple meaning. By going from right to left it describes the creation of a hole at point $\mathbf{r}'$ and the subsequent creation of an electron at the same point in real space, then some interaction taking place (due to the δ-function term), then first annihilation of the electron at $\mathbf{r}$, and then the annihilation of the hole at $\mathbf{r}$ (to avoid confusion, note that the annihilation of an electron is equivalent to the creation of a hole). Using the standard representation of the δ-function term

$$\delta \left(\frac{H_{cr} - \epsilon_0}{\hbar} - \omega \right) = \hbar\delta \left(\epsilon_0 + \hbar\omega - H_{cr} \right) = -\frac{\hbar}{\pi}\Im \left[\frac{1}{\epsilon_0 + \hbar\omega - H_{cr} + i0} \right], \tag{7.71}$$

where $\Im$ denotes the imaginary part, and comparing (7.71) with (7.70) we can *approximately* represent (7.70) in the form of a product of the Green's function

$$\mathcal{G}(\mathbf{r}, \mathbf{r}', \hbar\omega) = \langle 0|\hat{\Psi}(\mathbf{r})\frac{1}{\epsilon_0 + \hbar\omega - H_{cr} + i0}\hat{\Psi}^\dagger(\mathbf{r}')|0\rangle \tag{7.72}$$

of the electron ejected into the continuum of excited states, and two wave functions of the hole $\phi_i(\mathbf{r}') = \hat{\Psi}^\dagger(\mathbf{r}')|0\rangle$ and $\phi_i^*(\mathbf{r}) = \langle 0|\hat{\Psi}(\mathbf{r})$ created in one of the inner atomic shells by the impact with the incident high-energy electron. Having used all these approximations we arrive at (Vvedensky, 1992)

$$S(\mathbf{q}, \mathbf{q}', \omega) = -\frac{\hbar}{\pi} \int \exp(-i\mathbf{q} \cdot \mathbf{r} + i\mathbf{q}' \cdot \mathbf{r}')\phi_i^*(\mathbf{r})\Im[\mathcal{G}(\mathbf{r}, \mathbf{r}', \hbar\omega)]\phi_i(\mathbf{r}')d\mathbf{r} d\mathbf{r}'. \tag{7.73}$$

The product $\phi_i^*(\mathbf{r})\mathcal{G}(\mathbf{r}, \mathbf{r}', \hbar\omega)\phi_i(\mathbf{r}')$ equals the probability amplitude of a complex process involving the creation of a hole in the inner-shell atomic state $\phi_i(\mathbf{r}')$, the propagation of the ejected electron between $\mathbf{r}'$ and $\mathbf{r}$ as described by $\mathcal{G}(\mathbf{r}, \mathbf{r}', \hbar\omega)$, and the subsequent recombination of this electron with the hole in the state $\phi_i^*(\mathbf{r})$.

Note that despite its seemingly transparent meaning, the range of validity of the approximation (7.73) is fairly limited. Formula (7.70) assumes that the excitation (i.e. an electron above the Fermi energy) is created in the ground state of the system. This approximation is in fact more a matter of convenience rather than of mathematical accuracy and consistency because by adopting this point of view it is possible to comply with the definition of the one-particle Green's function (7.72). The latter is useful since there exists powerful mathematical machinery developed for the calculation of one-particle Green's functions. However, in fact the excited electron propagates between $\mathbf{r}'$ and $\mathbf{r}$ in the Fermi sea *containing a hole in one of the atomic core states*. Formula (7.73) neglects effects resulting from the interaction between the excited electron and the hole. In some materials (e.g. in compounds containing ions of transition metals or rare earth elements where the interaction between electrons is particularly strong) the role played by these effects is very substantial (Kanamori and Kotani, 1988, Vvedensky, 1992). Electron–hole interactions manifest themselves in the energy loss spectra through the appearance of additional peaks, or satellites, the origin of which would have remained a complete mystery if the presence of the interaction between the excited electron and the hole had been neglected. The problem of how to assess *quantitatively* the part played by electron–hole interactions remains one of the central problems in the field of electron energy loss spectroscopy (Kanamori and Kotani, 1988).

Within the range of validity of eqn (7.73) it is the energy-dependent one-electron Green's function $\mathcal{G}(\mathbf{r}, \mathbf{r}', \hbar\omega)$ that describes the fine structure of the electron energy loss spectrum in the interval of energies corresponding to the ionization of one of the inner electronic shells. The imaginary part of the diagonal element of this Green's function

$$-\frac{1}{\pi}\Im\mathcal{G}(\mathbf{r}, \mathbf{r}, \hbar\omega) \tag{7.74}$$

equals the $\mathbf{r}$-dependent density of unoccupied electronic states at point $\mathbf{r}$ and at energy $\hbar\omega - I_i$, where I_i is the potential of ionization of the i-th shell. For example, in the free-electron approximation we obtain the following expression:

$$\mathcal{G}(\mathbf{r}, \mathbf{r}', \hbar\omega) = -\frac{m}{2\pi\hbar^2}\frac{\exp(i\sqrt{2m(\hbar\omega - I_i)/\hbar^2}|\mathbf{r} - \mathbf{r}'|)}{|\mathbf{r} - \mathbf{r}'|},$$

the combination of which with (7.74) gives rise to a conventional result (see §149 of (Landau and Lifshitz, 1977)) that $-\Im\mathcal{G}(\mathbf{r}, \mathbf{r}, \hbar\omega) \sim \sqrt{\hbar\omega - I_i}$.

In simple metals many-body interactions between the excited electron and valence electrons give rise to the formation of a cloud of electron–hole pairs in the vicinity of the Fermi energy. In this case it is not the interaction between the excited electron and the hole created in the inner shell, but rather the interaction between the excited electron and the *valence* electrons in the solid (Vvedensky, 1992). The resulting effect of the formation of the cloud of electron–hole pairs is that in simple metals the form factor diverges as a function of ω at the threshold $\hbar\omega = I_i$ (Mahan, 1990).

Formula (7.74) shows that the spectrum of energy losses above the ionization threshold (assuming that effects of electron–electron interactions may be neglected) is proportional to the density of unoccupied one-particle states above the Fermi energy. In the dipole approximation, which corresponds to the linear expansion of exponential factors in (7.68) $(\exp(i\mathbf{q} \cdot \mathbf{r}_j) = 1 + i\mathbf{q} \cdot \mathbf{r}_j + ...)$, we arrive at the following expression for the form factor:

$$S(\mathbf{q}, \mathbf{q}', \omega) = -\frac{\hbar}{\pi} \int \phi_i^*(\mathbf{r})(\mathbf{q} \cdot \mathbf{r})\Im\mathcal{G}(\mathbf{r}, \mathbf{r}', \hbar\omega)(\mathbf{q}' \cdot \mathbf{r}')\phi_i(\mathbf{r}')d\mathbf{r}d\mathbf{r}'. \qquad (7.75)$$

If the core wave function is characterized by orbital momentum l, the only possible final states are those which are characterized by momenta $l \pm 1$. The form factor (7.75) therefore turns out to be proportional to the density of empty electronic states corresponding to momenta $l + 1$ and $l - 1$ (Egerton, 1986, Weng and Rez, 1989, Weng et al., 1989a, Weng et al., 1989b). The dipole approximation is often used in calculations of transition matrix elements where it proves to be very accurate, see (Saldin and Yao, 1990, Saldin and Ueda, 1992) and also (Ueda and Saldin, 1992). Among methods which do not rely on the dipole approximation we would like to mention the exactly solvable hydrogen model of inner-shell ionization first investigated by Maslen and Rossouw in (Maslen, 1983, Maslen and Rossouw, 1983). Further progress of this work is described in (Maslen and Rossouw, 1984, Rossouw and Maslen, 1984), in a monograph (Egerton, 1996) and in papers by Oxley and Allen (Oxley and Allen, 2000, Oxley and Allen, 2001).

To compute the momentum-resolved density of unoccupied electronic states above the Fermi energy it is necessary to use accurate *ab initio* methods. For example, wave functions and densities of the effective one-particle electronic states can be found numerically using computer codes developed in recent years for carrying out the total energy minimization and for the comparison of energies of various equilibrium crystal structures (Fuggle and Inglesfield, 1992, Rez et al., 1995, Rez et al., 1998). Calculations are usually performed using the local density approximation of density functional theory or some of its modifications (e.g. the generalized gradient approximation or the GW approximation). In many cases *ab initio* calculations reproduce the observed spectral features with a remarkable degree of accuracy (Rez et al., 1995). However in some cases, particularly where the spatially localized d or f states are involved in the inelastic transition, there is a substantial degree of uncertainty associated with the interpretation of the spectroscopic information, see (Kanamori and Kotani, 1988, Hüfner, 1994) and (Dudarev et al., 1998). The nature of this uncertainty can be traced back to the approximations involved in the transformation of (7.70) to (7.73).

7.6.3 *The extended energy loss fine structure (EXELFS)*

If the energy of the excited electron exceeds the ionization threshold by approximately 50 eV the conventional interpretation of energy loss spectra in terms of the density of one-particle electronic states above the Fermi energy becomes unreliable. To understand why our approach needs to be modified in the region 50 eV $\leq \hbar\omega - I \leq$ 1000 eV, where $\hbar\omega$ is the energy lost in an inelastic collision and

I is the ionization threshold, let us consider how the parameters characterizing the motion of the excited electron vary as a function of its energy.

First of all we note that as we move up the energy scale the band gaps in the spectrum of one-electron states become progressively narrower. The prime reason for this is that the Fourier components of the potential of the crystal are decreasing functions of the modulus of the reciprocal lattice vector $\mathbf{g}$. In addition the Fourier-components of the potential contain the temperature-dependent Debye–Waller factor $\exp[-\langle(\mathbf{u}\cdot\mathbf{g})^2\rangle/2]$, which suppresses all the diffraction effects (including the formation of band gaps) associated with high-order reciprocal lattice vectors $\mathbf{g}$. As band gaps become progressively narrower, it becomes necessary to take into account a quantity that we have so far disregarded in our discussion of the meaning of the one-particle Green's function $\mathcal{G}(\mathbf{r},\mathbf{r},\hbar\omega)$ of the excited electron. This important quantity is the *imaginary part* of the self-energy of the electron.

What is the self-energy of the electron and why is its imaginary part so important? To answer this question let us forget for a moment about all the complexity of the real many-body problem of electron energy loss spectra and consider a simple case where the electron *does not* interact with the environment. Since there is no interaction with other particles in the solid the Hamiltonian $\hat{H}$, which describes the electron, depends only on the coordinates of the electron. By solving the Schrödinger equation $(E-\hat{H})\psi=0$ we find the energies E_n and eigenfunctions ψ_n of this equation and calculate the Green's function of the electron as

$$G(\mathbf{r},\mathbf{r}',E)=\sum_n \psi_n(\mathbf{r})\frac{1}{E-E_n+i0}\psi_n^*(\mathbf{r}'). \tag{7.76}$$

The fact that we have been able to represent Green's function of the electron in this form crucially depends on the fact that we know the eigenfunctions and eigenvalues of the Schrödinger equation describing the motion of this electron.

In principle there is also an alternative way of finding $G(\mathbf{r},\mathbf{r}',E)$, which avoids carrying out summation over all the eigenstates. Taking into account that $\psi_n(\mathbf{r})$ satisfies the equation

$$-\frac{\hbar^2}{2m}\nabla^2\psi_n(\mathbf{r})+V(\mathbf{r})\psi_n(\mathbf{r})=E_n\psi_n(\mathbf{r}),$$

we obtain that the Green's function is a solution of the following inhomogeneous equation:

$$\left(E+\frac{\hbar^2}{2m}\nabla^2-V(\mathbf{r})\right)G(\mathbf{r},\mathbf{r}',E)=\delta(\mathbf{r}-\mathbf{r}'), \tag{7.77}$$

the solution of which can be represented in the form (7.76).

If we now take account of the fact that the electron actually does interact with other particles in the system, then we will have to modify the latter equation.

This modification, which takes account of the motion of an electron being affected by the motion of other electrons, has the form (see (Fulde, 1995), page 216)

$$\left(E + \frac{\hbar^2}{2m}\nabla^2 - V(\mathbf{r})\right) G(\mathbf{r}, \mathbf{r}', E) - \int d\mathbf{r}'' \Sigma(\mathbf{r}, \mathbf{r}'', E) G(\mathbf{r}'', \mathbf{r}', E) = \delta(\mathbf{r} - \mathbf{r}').$$

$$(7.78)$$

where $\Sigma(\mathbf{r}, \mathbf{r}'', E)$ is the self-energy of the electron.

To find the Green's function of the excited electron in the energy range 50 eV $\leq E \leq 1000$ eV, it turns out to be more convenient to use an expansion based on the quantum-mechanical theory of multiple scattering, see e.g. (Vvedensky, 1992) and (Durham, 1988), rather than to solve eqn (7.78).

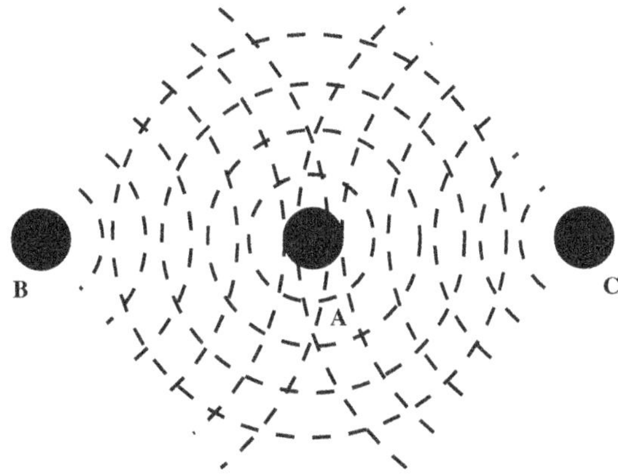

FIG. 7.4. Schematic illustration of the origin of extended energy loss fine structure (EXELFS) of ionization spectra. The ejected electron described by $\mathcal{G}_0(\mathbf{r}, \mathbf{r}', \hbar\omega)$ and propagating from atom A is scattered back by neighbouring atoms B and C. The sum of the initial and backscattered waves equals $\mathcal{G}(\mathbf{r}, \mathbf{r}', \hbar\omega)$. The imaginary part of this Green's function exhibits weak oscillations due to the interference between the initial and backscattered waves.

According to eqn (7.73), the energy dependence of the cross-section of inelastic scattering is determined by the imaginary part of the Green's function $\mathcal{G}(\mathbf{r}, \mathbf{r}', \hbar\omega)$, where the initial and the final coordinates $\mathbf{r}'$ and $\mathbf{r}$ lie in the region which is close to the centre of the atom. This follows from the fact that $\phi_i(\mathbf{r}')$ and $\phi_i^*(\mathbf{r})$ describe hole states localized in one of the core electronic shells of the atom. At the same time, propagation of the electron from $\mathbf{r}'$ to $\mathbf{r}$ does not necessarily involve trajectories that lie entirely in the atomic core. For example the electron may travel from $\mathbf{r}'$ to $\mathbf{r}$ by escaping into the space surrounding the original atom and returning back following a collision with one or several atoms of the first or second coordination shells. Formally the process can be described by an infinite series (Vvedensky, 1992)

$$\mathcal{G} = \mathcal{G}_0 + \mathcal{G}_0 \sum_\alpha t_\alpha \mathcal{G}_0 + \mathcal{G}_0 \sum_\alpha t_\alpha \mathcal{G}_0 \sum_{\beta \neq \alpha} t_\beta \mathcal{G}_0 + ..., \qquad (7.79)$$

where t_α is the so-called t-matrix, which is related to the amplitude of scattering via $f = -(m/2\pi\hbar^2)t$. The real-space representation of (7.79) is

$$\mathcal{G}(\mathbf{r}, \mathbf{r}', \hbar\omega) = -\frac{m}{2\pi\hbar^2} \frac{\exp(i\kappa|\mathbf{r} - \mathbf{r}'|)}{|\mathbf{r} - \mathbf{r}'|}$$
$$- \frac{m}{2\pi\hbar^2} \sum_\alpha \int d\mathbf{R} d\mathbf{R}' \frac{\exp(i\kappa|\mathbf{r} - \mathbf{R}|)}{|\mathbf{r} - \mathbf{R}|} f_\alpha(\mathbf{R}, \mathbf{R}') \frac{\exp(i\kappa|\mathbf{R}' - \mathbf{r}'|)}{|\mathbf{R} - \mathbf{r}'|} + ... \qquad (7.80)$$

where $\kappa = \sqrt{2m(\hbar\omega - I_i)/\hbar^2}$. In (7.80) $\mathbf{r}$ and $\mathbf{r}'$ are located near the origin while the integration over $\mathbf{R}$ and $\mathbf{R}'$ spans the region surrounding the centre of atom situated at $\mathbf{R}_\alpha$. Since the distance between atoms is normally several times larger than the effective atomic size, we have $|\mathbf{R} - \mathbf{R}_\alpha| \ll |\mathbf{r} - \mathbf{R}_\alpha|$. Using this inequality we approximate the arguments of the exponential functions in (7.80) as

$$|\mathbf{r} - \mathbf{R}| = |(\mathbf{r} - \mathbf{R}_\alpha) - (\mathbf{R} - \mathbf{R}_\alpha)| \approx |\mathbf{r} - \mathbf{R}_\alpha| - \frac{\mathbf{r} - \mathbf{R}_\alpha}{|\mathbf{r} - \mathbf{R}_\alpha|} \cdot (\mathbf{R} - \mathbf{R}_\alpha).$$

The second term of (7.80) becomes

$$\mathcal{G}^{(EXELFS)}(\mathbf{r}, \mathbf{r}', \hbar\omega) = -\frac{m}{2\pi\hbar^2} \sum_\alpha \frac{\exp(i\kappa|\mathbf{r} - \mathbf{R}_\alpha|)}{|\mathbf{r} - \mathbf{R}_\alpha|} \frac{\exp(i\kappa|\mathbf{R}_\alpha - \mathbf{r}'|)}{|\mathbf{R}_\alpha - \mathbf{r}'|}$$
$$\times f_\alpha\left(\kappa \frac{\mathbf{r} - \mathbf{R}_\alpha}{|\mathbf{r} - \mathbf{R}_\alpha|}, \kappa \frac{\mathbf{R}_\alpha - \mathbf{r}'}{|\mathbf{R}_\alpha - \mathbf{r}'|}\right). \qquad (7.81)$$

Note that vector $\mathbf{R}_\alpha - \mathbf{r}' \approx \mathbf{R}_\alpha$ points from the original atom to the atom from which the excited electron is reflected, and therefore the amplitude of scattering $f(-\kappa\mathbf{R}_\alpha/|\mathbf{R}_\alpha|, \kappa\mathbf{R}_\alpha/|\mathbf{R}_\alpha|)$ equals the amplitude of scattering through the angle $\vartheta = \pi$. Using (7.81), we find

$$-\Im\mathcal{G}^{(EXELFS)}(\mathbf{r}, \mathbf{r}', \hbar\omega) = -\frac{m}{2\pi\hbar^2} \sum_\alpha \frac{\sin(2\kappa R_\alpha + \delta(\kappa))}{R_\alpha^2} |f(\pi)| \exp(-2R_\alpha/\lambda).$$
$$(7.82)$$

Here we have also introduced a factor that takes into account the effect of attenuation of the electron wave propagating from the original atom to atom α and back. This attenuation is associated with the excitation of plasmons and electron–hole pairs. In (7.82) we have also included the phase $\delta(\kappa)$, which comes from the interaction of the electron with the potential of the atom situated at the origin. The presence of this phase factor shows that the range of validity of the free-electron approximation used for the evaluation of $\mathcal{G}_0(\mathbf{r}, \mathbf{r}', \hbar\omega)$ in (7.81) is limited. Analysis of experimental data shows that it is essential to take the phase factor $\delta(\kappa)$ into account in formula (7.82), since the dependence of this phase factor on κ introduces a systematic error into the procedure of determination of interatomic distances based on the Fourier deconvolution of (7.82). A more detailed discussion of the approximations leading to the derivation of equations similar to (7.82) can be found in (Stern, 1988).

The extended fine structure of energy loss spectra was first observed by Leapman and Cosslett (Leapman and Cosslett, 1976) who used the oscillating component of the energy loss spectra above the aluminium K edge to estimate the interatomic distances in the first coordination shell in pure aluminium and in aluminium oxide Al_2O_3. Considerable improvements in the spectrum acquisition techniques and the development of computer software for the analysis of data have since transformed EXELFS into a reliable method of investigation of short-range order in crystalline and amorphous materials, and also at surfaces and interfaces (Crescenzi, 1985). A good review describing many interesting applications of the method was given by Sarikaya et al. in (Sarikaya et al., 1996).

7.7 Summary

In this chapter we described how to evaluate the cross-section of diffuse scattering of high-energy electrons using the distorted wave Born approximation (DWBA). This approximation takes into account dynamical diffraction of electrons and scattering by defects, thermal fluctuations, and electron energy losses. Several examples considered in this chapter include the treatment of thermal diffuse scattering, scattering by plasmons and inner shell excitations, and the extended electron energy loss fine structure (EXELFS).

8

DIFFUSE AND INELASTIC SCATTERING – MULTIPLE SCATTERING EFFECTS

8.1 Introduction

In the previous chapter we showed that fluctuations of the crystal potential associated with thermal vibrations of atomic nuclei, randomly distributed lattice defects, and electronic excitations give rise to diffuse scattering of high-energy electrons. Using perturbation theory we derived an explicit expression for the amplitude of scattering of electrons by the fluctuation part of the potential. This expression takes into account the fact that the cross-section of diffuse scattering is affected by the presence of the average periodic potential of the crystal.

Generalizing the approach developed in the previous chapter, we can now give a classification of processes of scattering that occur when a high-energy electron propagates through a thin crystalline specimen. Diffraction is the most important of these processes in the sense that it is observed even in the case where incident electrons are scattered by a very thin crystal. Diffraction is a coherent process and in order to evaluate the relevant amplitude of scattering it is sufficient to solve the Schrödinger equation describing the interaction of the high-energy electron with atoms forming the crystal lattice.

As the thickness of the specimen increases we start to observe effects associated with diffuse scattering by spatial and temporal fluctuations of the interaction potential. Gradually scattering by fluctuations becomes dominant and at some point we find that almost every electron propagating through the specimen takes part in at least one event of inelastic or diffuse scattering. Backscattering of high-energy electrons provides a particularly striking example of the case where *all* the electrons scattered through angles greater than 90° undergo scattering by fluctuations.

Considering the limiting case of a thick crystal where scattering by fluctuations dominates, we need to recognize the important part that dynamical diffraction continues to play in the process of propagation of electrons through the material. Indeed, the fact that in a thick specimen high-energy electrons undergo multiple diffuse scattering does not mean that diffraction effects vanish entirely. On the contrary diffraction remains present for as long as the specimen retains its *average* periodicity. For example diffraction of diffusely scattered electrons gives rise to the formation of Kikuchi lines, bands, rings, and parabolas that can be readily seen in electron diffraction patterns observed in transmission, reflection, and backscattering geometries, as was discussed in Chapter 6.

We therefore face the problem of developing an approach that would make

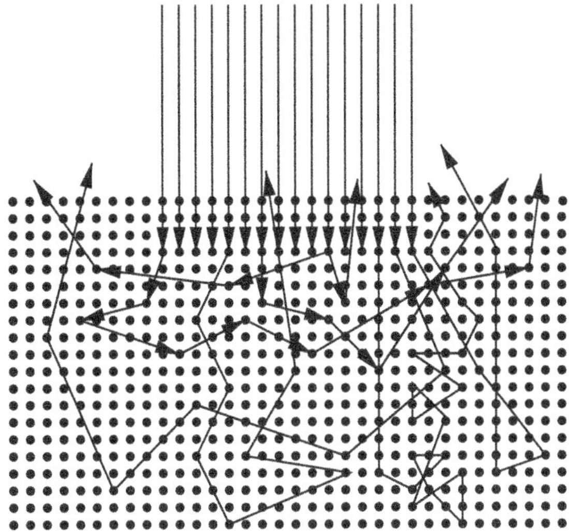

FIG. 8.1. Sketch illustrating why multiple scattering plays the dominant part in the process of backscattering of high-energy electrons from a crystalline specimen. The representation of the lattice by a regular array of atoms is an idealization. The majority of electrons are backscattered by fluctuations of the interaction potential associated with random thermal displacements of atoms from lattice sites.

it possible to evaluate the distribution of high-energy electrons over angle of scattering and energy, no matter how large is the volume of the material where electrons undergo diffraction and diffuse scattering. Ultimately the approach should be suitable for the treatment of electron backscattering in the geometry shown in Fig. 8.1. From the point of view of the perturbation treatment described in the previous chapter, the electron backscattering case proves to be the most difficult one. Indeed, in the backscattering geometry electrons are scattered by a specimen, the extent of which in the direction of the incident beam is semi-infinite. In this case it is *multiple* rather than *single* scattering by fluctuations that gives the dominant contribution to the observed distribution of backscattered electrons over energy and solid angle.

There are several fundamental issues that we encounter in a treatment combining diffraction and multiple diffuse scattering. These issues include the notion of coherence of the wave field of high-energy electrons and concepts of the density matrix and the quantum kinetic equation describing the evolution of the density matrix. It was not necessary to introduce these concepts in the previous chapter since in the case where diffuse scattering by fluctuations was weak and it was possible to treat it as a perturbation, we were able to use the distorted wave Born approximation (DWBA) and to find an explicit expression for the amplitude of scattering. However, in the case where we are interested in evaluating the cross-section of scattering by a fluctuating crystalline *medium*, the treatment of the problem requires using new approximations and new mathematical concepts.

We start by identifying the range of validity of the perturbation treatment of diffuse scattering developed in the previous chapter. We show that it is not only the energy of the electron and the strength of the interaction potential that have to satisfy certain conditions in order to justify the use of the DWBA. The size of the specimen is also an important parameter entering the condition of validity of the perturbation approach. The eventual breakdown of the perturbation treatment of diffuse scattering in thick specimens is not associated with the fact that the strength of the potential of interaction of electrons with fluctuations reaches a certain threshold. Instead it is the effect of *multiple* scattering by individual, relatively weak fluctuations that invalidates the DWBA. In other words for thin specimens the DWBA provides a sufficiently accurate way of calculating the cross-section of diffuse scattering. For thicker specimens the DWBA needs to be corrected to take effects of absorption of high-energy electrons into account. For even thicker specimens the DWBA breaks down entirely and the only approach that can still be used in that case requires solving the kinetic equation for the density matrix of high-energy electrons.

8.2   Breakdown of the DWBA and the optical potential model

We start by investigating the range of validity of the DWBA (7.14) and (7.32) derived in the previous chapter. Evidently, the perturbation treatment of diffuse scattering is only justified if the fluctuating part of the interaction potential is sufficiently weak. This requirement gives rise to condition (7.15). However, this condition does not specify whether or not the perturbation treatment of fluctuations is applicable to the entire crystal, or in other words whether effects of multiple scattering by fluctuations are significant. Now we consider this question in more detail.

We start from formula (7.14). The absolute value of the amplitude of diffuse scattering can be estimated as

$$|f_{\text{diffuse}}| \sim \left(\frac{m}{2\pi\hbar^2}\right) |\delta V| a^3, \tag{8.1}$$

where a is the characteristic size of the region where fluctuations of the potential are correlated. For example in the case of phonon diffuse scattering treated using the Einstein model of thermal vibrations (where it is assumed that thermal displacements of individual atoms are entirely independent) the correlation length a approximately equals the radius of an atom. The cross-section of diffuse scattering can be estimated as $\sigma_{\text{diffuse}} \approx 4\pi |f_{\text{diffuse}}|^2 \sim (m|\delta V| a^3/\hbar^2)^2/\pi$. The total cross-section of diffuse scattering of electrons by the entire crystal is equal to

$$\sigma_{\text{diffuse}}^{(\text{tot})} = n\sigma_{\text{diffuse}} A L, \tag{8.2}$$

where A is the surface area, L is the thickness of the specimen, and n is the volume density of atoms. We may treat diffuse scattering using a perturbation approach only if the specimen is nearly transparent for the incident high-energy electrons, or in other words if

$$\sigma_{\text{diffuse}}^{(\text{tot})} \ll A.$$

This is equivalent to the requirement that

$$L \ll (n\sigma_{\text{diffuse}})^{-1}, \tag{8.3}$$

stating that the thickness of the specimen must not exceed the mean free path $L_D = (n\sigma_{\text{diffuse}})^{-1}$, which is the average distance that an electron travels between successive events of scattering by fluctuations.

Why is it important to ensure that conditions of validity of DWBA are satisfied? The main reason is that unless we comply with condition (8.3), the calculated cross-sections of diffraction and diffuse scattering will not be compatible with the principle of conservation of the total probability. Indeed, since in the perturbation treatment of diffuse scattering we do not take into account the fact that the intensity of elastically scattered electrons decreases in order to compensate for the increase of intensity of diffusely scattered electrons, the total calculated intensity increases as $I = I_{\text{el}} + I_{\text{diffuse}} \sim I_0(1 + L/L_D)$. This shows that the perturbation treatment of diffuse scattering becomes unreliable for $L \sim L_D$.

Can the accuracy of the treatment be improved by taking into account the effective absorption of high-energy electrons (e.g. by introducing the concept of the optical potential)? By taking the effects of absorption into account it is indeed possible to improve the accuracy of the treatment of *diffraction* of electrons by relatively thick crystalline specimens and to explain some important experimental observations like the occurrence of anomalous absorption of electrons (Hirsch et al., 1977). But does the use of the optical potential improve the accuracy of calculation of the *total* intensity integrated over the entire solid angle of scattering? Unfortunately the improvement is fairly marginal. Indeed, by using the optical potential model we find that the intensity of the transmitted beam I_{el} attenuates exponentially as a function of thickness L of the crystal as $I_{\text{el}} = I_0 \exp(-L/L_D)$. A similar exponential factor also appears in the expression for the cross-section of diffuse scattering. For the total intensity of scattering we find that it now varies as a function of the thickness of the specimen L as $I = I_0 \exp(-L/L_D)(1 + L/L_D)$. Although in the limit of small thickness $L \ll L_D$ the total intensity is conserved better than in the case where absorption is not taken into account at all (the quantity $|(I/I_0) - 1|$ is now of the order of $\sim (L/L_D)^2$ instead of being of the order of $\sim (L/L_D)$), in the limit of large L the accuracy still remains poor.

What is the origin of deficiency of our approach? The main point is that so far we have been trying to treat diffuse scattering starting from the perturbation theory limit. This is neither the only possible nor the most adequate way of addressing the issue. In the case of scattering by a relatively thick specimen it is difficult to say what diffuse scattering events (for example the ones occurring near the entrance surface of the crystal or those occurring in the centre of the specimen, or those taking place near the exit surface) give the dominant contribution to the distribution of intensity observed in a diffraction pattern. It is logical to expect that a consistent approach to the problem of diffuse scattering must treat all these events on an equal footing, and it is the solution of the problem rather than our a priori guess that determines the relative contribution of diffuse scattering events occurring in various parts of the specimen to the observed electron diffraction pattern.

In what follows we describe how the treatment of diffuse scattering can be extended beyond the DWBA. This requires developing a self-consistent approach to multiple scattering based on the kinetic equation for the density matrix. We start by introducing the notion of coherence of the wave field of electrons and also the entity that quantifies this notion, namely the density matrix of high-energy electrons. We then derive a kinetic equation that describes the evolution of the density matrix in the process of multiple scattering. In what follows we consider how the density matrix method can be applied to treat multiple scattering of electrons by plasmon excitations, the formation of Kikuchi patterns, the problem of electron channelling and backscattering, and also the problem of X-ray and Auger electron production in a crystalline environment.

8.3 Diffraction and multiple incoherent scattering of electrons

The question about how to find an equation that would on one hand be compatible with the wave equation and on the other hand would describe multiple incoherent diffuse scattering was first raised (Barabanenkov and Finkelberg, 1968, Barabanenkov, 1968, Barabanenkov, 1969, Frisch, 1968, Ichimaru, 1978) in the theory of propagation of electromagnetic and radio waves through the atmosphere and in the treatment of scattering of conduction electrons by impurities in alloys (Edwards, 1958, Abrikosov et al., 1975). It was discovered that the treatment of multiple scattering of waves requires investigating the properties of the so-called *mutual coherence function* $\rho(\mathbf{r}, \mathbf{r}')$. This function is defined as the product of two wave amplitudes $\psi(\mathbf{r})$ and $\psi^*(\mathbf{r}')$ taken at points $\mathbf{r}$ and $\mathbf{r}'$ and averaged over the random distribution of atoms in the medium,

$$\rho(\mathbf{r}, \mathbf{r}') = \langle \psi(\mathbf{r})\psi^*(\mathbf{r}') \rangle, \tag{8.4}$$

where angle brackets $\langle ... \rangle$ denote averaging over atomic coordinates. In quantum mechanics the quantity defined by eqn (8.4) is called the *density matrix* (Blum, 1981). The meaning of $\rho(\mathbf{r}, \mathbf{r}')$ is relatively simple: its diagonal elements $\rho(\mathbf{r}, \mathbf{r})$ give the average probability of finding an electron at a point $\mathbf{r}$, while its off-diagonal elements $\rho(\mathbf{r}, \mathbf{r}')$ evaluated for $\mathbf{r} \neq \mathbf{r}'$ represent the measure of mutual *coherence* of the wave field of electrons at these two points.

Normally the off-diagonal elements of the density matrix $\rho(\mathbf{r}, \mathbf{r}')$ vanish in the limit $|\mathbf{r}\text{-}\mathbf{r}'| \to \infty$. However, in some cases these elements remain finite even if the distance between the two points $\mathbf{r}$ and $\mathbf{r}'$ is very large. For example, in the case where the state of the electron is described by a plane wave $\exp(i\mathbf{k} \cdot \mathbf{r})$

we find that the density matrix has the form $\rho(\mathbf{r}, \mathbf{r}') = \exp[i\mathbf{k} \cdot (\mathbf{r} - \mathbf{r}')]$. Here the off-diagonal elements satisfy condition $|\rho(\mathbf{r}, \mathbf{r}')| = 1$ independently of the distance $|\mathbf{r} - \mathbf{r}'|$ between points $\mathbf{r}$ and $\mathbf{r}'$. In other words, the wave field of the electron at point $\mathbf{r}$ remains entirely coherent with the wave field of the electron at any other point $\mathbf{r}'$ irrespectively of the distance between the points.

Furthermore, it can be shown that if the interaction potential contains no randomly fluctuating component, the wave field of electrons scattered by this potential is also described by a density matrix, the off-diagonal elements of which do not vanish in the limit $|\mathbf{r} - \mathbf{r}'| \to \infty$. In the absence of scattering by fluctuations electrons remain entirely coherent everywhere in space. Scattering by fluctuations destroys coherence and the off-diagonal elements of the density matrix $\rho(\mathbf{r}, \mathbf{r}')$ that describe electrons undergoing diffuse scattering vanish in the limit $|\mathbf{r} - \mathbf{r}'| \to \infty$. If the specimen is spatially homogeneous (this case corresponds to $V(\mathbf{r}) = V_0(\mathbf{r}) + \delta V(\mathbf{r})$, where $V_0(\mathbf{r}) = 0$ and where the correlation function $\langle \delta V(\mathbf{r}) \delta V(\mathbf{r}') \rangle$ depends only on the distance $|\mathbf{r} - \mathbf{r}'|$ between $\mathbf{r}$ and $\mathbf{r}'$), the density matrix can be represented in the form (Frisch, 1968, Barabanenkov and Finkelberg, 1968, Barabanenkov, 1968, Barabanenkov, 1969)

$$\rho(\mathbf{r}, \mathbf{r}') = \Phi[\mathbf{r} - \mathbf{r}'; (\mathbf{r} + \mathbf{r}')/2], \tag{8.5}$$

where the rate of variation of Φ as a function of $\mathbf{r} - \mathbf{r}'$ is many times the rate of variation of Φ as a function of $(\mathbf{r} + \mathbf{r}')/2$. The Fourier transform of $\Phi[\mathbf{r} - \mathbf{r}'; (\mathbf{r} + \mathbf{r}')/2]$ with respect to the variable $\mathbf{r} - \mathbf{r}'$ satisfies an equation that is similar to the Boltzmann transport equation (Lifshitz and Pitaevskii, 1979). Solutions of the transport equation are well known and this makes the approximate representation (8.5) suitable for investigating a broad range of problems of scattering of waves by random media. The drawback associated with the representation (8.5) is that it is difficult to set up boundary conditions for the function Φ, since the transformation of coordinates $\mathbf{r}$ and $\mathbf{r}'$ alters the shape of surfaces defined in the $\mathbf{r}$ and $\mathbf{r}'$ spaces (Gorodnichev et al., 1987). For example the representation (8.5) is not applicable to the treatment of electron backscattering, where taking into account appropriate boundary conditions is essential. Moreover, it is evident that the form (8.5) is unsuitable for treating scattering by a *periodic* medium. Therefore in order to evaluate the cross-section of diffuse scattering of electrons by a crystal, it is necessary to derive and to solve an equation describing the evolution of the density matrix $\rho(\mathbf{r}, \mathbf{r}')$ taken 'as it is' with no transformation applied to either of the two coordinates $\mathbf{r}$ and $\mathbf{r}'$. A quantum kinetic equation for the density matrix of high-energy electrons satisfying this requirement and describing diffraction, multiple diffuse scattering and energy losses of electrons in a crystalline material was derived by Dudarev et al. (Dudarev et al., 1993a).

8.4 Kinetic equation for the density matrix

We now outline the derivation of the kinetic equation for the density matrix and explain the meaning of approximations that we adopt in order to develop a tractable treatment of diffuse scattering of electrons by a crystalline solid.

We start by introducing an important new Green's function, which generalizes eqn (6.50) that we used previously in the treatment of resonance diffraction of electrons from a crystal surface. This new Green's function is defined by the equation

$$\left[E + \frac{\hbar^2}{2m}\nabla^2 - V_0(\mathbf{r}) \right] G_0(\mathbf{r}, \mathbf{r}', E) = \delta(\mathbf{r} - \mathbf{r}'). \tag{8.6}$$

This function gives the probability amplitude of finding an electron at point $\mathbf{r}$ that was emitted by a point source situated at $\mathbf{r}'$ and which propagated from point $\mathbf{r}'$ to point $\mathbf{r}$ in the periodic potential field $V_0(\mathbf{r})$. The potential $V_0(\mathbf{r})$ entering eqn (8.6) has no imaginary part, and therefore no absorption effects are included in the definition of the Green's function $G_0(\mathbf{r}, \mathbf{r}', E)$.

We now include the fluctuating part of the interaction potential and investigate how this affects the Green's function $\Gamma(\mathbf{r}, \mathbf{r}', E)$. This function describes the propagation of the electron from $\mathbf{r}'$ to $\mathbf{r}$ in the *actual* (not averaged) potential field, and satisfies the equation (here we assume that the fluctuating part of the potential is independent of time)

$$\left[E + \frac{\hbar^2}{2m}\nabla^2 - V_0(\mathbf{r}) - \delta V(\mathbf{r}) \right] \Gamma(\mathbf{r}, \mathbf{r}', E) = \delta(\mathbf{r} - \mathbf{r}'). \tag{8.7}$$

Is there a connection between the two Green's functions (8.7) and (8.6)? These two functions are related by an integral equation that we now derive by following a simple line of argument. In the limit $\delta V(\mathbf{r}) = 0$ the two functions $\Gamma(\mathbf{r}, \mathbf{r}', E)$ and $G_0(\mathbf{r}, \mathbf{r}', E)$ are identical. The presence of the fluctuating part of the potential $\delta V(\mathbf{r})$ gives rise to scattering, and this affects the way in which the electron propagates from $\mathbf{r}'$ to $\mathbf{r}$. Scattering occurs only at points $\mathbf{r}''$, where $\delta V(\mathbf{r}'') \neq 0$, and can be taken into account by considering processes where the electron first travels from $\mathbf{r}'$ to $\mathbf{r}''$, then undergoes scattering by fluctuations, and then propagates without further scattering by fluctuations from $\mathbf{r}''$ to $\mathbf{r}$. This process is described by the following equation:

$$\Gamma(\mathbf{r}, \mathbf{r}', E) = G_0(\mathbf{r}, \mathbf{r}', E) + \int d\mathbf{r}'' G_0(\mathbf{r}, \mathbf{r}'', E)\delta V(\mathbf{r}'')\Gamma(\mathbf{r}'', \mathbf{r}', E). \tag{8.8}$$

At first glance it is not immediately obvious why the first of the two Green's functions entering the integral term in this equation (this integral term describes scattering by fluctuations) is taken as $G_0(\mathbf{r}, \mathbf{r}'', E)$ rather than as $\Gamma(\mathbf{r}, \mathbf{r}'', E)$, i.e. why propagation of the electron from $\mathbf{r}''$ to $\mathbf{r}$ does not *seem* to be affected by the presence of the fluctuating part of the potential δV. The fact that our bookkeeping is in a good order and eqn (8.8) is correct can be proved by showing that $\Gamma(\mathbf{r}'', \mathbf{r}', E)$ satisfying eqn (8.8) also satisfies eqn (8.7). To show this, we substitute $\Gamma(\mathbf{r}'', \mathbf{r}', E)$, taken from eqn (8.8), into eqn (8.7), namely

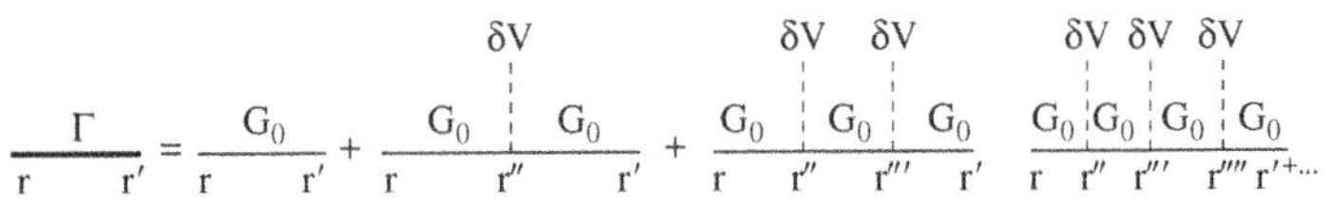

$$\underline{}_{\substack{\Gamma \\ r \qquad r'}} = \underline{}_{\substack{G_0 \\ r \qquad r'}} + \underset{r \qquad r'' \qquad r'}{\overset{\delta V}{G_0 \; G_0}} + \underset{r \quad r'' \quad r''' \quad r'}{\overset{\delta V \quad \delta V}{G_0 \; G_0 \; G_0}} + \underset{r \; r'' \; r''' \; r'''' \; r'+\dots}{\overset{\delta V \; \delta V \; \delta V}{G_0 \, G_0 \, G_0 \, G_0}}$$

FIG. 8.2. Graphical representation of the expansion (8.10) of the Green's function $\Gamma(\mathbf{r}, \mathbf{r}')$ in a power series in the fluctuating part of the interaction potential δV.

$$\left[E + \frac{\hbar^2}{2m} \nabla^2 - V_0(\mathbf{r}) \right] \Gamma(\mathbf{r}, \mathbf{r}', E) = \left[E + \frac{\hbar^2}{2m} \nabla^2 - V_0(\mathbf{r}) \right] G_0(\mathbf{r}, \mathbf{r}', E)$$

$$+ \left[E + \frac{\hbar^2}{2m} \nabla^2 - V_0(\mathbf{r}) \right] \int d\mathbf{r}'' G_0(\mathbf{r}, \mathbf{r}'', E) \delta V(\mathbf{r}'') \Gamma(\mathbf{r}'', \mathbf{r}', E) = \delta(\mathbf{r} - \mathbf{r}')$$

$$+ \int d\mathbf{r}'' \delta(\mathbf{r} - \mathbf{r}'') \delta V(\mathbf{r}'') \Gamma(\mathbf{r}'', \mathbf{r}', E) = \delta(\mathbf{r} - \mathbf{r}') + \delta V(\mathbf{r}) \Gamma(\mathbf{r}, \mathbf{r}', E). \qquad (8.9)$$

We see that the function $\Gamma(\mathbf{r}, \mathbf{r}', E)$ satisfying eqn (8.8) also satisfies (8.7). This proves the equivalence of the two equations, and simultaneously the correctness of the description of scattering by fluctuations given by the integral equation (8.8).

The advantage of using eqn (8.8) lies in the fact that this integral equation can be solved by iteration, or in other words by a procedure that describes scattering very literally by following how the electron emerges from $\mathbf{r}'$, goes to $\mathbf{r}''$, then to $\mathbf{r}'''$, etc., undergoing scattering by fluctuations at all the intermediate points. By substituting G_0 into the integral term on the right-hand side of (8.8), then resubstituting the results into the integral term again and again, we arrive at

$$\Gamma(\mathbf{r}, \mathbf{r}') = G_0(\mathbf{r}, \mathbf{r}') + \int d\mathbf{r}'' G_0(\mathbf{r}, \mathbf{r}'') \delta V(\mathbf{r}'') G_0(\mathbf{r}'', \mathbf{r}')$$

$$+ \int d\mathbf{r}'' \int d\mathbf{r}''' G_0(\mathbf{r}, \mathbf{r}'') \delta V(\mathbf{r}'') G_0(\mathbf{r}'', \mathbf{r}''') \delta V(\mathbf{r}''') G_0(\mathbf{r}''', \mathbf{r}')$$

$$+ \int d\mathbf{r}'' \int d\mathbf{r}''' \int d\mathbf{r}'''' G_0(\mathbf{r}, \mathbf{r}'') \delta V(\mathbf{r}'') G_0(\mathbf{r}'', \mathbf{r}''') \delta V(\mathbf{r}''')$$

$$\times G_0(\mathbf{r}''', \mathbf{r}'''') \delta V(\mathbf{r}'''') G_0(\mathbf{r}'''', \mathbf{r}') + \dots . \qquad (8.10)$$

In a schematic graphical form this expansion corresponds to the series of diagrams shown in Fig. 8.2. It is easy to establish a relation between the solid and the dashed lines shown in Fig. 8.2 and the relevant terms entering expansion (8.10). We use this graphical representation of the various terms in (8.10) because in the subsequent treatment of multiple scattering of electrons by fluctuations we need to be able to operate with higher order terms of the expansion (8.10), where it is impractical to use analytical formulae similar to (8.10).

Calculating the cross-section of scattering requires averaging the product of two amplitudes $\langle d\sigma / do \rangle = \langle f(\mathbf{k}, \mathbf{k}_0) f^*(\mathbf{k}, \mathbf{k}_0) \rangle = \langle f(\mathbf{k}, \mathbf{k}_0) f^\dagger(\mathbf{k}_0, \mathbf{k}) \rangle$. Since each amplitude is related to the Green's function Γ by a linear transformation, we now need to find out how to average a *product* of two series of the form (8.10). It is clear that by multiplying the two series and by averaging them we will obtain a vast number of various terms, each describing a particular process of scattering by fluctuations as well as diffraction by the average potential $V_0(\mathbf{r})$ (the latter is included in the definition (8.6) of the Green's function Γ and is therefore taken into account *exactly*). Our objective is therefore first to classify various terms and then to select those whose contribution to the cross-section of scattering is dominant. The approximation that we now adopt is in some sense equivalent to the Born approximation for scattering by the fluctuating part of the potential. We assume that the correlation radius of fluctuations r_c is relatively small and that we can neglect terms associated with the three-particle, four-particle, and further high-order correlation functions (Barabanenkov and Finkelberg (Barabanenkov and Finkelberg, 1968) discussed how to include these terms in the kinetic equation describing scattering of waves by a random medium). This is equivalent to stating that $\langle \delta V(\mathbf{r}_1)\delta V(\mathbf{r}_2)\delta V(\mathbf{r}_3)\rangle = 0$ and that higher order terms can be expressed as products of the second-order terms. This approximation leads to a sequence of terms representing $\langle \Gamma\Gamma^\dagger\rangle$, some of which are shown in Fig. 8.3. Neglecting the contribution of diagrams containing crossing lines (Gorodnichev et al., 1990) and retaining only the so-called 'ladder diagrams', we arrive at the following equation for $\langle \Gamma\Gamma^\dagger\rangle$,

$$\langle \Gamma\Gamma^\dagger\rangle = \langle \Gamma\rangle\langle \Gamma^\dagger\rangle + \langle \Gamma\rangle\langle \delta V \langle \Gamma\Gamma^\dagger\rangle \delta V\rangle\langle \Gamma^\dagger\rangle. \tag{8.11}$$

Writing all the arguments of Green's functions entering this equation in an explicit form, we arrive at

$$\langle \Gamma(\mathbf{r}, \mathbf{R}, E)\Gamma^*(\mathbf{r}', \mathbf{R}', E)\rangle = \langle \Gamma(\mathbf{r}, \mathbf{R}, E)\rangle\langle \Gamma^*(\mathbf{r}', \mathbf{R}', E)\rangle \tag{8.12}$$

$$+ \int\int d\mathbf{x} d\mathbf{x}' \langle \Gamma(\mathbf{r}, \mathbf{x}, E)\rangle\langle \Gamma^*(\mathbf{r}', \mathbf{x}', E)\rangle\langle \delta V(\mathbf{x})\delta V(\mathbf{x}')\rangle\langle \Gamma(\mathbf{x}, \mathbf{R}, E)\Gamma^*(\mathbf{x}', \mathbf{R}', E)\rangle.$$

This is a quantum-mechanical kinetic equation that describes successive interactions of the incident electron with statistical fluctuations of the potential. Propagation of the electron between successive events of scattering by fluctuations is described by the product of two *average* Green's functions $\langle \Gamma\rangle\langle \Gamma^\dagger\rangle$. The condition of validity of the kinetic equation (8.12) is to a certain extent similar to the condition of validity of the DWBA (in fact, eqn (8.12) describes a sequence of events of scattering by fluctuations, where each event is described by the DWBA).

In order to obtain an equation for the density matrix of high-energy electrons we note that the wave function $\psi(\mathbf{r})$ of the electron is related to the Green's function $\Gamma(\mathbf{r}, \mathbf{r}', E)$ via

$$\psi(\mathbf{r}) = \int d\mathbf{r}' \Gamma(\mathbf{r}, \mathbf{r}', E)S(\mathbf{r}', E), \tag{8.13}$$

where $S(\mathbf{r}', E)$ is the amplitude of sources of electrons. By multiplying (8.12) by $S(\mathbf{R}, E)S^*(\mathbf{R}', E)$ and integrating over the coordinates $\mathbf{R}$ and $\mathbf{R}'$, and also by

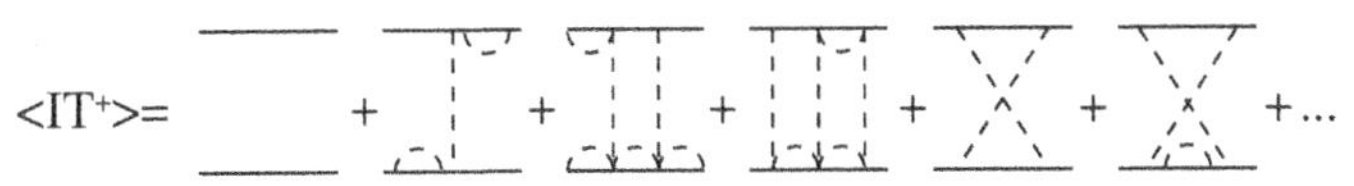

<IT⁺>= $\quad$ + $\quad$ + $\quad$ + $\quad$ + $\quad$ + ...

FIG. 8.3. Graphical representation of the terms arising from the multiplication and term-by-term averaging of the two series $\Gamma\Gamma^\dagger$, where each series represents an expansion of Γ and $\Gamma^\dagger$ in a power series in the fluctuating part of the interaction potential δV.

taking into account the definition of the density matrix (8.4), we obtain a general form of the kinetic equation describing the evolution of the density matrix of high-energy electrons

$$\rho(\mathbf{r}, \mathbf{r}', E) = \rho_0(\mathbf{r}, \mathbf{r}', E) + \int\int d\mathbf{x}d\mathbf{x}' G(\mathbf{r}, \mathbf{x}, E)G^*(\mathbf{r}', \mathbf{x}', E)$$

$$\times \int d\omega\, \bar{s}(\mathbf{x}', \mathbf{x}, \omega)\rho(\mathbf{x}, \mathbf{x}', E + \hbar\omega). \tag{8.14}$$

Here $\rho_0(\mathbf{r}, \mathbf{r}', E)$ is the density matrix of incident electrons, $G(\mathbf{r}, \mathbf{x}, E)$ is the *average* Green's function $G(\mathbf{r}, \mathbf{x}, E) = \langle\Gamma(\mathbf{r}, \mathbf{x}, E)\rangle$ describing propagation of an electron from $\mathbf{x}$ to $\mathbf{r}$ in the crystal. In eqn (8.14) $\bar{s}(\mathbf{x}', \mathbf{x}, \omega)$ is the mixed dynamic form factor introduced in the previous chapter and describing interactions of the incident electron with phonon and electronic excitations, and also scattering by aperiodic variations of the potential associated with randomly distributed defects. The average Green's function entering (8.14) satisfies the equation

$$\left[E + \frac{\hbar^2}{2m}\nabla^2 - V_0(\mathbf{r})\right] G(\mathbf{r}, \mathbf{r}', E)$$

$$- \int d\mathbf{x} \int d\omega\, \bar{s}(\mathbf{r}, \mathbf{x}, \omega)G_0(\mathbf{r}, \mathbf{x}, E - \hbar\omega)G(\mathbf{x}, \mathbf{r}', E) = \delta(\mathbf{r} - \mathbf{r}'). \tag{8.15}$$

Equations (8.14) and (8.15) also remain valid in the general case where the fluctuating part of the potential is both spatially and time dependent. They describe in a self-consistent way the evolution of the one-particle density matrix of high-energy electrons propagating through the crystal and undergoing diffraction and multiple diffuse scattering.

Using eqn (8.14) and following the argument leading to eqn (7.35) we find that the differential cross-section of *multiple* inelastic (diffuse) scattering is given by

$$\left(\frac{d^2\sigma}{dodE}\right)_{\text{inel}} = \frac{k}{k_0}\left(\frac{m}{2\pi\hbar^2}\right)^2\frac{1}{\hbar}$$

$$\times \int\int d\mathbf{r}d\mathbf{r}'\psi_{-\mathbf{k}}(\mathbf{r})\left[\psi_{-\mathbf{k}}(\mathbf{r}')\right]^* \int d\omega\, \bar{s}(\mathbf{r}', \mathbf{r}, \omega)\rho(\mathbf{r}, \mathbf{r}', E + \hbar\omega).$$

For example, in the case of single inelastic (diffuse) scattering and a partially coherent incident beam the cross-section of scattering has the form

$$\left(\frac{d^2\sigma}{do\,dE}\right)_{\text{inel}} = \frac{k}{k_0}\left(\frac{m}{2\pi\hbar^2}\right)^2 \frac{1}{\hbar}$$

$$\times \int\int d\mathbf{r}\,d\mathbf{r}'\,\psi_{-\mathbf{k}}(\mathbf{r})\,[\psi_{-\mathbf{k}}(\mathbf{r}')]^* \int d\omega\,\bar{s}\,(\mathbf{r}',\mathbf{r},\omega)\,\rho_0(\mathbf{r},\mathbf{r}',E+\hbar\omega),$$

where $\rho_0(\mathbf{r},\mathbf{r}',E)$ is the density matrix of the incident beam of electrons and the function $\psi_{-\mathbf{k}}(\mathbf{r})$ is related to the Green's function (8.15) in the same way as the function $\psi_{-\mathbf{k}}^{(0)}(\mathbf{r})$ is related to the Green's function $\mathcal{G}(\mathbf{r}',\mathbf{r})$ via eqn (7.13).

8.5 Loss of coherence due to multiple scattering by plasmons

The density matrix of the electron, or in other words the quantity the off-diagonal elements of which describe the *coherence* of the wave field of the electron in the crystal, satisfies a self-consistent equation (8.14). We shall see below that this quantity plays a central part in the theory of diffuse scattering. Formally, in every event of diffuse scattering the density matrix of the electron $\rho(\mathbf{x},\mathbf{x}',E+\hbar\omega)$ is convoluted with the mixed dynamic form factor $\bar{s}(\mathbf{x}',\mathbf{x},\omega)$. As a result both diagonal and off-diagonal elements of ρ, taken in the coordinate representation, contribute to the observed cross-section of scattering. In general it is not possible to find a closed equation for the *intensity* $\rho(\mathbf{x},\mathbf{x},E)$, i.e. an equation involving only the *diagonal* elements of the density matrix. In other words, it is inevitable that the full treatment of diffuse scattering requires solving eqn (8.14) for the density matrix considered as a complex function of two independent variables $\mathbf{x}$ and $\mathbf{x}'$. This poses a formidable mathematical problem and at the same time highlights the need to develop approximate methods for solving the kinetic equation. An important point associated with finding a solution of eqn (8.14) consists in investigating the difference in the behaviour of diagonal and off-diagonal elements of the density matrix.

In this section we study the evolution of diagonal and off-diagonal elements of the density matrix using a relatively simple example, where electrons propagate through a crystalline specimen in the absence of thermal vibrations and where diffuse scattering is associated only with plasmon excitations. This case may be considered as an approximate model describing the diffraction of electrons by a single crystal of silicon, where the fact that the Debye temperature of the material is high and the potential of interaction of electrons with an individual atom is relatively weak makes plasmon excitations the dominant channel of diffuse scattering.

Since the pioneering work by Howie (Howie, 1963) the question of how diffuse scattering affects the coherence of high-energy electrons propagating through crystals remained the subject of active experimental and theoretical research, see for example (Young and Rez, 1975, Rossouw and Whelan, 1981) and also the more recent studies (Wang, 1995, Allen and Josefsson, 1995). On the one hand the delocalization of the mixed dynamic form factor of plasmon excitations (7.67) makes it possible to consider plasmon scattering to some extent as coherent (Ajika et al., 1985, Cowley, 1988). On the other hand experimental observations (Bakenfelder et al., 1990) show that by eliminating the contribution of electron–electron inelastic collisions by energy filtering it is possible to improve substantially the contrast of electron microscope images. Experimental observations described in (Bakenfelder et al., 1990) suggest that collective electron excitations may indeed strongly affect the coherence of high-energy electrons propagating through the crystal.

The cross-section of scattering of high-energy electrons by plasmons (7.66) is anisotropic and electrons are predominantly scattered in the forward direction. The angular divergence of the electron beam resulting from plasmon scattering is fairly small, and it is this point that is often quoted as the origin of the observed 'preservation of contrast' of electron microscope images occurring in the presence of plasmon scattering (Ajika et al., 1985). Nevertheless, the finite angular width of the electron beam results in electrons propagating along different trajectories and undergoing diffraction under slightly different Bragg conditions, and thereby acquiring different phases. Repeated plasmon scattering gives rise to random mixing of phases and to the gradual loss of coherence of the wave field of high-energy electrons.

The picture that we intend to describe can be qualitatively understood as follows. After the first event of plasmon scattering the wave function of the fast electron can be represented in the form of a set of waves propagating in slightly different directions and therefore characterized by slightly different excitation errors (Howie, 1963, Ajika et al., 1985). Although the difference between the excitation errors of any two waves may be small, the phase difference between them increases linearly with the path length. An estimate based on the known form of the cross-section of electron-electron scattering (7.66) shows that the difference between phases of two inelastically scattered waves can reach π for a path length of the order of the inelastic mean free path l_{e-e} for plasmon and valence electron excitations. This estimate shows that multiple plasmon scattering averages out the phase factors describing interference between quantum states of high-energy electrons propagating through the crystal.

We may therefore expect that small-angle plasmon scattering should give rise to the rapid disappearance of thickness oscillations of cross-sections of processes depending on the distribution of the density of electrons at the lattice sites: for example, the cross-sections of secondary electron emission and high-angle backscattering. The transition from the oscillating to a monotonic behaviour of cross-sections considered as a function of the thickness of the crystal is often referred to as the transition from the 'dependent' to the 'independent' Bloch wave models (Hagemann and Reimer, 1979). The loss of coherence resulting from small-angle plasmon scattering represents the primary cause of this transition.

To follow the evolution of coherence of the wave field of electrons we use the kinetic equation for the density matrix $\rho(\mathbf{r}, \mathbf{r}', E)$, of which the diagonal elements, $\rho(\mathbf{r}, \mathbf{r}, E)$, give the probability $P(\mathbf{r}, E)$ of finding an electron with total energy E at a point $\mathbf{r}$ in the crystal. It is often important to know how $P(\mathbf{r}, E)$ varies as a function of coordinate z in the direction of the incident beam since

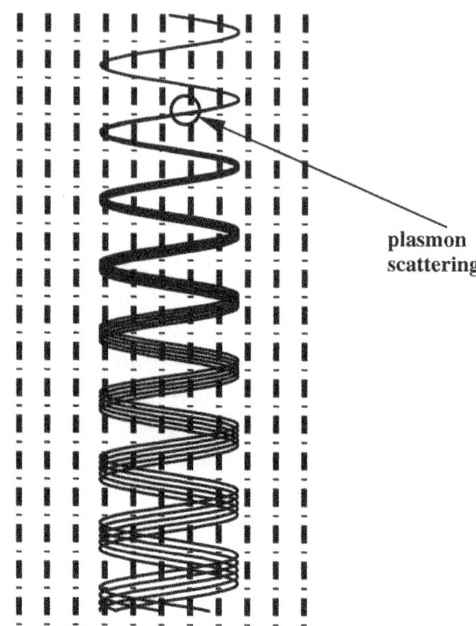

FIG. 8.4. Sketch illustrating how scattering of high-energy electrons by plasmons gives rise to the gradual decoherence of the wave field of electrons. The incident electron initially performs 'wavy' motion in the crystal lattice described by the oscillating solution of the Howie-Whelan equations (Hirsch et al., 1977). The finite angular width of the beam associated with plasmon scattering gives rise to the gradual accumulation of phase difference between electrons following slightly different trajectories in the crystal.

this quantity determines the yield of secondary electron emission and high-angle backscattering.

Consider the geometry of transmission electron diffraction shown in Fig. 8.4. Assuming that diffraction involves only reciprocal lattice vectors of the zero-order Laue zone, we write the density matrix of the high-energy electron in the form of a linear combination of Bloch waves with slowly varying amplitudes,

$$\rho(\mathbf{r}, \mathbf{r}', E) = \sum_{j,j'} \int_{BZ} \frac{d^2 q}{(2\pi)^2} \rho_{jj'}(\mathbf{q}, z, z', E) b_j(\mathbf{R}, \mathbf{q}) b_{j'}^*(\mathbf{R}', \mathbf{q}) \qquad (8.16)$$

$$\times \exp\left[i\sqrt{\frac{2mE}{\hbar^2}}(z - z') - i\frac{z E_j(\mathbf{q})}{\hbar v} + i\frac{z' E_{j'}(\mathbf{q})}{\hbar v} - \frac{z \mu_j(\mathbf{q})}{\hbar v} - \frac{z' \mu_{j'}(\mathbf{q})}{\hbar v} \right],$$

where $v = \sqrt{2E/m}$ is the velocity of the fast electron and the summation is performed over different branches j, j' of the dispersion surface. Integration over $\mathbf{q} = (q_x, q_y)$ is performed over the two-dimensional Brillouin zone (BZ). $b_j(\mathbf{R}, \mathbf{q}) = b_j(x, y, \mathbf{q})$ is a two-dimensional Bloch wave function describing the transverse motion of the electron in the *projected* two-dimensional periodic potential $V_0(x, y)$. This Bloch wave function satisfies the 'transverse' Schrödinger equation, with transverse energy $E_j(\mathbf{q})$,

$$-\frac{\hbar^2}{2m}\nabla_{\mathbf{R}}^2 b_j(\mathbf{R}, \mathbf{q}) + V_0(\mathbf{R})b_j(\mathbf{R}, \mathbf{q}) = E_j(\mathbf{q})b_j(\mathbf{R}, \mathbf{q}). \tag{8.17}$$

In eqn (8.16) $\mu_j(\mathbf{q})$ is the coefficient of absorption of the j-th Bloch state describing the attenuation of this state due to large-angle phonon scattering. Following the method described in (Dudarev et al., 1993a) we find a closed equation for the elements of the density matrix $\rho_{jj'}(\mathbf{q}, z)$ integrated over the energy spectrum of the incident electrons,

$$\rho_{jj'}(\mathbf{q}, z) = \int\limits_{0}^{E_{max}} \rho_{jj'}(\mathbf{q}, z, z, E)dE. \tag{8.18}$$

Considering only the small-angle *intrabranch scattering* of electrons by plasmons (Howie, 1963), we find that the density matrix satisfies the differential equation

$$l_{e-e}\frac{d}{dz}\rho_{jj'}(\mathbf{q}, z) = -\rho_{jj'}(\mathbf{q}, z) + \frac{1}{2\pi\ln\left(\frac{Q_c v}{\omega_{pl}}\right)}\int\limits_{|\mathbf{Q}|\leq Q_c}\frac{d^2Q}{Q^2 + (\omega_{pl}/v)^2} \tag{8.19}$$

$$\times\rho_{jj'}(\mathbf{q} + \mathbf{Q}, z)\exp\left(i\frac{z}{\hbar v}[E_j(\mathbf{q}) - E_{j'}(\mathbf{q}) + E_{j'}(\mathbf{q} + \mathbf{Q}) - E_j(\mathbf{q} + \mathbf{Q})]\right).$$

In this equation $Q_c = \omega_{pl}/v$ is the plasmon cut-off vector, and the inelastic mean free path l_{e-e} is given by the expression

$$l_{e-e} = \frac{4\pi\epsilon_0\hbar v^2}{e^2\omega_{pl}}\left[\ln(vQ_c/\omega_{pl})\right]^{-1}. \tag{8.20}$$

In (8.19) we assumed that vectors $\mathbf{q}$ and $\mathbf{q} + \mathbf{Q}$ lie in the first Brillouin zone.

Kinetic equations for the density matrix of fast charged particles propagating through a thin crystal were first derived by Kagan and Kononets for the case of channelling of protons (Kagan and Kononets, 1973) and by Rez for the case of electron diffraction (Rez, 1978). These authors used the similarity between the time-independent Schrödinger equation describing propagation of a fast electron through a solid and the time-dependent Schrödinger equation. A more general integral form of the quantum kinetic equation, suitable also for the treatment of electron backscattering and reflection electron diffraction, was derived later (Dudarev et al., 1993a).

Equation (8.19) illustrates a remarkable difference between the evolution of diagonal ($j = j'$) and off-diagonal ($j \neq j'$) elements of the density matrix in

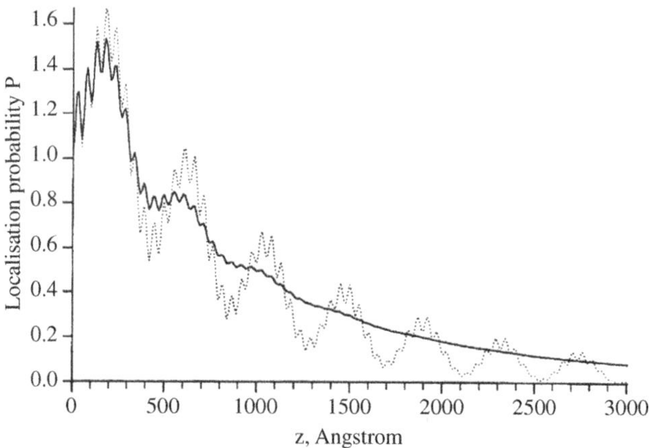

FIG. 8.5. Probability of finding an electron at point $\mathbf{R} = 0$ in the plane normal to the direction of the incident beam. This probability is plotted as a function of coordinate z in the direction of the beam. Point $\mathbf{R} = 0$ corresponds to a minimum of the projected potential $V_0(\mathbf{R})$. The two curves illustrate the difference between results obtained using the conventional dependent Bloch wave model (where no loss of coherence due to plasmon scattering is assumed, see dotted line) and the solution of the kinetic equation for the density matrix (solid line). Calculations were carried out for the Si 220 systematic row of reflections for the angle of incidence $\theta_0 = 17.12$ mrad and the primary beam energy of 30 keV. The absolute temperature $T = 293$ K. From (Dudarev et al., 1992a, Dudarev et al., 1993d).

the presence of multiple inelastic scattering. For example, the equation for the diagonal elements

$$l_{e-e}\frac{d}{dz}\rho_{jj}(\mathbf{q}, z) = -\rho_{jj}(\mathbf{q}, z) + \frac{1}{2\pi \ln\left(\frac{Q_c v}{\omega_{pl}}\right)} \int\limits_{|\mathbf{Q}| \leq Q_c} \frac{d^2 Q}{Q^2 + (\omega_{pl}/v)^2}\rho_{jj}(\mathbf{q} + \mathbf{Q}, z)$$

$$(8.21)$$

contains no phase factors. By integrating this equation over $\mathbf{q}$ we find that the total probability $\sum_j \int d^2 q \rho_{jj}(\mathbf{q}, z)$ is conserved

$$\frac{d}{dz}\left[\sum_j \int d^2 q \rho_{jj}(\mathbf{q}, z)\right] = 0.$$

No similar condition of conservation exists for the off-diagonal elements of the density matrix. This fact illustrates a fundamental feature of dissipative evolution of a quantum-mechanical system interacting with a random external environment, namely that the off-diagonal elements of the density matrix gradually attenuate owing to destructive interference between states of the system evolving along a group of slightly different trajectories in the parameter space. In the case of scattering of electrons by plasmons the phase factors enter the right-hand side of eqn (8.19). These phase factors affect only the off-diagonal elements of the density matrix taken in the Bloch state representation, and the significance of these factors depends on the law of *dispersion* of the relevant Bloch states. In the limiting case where the function $E_j(\mathbf{q})$ is independent of $\mathbf{q}$ (this case describes tightly bound Bloch states as shown in Fig. 6.3) the phase factors in eqn (8.19) vanish and the wave field of electrons retains its coherence even in the presence of plasmon scattering. In practice Bloch states always exhibit some dispersion, and the difference $E_{j'}(\mathbf{q} + \mathbf{Q}) - E_j(\mathbf{q} + \mathbf{Q})$ between the energies of any two Bloch states corresponding to the transverse momentum $\mathbf{q} + \mathbf{Q}$ is never exactly equal to the difference $E_{j'}(\mathbf{q}) - E_j(\mathbf{q})$ between energies of the same pair of states corresponding to the transverse momentum $\mathbf{q}$.

In the case where the angular width of the beam of electrons incident on the surface of the specimen is small, the initial condition for eqn (8.19) has the form

$$\rho_{jj'}(\mathbf{q}, 0) = (2\pi)^2 \delta(\mathbf{q} - \mathbf{q}_0) \rho_{jj'}^{(0)}(\mathbf{q}_0). \tag{8.22}$$

Here $\mathbf{q}_0$ is the projection of the momentum $\mathbf{k}_0$ of incident electrons on the (x, y) plane. The matrix elements $\rho_{jj'}^{(0)}(\mathbf{q}_0)$ are defined as products of the coefficients in the expansion of the function $\exp(i\mathbf{q}_0 \cdot \mathbf{R})$ in the form of a linear combination of Bloch functions $b_j(\mathbf{R}, \mathbf{q})$ (Kagan and Kononets, 1973).

If the solution of kinetic equation (8.19) is known, the probability of finding the electron at a point $\mathbf{r}$ is given by

$$P(\mathbf{r}) = \int \rho(\mathbf{R}, \mathbf{R}, z, z, E)dE = \sum_{j,j'} \int \frac{d^2q}{(2\pi)^2} \rho_{jj'}(\mathbf{q}, z) b_j(\mathbf{R}, \mathbf{q}) b_{j'}^*(\mathbf{R}, \mathbf{q})$$

$$\times \exp\left\{ -i\frac{z}{\hbar v}[E_j(\mathbf{q}) - E_{j'}(\mathbf{q})] - \frac{z}{\hbar v}[\mu_j(\mathbf{q}) + \mu_{j'}(\mathbf{q})] \right\}. \tag{8.23}$$

The cross-section of secondary emission and high-angle backscattering is approximately proportional to this probability taken at crystal lattice sites $\mathbf{r} = \mathbf{r}_a$. Neglecting the angular divergence of the beam due to electron–electron inelastic scattering, we find that the solution of (8.19) is given by $\rho_{jj'}(\mathbf{q}, z) = \rho_{jj'}(\mathbf{q}, 0)$, and eqn (8.23) reduces to the well-known expression of the dependent Bloch wave model (Hirsch et al., 1962, Cherns et al., 1973, Hagemann and Reimer, 1979).

The probability $P(0, z)$ of finding an electron at $\mathbf{R} = 0$ (corresponding to one of the minima of the projected potential $V_0(\mathbf{R})$) is shown as a function of z in Fig. 8.5 by the dotted line. This probability exhibits oscillating behaviour. The oscillations are associated with the off-diagonal elements of the density matrix $\rho_{jj'}(\mathbf{q}, z)$ that describe coherence between Bloch states of the electron propagating through the crystal. The monotonic attenuation of the functions shown in Fig. 8.5 is associated with large-angle phonon scattering.

To investigate how small-angle plasmon scattering affects the off-diagonal elements of the density matrix $\rho_{jj'}(\mathbf{q}, z)$, we have solved the kinetic equation

FIG. 8.6. Energy-unfiltered convergent beam electron diffraction pattern of the Si 000 and Si 220 reflections taken in the transmission diffraction geometry. The thickness of the crystal is 2.1 extinction distances for the 220 reflection, the absolute temperature is approximately 100 K, and the energy of the incident electrons is 80 keV.

(8.19) numerically. In the systematic diffraction case studied here the energy of a particular Bloch state depends only on the x projection of the quasimomentum, $E_j(\mathbf{q}) = E_j(q_x) + \hbar^2 q_y^2/2m$. This makes it possible to approximate the integral term in (8.19) by a finite sum over several hundred points in the Brillouin zone. The localization probability $P(0, z)$ calculated using numerical solutions of the kinetic equation is shown in Fig. 8.5 by the solid line. Oscillations of the localization probability shown by the solid curve vanish almost entirely for $z \geq 1000$ Å. By comparing this spatial scale with the inelastic mean free path (8.21) $l_{e-e} = 480$ Å for silicon at $E = 30$ keV, we find that Bloch states forming the wave field of the high-energy electron become mutually incoherent after just two successive events of scattering by plasmon excitations.

A comparison of the cross-sections of various types of inelastic excitations (Radi, 1970) shows that in many cases the inelastic mean free path of high-energy electrons for electron excitations is substantially smaller than the mean free path of phonon scattering. For instance in silicon at room temperature the average inelastic mean free path of phonon scattering is four times the value of l_{e-e} calculated using eqn (8.21). Therefore in many cases small-angle plasmon scattering represents the dominant mechanism giving rise to the loss of coherence of high-energy electrons propagating through a crystalline specimen.

Another illustration of the loss of coherence due to multiple small-angle scattering follows from the analysis of distributions of intensity in energy-unfiltered CBED patterns (Marthinsen et al. (Marthinsen et al., 1994) investigated a sim-

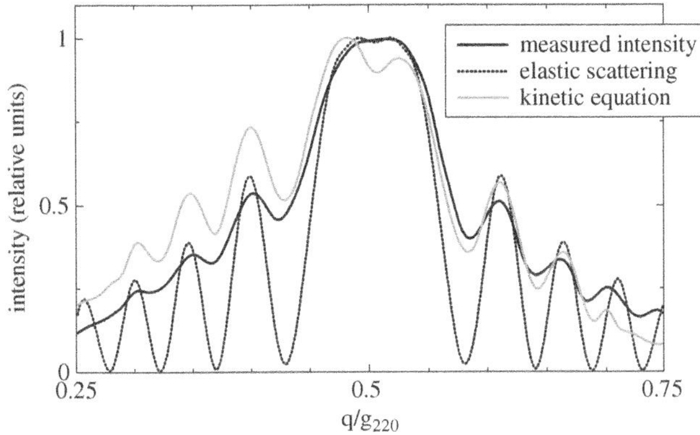

FIG. 8.7. Experimentally observed (solid line) and theoretically simulated profiles of intensity in the dark-field CBED pattern of the Si 220 reflection. The dotted curve was calculated using the 11-beam approximation of the dynamical diffraction theory taking into account only elastically scattered electrons. The grey curve represents the solution of the kinetic equation of the density matrix that takes into account multiple scattering of electrons by plasmons. The thickness of the crystal $L = 2340$ Å.

ilar problem using a somewhat different theoretical approach). For example, an experimental CBED pattern corresponding to the 110 systematic row of reciprocal lattice vectors of an Si single crystal obtained using a Philips CM12 analytical electron microscope at an accelerating voltage of 80 keV and a specimen temperature of approximately 100 K is shown in Fig. 8.6. Observations were made for several different values of the thickness of the specimen, and in each case the distribution of intensity across the 220 disc was digitized using a microdensitometer. The thickness of the crystal was determined by measuring the positions of intensity maxima and minima, and in the case shown in Fig. 8.6 the thickness was found to be 2860 ± 20 Å.

A densitometer trace across the 220 CBED disc of Fig. 8.6 is shown by the solid line in Fig. 8.7. The dotted line in Fig. 8.7 is the dark-field rocking curve simulated using conventional dynamical diffraction theory (Hirsch et al., 1977), where the effects of inelastic scattering are taken into account by an effective absorption of electrons. Dynamical diffraction calculations predict highly symmetrical profiles even in the limit of large thickness of the specimen. The light grey curve is the dark-field intensity profile evaluated by integrating the kinetic equation for the density matrix (8.19). The comparison of measured and calculated profiles given in Fig. 8.7 shows that the approach based on the kinetic equation for the density matrix agrees with experimental observations much better than calculations based on the conventional dynamical diffraction theory. The density matrix solution reproduces correctly the asymmetry of the profile first noted by Howie (Howie, 1963). This is of course not surprising since the kinetic equation describes inelastic scattering that actually occurs in a diffraction experiment, while the conventional dynamical diffraction treatment is unsuitable for even addressing the issue. What is significant, however, is that we have been able to develop a tractable approach to finding solutions of the kinetic equation, and this opens the possibility of applying the method to other interesting cases including the examples considered below.

8.6 Diffraction of diffusely scattered electrons: the formation of Kikuchi lines and bands

In this section we consider a very interesting application of the theory of diffuse scattering of high-energy electrons, namely the problem of formation of Kikuchi lines and bands shown in Figs 8.8 and 8.9. This topic is generally considered to be among the difficult problems of electron diffraction, and classic monographs such as those by Hirsch et al. (Hirsch et al., 1977) and Cowley (Cowley, 1990) provide only a qualitative description of the effect.

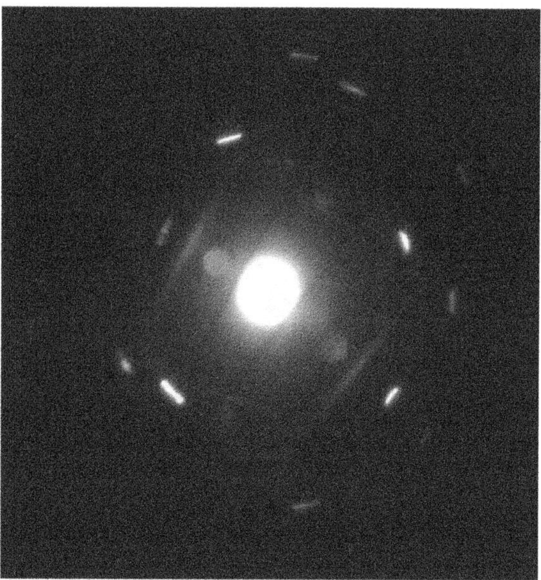

FIG. 8.8. Kikuchi pattern showing pairs of dark (near the centre of the pattern) and bright (on the periphery of the pattern) Kikuchi lines. This pattern was observed using a single crystal of silicon and 200 keV electrons. The direction of the incident electron beam is near the [110] zone axis.

We will show below that the treatment of Kikuchi lines and bands does not require going beyond a fairly elementary calculation based on the DWBA (7.35). Some more subtle aspects of the theory of Kikuchi patterns, involving, for example, the treatment of inversion of the contrast of Kikuchi bands occurring

FIG. 8.9. Electron diffraction pattern showing Kikuchi bands. The pattern was observed using 200 keV electrons and a relatively thick single crystal of silicon. Note that the intensity is higher in the central part of each Kikuchi band.

in the case of relatively thick specimens, require taking into account the effects of absorption and multiple diffuse scattering of electrons.

The first theoretical treatment of Kikuchi patterns based on the single inelastic scattering approximation was developed by Kainuma (Kainuma, 1955, Kainuma et al., 1955). Rossouw and Bursill (Rossouw and Bursill, 1985) gave an excellent example of application of this approximation to the quantitative investigation of the distribution of intensity in electron diffraction patterns of a rutile (TiO$_2$) single crystal. Chukhovskii et al. (Chukhovskii et al., 1973) showed that the reversal of contrast of Kikuchi bands occurring in the case of relatively thick crystalline specimens (Hall, 1970) is caused by absorption effects. Høier (Høier, 1973) developed the first *multiple scattering* treatment of Kikuchi patterns and explained why the reversal of contrast occurs first in the centre of the pattern and later at the edges of the pattern. The subsequent development of the density matrix approach (Dudarev and Ryazanov, 1988) was largely stimulated by the need to find a more transparent treatment of the same problem.

Here we investigate the formation of Kikuchi patterns in the case of diffuse scattering of electrons by a thin crystal. In eqns (7.35) and (7.34) we neglect energy losses and write the cross-section of diffuse scattering in the form

$$\left(\frac{d\sigma}{do}\right) = \left(\frac{m}{2\pi\hbar^2}\right)^2 \int d\mathbf{r}\,d\mathbf{r}'\,\psi^{(0)}_{-\mathbf{k}}(\mathbf{r}) \left[\psi^{(0)}_{-\mathbf{k}}(\mathbf{r}')\right]^* \psi^{(0)}_{\mathbf{k}_0}(\mathbf{r}) \left[\psi^{(0)}_{\mathbf{k}_0}(\mathbf{r}')\right]^* \langle \delta V(\mathbf{r})\delta V(\mathbf{r}')\rangle,$$

$$(8.24)$$

where δV is the fluctuating part of the potential associated with random thermal displacements of atoms from their equilibrium positions in the crystal lattice

$$\delta V(\mathbf{r}) = \sum_n [\varphi(\mathbf{r} - \mathbf{R}_n - \mathbf{u}_n) - \langle \varphi(\mathbf{r} - \mathbf{R}_n - \mathbf{u}_n)\rangle]. \qquad (8.25)$$

In this equation the summation over n is performed over all atoms in the crystal. We assume that the atoms are identical and that their thermal displacements are uncorrelated. The wave functions $\psi_{\mathbf{k}_0}^{(0)}(\mathbf{r})$ and $\psi_{-\mathbf{k}}^{(0)}(\mathbf{r})$ describe electrons that are incident on the specimen from directions $\mathbf{k}_0$ and $-\mathbf{k}$, respectively. Approximation (8.25) neglects the fact that atoms actually perform thermal *motion* and treats $\mathbf{u}_n$ as static random displacements. This approximation is justified by the fact that propagation of a high-energy electron through the crystal is a very fast process occurring on a much shorter time scale than the time scale characterizing the thermal motion of atoms.

In the treatment of Kikuchi patterns we focus our attention on the diffraction of diffusely scattered electrons. In this case in formula (8.24) we take the wave function of incident electrons in the form of a plane wave $\psi_{\mathbf{k}_0}^{(0)}(\mathbf{r}) = \exp(i\mathbf{k}_0 \cdot \mathbf{r})$, which we assume is not affected by the average periodic potential of the crystal. The wave function $\psi_{-\mathbf{k}}^{(0)}(\mathbf{r})$ is calculated using eqns (7.22)–(7.25), namely

$$\psi_{-\mathbf{k}}(\mathbf{r}) = \alpha^{(1)}b^{(1)}(\mathbf{k}^{(1)},\mathbf{r}) + \alpha^{(2)}b^{(2)}(\mathbf{k}^{(2)},\mathbf{r}), \tag{8.26}$$

where the wave vectors $\mathbf{k}^{(1)}$ and $\mathbf{k}^{(2)}$ of Bloch states $b^{(1)}$ and $b^{(2)}$ of diffusely scattered electrons are given by

$$\left[k_z^j\right]^2 = (-\mathbf{k})_z^2 + \frac{1}{2}\Big[(-\mathbf{k}_t)^2 - (-\mathbf{k}_t + \mathbf{g})^2 \tag{8.27}$$
$$+ (-1)^{j+1}\sqrt{[(-\mathbf{k}_t)^2 - (-\mathbf{k}_t + \mathbf{g})^2]^2 + 16m^2|V_{\mathbf{g}}|^2/\hbar^4}\Big].$$

The wave function $\psi_{-\mathbf{k}}^{(0)}(\mathbf{r})$ of diffusely scattered electrons is

$$\psi_{-\mathbf{k}}^{(0)}(\mathbf{r}) = \exp(-i\mathbf{k} \cdot \mathbf{r}) \exp\left[i\frac{m|V_{\mathbf{g}}|}{k_z\hbar^2}w(L-z)\right]$$
$$\times \left\{\cos\left[\frac{m|V_{\mathbf{g}}|}{k_z\hbar^2}(L-z)\sqrt{1+w^2}\right] - i\frac{w}{\sqrt{1+w^2}}\sin\left[\frac{m|V_{\mathbf{g}}|}{k_z\hbar^2}(L-z)\sqrt{1+w^2}\right]\right\}$$
$$+ \exp[i(-\mathbf{k}+\mathbf{g})\cdot\mathbf{r}]\exp\left[i\frac{m|V_{\mathbf{g}}|}{k_z\hbar^2}w(L-z)\right]$$
$$\times \frac{iw}{\sqrt{1+w^2}}\sin\left[\frac{m|V_{\mathbf{g}}|}{k_z\hbar^2}(L-z)\sqrt{1+w^2}\right]. \tag{8.28}$$

Here L is the thickness of the crystal and the deviation parameter w refers to the wave vector of *scattered* electrons $\mathbf{k}$ via

$$w = \frac{\hbar^2}{4m|V_{\mathbf{g}}|}[(-\mathbf{k}_t)^2 - (-\mathbf{k}_t + \mathbf{g})^2]. \tag{8.29}$$

Substituting (8.28) into (8.24) and neglecting the oscillating terms, we find

$$\left(\frac{d\sigma}{do}\right)_{\text{diffuse}} = \mathcal{I}(\mathbf{k} - \mathbf{k}_0; \mathbf{k} - \mathbf{k}_0) \tag{8.30}$$

$$+ \frac{1}{2(1 + w^2)}[\mathcal{I}(\mathbf{k} - \mathbf{k}_0 - \mathbf{g}; \mathbf{k} - \mathbf{k}_0 - \mathbf{g}) - \mathcal{I}(\mathbf{k} - \mathbf{k}_0; \mathbf{k} - \mathbf{k}_0)]$$

$$- \frac{w}{2(1 + w^2)}[\mathcal{I}(\mathbf{k} - \mathbf{k}_0; \mathbf{k} - \mathbf{k}_0 - \mathbf{g}) + \mathcal{I}(\mathbf{k} - \mathbf{k}_0 - \mathbf{g}; \mathbf{k} - \mathbf{k}_0)],$$

where

$$\mathcal{I}(\mathbf{q}; \mathbf{q}') = \left(\frac{m}{2\pi\hbar^2}\right)^2 \sum_n \int d\mathbf{r} d\mathbf{r}' \exp(-i\mathbf{q} \cdot \mathbf{r} + i\mathbf{q}' \cdot \mathbf{r}') \times \tag{8.31}$$

$$[\langle\varphi(\mathbf{r} - \mathbf{R}_n - \mathbf{u}_n)\varphi(\mathbf{r}' - \mathbf{R}_n - \mathbf{u}_n)\rangle - \langle\varphi(\mathbf{r} - \mathbf{R}_n - \mathbf{u}_n)\rangle\langle\varphi(\mathbf{r}' - \mathbf{R}_n - \mathbf{u}_n)\rangle].$$

Expressions (8.30) and (8.31) describing the distribution of intensity in electron Kikuchi patterns were first derived in (Kainuma, 1955, Kainuma et al., 1955). The first term in eqn (8.30) gives the cross-section of diffuse scattering of electrons in the absence of dynamical diffraction effects, the second term describes a complementary pair of Kikuchi lines, and the last term describes a Kikuchi band.

Equation (8.30) makes it easy to understand the nature of Kikuchi lines and bands. The formation of Kikuchi lines is associated entirely with the fact that the cross-section of scattering of high-energy electrons by atoms is anisotropic and that factor $\mathcal{I}(\mathbf{q}, \mathbf{q})$ decreases rapidly as a function of $|\mathbf{q}|$ for $|\mathbf{q}| > a^{-1}$, where a is the effective radius of the potential of an atom. For example for $\mathbf{k} \approx \mathbf{k}_0$ we find that $\mathcal{I}(\mathbf{k} - \mathbf{k}_0; \mathbf{k} - \mathbf{k}_0) \gg \mathcal{I}(\mathbf{k} - \mathbf{k}_0 - \mathbf{g}; \mathbf{k} - \mathbf{k}_0 - \mathbf{g})$ and the second term in eqn (8.30) describes an inverted Lorenzian profile centred around $w = 0$. Furthermore, since parameter w defined by eqn (8.29) depends only on the component of the wave vector $\mathbf{k}$ parallel to $\mathbf{g}$, in a diffraction pattern the inverted Lorenzian profile is seen as a narrow dark line (in Fig. 8.8 these narrow dark lines are seen close to the centre of the diffraction pattern). On the other hand, for wave vectors $\mathbf{k} \approx \mathbf{k}_0 + \mathbf{g}$ we have $\mathcal{I}(\mathbf{k} - \mathbf{k}_0; \mathbf{k} - \mathbf{k}_0) \ll \mathcal{I}(\mathbf{k} - \mathbf{k}_0 - \mathbf{g}; \mathbf{k} - \mathbf{k}_0 - \mathbf{g})$, and the second term in (8.30) gives rise to a positive Lorenzian profile describing a bright line running parallel to the dark line further away from the centre of the diffraction pattern (this can also be seen in Fig. 8.8). This analysis shows that Kikuchi lines are formed simply as a result of transfer of diffuse intensity from the central region in the diffraction pattern where diffuse intensity is higher, to the edges of the pattern where diffuse intensity is low.

The mechanism of formation of Kikuchi bands (described by the last term in eqn (8.30) and seen in Fig. 8.9) is somewhat more subtle. Consider the limit of purely isotropic scattering by fluctuations, where $\mathcal{I}(\mathbf{q}, \mathbf{q})$ =const. In this case the second term in eqn (8.30) vanishes. This is not surprising of course, since in the case where the cross-section of diffuse scattering is isotropic the intensity of scattering in the directions $\mathbf{k} \approx \mathbf{k}_0$ and $\mathbf{k} \approx \mathbf{k}_0 + \mathbf{g}$ is the same, and Kikuchi lines do not form. However, the last term in eqn (8.30) does not vanish even in the limit of isotropic scattering. This term describes a feature that is asymmetric with respect to the exact Bragg condition $w = 0$. To understand the meaning of this term, imagine a pair of lines in a diffraction pattern where each line corresponds to the exact Bragg condition for reciprocal lattice vectors $\pm\mathbf{g}$. Equation (8.30) states that between these two lines (where $w < 0$) the diffuse intensity is higher than in the region where $w > 0$. The resulting band-like features (i.e. bands of enhanced diffuse intensity seen in Fig. 8.9 and enclosed between imaginary pairs of lines corresponding to the exact Bragg conditions for reciprocal lattice vectors $\pm\mathbf{g}$) seen in diffuse electron diffraction patterns are called the Kikuchi bands. The origin of Kikuchi bands reflects the difference between real-space distributions of the density of electrons in Bloch waves of type 1 and 2 forming the wave function of diffusely scattered electrons (8.26), (8.28). Indeed, the probability of finding a scattered electron at a lattice site $\mathbf{R}_n$ equals

$$P_{-\mathbf{k}}(\mathbf{R}_n) = \left| a^{(1)} b^{(1)}(\mathbf{k}^{(1)}, \mathbf{R}_n) + a^{(2)} b^{(2)}(\mathbf{k}^{(2)}, \mathbf{R}_n) \right|^2 \tag{8.32}$$

$$= 1 - \frac{w}{1 + w^2} + \frac{w}{1 + w^2} \cos\left(\frac{2m|V_{\mathbf{g}}|}{k_{0z}\hbar^2}(L - z_n)\sqrt{1 + w^2} \right).$$

and this probability is higher for directions of scattering corresponding to $w < 0$ (the inner part of a Kikuchi band) than for directions of scattering corresponding to $w > 0$ (the outer part of a Kikuchi band). The asymmetry of the localization probability (8.32) is associated with the asymmetry of the real-space density distribution of the two Bloch waves forming the solution (8.26). Bloch wave 1, which is predominantly localized on lattice sites, has higher excitation amplitude $a^{(1)}$ in the region $w < 0$. Bloch wave 2, which is predominantly localized in the space between the atoms, has higher excitation amplitude $a^{(2)}$ in the region $w > 0$. Fluctuations of the interaction potential giving rise to diffuse scattering of electrons are mainly localized in the vicinity of lattice sites. As a result the intensity of diffuse scattering is higher for directions of scattering where the overlap between the wave function of scattered electrons $\psi_{-\mathbf{k}}(\mathbf{r})$ and atomic potentials is greater. This explains why the diffuse intensity is higher in the inner part of Kikuchi bands and why it is lower in the outer part of the band.

In the case of scattering by thick crystals the above arguments need to be modified in order to take into account absorption of both incident and diffusely scattered electrons (Chukhovskii et al., 1973). For the incident electrons absorption effects can be taken into account by multiplying the incident plane wave by an appropriate exponential factor

$$\psi_{\mathbf{k}_0}^{(0)}(\mathbf{r}) = \exp(i\mathbf{k}_0 \cdot \mathbf{r}) \exp(-\mu_0 z/2). \tag{8.33}$$

For the diffusely scattered electrons we must take into account the fact that the two Bloch states forming the wave function of the scattered electron (8.26) are characterized by *different* rates of absorption. We write

$$\psi_{-\mathbf{k}}(\mathbf{r}) = a^{(1)} b^{(1)}(\mathbf{k}^{(1)}, \mathbf{r}) \exp(-\mu^{(1)}[L - z]/2)$$
$$+ a^{(2)} b^{(2)}(\mathbf{k}^{(2)}, \mathbf{r}) \exp(-\mu^{(2)}[L - z]/2), \tag{8.34}$$

where $\mu^{(j)} = \mu_0 + (-1)^{j+1}\mu_1/\sqrt{1+w^2}$ and μ_1 is the anomalous absorption coefficient (Hirsch et al., 1977). Since $\mu^{(1)}$ is greater than $\mu^{(2)}$, in thick crystals the contribution of Bloch wave 1 to the distribution of intensity in the diffraction pattern is reduced in comparison with the contribution of Bloch wave 2. But we have just seen above that the probability of diffuse scattering into Bloch wave 1 is greater than the probability of scattering into Bloch wave 2. This gives rise to a complex picture of competition between scattering into different Bloch states and absorption of electrons occupying those states. As the thickness of the crystal increases, absorption effects become dominant and we observe how the contrast of Kikuchi bands changes from being initially bright in the middle and dark at the edges of a band to dark in the middle (due to the attenuation of Bloch state 1) and relatively bright at the edges (due to the relatively low rate of absorption of Bloch state 2) of a band. The presence of multiple diffuse scattering complicates this picture further, resulting in the distribution of diffuse intensity where Kikuchi bands have 'inverted' contrast in the centre of the diffraction pattern and normal contrast in the outer regions of the pattern (Høier, 1973, Dudarev and Ryazanov, 1988).

Concluding this section, we note a similarity between Kikuchi lines and bands observed in electron diffraction, and Kossel lines observed in experiments on diffuse scattering of neutrons. The treatment of Kossel lines requires taking into account both the potential and resonance scattering of neutrons by the atomic nuclei (Dudarev, 1990).

8.7 Kikuchi patterns in electron backscattering

The occurrence of Kikuchi lines and bands in electron diffraction patterns is a generic phenomenon associated with dynamical diffraction of diffusely scattered electrons. Kikuchi patterns are observed not only in the transmission geometry of diffraction but also in electron backscattering. In this section we show how to find the distribution of intensity in Kikuchi patterns observed in the backscattering geometry (Dudarev and Peng, 1991). For a more extensive review of the subject the interested reader may benefit from consulting comprehensive review paper by Baba-Kishi (Baba-Kishi, 2002).

The cross-section of diffuse scattering (8.24) can be written in a compact form as

$$\left(\frac{d\sigma}{do}\right)_{\text{diffuse}} = \sum_n \langle |\langle \psi_{-\mathbf{k}}|f_n - \langle f_n\rangle_T |\psi_{\mathbf{k}_0}\rangle|^2 \rangle_T, \tag{8.35}$$

where operation $\langle...\rangle_T$ denotes averaging over thermal atomic displacements $\mathbf{u}_n$ and $\mathbf{k}_0$ and $\mathbf{k}$ are the wave vectors of the incident and diffusely scattered electrons. f_n is the amplitude of scattering by the n-th atom in the crystal evaluated in the Born approximation

$$f_n(\mathbf{r}) = -\frac{m}{2\pi\hbar^2}\varphi_n(\mathbf{r} - \mathbf{R}_n - \mathbf{u}_n).$$

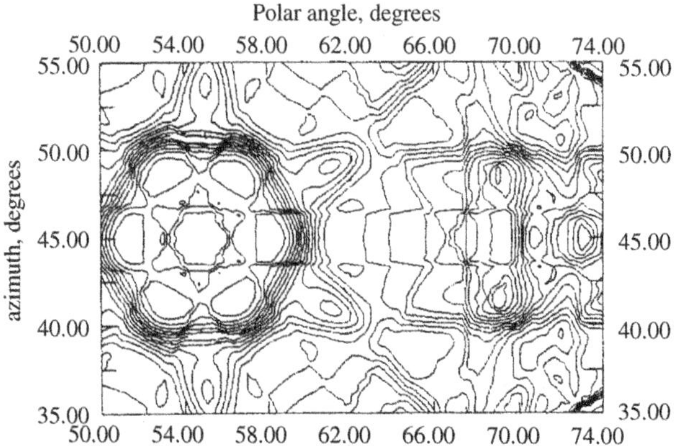

FIG. 8.10. Contour map of the intensity of electrons backscattered from a single crystal of aluminium calculated numerically for a range of azimuthal and polar angles in the vicinity of the $\langle 111 \rangle$ direction. The interval between contour lines is 0.05 and the intensity reaches its maximum value of 1.61 (relative to background) in the vicinity of the point $\phi = 45°$ and $\theta = 54.73°$. From (Dudarev and Peng, 1991).

In principle the wave functions of both the incident and diffusely scattered electrons are affected by dynamical diffraction. However, as before here we assume that the direction of incidence is chosen in such a way that diffraction of the *incident* electrons can be neglected. In the case where the crystal occupies the region $z > 0$, we take the wave function of the incident electrons in the form

$$\psi_{\mathbf{k}_0}(\mathbf{r}) = \exp\left(i\mathbf{k}_0 \cdot \mathbf{r} - \frac{z}{2\Lambda \cos \alpha_0}\right), \tag{8.36}$$

where α_0 denotes the angle of incidence and Λ is the effective length of attenuation of the primary beam (Seah, 1986).

The wave function of backscattered electrons can be found in the following way. The DWBA (7.14) states that this wave function represents the wave field corresponding to the inverted scattered plane wave $\exp(-i\mathbf{k} \cdot \mathbf{r})$. Since we are interested in the angular distribution of back-scattered electrons, we must now take into account dynamical diffraction effects. The wave function of backscattered electrons has the form of a linear combination of diffracted beams with slowly varying amplitudes

$$\psi_{-\mathbf{k}}(\mathbf{r}) = \exp(-i\mathbf{k} \cdot \mathbf{r}) \sum_h \phi_h(z, -\mathbf{k}) \exp(i\mathbf{g}_h \cdot \mathbf{r}), \tag{8.37}$$

where amplitudes $\phi_h(z, -\mathbf{k})$ satisfy the following system of coupled differential equations:

$$v_h \frac{d}{dz}\phi_h + \frac{i}{\hbar v}(\epsilon_h - \epsilon_0)\phi_h + \frac{1}{2\Lambda}\phi_h = -\frac{i}{\hbar v} \sum_s V_{hs}\phi_s, \tag{8.38}$$

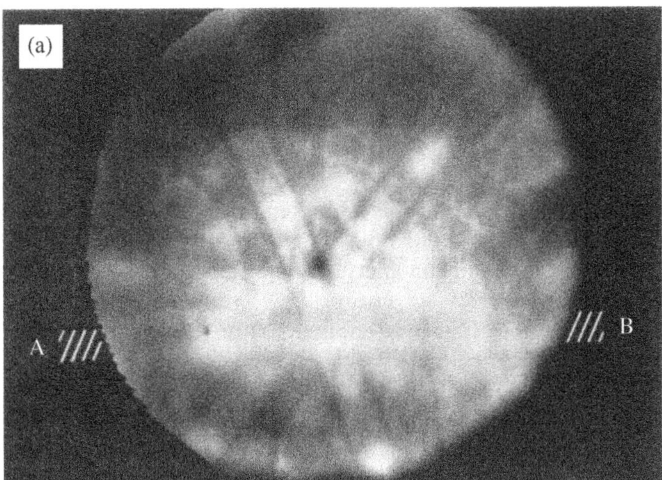

FIG. 8.11. Medium-energy electron diffraction patterns observed for the Si(111)7 × 7 surface using a spherical mirror analyser. The energy of incident electrons is 3 keV. From (Daimon and Ino, 1989).

where $\nu_h = (-\mathbf{k} + \mathbf{g}_h)_z/k > 0$, v is the velocity of electrons, and $\epsilon_h = \hbar^2(-\mathbf{k} + \mathbf{g}_h)^2/2m$. V_{hs} are the Fourier components of the average crystal potential $V_0(\mathbf{r})$, namely

$$V_{hs} = \frac{1}{\Omega_0} \int_{\Omega_0} d\mathbf{r} V_0(\mathbf{r}) \exp[i(\mathbf{g}_s - \mathbf{g}_h) \cdot \mathbf{r}], \tag{8.39}$$

and Ω_0 is the volume of a unit cell.

The boundary conditions for the amplitudes $\phi_n(z, -\mathbf{k})$ at the surface of the crystal $z = 0$ are $\phi_h(0, -\mathbf{k}) = 1$ for $h = 0$ and $\phi_h(0, -\mathbf{k}) = 0$ for $h \neq 0$. By substituting (8.36) and (8.37) into (8.35), we find that the angular distribution of backscattered electrons is given by

$$\left(\frac{d\sigma}{do}\right)_{\text{diffuse}} = \frac{\Lambda}{\Omega_0} \sum_{h,l} I_{hl}\sigma_{lh}(\mathbf{k}, \mathbf{k}_0), \tag{8.40}$$

where

$$I_{hl} = \int_0^\infty \frac{dz}{\Lambda} \exp\left(-\frac{z}{\Lambda \cos\alpha_0}\right) \phi_h(z, -\mathbf{k})\phi_l^*(z, -\mathbf{k}), \tag{8.41}$$

and

$$\sigma_{lh}(\mathbf{k}, \mathbf{k}_0) = \sum_n f_n(\mathbf{k} - \mathbf{k}_0 - \mathbf{g}_h)f_n^*(\mathbf{k} - \mathbf{k}_0 - \mathbf{g}_l) \left[\exp\left(-\langle[(\mathbf{g}_h - \mathbf{g}_l) \cdot \mathbf{u}_n]^2/2\rangle_T\right)\right.$$
$$\left. - \exp\left(-\langle[(\mathbf{k} - \mathbf{k}_0 - \mathbf{g}_l) \cdot \mathbf{u}_n]^2/2\rangle_T\right) \exp\left(-\langle[(\mathbf{k} - \mathbf{k}_0 - \mathbf{g}_h) \cdot \mathbf{u}_n]^2/2\rangle_T\right)\right], \tag{8.42}$$

where summation over n is performed over atoms in a unit cell. By integrating the differential equations (8.38), we find that the matrix elements I_{hl} satisfy a system of *algebraic* equations of the form

$$-\frac{\delta_{h0}\delta_{l0}}{\Lambda} + \frac{1}{\Lambda\cos\alpha_0}I_{hl} + \frac{1}{2\Lambda}\left(\frac{1}{\nu_h} + \frac{1}{\nu_l}\right)I_{hl} = -\frac{i}{\hbar v}\sum_s\left(\frac{H_{hs}}{\nu_h}I_{sl} - I_{hs}\frac{H_{sl}}{\nu_l}\right),$$

$$(8.43)$$

where $H_{hl} = (\epsilon_h - \epsilon_0)\delta_{hl} + V_{hl}$.

Equations (8.40) and (8.43) solve the problem of calculation of the angular distribution of electrons backscattered from a crystalline specimen, provided that single diffuse scattering gives the dominant contribution to the cross-section of scattering.

In the case where the direction of scattering $\mathbf{k}$ lies far from any crystallographic direction of high symmetry, the matrix elements I_{hl} are given by

$$I_{hl} = \delta_{h0}\delta_{l0}\frac{\cos\alpha\cos\alpha_0}{\cos\alpha + \cos\alpha_0},$$

and the cross-section of backscattering has the form

$$\left(\frac{d\sigma}{do}\right)_{\text{diffuse}} = \frac{\Lambda}{\Omega_0}\frac{\cos\alpha\cos\alpha_0}{\cos\alpha + \cos\alpha_0}\sigma_{00}(\mathbf{k},\mathbf{k}_0). \qquad (8.44)$$

This expression neglects dynamical diffraction effects and describes a featureless distribution of backscattered intensity. To investigate the part played by dynamical diffraction of diffusely scattered electrons we need to solve the system of algebraic equations (8.43) and to evaluate matrix elements I_{hl}. Figure 8.10 shows a contour map of the intensity backscattered from a single crystal of aluminium in the vicinity of the $\langle 111 \rangle$ direction.

In Fig. 8.10 we see a Kikuchi band situated in the range of azimuthal angles $40° < \phi < 50°$, and also a part of a ring Kikuchi pattern associated with bulk resonance scattering of electrons. An experimentally observed backscattering diffraction pattern (Daimon and Ino, 1989) is shown in Fig. 8.11. Kikuchi bands and rings are seen in both simulated and experimentally observed patterns. This illustrates the significant part played by many-beam diffraction effects in electron backscattering.

By observing the variation of contrast and orientation of electron backscattering Kikuchi diffraction patterns as a function of the position of the incident electron beam on the surface of a crystalline sample it is possible to determine the orientation of individual grains, and make grain boundaries visible in a scanning electron microscope (Dingley and Randle, 1992, Dingley and Field, 1997, Baba-Kishi, 1998).

8.8   Multiple diffuse scattering: an exact solution of the backscattering problem

The treatment of diffuse scattering of electrons that we have developed so far is based either on the small-angle scattering approximation (cf. the case of scattering by plasmons), or on the single diffuse scattering approximation used in the treatment of Kikuchi patterns. In this section we investigate cases where the kinetic equation for the density matrix can be solved exactly by taking into account the full sequence of events of diffuse scattering occurring in the crystal from the moment when the incident electron enters the specimen until the moment when it emerges back into the vacuum.

The solution of the problem given below is based on the method developed in (Dudarev, 1988, Dudarev, 1997). Consider a lattice of periodically distributed potential centres

$$V(\mathbf{r}) = \sum_n \varphi_n(\mathbf{r} - \mathbf{R}_n), \qquad n = 1, 2, ..., N. \tag{8.45}$$

We assume that the effective size r_0 of atom n is many times smaller than the wavelength $\lambda_0 = 2\pi/k_0$, i.e. $r_0 \ll \lambda_0$ (note that this is only a model assumption simplifying the mathematical treatment of the problem), and that each of the atomic potentials φ_n satisfies the condition of validity of the Born approximation

$$|\varphi_n| \ll \frac{\hbar^2}{mr_0^2}.$$

As before we write $V(\mathbf{r})$ as a sum of the average periodic potential

$$V_0(\mathbf{r}) = \sum_n \langle \varphi_n(\mathbf{r} - \mathbf{R}_n) \rangle \tag{8.46}$$

and its fluctuating part

$$\delta V(\mathbf{r}) = V(\mathbf{r}) - V_0(\mathbf{r}) = \sum_n \delta\varphi_n(\mathbf{r} - \mathbf{R}_n). \tag{8.47}$$

The average Green's function $G(\mathbf{r}, \mathbf{r}')$ of the electron satisfies the equation

$$\left[\frac{\hbar^2 k^2}{2m} + \frac{\hbar^2}{2m} \frac{\partial^2}{\partial \mathbf{r}^2} - V_0(\mathbf{r}) \right] G(\mathbf{r}, \mathbf{r}')$$

$$- \sum_n \int d\mathbf{r}'' \langle \delta\varphi_n(\mathbf{r} - \mathbf{R}_n) \delta\varphi_n(\mathbf{r}'' - \mathbf{R}_n) \rangle G_0(\mathbf{r}, \mathbf{r}'') G(\mathbf{r}'', \mathbf{r}') = \delta(\mathbf{r} - \mathbf{r}'). \tag{8.48}$$

Since $r_0 \ll \lambda_0$, we expand $G_0(\mathbf{r}, \mathbf{r}'')$ in a power series in $kr_0 \ll 1$

$$G_0(\mathbf{r}, \mathbf{r}'') = -\frac{m}{2\pi\hbar^2} \frac{\exp(ik|\mathbf{r} - \mathbf{r}''|)}{|\mathbf{r} - \mathbf{r}''|} \approx -\frac{m}{2\pi\hbar^2} \frac{1}{|\mathbf{r} - \mathbf{r}''|} - \frac{imk}{2\pi\hbar^2}, \tag{8.49}$$

and approximate $G(\mathbf{r}'', \mathbf{r}')$ in the fourth term on the left-hand side of eqn (8.48) by $G(\mathbf{r}, \mathbf{r}')$. This approximation is equivalent to introducing a *local* optical potential

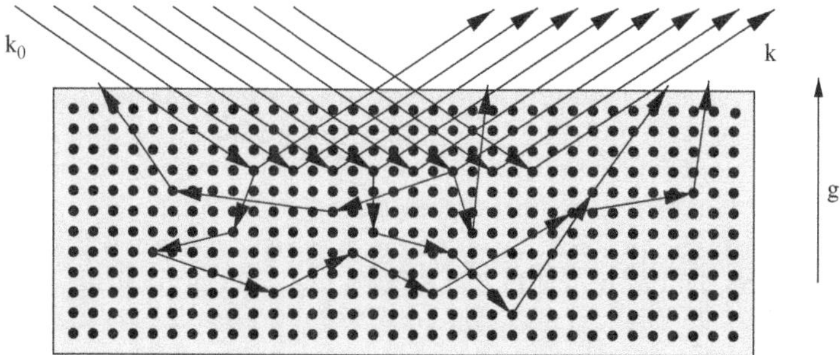

FIG. 8.12. Sketch illustrating the geometry of diffraction and the two types of processes contributing to the observed total cross-section of scattering. Electrons incident on the crystal may either be diffracted back into the vacuum or be scattered back into the vacuum after several successive events of diffuse scattering.

$$V_{eff}(\mathbf{r}) = \sum_n \left[\langle \varphi_n(\mathbf{r} - \mathbf{R}_n) \rangle - \frac{m}{2\pi\hbar^2} \int d\mathbf{r}'' \frac{\langle \delta\varphi_n(\mathbf{r} - \mathbf{R}_n)\delta\varphi_n(\mathbf{r}'' - \mathbf{R}_n) \rangle}{|\mathbf{r} - \mathbf{r}''|} \right]$$
$$- \frac{imk}{2\pi\hbar^2} \sum_n \left[\int d\mathbf{r}'' \langle \delta\varphi_n(\mathbf{r} - \mathbf{R}_n)\delta\varphi_n(\mathbf{r}'' - \mathbf{R}_n) \rangle \right]. \tag{8.50}$$

Equation (8.50) shows that fluctuations not only give rise to the imaginary part of the effective potential, but also renormalize the real part of this potential (the second term in square brackets in the first line of (8.50)). Assuming that scattering centres form a simple cubic lattice, we find the Fourier components of the optical potential

$$V_{eff}(\mathbf{g}) = V'_{eff}(\mathbf{g}) + iV''_{eff}(\mathbf{g}) = \frac{1}{\Omega_0} \int_{\Omega_0} d\mathbf{r} \exp(-i\mathbf{g} \cdot \mathbf{r}) V_{eff}(\mathbf{r}). \tag{8.51}$$

Since $|\mathbf{g}|r_0 \ll 1$ these quantities are practically independent of $\mathbf{g}$, and in what follows they will be referred to simply as V_{eff}.

Depending on the orientation of the reciprocal lattice vector $\mathbf{g}$ involved in the diffraction process it is common to distinguish between the two geometries of diffraction: (1) the Laue geometry corresponding to the case where $\mathbf{g}$ is parallel to the surface of the crystal and (2) the Bragg geometry where $\mathbf{g}$ is normal to the surface. In what follows we consider the Bragg diffraction case shown in Fig. 8.12. In the two-beam approximation the wave function $\psi_{\mathbf{k}_0}(\mathbf{r})$ is a superposition of two plane waves with slowly varying amplitudes

$$\psi_{\mathbf{k}_0}(\mathbf{r}) = \alpha(z) \exp(i\mathbf{k}_0 \cdot \mathbf{r}) + \beta(z) \exp[i(\mathbf{k}_0 + \mathbf{g}) \cdot \mathbf{r}], \tag{8.52}$$

where (see (Dudarev, 1997) for details of the derivation)

$$\alpha(z) = \exp\left[\frac{i}{2}\frac{m\epsilon_0}{\hbar^2 k_{0z}}z\left(1 - \sqrt{1 + \frac{4V_{eff}}{\epsilon_0}}\right)\right], \tag{8.53}$$

$$\beta(z) = \left[\frac{\epsilon_0}{2V_{eff}}\left(\sqrt{1 + \frac{4V_{eff}}{\epsilon_0}} - 1\right) - 1\right]\exp\left[\frac{i}{2}\frac{m\epsilon_0}{\hbar^2 k_{0z}}z\left(1 - \sqrt{1 + \frac{4V_{eff}}{\epsilon_0}}\right)\right],$$

and $\epsilon_0 = (\hbar^2/2m)[(\mathbf{k}_0 + \mathbf{g})^2 - \mathbf{k}_0^2]$.

The kinetic equation for the density matrix describing multiple incoherent scattering by a periodic system of fluctuating scattering centres has the form

$$\rho(\mathbf{r}, \mathbf{r}') = \psi_{\mathbf{k}_0}(\mathbf{r})\psi_{\mathbf{k}_0}^*(\mathbf{r}') + \frac{\sigma_{\rm el}}{4\pi}\left(\frac{2\pi\hbar^2}{m}\right)^2\sum_n G(\mathbf{r}, \mathbf{R}_n)G^*(\mathbf{r}', \mathbf{R}_n)\rho(\mathbf{R}_n, \mathbf{R}_n),$$

$$\tag{8.54}$$

where the cross-section of elastic diffuse scattering by a single centre is given by

$$\sigma_{\rm el} = 4\pi\left(\frac{m}{2\pi\hbar^2}\right)^2\int d\mathbf{r}d\mathbf{r}''\langle\delta\varphi_n(\mathbf{r} - \mathbf{R}_n)\delta\varphi_n(\mathbf{r}'' - \mathbf{R}_n)\rangle.$$

Comparing this with the definition of the optical potential (8.50), we see that the imaginary part of its Fourier component $V_{eff}''(\mathbf{g})$ equals $-\hbar v_0\sigma_{\rm el}/2$, where $v_0 = \hbar k_0/m$ is the velocity of the electron. Since $V_{eff}''(\mathbf{g})$ is a quantity proportional to the imaginary part of the amplitude of scattering by a single atom, it must also include the channel of scattering associated with absorption of particles (Dudarev, 1988). Taking this into account, we generalize the above formula and find that $V_{eff}''(\mathbf{g}) = -\hbar v_0\sigma_{tot}/2$, where $\sigma_{tot} = \sigma_{\rm el} + \sigma_{abs}$.

Returning to eqn (8.54) we note that this equation can be written in the form of a closed equation for the density of electrons *at the lattice sites* $P(\mathbf{R}_n) = \rho(\mathbf{R}_n, \mathbf{R}_n)$, namely

$$P(\mathbf{R}_n) = |\psi_{\mathbf{k}_0}(\mathbf{R}_n)|^2 + \frac{\sigma_{\rm el}}{4\pi}\left(\frac{2\pi\hbar^2}{m}\right)^2\sum_{n'}|G(\mathbf{R}_n, \mathbf{R}_{n'})|^2 P(\mathbf{R}_{n'}). \tag{8.55}$$

Since $\exp(i\mathbf{g}\cdot\mathbf{R}_n) = 1$ for all lattice sites, the density of electrons considered as a function of $\mathbf{R}_n$ varies slowly from one lattice site to another. For example, the density of electrons associated with the incident beam equals

$$|\psi_{\mathbf{k}_0}(\mathbf{R}_n)|^2 = \left|\frac{\epsilon_0}{2V_{eff}}\left(\sqrt{1 + \frac{4V_{eff}}{\epsilon_0}} - 1\right)\right|^2\exp\left(\frac{m\epsilon_0 z_n}{\hbar^2 k_{0z}}\Im\sqrt{1 + \frac{4V_{eff}}{\epsilon_0}}\right),$$

$$\tag{8.56}$$

where the choice of the branch of the complex-valued function $\sqrt{z}$ is determined by the condition $\Im\sqrt{z} > 0$. The characteristic scale of variation of this function from one lattice site to another is many times the period of the crystal lattice. A similar statement also applies to the average Green's function $G(\mathbf{R}_n, \mathbf{R}_{n'})$. This fact makes it possible to replace summation over lattice sites n in eqn (8.55) by integration

$$\sum_{n'} \rightarrow \frac{1}{\Omega_0} \int d\mathbf{R}_{n'}. \tag{8.57}$$

This shows that the problem of scattering by a lattice of fluctuating centres is mathematically equivalent to the problem of scattering by a random distribution of atoms (Gorodnichev et al., 1987, Gorodnichev et al., 1990). Following the same line of argument, we arrive at an explicit expression for the differential cross-section of diffuse scattering per unit surface area:

$$\left(\frac{d\sigma}{do} \right) = \frac{\sigma_{el}}{4\pi\sigma_{tot}} \left| \frac{\epsilon_0}{2V_{eff}} \left[\sqrt{1 + \frac{4V_{eff}}{\epsilon_0}} - 1 \right] \right|^2 \left| \frac{\epsilon}{2V_{eff}} \left[\sqrt{1 + \frac{4V_{eff}}{\epsilon}} - 1 \right] \right|^2$$

$$\times \left\{ \left[\cos\theta_0 \Re \sqrt{1 + \frac{4V_{eff}}{\epsilon_0}} \right]^{-1} + \left[\cos\theta \Re \sqrt{1 + \frac{4V_{eff}}{\epsilon}} \right]^{-1} \right\}^{-1} \tag{8.58}$$

$$\times H\left(\cos\theta_0 \Re \sqrt{1 + \frac{4V_{eff}}{\epsilon_0}}, \frac{\sigma_{el}}{\sigma_{tot}} \right) H\left(\cos\theta \Re \sqrt{1 + \frac{4V_{eff}}{\epsilon}}, \frac{\sigma_{el}}{\sigma_{tot}} \right),$$

where $H(\mu, \omega)$ is the Chandrasekhar function, the angle of incidence θ_0 equals $\theta_0 = \cos^{-1}(k_{0z}/k)$ and the angle of scattering is given by $\theta = \cos^{-1}(-\mathbf{k}_z/k_0)$. Given the complexity of the problem of multiple diffuse scattering, the final expression for the cross-section (8.58) looks very simple indeed.

Equation (8.58) shows how the cross-section of diffuse scattering depends on the direction of incidence (characterized by the two parameters ϵ_0 and θ_0) and on the direction of scattering (characterized by ϵ and θ). To evaluate the *total* cross-section of diffuse scattering, we integrate (8.58) over all directions of the vector $\mathbf{k}$ satisfying condition $\mathbf{k}_z < 0$. The simplest way of evaluating the integral over $d\phi$ and $d(\cos\theta)$ consists in employing a useful property of the Chandrasekhar function (Dudarev, 1988) (here μ denotes $\cos\theta$)

$$\frac{\sigma_1}{2\sigma_2} \int_0^1 \frac{\mu \, d\mu}{\mu_0 + \mu} H\left(\mu, \frac{\sigma_1}{\sigma_2} \right) H\left(\mu_0, \frac{\sigma_1}{\sigma_2} \right) = 1 - \sqrt{1 - \frac{\sigma_1}{\sigma_2}} H\left(\mu_0, \frac{\sigma_1}{\sigma_2} \right). \tag{8.59}$$

Comparing (8.58) and (8.59) we find the part of the cross-section of backscattering associated with multiple scattering by fluctuations

$$\sigma_{\text{diff}} = |(\epsilon_0/2V_{eff})[(1 + 4V_{eff}/\epsilon_0)^{1/2} - 1]|^2 \cos\theta_0 \Re(1 + 4V_{eff}/\epsilon_0)^{1/2}$$

$$\times \left[1 - \sqrt{1 - \frac{\sigma_{el}}{\sigma_{tot}}} H\left(\cos\theta_0 \Re(1 + 4V_{eff}/\epsilon_0)^{1/2}, \frac{\sigma_{el}}{\sigma_{tot}} \right) \right]. \tag{8.60}$$

To find the total cross-section we add to (8.60) the cross-section of Bragg scattering (i.e. the intensity of the beam diffracted by the atomic planes parallel to

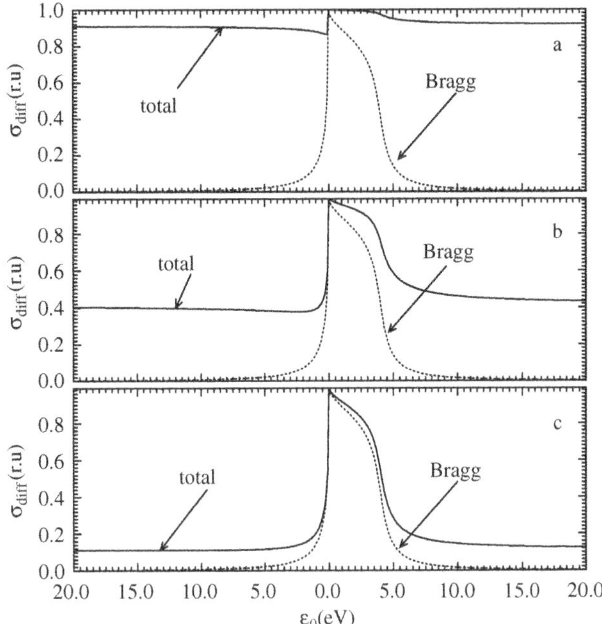

FIG. 8.13. Cross-sections of diffuse and Bragg scattering evaluated for three different values of the ratio σ_{el}/σ_{tot} as a function of parameter ϵ_0 characterizing the deviation of the wave vector of the incident beam from the exact Bragg condition. Results shown in this figure indicate that in the case where the real absorption is weak (i.e. in the case where an event of diffuse scattering gives rise only to the change of direction of motion of the electron), it is multiple scattering by fluctuations that gives the dominant contribution to the total cross-section of backscattering. Values of the ratio σ_{el}/σ_{tot} used in the calculations are: (a) 0.999, (b) 0.90, and (c) 0.50.

the surface of the crystal) $\sigma_{\text{Bragg}} = \cos\theta_0 |\beta_0|^2$. After some algebra we find an explicit expression for the total cross-section of backscattering

$$[\cos\theta_0]^{-1}\sigma = 1 - \Re(1 + 4V_{eff}/\epsilon_0)^{1/2}|(\epsilon_0/2V_{eff})[(1 + 4V_{eff}/\epsilon_0)^{1/2} - 1]|^2$$

$$\times \sqrt{1 - \frac{\sigma_{el}}{\sigma_{tot}}H\left(\cos\theta\Re(1 + 4V_{eff}/\epsilon_0)^{1/2}, \frac{\sigma_{el}}{\sigma_{tot}}\right)}. \tag{8.61}$$

This cross-section is normalized to the flux density $\cos\theta_0$ of the incident beam of electrons.

The first important conclusion that follows from the analysis of equation (8.61) is that our treatment of multiple diffuse scattering is consistent with the principle of conservation of current. Indeed, in the limiting case of zero absorption $\sigma_{el} = \sigma_{tot}$ the total normalized cross-section of backscattering (8.61) is *independent* of the direction of incidence and is equal to one. This result has a simple meaning: in the case where there is no *real* absorption, any incident particle will sooner or later emerge from the crystal back into the vacuum and will therefore contribute to the total cross-section of backscattering. Note that this comment only refers to the presence (or absence) of *real* absorption and is not relevant to the magnitude of the imaginary part of the optical potential. The latter quantity consists of two parts, one describing the real absorption of particles and the other part associated with scattering by fluctuations that do not result in any 'real' absorption of electrons. The 'real' absorption of electrons is in fact associated with energy losses (the treatment of which requires using time-dependent Green's functions and lies outside the scope of our present discussion). As electrons lose energy their velocity decreases, until they stop at some point inside the crystal giving no contribution to the cross-section of backscattering.

Figure 8.13 illustrates how absorption effects affect the total cross-section of backscattering in the vicinity of the Bragg condition. If the real absorption of electrons (resulting from electron energy losses) is absent, the total cross-section of backscattering is independent of the angle of incidence. As the cross-section of real absorption σ_{abs} increases (provided that the total cross-section of scattering by a single atom remains constant), the contribution of Bragg diffraction to the total cross-section of backscattering becomes dominant. In the limiting case where the 'real' absorption of electrons represents the only process contributing to the imaginary part of the potential, Bragg diffraction remains the only channel of scattering contributing to the total cross-section of electron backscattering (8.61).

In summary, in this section we have investigated how to describe diffuse scattering of electrons by various statistically disordered systems. We have found that provided that certain conditions are satisfied, a consistent approach to multiple diffuse scattering of electrons can be developed leading to the quantum kinetic equation for the density matrix. In a number of cases this equation can be solved analytically. Two such examples were considered above, and in both cases we found explicit expressions for the differential *and* total cross-sections of backscattering.

8.9 Electron channelling patterns and channelling imaging of crystal defects

Electron channelling (Coates, 1967, Booker et al., 1967) is another interesting phenomenon where diffraction and diffuse scattering produce a combined effect widely used in scanning electron microscopy (SEM) for imaging defects in crystalline materials. In practical terms electron channelling patterns observed in SEM represent two-dimensional maps (Joy et al., 1982) showing how the coefficient of electron backscattering varies as a function of azimuthal ϕ_0 and polar θ_0 angles of incidence. In fact almost any observed signal may be used in order to produce a two-dimensional electron channelling pattern (Joy et al., 1982). For example instead of looking at the variation of the total coefficient of backscattering we may observe the intensity of backscattering in a given direction, or the rate of Auger electron production. In each case the observed variation of the signal treated as a function of θ_0 and ϕ_0 reflects the local symmetry of the lattice in the vicinity of the point where the incident beam of electrons illuminates the surface of the specimen.

In this section we develop a method for computing the contrast of electron channelling patterns. We achieve this by solving the kinetic equation for the density matrix and by considering how the coefficient of electron backscattering varies as a function of the polar and azimuthal angles of incidence. In this solution we take into account both thermal diffuse scattering and energy losses of high-energy electrons in the crystal.

We adopt the point of view that since we intend to evaluate the distribution of backscattered electrons for a broad range of angles of incidence, we should not expect that an analytical treatment covering all the aspects of the problem can possibly be developed. We shall therefore attempt to design a numerical approach to solving the kinetic equation for the density matrix. We start by stating that in order to evaluate the cross-section of electron backscattering as a function of orientation of the incident beam it is necessary to solve an *inhomogeneous* transport equation describing multiple thermal diffuse scattering and energy losses of electrons. The source term in the transport equation can be found following the derivation given in Appendix A of (Dudarev et al., 1995b). The idea that it is the *inhomogeneous* transport equation that provides a suitable framework for treating backscattering of electrons from a crystalline specimen was proposed by Spencer and Humphreys (Spencer and Humphreys, 1980), who generalized the treatment of multiple scattering of electrons in a crystal developed by Hirsch and Humphreys (Hirsch and Humphreys, 1970). A good though somewhat simplified description of processes giving rise to the formation of electron channelling patterns is given in the paper by Reimer et al. (Reimer et al., 1971). The inhomogeneous transport equation that we intend to solve in order to find the angular distribution and the energy spectrum of backscattered electrons is

$$
\mathbf{n}\frac{\partial}{\partial \mathbf{r}}N(\mathbf{n}, E, \mathbf{r}) = \frac{1}{\Omega_0}\int do_{\mathbf{n}'}\int d\epsilon \frac{d^2\sigma}{do_{\mathbf{n}'}d\epsilon}(\mathbf{n}, E|\mathbf{n}', E+\epsilon)N(\mathbf{n}', E+\epsilon, \mathbf{r})
$$

$$
-\frac{1}{\Omega_0}\int do_{\mathbf{n}'}\int d\epsilon \frac{d^2\sigma}{do_{\mathbf{n}'}d\epsilon}(\mathbf{n}', E-\epsilon|\mathbf{n}, E)N(\mathbf{n}, E, \mathbf{r}) + Q(\mathbf{n}, E, \mathbf{r}). \qquad (8.62)
$$

Here $Q(\mathbf{n}, E, \mathbf{r})$ is the source term that describes electrons that undergo *single* scattering by thermal fluctuations of the crystal potential. Note that eqn (8.62) is incomplete. It *does not* describe electrons that have *not* undergone scattering by phonons and that propagate through the crystal retaining their coherence with the incident beam. To find the density matrix of these *coherently* scattered electrons we need to solve the usual set of equations describing many-beam dynamical diffraction by the periodic potential of the crystal. The source function entering the right-hand side of the transport equation (8.62) has the form

$$Q(\mathbf{n}, E, \mathbf{r}) = v \left(\frac{m}{2\pi\hbar^2}\right)^2 \delta(E - E_0) \frac{1}{\Omega_0} \sum_{j,h,l} \phi_h(\mathbf{r}) \phi_l^*(\mathbf{r}) \exp[i(\mathbf{G}_h - \mathbf{G}_l) \cdot \mathbf{r}_j]$$

$$\times \varphi_j(\mathbf{k}_0 + \mathbf{G}_h - \mathbf{k}) \varphi_j(\mathbf{k} - \mathbf{k}_0 - \mathbf{G}_l) \{\exp[-M_j(\mathbf{G}_l - \mathbf{G}_h)]$$

$$- \exp[-M_j(\mathbf{k}_0 + \mathbf{G}_h - \mathbf{k}) - M_j(\mathbf{k}_0 + \mathbf{G}_l - \mathbf{k})]\}, \tag{8.63}$$

where summation over j is performed over atoms in a unit cell, E_0 is the energy of the incident electrons, Ω_0 is the volume of a unit cell, v is the velocity of electrons and functions $\phi_h(\mathbf{x})$ entering (8.63) satisfy equations

$$i\frac{\hbar^2}{m}(\mathbf{k}_0 + \mathbf{G}_h) \cdot \frac{\partial}{\partial \mathbf{x}} \phi_h(\mathbf{x}) = (\epsilon_h - \epsilon_0)\phi_h(\mathbf{x}) + \sum_t \left(U_{ht} - \frac{i}{2}\gamma_{ht}\right) \phi_t(\mathbf{x}). \tag{8.64}$$

Together eqns (8.63) and (8.64) satisfy the conservation law

$$\int d\mathbf{r} \int dE \int do_{\mathbf{n}} Q(\mathbf{n}, E, \mathbf{r}) = -\int_S d\mathbf{S} \left[\frac{\hbar}{m} \sum_h (\mathbf{k}_0 + \mathbf{G}_h) |\phi_h(\mathbf{r})|^2\right], \tag{8.65}$$

which shows that the total flux generated by the source function $Q(\mathbf{n}, E, \mathbf{r})$ in the transport equation (8.62) equals the flux of electrons leaving the diffraction channels of scattering described by eqns (8.64).

In the case where (1) electron-electron interactions are treated as effective absorption and (2) diffraction of incident electrons is neglected, eqn (8.62) acquires a particularly simple form (this is of course an oversimplification that we use to illustrate how this equation can be solved numerically)

$$\cos\theta \frac{\partial}{\partial z} N(\cos\theta, \phi, z) = \frac{1}{\Omega_0} \int_{-\pi}^{\pi} d\phi' \int_0^{\pi} d\theta' \sin\theta' \, (d\sigma[\psi']/do')_{\mathrm{el}} \, N(\cos\theta', \phi', z)$$

$$- \frac{1}{\Omega_0}\sigma_{tot} N(\cos\theta, \phi, z) + \frac{1}{\Omega_0} (d\sigma[\psi_0]/do)_{\mathrm{el}} \exp(-n\sigma_{tot}z/\cos\theta_0), \tag{8.66}$$

where $\sigma_{tot} = \sigma_{\mathrm{el}} + \sigma_{\mathrm{inel}}$. In (8.66) ψ' is the angle between directions of propagation of the electron before and after an elastic collision,

$$\cos\psi' = \sin\theta \sin\theta' \cos(\phi - \phi') + \cos\theta \cos\theta', \tag{8.67}$$

and ψ_0 is given by (8.67) with θ' replaced by θ_0 and ϕ' replaced by ϕ_0, where θ_0 and ϕ_0 are the polar and azimuthal angles of incidence. The boundary condition for (8.67) at $z = 0$ is

$$N(\cos\theta, \phi, z = 0) = 0 \quad \text{for} \quad \cos\theta > 0, \tag{8.68}$$

and we are interested in finding the coefficient of backscattering

$$R_{tot} = \int_{-\pi}^{\pi} d\phi \int_{\pi/2}^{\pi} d\theta \sin\theta |\cos\theta| N(\cos\theta, \phi, z = 0). \tag{8.69}$$

To solve eqn (8.66) numerically we introduce a dimensionless variable $\tau = \sigma_{tot}z/\Omega_0$ and approximate the integral term by an L-point (where L equals 48 or 96) Gaussian quadrature. We find

$$\mu_j \frac{\partial}{\partial \tau} N(\mu_j, \tau) = \frac{1}{w_{tot}} \sum_{j'=1}^{L} \Phi(\mu_j, \mu_{j'}) N(\mu_{j'}, \tau) W_{j'} - N(\mu_j, \tau)$$

$$+ \frac{1}{w_{tot}} \Phi(\mu_j, \mu_0) \exp(-\tau/\mu_0), \qquad (8.70)$$

where $w_{tot} = \sigma_{tot}/\Omega_0$, $\mu = \cos\theta$, $\mu_0 = \cos\theta_0$, and

$$\Phi(\mu, \mu') = \frac{1}{\Omega_0} \int_{-\pi}^{\pi} d\phi \, d\sigma_{el} \left[\cos^{-1}\{(1 - \mu^2)^{1/2}(1 - \mu'^2)^{1/2} \cos\phi + \mu\mu'\} \right] / do'.$$

$$(8.71)$$

The system of linear differential equations (8.70) approximates the transport equation (the transport equation is recovered in the limit $L \to \infty$), and it can be solved by finding the eigenvalues and eigenvectors of matrix $\Pi_{jj'} = (w_{tot}\mu_j)^{-1}[\Phi(\mu_j, \mu_{j'})W_{j'} - \delta_{jj'}]$. Since the flux of electrons considered as a function of coordinate z in the direction normal to the surface vanishes in the limit $z \to \infty$, the solution of (8.70) contains no terms diverging in the limit $z \to \infty$. This condition makes it possible to find an explicit expression for the coefficient of backscattering

$$R_{tot} = \sum_{j=1}^{L} \sum_{m=1}^{L} |\mu_j| W_j X_{jm} C_m(0), \qquad (8.72)$$

where W_j denote Gaussian weights, X_{jm} are the eigenvectors of $\Pi_{jj'}$, and $C_m(0)$ are the coefficients involved in the expansion

$$N(\mu_j, \tau) = \sum_{m=1}^{L} X_{jm} C_m(\tau) \exp(p_m \tau). \qquad (8.73)$$

The numerical implementation of the procedure described above is straightforward (Dudarev and Whelan, 1994c). A comparison of total coefficients of quasielastic backscattering calculated using eqns (8.66)–(8.73) with experimental data shows that the theoretical and experimental values agree within 10%. For example for polycrystalline copper at $E_0 = 1$ keV we find $R = 3.3\%$, whereas the experimentally observed coefficient of quasielastic backscattering equals 3.4% (Schmid et al., 1983).

Regarding the proper treatment of electron energy losses, surprisingly it is a fact that *energy losses* are entirely responsible for the observed variation of the *total* coefficient of electron backscattering considered as a function of angles of incidence. It is a simple matter to prove the correctness of this statement. Let us consider a model where electrons are scattered in all directions by thermal fluctuations of the potential, but where there are no energy losses. In this case, given a sufficiently long interval of time, any electron incident on the surface will eventually approach it from the inside of the crystal and cross it, leaving the crystal for ever. In other words, provided that the beam of incident electrons remains stationary, at any given moment of time the number of particles entering the crystal will be equal to the number of particles leaving it. This conclusion evidently does not depend on the choice of the direction of incidence. This shows that in the absence of energy losses the total coefficient of backscattering remains constant $R_{tot} = 1$ and does not depend on the orientation of the incident beam.

A full-scale treatment of the problem of electron backscattering requires taking into account both incoherent quasielastic scattering of electrons by thermal fluctuations of the potential *and* electron energy losses. The transport equation now has the form

$$\cos\theta \frac{\partial}{\partial z} N(\cos\theta, \phi, E, z) = \int_{-\pi}^{\pi} d\phi' \int_{0}^{\pi} d\theta' \sin\theta' w_{ph}(\cos\psi', E) N(\cos\theta', \phi', E, z)$$

$$+ \frac{\partial}{\partial E} \left[\bar{\epsilon}(E) N(\cos\theta, \phi, E, z) \right] - w_{ph}^{(tot)}(E) N(\cos\theta, \phi, E, z) + Q(\cos\theta, \phi, E, z),$$

$$(8.74)$$

where the subscript '*ph*' refers to the cross-section of phonon scattering as opposed to the cross-section of energy losses represented by the second term on the right-hand side of eqn (8.74). Energy losses in this equation are described by the continuous slowing-down approximation. In this approximation the rate of energy losses is assumed to be independent of the direction of motion of the electron and the amount of energy lost by an electron is proportional to the total length of the trajectory of the electron inside the crystal. The boundary condition for eqn (8.74) has the form

$$N(\mu, E, z = 0) = 0 \quad \text{for} \quad 0 < \mu < 1, \tag{8.75}$$

and we are interested in finding the distribution of electrons $N(\mu, E, z = 0)$ in the region of negative values of μ where $-1 < \mu < 0$.

Equation (8.74) can be solved using the supermatrix algorithm developed by Fathers and Rez (Fathers and Rez, 1979, Fathers and Rez, 1984). This algorithm is based on the observation that in eqn (8.74) the term describing energy losses can be approximated by a finite difference

$$\frac{\partial}{\partial E} \left[\bar{\epsilon}(E) N(\cos\theta, \phi, E, z) \right] \approx \frac{\bar{\epsilon}(E_{i-1})}{\Delta E} N(\cos\theta, \phi, E_{i-1}, z)$$

$$- \frac{\bar{\epsilon}(E_i)}{\Delta E} N(\cos\theta, \phi, E_i, z), \tag{8.76}$$

and the entire energy spectrum is described by a smooth curve interpolated through a finite set of points $E_0 > E_1 > E_2....$ Energy losses are approximated

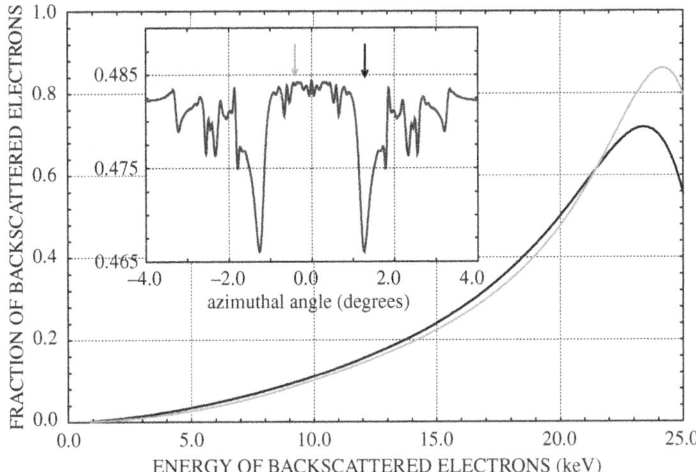

FIG. 8.14. Energy loss spectra of electrons backscattered from a silicon single crystal. The spectrum shown in grey corresponds to the direction of incidence where the total coefficient of backscattering is a maximum. The spectrum shown in black corresponds to the direction of incidence where the total coefficient of backscattering is a minimum. The total area under the curves gives the total coefficient of backscattering for a given direction of incidence. The energy E_0 of the incident electrons equals 25 keV and the angle of incidence θ_0 is equal to $69°$. The dependence of the total coefficient of backscattering on the azimuthal angle ϕ_0 is shown in the inset.

by a series of successive transitions of electrons from higher to lower energy groups. In practice solving eqn (8.74) involves additional algebraic manipulations: for example expanding $N(\cos\theta, \phi, E, z)$ as a Fourier series in ϕ. We find the solution by successively diagonalizing matrices describing phonon scattering of electrons *at a given energy* and by recursively evaluating angular distributions of backscattered electrons at points E_0, E_1, E_2... of the energy spectrum.

An interesting question that we are now able to investigate is how the shape of the energy spectrum of backscattered electrons depends on diffraction conditions for the incident beam, i.e. how the spectrum varies as a function of the mutual orientation of the incident beam and the crystal lattice. Figure 8.14 shows the energy spectrum of electrons backscattered from a silicon single crystal. The maximum of the spectrum is shifted down slightly from the energy E_0 of the incident electrons, and this is a remarkable feature of the solution obtained using the supermatrix algorithm (this maximum can indeed be readily observed experimentally, see e.g. an experimental investigation of the effect by Wolf et al. (Wolf et al., 1970)). Solutions found using less accurate approximations (Sandstrom et al., 1974) cannot explain the origin of this maximum.

By plotting the solution of the transport equation as a function of polar and azimuthal angles of incidence we can now simulate two-dimensional electron

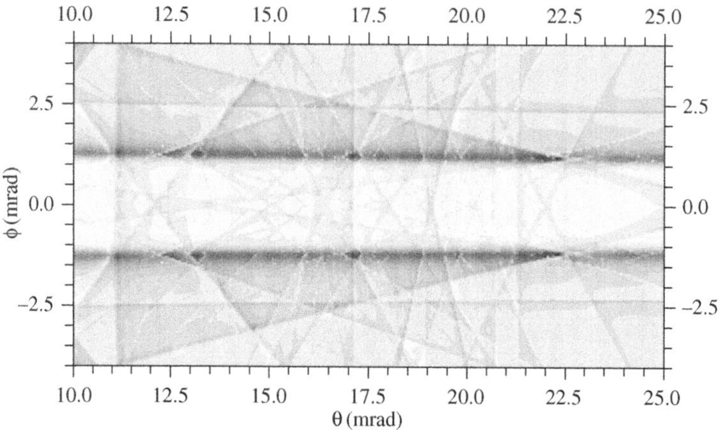

FIG. 8.15. Distribution of intensity in an electron channelling pattern simulated for crystalline silicon using the numerical approach described in the text. The channelling band running through the centre of the pattern corresponds to the 220 diffraction reflection. ϕ and θ are the azimuthal and polar angles of incidence.

channelling patterns similar to the one shown in Fig. 8.15.

Another experimentally observed effect, the origin of which we can now understand using eqn (8.74), is the phenomenon of the contrast reversal of channelling patterns. This effect was discovered by Ichinokawa et al. (Ichinokawa et al., 1974), who showed that the *sign* of the contrast of electron channelling patterns depends on the position of the detector of backscattered electrons. Solutions of the transport equations investigated by (Dudarev et al., 1995b) show that the origin of this phenomenon is associated with variations of the shape of the *angular distribution* of electrons backscattered from a crystal at a grazing incidence considered as a function of the direction of incidence. To explain the effect it is necessary to consider the interplay between anomalous transmission effects, energy losses, and *multiple* thermal diffuse scattering of electrons. By solving the transport equation (8.74) numerically it is possible to address and to explain all the features observed experimentally (Dudarev et al., 1995b).

The fact that the coefficient of electron backscattering is sensitive to the orientation of the incident beam makes it possible to obtain images of defects situated near the surface of the crystal (Wilkinson and Hirsch, 1997). The idea that channelling of electrons could be used for electron microscope imaging of defects was proposed by Booker et al. (Booker et al., 1967). The first simulations of images of lattice defects were performed by Clarke and Howie (Clarke and Howie, 1971) and by Spencer et al. (Spencer et al., 1972). Later Clarke (Clarke, 1971) attempted to observe crystal defects in SEM using backscattered electrons. To improve the contrast of images, Morin et al. (Morin et al., 1979) performed a series of experiments that showed that the quality of electron channelling im-

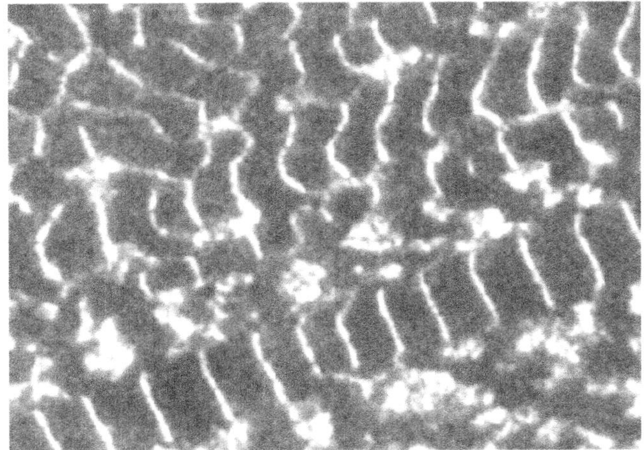

FIG. 8.16. Electron channelling contrast image of a fatigued copper single crystal at saturation. The typical ladder structure of persistent slip bands consisting of sets of dislocation walls is seen in this image. The image was obtained using a JEOL 6300 scanning electron microscope operated at 30 keV at a magnification of 3000 times and using the outer dark edge of the 220 channelling band. (Courtesy of J. Ahmed.)

ages can be improved significantly by using energy filtering and observing only quasielastically scattered electrons. More recently the Oxford group led by P. B. Hirsch (Czernuszka et al., 1990, Wilkinson et al., 1993, Ahmed et al., 1997) demonstrated the possibility of obtaining high-quality electron channelling contrast images (ECCI) of defects using commercial SEM. This has been achieved by using a more sensitive detector of backscattered electrons and inclined geometry of scattering. An example of an electron channelling contrast image of fatigue dislocation structures (persistent slip bands) in crystalline copper is shown in Fig. 8.16.

By examining the simulated channelling pattern shown in Fig. 8.15 it is easy to understand the mechanism of formation of channelling images of defects. Indeed, the electron channelling pattern plotted in Fig. 8.15 shows that in some regions of the pattern the *dispersion* of the observed signal, i.e. the local rate of variation of channelling contrast considered as a function of either θ or ϕ, is very high. For example this occurs for the directions of incidence that are close to the edges of channelling bands (i.e. near the deep dark lines running across the channelling pattern). If the direction of incidence corresponds to a region where the dispersion of the backscattering signal is substantial, a small local change in the orientation of crystallographic atomic planes associated with the presence of a defect will give rise to a substantial variation of the observed signal. By scanning a reasonably well focussed electron beam over a region of the surface while keeping the angles of incidence constant, it is possible to obtain an image showing the spatial distribution of deformation in the crystal.

Since the average level of contrast of channelling patterns is relatively low (the difference between intensities of bright and dark areas in Fig. 8.15 does not typically exceed 10%), channelling contrast images can be simulated using a perturbative approach proposed by Howie (Howie, 1974).

To simulate a channelling contrast image of a defect we need to evaluate how the coefficient of backscattering varies owing to local bending of atomic planes in the vicinity of the defect. The formal treatment of the problem involves transforming the transport equation into the integral form, introducing the suitable Green's function of this equation, and solving it iteratively. The first-order term of the series has the form

$$\delta R(\cos\theta, \phi, E) = R(\cos\theta, \phi, E) - R_{\mathrm{rand}}(\cos\theta, \phi, E) \tag{8.77}$$

$$\sim \int dS \int_0^\infty dz\, z \left[\sum_{h,l} \Phi_h(\mathbf{r}) \Phi_l^*(\mathbf{r}) \frac{\gamma_{lh}}{\hbar} - v_0 w_{\mathrm{el}}^{(tot)}(E_0) \exp\left(-\frac{w_{\mathrm{el}}^{(tot)}(E_0) z}{\cos\theta_0} \right) \right].$$

Simulations of electron channelling contrast images of dislocations carried out using this equation make it possible to interpret various features seen in images observed experimentally (Dudarev, 1999).

8.10 Diffraction effects in inner-shell ionization, X-ray, and Auger electron production

The fact that diffraction of electrons in a crystal gives rise to the formation of Bloch states, where the distribution of density of electrons is spatially heterogeneous and periodic, also gives rise to a strongly pronounced dependence of the cross-section of various inelastic interactions on the orientation of the incident electron beam. This was first pointed out by Hirsch, Howie and Whelan (Hirsch et al., 1962) who noted that the formation of two Bloch states, one localized between the atomic planes and the other at the lattice sites, should give rise to a dependence of the intensity of characteristic X-ray emission on the angle of incidence of the electron beam. Experimentally the effect was discovered by Duncumb (Duncumb, 1962) soon after the theoretical prediction (Hirsch et al., 1962) became known. A thorough experimental investigation of the effect was later performed by Hall (Hall, 1966) who confirmed the validity of the theoretical treatment based on the dynamical theory of electron diffraction. A remarkably clear treatment of characteristic X-ray production in thin crystals illuminated by high-energy electrons was given by Cherns et al. (Cherns et al., 1973). The fact that the shape of the curve describing how the intensity of characteristic X-ray emission varies as a function of the angle of incidence is sensitive to the location of impurity atoms in the crystal lattice was demonstrated by Spence and Taftø (Spence and Taftø, 1983, Taftø and Spence, 1982, Spence and Kim, 1988). The subsequent work by Allen and Rossouw and their colleagues (Allen and Rossouw, 1993b, Allen et al., 1994,

Josefsson et al., 1994, Oxley and Allen, 2000, Oxley and Allen, 2001) provided the much needed theoretical foundation for quantitative applications of the method.

Another case where cross-sections of inelastic processes exhibit strong dependence on the orientation of the incident beam is the case of Auger electron emission from crystalline surfaces investigated by Andersen and Howie (Andersen and Howie, 1975) and by Gomoyunova and her colleagues in St. Petersburg (Gomoyunova et al., 1982, Gomoyunova et al., 1990). A comprehensive review of this subject is given by Chambers (Chambers, 1992). Since characteristic X-ray emission and Auger electron emission result from ionization of inner electronic shells of atoms by the incident high-energy electrons, we treat both cases using a unified approach. In this way we will be able to investigate some subtle aspects of inelastic scattering occurring in a crystalline environment, e.g. the effects of delocalization in electron impact ionization (Allen and Rossouw, 1993a). Our approach is based on the observation that once the solution of the kinetic equation describing the (partially coherent) wave field of high-energy electrons is found, the cross-section of characteristic X-ray or Auger electron emission can be evaluated simply by finding the rate of inelastic transitions occurring in a given atom exposed to the *known* wave field of high-energy electrons.

A non-perturbative expression for the rate of inelastic transitions has the form (Goldberger and Watson, 1964)

$$w = \frac{2\pi}{\hbar} \sum_{j,j'} \langle f|\hat{T}|j\rangle \rho_{jj'} \langle j'|\hat{T}|f\rangle \delta(\epsilon_f - \epsilon_0), \tag{8.78}$$

where ϵ_0 is the energy of the initial state of the system (in our case it is the sum of energies of the incident high-energy electron and the crystal) and ϵ_f is the energy of the final state of the system (in the final state an electronic shell of one of the atoms is ionized as a result of interaction with the incident high-energy electron). In eqn (8.78) $\hat{T}$ is the scattering matrix operator that describes how the incident high-energy electron interacts with the atomic electron occupying an inner electronic shell.

Using the real-space representation of the density matrix, we write the rate of ionization of a given atom a in the form

$$w_a = \frac{2\pi}{\hbar} \int_0^{E_0} dE \int_0^{E_0} dE' \int d\mathbf{r} \int d\mathbf{r}' \int d\mathbf{r}_1 \sum_{\nu,\nu'} \sum_f \delta(\varepsilon_f + E' - \varepsilon_0 - E_0)$$

$$\times \frac{1}{\mathcal{Z}} \exp\left(-\frac{\mathcal{E}_\nu}{k_B T}\right) \langle \nu', f|\hat{T}_a(\mathbf{r}_1, \mathbf{r})|\nu, 0\rangle \rho(\mathbf{r}, \mathbf{r}', E)\langle \nu, 0|\hat{T}_a^\dagger(\mathbf{r}', \mathbf{r}_1)|f, \nu'\rangle. \tag{8.79}$$

In this equation $\mathcal{Z} = \sum_\nu \exp\left(-\mathcal{E}_\nu/k_B T\right)$ is the partition function for the vibrational motion of atoms, ν and ν' are sets of quantum numbers characterizing the initial and final states of the phonon subsystem of the crystal, and ε_f and

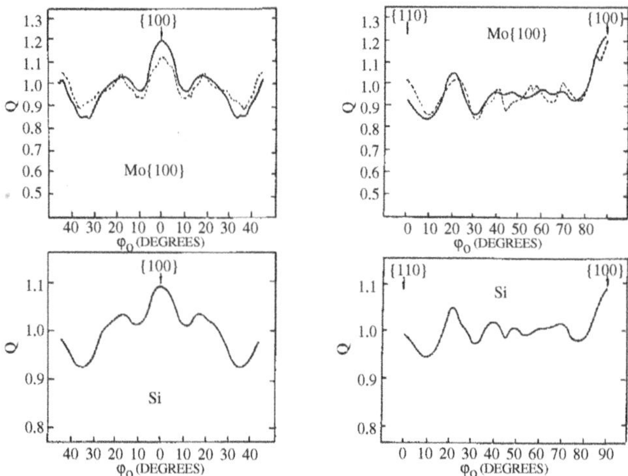

FIG. 8.17. Azimuthal dependence of the rate of $L_{2,3}VV$ Auger electron yield from the Mo(100) and Mo(110) surfaces (top) and from a submonolayer film of silicon atoms deposited on the two surfaces. Experimentally observed quantities are shown by solid lines. Dashed lines shows the predicted variation of the rate of Auger electron emission evaluated using eqn (8.85). From (Gomoyunova et al., 1990).

ε_0 are the energies of the final and initial states of ionization of atom a. Integration over $\mathbf{r}_1$ is performed over the coordinate of the high-energy electron after the inelastic collision. To find the rate of characteristic X-ray emission we need to add contributions (8.79) from all atoms in the crystal. In the case of Auger electron emission contributions of individual atoms need to be weighted with the exponential factor

$$I_{\text{Auger}} \sim \sum_a \exp(-z_a/\Lambda)w_a, \tag{8.80}$$

taking into account the attenuation of the Auger signal as a function of distance z_a from the surface of the crystal. Λ is the effective escape depth of Auger electrons.

Using eqn (8.79) we find how the rate of Auger emission depends on the orientation of the incident beam of electrons (Gomoyunova et al., 1990). A similar problem of low impact parameter ionization of atoms in a thin crystal has recently been investigated by Rossouw et al. (Rossouw et al., 1998). In the case of Auger electron emission from a crystal surface we need to find the rate of ionization of atoms situated at relatively small depths (of the order of a few nanometres) below the surface. The wave field of high-energy electrons near the surface consists of two components. The first component describes electrons directly incident on the surface of the crystal while the second component represents electrons that underwent multiple scattering in the crystal bulk and form the flux of primary electrons backscattered from the crystal, namely

$$\rho(\mathbf{r}, \mathbf{r}', E) = \delta(E - E_0) \sum_{h,l} \phi_h(z) \phi_l^*(z') \exp[i(\mathbf{k}_0 + \mathbf{g}_h) \cdot \mathbf{r} - i(\mathbf{k}_0 + \mathbf{g}_l) \cdot \mathbf{r}']$$

$$+ \int_{\mathbf{n}_z < 0} d^2 n\, C(\mathbf{n}, E) \exp[i k \mathbf{n} \cdot (\mathbf{r} - \mathbf{r}')] \tag{8.81}$$

The first term in this equation is proportional to the direct product of wave functions $\psi_{\mathbf{k}_0}^{(0)}(\mathbf{r})$ of electrons undergoing dynamical diffraction in the crystal

$$\psi_{\mathbf{k}_0}^{(0)}(\mathbf{r}) = \sum_h \phi_h(z) \exp[i(\mathbf{k}_0 + \mathbf{g}_h) \cdot \mathbf{r}], \tag{8.82}$$

where summation is performed over reciprocal lattice vectors $\mathbf{g}_h$. Taking into account the fact that the elements of scattering matrix $\hat{\mathcal{T}}$ depend on the coordinates of atom a via

$$\int d\mathbf{r} d\mathbf{r}' \exp(i\mathbf{k}' \cdot \mathbf{r}' - i\mathbf{k} \cdot \mathbf{r}) \mathcal{T}_a(\mathbf{r}, \mathbf{r}') = \mathcal{T}_a(\mathbf{k}, \mathbf{k}') \exp[i(\mathbf{k}' - \mathbf{k}) \cdot (\mathbf{R}_a + \mathbf{u}_a)], \tag{8.83}$$

where $\mathbf{u}_a$ is the thermal displacement of atom a from its equilibrium position $\mathbf{R}_a$, and that summation over initial and final states of the phonon subsystem of the crystal can be performed using the formula

$$\frac{1}{\mathcal{Z}} \sum_{\nu,\nu'} \exp\left(-\frac{\mathcal{E}_\nu}{k_B T}\right) \langle \nu' | e^{i\mathbf{k} \cdot \mathbf{u}_a} | \nu \rangle \langle \nu | e^{-i\mathbf{k}' \cdot \mathbf{u}_a} | \nu' \rangle = \exp\left[-\frac{1}{2} M_a(\mathbf{k} - \mathbf{k}')\right], \tag{8.84}$$

where $M_a(\mathbf{k} - \mathbf{k}')$ is the Debye–Waller factor for atom a, we find an explicit expression of the rate of Auger emission from the crystal surface for a given orientation of the incident beam

$$I_{\mathrm{Auger}} = \sum_a e^{-z_a/\Lambda} \sum_{h,l} \phi_h(z_a) \phi_l^*(z_a) e^{i(\mathbf{g}_h - \mathbf{g}_l) \cdot \mathbf{R}_a} e^{-\frac{1}{2} M_a(\mathbf{g}_h - \mathbf{g}_l)} \Gamma_{lh}^{(a)}(E_0)$$

$$+ \sum_a \exp(-z_a/\Lambda) \int_{\mathbf{n}_z < 0} d^2 n \int dE\, C(\mathbf{n}, E) \Gamma_{00}^{(a)}(E). \tag{8.85}$$

In this equation the quantities

$$\Gamma_{lh}^{(a)}(E) = \int \frac{d\mathbf{k}'}{(2\pi)^2} \sum_f \delta\left(\varepsilon_f - \varepsilon_0 + \frac{(\mathbf{k}')^2}{2m} - E\right)$$

$$\times \langle 0 | \mathcal{T}_a^\dagger(\mathbf{k}_0 + \mathbf{g}_l; \mathbf{k}') | f \rangle \langle f | \mathcal{T}_a(\mathbf{k}'; \mathbf{k}_0 + \mathbf{g}_h) | 0 \rangle \tag{8.86}$$

are the *generalized* ionization cross-sections.

Maslen and Rossouw (Maslen and Rossouw, 1983, Maslen and Rossouw, 1984, Rossouw and Maslen, 1984) investigated these cross-sections and explained why it is natural that the rates of inelastic processes occurring in a crystalline environment are expressed not only in terms of 'ordinary' cross-sections of scattering

Γ_{00} but also in terms of the somewhat unusual quantities of the form (8.86), where $h \neq l$.

Equation (8.85) can be readily applied to the analysis of experimental observations (Gomoyunova et al., 1990). In the problem of Auger electron emission from a crystalline specimen the dependence of the observed signal on the orientation of the incident beam comes from (1) dynamical diffraction of incident electrons and (2) the dependence of the flux of backscattered electrons on the direction of incidence. Note that point (2) is in fact identical to the problem of formation of channelling patterns discussed earlier in this chapter. It is possible to separate contributions (1) and (2). Experimentally this can be achieved by observing the rate of Auger emission from a very thin layer of foreign atoms adsorbed on the surface (Gomoyunova et al., 1990). Since these atoms are illuminated by the *spatially homogeneous* direct flux of incident electrons, the orientational dependence of the Auger signal from those atoms comes only from the variation of the coefficient of backscattering. Figure 8.17 illustrates this point in detail. In the experiment performed by (Gomoyunova et al., 1990) the authors observed the variation of the intensity of Auger emission from molybdenum atoms of the crystalline substrate and from a very thin layer of silicon atoms deposited on the surface. The fact that the rate of Auger emission from silicon atoms showed notable dependence on the orientation of the incident beam confirms the significance of the contribution to the cross-section of ionization associated with electrons backscattered from the crystalline substrate.

Probably the most interesting phenomenon associated with the excitation of inner electronic shells by high-energy electrons in a crystalline environment is related to the delocalization of impact ionization of atoms. The term 'delocalization' is used here to describe the fact that the probability of ionization of an atom at point $\mathbf{R}_a$ is *not* proportional to the density $\rho(\mathbf{R}_a, \mathbf{R}_a)$ of the incident high-energy electrons at this point (the 'local' approximation assuming that the probability of ionization of an atom is proportional to the density of electrons $|\psi(\mathbf{R}_a)|^2$ was used in early papers on X-ray production in thin crystals (Hall, 1966, Cherns et al., 1973)). On one hand, delocalization effects result from the thermal motion of atoms in the crystal, and this is reflected in the presence of the Debye–Waller factors in the first term on the right-hand side of eqn (8.85). On the other hand (and this is far more important), delocalization is associated with the long range of the Coulomb interaction between the incident high-energy electron and the electron occupying an inner electronic shell of an atom. For example an incident high-energy electron moving at velocity v can create a hole in an electronic shell characterized by the binding energy E_b at a distance $b \sim v/E_b$. It is a simple matter to see that b may be many times the lattice constant. For example, in the case of ionization of the $M_{4,5}$ electronic shell of a molybdenum atom ($E_b = 230$ eV) by an incident 200 keV electron we find that $b \sim 8$ Å. Formally the Coulomb delocalization of the ionization process is manifested in the fact that in eqn (8.85) $|\Gamma_{lh}| < |\Gamma_{00}|$.

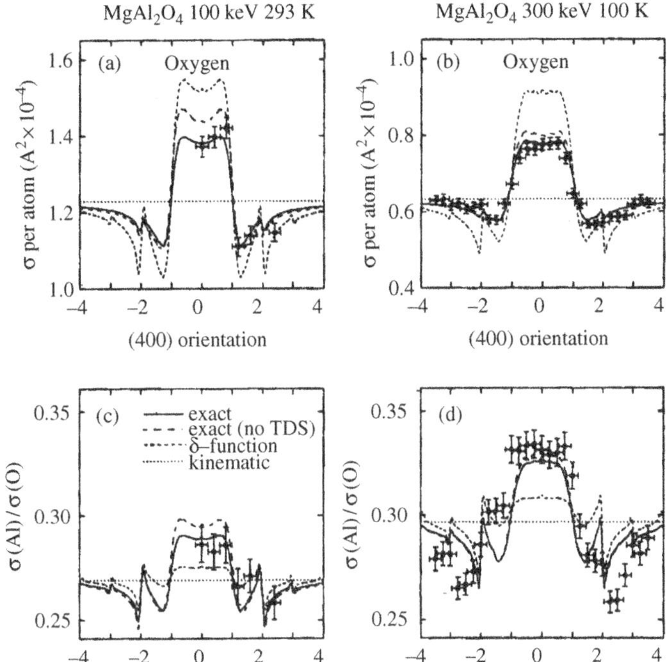

FIG. 8.18. Curves showing how the cross-sections of ionization of inner electronic shells of various atoms in MgAl$_2$O$_4$ vary as a function of the angle between the direction of incidence and the (400) crystallographic planes. An abscissa value of unity indicates that g_{400} is in the exact Bragg orientation. The curves labelled 'exact, no TDS' retain the effect of delocalization of impact ionization and those labelled 'δ-function' assume a point-like ionization and include the effect of thermal motion of atoms in the crystal. From (Allen et al., 1994).

Allen et al. (Allen et al., 1994) performed a comprehensive study of the effect of delocalization of ionization of atoms in a crystalline environment on the rate of characteristic X-ray emission. By investigating how the shape of profiles of the rate of X-ray emission depends on the orientation of the incident beam of electrons and the assumed degree of delocalization of the process of impact ionization Allen et al. (Allen et al., 1994) were able to prove that delocalization plays a very significant part and must be taken into account in order to interpret experimental observations quantitatively. The very good agreement achieved between the predicted and experimental curves shown in Fig. 8.18 shows that at the present level of accuracy of experimental techniques and theoretical understanding, spectroscopical observations performed using crystalline samples can provide valuable information about the atomic structure of natural and technological materials.

8.11　Summary

In this chapter we have introduced the density matrix approach to dynamical electron diffraction. The method provides a suitable foundation for the treatment of coherence of electrons and effects of multiple diffuse scattering. We have discussed several applications of the method including the treatment of multiple scattering by plasmons, the formation of Kikuchi patterns, the problem of electron backscattering, and inner-shell impact ionization. We have shown that all these seemingly difficult issues can in fact be treated using a fairly simple and straightforward approach.

9

CRYSTAL AND DIFFRACTION SYMMETRY

9.1 Introduction

Convergent-beam electron diffraction (CBED) provides the most convenient way of determining the symmetry of crystalline materials, and this method has acquired great popularity in the electron diffraction community. In an electron microscope CBED patterns are obtained using a method analogous to the case of critical illumination in optics, where the electron source is (approximately) conjugate to the specimen. The main advantages of the CBED method resulting from the use of a small electron probe (10 Å– 100 Å) are: (1) there is negligible thickness variation within the illuminated region; (2) any significant buckling of the specimen is unlikely to occur within the illuminated region during observation; (3) the variation of intensity as a function of the angle of incidence in the CBED discs contains important information about the structure of the crystal; (4) a large amount of diffraction data exceeding millions of data points can be collected in parallel in a few seconds. A CBED pattern is therefore better defined and more suited for comparison with theoretical calculations (Spence and Zuo, 1992, Steeds, 1983) than a conventional selected area electron diffraction (SAED) pattern.

The description of CBED patterns obtained in a conventional transmission electron microscope (CTEM) is normally based on the assumption that the sample is illuminated by an *extended fully incoherent* source (see Fig. 1.4). Each point on the effective source disc CA (the condenser lens aperture) defines an independent source of electrons, i.e. the direction $\mathbf{k}_0$ of a plane wave incident on the specimen. Each of these source points (such as Q) is conjugate to a set of points Q_i in each diffraction disc (Spence and Zuo, 1992). Imaging conditions in a field-emission gun (FEG) electron microscope are, however, better approximated by the assumption that the electron source is coherent. Fortunately in the case of diffraction by a perfect crystal, the intensity distribution within each individual disc of a CBED pattern obtained using an FEG electron microscope is identical to that obtained using a CTEM or an incoherent electron source. In the overlapping regions between the discs the intensities are, however, different for the two types of illumination, i.e. coherent and incoherent, and in principle these regions provide information about the phase relationship between adjacent diffracted beams (Cowley, 1990). In this chapter we will consider diffraction by a perfect crystal, and no distinction will be made between coherent and incoherent electron sources.

The amplitudes and intensities of diffracted beams in general depend in a

311

complex way on the crystal thickness and the orientation of the incident beam of electrons. For some crystals there are certain incident beam directions for which the variation of beam intensities with thickness is described by simple expressions, but these cases are rare. The only generally valid statements that can be made about the observable intensities are those related to the symmetries of the diffraction pattern. Symmetry information obtained from dynamical scattering of electrons is fundamentally different in nature from that obtained from kinematical scattering. The diffraction pattern symmetries observed in the transmission case reflect the symmetry of the wave function emerging from the bottom face of the crystal. The symmetry of this wave function reflects the whole history of propagation of the electron wave through the crystal. The observed symmetries therefore describe the illuminated part of the crystal, rather than the symmetry of an ideal crystal lattice. For example the symmetry of the diffraction pattern of a wedge-shaped crystal is usually lower than the symmetry of a rectangular slab. Similarly CBED patterns obtained from a rectangular slab crystal will show lower symmetry if the top face of the slab is not perpendicular to the incident electron beam. The type of surface termination may also affect the overall observed symmetry of the crystal. For example if the unit cell contains several distinct layers of atoms, then the type of surface termination at one or another layer of atoms will give rise to an integral or non-integral number of unit cells in the incident beam direction affecting the symmetry of the observed CBED pattern.

In this chapter we will consider how CBED patterns can be used for the determination of the symmetry of crystalline samples. The procedure for crystal symmetry group identification described below is largely based on the work by Gjønnes and Moodie (Gjønnes and Moodie, 1965), Goodman (Goodman, 1975), Buxton et al. (Buxton et al., 1976), and on a series of papers by Tanaka et al. (Tanaka et al., 1983a, Tanaka et al., 1983b). The automation of this procedure is based on the early proposal by Peng and Li (Peng and Li, 1992) and a more recent genetic algorithm implementation of the method (Hu et al., 2000b).

9.2 Representation of symmetry

The symmetry relationships of a crystal are most conveniently represented by using *crystal projections* where each face of the crystal is represented by a point, and where points are related to each other by certain rules depending on the angles involved in the construction of a projection.

To construct a projection we draw surface normals to all the faces of the crystal. A *spherical projection* corresponds to the case where a sphere is drawn around the crystal, and where the centre of the sphere is located at the point of intersection of the surface normals (see Fig. 9.1a). Each point of intersection of the normals with the sphere is marked by a dot (the pole of the corresponding face).

The spherical projection is three dimensional. In electron diffraction, however, it is more desirable to represent the projection on a two-dimensional sheet of

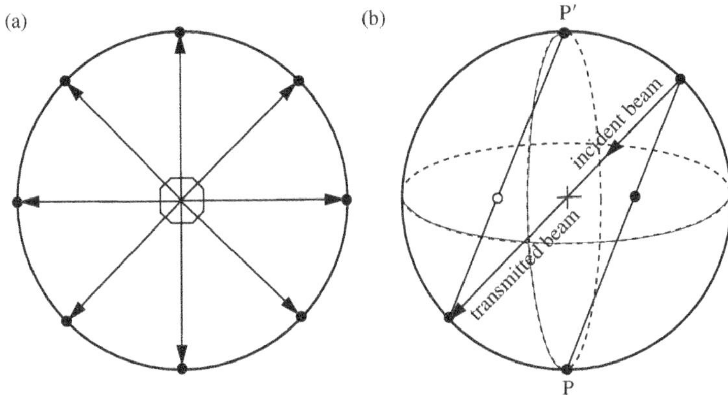

FIG. 9.1. Schematic diagram showing the construction of (a) spherical projection and (b) stereographic projection.

paper. The method most widely used in crystallography is the *stereographic projection*, the construction of which is illustrated in Fig. 9.1b. We imagine that a plane sheet of paper passes horizontally through the centre of the spherical projection. The intersection of the surface of the sphere with this plane forms the primitive circle. Each point on the upper half of the spherical projection (such as the point resulting from the incident beam shown in Fig. 9.1b) is then projected onto the plane of the paper by joining it to the south (bottom) pole P of the sphere, where the point of intersection is marked on the paper by a small dot. Points on the lower half of the spherical projection are projected upwards to the north (top) pole P' on the sphere diametrally opposite to P. To distinguish between the two projections, a point on the lower half of the sphere is marked on the paper by a small open circle instead of by a dot.

The symmetry relations in a diffraction pattern, in particular in a CBED pattern, may be conveniently represented (Buxton et al., 1976) by the stereographic projection, where the crystal zone axis of interest is marked by a cross at the centre of the projection (Fig. 9.2). An extended ideally incoherent source is usually assumed for CBED experiments performed in a CTEM. Each dot on the upper (northern) hemisphere within the incident electron beam cone defines the direction of a plane wave incident on the specimen, and each of these source points (such as Q) is conjugate to a set of points Q' shown in Fig. 9.2, one point in each diffraction disc.

9.3 The reciprocity principle

A good understanding of the reciprocity principle is important for understanding the procedure of symmetry determination described below. The principle of reciprocity was first used by von Laue in the context of electron diffraction to interpret the contrast of Kikuchi lines observed in electron diffraction patterns. The reciprocity theorem then found many other applications in electron

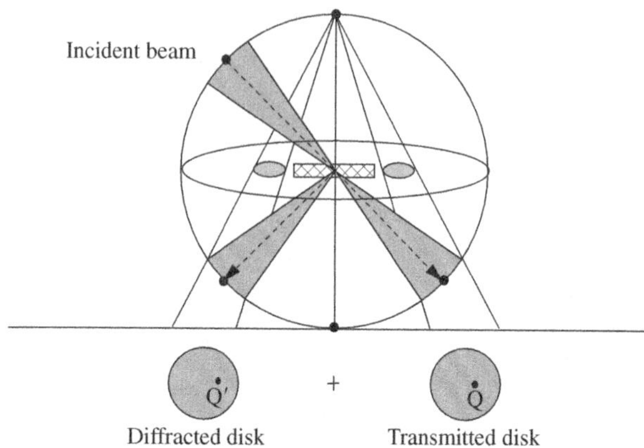

FIG. 9.2. Schematic diagram illustrating the stereographic projection widely used for investigating the symmetry of CBED patterns.

diffraction, including the interpretation of STEM image contrast (Cowley, 1969) and CBED pattern symmetries (Buxton et al., 1976). The validity of the theorem was proved in Chapter 3 by using the symmetry of the free-space Green's function, and here we simply restate this theorem: *The amplitude $\Psi(\mathbf{r}, \mathbf{r}_A)$ of radiation at point $\mathbf{r}$ due to a source at point $\mathbf{r}_A$ in the presence of arbitrary potential $V(\mathbf{r})$ is identical to the amplitude $\Psi(\mathbf{r}_A, \mathbf{r})$ at $\mathbf{r}_A$ produced by the same source placed at point $\mathbf{r}$, i.e. $\Psi(\mathbf{r}, \mathbf{r}_A) = \Psi(\mathbf{r}_A, \mathbf{r})$.*

For electron waves this principle can be generalized to include a vector scattering field such as the magnetic field of an electron lens or a deflection coil if the reversal of the direction of the electron beam is accompanied by the reversal of the direction of the magnetic field. The same relationship in terms of wave intensities rather than amplitudes can be applied to inelastic scattering, provided that the energy losses associated with inelastic scattering are sufficiently small (Pogany and Turner, 1968).

9.4 Symmetry elements and their identification

In CBED we are mainly interested in the problem of scattering by a thin crystalline film, where the surfaces may be considered as being parallel to each other and where we may assume that the crystal contains no defects. The symmetry of a thin crystalline sample is described in terms of (1) one or more planes of reflection symmetry, symbolized by m (*mirror*); (2) *rotation axes*, symbolized in terms of their degree 1,2,3,4, or 6; (3) axes of *rotary inversion* (or *inversion axes*), symbolized also in terms of their degree, $\bar{1}, \bar{2}, \bar{3}, \bar{4}$, or $\bar{6}$. The inversion axes are compound symmetry elements. The operation of rotation through the angle indicated by for example the degree 4 is carried out first, followed by inversion through a centre, and is indicated by the bar above the degree of rotation, e.g. $\bar{4}$. Based on these symmetry elements 32 crystal point groups may be constructed in a systematic way (Phillips, 1971). The principal axis (normal to the paper in a projection) is first set down. It may be either a rotation axis X or an inversion axis $\overline{X}$. If there is a reflection plane normal to this axis, the symbol m is added as X/m (rotation axis normal to the plane of symmetry), while a reflection plane through the axis is written without the stroke as $X\text{m}$ or $\overline{X}\text{m}$ (rotation or inversion axis with a vertical plane of symmetry). If both kinds of plane are present the symbol is X/mm. A horizontal diad axis (normal to the principal axis) is indicated by adding another figure such as $X2$, denoting a principle axis X normal to the projection and a two-fold horizontal axis normal to the principal axis.

An ideal specimen may have a maximum of ten symmetry elements: 1, 2, 3, 4, 6, m, i, m', 2' and $\overline{4}$. In relation to how they affect diffraction intensities, these symmetry elements may be divided into two categories, type I and type II. The type I (or two-dimensional) symmetries are those leaving both the crystal and the z axis (nearly parallel to the incident beam) unchanged. These symmetries include the rotation axes, 1, 2, 3, 4, 6, and mirror planes, m, parallel to the surface normal. These symmetry operations are also called vertical symmetry operations, since they leave the z axis of the specimen unchanged. The type II (or three-dimensional) symmetries are those leaving the crystal unchanged except for the inversion of the z axis. These symmetries include the horizontal mirror plane m' (which is parallel to the surface of the specimen), the horizontal two-fold axis 2', the four-fold rotary inversion axis $\overline{4}$ parallel to the surface normal, and the centre of inversion i. The type II symmetry operations are also called horizontal symmetry operations since they invert the z axis.

A CBED pattern taken for an exact zone axis incidence is called the zone axis pattern (ZAP). A ZAP contains a bright-field pattern (BP) and a whole pattern (WP). A BP usually refers to the CBED pattern appearing in the bright-field (transmitted) disc, while a WP consists of the transmitted disc plus all the diffracted discs. The six type I symmetry elements 1, 2, 3, 4, 6, and m may result in one-, two-, three-, four-, and six-fold rotation and a mirror m_v symmetries in the WP, where the subscript v means that the mirror plane is a vertical one as opposed to the horizontal mirror plane discussed below.

While the effects of symmetry operations of type I on electron diffraction patterns are evident, i.e. the diffraction patterns show the same symmetry (a formal derivation will be given in the next section), the effect of symmetry operations of type II is less obvious. Nevertheless, their effect may be elucidated using the reciprocity theorem.

To illustrate how the symmetry affects the appearance of electron diffraction patterns, we first consider the case shown in Fig. 9.3 for a crystal with a horizontal mirror plane (Steeds, 1983). We start from Fig. 9.3a. Application of the horizontal mirror m' to this geometry generates Fig. 9.3b. Applying the reciprocity principle, we interchange the source and the observer to obtain Fig. 9.3c. Comparing Figs 9.3a and 9.3c, we see that the transmitted electron beam intensity satisfies the following relationship:

$$I_0(\theta) = I_0(-\theta). \tag{9.1}$$

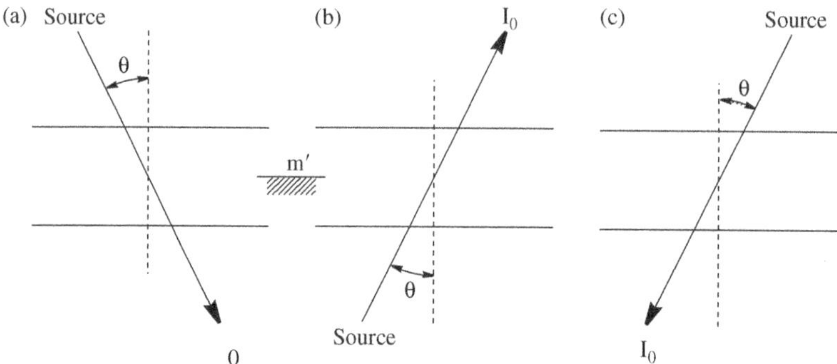

FIG. 9.3. Schematic diagram showing the effect of a horizontal mirror plane on the transmitted beam intensity.

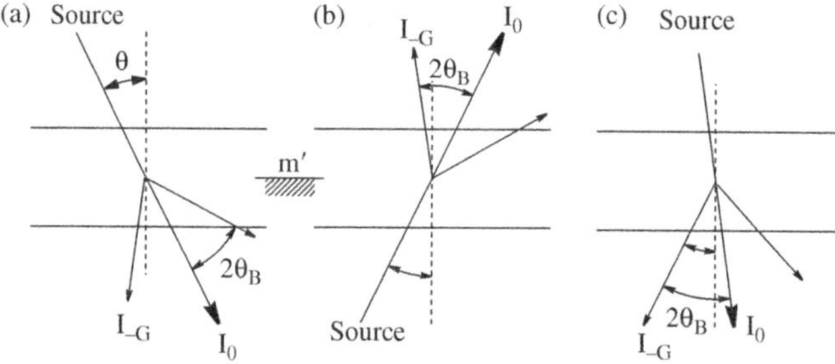

FIG. 9.4. Schematic diagram showing the effect of a horizontal mirror plane symmetry operation on the diffracted beam intensity.

It should be noted that since the above argument is independent of the azimuthal angle about the direction of incidence, the two-dimensional symmetry expected in the direct beam may therefore be generated by a rotation of 180 ° about the surface normal. The presence of a horizontal mirror plane therefore results in a two-fold rotation symmetry in the direct beam disc or BP.

To extend the argument to the diffracted beam or dark-field pattern (DP), we now consider the diffraction geometry shown in Fig. 9.4a, which is similar to Fig. 9.3a except that the observer is now placed at $-\mathbf{G}$. The horizontal mirror symmetry operation m′ results in Fig. 9.4b, and the application of the reciprocity theorem gives Fig. 9.4c. By comparing Figs 9.4a and 9.4c we find a relationship between diffracted beam intensities for the incident beam along θ (Fig. 9.4a) and $2\theta_B - \theta$ (Fig. 9.4c)

$$I_{-G}(\theta) = I_{-G}(2\theta_B - \theta), \tag{9.2}$$

where θ_B is the Bragg angle. Denoting $\theta = \phi + \theta_B$ and substituting this into eqn (9.2), we find

$$I_{-G}(\theta_B + \phi) = I_{-G}(\theta_B - \phi). \tag{9.3}$$

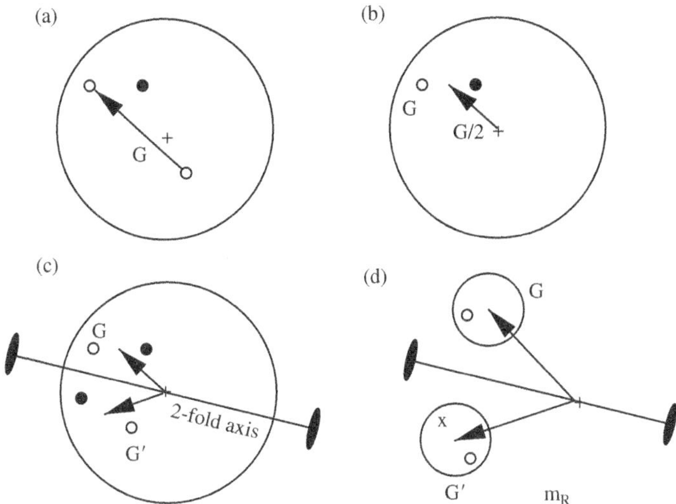

FIG. 9.5. Stereographic projections illustrating how a horizontal 2-fold rotation axis affects the CBED pattern symmetry.

By considering beams incident at other azimuthal angles for the same polar angle θ in Fig. 9.4a, we see that the CBED disc at $-\mathbf{G}$ will have two-fold symmetry about the point $\theta = \theta_B$. In other words if the Bragg condition for the $\mathbf{G}$-th diffracted disc is satisfied, the dark-field disc shows two-fold rotation symmetry about the Bragg condition at $\mathbf{K} = -\mathbf{G}/2$.

The way in which the symmetry of the specimen determines the symmetry of a convergent beam electron diffraction pattern can also be analysed using the stereographic projection as illustrated in Figs. 9.5 (a)-(d) for the case of a horizontal 2-fold axis (Tanaka, 1989). The cross at the centre of each diagram represents the zone axis of the specimen. In Fig. 9.5(a) the black dot represents the incident beam, and the circle diametrally opposite represents the directly transmitted beam. The direction of the diffracted beam is represented by a circle displaced by the diffraction vector $\mathbf{G}$ from the transmitted beam circle. The angles of scattering are assumed to be small so that the diagrams represent an angular region close to the zone axis. The angular displacement of twice the Bragg angle between transmitted and diffracted beams can then be represented by the vector $\mathbf{G}$. In Fig. 9.5 the primitive circle of the stereogram is represented only schematically. In reality the radius of the primitive circle would be about two orders of magnitude greater than $|\mathbf{G}|$. In Fig. 9.5(b) the circle representing the directly transmitted beam is omitted. The mid-point of the line joining the black dot and circle G is at a vector position $\mathbf{G}/2$ from the zone axis. Both the dot and the circle G represent beams travelling downwards. The operation of a horizontal 2-fold axis converts these beams into upward travelling beams, the

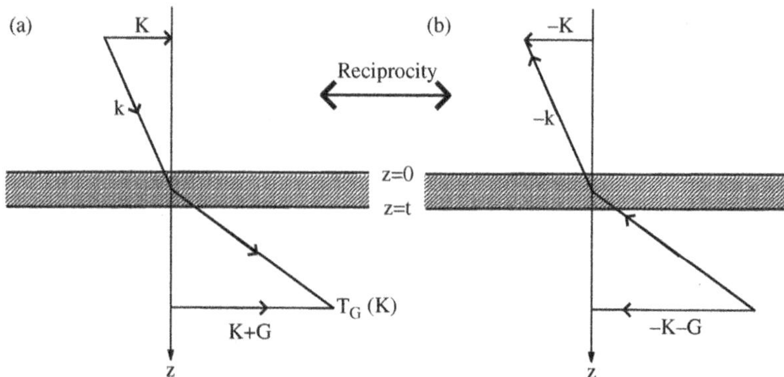

FIG. 9.6. Schematic diagram showing (a) a beam of electrons incident on a thin crystal and (b) reciprocity interchanging of the source and observer and reversing the direction of the incident electron beam. In (a) the wave field is referred to as $\psi_K^{(+)}$ while that shown in (b) is referred to as $\psi_{-K-G}^{(-)}$.

directions of which are denoted by the dot (diffracted beam) and the circle G' (incident beam) in the lower half of Fig. 9.5(c). However, reversing the directions of these beams and using the reciprocity priciple (section 9.3), makes the dot represent an incident beam and the circle represent a diffracted beam $\mathbf{G}'$, both travelling downwards. If a convergent incident beam of uniform angular intensity distribution is employed, so that the intensities of the incident beams represented by dots in 9.5(c) are equal, then the diffracted beam intensities at $\mathbf{G}$ and $\mathbf{G}'$ will also be equal. Conventionally the incident beam dots and the primitive circle are then omitted, and circles centred at $\mathbf{G}/2$ and $\mathbf{G}'/2$ are drawn to represent the convergent diffracted beam discs related by the 2-fold axis as shown in Fig. 9.5(d). The symmetry of Fig. 9.5(d) is denoted by the symbol m_R. m denotes a vertical mirror plane containing the horizontal 2-fold axis, which relates the circle in the disc G to the point marked $\times$ in the disc G'. The subscript R denotes a rotation of 180° about the centre of the disc G', which transfers the point $\times$ to the circle G'.

9.5   Diffraction symmetry – a formal derivation

9.5.1   *Basic solutions and relations*

In this section we will provide a formal derivation of the effects of diffraction symmetry to illustrate how point and space group symmetries affect the symmetry of CBED patterns. We assume that the beam of high-energy electrons incident on a thin crystalline sample shown in Fig. 9.6a may be represented by a plane wave $\exp(i\mathbf{k} \cdot \mathbf{r})$, and the amplitude of the wave diffracted into direction $\mathbf{K} + \mathbf{G}$ is given by the transmission coefficient $\mathcal{T}_\mathbf{G}(\mathbf{K})$. Here as before we use capital letters to represent components of vectors in the (x, y) plane so that $\mathbf{r} = (\mathbf{X}, z)$, $\mathbf{k} = (\mathbf{K}, k_z)$, and $\mathbf{g} = (\mathbf{G}, g_z)$.

Let

$$\Gamma_G(\mathbf{K}) = \sqrt{k^2 - (\mathbf{K} + \mathbf{G})^2}, \quad \text{if } k^2 \geq (\mathbf{K}+\mathbf{G})^2,$$
$$= i\sqrt{(\mathbf{K} + \mathbf{G})^2 - k^2} \quad \text{otherwise.} \tag{9.4}$$

A plane wave $\exp(i\mathbf{k} \cdot \mathbf{r}) = \exp\left\{i\mathbf{K} \cdot \mathbf{X} + i\Gamma_0(\mathbf{K})z\right\}$ represents an electron beam incident on the top surface of the crystal. This case is referred to by the superscript $(+)$ as opposed to $(-)$, which denotes a beam incident on the bottom surface of the crystal from below the crystal. The electron wave function below the crystal for $z \geq t$ is given by

$$\psi_k^{(+)}(\mathbf{r}) = \sum_G \mathcal{T}_G^{(+)}(\mathbf{K}) \exp\left\{i(\mathbf{K} + \mathbf{G}) \cdot \mathbf{X} + i\Gamma_G(\mathbf{K})z\right\}. \tag{9.5}$$

The wave field above the crystal ($z < 0$) has the form

$$\psi_k^{(+)}(\mathbf{r}) = \exp\left\{i\mathbf{K}\cdot\mathbf{X} + i\Gamma_0(\mathbf{K})z\right\} + \sum_G \mathcal{R}_G^{(+)}(\mathbf{K}) \exp\left\{i(\mathbf{K}+\mathbf{G})\cdot\mathbf{X} - i\Gamma_G(\mathbf{K})z\right\}. \tag{9.6}$$

The incident beam is included explicitly in this expression which also contains backscattered waves with amplitudes $\mathcal{R}_G$. Similarly, if the electron beam is incident from below the crystal we find for $z < 0$

$$\psi_k^{(-)}(\mathbf{r}) = \sum_G \mathcal{T}_G^{(-)}(\mathbf{K}) \exp\left\{i(\mathbf{K} + \mathbf{G}) \cdot \mathbf{X} - i\Gamma_G(\mathbf{K})z\right\}, \tag{9.7}$$

For $z \geq t$ we write

$$\psi_k^{(-)}(\mathbf{r}) = \exp\left\{i\mathbf{K}\cdot\mathbf{X} - i\Gamma_0(\mathbf{K})z\right\} + \sum_G \mathcal{R}_G^{(-)}(\mathbf{K}) \exp\left\{i(\mathbf{K}+\mathbf{G})\cdot\mathbf{X} + i\Gamma_G(\mathbf{K})z\right\}. \tag{9.8}$$

The solutions $\psi_k^{(+)}$ and $\psi_k^{(-)}$ are related by the reciprocity principle, which follows from the identity (Bilhorn et al., 1964)

$$\int_{plate} d\mathbf{r} \left[\psi_k^{(+)}(\mathbf{r})E_k\psi_{k'}^{(-)}(\mathbf{r}) - \psi_{k'}^{(-)}(\mathbf{r})E_k\psi_k^{(+)}(\mathbf{r})\right] = 0.$$

By using the Schrödinger equation

$$H\psi_k^{(\pm)}(\mathbf{r}) = \left[-\frac{\hbar^2}{2m}\nabla^2 + V(\mathbf{r})\right]\psi_k^{(\pm)}(\mathbf{r}) = E_k\psi_k^{(\pm)}(\mathbf{r}) \tag{9.9}$$

and the expressions (9.5)–(9.8), we find that (Buxton et al., 1976)

$$\Gamma_G(\mathbf{K})\mathcal{T}_G^{(+)}(\mathbf{K}) = \Gamma_G(-\mathbf{K} - \mathbf{G})\mathcal{T}_G^{(-)}(-\mathbf{K} - \mathbf{G}). \tag{9.10}$$

In other words the reciprocity principle requires that we transpose the source and the observer in Fig. 9.6a and reverse the direction of the electron beam as shown in Fig. 9.6b to satisfy eqn (9.10).

9.5.2 *Effect of the space group symmetry*

We now consider the effect of a coordinate transformation $\mathcal{S}$, made up of a rotation or reflection R and a translation $\boldsymbol{\nu}$,

$$\mathcal{S}\mathbf{r} = \mathbf{r}' = R\mathbf{r} + \boldsymbol{\nu}. \tag{9.11}$$

This may be most conveniently represented by the *Seitz* symbol

$$\mathcal{S} = \{R|\boldsymbol{\nu}\}. \tag{9.12}$$

The result of two successive transformations $\mathcal{S}_1$ followed by $\mathcal{S}_2$ is given by

$$\mathcal{S}_1\mathcal{S}_2 = \{R_2|\boldsymbol{\nu}_2\}\{R_1|\boldsymbol{\nu}_1\} = \{R_2 R_1|R_2\boldsymbol{\nu}_1 + \boldsymbol{\nu}_2\}.$$

In particular the inverse transformation has the form

$$\mathcal{S}^{-1} = \{R|\boldsymbol{\nu}\}^{-1} = \{R^{-1}| - R^{-1}\boldsymbol{\nu}\}. \tag{9.13}$$

If $\mathcal{S}$ is a symmetry operator that leaves the Hamiltonian $H = -(\hbar^2/2m)\nabla^2 + V$ invariant, i. e.

$$\mathcal{S}H\mathcal{S}^{-1} = H,$$

it follows from the properties of the Seitz symbols (9.11) that if $\psi_k(\mathbf{r})$ is a solution of the Schrödinger equation (9.9) then so is

$$\mathcal{S}\psi_k(\mathbf{r}) = \psi_k(\mathcal{S}^{-1}\mathbf{r}) = \psi_k[R^{-1}(\mathbf{r} - \boldsymbol{\nu})], \tag{9.14}$$

i.e.

$$H(\mathcal{S}\psi_k) = \mathcal{S}(H\psi_k) = E_k(\mathcal{S}\psi_k).$$

It can be readily proved (see e.g. Chapter 6 of (Altmann, 1991)) that a general space group operator $\mathcal{S} = \{R|\boldsymbol{\nu}\}$, when acting on a Bloch wave function associated with the vector $\mathbf{k}$, transforms it into another Bloch wave characterized by vector $R\mathbf{k}$

$$\mathcal{S}\psi_k(\mathbf{r}) = A\psi_{R\mathbf{k}}(\mathbf{r}),$$

where A is a constant phase factor independent of $\mathbf{r}$. It should be noted that not all the symmetry operations of the crystal space group leave the Hamiltonian (9.9) invariant. Among all the crystal space group symmetry operations only those that leave the surfaces of the sample unchanged should be regarded as valid $\mathcal{S}$ leaving the Hamiltonian invariant (Goodman, 1975).

The two types of symmetries introduced earlier may be defined mathematically as follows. If the z axis is chosen perpendicular to the surface of the specimen as in Fig. 9.6a, it follows that only those symmetry operations that can be written in partitioned form

$$R = \begin{pmatrix} \mathcal{R} & 0 \\ 0 & 1 \end{pmatrix} \quad \text{with } \boldsymbol{\nu} = (\mathbf{V}, 0), \tag{9.15}$$

and

$$R' = \begin{pmatrix} \mathcal{R}' & 0 \\ 0 & -1 \end{pmatrix} \quad \text{with } \boldsymbol{\nu} = (\mathbf{V}, t), \tag{9.16}$$

leave the Hamiltonian invariant. The former are type I symmetry operations which leave the z-axis invariant, while the latter are type II symmetry operations, which change z into $t - z$, where t is the thickness of the crystal plate.

There are two fundamental relations from which we can deduce information about the symmetry of diffraction patterns. For the type I symmetry operations (9.15) we have

$$\mathcal{T}_G^{(+)}(\mathbf{K}) = \exp\{i\mathcal{R}\mathbf{G} \cdot \mathbf{V}\}\mathcal{T}_{\mathcal{R}G}^{(+)}(\mathcal{R}\mathbf{K}), \tag{9.17}$$

and for the type II symmetry operation (9.16)

$$\Gamma_G(\mathbf{K})\mathcal{T}_G^{(+)}(\mathbf{K})\exp\{i\Gamma_G(\mathbf{K})t\} = \exp\{i\mathcal{R}'\mathbf{G} \cdot \mathbf{V}'\}\Gamma_{\mathcal{R}'G}[-\mathcal{R}'(\mathbf{K}+\mathbf{G})] \tag{9.18}$$
$$\times \mathcal{T}_{\mathcal{R}'G}^{(+)}[-\mathcal{R}'(\mathbf{K}+\mathbf{G})]\exp\{i\Gamma_{\mathcal{R}'G}[-\mathcal{R}'(\mathbf{K}+\mathbf{G})]t\}.$$

Here we will consider how to use these relations to obtain information about the symmetry of diffraction patterns. We defer the derivation of these equations to section 9.5.4. Since in electron diffraction we normally observe scattered *intensities*, we ignore the phase factors entering the above equations. In the case of small-angle scattering typical for THEED, $k \gg |\mathbf{K}+\mathbf{G}|$, we approximate $\Gamma_G(\mathbf{K})$ by k, i.e.

$$\Gamma_G(\mathbf{K}) = \sqrt{k^2 - (\mathbf{K}+\mathbf{G})^2} \approx k\left[1 - \frac{(\mathbf{K}+\mathbf{G})^2}{2k^2}\right] \approx k.$$

Eqns (9.17) and (9.18) then become

$$\mathcal{T}_G(\mathbf{K}) = \mathcal{T}_{\mathcal{R}G}(\mathcal{R}\mathbf{K}), \tag{9.19}$$
$$\mathcal{T}_G(\mathbf{K}) = \mathcal{T}_{\mathcal{R}'G}[-\mathcal{R}'(\mathbf{K}+\mathbf{G})]. \tag{9.20}$$

Here we have omitted the $(+)$ sign since the above equation applies to both cases where electrons are incident on the top or the bottom surface of the specimen.

First we consider eqn (9.19). This relation states that the intensity of scattering in the direction $\mathbf{K}$ in the G-th disc, $I_G(\mathbf{K}) = |\mathcal{T}_G(\mathbf{K})|^2$, is the same as that corresponding to the direction $\mathcal{R}\mathbf{K}$ in the disc $\mathcal{R}G$. Since $\mathcal{R}$ stands for a rotation or reflection in the (x, y) plane of the disc parallel to the surface of the crystal plate, it has to belong to the 10 two-dimensional crystallographic point groups. Eqn (9.19) alone then gives 10 groups of symmetry relations among the diffracted beam intensities, and these relations can be described by the conventional 10 two-dimensional crystallographic point groups. In particular, in the case where $G = 0$ (the transmitted disc), we have

$$\mathcal{T}_0(\mathbf{K}) = \mathcal{T}_0(\mathcal{R}\mathbf{K}),$$

i.e. the bright-field disc or BP exhibits the rotation or reflection symmetry of the specimen.

We now consider the type II symmetry operations. In eqn (9.20) the rotation or reflection $\mathcal{R}'$ must also belong to the set of two-dimensional crystal point groups. We can make this more transparent by redefining the incident electron wave vector $\mathbf{K}$ with respect to the Bragg condition $-\mathbf{G}/2$ (in the case where $\mathbf{K} = -\mathbf{G}/2$ the Bragg condition is satisfied exactly), so that in this new coordinate system the incident electron wave vector $\mathbf{Q}$ is related to the old $\mathbf{K}$ via

$$\mathbf{Q} = \mathbf{K} + \mathbf{G}/2, \tag{9.21}$$

and in this new coordinate system $\mathbf{Q} = 0$ corresponds to the exact Bragg condition. Substituting this relation into eqn (9.19) for the type I symmetry, we find that

$$\mathcal{T}_G(\mathbf{Q}) = \mathcal{T}_G(\mathbf{K}) = \mathcal{T}_G\left(-\frac{\mathbf{G}}{2} + \mathbf{Q}\right) = \mathcal{T}_{\mathcal{R}G}(\mathcal{R}\mathbf{K})$$

$$= \mathcal{T}_{\mathcal{R}G}\left[-\frac{\mathcal{R}\mathbf{G}}{2} + \mathcal{R}\mathbf{Q}\right] = \mathcal{T}_{\mathcal{R}G}(\mathcal{R}\mathbf{Q}),$$

i.e. that the fundamental symmetry relation for the type I symmetry is unchanged by this transformation. Similarly for the type II symmetry we have from equation (9.20)

$$\mathcal{T}_G(\mathbf{K}) = \mathcal{T}_{\mathcal{R}'G}[-\mathcal{R}'(\mathbf{K} + \mathbf{G})]$$

$$= \mathcal{T}_{\mathcal{R}'G}[-\mathcal{R}'(-\frac{1}{2}\mathbf{G} + \mathbf{Q} + \mathbf{G})] = \mathcal{T}_{\mathcal{R}'G}[-\frac{1}{2}\mathcal{R}'\mathbf{G} + \mathcal{R}'(-\mathbf{Q})].$$

However, from the above definition we see that

$$\mathcal{T}_{\mathcal{R}'G}(\mathcal{R}\mathbf{Q}) = \mathcal{T}_{\mathcal{R}'G}\left(-\frac{1}{2}\mathcal{R}'\mathbf{G} + \mathcal{R}'\mathbf{Q}\right),$$

and we arrive at a greatly simplified form of the second fundamental symmetry relation (9.20)

$$\mathcal{T}_G(\mathbf{Q}) = \mathcal{T}_{\mathcal{R}'G}(-\mathcal{R}'\mathbf{Q}). \tag{9.22}$$

This relation means that while the reciprocal lattice vector $\mathbf{G}$ is rotated (or reflected) by $\mathcal{R}'$ into $\mathcal{R}'G$, the new vector $\mathbf{Q}$ is rotated by $\mathcal{R}'$ with respect to the origin of the reciprocal lattice (or zone axis) to become $\mathcal{R}'\mathbf{Q}$, and is then inverted through the Bragg position $-\mathbf{G}/2$ to become $-\mathcal{R}'\mathbf{Q}$. The simplest example of relation (9.22) is given by the case where $\mathcal{R}'$ is just the identity operation so that $\mathcal{T}_G(\mathbf{Q}) = \mathcal{T}_G(-\mathbf{Q})$. This is 1_R symmetry which is expected for the G-th dark-field disc when a horizontal mirror m' is present, regardless of whether or not the crystal has true inversion symmetry. Here the subscript R has been used to remind us that in deriving the diffraction symmetry the reciprocity principle has been used, and that we need to invert the point $\mathbf{Q}$ through the Bragg position after performing the ordinary symmetry operation $\mathcal{R}'$ on this vector. This is

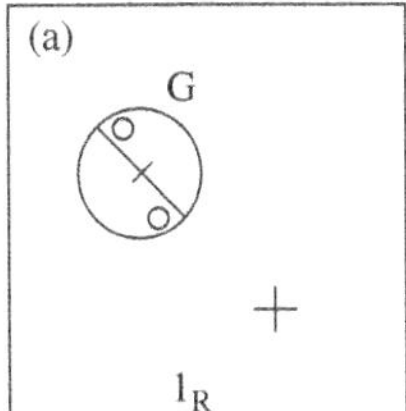
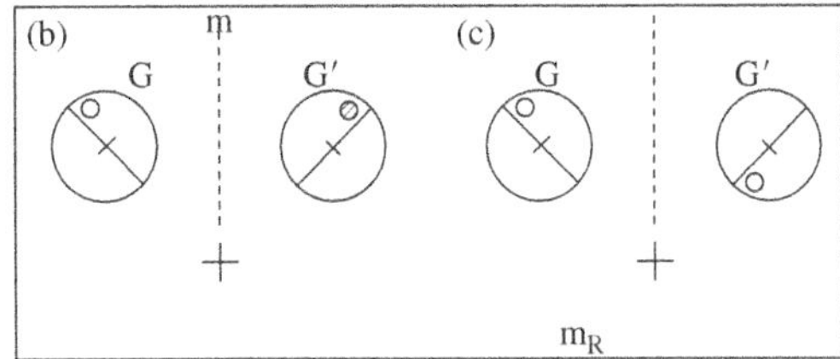

FIG. 9.7. Stereograms showing (a) the 1_R symmetry and (c) the m_R symmetry. (b) is an intermediate stage, showing the effect of a simple m symmetry on $\mathbf{G}$ (that changes G into G') and on the scattering vector $\mathbf{Q}$ (denoted as an open circle before the operation and a shadowed circle after the operation).

equivalent to a rotation of $180°$ about the Bragg position, and therefore the subscript R can also be thought of as denoting this rotation.

Figure 9.7a shows a stereogram illustrating the 1_R symmetry relation, where for the 1_R case only the internal symmetry is present within the dark-field disc (represented as the large circle centred at the Bragg position $\mathbf{K} = -\mathbf{G}/2$). In the previous section we discussed the CBED symmetries resulting from a horizontal two-fold axis $2'$. For this symmetry operation, the relevant $\mathcal{R}'$ operation is a mirror, as shown in Figs 9.7b and 9.7c. This mirror operation changes the G-th disc into the G'-th disc, and the scattering vector $\mathbf{Q}$ (small open circle in Fig. 9.7b) into $\mathbf{Q}'$ (small shadowed circle in Fig. 9.7b). The further inversion through the Bragg position at $-\mathbf{G}'/2$ then gives the m_R symmetry illustrated in Fig. 9.7c.

The expected diffraction symmetry resulting from other type II symmetry operations i and $\bar{4}$ is illustrated in Fig. 9.8, together with that resulting from the type I symmetry 4. For the case of inversion the corresponding $\mathcal{R}'$ is the two-fold rotation about the crystal zone axis (denoted in the figure by the cross). A direct application of this two-fold rotation changes the incident wave vector $\mathbf{Q}$ in the G-th disc (denoted as a small open circle) into a shadowed circle (see Fig. 9.8a) in the $-G$-th disc, and further inversion through the Bragg position for this $-G$-th disc then gives the 2_R symmetry between the two open circles. The $\bar{4}$ operation, on the other hand, may be regarded as the repetition of a $90°$ rotation about the vertical axis with each rotation being followed by reflection in a horizontal mirror m'. The expected diffraction symmetry for the vertical four-fold rotation axis is shown in Fig. 9.8b, and further symmetry operation by the horizontal mirror converts the corresponding shadowed circles of Fig. 9.8c into open circles. The symmetry amongst the open circles is denoted by 4_R.

9.5.3 *Diffraction groups and the symmetry of CBED patterns*

The symmetry relations between the intensities of diffracted beams form groups, and these groups are called *diffraction groups*. It may be readily shown that direct application of the fundamental symmetry relation (9.19) to the type I or vertical symmetry elements 1, 2, 3, 4, 6, and m results in 10 diffraction groups, denoted

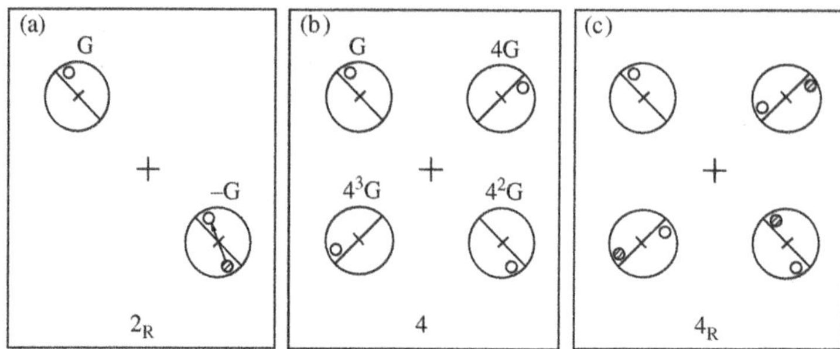

FIG. 9.8. Stereograms showing (a) the 2_R symmetry, (b) the 4 symmetry, and (c) the 4_R symmetry.

by 1, 2, 3, 4, 6, 1m, 2mm, 3m, 4mm, and 6mm. These 10 diffraction groups are isomorphic to the 10 two-dimensional crystallographic point groups shown in Fig. 9.9. On the left of each group this figure shows the equivalent positions, and on the right are the symmetry operations. The equivalent positions are two-dimensional and are therefore represented by points on the primitive circle.

In the preceding sections we showed that the application of the second fundamental symmetry relation (9.22) to the four type II or horizontal symmetry operations i, 2′, m′ and $\overline{4}$ results in four new diffraction groups 1_R, 2_R, m_R, and 4_R. By combining these four new diffraction groups with the 10 two-dimensional diffraction groups we obtain 31 diffraction groups, and a complete stereographic representation of all the 31 diffraction groups was given by Buxton et al. in their Table 1 (Buxton et al., 1976). In many cases the two-dimensional pattern symmetries of the transmitted disc (BP symmetry) and of the whole pattern (WP symmetry) may uniquely determine the diffraction groups, and these BP and WP symmetries may be readily found from the diagrams of the diffraction groups given by Buxton et al. (Buxton et al., 1976). The BP symmetry may be explicitly found by superimposing all the large discs in the diagram of the diffraction group at its centre in the form of a single bright-field disc. The symmetry of this disc is the BP symmetry. Figure 9.10a shows the group 4_R, and the corresponding bright-field symmetry is determined by superimposing the four dark-field discs in Fig. 9.10b. This pattern has the point group symmetry 4. On the other hand, to determine the WP symmetry all the small circles on the inner halves of the discs representing features below the Bragg angle should be ignored. The corresponding WP symmetry of the 4_R group is therefore 2 as shown in Fig. 9.10c. In Table 9.1 we list all the expected BP, WP, DP symmetries as well as those corresponding to pairs of $\pm G$ discs for all the 31 diffraction groups.

Since the point group symmetry of the specimen slab must be one of the 31 point groups of diperiodic plane figures, it is no surprise that the diffraction groups are isomorphic to them and therefore to the 31 Shubnikov point groups of coloured plane figures (Shubnikov and Belov, 1964). The diffraction groups were labelled by analogy with the Shubnikov groups (Buxton et al., 1976) so that the

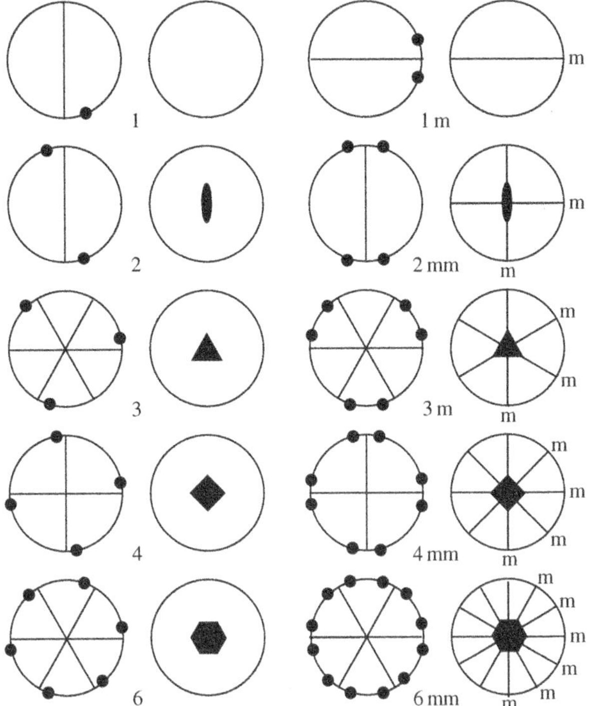

FIG. 9.9. Stereograms of the 10 two-dimensional point groups. Each group is denoted by the full Hermann–Mauguin or 'International' notation (this notation is recommended by the International Tables for Crystallography). m denotes a horizontal mirror line. Note that the same symmetry elements (with m denoting a vertical mirror plane) denote the 10 polar three-dimensional point groups.

isomorphism is obvious, e.g. 4_R is isomorphic to the Shubnikov group $4'$.

It should be noted that in a BP there are not only two-dimensional but also three-dimensional symmetries. Because of this the BP sometimes exhibits a symmetry that is higher than that of the WP. The symmetry expected for a BP as a result of a three-dimensional symmetry element (m', i, $2'$, and $\bar{4}$) may be obtained by moving the centres of dark-field discs in Figs. 9.8 and 9.7 onto the zone axis. This procedure gives symmetries 2, 1, m, and 4 in the BP, instead of 1_R, 2_R, m_R, and 4_R in DPs. The fact that the inversion operation i results only in a 1 symmetry in the BP indicates that the BP cannot be used for deciding whether or not a specimen has an inversion centre. When the projected potential approximation is applicable, i.e. if the variation of the potential along the zone axis can be ignored, a horizontal mirror plane half-way through the sample is added resulting in only 10 distinct diffraction groups isomorphic to the 10 grey Shubnikov groups (Shubnikov and Belov, 1964). These 10 diffraction groups may be used to distinguish between all but the 4 and $\bar{4}$ diffraction groups.

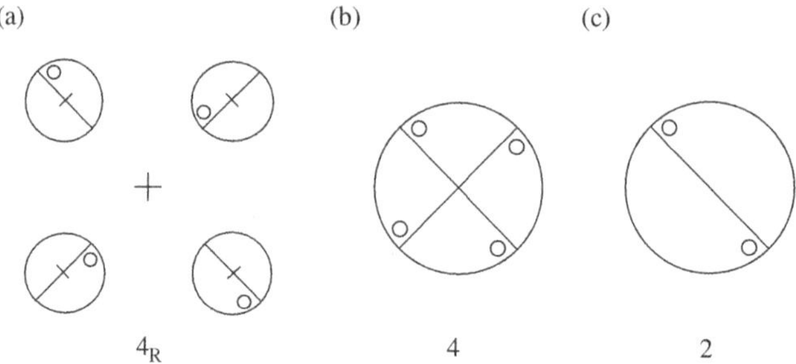

FIG. 9.10. (a) Diffraction group 4_R, (b) the corresponding BP symmetry, and (c) the WP symmetry.

9.5.4 *Derivation of the fundamental symmetry relations*

We now consider how the space group symmetry operations transform the electron wave function and derive the symmetry relations between diffracted beams. Applying an operator $\mathcal{S} = \{R|\boldsymbol{\nu}\}$ with R of the form (9.15) to the wave function $\psi_k^{(+)}$, we find from (9.6) that for a beam of electrons incident on the top surface of the crystal in the region $z < 0$

$$\mathcal{S}\psi_k^{(+)}(\mathbf{r}) = \exp\{i\mathbf{k} \cdot \mathcal{S}^{-1}\mathbf{r}\} + \sum_G \mathcal{R}_G^{(+)}(\mathbf{K}) \exp\left\{ i[(\mathbf{K} + \mathbf{G}) - \Gamma_G(\mathbf{K})\hat{\mathbf{z}}] \cdot \mathcal{S}^{-1}\mathbf{r}\right\},$$

where $\hat{\mathbf{z}}$ is the unit vector along the z axis. Since a point group (rotation or reflection) operator R must not affect the scalar product $\mathbf{k}\cdot\mathbf{r}$, i.e. $R\mathbf{k}\cdot R\mathbf{r} = \mathbf{k}\cdot\mathbf{r}$, we find

$$\mathbf{k} \cdot R^{-1}\mathbf{r} = \{R\mathbf{k}\} \cdot \{R(R^{-1}\mathbf{r})\} = R\mathbf{k} \cdot \mathbf{r}.$$

More generally, for any function

$$Rf(\mathbf{s} \cdot \mathbf{r}) = f(\mathbf{s} \cdot R^{-1}\mathbf{r}) = f(R\mathbf{s} \cdot \mathbf{r}).$$

Recalling that $\boldsymbol{\nu} = (\mathbf{V}, 0)$ we have

$$\begin{aligned}
\mathbf{k} \cdot \mathcal{S}^{-1}\mathbf{r} &= \mathbf{k} \cdot (R^{-1}\mathbf{r} - R^{-1}\boldsymbol{\nu}) = R\mathbf{k} \cdot (\mathbf{r} - \boldsymbol{\nu}) \\
&= R\{\mathbf{K} + \Gamma_0(\mathbf{K})\hat{\mathbf{z}}\}(\mathbf{X} - \mathbf{V} + z\hat{\mathbf{z}}) = [\mathcal{R}\mathbf{K} + \Gamma_0(\mathcal{R}\mathbf{K})\hat{\mathbf{z}}](\mathbf{X} - \mathbf{V} + z\hat{\mathbf{z}}) \\
&= \mathcal{R}\mathbf{K} \cdot (\mathbf{X} - \mathbf{V}) + \Gamma_0(\mathcal{R}\mathbf{K})z, \tag{9.23}
\end{aligned}$$

and

$$\begin{aligned}
[(\mathbf{K} + \mathbf{G}) - \Gamma_G(\mathbf{K})\hat{\mathbf{z}}]\mathcal{S}^{-1}\mathbf{r} &= [(\mathbf{K} + \mathbf{G}) - \sqrt{k^2 - (\mathbf{K} + \mathbf{G})^2}\hat{\mathbf{z}}](R^{-1}\mathbf{r} - R^{-1}\boldsymbol{\nu}) \\
&= R[(\mathbf{K} + \mathbf{G}) - \sqrt{k^2 - (\mathbf{K} + \mathbf{G})^2}\hat{\mathbf{z}}](\mathbf{r} - \mathbf{V}) \\
&= [\mathcal{R}(\mathbf{K} + \mathbf{G}) - \sqrt{k^2 - (\mathcal{R}\mathbf{K} + \mathcal{R}\mathbf{G})^2}\hat{\mathbf{z}}](\mathbf{r} - \mathbf{V}) \\
&= [\mathcal{R}(\mathbf{K} + \mathbf{G}) - \Gamma_{\mathcal{R}\mathbf{G}}(\mathcal{R}\mathbf{K})\hat{\mathbf{z}}](\mathbf{X} - \mathbf{V} + z\hat{\mathbf{z}}) \\
&= \mathcal{R}(\mathbf{K} + \mathbf{G}) \cdot (\mathbf{X} - \mathbf{V}) - \Gamma_{\mathcal{R}G}(\mathcal{R}\mathbf{K})z. \tag{9.24}
\end{aligned}$$

Table 9.1 Symmetries of bright-field patterns (BP), whole patterns (WP), dark-field patterns, (DP) and positive and negative dark-field patterns ±DPs and corresponding diffraction groups

BP	WP	DP	±DPs	Diffraction groups
1	1		1	1
			2_R	2_R
2	1			1_R
	2	1		2
		2		21_R
3	3		1	3
			2_R	6_R
4	2			4_R
	4	1		4
		2		41_R
6	3			31_R
	6	1		6
		2		61_R
m	1			m_R
	m		1	m
			2_R	$2_R mm_R$
2mm	2			$2m_R m_R$
	m			$2_R mm_R$
	2mm	1(m)		2mm
		2(mm)		$2mm1_R$
3m	3			$3m_R$
	3m	1		3m
			2_R	$6_R mm_R$
4mm	4			$4m_R m_R$
	2mm			$4_R mm_R$
	4mm	1(m)		4mm
		2(mm)		$4mm1_R$
6mm	6			$6m_R m_R$
	3m			$3m1_R$
	6mm	1(m)		6mm
		2(mm)		$6mm1_R$

Substitution of relations (9.23) and (9.24) back into the expression for $\mathcal{S}\psi_k^{(+)}$ gives

$$\mathcal{S}\psi_k^{(+)}(\mathbf{r}) = \exp\left\{i\mathcal{R}\mathbf{K}\cdot(\mathbf{X}-\mathbf{V})+i\Gamma_0(\mathcal{R}\mathbf{K})z\right\} \tag{9.25}$$

$$+ \sum_G \mathcal{R}_G^{(+)}(\mathbf{K})\exp\left\{i\mathcal{R}(\mathbf{K}+\mathbf{G})\cdot(\mathbf{X}-\mathbf{V})-i\Gamma_{\mathcal{R}G}(\mathcal{R}\mathbf{K})z\right\}$$

$$= \exp\left\{i\mathcal{R}\mathbf{K}\cdot(\mathbf{X}-\mathbf{V})+i\Gamma_0(\mathcal{R}\mathbf{K})z\right\}$$

$$+ \sum_{G'} \mathcal{R}_G^{(+)}(\mathbf{K})\exp\left\{-i(\mathcal{R}\mathbf{K}+\mathbf{G}')\cdot\mathbf{V}+i(\mathcal{R}\mathbf{K}+\mathbf{G}')\cdot\mathbf{X}-i\Gamma_{G'}(\mathcal{R}\mathbf{K})z\right\},$$

where $\mathbf{G}' = \mathcal{R}\mathbf{G}$. On the other hand, for a beam of electrons incident from above the crystal along the direction $\mathcal{R}\mathbf{K}$ we have

$$\psi_{\mathcal{R}\mathbf{K}}^{(+)}(\mathbf{r}) = \exp\left\{i\mathcal{R}\mathbf{K}\cdot\mathbf{X}+i\Gamma_0(\mathcal{R}\mathbf{K})z\right\}$$

$$+ \sum_{G'} \mathcal{R}_{G'}^{(+)}(\mathcal{R}\mathbf{K})\exp\left\{i(\mathcal{R}\mathbf{K}+\mathbf{G}')\cdot\mathbf{X}-i\Gamma_{G'}(\mathcal{R}\mathbf{K})z\right\}. \tag{9.26}$$

Since the solution of the Schrödinger equation for an incident plane wave is unique, comparison of the above two expressions (9.25) and (9.26) shows that

$$\mathcal{S}\psi_k^{(+)}(\mathbf{r}) = \exp\{-i\mathcal{R}\mathbf{K}\cdot\mathbf{V}\}\psi_{\mathcal{R}\mathbf{K}}^{(+)}(\mathbf{r}), \tag{9.27}$$

and for the reflected beam amplitudes

$$\mathcal{R}_G^{(+)}(\mathbf{K}) = \exp\{i\mathcal{R}\mathbf{G}\cdot\mathbf{V}\}\mathcal{R}_{\mathcal{R}G}^{(+)}(\mathcal{R}\mathbf{K}). \tag{9.28}$$

Applying a type I or vertical symmetry operation to $\psi_k^{(+)}$ below the crystal in the region $z > t$, we find

$$\mathcal{S}\psi_k^{(+)}(\mathbf{r}) = \sum_G \mathcal{T}_G^{(+)}(\mathbf{K})\exp\left\{i[(\mathbf{K}+\mathbf{G})+\Gamma_G(\mathbf{K}\hat{\mathbf{z}})](\mathcal{S}^{-1}\mathbf{r})\right\}$$

$$= \sum_G \mathcal{T}_G^{(+)}(\mathbf{K})\exp\left\{i\mathcal{R}[(\mathbf{K}+\mathbf{G})+\Gamma_G(\mathbf{K})\hat{\mathbf{z}}](\mathbf{X}-\mathbf{V}+z\hat{\mathbf{z}})\right\}$$

$$= \sum_G \mathcal{T}_G^{(+)}(\mathbf{K})\exp\left\{i(\mathcal{R}\mathbf{K}+\mathcal{R}\mathbf{G})\cdot(\mathbf{X}-\mathbf{V})+i\Gamma_{\mathcal{R}G}(\mathcal{R}\mathbf{K})z\right\}.$$

But for a plane wave incident along $\mathcal{R}\mathbf{K}$ we have

$$\psi_{\mathcal{R}\mathbf{K}}^{(+)}(\mathbf{r}) = \sum_{G'} \mathcal{T}_{G'}^{(+)}(\mathcal{R}\mathbf{K})\exp\left\{i(\mathcal{R}\mathbf{K}+\mathbf{G}')\cdot\mathbf{X}+i\Gamma_{G'}(\mathcal{R}\mathbf{K})z\right\},$$

where $\mathbf{G}' = \mathcal{R}\mathbf{G}$. By comparing the above two expressions we obtain relation (9.27) and

$$\mathcal{T}_G^{(+)}(\mathbf{K}) = \exp\{i\mathcal{R}\mathbf{G}\cdot\mathbf{V}\}\mathcal{T}_{\mathcal{R}G}^{(+)}(\mathcal{R}\mathbf{K}), \tag{9.29}$$

and this is exactly the first of the fundamental symmetry relations (9.17).

Similarly for a horizontal symmetry operation $\mathcal{S}' = \{\mathcal{R}'|\boldsymbol{\nu}'\}$ of the form (9.16) we have

$$\mathcal{S}'\psi_k^{(+)}(\mathbf{r}) = \exp\left\{-i\mathcal{R}'\mathbf{K}\cdot\mathbf{V} + i\Gamma_0(\mathcal{R}'\mathbf{K})t\right\}\psi_{\mathcal{R}'\mathbf{K}}^{(-)}(\mathbf{r}). \qquad (9.30)$$

Applying $\mathcal{S}'$ to the wave function $\psi_k^{(+)}(\mathbf{r})$ below the crystal, we find

$$\begin{aligned}
\mathcal{S}'\psi_k^{(+)} &= \sum_G \mathcal{T}_G^{(+)}(\mathbf{K})\exp\left\{i[(\mathbf{K}+\mathbf{G})+\Gamma_G(\mathbf{K})\hat{\mathbf{z}}](\mathcal{S}')^{-1}\mathbf{r}\right\} \\
&= \sum_G \mathcal{T}_G^{(+)}(\mathbf{K})\exp\left\{iR'[(\mathbf{K}+\mathbf{G})+\Gamma_G(\mathbf{K})\hat{\mathbf{z}}](\mathbf{r}-\mathbf{V}-t\hat{\mathbf{z}})\right\} \\
&= \sum_G \left\{\mathcal{T}_G^{(+)}(\mathbf{K})\exp\left[-i(\mathcal{R}'\mathbf{K}+\mathbf{G}')\cdot\mathbf{V}+i\Gamma_{G'}(\mathcal{R}'\mathbf{K})t\right]\right\} \\
&\quad \times \exp\left\{i(\mathcal{R}'\mathbf{K}+\mathbf{G}')\cdot\mathbf{X}-i\Gamma_{G'}(\mathcal{R}'\mathbf{K})z\right\}.
\end{aligned}$$

On the other hand, for an electron beam incident on the bottom surface of the crystal from below we have

$$\psi_{\mathcal{R}'\mathbf{K}}^{(-)}(\mathbf{r}) = \sum_{G'} \mathcal{T}_{\mathcal{R}'G}^{(-)}(\mathcal{R}'\mathbf{K})\exp\left\{i(\mathcal{R}'\mathbf{K}+\mathbf{G}')\cdot\mathbf{X}-i\Gamma_{G'}(\mathcal{R}'\mathbf{K})z\right\},$$

where $\mathbf{G}' = \mathcal{R}'\mathbf{G}$. Using (9.30) and the expressions for $\mathcal{S}'\psi_k^{(+)}$ and $\psi_{\mathcal{R}'\mathbf{K}}^{(-)}$ above the crystal, we then obtain for a vertical symmetry operation $\mathcal{S}' = \{R'|\boldsymbol{\nu}'\}$

$$\mathcal{T}_G^{(+)}(\mathbf{K}) = \exp\left\{i\mathcal{R}'\mathbf{G}\cdot\mathbf{V}'\right\}\exp\left\{i[\Gamma_0(\mathcal{R}'\mathbf{K})-\Gamma_{\mathcal{R}'G}(\mathcal{R}'\mathbf{K})]t\right\}\mathcal{T}_{\mathcal{R}'G}^{(-)}(\mathcal{R}'\mathbf{K}). \quad (9.31)$$

Taking into account that $\Gamma_0(\mathcal{R}'\mathbf{K}) = \Gamma_0(\mathbf{K})$ and $\Gamma_{\mathcal{R}'G}(\mathcal{R}'\mathbf{K}) = \Gamma_G(\mathbf{K})$, the above equation becomes

$$\mathcal{T}_G^{(+)}(\mathbf{K}) = \exp\left\{i\mathcal{R}'\mathbf{G}\cdot\mathbf{V}'\right\}\exp\{i[\Gamma_0(\mathbf{K})-\Gamma_G(\mathbf{K})]t\}\mathcal{T}_{\mathcal{R}'G}^{(-)}(\mathcal{R}'\mathbf{K}), \qquad (9.32)$$

and this is eqn (2.21) of Buxton *et al.* (Buxton et al., 1976) (note that in that paper the prime on the vector $\mathbf{V}'$ is missing). This formula relates the diffracted beam amplitudes for the electron incident from above the crystal to those for the beam incident from below. Writing $\mathcal{R}'\mathbf{K}' = -\mathbf{K} - \mathbf{G}$ and $\mathbf{G} = \mathcal{R}'\mathbf{G}'$, and using the reciprocity relation (9.10), we find

$$\Gamma_{\mathcal{R}'G'}[-\mathcal{R}'(\mathbf{K}'+\mathbf{G}')]\mathcal{T}_{\mathcal{R}'G'}^{(+)}[-\mathcal{R}'(\mathbf{K}'+\mathbf{G}')] = \Gamma_{\mathcal{R}'G'}(\mathcal{R}'\mathbf{K}')\mathcal{T}_{\mathcal{R}'G'}^{(-)}(\mathcal{R}'\mathbf{K}').$$

Multiplying eqn (9.32) from the left by $\Gamma_{\mathcal{R}'\mathbf{G}'}(\mathcal{R}'\mathbf{K}')$, substituting the above form of the reciprocity relation into (9.32), and noting that $\Gamma_{\mathcal{R}'\mathbf{G}}(\mathcal{R}'\mathbf{K}) = \Gamma_G(\mathbf{K})$ and $\Gamma_0(\mathcal{R}'\mathbf{K}) = \Gamma_{\mathcal{R}'G}[-\mathcal{R}'(\mathbf{K}+\mathbf{G})]$, we finally arrive at a relation among diffracted

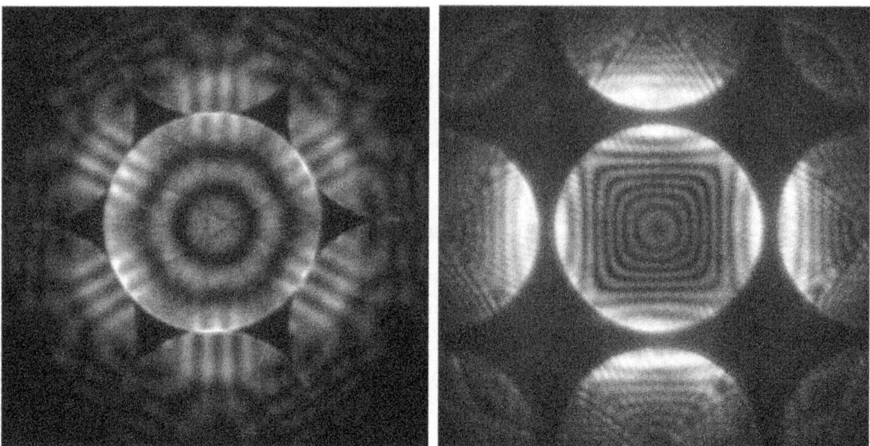

FIG. 9.11. Energy-filtered CBED patterns obtained from (left) $\langle 111 \rangle$ and (right) $\langle 100 \rangle$ zone axes of a silicon single crystal.

beam amplitudes valid in the case where the electron beam is incident on the top surface of the crystal

$$\Gamma_G(\mathbf{K})\mathcal{T}_G^{(+)}(\mathbf{K}) \exp\left\{i\Gamma_G(\mathbf{K})t\right\} = \exp\left\{i\mathcal{R}'\mathbf{G}\cdot\mathbf{V}'\right\}\Gamma_{\mathcal{R}'G}[-\mathcal{R}'(\mathbf{K}+\mathbf{G})]$$
$$\times \mathcal{T}_{\mathcal{R}'G}^{(+)}[-\mathcal{R}'(\mathbf{K}+\mathbf{G})] \exp\left\{i\Gamma_{\mathcal{R}'G}[-\mathcal{R}'(\mathbf{K}+\mathbf{G})]t\right\}.$$

This is our second fundamental symmetry relation (9.18). By combining it with the first fundamental relation (9.29) we can identify all the symmetry elements characterizing the diffraction pattern.

9.6　Crystal point group determination

In the preceding sections we have shown that a perfect crystal slab, with its normal parallel to the axis of the electron microscope, may have 10 symmetry elements, of which six are type I symmetry elements and the remaining four are type II symmetry elements. Direct application of the six type I symmetry elements gives rise to ten groups of symmetry, while the addition of the type II symmetry elements result in four more groups of symmetry, or diffraction groups. A total of 31 diffraction groups can be constructed by combining these diffraction groups consisting of both the type I and type II symmetry elements. A complete graphical representation of the 31 diffraction groups was given by Buxton et al. (Buxton et al., 1976), where each diagram represents the symmetry relations between the dark-field CBED discs. Relations between the 31 diffraction groups and the 32 crystal point groups have also been tabulated by Buxton et al. and their table is reproduced in Table 9.2.

The point group of a crystal may in principle be determined from any of the four types of CBED patterns: BP, WP, DP, and a pair of dark-field patterns

Table 9.2 Relation between the diffraction groups (rows) and the crystal point groups (columns)

	1	$\bar{1}$	2	m	2/m	222	mm2	mmm	4	$\bar{4}$	4/m	422	4mm	$\bar{4}2m$	4/mmm	3	$\bar{3}$	32	3m	$\bar{3}m$	6	$\bar{6}$	6/m	622	6mm	$\bar{6}m2$	6/mmm	23	m3	432	$\bar{4}3m$	m3m
$6mm1_R$																											X					
$3m1_R$																										X						
$6mm$																									X							
$6m_Rm_R$																								X								
61_R																							X									
31_R																						X										
6																					X											
6_Rmm_R																				X												X
$3m$																			X												X	
$3m_R$																		X												X		
6_R																	X												X			
3																X												X				
$4mm1_R$															X																	X
4_Rmm_R														X																	X	
$4mm$													X																			
$4m_Rm_R$												X																		X		
41_R											X																					
4_R										X																						
4									X																							
$2mm1_R$								X							X												X		X			X
2_Rmm_R					X						X									X			X									
$2mm$							X																			X						
$2m_Rm_R$						X						X		X										X				X		X		
$m1_R$							X						X	X											X						X	
m				X															X			X										
m_R			X						X	X											X											
21_R					X																											
2_R		X			X			X			X				X		X			X			X				X		X			X
2			X															X														
1_R				X																												
1	X		X	X		X	X		X	X		X	X	X		X		X	X		X	X		X	X	X		X		X	X	

($\pm$DP) with opposite indices (Tanaka, 1989). We now proceed to explain the procedure for point group determination with a worked example for a single crystal of silicon. Figure 9.11a shows a $\langle 111 \rangle$ zone axis CBED pattern obtained from a silicon single crystal. The BP has a three-fold rotation axis and three mirror planes, or 3m symmetry, and the WP has the same symmetry. In Table 9.1 are listed all the expected symmetries of BP, WP, DP, and $\pm$DP and the corresponding diffraction groups. By consulting this table we see that only two diffraction groups, 3m and 6_Rmm$_R$, are compatible with the observed BP and WP symmetry of Fig. 9.11a. Relations between the diffraction groups and the crystal point groups are listed in Table 9.2 (Buxton et al., 1976). From this table we see that there are two point groups $\bar{3}$m and m3m related to the diffraction group 6_Rmm$_R$, and another two 3m and $\bar{4}$3m related to the diffraction group 3m. To determine which of the two point groups is the correct one, another zone axis CBED was taken and is shown in Fig. 9.11b. Both the BP and WP show the 4mm symmetry. From Table 9.1 we find that the symmetries of this pattern are compatible with diffraction groups 4mm and 4mm1_R. From Table 9.2 we find that the compatible crystal point groups are 4mm (which gives rise to diffraction group 4mm, 4/mmm, and m3m (both resulting in diffraction group 4mm1_R).

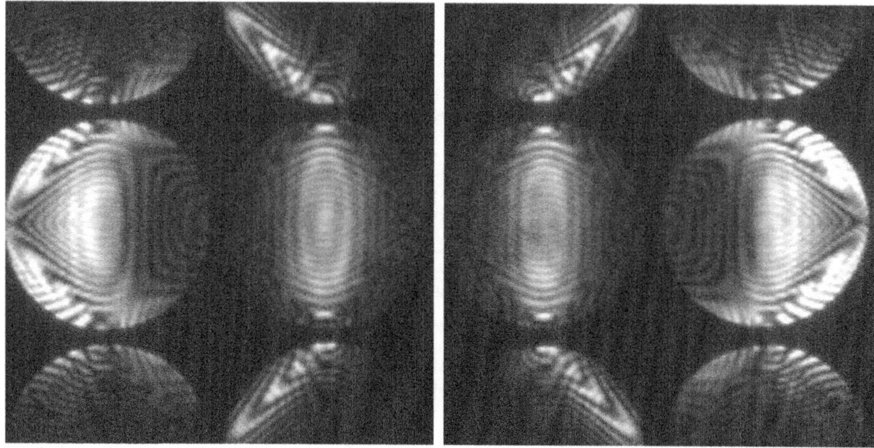

FIG. 9.12. A pair of 200 dark-field CBED $\pm G$ patterns obtained from a single crystal of silicon near the $\langle 100 \rangle$ zone axis.

Since both patterns 9.11a and 9.11b were obtained from the same crystal, the true crystal point group must be the one compatible with both patterns. From the above analysis we see that the only point group compatible both with the BP and the WP symmetries of Figs 9.11a and 9.11b is m3m. The point group of silicon is therefore uniquely determined to be m3m.

Often the BP and WP symmetries alone are sufficient for the determination of the corresponding diffraction group. If these symmetries do not identify the diffraction group uniquely, the DP symmetry (i.e. the symmetry of one dark-field disc, and a comparison of the $\pm G$ discs) may help in discriminating between different groups. Figure 9.12 shows an example of a pair of 040 and 0$\bar{4}$0 DPs, while in Table 9.1 are listed all of the expected symmetries of DP and $\pm G$ DPs. For example, this table may be used to discriminate between groups 6 and 61_R by utilizing the DP symmetry, since the former is compatible only with the 1 DP symmetry, while the later is compatible with 2_R.

9.7 Crystal space group determination

For a crystal slab bounded by parallel surfaces, there are three possible space group symmetry operations: (1) a horizontal screw axis $2_1'$, (2) a vertical glide plane g with a horizontal glide vector, and (3) a horizontal glide plane g'. From these elements combined with the 10 symmetry elements of the point groups we can generate 80 space groups.

In X-ray and neutron diffraction, kinematically forbidden reflections are used in order to determine the crystal space group. In electron diffraction kinematically forbidden reflections due to screw axes and glide planes still appear owing to dynamical diffraction effects. However, Gjønnes and Moodie in 1965 showed that these reflections may still vanish for certain crystal orientations with respect to the incident beam. These extinction lines are often called G–M lines after Gjønnes and Moodie (Gjønnes and Moodie, 1965).

9.7.1 *Formation of G–M lines*

9.7.1.1 *Systematic extinction for a horizontal screw* We first consider the systematic kinematic extinction due to the $2'_1$ screw axis parallel to the y direction (see Fig. 9.13). In this case atoms occur in pairs with fractional atomic coordinates (x, y, z) and $(\overline{x}, y + 1/2, \overline{z})$. Assuming that the crystal contains a total of N atoms, the crystal structure factor for the (hkl) reflection is given by

$$F(hkl) = \sum_{j=1}^{N/2} f_j \{ \exp[-2\pi i(hx_j + ky_j + lz_j)] + \exp[-2\pi i(-hx_j + ky_j - lz_j) - \pi i k] \}.$$

For even k we find

$$F(hkl) = \sum_{j=1}^{N/2} 2f_j \exp(-2\pi i k y_j) \cos[2\pi(hx_j + lz_j)], \tag{9.33}$$

and for odd k we have

$$F(hkl) = \sum_{j=1}^{N/2} 2i f_j \exp(-2\pi i k y_j) \sin[2\pi(hx_j + lz_j)]. \tag{9.34}$$

Equation (9.34) shows that systematic absences occur for $h = l = 0$ if k is odd, i.e.

$$F(0k0) = 0. \tag{9.35}$$

9.7.1.2 *Systematic absences for a vertical glide plane* We now consider a vertical glide plane with a horizontal glide vector, e.g. a b-glide plane perpendicular to the x axis. Atoms now form pairs with coordinates (x, y, z) and $(\overline{x}, y + 1/2, z)$. This gives

$$F(hkl) = \sum_{j=1}^{N/2} f_j \{ \exp[-2\pi i(hx_j + ky_j + lz_j)] + \exp[-2\pi i(-hx_j + ky_j + lz_j) - \pi i k] \}.$$

For even k the above equation reduces to

$$F(hkl) = \sum_{j=1}^{N/2} 2f_j \exp[-2\pi i(ky_j + lz_j)] \cos(2\pi hx_j), \tag{9.36}$$

and for odd k

$$F(hkl) = \sum_{j=1}^{N/2} 2i f_j \exp[-2\pi i(ky_j + lz_j)] \sin(2\pi hx_j). \tag{9.37}$$

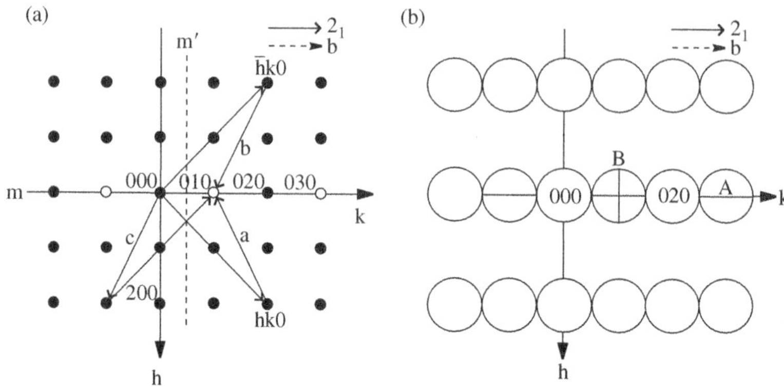

FIG. 9.13. Generation of the G–M lines in kinematically forbidden reflections due to a b-glide plane and the 2_1 screw axis: (a) shows three diffraction paths a, b and c and (b) the formation of A and B G–M lines in forbidden reflections. In (b) the sample has been tilted in such a way that the first-order (010) reflection satisfies the exact Bragg condition.

From these equations we can see that systematic absences occur for odd k if $h = 0$, i.e.

$$F(0kl) = 0. \tag{9.38}$$

9.7.1.3 *Formation of G–M lines* Figure 9.13 shows multiple scattering paths giving rise to a kinematically forbidden reflection $0k0$, where k is odd. This reflection is forbidden owing to a b-glide plane perpendicular to the x axis and/or a $2'_1$ screw axis in the b-direction. According to the multislice method, every dynamical scattering path pertinent to a reflection $\mathbf{h} = (hkl)$ can be expressed as the following product:

$$F(\mathbf{h}_1)F(\mathbf{h}_2), ..., F(\mathbf{h}_n)Z(\zeta, \zeta_1, ..., \zeta_{n-1}), \tag{9.39}$$

where ζ_r is the excitation error of the reflection $\mathbf{h}_r = \sum_{j=1}^{r} \mathbf{h}_j$ and $\mathbf{h} = \mathbf{h}_n$, and $Z(\zeta, ...)$ is a geometrical factor that depends only on these excitation errors. Let us first consider a dynamical diffraction path 'a' for the 010 forbidden reflection. The path 'b' is geometrically equivalent to path 'a' with respect to the glide plane and the $2'_1$ screw axis. From eqns (9.33)–(9.37) for both the glide plane and the $2'_1$ screw axis we find the following relations between the crystal structure factors:

$$F(h, k, 0) = F(\bar{h}, k, 0) \quad \text{for } k = 2n$$

$$F(h, k, 0) = -F(\bar{h}, k, 0) \quad \text{for } k = 2n + 1,$$

i.e. the structure factor of a reflection $hk0$ located in the upper half of the diagram and that of a reflection $\bar{h}, k, 0$ located in the lower half have the same phase for reflections with even k, but have opposite phases for reflections with odd k.

Since a multiple scattering path giving rise to reflection $(0k0)$ with odd k always contains an odd number of reflections, the following equations are satisfied:

$$F(h_1, k_1)F(h_2, k_2), ..., F(h_n, k_n)(\text{for path a})$$
$$= -F(\overline{h}_1, k_1)F(\overline{h}_2, k_2), ..., F(\overline{h}_n, k_n)(\text{for path b}),$$

where $\sum_{i=1}^{n} h_i = 0$ and $\sum_{i=1}^{n} k_i = k$, and $k =$odd. The function $Z(\zeta, ...)$ involving the excitation errors is omitted here because this function remains the same for paths 'a' and 'b' if the projection of the Laue point along the zone axis lies on the k axis. Since the waves following paths 'a' and 'b' have the same amplitudes but opposite signs, these two waves cancel each other in the $0k0$ discs ($k =$odd), giving rise to the horizontal dark lines A in the forbidden discs shown in Fig. 9.13. The absence lines A run along the direction of the 2_1 screw axis or the glide translation.

We now consider path 'c' shown in Fig. 9.13. In path 'c' the reflections are arranged in the reverse order in comparison with those in path 'b'. When the 010 reflection is in the exact Bragg position, the two paths 'a' and 'c' are symmetric with respect to the bisector of the 010 vector (denoted by the dashed line m' in Fig. 9.13), and the waves propagating along these paths have the same amplitude but opposite signs, i.e.

$$F(h_1, k_1)F(h_2, k_2), ..., F(h_n, k_n)(\text{for path a}) \tag{9.40}$$
$$= -F(\overline{h}_n, k_n), ..., F(\overline{h}_2, k_2)F(\overline{h}_1, k_1)(\text{for path c}). \tag{9.41}$$

As a result these two waves cancel out each other on the (010) disc, resulting in a vertical dark line B seen in this disc in Fig. 9.13. The absence line B is perpendicular to line A. Under the ZOLZ approximation ($l = 0$), the glide plane and the $2_1'$ screw axis produce the same dynamical extinction lines A and B. We shall call these lines A_2 and B_2 G–M lines, where the subscript 2 refers to the interaction occurring in two dimensions.

When interactions due to HOLZ reflections for which $l \neq 0$ are involved, the glide plane produces only G–M lines of the type A, and the $2_1'$ screw axis produces only the G–M lines of the type B. This can be demonstrated by using eqns (9.33) and (9.34). For the $2_1'$ screw axis we have

$$F(hkl) = (-1)^k F(\overline{h}k\overline{l}), \tag{9.42}$$

and for the glide plane we find

$$F(hkl) = (-1)^k F(\overline{h}kl). \tag{9.43}$$

In the case of the glide plane the two waves following paths 'a' and 'b' have opposite signs and form the G–M lines A, whereas the G–M lines B are not produced because eqn (9.42) holds only for the 2_1 screw axis. In the case of a $2_1'$ screw axis only the waves following the paths 'a' and 'c' have opposite signs

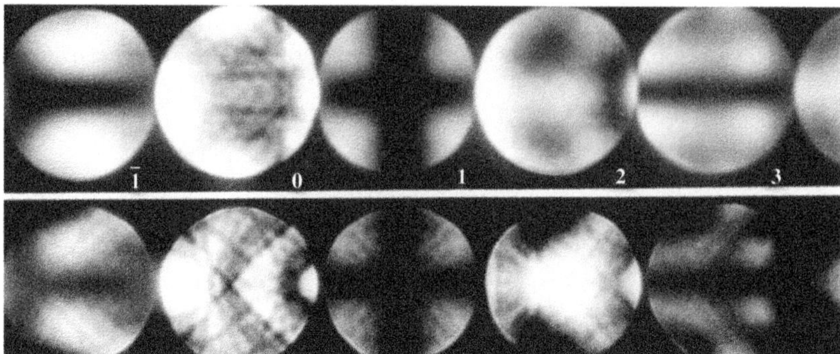

FIG. 9.14. CBED patterns obtained from thin (top) and thick (bottom) areas of a (001) FeS$_2$ film. From (Tanaka and Terauchi, 1985).

(see eqn (9.42)), forming the G–M lines of the type B only. It is appropriate to call these G–M lines A_3 and B_3 G–M lines, since they are associated with three-dimensional interactions. It was also predicted (Gjønnes and Moodie, 1965) that a horizontal glide plane g' gives a dark spot at the crossing point between lines A and B in Fig. 9.13. This spot results from the destructive interference between waves following the paths 'b' and 'c'.

9.7.2 *Identification of G–M lines*

The horizontal screw axis $2_1'$ and the vertical glide plane g can be identified in the following way. If the three-dimensional G–M lines A_3 and B_3 are observed separately from the two-dimensional G–M lines A_2 and B_2, it is obvious that the observed G–M lines are associated either with a screw axis or with a glide plane. However, it is not easy to observe the three-dimensional G–M lines since broad two-dimensional G–M lines are also present in the diffraction pattern. The presence of the three-dimensional G–M lines can be established by inspecting the symmetries of fine HOLZ lines in forbidden reflections instead of the direct observation of the G–M lines. In other words if the HOLZ lines form the three-dimensional G–M lines, the HOLZ lines are symmetric with respect to the two-dimensional G–M lines, and vice versa. In the case where the HOLZ lines are symmetric about A_2 G–M lines the specimen crystal has a glide plane. In the case where the HOLZ lines are symmetric with respect to B_2 G–M lines, a 2_1 screw axis is present.

Figure 9.14 shows CBED patterns taken at the 010 Bragg orientation from thin (top) and thick (bottom) areas of an FeS$_2$ sample. The space group of this crystal is $P \dfrac{2_1}{a} \bar{3}$, and the electron beam is incident near the [001] zone axis. In the odd-order discs of Fig. 9.14 (top) we can see broad G–M lines resulting from the two-dimensional interactions. On the other hand the fine HOLZ lines due to the three-dimensional effects are clearly seen in the bottom pattern of Fig. 9.14. The HOLZ lines are symmetric with respect to both the A_2 and B_2 G–M lines in the 010 disc. This fact proves the presence of the A_3 and B_3 G–M lines, or the presence of a glide plane and a $2'_1$ screw axis, and this agrees with the space group of FeS_2.

9.7.3 *Space group determination*

Tanaka et al. (Tanaka et al., 1983b) carried out a detailed analysis of how the G–M line rules originally given by Gjønnes and Moodie (Gjønnes and Moodie, 1965) are modified in the case where several crystal symmetry elements giving rise to G–M lines coexist, and where the symmetry elements are combined with various lattice types. Using these results the G–M lines A_2, A_3, B_2 and B_3 expected for various crystal orientations for all the space groups have been tabulated (Tanaka and Terauchi, 1985). By consulting the tables 181 space groups out of 230 can be identified directly by using G–M lines.

We now consider an example illustrating the procedure for determination of the space group of a crystal. Consider the magnetic superconductor $ErRh_4B_4$. The room temperature phase of $ErRh_4B_4$ belongs to the tetragonal crystal system and has lattice spacings $a = 0.5292$ nm and $c = 0.7375$ nm. The point group of this material has been identified as 4/mmm following the procedure discussed above. Figure 9.15a shows the CBED pattern taken for the [100] zone axis incidence at the 030 Bragg setting. The A_2 and B_2 G–M lines are seen in the 030 disc. The three-dimensional G–M lines are not observed here since the HOLZ interactions are weak. By consulting the [100] column of Table 9.3 (this column is taken from Table 10 of Tanaka et al. (Tanaka et al., 1983b)), we find that there are four space groups that are compatible with the point group 4/mmm and the A_2 and B_2 G–M lines appearing in Fig. 9.15a. These are the space groups numbered 129, 130, 137, and 138. Figure 9.15b shows a CBED pattern taken with a $[0kl]$ incidence at the 300 Bragg setting. The A_2 and B_2 G–M lines are seen in the 300 disc. However, no A-type G–M lines are seen in the 013 disc. By consulting the $[0kl]$ column of Table 9.3 only two space groups with numbers 129 and 137 remain on our list. Finally we examine the CBED pattern taken for the $[hhl]$ incidence (Fig. 9.15c). In the $11\overline{3}$ disc we observe an A_3 G–M line. By consulting the last column of Table 9.3 we see that the space group of the crystal is number 137.

9.8 Automated identification of CBED pattern symmetry

The CBED technique uses the dynamical nature of electron diffraction and should therefore be able to identify all the point groups (Buxton et al., 1976, Tanaka, 1989) as well as the majority of space groups (Gjønnes and Moodie, 1965, Steeds, 1983, Tanaka et al., 1983b). The number of structures determined independently by electron diffraction techniques is, however, by no means comparable with the number solved by X-ray diffraction techniques. This slow progress is largely due to the fact that in electron diffraction automated procedures are not widely used for identifying the symmetry elements seen in CBED patterns. In this section, we will describe a procedure based on a genetic algorithm and show

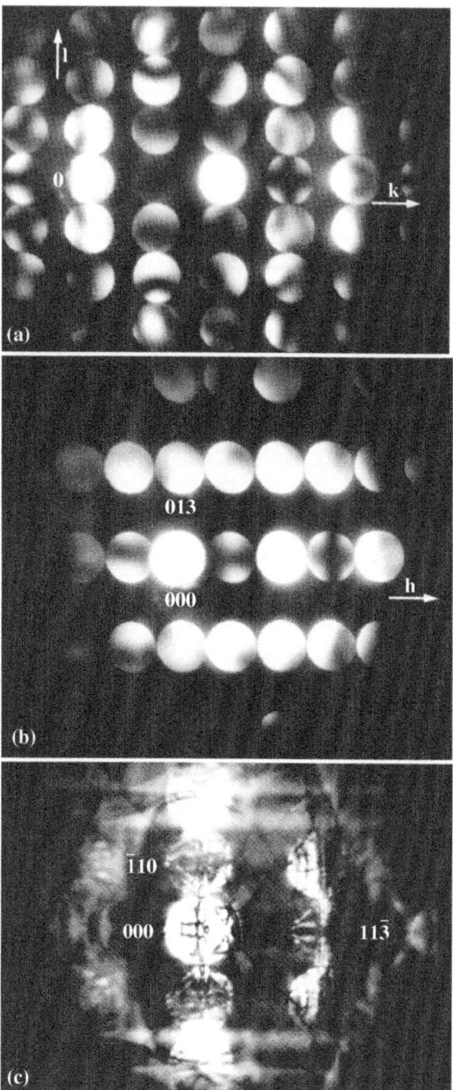

FIG. 9.15. (a) [100], (b) [0kl], and (c) [hhl] zone axis CBED patterns obtained from an ErRh$_4$B$_4$ sample. From (Tanaka and Terauchi, 1985).

that this procedure may be used efficiently for automated identification of the symmetry of CBED patterns and for the point group determination.

Although the goal of the automated identification of symmetry in CBED patterns may be formulated in a straightforward way as an optimization problem (Peng and Li, 1992, Vincent and Walsh, 1997), the development of a practical implementation of the optimization scheme has taken nearly a decade. Peng and

Table 9.3 The types of G–M line exhibited by four space groups belonging to point group 4/mmm

Space	Group	[100]	[0kl]	[hhl]
129	P4/nmm	A_2 B_2	A_2 B_2	
	P4/n2$_1$/m2/m	A_3 B_3	B_3	
130		A_2 B_2	A_2 B_2	
	P4/ncc	A_3 B_3	A_3	A_2 B_2
	P4/n2$_1$/c2/c		A_2 B_2	A_3
		A_3	B_3	
137	P4$_2$/nmc	A_2 B_2	A_2 B_2	A_2 B_2
	P4$_2$/n2$_1$/m2/c	A_3 B_3	B_3	A_3
138		A_2 B_2	A_2 B_2	
	P4$_2$/ncm	A_3 B_3	A_3	
	P4$_2$/n2$_1$/c2/m		A_2 B_2	
		A_3	B_3	

Li (Peng and Li, 1992) proposed that by defining a suitable error function it should be possible to identify automatically the presence of symmetry elements in a CBED pattern. Although they successfully found rotation axes in simulated CBED patterns, a direct application of this procedure to experimental CBED patterns failed owing to the poor quality of the digital CBED patterns available at that time. The situation greatly improved during the early 1990's when a new generation of field-emission gun (FEG) transmission electron microscopes became available. Using a Gatan Imaging Filter (GIF) system high-quality digital CBED patterns may be obtained from a subnanometre region, although in most applications such a small probe size is not required. Initial applications of the original scheme (Peng and Li, 1992) for automated identification of symmetry elements proved to be successful (Zou, 1996), but it was soon found that the error functions involved in the minimization scheme have many local maxima (see Fig. 9.16), and standard minimization procedures, such as the down-hill simplex method (Press et al., 1986), failed most of the time unless a fairly good guess about the position of the centre of symmetry or rotation axes is made before the procedure is applied. Later some global optimization procedures were implemented, and in this section we will briefly introduce these procedures and show that a genetic algorithm may be used for the efficient automated identification of symmetry elements in the CBED patterns (Hu et al., 2000b).

9.8.1 *Genetic algorithm — basic concepts and implementation*

In this section we will show how the genetic algorithm may be used for searching the symmetry exhibited by a CBED pattern. The genetic algorithm is based on the concept of Darwinian evolution and is an effective tool for solving complex global optimization problems (Holland, 1975, Jong, 1975). The algorithm is a stochastic search method consisting of the following steps that are similar

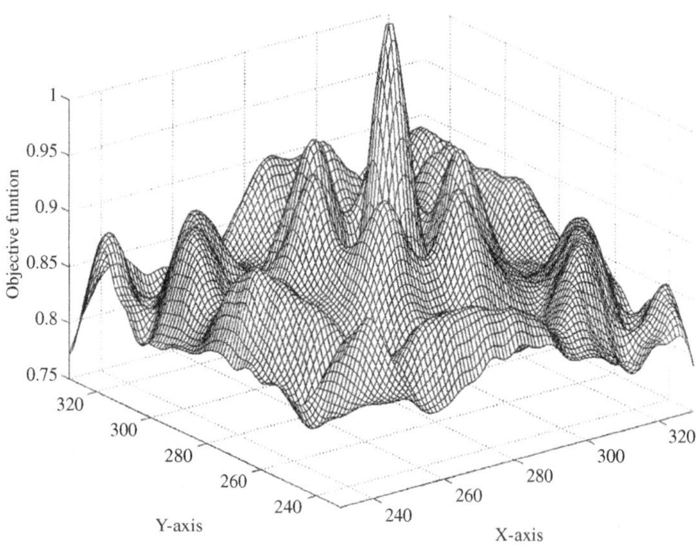

FIG. 9.16. Two-dimensional representation of the objective function $\mathcal{O}(x, y)$ for the experimental $\langle 111 \rangle$ zone axis CBED pattern shown in Fig. 9.17a and a three-fold rotation axis. From (Hu et al., 2000b).

to natural evolution: *selection, crossover, and mutation.* In an implementation of the algorithm a subset of possible solutions is first selected to simulate a population of individuals, where this population is maintained throughout the evolution. In each generation or iteration every individual is evaluated with respect to its fitness (related to our objective function value). The selection rule is that the higher the fitness of an individual (or a trial solution), the higher is the probability that the individual would be selected for survival and to produce off-spring. The selected individuals are then subjected to crossover to generate new individuals, and mutation is applied to the new individuals to produce a new generation. The 'schema theorem' or the 'building block hypothesis' (Holland, 1975) predicts that the fitness of individuals will improve in the course of evolution or iteration.

9.8.1.1 *Representation of the population* In classic genetic algorithms individuals are coded into binary strings. In our scheme every component of a trial solution is coded into a binary string, and all components are combined into a long string. The length of each component string depends on the range of possible solutions. Assuming for example that one component of a trial solution is x, with x being in the range (0-511), we represent it by a 9 bit binary string. For continuous optimization problems it was shown that a real floating-point number representation is sometimes more convenient (Masuda et al., 1993, Djurisic, 1998). In this representation every individual is represented by a real floating-point number. If one component of a trial solution is x, its representation is also x. For the quantitative evaluation of a CBED pattern it is convenient to use the real floating-point number representation, and typically 100 individuals or trial solutions were used to simulate the population. The first generation of individuals were generated randomly and distributed throughout the whole solution space spanned by the free parameters.

9.8.1.2 *Objective function and fitness*   The fitness of an individual is the only piece of knowledge that the genetic algorithm needs to solve the optimization problem. In this section we will first define an objective function $\mathcal{O}$ measuring the similarity between the original CBED pattern and the operated pattern related to the original one by a symmetry operation $\mathcal{G}$. In general, several free parameters are needed to define a symmetry operation. For example, to specify a two-dimensional rotation axis we need three free parameters, i.e. the coordinates of the rotation axis (x, y) and the angle of rotation θ associated with this axis. In the case of CBED patterns our goals are two-fold: (1) to test whether a particular symmetry exists in a CBED pattern, and if the symmetry does exist then (2) to find the free parameters which are needed to define uniquely the symmetry operation. If we define the objective function $\mathcal{O}$ to be a measure of the similarity between the original and the symmetry operated CBED pattern, we will then be able to define the optimization problem. Consider a symmetry-operation defined by n free parameters. If the symmetry is truly a symmetry of the CBED pattern, a prominent global maximum is expected to occur in the n-dimensional space spanned by the free parameters, where its location in the parameter space will define the symmetry operation. If no prominent global maximum exists, the symmetry operation is rejected. In a genetic algorithm the fitness level does not in general represent the objective function itself, but is normally related to it through a suitable transformation. Ultimately it is the fitness that a genetic algorithm aims to maximize. There are two main reasons why the value of the objective function is not used as a measure of fitness. The first reason is to avoid premature convergence, and the second is to separate more effectively individuals characterized by similar values of the objective function.

Consider the application of the method to CBED patterns. Each CBED pattern may be described by a two-dimensional array $f(i, j)$, where $(i, j) = \mathbf{X}$ are the coordinates of pixels forming the pattern and $f(i, j)$ is the intensity in a given pixel. The level of similarity between any two patterns $f(\mathbf{X})$ and $g(\mathbf{X})$ may in principle be described by a correlation function defined by

$$corr(\boldsymbol{\xi}) = \int g(\mathbf{X})h(\mathbf{X} + \boldsymbol{\xi})d\mathbf{X}, \tag{9.44}$$

or, in the discrete form,

$$corr(k, l) = \sum_{i,j} g(i, j)h(i + k, j + l).$$

For two identical patterns shifted with respect to each other by $\mathbf{X}_0 = (k_0, l_0)$, the correlation function $corr(k, l)$ will show a maximum around $\mathbf{X}_0$. The relative shift between the two similar patterns may therefore be readily found by calculating the two-dimensional map of $corr(\boldsymbol{\xi})$ and searching for the maximum in this map. Since this two-dimensional correlation map can be calculated using the fast Fourier transform (FFT), the correlation function method provides a conceptually simple and numerically efficient way of identifying the relative shift of two patterns.

In the problem of determination of a crystal point group, we are mainly concerned with finding the point symmetry elements rather than relative shifts between patterns. An appropriate function describing the similarity between two patterns is the inner product of the two patterns

$$(\mathbf{f}, \mathbf{g}) = \int f(\mathbf{X})g(\mathbf{X})d\mathbf{X}, \qquad (9.45)$$

where the pattern $f(\mathbf{X})$ is regarded as a vector $\mathbf{f}$, and the inner product between the two vectors is defined as above. Assuming that the pattern $g(\mathbf{X})$ is obtained by performing a translation $-\boldsymbol{\xi}$ on the original pattern $f(\mathbf{X})$, we then have $g(\mathbf{X}) = f(\mathbf{X}+\boldsymbol{\xi})$. Substitution of $g(\mathbf{X}) = f(\mathbf{X}+\boldsymbol{\xi})$ into the above expression for the inner product between $\mathbf{g}$ and $\mathbf{f}$ gives the usual expression for the correlation function (9.44). The inner product (9.45) can therefore be regarded as a generalized correlation function. By introducing a general symmetry operation $\mathcal{G}$ and applying it to $f(\mathbf{X})$ we obtain

$$g(\mathbf{X}) = \mathcal{G}f(\mathbf{X}) = f(\mathcal{G}^{-1}\mathbf{X}).$$

In an explicit form the inner product (9.45) between the original and operated patterns can be written as

$$(\mathbf{f}, \mathbf{g}) = (\mathbf{f}, \mathcal{G}\mathbf{f}) = \int f(\mathbf{X})f(\mathcal{G}^{-1}\mathbf{X})d\mathbf{X}, \qquad (9.46)$$

and in what follows we will take this inner product as our objective function for the symmetry operation $\mathcal{G}$. It should be noted, however, that the above definition for the inner product depends on the absolute intensities of the two patterns. Since here we are mainly interested in finding the similarity between the patterns rather than their absolute intensities, a normalized objective function $\mathcal{O}$ is more appropriate and may be defined as

$$\mathcal{O} = \frac{(\mathbf{f}, \mathbf{g})}{\sqrt{(\mathbf{f}, \mathbf{f})(\mathbf{g}, \mathbf{g})}} = \frac{(\mathbf{f}, \mathcal{G}\mathbf{f})}{\sqrt{(\mathbf{f}, \mathbf{f})(\mathcal{G}\mathbf{f}, \mathcal{G}\mathbf{f})}} = \frac{\sum_{i,j} f(i, j)g(i, j)}{\sqrt{[\sum_{ij} f^2(i, j)][\sum_{k,l} g^2(k, l)]}}. \qquad (9.47)$$

To understand the meaning of this objective function $\mathcal{O}$ we regard it as $\cos\alpha$, where α is the angle between the two vectors $\mathbf{f}$ and $\mathbf{g}$. Since the pattern intensities $f(i, j)$ and $g(i, j)$ take only positive values for all pixels, the objective function

$\mathcal{O}$ takes only positive values between 0 and 1. For the automated identification of symmetry in CBED patterns it is appropriate to use a positive fitness value in the range from 0 to 100. This can be achieved by using a linear transformation

$$f(\zeta) = a\mathcal{O}(\zeta) + b, \tag{9.48}$$

where ζ is a point in the parameter space, and a and b are two constants. Typically we use $a = 100$ and $b = -30$, but in all cases we keep the fitness values greater than zero.

9.8.1.3 *Selection*   The probability of an individual or a trial solution ζ_i being selected to produce offspring in the next generation is defined by

$$p(i) = \frac{f(\zeta_i)}{\sum_i f(\zeta_i)},$$

where $f(\zeta_i)$ is the fitness value of i-th individual. It may be readily verified that $\sum_j p(j) = 1$ and that $p(j)$ takes positive values in the range $(0 \rightarrow 1)$. For any individual, a random number r between 0 and 1 is generated. The individual is selected if $r \leq p(i)$, and it is rejected otherwise. In addition an elitist selection scheme is employed. In this scheme P_e percent of the new population is produced directly from individuals with the highest fitness values in the current generation without participating in the crossover. Typically we found that $P_e \approx 10\%$ gives good results. It should also be noted that the number of individuals to take part in the crossover should be even, so that pairs may be formed between these individuals.

9.8.1.4 *Crossover*   The operation of the crossover is the most important genetic operation, which is used to exchange subsets of elements between the two parental individuals. Various crossover operators have been proposed, e.g. the one-point crossover and the two-point crossover schemes. These are given in Table 9.4. In the one-point crossover scheme the last four digits (given in bold face and separated from the remaining part of the string by one point) of the parent strings are exchanged to produce offspring, while in the two-point crossover scheme the four digits in the middle (given in bold face and separated from the remainder of the string by two gaps) are exchanged.

Table 9.4 Illustration of the one- and two-point crossover operations

	Parents	Offspring
One-point	11010 **0101**	11010 **0110**
crossover	10111 **0110**	10111 **0101**
Two-point	110 **1001** 01	110 **1101** 01
crossover	101 **1101** 10	101 **1001** 10

In this work we used the real floating-point number representation and the linear crossover scheme. For any two individuals in the parent generation, ζ_1 and

ζ_2, offspring ζ_1' and ζ_2' are generated according to the following linear transformation:

$$\zeta_1' = \zeta_1 r + \zeta_2(1 - r)$$
$$\zeta_2' = \zeta_1(1 - r) + \zeta_2 r$$

where r is a random number taken from the range $(0,1)$.

9.8.1.5 *Mutation* The main role of the operation of mutation is to escape from being trapped in local extremes and to guarantee convergence to the global extreme. A random selection procedure is applied to all individuals, and among them P_m % are selected for mutation. In our program a value of 5% for P_m is used. For a selected individual a perturbation about its position in the solution space is applied. For the coordinates (x, y) of the rotation axis a new position is selected randomly within a circle (within a radius of about 5 pixels) around (x, y). To summarize the evolution from generation to generation, individuals are firstly divided into two groups, i.e. the elite group and the group formed by the remaining individuals which are subjected to the operation of crossover. The new individuals, after the operation of crossover but before mutation, consist therefore of two groups: (1) the elite group which is about 10% of the population (P_e=10%) and (2) the crossover group. The two groups of individuals are then subjected to a selection process in which about 5% are selected and subjected to the mutation operation. The individuals after the operation of mutation form a new generation.

9.8.1.6 *Termination* We found that a global maximum can usually be found after approximately 30 iterations. To further guarantee the convergence, a total of 40 iterations are defined as the termination condition in our implementation. A typical search for the symmetry elements in a CBED pattern takes approximately 3 minutes on a Pentium III personal computer.

9.8.2 *Identification of CBED pattern symmetry*

Figure 9.16 shows a two-dimensional map of the objective function $\mathcal{O}(\mathbf{X})$ for a three-fold rotation axis and the Si$\langle 111 \rangle$ zone axis experimental CBED pattern shown in Fig. 9.17a. The boundary of the map corresponds to the white box shown in the bright-field (BF) disc of Fig. 9.17a. This map shows that a prominent global maximum exists among many local maxima surrounding the global maximum, suggesting that a three-fold rotation symmetry is indeed present in the BF disc of Fig. 9.17a, and that the standard procedures, such as the downhill simplex methods, are not applicable for finding the global maximum. The coordinate system that we used to index the pixels is such that the x-axis points from left to right horizontally, and the y-axis points from top to bottom vertically. The $(0, 0)$ pixel is therefore in the top left corner of the pattern, and for a $512{\times}512$ pattern the centre of the pattern is at $(255, 255)$.

9.8.2.1 *Rotation axes* We first consider the identification of the rotation symmetry operation. For a pure rotation axis we need to define the position of the

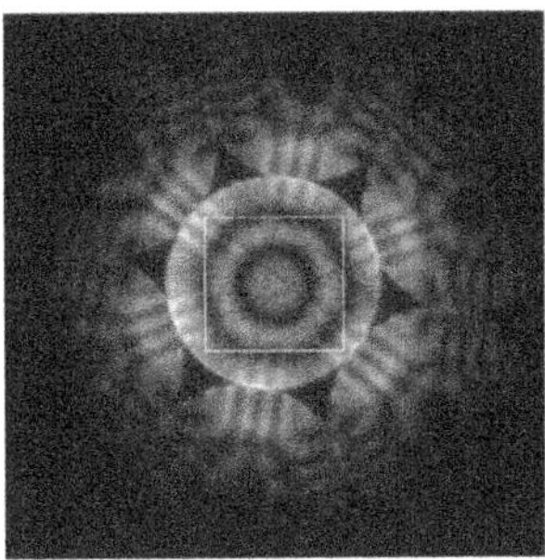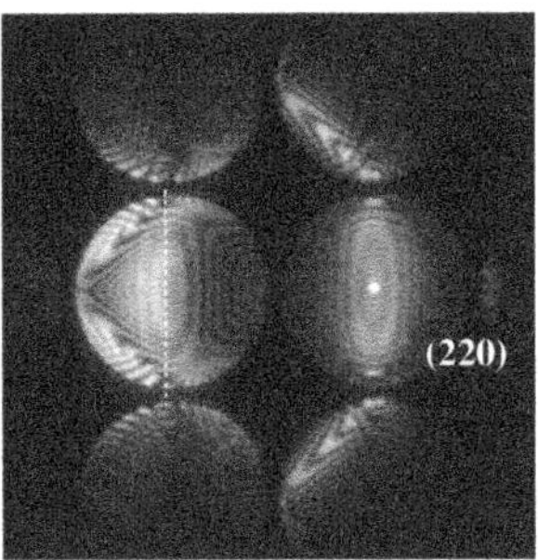

FIG. 9.17. Experimental CBED patterns obtained from a single crystal of silicon. Shown in (a, left) is a $\langle 111 \rangle$ zone axis pattern, in (b, middle) is a $\langle 001 \rangle$ zone axis pattern, and in (c, right) is a (220) DF pattern near the $\langle 001 \rangle$ zone axis.

axis as well as the rotation angle associated with the axis. Conventional crystallography tells us that the rotation angle takes only discrete values, i.e. $60°$, $90°$, $120°$ and $180°$. Among these angles, $180°$ and $120°$ appear most frequently. This is because if $90°$ or $60°$ rotations are the allowed symmetry operations, so are the $180°$ or $120°$ rotations. In searching for a rotation axis we start by restricting the angle to be either $120°$ or $180°$. Search is then performed in the two-dimensional space spanned by values of $\mathbf{x}$, the position of the axis, and in the angular space with only two possible values, i.e. $120°$ and $180°$. Once the position $\mathbf{x}_0$ for either a $120°$ or a $180°$ rotation axis is found, an angular plot of the objective function, i.e. an $\mathcal{O}(\theta)$ plot, is produced, from which other rotation axes may then be readily identified.

Figure 9.17a shows a $\langle 111 \rangle$ zone axis CBED pattern. Figure 9.17b is a $\langle 001 \rangle$ zone axis CBED pattern and Fig. 9.17c is a 220 dark-field (DF) pattern taken near the $\langle 001 \rangle$ zone axis. By restricting the solutions to be within the range of pixel numbers $100 < x, y < 400$ (pixels) for a 512×512 CBED pattern (all CBED patterns shown in Fig. 9.17 are 512×512), and θ to be either $120°$ or $180°$, a global maximum is found at $(x, y) = (279.4, 281.5)$ for Fig. 9.17a and the rotation angle of $120°$. By fixing the centre of rotation at this position, an angular plot of $\mathcal{O}(\theta)$ is evaluated and plotted in Fig. 9.18a. Two prominent maxima may be easily identified, one at $120°$ and the other at $240°$ respectively, both corresponding to a three-fold rotation axis. Three less pronounced peaks are also visible, at $60°$, $180°$, and $300°$ respectively. These peaks are consistent with a six-fold rotation axis, which is expected for an ideal two-dimensional crystal. Crystal potential variation along the $\langle 111 \rangle$ zone axis introduces a finite deviation from an ideal two-dimensional periodic potential. This deviation is usually small, and it manifests itself mainly in the form of fine dark lines in the BF disc. Figure 9.18a demonstrates that our objective function $\mathcal{O}$ is indeed a good indicator of the true crystal symmetry, i.e. it can pick up the symmetry associated with the weak dark lines in the BF disc.

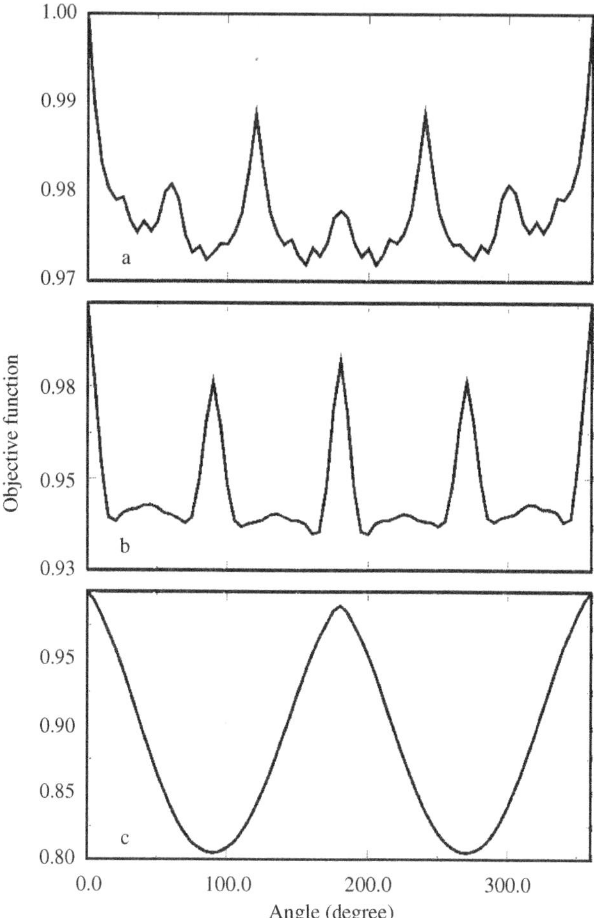

FIG. 9.18. Angular plots $\mathcal{O}(\theta)$ for rotation about a centre lying in (a) the BF disc of Fig. 9.17a, (b) the BF disc of Fig. 9.17b, and (c) the (220) dark field disc of Fig. 9.17c. From (Hu et al., 2000b).

In the case shown in Fig. 9.17b we find a global maximum at (219.6, 244.3) for a two-fold rotation axis. The corresponding $\mathcal{O}(\theta)$ plot is shown in Fig. 9.18b. Three prominent peaks can be readily identified at 90°, 180° and 270° respectively. These peaks correspond to a four-fold rotation axis, and clearly no other rotation symmetry exists in Fig. 9.17b.

Figure 9.17c may also be called a $+G$ pattern. The $+G$ disc (i.e. the 220 disc) may sometimes show rotational symmetry. The procedure for identifying symmetry in this DF disc is the same as that for a BF disc, and a global maximum is found at (312.5, 276.0) (as indicated in the figure by a white dot). The corresponding angular plot $\mathcal{O}(\theta)$ is shown in Fig. 9.18c, revealing clearly a two-fold

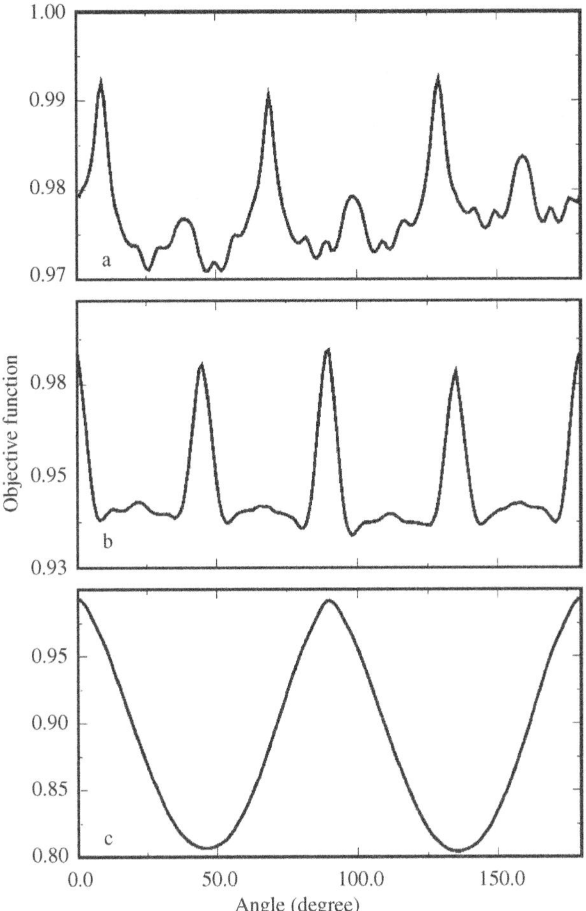

FIG. 9.19. Angular plots $\mathcal{O}(\theta)$ for a mirror operation for (a) Fig. 9.17a, (b) Fig. 9.17b, and (c) Fig. 9.17c respectively. From (Hu et al., 2000b).

rotation axis at $180°$.

9.8.2.2 *Mirror planes* For the identification of mirror planes we should consider two distinct alternatives. In the first case the mirror symmetry coexists with a rotation symmetry. More than one mirror plane will then be generated by rotations, and the number of mirror planes will then be equal to the order of the rotation axis or to half of it. For example, if a four-fold rotation symmetry exists in a CBED pattern, the number of mirror planes corresponding to that pattern may be equal to either four or two, where the first case corresponds to the diffraction group 4mm and the second case to the diffraction group 4m. Since the mirror planes pass through the rotation axis (assuming that it is a n-fold axis), the mirror reflection operation may then be specified uniquely by a single angle θ, and a one-dimensional search in the angular range $(0, \pi/n)$ will then be sufficient for identifying the mirror plane. Figure 9.19 shows three plots of $\mathcal{O}(\theta)$ for mirror operation. The three figures 9.19a, 9.19b and 9.19c correspond to the mirror planes passing through the three-fold rotation axis of Fig. 9.17a (in the BF disc), the four-fold rotation axis of Fig. 9.17b (in the BF disc), and the two-fold rotation axis of Fig. 9.17c (in the 220 DF disc) respectively. Three prominent peaks can be identified in Fig. 9.19a, corresponding to the presence of three mirror planes in Fig. 9.17a. The diffraction group is therefore 3m. Four prominent peaks are clearly visible in Fig. 9.19b, corresponding to four mirror planes in Fig. 9.17b and the diffraction group 4mm. On the other hand, the two prominent peaks seen in Fig. 9.19c correspond to the two perpendicular mirror planes in the (220) BF disc of Fig. 9.17c and the diffraction group 2mm.

The identification of a single mirror plane is more difficult than that of multiple mirror planes coexisting with the rotation axis. This is because a mirror plane needs at least three free parameters to specify it, i.e. a point (x, y) in the mirror plane and an angle θ that the mirror plane makes with, say, the horizontal x axis. In principle this does not present a conceptual problem since a search can be performed in the three-dimensional parameter space (x, y, θ). The procedure may, however, be greatly simplified if we note that the mirror must either be parallel to a predefined line (e.g. the white dotted line in the BF disc of Fig. 9.17c) or intersect this line. By placing the point (x, y) on this line the three-dimensional optimization problem can be reduced to a two-dimensional one. For example, a search made for the BF disc shown in Fig. 9.17c found a mirror plane defined by $\mathbf{X} = (276.0, \ 276.0)$, and the angle $\theta = 179°$ with respect to the horizontal x axis. A Windows-based program package SYMCBED has been developed based on the procedure outlined above, and it can in principle be applied to the determination of all the point groups (Hu et al., 2000a).

9.9 Summary

We have shown that for a perfect rectangular crystal bounded by parallel surfaces there is a maximum of 10 symmetry elements, of which six (1,2,3,4,6, and m) are type I or vertical symmetry elements leaving the z axis unchanged and the remaining four (i,m',2', and $\bar{4}$) are type II or horizontal symmetry elements reversing the direction of the z axis.

While a combination of the type I symmetry elements results in the 10 familiar two-dimensional point groups, a combination of both the type I and type II symmetry elements results in the total of 31 groups that are isomorphic to the 31 point groups of diperiodic plane figures and to the 31 Shubnikov point groups of coloured plane figures. These symmetry operations impose certain conditions on the symmetry of intensity distributions in the CBED discs, giving rise to 31 diffraction groups describing the symmetry of electron diffraction patterns.

Using relations between the symmetries of CBED patterns, diffraction groups, and crystal point groups, we have shown how all the crystal point groups may be unambiguously determined by analysing one or a few CBED patterns. An automated procedure and a Windows-based program can identify symmetry elements of a CBED pattern and automatically determine the point group of the crystal.

All the diffraction symmetry relations established in this chapter are based on the use of the reciprocity principle. This means that the treatment given in this chapter refers only to diffraction patterns formed by elastically scattered electrons. In practice only energy-filtered CBED patterns or very thin specimens should be used when applying the procedure described above to the determination of symmetry of crystalline samples.

10

PERTURBATION METHODS AND TENSOR THEORY

10.1 Introduction

A very important class of theoretical methods widely used in high-energy electron diffraction is based on the use of perturbation approaches. In the early days of the development of electron diffraction theory, high-speed computers were not as readily accessible as today. Much of the early effort was therefore focussed on finding analytical solutions of few-beam cases, while errors associated with weaker beams and diffuse and inelastic scattering were treated using perturbation methods. The first perturbation treatment of electron diffraction aiming at the evaluation of the effect of weaker beams on the strong beams was developed by Bethe (Bethe, 1928). His method is still widely used in quantitative electron diffraction calculations and is usually referred to as the 'Bethe potentials' method. Since the early 1960s perturbation methods of various forms have been formulated and used for taking account of effects like inelastic diffuse scattering (Hashimoto et al., 1962, Hirsch et al., 1977), and HOLZ diffraction (Bird, 1989). They have also been successfully applied to crystal structure determination (Vincent et al., 1984) and crystal structure factor refinement (Zuo, 1991, Bird and Saunders, 1992). A formal mathematical expression for the first-order partial derivatives of the scattering matrix has been obtained by Speer et al. (Speer et al., 1990). In this chapter we will briefly review perturbation methods, presenting the results in a particularly simple tensor form and outlining possible applications of tensor theory to the retrieval of information about crystal structure from experimental observations.

Perturbation theory assumes that the crystal potential may be represented by a sum of two terms

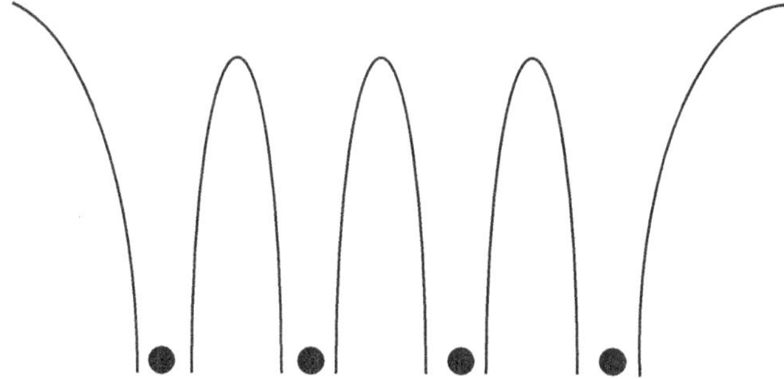

FIG. 10.1. Schematic diagram showing variation of the potential along a line of atoms.

350

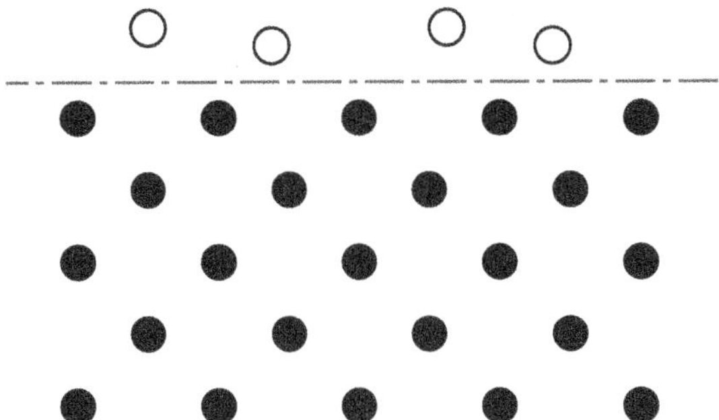

FIG. 10.2. Schematic diagram showing a reconstructed surface layer and the bulk crystal beneath the surface.

$$V(\mathbf{r}) = V_0(\mathbf{r}) + \Delta V(\mathbf{r}), \qquad (10.1)$$

where $V_0(\mathbf{r})$ is the potential for which the solution of the Schrödinger equation is known, and $\Delta V(\mathbf{r})$ is the difference between the real potential and $V_0(\mathbf{r})$. The key point in any perturbation scheme is to choose $V_0(\mathbf{r})$ in such a way that $\Delta V(\mathbf{r})$ introduces only a weak perturbation in the motion of the electron in the potential $V_0(\mathbf{r})$. Figures 10.1–10.4 give examples of some of the systems of practical importance where we can make such a choice.

A very important application of electron diffraction concerns the determination of the valence electron redistribution occurring when neutral atoms are brought together to form a solid. Figure 10.1 shows the one-dimensional distribution of a potential formed by placing neutral atoms together, and this potential represents an obvious choice for $V_0(\mathbf{r})$. Charge redistribution associated with the formation of chemical bonds in a crystal introduces only a weak perturbation $\Delta V(\mathbf{r})$ of the total potential, and this perturbation is typically smaller than 1% of $V_0(\mathbf{r})$, see e.g. (Spence and Zuo, 1992, Saunders et al., 1995, Høier et al., 1993, Deininger et al., 1994). While scattering by $V_0(\mathbf{r})$ may be very strong and should be dealt with by using the full dynamical theory, additional scattering by $\Delta V(\mathbf{r})$ is sufficiently weak so that the perturbation treatment can be applied. We will show later that second-order perturbation theory works well in this case, and provides an excellent basis for solving the inverse problem for CBED patterns (Peng and Zuo, 1995).

We now consider the problem of the determination of coordinates of atoms in a crystal. We may identify at least three important areas of interest. The first area is concerned with scattering by a surface layer shown schematically in Fig. 10.2. Atoms in the surface layer usually have positions that differ from those of the bulk crystal (Zangwill, 1988), but nevertheless we may take $V_0(\mathbf{r})$ to be that of a bulk terminated surface. Since the positions of surface atoms usually differ from those of bulk atoms by only a small amount, and the thickness of the surface layer is typically less than 10 Å, $\Delta V(\mathbf{r})$ resulting from surface reconstruction or relaxation is usually weak. Perturbation theory may then be used for treating

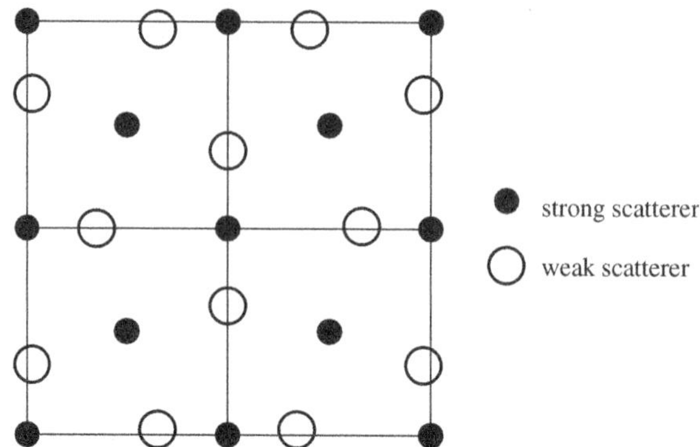

FIG. 10.3. Schematic diagram showing a crystal composed of heavy and light atoms.

scattering by the potential $\Delta V(\mathbf{r})$ localized near the surface, while a full dynamical treatment is needed for dealing with scattering by the reference structure, i.e. the bulk terminated surface and the bulk part of the sample beneath the surface (Peng and Dudarev, 1993a, Peng and Dudarev, 1993b).

The second area concerns systems composed of heavy and light atoms. Using techniques such as CBED and HREM (Spence, 1988) we may determine the space and point groups and the lattice type of the crystal (see Chapter 9 and (Gjønnes and Moodie, 1965, Buxton et al., 1976, Tanaka et al., 1983a)). In this case we may place heavy atoms at the lattice sites and assign arbitrary positions to the light atoms to obtain $V_0(\mathbf{r})$ for the system (Fig. 10.3). The difference $\Delta V(\mathbf{r})$ now results only from the inaccuracy in the assumed atomic coordinates of the light atoms, and in many cases it is reasonable to treat $\Delta V(\mathbf{r})$ as a perturbation (Peng, 1995).

The third area concerns the refinement of an almost known structure. A reference structure may be obtained using, say, HREM or a combination of HREM and direct methods (Li, 1994). Errors resulting from multiple scattering and finite crystal thickness may then be treated perturbatively.

No crystal is perfect. There are various kinds of defects in a real crystal. Figure 10.4 is a schematic diagram showing several atoms that are displaced from their ideal positions because of the presence of a point defect. For this type of system we may define $V_0(\mathbf{r})$ as that due to the average structure and regard the difference potential $\Delta V(\mathbf{r})$ as a perturbation.

The aim of developing perturbation methods is two-fold. Firstly, we need efficient procedures for evaluating the amplitudes of dynamically diffracted beams of electrons. For this reason in this chapter we will discuss only the first-order and second-order perturbation methods. Although in principle higher order perturbation theory may be readily developed following a similar procedure as outlined in standard textbooks on quantum mechanics, these higher order perturbation methods are numerically less efficient than *exact* numerical solutions and are therefore not useful for practical purposes. Secondly, perturbation expressions

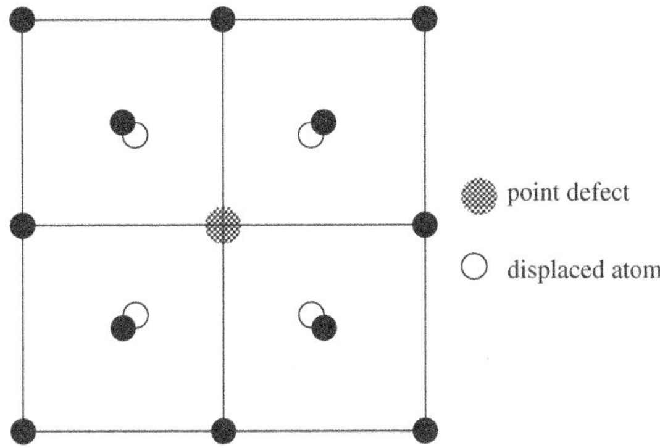

FIG. 10.4. Schematic diagram showing lattice distortion around a point defect.

are analytical. These analytical expressions may in principle be used for gaining deeper understanding of dynamical electron diffraction processes, and, as will be shown in this chapter, for crystal structure determination.

10.2 Perturbation treatment of a periodic structure

In the case of a periodic structure it is convenient to use the Bloch wave method. We start with the fundamental equations of dynamical electron diffraction or the Bethe equations

$$[K^2 - (\mathbf{k}^{(j)} + \mathbf{g})^2]C_g^{(j)} + \sum_{h \neq g} U_{gh}C_h^{(j)} = 0. \tag{10.2}$$

Here the Fourier components of the potential U_{gh} are related to V_{gh} via $U_{gh} = -2mV_{gh}/\hbar^2$, $\mathbf{K}$ is the electron wave vector derived from the incident electron wave vector $\mathbf{k}_0$ after correction for the inner potential U_0, i.e. $K^2 = k_0^2 + U_0$, and the j-th electron wave vector $\mathbf{k}^{(j)}$ inside the crystal is

$$\mathbf{k}^{(j)} = \mathbf{K} + \gamma^{(j)}\mathbf{n}, \tag{10.3}$$

where $\mathbf{n}$ is a unit vector inward normal to the surface. In THEED geometry we may safely neglect backscattering effects that are associated with the γ^2 term in eqn (10.2) (resulting from the substitution of (10.3) into this equation) and obtain a first-order eigensystem (Humphreys, 1979, Peng and Whelan, 1990b)

$$2K_z S_g C_g^{(j)} + \sum_{h \neq g} U_{gh}C_h^{(j)} = 2K_z(1 + g_z/K_z)\gamma^{(j)}C_g^{(j)}. \tag{10.4}$$

Here S_g is the usual excitation error for reflection $\mathbf{g}$ measured in the z direction. Neglecting a term in S_g^2, we find

$$2K_z S_g \approx K^2 - (\mathbf{K} + \mathbf{g})^2, \tag{10.5}$$

and the g-th diffracted beam amplitude is given by

$$\mathcal{F}_g = \sum_j \alpha^{(j)} C_g^{(j)} \exp(i\gamma^{(j)} z),$$

where $\alpha^{(j)}$ is the excitation amplitude of the j-th Bloch wave and $C_g^{(j)}$ is the corresponding eigenvector. The aim of the perturbation method is to find changes in the eigenvalues $\Delta\gamma^{(j)}$ and eigenvectors $\Delta C_g^{(j)}$ of the system when a small perturbation is applied. In what follows we consider a second-order perturbation treatment suitable for the case of a general non-Hermitian eigensystem.

10.2.1 *Bloch waves, left-hand, and right-hand eigenvectors*

We see that the form of eqn (10.4) is not particularly symmetric. For convenience we transform the eigenvector $C_g^{(j)}$ in the following way:

$$B_g^{(j)} = \left(1 + \frac{g_z}{K_z}\right)^{1/2} C_g^{(j)},$$

and call B_g the *right-hand eigenvector*. Substituting the above definition into eqn (10.4), we obtain a linear eigensystem

$$\frac{2K_z S_g B_g^{(j)}}{(1 + g_z/K_z)} + \sum_{h \neq g} \frac{U_{gh} B_h^{(j)}}{\sqrt{1 + g_z/K_z}\sqrt{1 + h_z/K_z}} = 2K_z \gamma^{(j)} B_g^{(j)}.$$

In the case where absorption effects are neglected the matrix involved in the above equation is Hermitian (indeed, if the potential $U(\mathbf{r})$ is a real function then $U_{gh} = U_{hg}^*$). In general it is easier to solve this more symmetric matrix equation than eqn (10.4). In a general diffraction case involving an absorbing crystal, the above eigensystem is a non-Hermitian one. In what follows we develop a general perturbation procedure suitable for a general non-Hermitian eigensystem.

In matrix notation the above equation may be rewritten as

$$(\mathbf{S} + \mathbf{U})\mathbf{B} = \mathbf{B}\mathbf{\Upsilon}, \tag{10.6}$$

where matrices $\mathbf{S}$ and $\mathbf{\Upsilon}$ are diagonal

$$\{\mathbf{S}\}_{gg} = \frac{2K_z S_g}{(1 + g_z/K_z)}, \quad \{\mathbf{\Upsilon}\}_{jj} = 2K_z \gamma^{(j)},$$

and the elements of matrices $\mathbf{U}$ and $\mathbf{B}$ are given by

$$\{\mathbf{U}\}_{gh} = \frac{U_{gh}}{\sqrt{1 + g_z/K_z}\sqrt{1 + h_z/K_z}},$$

and

$$\{\mathbf{B}\}_{gj} = B_g^{(j)}.$$

Similarly we can define a set of *left-hand eigenvectors* $\overline{B}_g$, satisfying

$$\frac{2K_z S_g \overline{B}_g^{(j)}}{(1 + g_z/K_z)} + \sum_{h \neq g} \frac{\overline{B}_h^{(j)} U_{hg}}{\sqrt{1 + g_z/K_z}\sqrt{1 + h_z/K_z}} = 2K_z \gamma^{(j)} \overline{B}_g^{(j)}.$$

In matrix notation we can rewrite the above equation as

$$\overline{\mathbf{B}}(\mathbf{S} + \mathbf{U}) = \boldsymbol{\Upsilon}\overline{\mathbf{B}}, \tag{10.7}$$

with

$$\{\overline{\mathbf{B}}\}_{jg} = \overline{B}_g^{(j)}.$$

By multiplying eqn (10.6) by $\mathbf{B}^{-1}$ first from the left and then from the right, we obtain

$$\mathbf{B}^{-1}(\mathbf{S} + \mathbf{U}) = \boldsymbol{\Upsilon}\mathbf{B}^{-1}. \tag{10.8}$$

Comparing eqn (10.8) with eqn (10.7), we see that $\overline{\mathbf{B}} = \mathbf{B}^{-1}$, leading to the following orthogonality relations: $\overline{\mathbf{B}} \cdot \mathbf{B} = \mathbf{I}$ and $\mathbf{B} \cdot \overline{\mathbf{B}} = \mathbf{I}$. In explicit form we have

$$\sum_j B_g^{(j)} \overline{B}_h^{(j)} = \delta_{hg}, \quad \sum_g B_g^{(j)} \overline{B}_g^{(j')} = \delta_{jj'}. \tag{10.9}$$

10.2.2 *Non-degenerate perturbation theory*

The assumption that $\Delta U(\mathbf{r})$ is small suggests that both the perturbed eigenfunctions and eigenvalues may be expanded in a power series in $\Delta U(\mathbf{r})$. Retaining terms up to the second order in $\Delta U(\mathbf{r})$, we have

$$\left.\begin{aligned}
\mathbf{U}(\mathbf{r}) &= \mathbf{U}_0(\mathbf{r}) + \lambda \Delta \mathbf{U}(\mathbf{r}) \\
\mathbf{B} &= \mathbf{B}_0 + \lambda \mathbf{B}_1 + \lambda^2 \mathbf{B}_2 \\
\boldsymbol{\Upsilon} &= \boldsymbol{\Upsilon}_0 + \lambda \boldsymbol{\Upsilon}_1 + \lambda^2 \boldsymbol{\Upsilon}_2
\end{aligned}\right\} \tag{10.10}$$

where the parameter λ has been chosen in such a way that the limiting form of eqn (10.6), corresponding to $\lambda \to 0$,

$$(\mathbf{S} + \mathbf{U}_0)\mathbf{B}_0 = \mathbf{B}_0 \boldsymbol{\Upsilon}_0, \tag{10.11}$$

is solvable. We will call this equation the reference structure equation. Substituting eqn (10.10) into eqn (10.6), we have

$$(\mathbf{S} + \mathbf{U}_0 + \lambda \Delta \mathbf{U})(\mathbf{B}_0 + \lambda \mathbf{B}_1 + \lambda^2 \mathbf{B}_2) = (\mathbf{B}_0 + \lambda \mathbf{B}_1 + \lambda^2 \mathbf{B}_2)(\boldsymbol{\Upsilon}_0 + \lambda \boldsymbol{\Upsilon}_1 + \lambda^2 \boldsymbol{\Upsilon}_2).$$

From the terms linear in λ we find

$$(\mathbf{S} + \mathbf{U}_0)\mathbf{B}_1 + \Delta \mathbf{U}\mathbf{B}_0 = \mathbf{B}_0 \boldsymbol{\Upsilon}_1 + \mathbf{B}_1 \boldsymbol{\Upsilon}_0, \tag{10.12}$$

. The terms proportional to λ^2 give

$$(\mathbf{S} + \mathbf{U}_0)\mathbf{B}_2 + \Delta \mathbf{U}\mathbf{B}_1 = \mathbf{B}_0 \boldsymbol{\Upsilon}_2 + \mathbf{B}_1 \boldsymbol{\Upsilon}_1 + \mathbf{B}_2 \boldsymbol{\Upsilon}_0. \tag{10.13}$$

We will show below how the quantities entering the above equations may be expressed in terms of the reference structure.

10.2.3 *First-order perturbation*

We first consider eqn (10.12). By introducing a matrix $\boldsymbol{\alpha}_1$ which we call the matrix of coefficients, we write

$$\mathbf{B}_1 = \mathbf{B}_0 \boldsymbol{\alpha}_1. \tag{10.14}$$

By using eqn (10.11) for the reference structure, we transform eqn (10.12) as follows:

$$\mathbf{B}_0(\boldsymbol{\Upsilon}_0\boldsymbol{\alpha}_1 - \boldsymbol{\alpha}_1\boldsymbol{\Upsilon}_0) + \Delta\mathbf{U}\mathbf{B}_0 = \mathbf{B}_0\boldsymbol{\Upsilon}_1.$$

By multiplying both side of the above equation by $\overline{\mathbf{B}}_0$, we find

$$(\overline{\mathbf{B}}_0\mathbf{B}_0)(\boldsymbol{\Upsilon}_0\boldsymbol{\alpha}_1 - \boldsymbol{\alpha}_1\boldsymbol{\Upsilon}_0) + \overline{\mathbf{B}}_0\Delta\mathbf{U}\mathbf{B}_0 = (\overline{\mathbf{B}}_0\mathbf{B}_0)\boldsymbol{\Upsilon}_1.$$

For the diagonal terms $j = j'$ we have

$$\{\boldsymbol{\Upsilon}_1\}_{jj} = \{\overline{\mathbf{B}}_0\Delta\mathbf{U}\mathbf{B}_0\}_{jj}.$$

Explicitly

$$\gamma_1^{(j)} = \frac{1}{2K_z}U^{(jj)}, \tag{10.15}$$

where

$$U^{(jj')} = \sum_{g,h} \frac{\overline{B}_{0g}^{(j)}\Delta U_{gh}B_{0h}^{(j')}}{\sqrt{1+g_z/K_z}\sqrt{1+h_z/K_z}}.$$

For the off-diagonal terms $j \neq j'$:

$$(\{\boldsymbol{\Upsilon}_0\}_{jj} - \{\boldsymbol{\Upsilon}_0\}_{j'j'})\{\boldsymbol{\alpha}_1\}_{jj'} = -\{\overline{\mathbf{B}}_0\Delta\mathbf{U}\mathbf{B}_0\}_{jj'} = -U^{(jj')},$$

where for a non-degenerate case, $\gamma_0^{(j)} \neq \gamma_0^{(j')}$, we find

$$\{\boldsymbol{\alpha}_1\}_{jj'} = -\frac{1}{2K_z}\frac{U^{(jj')}}{\gamma_0^{(j)} - \gamma_0^{(j')}}.$$

Similarly we define

$$\overline{\mathbf{B}}_1 = \overline{\boldsymbol{\alpha}}_1\overline{\mathbf{B}}_0, \tag{10.16}$$

and obtain the off-diagonal terms as

$$\{\overline{\boldsymbol{\alpha}}_1\}_{jj'} = \frac{1}{2K_z}\frac{U^{(jj')}}{\gamma_0^{(j)} - \gamma_0^{(j')}}.$$

The diagonal elements $\{\boldsymbol{\alpha}_1\}_{jj}$ and $\{\overline{\boldsymbol{\alpha}}_1\}_{jj}$ can be found from the normalization condition

$$(\mathbf{B}_0 + \lambda\mathbf{B}_1)(\overline{\mathbf{B}}_0 + \lambda\overline{\mathbf{B}}_1) = \mathbf{I}.$$

By neglecting the second-order term, we see that the above condition requires that $\overline{\boldsymbol{\alpha}}_{1_{jj}} + \boldsymbol{\alpha}_{1_{jj}} = 0$. This can be satisfied if we choose

$$\{\overline{\alpha}_1\}_{jj} = \{\alpha_1\}_{jj} = 0.$$

Using the definition (10.14), we find

$$B_{1g}^{(j)} = \sum_{j'} B_{0g}^{(j')} \{\alpha_1\}_{j'j} = \frac{1}{2K_z} \sum_{j' \neq j} \frac{B_{0g}^{(j')} U^{(j'j)}}{\gamma_0^{(j)} - \gamma_0^{(j')}}, \tag{10.17}$$

and

$$\overline{B}_{1g}^{(j)} = \sum_{j' \neq j} \{\alpha_1\}_{jj'} \overline{B}_{0g}^{(j')} = \frac{1}{2K_z} \sum_{j' \neq j} \frac{\overline{B}_{0g}^{(j')} U^{(jj')}}{\gamma_0^{(j)} - \gamma_0^{(j')}}. \tag{10.18}$$

10.2.4 *Second-order perturbation*

We now consider the second-order equation (10.13). By introducing

$$\mathbf{B}_2 = \mathbf{B}_0 \boldsymbol{\alpha}_2, \tag{10.19}$$

using (10.11), and multiplying both sides of (10.13) by $\overline{\mathbf{B}}_0$, we find that

$$(\overline{\mathbf{B}}_0 \mathbf{B}_0)(\boldsymbol{\Upsilon}_0 \boldsymbol{\alpha}_2 - \boldsymbol{\alpha}_2 \boldsymbol{\Upsilon}_0) + \overline{\mathbf{B}}_0 \Delta \mathbf{U} \mathbf{B}_1 = (\overline{\mathbf{B}}_0 \mathbf{B}_0)(\boldsymbol{\Upsilon}_2 + \boldsymbol{\alpha}_1 \boldsymbol{\Upsilon}_1).$$

For the diagonal terms $j = j'$ we have $\{\boldsymbol{\alpha}_1\}_{jj} = 0$ and therefore

$$\{\boldsymbol{\Upsilon}_2\}_{jj} = \{(\overline{\mathbf{B}}_0 \Delta \mathbf{U} \mathbf{B}_1)\}_{jj},$$

or explicitly

$$\gamma_2^{(j)} = \frac{1}{2K_z} \sum_{j'} \{\overline{\mathbf{B}}_0 \Delta \mathbf{U} \mathbf{B}_0\}_{jj'} \{\alpha_1\}_{j'j} = \left(\frac{1}{2K_z}\right)^2 \sum_{j' \neq j} \frac{U^{(jj')} U^{(j'j)}}{\gamma_0^{(j)} - \gamma_0^{(j')}}. \tag{10.20}$$

For the off-diagonal terms $j' \neq j$ we find

$$\left(\{\boldsymbol{\Upsilon}_0\}_{jj} - \{\boldsymbol{\Upsilon}_0\}_{j'j'}\right) \{\alpha_2\}_{jj'} = -\{\overline{\mathbf{B}}_0 \Delta \mathbf{U} \mathbf{B}_1\}_{jj'} + \{\alpha_1\}_{jj'} \{\boldsymbol{\Upsilon}_1\}_{j'j'}.$$

Substitution of eqns (10.17) and (10.15) into the above equation gives

$$\{\alpha_2\}_{jj'} = -\left(\frac{1}{2K_z}\right)^2 \frac{1}{\gamma_0^{(j)} - \gamma_0^{(j')}} \left\{ \frac{U^{(jj')} U^{(j'j')}}{\gamma_0^{(j)} - \gamma_0^{(j')}} + \sum_{i \neq j'} \frac{U^{(ji)} U^{(ij')}}{\gamma_0^{(j')} - \gamma_0^{(i)}} \right\}.$$

Carrying out similar calculations for the left-hand eigenvectors, we find that

$$\{\overline{\alpha}_2\}_{jj'} = -\left(\frac{1}{2K_z}\right)^2 \frac{1}{\gamma_0^{(j)} - \gamma_0^{(j')}} \left\{ \frac{U^{(jj)} U^{(jj')}}{\gamma_0^{(j)} - \gamma_0^{(j')}} - \sum_{i \neq j} \frac{U^{(ji)} U^{(ij')}}{\gamma_0^{(j)} - \gamma_0^{(i)}} \right\}.$$

The diagonal elements of $\{\alpha_2\}_{jj}$ and $\{\overline{\alpha}_2\}_{jj}$ can now be found by imposing the requirement that

$$(\mathbf{B}_0 + \lambda\mathbf{B}_1 + \lambda^2\mathbf{B}_2)(\overline{\mathbf{B}}_0 + \lambda\overline{\mathbf{B}}_1 + \lambda^2\overline{\mathbf{B}}_2) = \mathbf{I}.$$

Considering the terms proportional to λ^2, we arrive at the condition

$$\mathbf{B}_0\overline{\mathbf{B}}_2 + \mathbf{B}_1\overline{\mathbf{B}}_1 + \mathbf{B}_2\overline{\mathbf{B}}_0 = \mathbf{I},$$

which gives an equation for $\{\alpha_2\}_{jj}$ and $\{\overline{\alpha}_2\}_{jj}$:

$$\{\overline{\alpha}_2\}_{jj} + \{\alpha_2\}_{jj} + \left(\frac{1}{2K_z}\right)^2 \sum_{j'\neq j} \frac{U^{(jj')}U^{(j'j)}}{[\gamma_0^{(j)} - \gamma_0^{(j')}]^2} = 0.$$

This equation can be satisfied by choosing

$$\{\alpha_2\}_{jj} = \{\overline{\alpha}_2\}_{jj} = -\frac{1}{2}\left(\frac{1}{2K_z}\right)^2 \sum_{j'\neq j} \frac{U^{(jj')}U^{(j'j)}}{[\gamma_0^{(j)} - \gamma_0^{(j')}]^2}.$$

By combining this with the definition (10.19) we have

$$B_{2g}^{(j)} = \left(\frac{1}{2K_z}\right)^2 \sum_{j'\neq j} \frac{B_{0g}^{(j')}}{\gamma_0^{(j)} - \gamma_0^{(j')}}\left\{ -\frac{U^{(j'j)}U^{(jj)}}{\gamma_0^{(j)} - \gamma_0^{(j')}} - \sum_{i\neq j}\frac{U^{(j'i)}U^{(ij)}}{\gamma_0^{(j)} - \gamma_0^{(i)}}\right\}$$
$$- \frac{1}{2}\left(\frac{1}{2K_z}\right)^2 \sum_{j'\neq j} \frac{U^{(jj')}U^{(j'j)}}{[\gamma_0^{(j)} - \gamma_0^{(j')}]^2} B_{0g}^{(j)}. \tag{10.21}$$

Similarly we find an expression for $\overline{B}_{2g}^{(j)}$:

$$\overline{B}_{2g}^{(j)} = \left(\frac{1}{2K_z}\right)^2 \sum_{j'\neq j} \frac{\overline{B}_{0g}^{(j')}}{\gamma_0^{(j)} - \gamma_0^{(j')}}\left\{ -\frac{U^{(jj)}U^{(jj')}}{\gamma_0^{(j)} - \gamma_0^{(j')}} + \sum_{i\neq j}\frac{U^{(ji)}U^{(ij')}}{\gamma_0^{(j)} - \gamma_0^{(i)}}\right\}$$
$$- \frac{1}{2}\left(\frac{1}{2K_z}\right)^2 \sum_{j'\neq j} \frac{U^{(jj')}U^{(j'j)}}{[\gamma_0^{(j)} - \gamma_0^{(j')}]^2} \overline{B}_{0g}^{(j)}. \tag{10.22}$$

10.3 Tensor THEED

Although the expression for the perturbation correction terms $\mathbf{\Upsilon}_1, \mathbf{\Upsilon}_2, \mathbf{B}_1, \mathbf{B}_2, \overline{\mathbf{B}}_1$, and $\overline{\mathbf{B}}_2$ are somewhat complicated, in this section we will show that in fact their dependence on the perturbation $\Delta\mathbf{U}$ may be represented in a very simple matrix form which we will call tensor THEED by analogy with the well-developed technique of the tensor low energy electron diffraction (Rous, 1992).

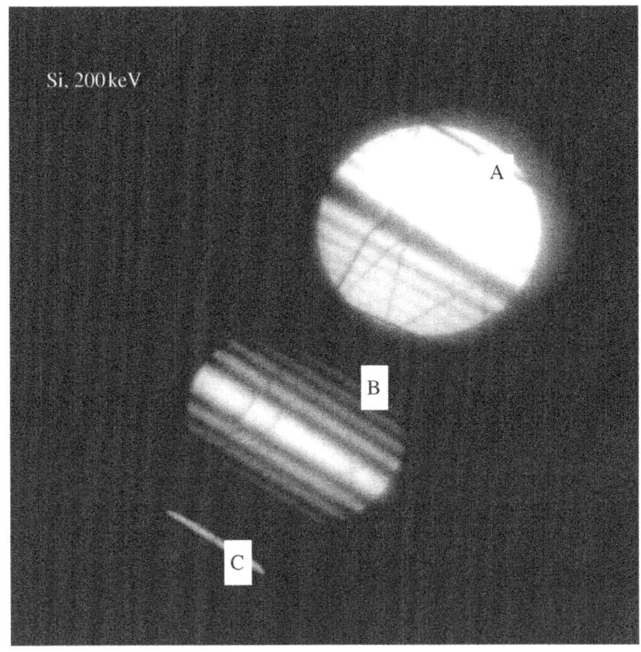

FIG. 10.5. Experimental CBED pattern from a single crystal of silicon.

Assuming that $\lambda = 1$ and writing $\Delta\mathbf{\Upsilon} = \mathbf{\Upsilon}_1 + \mathbf{\Upsilon}_2$, we drop the subscript zero for the reference structure, and then by using (10.15) and (10.20) we find that

$$\Delta\gamma^{(j)} = \frac{1}{2K_z}U^{(jj)} + \left(\frac{1}{2K_z}\right)^2 \sum_{j' \neq j} \frac{U^{(jj')}U^{(j'j)}}{\gamma^{(j)} - \gamma^{(j')}}.$$

Similarly by writing $\Delta\mathbf{B} = \mathbf{B}_1 + \mathbf{B}_2$ and using eqns (10.17) and (10.21) we find

$$\Delta B_g^{(j)} = \frac{1}{2K_z}\sum_{j' \neq j}\frac{B_g^{(j')}U^{(j'j)}}{\gamma^{(j)} - \gamma^{(j')}} + \left(\frac{1}{2K_z}\right)^2 \sum_{j' \neq j}\frac{B_g^{(j')}}{\gamma^{(j)} - \gamma^{(j')}}\left\{-\frac{U^{(j'j)}U^{(jj)}}{\gamma^{(j)} - \gamma^{(j')}}\right.$$

$$\left. - \sum_{i \neq j}\frac{U^{(j'i)}U^{(ij)}}{\gamma^{(j)} - \gamma^{(i)}}\right\} - \frac{1}{2}\left(\frac{1}{2K_z}\right)^2 \sum_{j' \neq j}\frac{U^{(jj')}U^{(j'j)}}{[\gamma^{(j)} - \gamma^{(j')}]^2}B_g^{(j)}.$$

For $\Delta\overline{\mathbf{B}} = \overline{\mathbf{B}}_1 + \overline{\mathbf{B}}_2$ we use eqns (10.18) and (10.22) and find that

$$\Delta\overline{B}_g^{(j)} = \frac{1}{2K_z}\sum_{j' \neq j}\frac{\overline{B}_g^{(j')}U^{(jj')}}{\gamma^{(j)} - \gamma^{(j')}} + \left(\frac{1}{2K_z}\right)^2 \sum_{j' \neq j}\frac{\overline{B}_g^{(j')}}{\gamma^{(j)} - \gamma^{(j')}}\left\{-\frac{U^{(jj)}U^{(jj')}}{\gamma^{(j)} - \gamma^{(j')}}\right.$$

$$\left. + \sum_{i \neq j}\frac{U^{(ji)}U^{(ij')}}{\gamma^{(j)} - \gamma^{(i)}}\right\} - \frac{1}{2}\left(\frac{1}{2K_z}\right)^2 \sum_{j' \neq j}\frac{U^{(jj')}U^{(j'j)}}{[\gamma^{(j)} - \gamma^{(j')}]^2}\overline{B}_g^{(j)}.$$

The above expressions may be represented in a compact tensor form. We introduce the notation $g - h = \ell$ or $h = g - \ell$, and from the definition of $U^{(jj')}$, we find that

$$U^{(jj')} = \sum_g \sum_h \frac{\overline{B}_g^{(j)} \Delta U_{gh} B_h^{(j')}}{\sqrt{1 + g_z/K_z}\sqrt{1 + h_z/K_z}} \tag{10.23}$$

$$= \sum_\ell \Big\{ \sum_g \frac{\overline{B}_g^{(j)} B_{g-\ell}^{(j')}}{\sqrt{1 + g_z/K_z}\sqrt{1 + (g-\ell)_z/K_z}} \Big\} \Delta U_\ell = 2K_z \sum_\ell \mathcal{U}_\ell^{(jj')} \cdot \Delta U_\ell,$$

where

$$\mathcal{U}_\ell^{(jj')} = \frac{1}{2K_z} \sum_g \frac{\overline{B}_g^{(j)} B_{g-\ell}^{(j')}}{\sqrt{1 + g_z/K_z}\sqrt{1 + (g_z - \ell_z)/K_z}}. \tag{10.24}$$

By substituting the definition (10.24) into the expressions for $\Delta \Upsilon$, $\Delta \mathbf{B}$, and $\Delta \overline{\mathbf{B}}$ we arrive at the following tensor expression for the correction to the eigenvalues

$$\Delta \gamma^{(j)} = {}^1\mathcal{U}^{(j)} \cdot \Delta \mathbf{U} + \Delta \mathbf{U} \cdot {}^2\mathcal{U}^{(j)} \cdot \Delta \mathbf{U}, \tag{10.25}$$

where

$$\{{}^1\mathcal{U}^{(j)}\}_\ell = \mathcal{U}_\ell^{(jj)}, \quad \{{}^2\mathcal{U}^{(j)}\}_{\ell k} = \sum_{j' \neq j} \frac{\mathcal{U}_\ell^{(jj')}\mathcal{U}_k^{(j'j)}}{\gamma_0^{(j)} - \gamma_0^{(j')}}.$$

The perturbation terms for the eigenvectors have the form

$$\Delta B_g^{(j)} = {}^1\mathcal{E}_g^{(j)} \cdot \Delta \mathbf{U} + \Delta \mathbf{U} \cdot {}^2\mathcal{E}_g^{(j)} \cdot \Delta \mathbf{U} \tag{10.26}$$

$$\Delta \overline{B}_g^{(j)} = {}^1\overline{\mathcal{E}}_g^{(j)} \cdot \Delta \mathbf{U} + \Delta \mathbf{U} \cdot {}^2\overline{\mathcal{E}}_g^{(j)} \cdot \Delta \mathbf{U}, \tag{10.27}$$

where

$$\{{}^1\mathcal{E}_g^{(j)}\}_\ell = \sum_{j' \neq j} \frac{B_g^{(j')}\mathcal{U}_\ell^{(j'j)}}{\gamma^{(j)} - \gamma^{(j')}}, \quad \{{}^1\overline{\mathcal{E}}_g^{(j)}\}_\ell = \sum_{j' \neq j} \frac{\overline{B}_g^{(j')}\mathcal{U}_\ell^{(jj')}}{\gamma^{(j)} - \gamma^{(j')}},$$

and

$$\{{}^2\mathcal{E}_g^{(j)}\}_{\ell k} = \sum_{j' \neq j} \frac{B_g^{(j')}}{\gamma^{(j)} - \gamma^{(j')}} \left\{ -\frac{\mathcal{U}_\ell^{(j'j)}\mathcal{U}_k^{(jj)}}{\gamma^{(j)} - \gamma^{(j')}} - \sum_{i \neq j} \frac{\mathcal{U}_\ell^{(j'i)}\mathcal{U}_k^{(ij)}}{\gamma^{(j)} - \gamma^{(i)}} \right\}$$

$$- \frac{1}{2} \sum_{j' \neq j} \frac{\mathcal{U}_\ell^{(jj')}\mathcal{U}_k^{(j'j)}}{[\gamma^{(j)} - \gamma^{(j')}]^2} B_g^{(j)}.$$

$$\{{}^2\overline{\mathcal{E}}_g^{(j)}\}_{\ell k} = \sum_{j' \neq j} \frac{\overline{B}_g^{(j')}}{\gamma^{(j)} - \gamma^{(j')}} \left\{ -\frac{\mathcal{U}_\ell^{(jj)}\mathcal{U}_k^{(jj')}}{\gamma^{(j)} - \gamma^{(j')}} + \sum_{i \neq j} \frac{\mathcal{U}_\ell^{(ji)}\mathcal{U}_k^{(ij')}}{\gamma^{(j)} - \gamma^{(i)}} \right\}$$

$$- \frac{1}{2} \sum_{j' \neq j} \frac{\mathcal{U}_\ell^{(jj')}\mathcal{U}_k^{(j'j)}}{[\gamma^{(j)} - \gamma^{(j')}]^2} \overline{B}_g^{(j)}.$$

The amplitude of the gth diffracted beam is given by

$$\mathcal{F}_g = \sum_j (\overline{B}_0^{(j)} + \Delta \overline{B}_0^{(j)})(B_g^{(j)} + \Delta B_g^{(j)}) \exp[i(\gamma^{(j)} + \Delta\gamma^{(j)})z]$$

$$= \sum_j (\overline{B}_0^{(j)} + {}^1 \overline{\mathcal{E}}_0^{(j)} \cdot \Delta \mathbf{U} + \Delta \mathbf{U} \cdot {}^2 \overline{\mathcal{E}}_0^{(j)} \cdot \Delta \mathbf{U})$$

$$\times (B_g^{(j)} + {}^1 \mathcal{E}_g^{(j)} \cdot \Delta \mathbf{U} + \Delta \mathbf{U} \cdot {}^2 \mathcal{E}_g^{(j)} \cdot \Delta \mathbf{U})$$

$$\times \exp\left[i(\gamma^{(j)} + {}^1 \mathcal{U}^{(j)} \cdot \Delta \mathbf{U} + \Delta \mathbf{U} \cdot {}^2 \mathcal{U}^{(j)} \cdot \Delta \mathbf{U})z\right]. \tag{10.28}$$

These expressions are the basis of the tensor theory of transmission HEED (Peng and Dudarev, 1993a, Peng and Dudarev, 1993b, Peng and Zuo, 1995), similar to the tensor theory of low-energy electron diffraction (LEED) (Rous, 1992, Pendry et al., 1988).

In addition to its simplicity, the tensor theory expression (10.28) for the diffracted beam amplitudes has the advantage of being extremely efficient numerically. Once a dynamical diffraction calculation has been performed for the reference structure and the relevant vectors and tensors $\mathcal{U}$ and $\mathcal{E}$ found, the numerical evaluation of the amplitudes of diffracted beams does not require going beyond simple multiplication that scales as $np(1 + p)$, where n is the number of Bloch waves characterized by appreciable values of excitation amplitudes and p is the number of variable structure factors. In a typical case we might have $n < 30$ and $p < 10$. The tensor theory is then at least two orders of magnitude faster than a full dynamical calculation that scales as N^3, where N is the total number of diffracted beams (typically $N > 100$ for the case of zone axis diffraction).

To examine how the tensor theory works in practice, we have investigated how the intensities of diffracted beams vary as a function of the structure factor. The calculations were carried out for the systematic diffraction geometry, and a typical CBED pattern taken for the geometry of this type is shown in Fig. 10.5. Figure 10.6 shows three curves illustrating how the intensity of the transmitted beams varies as a function of the U_{200} structure factor for the case of an MgO crystalline specimen and systematic diffraction geometry. Figures 10.6a-c correspond to the 000, 200 and 400 CBED discs respectively. Calculations were performed assuming that the beam energy is 100 keV and the thickness of the crystal is 1000 Å. The full dynamical calculation was performed retaining 92 beams in eqns (10.2), and also using first-order ('1st order') and second-order ('2nd order') tensor theory. It is seen that while the first-order tensor theory may deviate appreciably from the full dynamical theory, the second-order tensor theory agrees almost perfectly with the full dynamical calculations. It should be noted that the range of variation of the structure factor ΔU_{200} shown in the figure 10.6 exceeds 5% of its magnitude, i.e. $\Delta U_{200}/U_{200} \geq 5\%$, and this is comparable with the maximum possible change of a structure factor in a real material resulting from the redistribution of charge associated with the formation of chemical bonds between atoms.

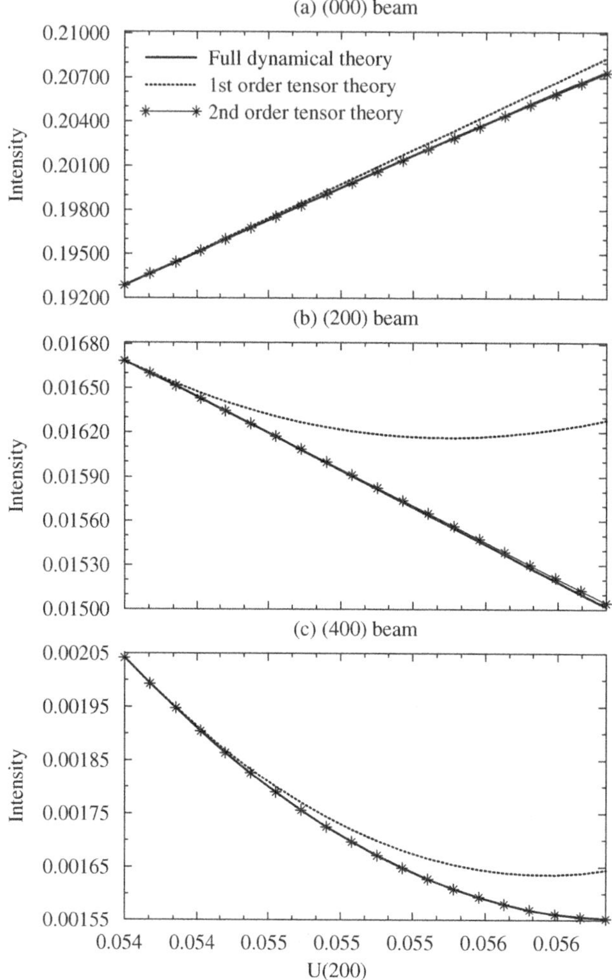

FIG. 10.6. Curves showing how the (a) 000, (b) 200, and (c) 400 diffracted beam intensities vary as a function of the U_{200} structure factor (measured in Å^{-2}). Calculations were performed for an MgO single crystal for a beam energy of 100 keV, and assuming systematic diffraction geometry.

10.4 Direct inversion of THEED data

One of the major challenges in electron microscopy is associated with the problem of inversion of dynamical electron diffraction data. By using a modern transmission electron microscope equipped with an energy-filtering facility, such as a Gatan Imaging Filter (GIF), diffraction data may be acquired extremely efficiently using the CBED technique. A large amount of diffraction data exceeding millions of data points may be collected in a few seconds from a region as small as a nanometre. In contrast to this, the retrieval of structural information from the experimental data takes much longer. The most successful procedure for the retrieval of structural information from CBED data has been developed by Spence and Zuo (Spence and Zuo, 1992), who used full dynamical electron diffraction theory. However, full scale dynamical diffraction calculations become very time consuming as the crystal structure becomes more complex. Because of this, so far only a few simple systems have been investigated using the Spence and Zuo approach. More efficient schemes have also been proposed for the direct inversion of the diffraction data, but these schemes have yet to be implemented experimentally (Spargo et al., 1994, Allen and Rossouw, 1993a, Rez, 1999, Allen et al., 1999, Spence, 1998, Spence et al., 1999). In this and the following sections we will show how the tensor theory of electron diffraction can be used to address the inversion problem for both the THEED and RHEED geometries.

As we briefly mentioned in Chapter 1, a quantitative electron diffraction study normally requires defining and minimizing a *merit function* that quantifies the agreement between a theoretical model and experimental observations (Spence and Zuo, 1992)

$$\chi^2 = \sum_i \frac{(I_{exp}^{(i)} - CI_{cal}^{(i)})^2}{\sigma_i^2}, \tag{10.29}$$

where C is a normalization constant. Information about the crystal structure can be obtained by adjusting parameters of the model, such as structure factors and atomic coordinates, until the minimum of the merit function is found. The adjustment process is therefore a problem of many-dimensional minimization. In general this problem is tedious and complicated, and there is no guarantee that the minimization procedure will be able to find the true minimum of (10.29).

In this section we will show that in the cases where the tensor theory discussed in the preceding sections is applicable, the general problem of minimization of χ^2 in the multidimensional parameter space may be reduced to a simple problem of matrix inversion, and structural information may be directly extracted from experimentally observed diffraction patterns. It must be emphasized, however, that the tensor theory is only an approximate method, and its validity is in general limited to relatively small (a few percent) variations of the charge density distribution and to atomic displacements not exceeding 0.2 $\overset{\circ}{A}$. In a general case where the degree of charge redistribution and/or atomic displacements are substantial, the tensor theory may be combined with a global search algorithm by dividing the parameter space into subspaces, defined in such a way that the tensor theory approach is valid in each subspace. While the global method may always be applied to choose between subspaces, in each subspace the tensor theory may be used to find the local minimum directly. In what follows we examine the validity of the tensor theory and discuss applications of the method to the direct inversion of diffraction data.

10.4.1　*Inversion of crystal structure factors*

We first discuss the problem of direct inversion of crystal structure factors. When a crystal is formed by bringing together neutral atoms, the real-space distribution of the density of valence electrons changes to reduce the total energy of the system. In this way chemical bonds are formed (Pauling, 1960, Pettifor, 1995). In the case of elastic electron diffraction the effective one-electron potential may be written as

$$V(\mathbf{r}) = \sum_i \phi_i(\mathbf{r} - \mathbf{r}_i) + \Delta V(\mathbf{r}) = V_0(\mathbf{r}) + \Delta V(\mathbf{r}), \qquad (10.30)$$

where the first term on the right-hand side represents the contribution from neutral atoms, and the second term results from the redistribution of charge density. An important feature of charge density distribution in a solid is that the first term in the above equation already includes a substantial amount of charge overlap. The additional contribution associated with the formation of anisotropic chemical bonds represented by the second term is very small compared with the first term. Typically this effect accounts for $\sim 0.01\%$ of the total charge in a covalent crystal. We therefore have a perfect case for the perturbation treatment of the effect of $\Delta V(\mathbf{r})$ on the diffracted beam amplitudes. Using eqns (10.25)–(10.27), the diffracted beam amplitude can be expanded as a power series in $\Delta \mathbf{U}$ (we recall here that the quantities $U(\mathbf{r})$ and $V(\mathbf{r})$ are related via $U(\mathbf{r}) = -2mV(\mathbf{r})/\hbar^2$. By retaining terms up to the second order we find

$$\mathcal{F}_g = \mathcal{F}_g^{(0)} + \mathbf{X}_g \cdot \Delta \mathbf{U} + \Delta \mathbf{U} \cdot \mathbf{A}_g \cdot \Delta \mathbf{U}, \qquad (10.31)$$

where

$$\mathbf{X}_g = \sum_j \{\overline{B}_0^{(j)} \, {}^1\!\boldsymbol{\mathcal{E}}_g^{(j)} + B_g^{(j)} \, {}^1\!\overline{\boldsymbol{\mathcal{E}}}_0^{(j)} + iz\overline{B}_0^{(j)} B_g^{(j)} \, {}^1\!\boldsymbol{\mathcal{U}}^{(j)}\} \exp(i\gamma^{(j)}z),$$

and

$$\begin{aligned}
\mathbf{A}_g = \sum_j \Big\{ &\overline{B}_0^{(j)} \, {}^2\!\boldsymbol{\mathcal{E}}_g^{(j)} + B_g^{(j)} \, {}^2\!\overline{\boldsymbol{\mathcal{E}}}_0^{(j)} + {}^1\!\overline{\boldsymbol{\mathcal{E}}}_0^{(j)} \, {}^1\!\boldsymbol{\mathcal{E}}_g^{(j)} \\
&+ iz[\overline{B}_0^{(j)} \, {}^1\!\boldsymbol{\mathcal{E}}_g^{(j)} + B_g^{(j)} \, {}^1\!\overline{\boldsymbol{\mathcal{E}}}_0^{(j)}]^1\!\boldsymbol{\mathcal{U}}^{(j)} \\
&+ \overline{B}_0^{(j)} B_g^{(j)} [iz \, {}^2\!\boldsymbol{\mathcal{U}}^{(j)} - \frac{z^2}{2} \, {}^1\!\boldsymbol{\mathcal{U}}^{(j)} \, {}^1\!\boldsymbol{\mathcal{U}}^{(j)}] \Big\} \exp(i\gamma^{(j)}z).
\end{aligned}$$

In what follows it is convenient to treat the real and imaginary parts of the Fourier coefficients of the perturbation potential ΔU_ℓ as separate parameters. By introducing the notation

$$a_{2\ell} = Re(\Delta U_\ell), \quad a_{2\ell+1} = Im(\Delta U_\ell); \quad X'_{2\ell} = X_\ell, \; X'_{2\ell+1} = iX_\ell;$$

$$A'_{2\ell,2k} = A_{\ell k}, \quad A'_{2\ell+1,2k} = iA_{\ell k}, \quad A'_{2\ell,2k+1} = iA_{\ell k}, \quad A'_{2\ell+1,2k+1} = -A_{\ell k},$$

we find

$$\mathcal{F}_g = \mathcal{F}_g^{(0)} + \mathbf{X}_g' \cdot \mathbf{a} + \mathbf{a} \cdot \mathbf{A}_g' \cdot \mathbf{a}, \tag{10.32}$$

where the components of the parameter vector $\mathbf{a}$ are now real. Since in a quantitative CBED investigation many hundreds of diffraction data points need to be analysed, it is convenient to eliminate the index of a CBED disc $\mathbf{g}$ and to replace it with a data point index i, i.e. to replace $\mathcal{F}_g^{(0)}$ with $\mathcal{F}_0^{(i)}$. By using expression (10.32), and by assuming that

$$|\Delta\mathcal{F}_g| = |\mathcal{F}_g - \mathcal{F}_g^{(0)}| \ll |\mathcal{F}_g^{(0)}|, \tag{10.33}$$

we find

$$I_{exp}^{(i)} - CI_{cal}^{(i)} = I_{exp}^{(i)} - C|\mathcal{F}_0^{(i)} + \Delta\mathcal{F}^{(i)}|^2 = \Delta I^{(i)} - \mathbf{Y}^{(i)} \cdot \mathbf{a} - \mathbf{a} \cdot \mathbf{D}^{(i)} \cdot \mathbf{a},$$

where

$$\Delta I^{(i)} = I_{exp}^{(i)} - C|\mathcal{F}_0^{(i)}|^2,$$

and

$$\{\mathbf{Y}^{(i)}\}_\ell = 2CRe(\mathcal{F}_0^{*(i)} X_\ell^{'(i)}), \quad \{\mathbf{D}^{(i)}\}_{\ell,k} = C\big[2Re(\mathcal{F}_0^{*(i)} A_{\ell k}^{'(i)}) + X_\ell^{'(i)} X_k^{'*(i)}\big].$$

The χ^2 function defined by (10.29) now becomes

$$\chi^2(\mathbf{a}) = \sum_i \frac{1}{\sigma_i^2}[I_{exp}^{(i)} - CI_{cal}^{(i)}]^2 \approx \sum_i \frac{1}{\sigma_i^2}[\Delta I^{(i)} - \mathbf{Y}^{(i)} \cdot \mathbf{a} - \mathbf{a} \cdot \mathbf{D}^{(i)} \cdot \mathbf{a}]^2. \tag{10.34}$$

10.4.1.1 *The linear model* In this model we retain only terms linear in $\mathbf{a}$ in the bracket of eqn (10.34), so that χ^2 reduces to

$$\chi^2(\mathbf{a}) = \sum_i \frac{1}{\sigma_i^2}[\Delta I^{(i)} - \sum_\ell Y_\ell^{(i)} a_\ell]^2. \tag{10.35}$$

At the minimum of χ^2 the first-order derivatives with respect to $\mathbf{a} = \{a_k\}$ must vanish,

$$(\nabla\chi^2)_{a_k} = 0 = \sum_i \frac{2}{\sigma_i^2}[\Delta I^{(i)} - \sum_\ell Y_\ell^{(i)} a_\ell]Y_k^{(i)}.$$

This leads to

$$\sum_i \frac{Y_k^{(i)}}{\sigma_i} \cdot \frac{\Delta I^{(i)}}{\sigma_i} = \sum_i \frac{Y_k^{(i)}}{\sigma_i} \cdot \sum_\ell \frac{Y_\ell^{(i)}}{\sigma_i} \cdot a_\ell, \tag{10.36}$$

for the entire set of parameters a_k. By defining the *design matrix* $\{\mathbf{M}\}_{i,\ell} = Y_\ell^{(i)}/\sigma_i$, its transpose $\{\mathbf{M}^T\}_{k,i} = Y_k^{(i)}/\sigma_i$, and a vector $\mathbf{b}$ with components $b_i = \Delta I^{(i)}/\sigma_i$, the normal equation (10.36) can be represented in the form

$$\mathbf{M}^T \cdot \mathbf{b} = (\mathbf{M}^T\mathbf{M}) \cdot \mathbf{a}.$$

Formally this matrix equation can be inverted giving rise to

$$\mathbf{a} = (\mathbf{M}^T\mathbf{M})^{-1}\mathbf{M}^T \cdot \mathbf{b}; \tag{10.37}$$

and the parameters $\{a_k\}$ of the model may be found by a direct inversion of the matrix $(\mathbf{M}^T\mathbf{M})$, which depends only on the reference structure.

10.4.1.2 *The quadratic model* We now consider the extension of the above linear model to include quadratic terms. If the parameters of the model are well chosen, so that function χ^2 is sufficiently close to its minimum, we may expand $\chi^2(\mathbf{a})$ in the vicinity of the set of parameters $\mathbf{a}_0$ characterizing the reference structure as

$$\chi^2(\mathbf{a}) = \chi^2(\mathbf{a}_0) + \nabla\chi^2(\mathbf{a}_0) \cdot (\mathbf{a} - \mathbf{a}_0) + \frac{1}{2}(\mathbf{a} - \mathbf{a}_0) \cdot \mathbf{H} \cdot (\mathbf{a} - \mathbf{a}_0), \qquad (10.38)$$

where $\nabla\chi^2(\mathbf{a})$ is the gradient of function χ^2 given by

$$\{\nabla\chi^2(\mathbf{a})\}_\ell = -\sum_i \frac{2}{\sigma_i^2}[\Delta I^{(i)} - \mathbf{Y}^{(i)} \cdot \mathbf{a} - \mathbf{a} \cdot \mathbf{D}^{(i)} \cdot \mathbf{a}][Y_\ell^{(i)} + \sum_k (D_{\ell k}^{(i)} + D_{k\ell}^{(i)})a_k],$$

$$(10.39)$$

and the matrix $\mathbf{H}$ is the *Hessian matrix*, the elements of which are given by

$$\begin{aligned}
\{\mathbf{H}\}_{\ell k} &= \frac{\partial^2 \chi^2}{\partial a_\ell \partial a_k} \\
&= \sum_i \frac{2}{\sigma_i^2}\{[Y_k^{(i)} + \sum_h (D_{kh}^{(i)} + D_{hk}^{(i)})a_h][Y_\ell^{(i)} + \sum_h (D_{\ell h}^{(i)} + D_{h\ell}^{(i)})a_h] \\
&\quad - [\Delta I^{(i)} - \mathbf{Y}^{(i)} \cdot \mathbf{a} - \mathbf{a} \cdot \mathbf{D}^{(i)}\mathbf{a}](D_{\ell k}^{(i)} + D_{k\ell}^{(i)})\}.
\end{aligned} \qquad (10.40)$$

For the true structure where $\chi^2(\mathbf{a})$ is minimum, the gradient of $\chi^2(\mathbf{a})$ must vanish, and we obtain

$$\nabla\chi^2(\mathbf{a}) = \nabla\chi^2(\mathbf{a}_0) + \mathbf{H} \cdot (\mathbf{a} - \mathbf{a}_0) = 0. \qquad (10.41)$$

The solution of this equation is given by

$$\mathbf{a} = \mathbf{a}_0 - \mathbf{H}^{-1} \cdot \nabla\chi^2(\mathbf{a}_0), \qquad (10.42)$$

and we see that again the model parameters $\{a_k\}$ are found by direct inversion of a matrix.

In the case where the second-order matrix $\mathbf{D}^{(i)}$ is small, eqns (10.39) and (10.40) for the reference structure corresponding to $\mathbf{a} = 0$ become

$$\frac{\partial\chi^2}{\partial a_\ell} \approx -\sum_i \frac{2}{\sigma_i^2}\Delta I^{(i)}Y_\ell^{(i)}, \qquad \frac{\partial^2\chi^2}{\partial a_\ell \partial a_k} \approx \sum_i \frac{2}{\sigma_i^2}[Y_k^{(i)}Y_\ell^{(i)}],$$

and eqn (10.41) reduces to

$$-\sum_i \frac{1}{\sigma_i^2}[\Delta I^{(i)} - \mathbf{Y}^{(i)} \cdot \mathbf{a}]Y_k^{(i)} = 0.$$

This equation is identical to the equation above eqn (10.36) and as expected, the quadratic model reduces to the linear model.

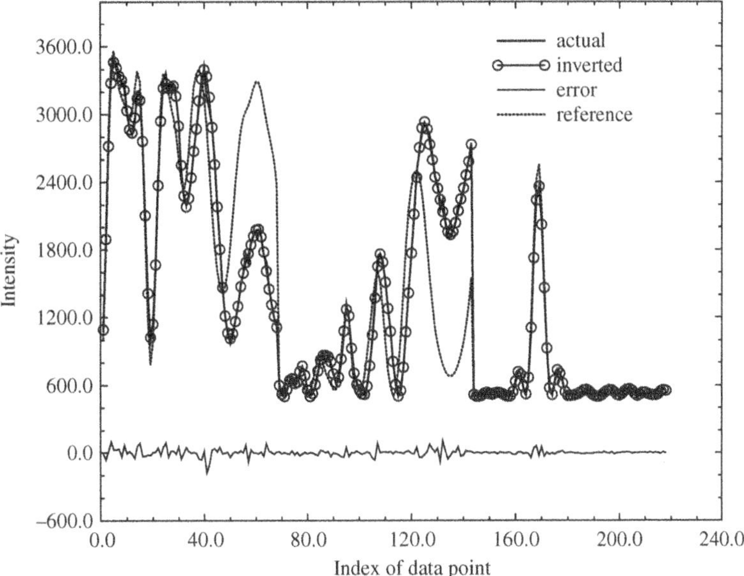

FIG. 10.7. Rocking curves calculated for the systematic diffraction geometry and for 100 keV beam energy. The curves correspond to the actual, reference and inverted structures. The lower part of the figure shows the curve representing the difference between the curves calculated for the actual and the inverted structures.

10.4.1.3 *A worked example*   The direct inversion schemes discussed above may be illustrated by performing an 'ideal experiment', the results of which are shown in Fig. 10.7 for a single crystal of MgO and 100 keV primary beam energy. The curves shown in the figure represent the transmitted and diffracted beam intensities as a function of the angle of incidence. These curves are usually called 'rocking curves'. The label on the horizontal axis is the data index i, and data points in the range between 1 and 68 represent the transmitted beam intensity, the points in the range between 69 and 143 represent the intensity of the 200 diffracted beam, and the points in the range between 145 and 218 correspond to the 400 CBED disc. The 'actual' rocking curve shown in the figure is calculated for a 'real structure' corresponding to complex $U_{200} = (0.058, 0.001)$ Å^{-2} and $U_{400} = (0.024, 0.00043)$ Å^{-2}, using full dynamical electron diffraction theory. The thickness of the crystal is 100 nm. We take this curve as the ideal 'experimental curve' and add Poisson-distributed noise to it to simulate the Poisson-distributed random errors normally introduced in experimental observations. The starting 'reference' curve is calculated for a reference structure that assumes that the crystal is composed of neutral atoms and is characterized by the following values of the structure factors $U_{200} = (0.054, 0.001)$ Å^{-2} and $U_{400} = (0.023, 0.0004)$ Å^{-2}. The two 'actual' and 'reference' curves shown in Fig. 10.7 are quite different, reflecting the high sensitivity of electron diffraction to the low order crystal structure factors and, through this, to the fine details of the valence charge distribution. The difference between the two curves gives $\Delta I^{(i)}$ for all the data points, and these data are then used to calculate the design matrix $\mathbf{M}$ defined by (10.36) and subsequently to invert the parameter vector $\mathbf{a}$ by using either the linear or quadratic models discussed in the preceding section.

The 'inverted' rocking curve is calculated for the set of parameter values determined by the inversion procedure. We find that a direct application of the linear inversion model gives rise to a noticeable difference between the 'actual' and 'inverted' curves, indicating that the linear model does not provide a satisfactory description of the process of scattering by $\Delta U(\mathbf{r})$. Nevertheless, the 'actual' and 'inverted' rocking curves are more similar than the 'actual' and the 'reference' curves. The χ^2 value characterizing the quality of the inversion procedure is 4.74. A new reference structure can now be defined as the reference structure, and the relevant matrices and vectors $\mathbf{M}$, $\mathbf{M}^T$, and $\mathbf{b}$ may now be evaluated for this new structure. Repeated applications of the linear inversion scheme gives rise to the curve shown in Fig. 10.7. The agreement between the 'actual' and 'inverted' rocking curves is almost perfect, where the residual value of χ^2 of 1.05 compares well with the ideal Poisson value of 1.0.

A similar procedure is used in the quadratic inversion scheme. The value of χ^2 resulting from the first application of the quadratic inversion model is 1.2, and the value returned from the second iteration step is 1.05, matching that of the linear model. The inverted values of the U_{200} and U_{400} structure factors are also similar. The accuracy is 0.5% for the real part and 15.0% for the imaginary part of U_g. The larger error in the value of the imaginary parts of U_g is partly due to the normalization constant C introduced in eqn (10.29). For example the uniform attenuation of the intensity of diffracted beams due to the imaginary part of the potential may be represented by this normalization constant C. In particular, any information about the value of the imaginary part of the mean inner potential U_0 is completely lost in eqn (10.29) in the THEED case.

The direct inversion schemes have also been applied to experimental CBED data (Peng and Zuo, 1995). Figure 10.8 shows an observed energy-filtered rocking curve (marked 'Experimental') and an inverted curve (marked 'Inverted') evaluated for an MgO single crystal by using the quadratic inversion scheme. The starting distribution of the potential $U_0(\mathbf{r})$ is constructed from electron atomic scattering factors of neutral atoms (Doyle and Turner, 1968, Peng et al., 1996e), and the absorption effects are included using the Einstein model of TDS scattering (Bird and King, 1990, Peng et al., 1996d). This gives the reference values for the structure factors of $U_{200} = (0.057334, 0.000791)$ Å^{-2} and $U_{400} = (0.024531, 0.000606)$ Å^{-2}. Since the diffraction data shown in Fig. 10.8 were obtained for the $2h,0,0$ systematic diffraction geometry, where the 400 beam satisfies the exact Bragg condition, the diffracted beam amplitudes are mostly

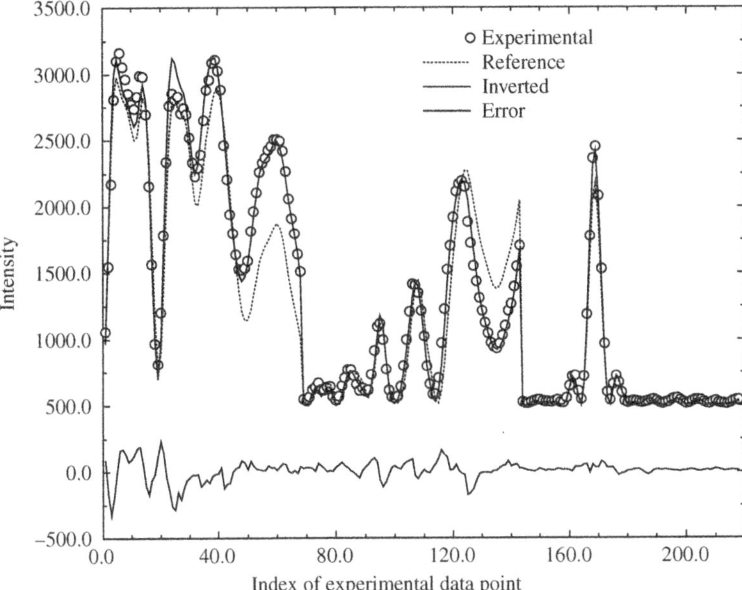

Fig. 10.8. Experimental, reference and inverted rocking curves measured and calculated for 100 keV primary beam energy for an MgO single crystal (after Peng and Zuo, (Peng and Zuo, 1995)). Calculations involved 94 beams and the thickness of the sample was found to be 95.5 nm.

sensitive to values of structure factors of the U_{2h00} type. The initial inversion of the experimental rocking curve was performed by using only four model parameters, namely the real and imaginary parts of the structure factors U_{200} and U_{400}, giving $U_{200} = (0.055109 \pm 0.000001, 0.000697 \pm 0.0000004)$ Å^{-2} and $U_{400} = (0.024211 \pm 0.000003, 0.000412 \pm 0.0000001)$ Å^{-2}. The residual value of the χ^2 function was found to be equal to 8.7. A newer reference structure was then constructed on the basis of the inverted model parameters, but it was found that further iterations did not improve the value of χ^2, indicating that that higher order corrections were small. On the other hand, a procedure involving a greater number of model parameters was found to improve the quality of the solution, progressively reducing the χ^2 values. In the case where nine model parameters were used, including the real and imaginary parts of U_{600} and U_{800} and the crystal thickness, the final value of χ^2 was found to converge to 8.2. However, the procedure leads to increased variances for the values of the U_{200} and U_{400} structure factors. The variance of the U_{600} factor is comparable with its value, and that for the U_{800} factor exceeds it. We should not, therefore, consider the values of these two factors as being reliable. Although the inclusion of these factors as parameters of the model does improve the fitting, the actual reason for the larger residual value of χ^2 is more likely to be associated with other factors, such as changes of experimental conditions during measurements. A more recent investigation of the case of silicon (Peng, 2000) showed that after a careful calibration and characterization of the instruments involved and the necessary deconvolution of the experimental data (Ren et al., 1997), it was possible to achieve $\chi^2 = 1.4$. This low value of the merit function shows that almost all the systematic errors were eliminated in the inversion procedure.

In the case of the linear and quadratic models, the set of model parameters obtained from the inversion scheme is unique. This follows from the fact that $\chi^2(\mathbf{a})$ is parabolic in the vicinity of its minimum $\mathbf{a}_0$, where the gradient $\nabla\chi^2(\mathbf{a}_0)$ vanishes. To a certain extent this conclusion remains valid even when the higher order terms are included, provided that the ratio $\Delta U/U_0$ remains small.

10.4.2 *Inversion of atomic coordinates*

The problem of the determination of positions of atoms in a crystal structure is closely related to the problem of inversion of crystal structure factors discussed in the previous section. Once the important crystal structure factors V_g are found we can find the atomic coordinates from a Fourier series

$$V(\mathbf{r}) = \sum_g V_g \exp(i\mathbf{g}\cdot\mathbf{r})$$

that reveals the atoms directly as maxima in the map of $V(\mathbf{r})$ (Spence, 1993). However, this approach proves to be not very efficient since a high resolution map requires knowing both amplitudes and phases of a large number of structure factors. In general, given that the crystal point and space groups are determined from the dynamical CBED patterns (Gjønnes and Moodie, 1965, Buxton et al., 1976, Tanaka et al., 1983a, Tanaka et al., 1983b), the number of unknown atomic coordinates is far smaller than the number of structure factors required to produce a high-resolution map. We explore below how changes of diffracted beam intensities considered as functions of the atomic coordinates can be used for crystal structure determination.

10.4.2.1 *Tensor expressions for the case of a thin film* Consider a thin film and assume that we can make a good guess about its basic crystal structure (Vincent et al., 1984, Tsuda and Tanaka, 1995). The task then is to refine the structure. Using our initial guess we define a reference structure characterized by a set of atomic coordinates $\{\mathbf{r}_i\}$. For this reference structure the structure factors (here we do not distinguish between the usual crystal structure factor F_g and U_g since the two quantities are related by a constant factor) are given by

$$U_g(ref) = -\frac{4\pi}{\Omega_0}\left(\frac{m}{m_0}\right)\sum_j f_j^{(e)}(s)\exp(-B_j s^2)\exp(-i\mathbf{g}\cdot\mathbf{r}_j),$$

where Ω_0 is the volume of a unit cell, $s = g/4\pi$, $f_j^{(e)}(s)$ is the atomic scattering factor, and B_j is the Debye–Waller (BW) factor of the j-th atom. By representing the set of atomic coordinates for the actual structure by $\{\mathbf{r}_j + \delta\mathbf{r}_j\}$, we find

$$U_g(act) = -\frac{4\pi}{\Omega_0}\left(\frac{m}{m_0}\right)\sum_j f_j^{(e)}(s)\exp(-B_j s^2)\exp(-i\mathbf{g}\cdot\mathbf{r}_j)\exp(-i\mathbf{g}\cdot\delta\mathbf{r}_j).$$

The difference between the structure factors is given by

$$\Delta U_g = U_g(act) - U_g(ref) = \sum_j U_{jg}\mathcal{S}_{jg}, \tag{10.43}$$

where

$$U_{jg} = -\frac{4\pi}{\Omega_0}\left(\frac{m}{m_0}\right)f_j^{(e)}(s)\exp(-B_j s^2)\exp(-i\mathbf{g}\cdot\mathbf{r}_j),$$

and

$$\mathcal{S}_{jg} = \exp(-i\mathbf{g}\cdot\delta\mathbf{r}_j) - 1. \tag{10.44}$$

Using the first-order tensor theory (10.28), we find

$$\begin{aligned}
\Delta\mathcal{F}_g &= \sum_j\left[(\overline{B}_0^{(j)} + \Delta\overline{B}_0^{(j)})(B_g^{(j)} + \Delta B_g^{(j)})e^{i\Delta\gamma^{(j)}z} - \overline{B}_0^{(j)}B_g^{(j)}\right]e^{i\gamma^{(j)}z} \\[2mm]
&\approx \sum_j\left\{\overline{B}_0^{(j)}B_g^{(j)}[e^{i\Delta\gamma^{(j)}z} - 1] + \Delta\overline{B}_0^{(j)}B_g^{(j)} + \overline{B}_0^{(j)}\Delta B_g^{(j)}\right\}e^{i\gamma^{(j)}z} \\[2mm]
&= \sum_j\left\{\overline{B}_0^{(j)}B_g^{(j)}[\exp(iU^{(jj)}z/2K_z) - 1]\right.\\[2mm]
&\quad\left. + \sum_j\sum_{j'\neq j}\left[\frac{\overline{B}_0^{(j')}U^{(jj')}B_g^{(j)}}{\gamma^{(j)} - \gamma^{(j')}} + \frac{\overline{B}_0^{(j)}U^{(j'j)}B_g^{(j')}}{\gamma^{(j)} - \gamma^{(j')}}\right]\right\}\exp(i\gamma^{(j)}z) \\[2mm]
&= \sum_j\overline{B}_0^{(j)}B_g^{(j)}[\exp(iU^{(jj)}z/2K_z) - 1]\exp(i\gamma^{(j)}z) \\[2mm]
&\quad + \sum_j\sum_{j'\neq j}\frac{\exp(i\gamma^{(j)}z) - \exp(i\gamma^{(j')})}{\gamma^{(j)} - \gamma^{(j')}}\overline{B}_0^{(j)}U^{(jj')}B_g^{(j')}.
\end{aligned}$$

This lengthy expression may be written in a compact tensor form

$$\Delta\mathcal{F}_g \approx \sum_{j,j'} T^{(jj')}\mathcal{X}^{(jj')}, \tag{10.45}$$

where

$$T^{(jj')} = \begin{cases} \overline{B}_0^{(j)}B_g^{(j)}\exp(i\gamma^{(j)}z) & \text{if } j = j' \\[3mm] \overline{B}_0^{(j)}B_g^{(j')}\cdot\dfrac{\exp(i\gamma^{(j)}z) - \exp(i\gamma^{(j')})}{\gamma^{(j)} - \gamma^{(j')}} & \text{if } j \neq j' \end{cases}$$

and

$$\mathcal{X}^{(jj')} = \begin{cases} \exp(iU^{(jj)}z/2K_z) - 1 & \text{if } j = j' \\ U^{(jj')} & \text{if } j \neq j' \end{cases} .$$

When the perturbation is weak, we can use the following expansion,

$$\exp(iU^{(jj')}z/2K_z) - 1 \approx iU^{(jj')}z/2K_z,$$

and obtain a linear tensor expression for the diffracted beam amplitudes

$$\Delta\mathcal{F}_g = \sum_\ell \mathcal{T}_\ell \cdot \Delta U_\ell, \tag{10.46}$$

where

$$\mathcal{T}_\ell = \frac{1}{2K_z} \sum_{j,j'} \frac{\exp(i\gamma^{(j)}z) - \exp(i\gamma^{(j')})}{\gamma^{(j)} - \gamma^{(j')}} \overline{B}_0^{(j')} B_g^{(j)} \mathcal{U}_\ell^{(jj')}.$$

By substituting (10.43) into (10.46), we arrive at

$$\Delta\mathcal{F}_g = \sum_\ell (\mathcal{T}_g)_\ell \sum_j U_{j\ell} \mathcal{S}_{j\ell} = \sum_{j,\ell} \left\{ U_{j\ell} (\mathcal{T}_g)_\ell \right\} \mathcal{S}_{j\ell} = \sum_{j,\ell} (\mathcal{M}_g)_{j\ell} \mathcal{S}_{j\ell}, \tag{10.47}$$

where

$$(\mathcal{M}_g)_{j\ell} = U_{j\ell} (\mathcal{T}_g)_\ell.$$

When the atomic displacements are small, we can expand $\mathcal{S}_{j\ell}$ as a polynomial series in the Cartesian coordinates of the displacements $\delta\mathbf{r}_j = \{\delta r_{jk}\}$ ($k = 1, 2, 3$), and find

$$\Delta\mathcal{F}_g = \sum_{j,\ell} \mathcal{M}_{j\ell} \sum_{k=1}^{3} \sum_{n=1}^{\infty} \frac{1}{n!} (i\ell_k)^n (-\delta r_{jk})^n = \sum_{j,k,n} \mathcal{T}_{jkn} (\delta r_{jk})^n, \tag{10.48}$$

where

$$\mathcal{T}_{jkn} = \sum_\ell \frac{1}{n!} (-i\ell_k)^n \mathcal{M}_{j\ell}.$$

In practice we have to terminate the series after a finite number of terms. If the summation is terminated at the n-th term, we obtain the n-th order tensor approximation. For example the first-order tensor expression is

$$\Delta\mathcal{F}_g = \sum_{j,k} \mathcal{T}_{jk1} \delta r_{jk}, \tag{10.49}$$

where $\Delta\mathcal{F}_g$ is proportional to the atomic displacements δr_{jk}.

10.4.2.2 *Tensor expressions for a surface structure* For surface structure determination we consider a model system consisting of a reconstructed surface layer and a bulk crystal slab of thickness t underneath the surface. For the surface layer we can use the tensor expression for $\Delta\mathcal{F}_g$ found above. When diffracted beams leave the surface layer their amplitudes are given by (10.47)

$$^s\mathcal{F}_g = {}^s\mathcal{F}_g(ref) + \sum_{j,\ell} ({}^s\mathcal{M}_g)_{j\ell}\mathcal{S}_{j\ell},$$

where the superscript s indicates that the relevant quantities are associated with the surface layer. Dynamical scattering processes occurring in the crystal bulk can be represented by the scattering matrix $^b\mathbf{M}$ (Peng and Whelan, 1991d). The amplitudes of diffracted beams at the bottom of the crystal slab are given by

$$^{s+b}\mathcal{F}_g = \sum_h {}^b M_{gh}\,{}^s\mathcal{F}_h = \sum_h {}^b M_{gh}\,{}^s\mathcal{F}_h(ref) + \sum_h {}^b M_{gh} \sum_{j,\ell} ({}^s\mathcal{M}_h)_{j\ell}\mathcal{S}_{j\ell},$$

where the superscript b denotes the bulk and $s+b$ denotes the entire (bulk plus surface) system. By introducing a new matrix

$$(^{s+b}\mathcal{M}_g)_{j\ell} = \sum_h {}^b M_{gh}\,(^s\mathcal{M}_h)_{j\ell},$$

we arrive at

$$\Delta\mathcal{F}_g = \sum_{j,\ell}\left\{\sum_h {}^b M_{gh}\,(^s\mathcal{M}_h)_{j\ell}\right\}\mathcal{S}_{j\ell} = \sum_{j,\ell} (^{s+b}\mathcal{M}_g)_{j\ell}\mathcal{S}_{j\ell}. \tag{10.50}$$

Again we arrive at an expression that has the required tensor form.

10.4.2.3 *Direct inversion of atomic coordinates* We have just shown that for both the thin film and the surface layer scattering we arrive at a tensor expression of the form

$$\Delta\mathcal{F} = \sum_{j,\ell} \mathcal{M}_{j\ell}\mathcal{S}_{j\ell}, \tag{10.51}$$

where for convenience we dropped the diffracted beam index g. In this section we will show not only that this tensor expression leads to substantial savings of computing time, but also that in some cases this expression may be used for the direct determination of crystal and surface structure. To a first order in δr, from eqn (10.44) we find

$$S_{jg} = \exp(-i\mathbf{g}\cdot\delta\mathbf{r}_j) - 1 \approx -i\sum_k^3 g_k\delta r_{jk},$$

and this leads to

$$\Delta\mathcal{F} = \sum_{j,h} \mathcal{M}_{jh} \sum_{k=1}^{3} (-ih_k \delta r_{jk}) = \sum_{\ell} \mathcal{T}_\ell \delta r_\ell \tag{10.52}$$

where $\ell = (j,k)$, the index k denotes one of the three orthogonal Cartesian axes x, y, z, and

$$\mathcal{T}_\ell = -i \sum_h h_k \mathcal{M}_{jh}.$$

Using the above tensor expression (10.52), the χ^2 function may be written as

$$\chi^2(\delta\mathbf{r}) = \sum_i \left[\frac{I^{(i)}_{exp} - CI^{(i)}_{cal}}{\sigma_i} \right]^2 \approx \sum_i \left[\frac{\Delta I^{(i)} - \sum_\ell \mathcal{N}_{i\ell}\delta r_\ell}{\sigma_i} \right]^2, \tag{10.53}$$

where

$$\Delta I^{(i)} = I^{(i)}_{exp} - C|\mathcal{F}_i(ref)|^2, \quad \mathcal{N}_{i\ell} = 2CRe\{\mathcal{F}_i^*(ref)\mathcal{T}_\ell^{(i)}\},$$

and i is the index of the data point.

The χ^2 function is minimum only if the first-order derivatives of the function with respect to all atomic displacements δr_ℓ vanish

$$\frac{\partial \chi^2}{\partial(\delta r_\ell)} = 0 = -\sum_i \frac{2}{\sigma_i^2} \left[\Delta I^{(i)} - \sum_{\ell'} \mathcal{N}_{i\ell'}\delta r_{\ell'} \right] \mathcal{N}_{i\ell}.$$

We can rewrite the above set of equations in a compact matrix form as

$$\mathcal{N}^T \cdot \mathbf{\Delta I} = (\mathcal{N}^T \cdot \mathcal{N}) \cdot \delta\mathbf{r}, \tag{10.54}$$

where $\delta\mathbf{r} = \{\delta x_1, \delta y_1, \delta z_1, ...\}$, $\mathbf{\Delta I} = \{\Delta I_i/\sigma_i\}$, $\{\mathcal{N}\}_{i\ell} = \{\mathcal{N}_{i\ell}/\sigma_i\}$ and $\{\mathcal{N}^T\}_{\ell i} = \{\mathcal{N}_{i\ell}/\sigma_i\}$. Formally the solution of this matrix equation can be written as

$$\delta\mathbf{r} = (\mathcal{N}^T \cdot \mathcal{N})^{-1}\mathcal{N}^T \cdot \mathbf{\Delta I}, \tag{10.55}$$

which shows that the atomic coordinates can be obtained by direct matrix inversion.

10.4.2.4 *Numerical examples* We now give two numerical examples illustrating how a structure can be determined by means of the tensor-theory-based direct schemes. The first example addresses the problem of the determination of the structure of the low temperature phase of $SrTiO_3$ (see Fig. 10.9). $SrTiO_3$ undergoes a phase transition from its high-temperature cubic form, where the space group symmetry is Pm3m, to a low temperature non-polar tetragonal form, where the space group symmetry is I4/mcm. This transformation occurs through the rotation of TiO_6 octahedra about a tetrad axis (Müller et al., 1968). Our aim is to determine the rotation angle of the TiO_6 octahedra in the low-temperature phase. Previously this structure was studied (Tsuda and Tanaka, 1995) using full dynamical theory and a trial-and-error method. A detailed structure model

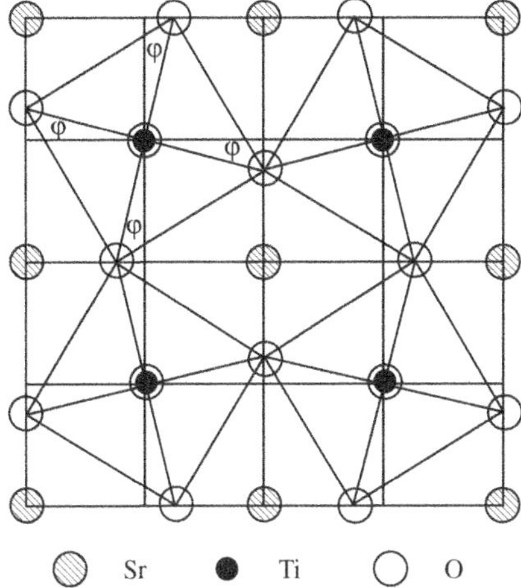

FIG. 10.9. Schematic diagram showing the [001] projected structure of an SrTiO$_3$ single crystal.

of the low temperature phase of SrTiO$_3$ was given in (Unoki and Sakudo, 1967), and the rotation angle ϕ of the TiO$_6$ octahedra was found to vary continuously from 0° at 103K to 2.1° at 4.2K.

Figure 10.10 shows curves of the variation of the transmitted beam intensity as a function of the angle ϕ of rotation of the TiO$_6$ octahedra. The curves were calculated for 100 keV primary beam energy, a [001] zone axis incidence and crystal thickness of 500 Å. For simplicity the same value of the Debye–Waller factor of 1.2 was used for all atoms in the crystal. The perturbation $\Delta U(\mathbf{r})$ in this case represents the difference between the potential of the low-temperature phase and that of the high-temperature phase. In Fig. 10.10a the solid curve was calculated using full dynamical theory, the dashed curve was calculated using first-order perturbation theory and the curve highlighted by circles was calculated using second-order perturbation theory. It is seen that while first-order perturbation theory works well for the entire range of possible rotation angles, second-order perturbation theory gives results that are almost identical to those obtained by using the full dynamical theory. Fig. 10.10b shows similar curves. The solid curve was calculated using the full dynamical theory, the dotted curve was calculated using first-order perturbation theory retaining only the term linear in the rotation angle ϕ in the expansion of the matrix $\mathcal{S}_{j,\ell}$, while the dot-dashed curve was calculated taking into account second-order terms. The first-order correction in the rotation angle ϕ (see the dashed curve) has a negligible effect on the transmitted beam intensity. This is because with the projected potential approximation the crystal structure is antisymmetric with respect to the rotation of the neighbouring oxygen atoms. The second-order correction (see the dot-dashed curve) works rather well. Over the entire range of

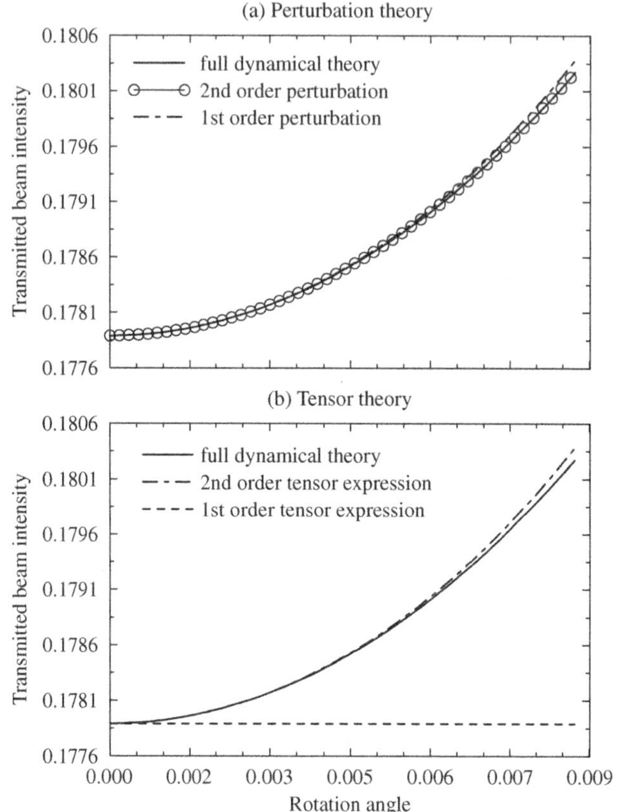

FIG. 10.10. Variations of the transmitted beam intensity as a function of the angle of rotation of the TiO_6 octahedra in an $SrTiO_3$ single crystal. The three curves in (a) were calculated using full dynamical theory and first-order and second-order perturbation theory, and in (b) the curves were calculated by using first-order perturbation theory based on first-order and second-order tensor expansions. The crystal thickness is 500 Å.

angles of rotation of the TiO_6 octahedra the second order correction agrees very well with full dynamical theory calculations.

We now consider a computer 'experiment' on the direct inversion of the rotation angle ϕ (see Fig. 10.11). Referring to Fig. 10.11, the 'actual' rocking curve was calculated for a rotation angle $\phi = 0.034$ rad, while the calculated 'reference' curve corresponds to a reference structure with $\phi = 0$. Since the first-order correction nearly vanishes, only one parameter has to be determined from the inversion scheme, i.e. the square of the rotation angle ϕ^2. The rocking curve obtained using the inversion procedure is shown in Fig. 10.11. This curve is seen to be practically indistinguishable from that of the ideal 'experimental' curve,

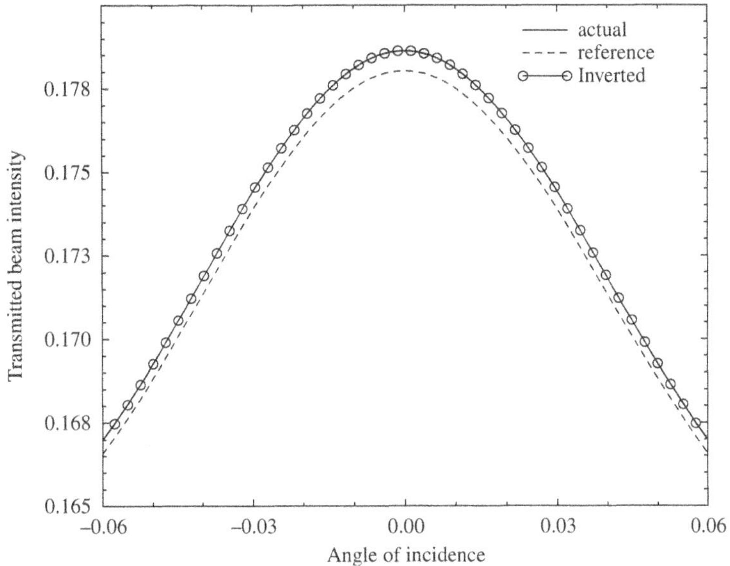

FIG. 10.11. Calculated and inverted transmitted beam rocking curves for 100 keV primary beam energy and a single crystal of SrTiO$_3$. The crystal thickness is 500 Å.

and the residual χ^2 value obtained by the inversion procedure is 5.107×10^{-5}, representing a perfect match between the 'actual' and the 'inverted' curves.

As our second numerical example we consider an Si(001) (2×1) reconstructed surface. The crystal consists of a 500 Å thick crystal slab and a reconstructed top surface layer of thickness 5.43 Å. Figure 10.12 shows the side view (along the [110] zone axis) of a bulk terminated ideal Si(001) surface structure. The surface atoms are known to relax in both the x and z directions (Yin and Cohen, 1981). Since in THEED geometry the amplitudes of diffracted beams are not sensitive to atomic displacements in the beam direction, we shall only consider atomic displacements along the x axis.

The intensity of the transmitted beam as a function of surface atomic relaxation is shown in Fig. 10.13. In this figure the zero of surface atom relaxation corresponds to the ideal bulk terminated surface structure, while unity corresponds to the Yin and Cohen surface atomic configuration. A total of eight surface atoms are involved in the relaxation, where the horizontal atomic displacements (along the x axis) are 0.573 Å, −1.038 Å, 0.093 Å, −0.115 Å, −0.007 Å, −0.034 Å, 0.061 Å and −0.060 Å respectively for atoms 1 to 4' of the Yin and Cohen model (see Fig. 10.12). In this figure the solid curve was calculated using full dynamical theory, while the dot dashed curve was calculated using first-order perturbation

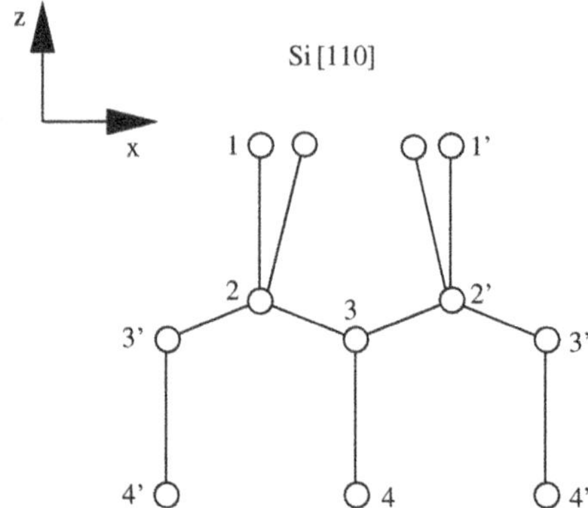

FIG. 10.12. Side view of a reconstructed Si(001) surface structure. The dimerization of the surface leads to the formation of strong bonds between the atoms 1 and 1'.

theory. This latter theory works well even for rather large atomic displacements, and this is consistent with the fact that in this case all the extinction distances are much greater than the thickness of the surface layer (5.43 Å) .

Figure 10.14 shows three curves of the transmitted beam intensity as a function of the surface atom relaxation parameter. These curves are similar to those given in Fig. 10.13, but now the dashed and dot–dashed curves in the figure were calculated using first-order and second-order tensor expressions respectively. This figure shows that both the first and the second order tensor expressions deviate from the result obtained using the full dynamical diffraction theory in the limit of large atomic relaxation. The range of validity of the second order tensor expression for the atomic coordinates is approximately given by 20% of the total relaxation towards the Yin and Cohen model, provided that we start from the ideal bulk terminated surface structure.

Although in the limit of large surface atom relaxation the tensor expression deviates from the dynamical diffraction theory calculation, in the limit of small relaxation the tensor expression is still adequate and direct inversion of structure is possible. Figure 10.15 shows two THEED rocking curves representing the 200 surface superlattice reflections from an Si(001) (2×1) surface. The 'actual' curve was calculated using the Yin and Cohen model (Yin and Cohen, 1981), while the 'reference' curve was calculated for a reference structure where two atoms (atoms 1 and 1' of Fig. 10.12) deviate from the actual structure by 0.1146Å and −0.2076 Å respectively. The inverted rocking curve obtained using the tensor inversion scheme is given in Fig. 10.15. The second-order tensor expressions were used. The inverted rocking curve is seen to be indistinguishable from that of the actual curve, confirming the validity of the tensor approach for a small perturbation of the crystal structure.

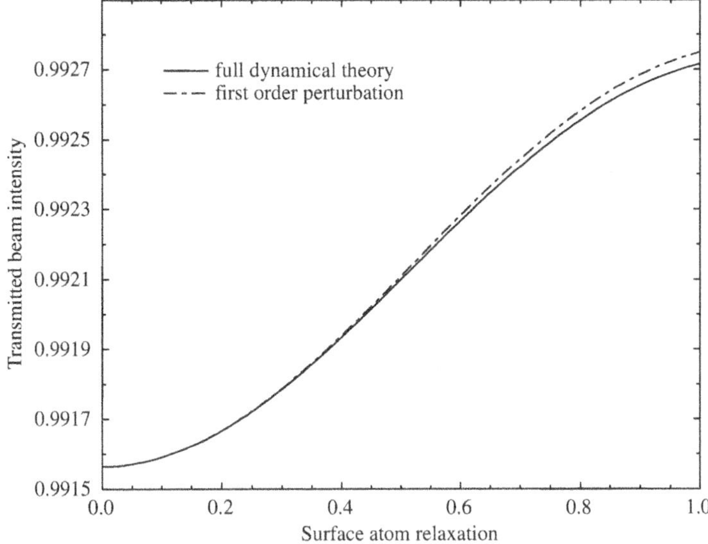

FIG. 10.13. Variations of the transmitted beam intensity as a function of the coordinate (see text) characterizing the relaxation of surface atoms. The curves shown in this figure were calculated for 100 keV primary beam energy for incidence along the [001] zone axis of the Si(2×1) surface.

10.5 Perturbation methods for non-periodic structures

In principle the perturbation methods developed for a periodic structure can also be used for the treatment of non-periodic objects. This can be achieved by using the periodic continuation method (Cowley, 1992), where we first define a giant unit cell containing defects or surfaces and then apply the Bloch-wave based method to an artificial crystal composed of the giant unit cells. In practice this approach is not very efficient and should be avoided whenever possible. In this section we consider perturbation methods that are particularly suited for treating non-periodic structures.

10.5.1 *The DWBA treatment of diffraction by a non-periodic structure*

We again start by separating the total potential $U(\mathbf{r})$ into two parts

$$U(\mathbf{r}) = U_0(\mathbf{r}) + \Delta U(\mathbf{r}). \tag{10.56}$$

We assume that we have a reference structure characterized by the potential $U_0(\mathbf{r})$, and that $\Delta U(\mathbf{r})$ introduces only a small perturbation. The motion of an electron in the potential $U_0(\mathbf{r})$ is described by the wave function $\psi_0(\mathbf{r})$ satisfying

$$[\nabla^2 + k_0^2 + U_0(\mathbf{r})]\psi_0(\mathbf{r}) = 0, \tag{10.57}$$

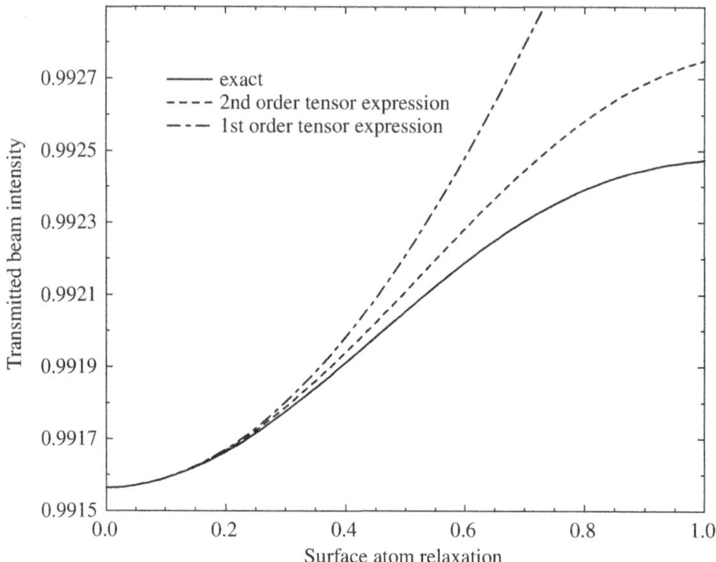

FIG. 10.14. Same as Fig. 10.13, except that the calculations were performed by using first- and the second-order tensor theory.

and the associated Green's function $G_U(\mathbf{r}, \mathbf{r}')$ satisfies

$$[\nabla^2 + k_0^2 + U_0(\mathbf{r})]G_U(\mathbf{r}, \mathbf{r}') = \delta(\mathbf{r} - \mathbf{r}'). \tag{10.58}$$

Our aim here is to find an approximate solution of the following inhomogeneous equation,

$$[k_0^2 + \nabla^2 + U_0(\mathbf{r})]\psi(\mathbf{r}) = -\Delta U(\mathbf{r})\psi(\mathbf{r}), \tag{10.59}$$

using the known wave function $\psi_0(\mathbf{r})$ and the Green's function $G_U(\mathbf{r}, \mathbf{r}')$. Formally the solution of (10.59) is given by

$$\psi(\mathbf{r}) = \psi_0(\mathbf{r}) - \int G_U(\mathbf{r}, \mathbf{r}')\Delta U(\mathbf{r}')\psi(\mathbf{r}')d\mathbf{r}', \tag{10.60}$$

and the validity of this expression can be easily verified by substituting it back into (10.59). The amplitude of scattering by the crystal is determined by the asymptotic form of the wave function in the region where $U(\mathbf{r}) = 0$. Let $\mathcal{F}_0(\mathbf{k})$ be the amplitude of scattering by potential $U_0(\mathbf{r})$ and $\mathcal{F}(\mathbf{k})$ be the amplitude of scattering by potential $U(\mathbf{r})$. For $\psi_0(\mathbf{r})$ we have

$$\psi_0(\mathbf{r}) \approx \exp(i\mathbf{k}_0 \cdot \mathbf{r}) + \frac{\exp(ikr)}{r}\mathcal{F}_0(\mathbf{k}), \tag{10.61}$$

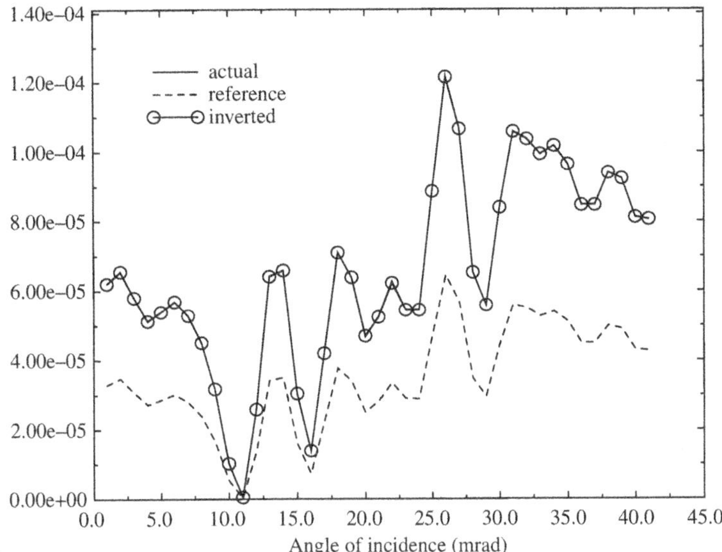

FIG. 10.15. Calculated and restored 200 superlattice diffracted beam rocking curves for a Si(001)-2×1 reconstructed surface.

and for $\psi(\mathbf{r})$

$$\psi(\mathbf{r}) \approx \exp(i\mathbf{k}_0 \cdot \mathbf{r}) + \frac{\exp(ikr)}{r}\mathcal{F}(\mathbf{k}). \tag{10.62}$$

Substitution of (10.61) and (10.62) into (10.60) gives

$$\mathcal{F}(\mathbf{k})\frac{\exp(ikr)}{r} = \mathcal{F}_0(\mathbf{k})\frac{\exp(ikr)}{r} - \int \mathsf{G}_U(\mathbf{r}, \mathbf{r}')\Delta U(\mathbf{r}')\psi(\mathbf{r}')d\mathbf{r}'. \tag{10.63}$$

To find an explicit expression for $\mathcal{F}(\mathbf{k})$ in terms of $\mathcal{F}_0(\mathbf{k})$ and $\Delta U(\mathbf{r})$, we perform a two-dimensional Fourier transform of both sides of eqn (10.63). For the Green's function we have

$$\mathsf{G}(\mathbf{r}, \mathbf{r}') = -\frac{1}{4\pi}\frac{\exp(ik|\mathbf{r} - \mathbf{r}'|)}{|\mathbf{r} - \mathbf{r}'|},$$

where by writing $\mathbf{k} = (\mathbf{Q}, k_z)$ and $\mathbf{r} = (\mathbf{X}, z)$, we find

$$\mathcal{F}(\mathbf{k}) \int \mathsf{G}(\mathbf{r}, 0) \exp(-i\mathbf{Q} \cdot \mathbf{X})d\mathbf{X} = \mathcal{F}_0(\mathbf{k}) \int \mathsf{G}(\mathbf{r}, 0) \exp(-i\mathbf{Q} \cdot \mathbf{X})d\mathbf{X}$$

$$-\frac{1}{4\pi}\int \left\{ \int \mathsf{G}_U(\mathbf{r}, \mathbf{r}') \exp(-i\mathbf{Q} \cdot \mathbf{X})d\mathbf{X} \right\}\Delta U(\mathbf{r}')\psi(\mathbf{r}')d\mathbf{r}'.$$

Using eqn (B.10) of Appendix B

$$\int \mathsf{G}(\mathbf{r})\exp(-i\mathbf{Q}\cdot\mathbf{X})d\mathbf{X} = -\frac{i}{2|k_z|}\exp(ik_z|z|),$$

the symmetry property of Green's function $\mathsf{G}_U(\mathbf{r},\mathbf{r}') = \mathsf{G}_U(\mathbf{r}',\mathbf{r})$ and eqn (B.18)

$$\int \mathsf{G}_U(\mathbf{r},\mathbf{r}')\exp(-i\mathbf{Q}\cdot\mathbf{X})d\mathbf{X} = -\frac{i}{2|k_z|}\exp(ik_z z)\psi_{-\mathbf{k}}(\mathbf{r}'),$$

we find

$$\mathcal{F}(\mathbf{k}) = \mathcal{F}_0(\mathbf{k}) + \frac{1}{4\pi}\int \psi_{-\mathbf{k}}(\mathbf{r}')\Delta U(\mathbf{r}')\psi(\mathbf{r}')d\mathbf{r}'. \tag{10.64}$$

Here we used the notation $\psi_{-\mathbf{k}}(\mathbf{r})$ to denote the electron wave function resulting from an incident wave of the form $\exp(-i\mathbf{k}\cdot\mathbf{r})$. In particular we have $\psi_0(\mathbf{r}) = \psi_{\mathbf{k}_0}(\mathbf{r})$.

An approximate solution of eqn (10.60) can be obtained by iteration. For the wave function we find

$$\psi(\mathbf{r}) = \psi_0(\mathbf{r}) - \int \mathsf{G}_U(\mathbf{r},\mathbf{r}')\Delta U(\mathbf{r}')\psi_0(\mathbf{r}')d\mathbf{r}' \tag{10.65}$$

$$+ \int \mathsf{G}_U(\mathbf{r},\mathbf{r}')\Delta U(\mathbf{r}')\mathsf{G}_U(\mathbf{r}',\mathbf{r}'')\Delta U(\mathbf{r}'')\psi_0(\mathbf{r}'')d\mathbf{r}'d\mathbf{r}'' + ...;$$

For the amplitude of scattering eqn (10.64) gives

$$\mathcal{F}(\mathbf{k},\mathbf{k}_0) = \mathcal{F}_0(\mathbf{k},\mathbf{k}_0) + \frac{1}{4\pi}\int \psi_{-\mathbf{k}}(\mathbf{r}')\Delta U(\mathbf{r}')\psi_{\mathbf{k}_0}(\mathbf{r}')d\mathbf{r}' \tag{10.66}$$

$$- \frac{1}{4\pi}\int \psi_{-\mathbf{k}}(\mathbf{r}')\Delta U(\mathbf{r}')\mathsf{G}_U(\mathbf{r}',\mathbf{r}'')\Delta U(\mathbf{r}'')\psi_{\mathbf{k}_0}(\mathbf{r}'')d\mathbf{r}'d\mathbf{r}'' + ...,$$

This is the distorted wave perturbation expansion in the power series of $\Delta U(\mathbf{r})$ (Dudarev et al., 1993a, Dudarev et al., 1993c). Here we have replaced the amplitude of scattering $\mathcal{F}(\mathbf{k})$ by $\mathcal{F}(\mathbf{k},\mathbf{k}_0)$ to emphasize that the amplitude depends on wave vectors of the incident and scattered waves. On the right-hand side of eqn (10.66), the first term denotes the amplitude of scattering of the high-energy electrons by potential $U_0(\mathbf{r})$. The second- and higher-order terms represent corrections due to single, double, and multiple scattering processes by the perturbation $\Delta U(\mathbf{r})$. Provided that the potential $U_0(\mathbf{r})$ is well chosen, in many applications the first order correction is sufficient, giving

$$\Delta\mathcal{F}(\mathbf{k},\mathbf{k}_0) = \mathcal{F}(\mathbf{k},\mathbf{k}_0) - \mathcal{F}_0(\mathbf{k},\mathbf{k}_0) \approx \frac{1}{4\pi}\int \psi_{-\mathbf{k}}(\mathbf{r}')\Delta U(\mathbf{r}')\psi_{\mathbf{k}_0}(\mathbf{r}')d\mathbf{r}'. \tag{10.67}$$

This is the distorted wave Born approximation (DWBA) that we discussed earlier in Chapter 7 in connection with the treatment of inelastic and diffuse scattering of electrons.

10.6 Tensor RHEED and the direct inversion of a surface structure

In the case of RHEED from a semi-infinite bulk crystal the variation of potential in the direction normal to the surface is not periodic and the DWBA may be used to investigate the effect of this perturbation on the observed RHEED intensities.

We assume that our reference structure characterized by a set of atomic coordinates $\{\mathbf{r}_i\}$ does not deviate significantly from the actual structure and that the distorted wave approximation applies. For the actual structure we have a set of atomic coordinates $\mathbf{r}_i(act) = \mathbf{r}_i + \delta\mathbf{r}_i$. The total potential can be partitioned as

$$U(\mathbf{r}) = -\frac{2m}{\hbar^2} \sum_j \varphi_j(\mathbf{r} - \mathbf{r}_j - \delta\mathbf{r}_j) = U_0(\mathbf{r}) + \Delta U(\mathbf{r}), \tag{10.68}$$

where

$$U_0(\mathbf{r}) = -\frac{2m}{\hbar^2} \sum_j \varphi_j(\mathbf{r} - \mathbf{r}_j), \tag{10.69}$$

and $\varphi(\mathbf{r})$ is the potential of an individual atom. Expanding eqn (10.68) we have

$$\Delta U(\mathbf{r}) = \sum_{j,k} \mathcal{T}_{jk}^{(1)} \delta r_{jk} + \sum_{j,k,l} \mathcal{T}_{jkl}^{(2)} \delta r_{jk} \delta r_{jl} + \sum_{j,k,l,m} \mathcal{T}_{jklm}^{(3)} \delta r_{jk} \delta r_{jl} \delta r_{jm} + ..., \tag{10.70}$$

where indices k, ℓ, m denote one of the three orthogonal Cartesian axes. The first two terms in eqn (10.70) are given by

$$\mathcal{T}_{jk}^{(1)} = -\frac{2m}{\hbar^2} \nabla_k \varphi(\mathbf{r} - \mathbf{r}_j), \quad \mathcal{T}_{jkl}^{(2)} = -\frac{2m}{\hbar^2} \frac{1}{2} \nabla_k \nabla_l \varphi(\mathbf{r} - \mathbf{r}_j) \tag{10.71}$$

and analytical expressions for these tensors can be found by using analytical expression for the electron atomic scattering factor.

By substituting (10.70) into (10.67), we find a tensor expression for the amplitude of scattering

$$\Delta\mathcal{F}(\mathbf{k}, \mathbf{k}_0) = \sum_{j,k} \mathcal{M}_{jk}^{(1)} \delta r_{jk} + \sum_{j,k,l} \mathcal{M}_{jkl}^{(2)} \delta r_{jk} \delta r_{jl} + ..., \tag{10.72}$$

where

$$\mathcal{M}^{(n)} = \frac{1}{4\pi} \int \psi_{-\mathbf{k}}(\mathbf{r}) \mathcal{T}^{(n)}(\mathbf{r}) \psi_{\mathbf{k}_0}(\mathbf{r}) d\mathbf{r}. \tag{10.73}$$

The above expression (10.72) is quite general, and it can be used in both the RHEED and the THEED geometries of diffraction for either periodic or aperiodic structures. If the reference structure is close to the actual structure, we can use the first-order tensor expression

$$\Delta\mathcal{F}(\mathbf{k}, \mathbf{k}_0) \approx \sum_{j,k} \mathcal{M}_{jk}^{(1)} \delta r_{jk}. \tag{10.74}$$

In this case the amplitudes of diffracted beams depend linearly on the atomic displacements.

It is often convenient to retain the two-dimensional periodicity of the crystal in the plane parallel to the surface and write

$$U(\mathbf{r}) = \sum_G U_G(z) \exp(i\mathbf{G} \cdot \mathbf{x}), \qquad (10.75)$$

where the $\mathbf{G}$'s form a set of two-dimensional reciprocal lattice vectors parallel to the surface. Applying Bloch's theorem to the two-dimensional periodic crystal, we find

$$\psi_{\mathbf{k}_0}(\mathbf{r}) = \exp(i\mathbf{k}_0 \cdot \mathbf{x}) \sum_G \phi_G(\mathbf{k}_0, z) \exp(i\mathbf{G} \cdot \mathbf{x}). \qquad (10.76)$$

Substitution of (10.75) and (10.76) into (10.67) gives

$$\Delta\mathcal{F}(\mathbf{k}, \mathbf{k}_0) \approx \frac{1}{4\pi} \sum_G \sum_{G'} \sum_{G''} \int d\mathbf{x} \exp[i(\mathbf{k}_0 - \mathbf{k} + \mathbf{G} + \mathbf{G}' + \mathbf{G}'') \cdot \mathbf{x}] \qquad (10.77)$$

$$\times \int dz \phi_G(\mathbf{k}_0, z) \Delta U_{G'}(z) \phi_{G''}(-\mathbf{k}, z)$$

$$= \pi \sum_G \sum_{G'} \sum_{G''} \delta(\mathbf{k}_{0t} - \mathbf{k}_t + \mathbf{G} + \mathbf{G}' + \mathbf{G}'') \int dz \phi_G(\mathbf{k}_0, z) \Delta U_{G'}(z) \phi_{G''}(-\mathbf{k}, z),$$

where the subscript t denotes the surface parallel component of a vector. In RHEED the wave function in the vacuum region has the form of a superposition of plane waves

$$\psi(\mathbf{r}) = \exp(i\mathbf{k}_0 \cdot \mathbf{r}) + \sum_G \mathcal{R}_G \exp[i(\mathbf{k}_0 + \mathbf{G}) \cdot \mathbf{x} - i\sqrt{k_0^2 - (\mathbf{k}_0 + \mathbf{G})_t^2}\, z]. \qquad (10.78)$$

In terms of $\mathcal{R}_G$ eqn (10.77) may be rewritten as

$$\Delta\mathcal{R}_G = \frac{i}{2\sqrt{k_0^2 - (\mathbf{k}_0 + \mathbf{G})_t^2}} \sum_{G'} \sum_{G''} \int dz \phi_{G'}(\mathbf{k}_0, z) \Delta U_{G-G'-G''}(z) \phi_{G''}(-\mathbf{k}_G, z).$$

$$(10.79)$$

As an illustration we consider the surface relaxation occurring on the Ni(001) $p(2\times2)$ surface (see Fig. 10.16 and (Rous, 1992)). We consider the RHEED geometry of diffraction. Figure 10.17 shows two curves which illustrate how the amplitude of the specular beam varies as a function of surface relaxation. The latter is defined as the displacement of the Ni atoms of the top surface layer in the surface normal direction, where the reference atomic positions are taken as those corresponding to the bulk terminated crystal. Positive values in the figure correspond to an inward surface relaxation. Calculations were performed for an angle of incidence of 58 mrad using the (exact) one-rod approximation of dynamical RHEED theory (Ichimiya, 1987, Peng and Whelan, 1991a) and the tensor approximation. The TDS effects were treated using the Einstein model (Hall and Hirsch, 1965b).

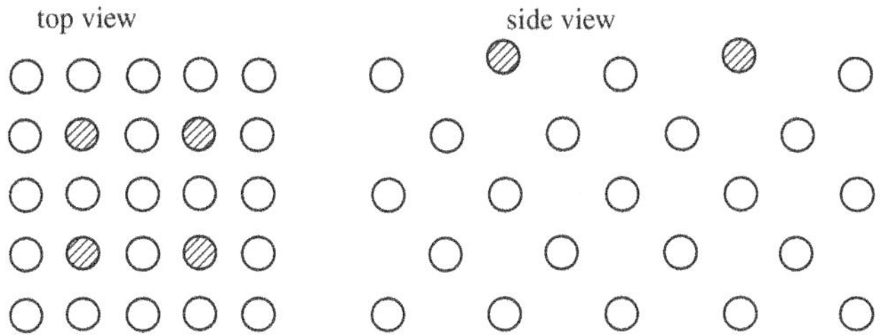

FIG. 10.16. Schematic diagrams showing the top and side views of the Ni(100) $p(2\times2)$ reconstructed surface. The shaded circles represent displaced atoms of the top surface layer (in the surface normal direction).

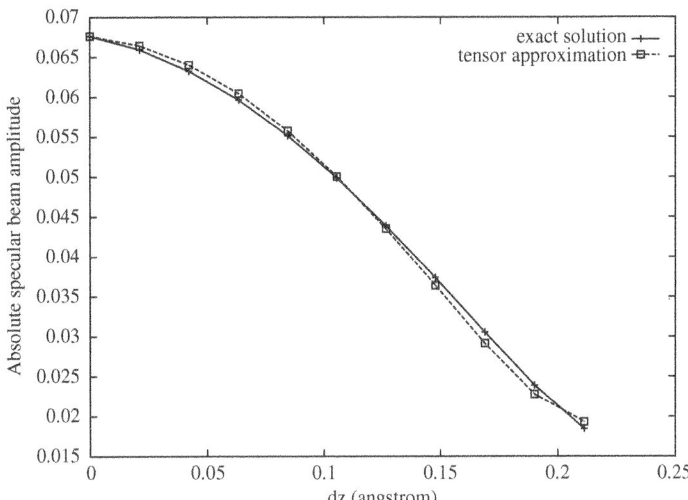

FIG. 10.17. Variation of the amplitude of the specular reflected beam as a function of the displacement of the top surface atom in the Ni(001) $p(2 \times 2)$ surface. The origin of the system of coordinates corresponds to the bulk terminated atom position, and the calculations were made for the angle of incidence 58 mrad and an electron energy of 12.5 keV.

The Debye–Waller factor of 0.16 Å^2 used in calculations corresponds to the temperature of 93 K (Radi, 1970). This figure shows that the tensor approximation works well over a fairly wide range of atomic displacements characterizing the surface relaxation.

By using the tensor expressions we can apply the inversion schemes discussed in the previous sections to determine directly the amplitude of surface relaxation. Fig. 10.18 shows two RHEED rocking curves calculated for the Ni(001) $p(2 \times 2)$ surface. The exact curve was calculated for the relaxed surface for dz=0.15 Å, while the reference curve was calculated for the bulk terminated reference surface.

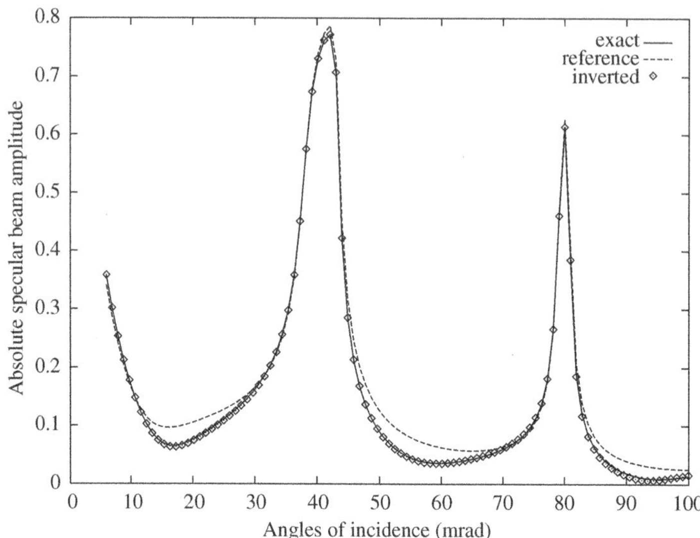

FIG. 10.18. The one-rod RHEED rocking curves calculated for the Ni(001) $p(2 \times 2)$ surface. Calculations were performed using eqn (10.79), and terms up to the third order were used for the evaluation of ΔU (Peng and Dudarev, 1993b).

Using the set of ΔI_i characterising the difference between the two curves and the tensor expressions for the reference structure, we solved the inverse problem of surface relaxation and calculated the inverted rocking curve shown in Fig. 10.18. We found that the inverted rocking curve is indistinguishable from the actual curve, again confirming the high accuracy of the tensor approach.

10.7 Summary

Perturbation theory can be successfully applied to a wide range of practical problems including diffuse and inelastic scattering and crystal and surface structure determination. The starting point of any perturbation treatment consists in choosing a reference structure such that the potential may be represented in the form of a sum of the potential of the known reference structure U_0 and a perturbation term ΔU

$$U(\mathbf{r}) = U_0(\mathbf{r}) + \Delta U(\mathbf{r}).$$

For a known reference structure we calculate the eigenvalues γ and eigenvectors B_g. Corrections due to the presence of the perturbation can be written in a simple tensor form

$$\Delta\gamma^{(j)} = {}^1\boldsymbol{\mathcal{U}}^{(j)} \cdot \Delta\mathbf{U} + \Delta\mathbf{U} \cdot {}^2\boldsymbol{\mathcal{U}}^{(j)} \cdot \Delta\mathbf{U},$$

$$\Delta B_g^{(j)} = {}^1\boldsymbol{\mathcal{E}}_g^{(j)} \cdot \Delta\mathbf{U} + \Delta\mathbf{U} \cdot {}^2\boldsymbol{\mathcal{E}}_g^{(j)} \cdot \Delta\mathbf{U},$$

$$\Delta\overline{B}_g^{(j)} = {}^1\overline{\boldsymbol{\mathcal{E}}}_g^{(j)} \cdot \Delta\mathbf{U} + \Delta\mathbf{U} \cdot {}^2\overline{\boldsymbol{\mathcal{E}}}_g^{(j)} \cdot \Delta\mathbf{U},$$

where the diffracted beam amplitudes are given by

$$\mathcal{F}_g = \sum_j (\overline{B}_0^{(j)} + \Delta\overline{B}_0^{(j)})(B_g^{(j)} + \Delta B_g^{(j)}) \exp[i(\gamma^{(j)} + \Delta\gamma^{(j)})z].$$

A full dynamical calculation of the amplitudes of diffracted beams (which scales as N^3 as a function of the total number N of excited electron beams) can now be reduced to a sequence of multiplications. A perturbation approach makes it possible to develop schemes for solving the inverse diffraction problem and for the direct determination of crystal structure factors and coordinates of atoms in bulk specimens and at surfaces.

11

DIGITAL ELECTRON MICROGRAPH RECORDING AND BASIC PROCESSING

11.1 Introduction

In this chapter we will discuss the recording and basic processing of digitized electron micrographs, including electron microscope images or diffraction patterns. The major advantage associated with using digitized images is that they may be readily processed with computers and compared with theoretical calculations.

The commonly used electron recording media include photographic film, TV cameras, charge-coupled device (CCD) cameras (Mooney et al., 1990), and image plates (IP) (Ogura et al., 1994). Ideally an electron recording device should be able to record electron images without degrading their resolution and adding appreciable noise, and to have a large number of pixels, wide dynamic range, and excellent linearity. An ideal electron recording device should also be of the on-line type, i.e. it should be able to display images simultaneously or within seconds of the moment when they are recorded. Photographic film and IP are off-line recording media and are not suitable for the majority of on-line applications.

CCD cameras surpass photographic film in dynamic range (typically more than 16000 counts for the slow-scan CCD (SSC) but only about 200 grey levels for film), linearity (typically better than 1% over the whole dynamic range for the SSC but poor in the case of photographic film), granularity (largely absent in the SSC but a significant source of noise in film), and sensitivity. A CCD camera is slightly worse than film in absolute resolution, but this can be compensated for by increasing the microscope magnification. It also has a smaller number of pixels than film (2048×2048 for the SSC compared with as high as 10000×10000 for film (Krivanek et al., 1991)). The main advantage of TV cameras is their speed of recording. The recorded images are, however, very noisy and are not suitable for quantitative applications. In the first three sections of this chapter we will discuss basic elements of the SSC camera and their characteristics..

The major drawback of digitizing a micrograph is that inevitably a finite number of pixels is involved in this process, and therefore the obtainable level of resolution is limited (for SSC the camera size is typically 1024×1024 pixels and for IP 3000×4000 pixels). On the one hand, to achieve the best possible resolution one should use as many pixels as possible when recording a micrograph digitally, but on the other hand by using more pixels or a larger camera size one inevitably needs a longer recording time and greater computer storage. Therefore from a practical point of view, it is desirable to use as few pixels as possible, but at the same time this should have as small an effect as possible on the level

388

of image resolution. In the last three sections of this chapter we will discuss all these questions, i.e. how to choose the digital camera size or the number of pixels for recording a digital micrograph and how to recover fine details of an originally continuous micrograph from its digital counterpart.

11.2 Basic features of CCDs

A CCD is a sensor based on a large crystalline silicon wafer. The wafer is divided into many square potential wells formed by a conducting gate structure situated on the top of a thin layer of silicon dioxide grown on the surface of the silicon wafer. Illumination by the incident electrons generates electron–hole pairs in the potential wells. The resulting charge is collected in each well and is read serially by an output amplifier. The readings are digitized using an analog-to-digital (A/D) converter, and the digitized data are transferred to and stored in a computer. Pictorially a CCD may be considered as a two-dimensional array of square elements (pixels) generating an image.

Although in principle incident electrons may be used to generate a distribution of induced charge in a CCD, in practice a scintillator is also involved in the process of recording an image. This is because a single 100 keV electron fired directly into a CCD would produce about 30000 electron–hole pairs so that only 17 primary electrons would generate more than 500000 thermalized electrons per pixel and saturate the CCD. For this level of electron dose the statistical noise would be of the order of $\sqrt{17}$ and the signal-to-noise ratio (SNR) would be of the order of 24%. For many TEM applications the image contrast lies in the range 5–20%, and a level of noise of the order of 24% would therefore obscure most of the features of the image. It is therefore essential to reduce the number of electron-hole pairs in a CCD chip. The best way of achieving this consists in using a scintillator such as a YAG (Yttrium Aluminium Garnet) crystal or a phosphor to convert the incident electrons into photons (Krivanek et al., 1991).

Slow-scan CCD sensors differ from the usual CCDs in several respects. Firstly, the area of pixels of an SSC sensor is larger, permitting more electrons to be stored in each pixel and producing a larger dynamic range. Secondly, the SSCs are optimized for operating at lower temperatures and slower readout speed. This minimizes the CCD dark current and readout noise and increases the sensitivity to a level where a single electron may be detected.

Figure 11.1 shows a sketch of an SSC camera for recording electron images described by Spence and Zuo (Spence and Zuo, 1988), and by Krivanek and Mooney (Krivanek and Mooney, 1993). This design uses a single-crystal YAG scintillator or a gadolinium oxysulphide phosphor to detect the incident electrons and convert the signal into photons. The photons generated by the scintillator are transferred to the CCD by a fiber-optics coupling. The silicon chip is usually cooled by a Peltier cooler to about -30°C to reduce the dark current generated by the CCD. The readout voltage of the CCD signal is digitized by a 12 bit A/D converter. The digitization is typically set so that one count corresponds to one or two electrons incident on a YAG scintillator, or 0.2–0.4 electrons incident on a

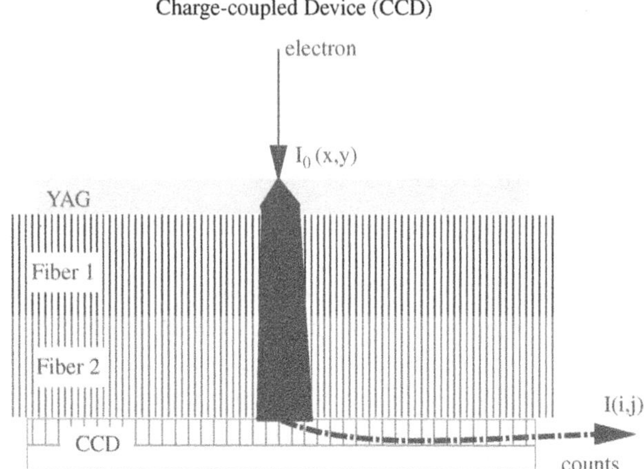

FIG. 11.1. Schematic diagram of a SSC camera for recording electron images.

gadolinium oxysulphide scintillator. This design illustrated in Fig. 11.1 requires an exposure time in the range between 0.1 s and 100 s.

11.3   A basic model of an SSC camera

The process of electron detection by an SSC camera may be divided into three separate stages (Daberkow et al., 1991, Ishizuka, 1993, Zuo, 1997): (1) the conversion of incident electrons into photons in the scintillator, (2) the transport of photons from the scintilator to the CCD array via a fiber-optic or lens coupling, and (3) conversion of photons into electrons and the readout of the resulting signal in the CCD.

Let $\overline{E}$ be the average energy of the electron absorbed in the scintillator, $\overline{E_{ph}}$ be the average energy of photons generated in the scintillator, and ϵ be the efficiency of the energy conversion of the incident electron into a photon. The average number of photons generated by an incident electron is given by

$$n_{ph} = \epsilon \overline{E} / \overline{E}_{ph}. \tag{11.1}$$

For a $Y_3Al_5O_{12}:Ce^{2+}$ (YAG) scintillator the average photon energy $\overline{E}_{ph} = 2.21$ eV and the energy conversion coefficient $\epsilon = 5\%$ (Daberkow et al., 1991). The absorption of a 100 keV electron produces 2262 photons in the scintillator. Among these photons only a fraction of the order of $\eta_L n_{ph}$ would reach the CCD chip, where η_L is the efficiency of the coupling system between the scintillator and the CCD. The number of electron–hole pairs produced by a photon in the CCD is described by the quantum efficiency coefficient η. An incident high-energy electron therefore produces an average number of electron–hole pairs $\overline{n}_e$ in the CCD given by

$$\overline{n}_e = \eta\eta_L n_{ph}. \tag{11.2}$$

Taking into account that for a typical fiber plate the efficiency of the coupling system η_L is of the order of 6%, and that for a THX-31156 CCD the quantum efficiency coefficient $\eta = 30\%$, we find $\overline{n}_e = 40$ for a 100 keV primary electron. For the readout process, assuming that each digital count from the A/D converter corresponds to S electron–hole pairs, we obtain an estimate of the overall gain

$$g = \overline{n}_e/S, \tag{11.3}$$

which is the average number of digital counts produced by each incident electron. The digitization of the readout voltage from the SSC by the A/D converter is typically adjusted in such a way that one count corresponds to one or two primary electrons incident on the YAG, i.e. S is of the order of 20 to 40.

The response of an SSC camera to a single incident electron is described by the point spread function (PSF) of the SSC camera. This function measures the lateral spread of a single electron signal caused by the angular spread of photons due to multiple scattering of photons in the scintillator and in the fibers. Mathematically these processes may be expressed in the form of a convolution

$$I(i,j) = gh(i,j) * I_0(i,j) + B(i,j) \tag{11.4}$$

$$= \int_{y_j-\Delta y/2}^{y_j+\Delta y/2} dy \int_{x_i-\Delta x/2}^{x_i+\Delta x/2} dx \left[g \int\int I_0(X,Y)h(X-x,Y-y)dXdY \right] + B(i,j),$$

where $I(i,j)$ is the SSC readout from a pixel (i,j), $I_0(x,y)$ is the intensity distribution at the entrance surface of the scintillator, Δx and Δy are the width and the height of the pixel (for a typical SSC $\Delta x = \Delta y$). $h(x,y)$ represents the PSF and $B(i,j)$ denotes the background noise of the SSC.

11.4   Main characteristics of an SSC camera

An SSC is characterized by (1) the sensitivity or the overall gain g, (2) the dynamic range, (3) the PSF $h(x,y)$, and (4) the detector quantum efficiency (DQE) (Herrmann and Krahl, 1984). In this section we will consider the meaning of these parameters.

11.4.1   *The overall gain*

As discussed in the preceding section, the overall gain factor g is given by the average number of digital counts produced by an incident electron. Alternatively the overall gain may be defined as the ratio of the average number of SSC output counts $\overline{I}$ to the average electron dose per pixel $\overline{N}_e$

$$g = \overline{I}/\overline{N}_e. \tag{11.5}$$

Experimentally the overall gain may be measured by fitting the observed $\overline{I} - \overline{N}_e$ curves, where $\overline{N}_e$ may be estimated on the basis of the exposure time

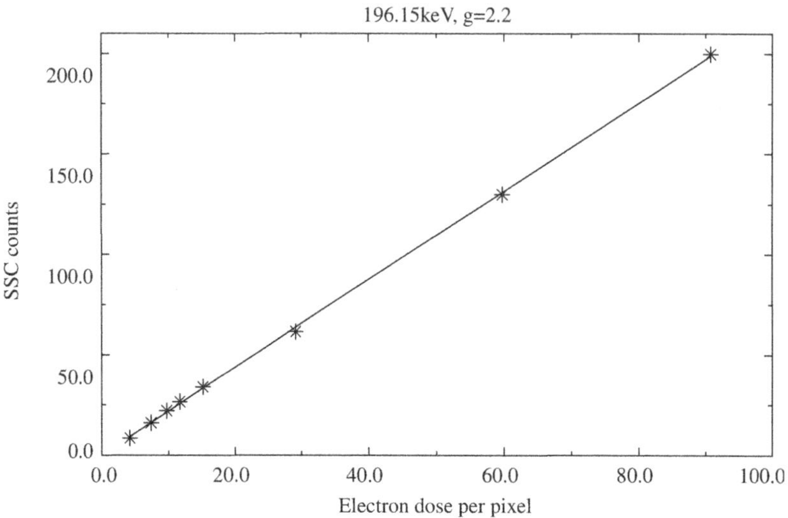

FIG. 11.2. SSC counts $\overline{I}$ vs incident electron dose per pixel $\overline{N}_e$ for an SSC.

given by the microcontroller of the electron microscope. Figure 11.2 gives an example of such measurements for an SSC fitted to a Philips CM200/FEG electron microscope. By fitting a straight line to these measurements we obtain a value for the overall gain g of 2.2. This means that on average we should expect to obtain 220 counts for every 100 incident electrons.

The overall gain depends on a number of factors and varies as a function of scintillator thickness and efficiency of coupling and conversion across the pixels of the SSC. Fluctuations of the gain factor may be approximately eliminated by means of the gain normalization operation

$$I(i,j) = \frac{I^{raw}(i,j) - I^{dark}(i,j)}{I^{gain}(i,j) - I^{dark-ref}(i,j)}, \tag{11.6}$$

where $I^{raw}(i,j)$ is the raw image, and $I^{dark}(i,j)$ is the dark current image recorded just before the acquisition of the raw image. The gain or reference image $I^{gain}(i,j)$ is acquired with uniform illumination, and $I^{dark-ref}(i,j)$ is the dark-current image recorded just before the acquisition of the gain image. Figure 11.3 shows the effect of gain normalization on the image obtained for uniform illumination. It is seen that while the raw image (Fig. 11.3a) shows many structural features associated with the structure of the YAG crystal and coupling fibers, these unwanted features are completely removed from the raw image through the gain normalization procedure. The resulting gain normalized image (Fig. 11.3d) shows a uniform intensity distribution superimposed on a noisy background, as is expected for uniform illumination.

FIG. 11.3. (a) Raw image, (b) gain or reference image, (c) dark-current image, and (d) gain normalized image.

11.4.2 *Resolution and the point spread function*

The resolution of an SSC camera is described by its point spread function (PSF), which is the intensity distribution recorded by the SSC in the case of a single incident electron. Since the spread of the signal in the scintillator and optical fibers is symmetric in all directions, the PSF is rotationally symmetric and may be described by a one-dimensional modulation transfer function (MTF) defined as

$$M(q) = |H(q, 0)|, \tag{11.7}$$

where

$$H(u, v) = \frac{1}{\Delta^2} \int \int h(x, y) \exp[-2\pi i(ux + vy)] dx dy \tag{11.8}$$

is the Fourier transform of the PSF $h(x, y)$, and Δ is the pixel size. In principle, the best level of resolution that can be achieved using a SSC camera is equal to the size of one pixel. The ideal MTF for an SSC is therefore given by the function

$$M(q) = \left| \frac{1}{\Delta^2} \int_{-\Delta/2}^{\Delta/2} \int_{-\Delta/2}^{\Delta/2} \exp(-2\pi i q x) dx dy \right| = \frac{\sin(\pi q \Delta)}{\pi q \Delta}. \tag{11.9}$$

Among the many experimental methods for measuring the MTF, the noise method (de Ruijter and Weiss, 1992) is the most convenient and widely used. The basic principle of this method is the following. The scintillator of the SSC is evenly illuminated by a wide electron beam at low magnification and high electron dose. The expected number of electrons incident on each pixel of the YAG scintillator remains approximately constant. The actually recorded number of electrons is a stochastic variable governed by Poisson statistics. In the limit of high electron dose the recorded image intensity is much higher than the dark current $B(i, j)$ and the latter may be safely neglected. The recorded image intensity is approximately given by

$$I(x, y) = gh(x, y) * I_0(x, y),\tag{11.10}$$

where $h(x, y)$ is the PSF and $I_0(x, y)$ is the intensity distribution of the incident electrons at the entrance surface of the scintillator. To a good approximation this intensity distribution may be regarded as white noise since the number of electrons incident on a pixel is independent of the number of electrons incident on any other pixel. Since the Fourier transform of white noise is a constant function, i.e. $\mathcal{F}\{I_0(x, y)\}$ =const, we obtain

$$H(u, v) = C \times \mathcal{F}\{I(x, y)\},\tag{11.11}$$

in which C is a constant which may be found by imposing the requirement $H(0, 0) = 1$, which is equivalent to the normalization condition imposed on the PSF

$$\int h(x, y)dxdy = 1.0.\tag{11.12}$$

Figure 11.4 shows the angular averaged cross-section of the experimental MTF measured for an SSC operated at 196.15 keV primary beam energy. Although the angular averaged one-dimensional MTF is characterized by the reduced level of noise compared with the two-dimensional raw experimental data, this function is still very noisy and not very well suited for the image deconvolution. In practice the experimental one-dimensional MTF is usually fitted by an analytical function suitable for subsequent manipulations like the deconvolution of electron microscope images and Fourier transformations. The most widely used function has the form

$$M(q) = \frac{a}{1 + \alpha q^2} + \frac{b}{1 + \beta q^2} + c.\tag{11.13}$$

The experimental curve shown in Fig. 11.4 is described by the following five fitting parameters $a = 0.1812$, $\alpha = 7985$, $b = 0.7666$, $\beta = 19.6225$, and $c = 0.0522$. The PSF of an SSC may be obtained simply by the Fourier transform of the one-dimensional MTF. Figure 11.5 shows a PSF obtained by the Fourier transform of the fitted curve of Fig. 11.4. The half-width of the PSF is approximately equal

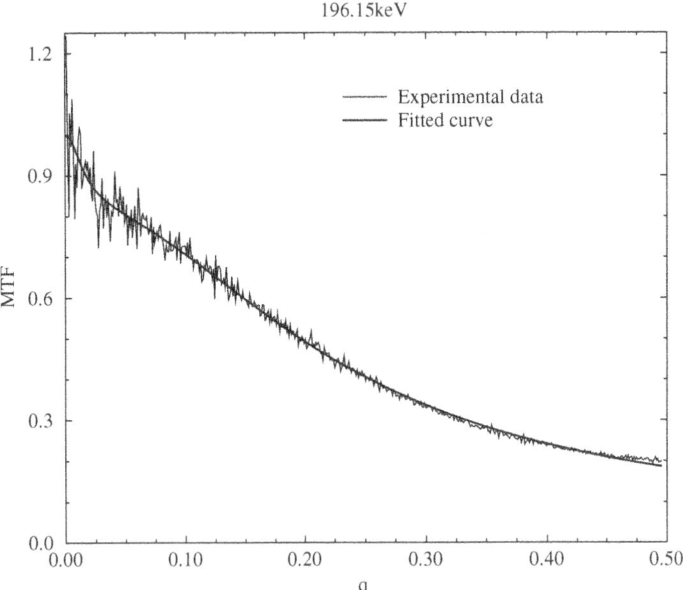

FIG. 11.4. Modulation transfer function (MTF) measured using the noise method. In the figure q is measured in units of $1/\Delta$, with Δ being the size of a pixel.

to two pixels, suggesting that for a point illumination the recorded digital image spreads over at least nine pixels. Fortunately this spread may be corrected using the image deconvolution technique (Ren et al., 1997).

11.4.3 *The detection quantum efficiency*

The noise performance of a detector is commonly characterized by the detection quantum efficiency (DQE), which is a parameter defined as the square of the ratio of the output and input signal-to-noise ratios (SNR)

$$\text{DQE} = \left(\frac{\text{SNR}_{out}}{\text{SNR}_{in}}\right)^2 = \frac{\overline{I}^2/\text{Var}[I]}{\overline{N_e}^2/\text{Var}[N_e]}, \tag{11.14}$$

where $\overline{N_e}$ is the averaged number of electrons incident on the surface of the CCD chip and $\text{Var}[N_e]$ is the corresponding variance, i.e. $\text{Var}[N_e] = \overline{(\delta N_e)^2}$ with $\delta N_e = N_e - \overline{N_e}$, and $\text{SNR}_{in} = \overline{N_e}/\sqrt{\overline{(\delta N_e)^2}}$. $\overline{I}$ is the averaged output number of electrons from the CCD, and $\text{Var}[I]$ is the corresponding variance. The DQE function is a measure of the noise amplification of the detector. For an ideal detector where every electron is detected with no additional noise added we have DQE=1.

Assuming a uniform illumination $I_0(i,j)$ with an average number of electrons per pixel $\overline{N_e}$, the intensity distribution on the entrance surface of the CCD chip

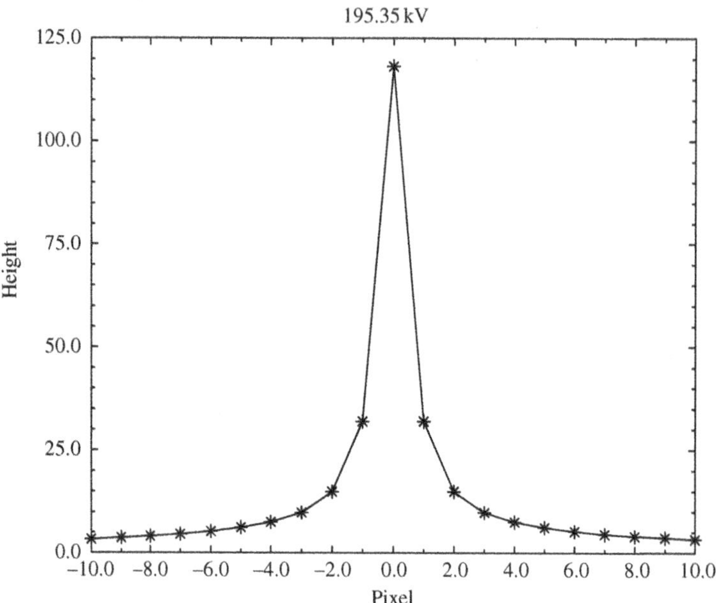

FIG. 11.5. Point spread function (PSF) for a SSC.

is given by

$$N_e(i,j) = \sum_{l,m} h(i-l, j-m) I_0(l,m). \tag{11.15}$$

According to the error propagation rule the variance in $N_e(i,j)$ is then given by

$$\mathrm{Var}[N_e(i,j)] = \sum_{l,m} \left\{ \frac{\partial N_e(i,j)}{\partial I_0(l,m)} \right\}^2 \mathrm{Var}[I_0(l,m)] = m\overline{N_e}, \tag{11.16}$$

where we have used the relation

$$\mathrm{Var}[I_0(i,j)] = \overline{N_e}, \tag{11.17}$$

owing to the fact that the incident electrons obey Poisson statistics, and

$$m = \sum_{l,m} \left\{ \frac{\partial N_e(i,j)}{\partial I_0(l,m)} \right\}^2 = \sum_{l',m'} h^2(l',m'). \tag{11.18}$$

Experimentally the factor m may be found using the MTF described in the preceding section and the two-dimensional Parseval theorem

$$m = \sum_{l,m} h_{l,m}^2 = \frac{1}{N^2} \sum_{u=0}^{N-1} \sum_{v=0}^{N-1} |H(u,v)|^2. \tag{11.19}$$

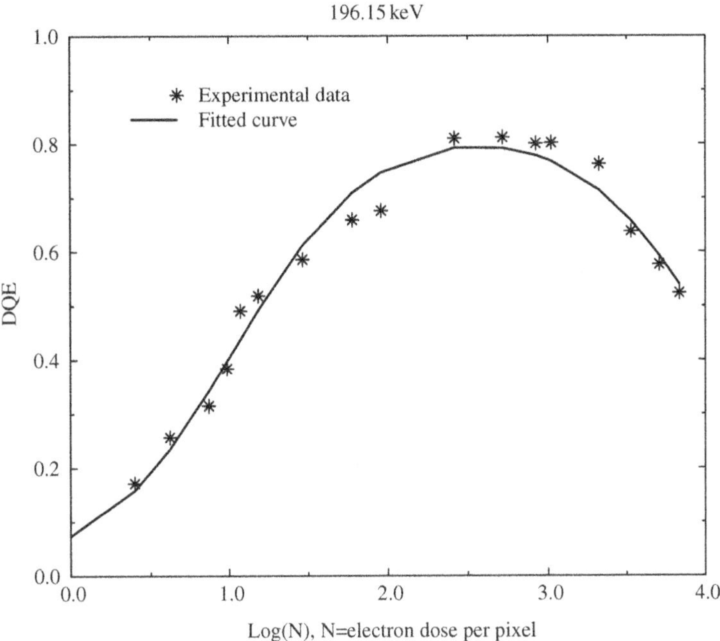

FIG. 11.6. Experimentally measured detection quantum efficiency (DQE) as a function of the number of electrons incident on each pixel of an SSC.

Substituting the expressions for $\mathrm{Var}[N_e]$ into eqn (11.14) and noticing that $g = \overline{I}/\overline{N_e}$, we obtain

$$\mathrm{DQE} = mg\overline{I}/\mathrm{Var}[I]. \tag{11.20}$$

It should be noted that DQE depends on the average number of output electrons. Experimentally for a given m and g we can estimate DQE for different $\overline{I}$ using noisy images, and this is the method used in quantitative CBED (Ren et al., 1997).

Figure 11.6 shows an experimentally measured dependence of DQE on the number of electrons incident on each pixel of the SSC attached to a Gatan Image Filter (GIF). This dependence may be fitted with a physical model based on the electron detection process

$$\mathrm{DQE}(\overline{I}) = \frac{1}{A + B\overline{I} + C/\overline{I}}, \tag{11.21}$$

and for the experimental curve shown in Fig. 11.6 the fitting gives $A = 1.186$, $B = 9.83 \times 10^{-5}$, and $C = 12.9$. This figure clearly shows that for either a very low electron dose (less than 100) or a very high dose (higher than 10000) the noise is greatly amplified. The electron dose range corresponding to the best performance of the SSC lies in the interval between a few hundred and a few thousand.

11.5 The sampling theorem

In this and in the following two sections we will consider how to choose the size of a digital camera for recording a digital micrograph, and how to recover fine details of the original object from the recorded micrograph. The answer to the first question is given by the sampling theorem, and in this section we will introduce the sampling theorem using a one-dimensional model. The extension to two dimensions can be made readily by adding one more variable. In the most common case the electron micrograph $I(x)$ is sampled at evenly spaced intervals. If Δ denotes such an interval, the digital micrograph is given by

$$I_n = I(n\Delta), \quad n = ..., -2, -1, 0, 1, 2, ... \tag{11.22}$$

In information theory the reciprocal of the interval Δ is called the sampling rate, and is related to a characteristic frequency f_c given by (Goodman, 1968)

$$f_c = 1/(2\Delta), \tag{11.23}$$

where f_c is called the Nyquist critical frequency (Bates and McDonnell, 1986). The sampling theorem states that if a continuous function $I(x)$, sampled at intervals Δ, happens to be bandwidth limited to frequencies smaller in magnitude than f_c, i.e. the Fourier transform of $I(x)$ equals zero for all $|f| > f_c$, then the function $I(x)$ is completely determined by its sample values I_n. In fact $I(x)$ is given explicitly by the formula

$$I(x) = \Delta \sum_{n=-\infty}^{\infty} I_n \frac{\sin[2\pi f_c(x - n\Delta)]}{\pi(x - n\Delta)}. \tag{11.24}$$

Since the transfer of information in an electron microscope is not perfect, all electron micrographs are bandwidth limited, i.e. all information is lost beyond the information limit of the electron microscope. The Nyquist critical frequency is then related to the resolution limit of the electron microscope u_{lim}, which normally does not exceed 1.0 Å^{-1}. When an objective aperture is used for recording an image, f_c equals the radius of the corresponding aperture in reciprocal space. It is a corollary of the sampling theorem that for all electron microscopes, provided that the micrograph is sampled with more than two points per $r_0 = 1/u_{lim}$, a continuous intensity distribution can be retrieved completely from its discrete digital counterpart. Alternatively if a given level of point resolution r_0 is to be achieved when recording an electron image, the image should be sampled with at least two pixels per r_0.

We now consider the consequences of undersampling, i.e. the situation where a continuous image is sampled with a frequency which is smaller than the Nyquist critical frequency. In this case the entire spectral density lying outside of the frequency range $-f_c < u < f_c$ is spuriously moved back into the range as illustrated in Fig. 11.7. This phenomenon is called aliasing. What happens is that frequencies from outside of the frequency range defined by f_c are aliased or falsely

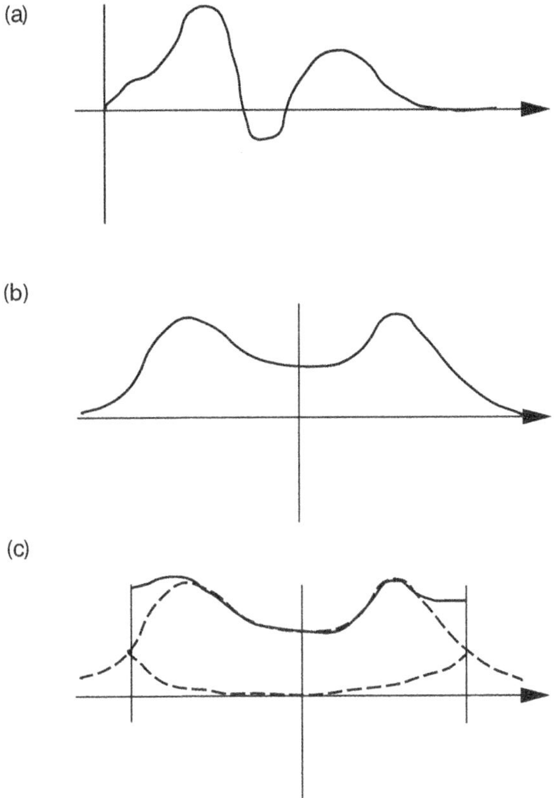

FIG. 11.7. Schematic diagrams showing (a) a continuous function sampled with an
interval Δ in real space; (b) the continuous Fourier transform of (a); (c) the discrete
Fourier transform of (a) and the effect of aliasing.

translated into that range by the very act of discrete sampling. This effect can be
eliminated either by sampling the continuous image at a rate sufficiently rapid
to satisfy the sampling theorem or by low-pass filtering the original image before
sampling (e.g. by using a smaller objective aperture).

11.6 Discrete and fast Fourier transform

11.6.1 *Discrete Fourier transform*

In this section we provide the mathematical background for much of the following
discussion of image processing. We will first evaluate the Fourier transform of
a function defined at a finite number of sampling points. The functions we are
interested in are periodic functions or functions that are defined in a finite interval
that can be treated as a single period of a periodic function. Figure 11.8 illustrates
the situation.

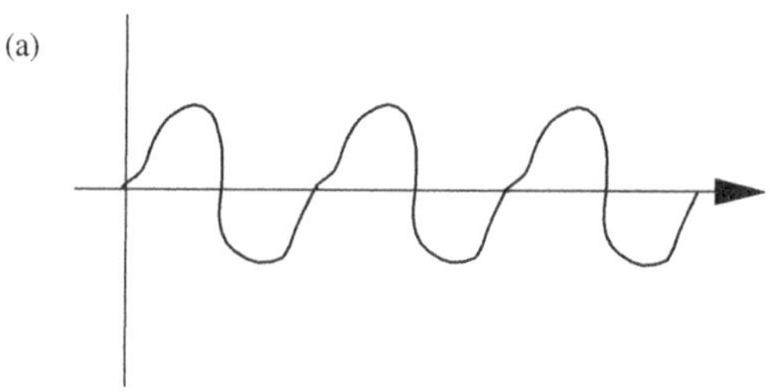

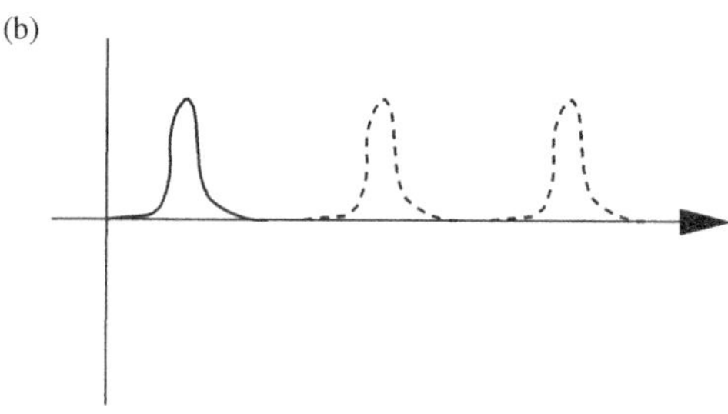

FIG. 11.8. Schematic diagrams showing (a) a periodic function and (b) an arbitrary periodic function generated from a function which is defined in a finite interval only.

Suppose that we have N consecutive sampled values of a one-dimensional image $I(x)$

$$I_k = I(x_k), \quad x_k = k\Delta, \quad k = 0, 1, 2, ... N - 1. \tag{11.25}$$

Since the functions we are interested in are either periodic functions or functions defined in a finite interval, the whole interval of x is assumed to be contained in the range of N points given above. We would like to evaluate the Fourier transform $J(u)$ at discrete points

$$u_n = \frac{n}{N\Delta}, \quad n = -\frac{N}{2}, ..., \frac{N}{2}. \tag{11.26}$$

The Fourier transform of a continuous function $I(x)$ is given by

$$J(u) = \int_{-\infty}^{\infty} I(x)\exp(-2\pi iux)dx, \tag{11.27}$$

which may be approximated using the trapezoidal rule, which expresses the integral as a discrete sum of terms of the form $\frac{1}{2}[I(x_i) + I(x_{i+1})]$ to yield

$$J(u_n) = \int_{-\infty}^{\infty} I(x)\exp(-2\pi iu_n x)dx \approx \frac{1}{2}\Delta \sum_{k=0}^{N-1}[I(x_k)+I(x_{k+1})]\exp(-2\pi ikn/N).$$

$$\tag{11.28}$$

Since the function $I(x)$ is periodic, i.e. $I(x_0) = I(x_N)$, the above expression becomes

$$J_n = \Delta \sum_{k=0}^{N-1} I_k \exp(-2\pi i k n/N), \tag{11.29}$$

and the summation in the above equation is called the discrete Fourier transform (DFT) of a function defined at N points of I_k. It should be pointed out that while the DFT is only an approximate form of the continuous Fourier transform, it has symmetry properties that are almost exactly the same as the properties of the continuous Fourier transform. A formula for the discrete inverse Fourier transform, which recovers the set I_k from the set J_n is

$$I_k = \frac{1}{N} \sum_{n=0}^{N-1} J_n \exp(2\pi i k n/N). \tag{11.30}$$

11.6.2 *Fast Fourier transform (FFT)*

The direct application of eqn (11.29) to the calculation of $\{J_n\}$ involves N^2 complex multiplications plus a smaller number of operations required to generate quantities like $\exp(-2\pi i k n/N)$. In other words it is an $O(N^2)$ process. A careful analysis reveals, however, that the DFT can be computed by performing only $O(N \log_2 N)$ operations with an algorithm called the Fast Fourier Transform (FFT). For an image of 1024×1024 pixels the difference between $N \log_2 N$ and N^2 is approximately 52428 times! To illustrate this point we rewrite eqn (11.29) as follows:

$$
\begin{aligned}
J_n &= \Delta \sum_{k=0}^{N-1} I_k \exp(-2\pi i k n/N) \\
&= \sum_{k=0}^{N/2-1} I_{2k} \exp(-2\pi i n(2k)/N) + \sum_{k=0}^{N/2-1} I_{2k+1} \exp[-2\pi i n(2k+1)/N] \\
&= \sum_{j=0}^{N/2-1} I_{2k} \exp[-2\pi i n j/(N/2)] + W^n \sum_{k=0}^{N/2-1} I_{2k+1} \exp[-2\pi i n k/(N/2)] \\
&= F_n^e + W^n F_n^o,
\end{aligned}
\tag{11.31}
$$

where $W = \exp(-2\pi i/N)$, F_n^e denotes the n-th component of the Fourier transform of $\{I_{2k}\}$, and F_n^o is the Fourier transform of $\{I_{2k+1}\}$. If N is an integer power of 2, the procedure can be repeated recursively to obtain F_n^{eo}, F_n^{eoo}, etc., until we have subdivided the data all the way down to transforms of length 1. The DFT is then a sum of $\log_2 N$ terms of the form $F_n^{eoeooee\ldots eoe}$, and the calculation of the total of N components of J_n is a process of order $O(N \log_2 N)$. There are many implementations of this algorithm for calculating the DFT, and a detailed account of the method is given in (Brigham, 1974) and (Press et al., 1986).

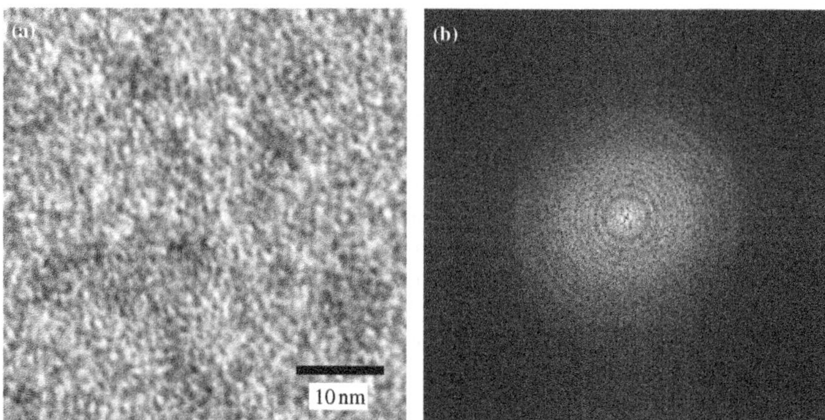

FIG. 11.9. Digital (a) HREM image and (b) corresponding diffractogram of an amorphous thin film. The HREM image was obtained for 200 keV primary beam energy, a defocus value of -549 Å, and it consists of 256×256 pixels.

11.7 Restoration of images

In this section we will discuss how to retrieve missing data points in a continuous image from its digital finite size counterpart (Peng, 1996). This procedure is in agreement with the sampling theorem, which states that in principle a continuous micrograph may be retrieved fully from the corresponding discrete digital micrograph provided that the bandwidth of the latter is limited and provided that the sampling rate satisfies the requirements of the sampling theorem.

Figure 11.9 shows a digital HREM image of an amorphous silicon thin film and also the corresponding diffractogram (DFT of the image). The HREM image was recorded using a GATAN slow-scan CCD camera using 256×256 pixels and a 200 keV primary beam. The diffractogram clearly shows that the real-space HREM image is bandwidth limited and that the sampling theorem is satisfied or, in other words, the diffractogram contains all frequencies of the spectrum up to the information limit or the Nyquist frequency. The resolution of the digital image and the diffractogram shown in Fig. 11.9 is rather moderate. In what follows we will discuss how to generate more data points, and how to add them to the original digital micrographs to improve the resolution.

11.7.1 *Generation of data points in reciprocal space*

We first discuss the generation of data points in reciprocal space. For simplicity we consider only a one-dimensional image $I(x)$. Extension to two dimensions can be made readily by adding one more variable. If the real-space image is sampled at evenly spaced points we have a set of sampled values $\{I(x_n),\ x_n = n\Delta\}$, where n is an integer. In reciprocal space the sampling interval is given by $\Delta u = 1/(N\Delta)$, where N is the total number of the sampled values. Usually the DFT is performed using discrete values $u_n = n/(N\Delta)$, with $n = -N/2, ..., N/2$. The

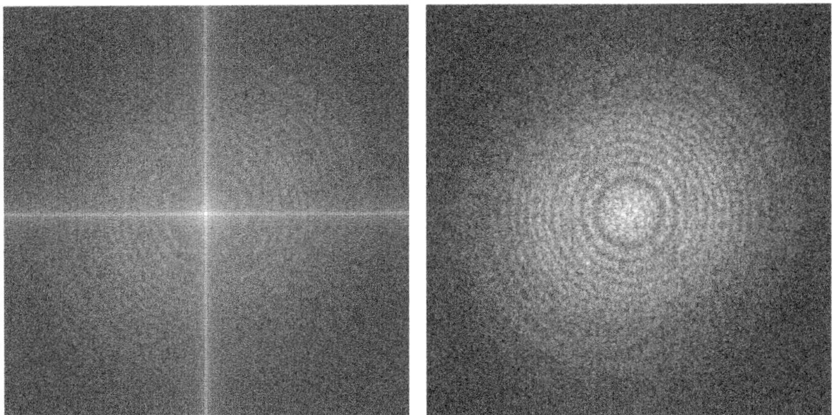

FIG. 11.10. High-resolution diffractograms generated using four repeated FFTs of Fig. 11.9a (a, left) without and (b, right) with the mean intensity of the original HREM image subtracted.

DFT of the electron image is given by (11.29) that maps N values of $I(x_k)$ to N values of $J(u_n)$. While it is true that with N values of the input, we will only be able to produce no more than N independent values of the output, we can nevertheless generate *more* than N values of the output, although not all of them will be independent. In fact, for any spatial frequency u we can always write $u = u_n + \delta u$ with $\delta u < 1/(N\Delta)$ to obtain

$$J(u_n + \delta u) = \Delta \sum_{k=0}^{N-1} I(x_k) \exp(-2\pi i u x_k)$$

$$= \Delta \sum_{k=0}^{N-1} \left\{ I(x_k) \exp(-2\pi i \delta u x_k) \right\} \exp(-2\pi i k n / N). \quad (11.32)$$

Comparing this expression with eqn (11.29) we see immediately that in order to obtain another set of values of $J(u)$ at points displaced from the normal set of $\{u_n\}$ by δu, all that we need to do is first multiply the original real-space image by a phase function to obtain $I(x) \exp(-2\pi i \delta u x)$, and then to Fourier-transform this new function to obtain $\{J(u_n + \delta u)\}$. In this way the data points in the reciprocal space can be increased indefinitely using repeated applications of the FFT.

Figure 11.10 shows two diffractograms generated from Fig. 11.9a using four repeated applications of the FFT. It is seen that while Fig. 11.10a contains more detail than Fig. 11.9b, this diffractogram is dominated by a pair of bright crossed lines. These lines result from the DFT of a constant term. To illustrate this point we write $I(x)$ as

$$I(x) = I'(x) + I_0, \quad (11.33)$$

where I_0 is the mean intensity of $I(x)$. The DFT of $I(x)$ is given by

$$J(u) = \Delta \sum_{k=0}^{N-1} \{I'(x_k) + I_0\} \exp(-2\pi i u x_k)$$

$$= J'(u) + \Delta I_0 \exp(-\pi i u \Delta) \frac{\sin[\pi i u (N-1)\Delta]}{\sin(\pi i u \Delta)}, \qquad (11.34)$$

where $J'(u)$ is the DFT of $I'(x)$ and the second term on the right hand side gives rise to one of the bright crossed lines seen in Fig. 11.10a. To remove this bright cross in Fig. 11.10b we show the calculated DFT of $I(x) - I_0$ rather than that of $I(x)$. The bright cross has now disappeared from the diffractogram, which shows features typical of the white-noise characteristics of an amorphous material.

11.7.2 *Generation of data points in real space*

While it is possible to use (11.24) to generate data points in real space, this operation proves to be an $O(mN^2)$ process if we want to add mN more data points to the original digital micrograph. A more efficient scheme follows directly from the consideration given in the previous section. For any x in real space we can always write $x = \delta x + x_k$ with $\delta x < \Delta$. The direct inverse DFT then gives

$$I(\delta x + x_k) = \frac{1}{N} \sum_{n=0}^{N-1} J(u_n) \exp[2\pi i(\delta x + k\Delta)u_n] \qquad (11.35)$$

$$= \frac{1}{N} \sum_{n=0}^{N-1} \Big\{ J(u_n) \exp[2\pi i \delta x n/(N\Delta)] \Big\} \exp[2\pi i k n/N],$$

and this equation shows that $I(x)$ can be obtained simply from the inverse DFT of $J(u_n) \exp[2\pi i \delta x n/(N\Delta)]$. Since the inverse DFT can be obtained by using the FFT algorithm, the operation of adding mN more data points is then an $O(mN \log_2 N)$ process. This scheme is therefore $N/\log_2 N$ times faster than an approach based on formula (11.24). For an image of 1024×1024 pixels the improvement in computational speed is nearly 53000-fold.

Figure 11.11 shows (a) an original 64×64 and (b) a generated 512×512 digital HREM image of an amorphous silicon thin film. Figure 11.11b was obtained by using the multiple FFT algorithm that added 64 times more data points to the original digital micrograph. This figure clearly shows some fine detail that was not visible in the original micrograph shown in Fig. 11.11a.

11.8 Summary

The slow-scan CCD camera is a digital electron recording device that has excellent linearity and allows on-line recording of electron images. A slow-scan CCD camera is characterized by its overall gain factor g, which is the average number of counts produced by each incident electron, by its point spread function, which

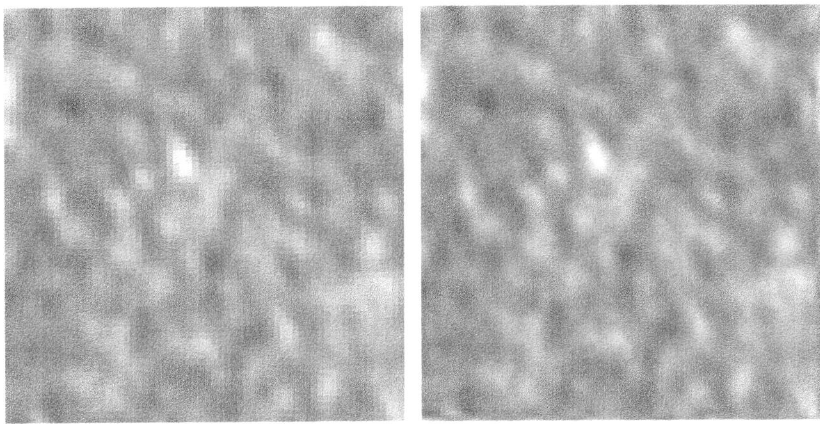

FIG. 11.11. HREM image of amorphous silicon with (a, left) 64×64 pixels, and (b, right) 512×512 pixels.

limits the resolution of the camera, and by its detector quantum efficiency (DQE) which characterizes the camera's ability to generate noise-free images. These parameters depend on the type of the camera and may be measured experimentally following the procedure described in this chapter.

The minimum number of pixels that should be used to record a digital micrograph is two pixels per r_0, where r_0 is equal either to the required level of resolution in real space or to the inverse of the bandwidth or the cut-off of the Fourier spectrum of the original image. For digital micrographs recorded with this condition satisfied, all fine detail of the original image may be recovered from the digital micrograph.

12

IMAGE FORMATION AND THE RETRIEVAL OF THE ELECTRON WAVE FUNCTION

12.1 Introduction

In this chapter we discuss the process of image formation in a transmission electron microscope. For electron microscopes fitted with either a tungsten or LaB_6 filament the illumination may be regarded as perfectly incoherent, while for field-emission gun (FEG) electron microscopes the illumination is partially coherent. In the second section of this chapter we introduce the concept of coherence in the context of electron illumination. In the third section we then proceed with a discussion of how images are formed in both a conventional electron microscope and an FEG electron microscope. In the final section we discuss various approaches to image deconvolution in order to retrieve the wave function of high-energy electrons at the exit surface of the sample.

12.2 Electron source and coherence

In an electron microscope an effective electron source can be defined to lie in the exit pupil of the second condenser lens, which is usually taken as coinciding with the illumination aperture. The effective source is an imaginary electron emitter filling the illumination aperture. The effective source of electrons is partially coherent, while real electron sources emit electrons incoherently. The problem of partial coherence results from two physical features of real electron guns, namely the finite energy spread and hence the spread of wavelength, and the finite emitting area.

12.2.1 *Partial coherence and the complex degree of coherence*

The electron wave field generated jointly at a certain point by several different sources is equal to the superposition

$$\psi = \psi_1 + \psi_2 + \psi_3 + ..., \tag{12.1}$$

of wave fields $\psi_1, \psi_2, \psi_3, ...$ generated at this point separately by each source. In the simplest case two sources generate two plane waves characterized by the same frequency ω. The wave functions ψ_1 and ψ_2 in this case have the form

$$\psi_1 = A_1 \exp[i(\mathbf{k}_1 \cdot \mathbf{r} - \omega t + \phi_1)], \quad \psi_2 = A_2 \exp[i(\mathbf{k}_2 \cdot \mathbf{r} - \omega t + \phi_2)], \tag{12.2}$$

where the phases ϕ_1 and ϕ_2 are introduced to allow for any phase difference between the sources of the two waves. If the phase difference $\phi_1 - \phi_2$ is constant,

406

the two sources are said to be *mutually coherent*. If the sources of the two waves are *mutually incoherent*, then the quantity $\phi_1 - \phi_2$ varies in a random fashion as a function of time.

The superposition of two plane waves results in an intensity distribution

$$I = \psi\psi^* = |\psi_1 + \psi_2|^2 = A_1^2 + A_2^2 + 2A_1 A_2 \cos\phi,$$

where

$$\phi = (\mathbf{k}_1 - \mathbf{k}_2) \cdot \mathbf{r} + \phi_1 - \phi_2.$$

In the actual case of interference of two or more waves, the amplitudes and phases of the waves usually vary with time in a random fashion. It is usually more convenient to define a time-averaged intensity

$$I = \langle \psi\psi^* \rangle = \langle A_1^2 + A_2^2 + 2A_1 A_2 \cos[(\mathbf{k}_1 - \mathbf{k}_2) \cdot \mathbf{r} + (\phi_1 - \phi_2)]\rangle. \tag{12.3}$$

We now consider a special case where the two waves originate from the same common source. The phases of the two waves differ because of the difference between their optical paths. Let us call t the time required for one wave to follow path 1 and $t + \tau$ the time required for the other to follow path 2. The interference term in eqn (12.3) may be written as

$$\Gamma_{12}(\tau) = \langle \psi_1(t)\psi_2^*(t + \tau)\rangle. \tag{12.4}$$

The function $\Gamma_{12}(\tau)$ is called the *mutual coherence function* or the *correlation function* of the two waves ψ_1 and ψ_2, and in particular the function

$$\Gamma_{11}(\tau) = \langle \psi_1(t)\psi_1^*(t + \tau)\rangle$$

is known as the *autocorrelation function* or the *self-correlation function*.

In practice it is convenient to use a normalized correlation function called the *complex degree of partial coherence*

$$\gamma_{12}(\tau) = \frac{\Gamma_{12}(\tau)}{\sqrt{\Gamma_{11}(0)\Gamma_{22}(0)}}.$$

The effect of this normalization is to ensure that $|\gamma_{12}|$ lies between zero and unity (Born and Wolf, 1980), i.e. $|\gamma_{12}| \leq 1$. In terms of $|\gamma_{12}(\tau)|$ we have the following types of coherence:

$$|\gamma_{12}(\tau)| = \begin{cases} 1 & \text{complete coherence,} \\ <1 & \text{partial coherence,} \\ 0 & \text{complete incoherence.} \end{cases} \tag{12.5}$$

In a pattern of interference fringes the intensity varies between two limits:

$$I_{max} = I_1 + I_2 + 2\sqrt{I_1 I_2}|\gamma_{12}|, \quad I_{min} = I_1 + I_2 - 2\sqrt{I_1 I_2}|\gamma_{12}|.$$

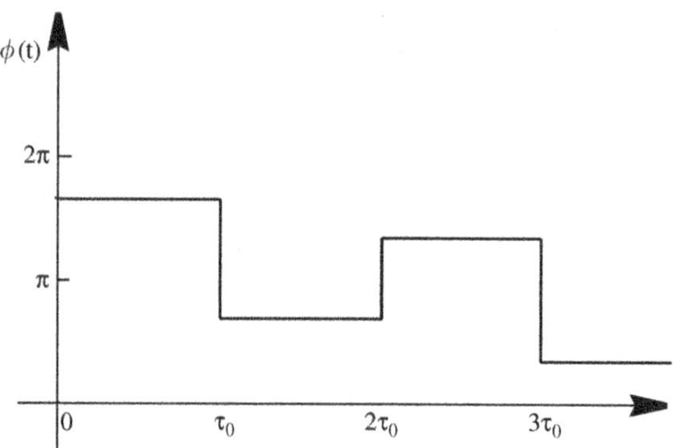

FIG. 12.1. Phase angle $\phi(t)$ changes randomly in the range 0 to 2π after each time interval equal to the coherence time τ_0.

The *fringe visibility* ν is defined as the ratio

$$\nu = \frac{I_{max} - I_{min}}{I_{max} + I_{min}} = \frac{2\sqrt{I_1 I_2}}{I_1 + I_2}|\gamma_{12}|, \tag{12.6}$$

i.e. the fringe visibility is proportional to the modulus of the degree of partial coherence.

12.2.2 Temporal coherence

The concept of temporal coherence is associated with the spread of energy of the electrons or the spread of electron wavelength. To treat this problem we first consider a quasi-monochromatic wave

$$\psi(t) = A \exp[i(\mathbf{k} \cdot \mathbf{r} - \omega t)] \exp[i\phi(t)], \tag{12.7}$$

where the phase $\phi(t)$ is a random step function of duration τ_0, usually called the coherence time. This phase changes randomly in the range 0 to 2π as shown in Fig. 12.1.

Assuming that $|\psi_1| = |\psi_2| = A$, we obtain

$$\gamma_{12}(\tau) = \langle \exp(i\omega\tau) \exp\{i[\phi(t) - \phi(t + \tau)]\}\rangle$$

$$= \exp(i\omega\tau) \lim_{T\to\infty} \frac{1}{T} \int_0^T \exp\{i[\phi(t) - \phi(t + \tau)]\}dt.$$

First consider the case when $\tau > \tau_0$. The integral above can then be expressed as the sum of integrals over each successive interval of time τ_0. For a particular interval the phase difference $\Delta\ (= \phi(t) - \phi(t + \tau))$ will be constant for $\phi(t)$ varying as in Fig. 12.1. Δ will vary randomly from interval to interval, and therefore the sum of contributions for all intervals will average to zero (random phase approximation), i.e.

$$\int_0^\infty \exp\{i[\phi(t) - \phi(t + \tau)]\}dt = 0, \text{ for } \tau > \tau_0.$$

On the other hand, if $\tau < \tau_0$, then for the first time interval $0 < t < \tau_0$, we have

$$\phi(t) - \phi(t + \tau) = \begin{cases} 0 & \text{if } 0 < t < \tau_0 - \tau \\ \Delta & \text{if } \tau_0 - \tau < t < \tau_0, \end{cases}$$

where Δ is a random phase difference between 0 and 2π. For the first interval, we have

$$\frac{1}{\tau_0}\int_0^{\tau_0} \exp\{i[\phi(t) - \phi(t + \tau)]\}dt = \frac{\tau_0 - \tau}{\tau_0} + \frac{\tau}{\tau_0}\exp(i\Delta).$$

For all subsequent intervals, using the same argument, we may obtain the same result. Since Δ is random, the whole integral averages to the first term. We therefore obtain for the partial coherence function

$$\gamma_{12} = \begin{cases} (1 - \tau/\tau_0)\exp(i\omega\tau) & \text{if } \tau < \tau_0, \\ 0 & \tau \geq \tau_0. \end{cases} \tag{12.8}$$

Evidently the fringe visibility $|\gamma|_{12}$ decreases linearly as a function of time, and drops to zero if τ exceeds the coherence time τ_0. This means that the path difference between the two beams must not exceed the value $v\tau_0 = l_c$ in order to obtain interference fringes, where v is the velocity of the electron. The quantity l_c is called the *coherence length*. It is in fact the length of an uninterrupted wave train.

We now investigate the relationship between the frequency spread, or the line width and the coherence of an electron source. In particular we will consider a case where the wave function $\psi(t)$ represents a single wave train of finite duration τ_0. The temporal variation of this wave train is given by the function

$$\psi(t) = \begin{cases} \exp(-i\omega_0 t) & \text{if } -\tau_0/2 < t < \tau_0/2 \\ 0 & \text{otherwise.} \end{cases} \tag{12.9}$$

Taking the Fourier transform we arrive at

$$g(\omega) = \frac{1}{2\pi}\int_{-\infty}^{+\infty} \psi(t)\exp(i\omega t)dt = \frac{1}{\pi} \cdot \frac{\sin[(\omega - \omega_0)\tau_0/2]}{\omega - \omega_0}. \tag{12.10}$$

We see that the spectral distribution is maximum for $\omega = \omega_0$ and that it drops to zero for $\omega = \omega_0 \pm 2\pi/\tau_0$. The 'width' $\Delta\omega$ of the frequency distribution is given by

$$\Delta\omega = 2\pi/\tau_0, \quad \text{or} \quad \delta\nu = 1/\tau_0.$$

Now if we consider a sequence of wave trains, and if the pulses are not all of the same duration, i.e. if τ_0 varies from pulse to pulse, then we can think of an average time $\langle\tau_0\rangle$. The precise form of the spectral distribution is different from

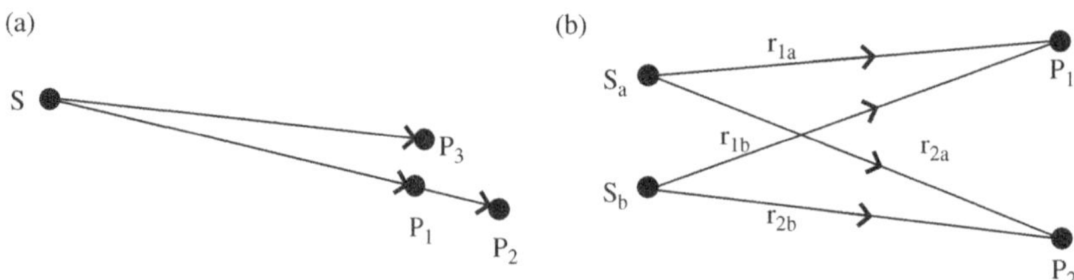

FIG. 12.2. Schematic diagram showing longitudinal and transverse spatial coherence
of (a) one source of electrons and (b) two sources of electrons.

that of the single pulse, but the width of the corresponding frequency spectrum
is still proportional to $\langle \tau_0 \rangle^{-1}$. On the other hand, if a spectral source has a line
width $\Delta \nu$, then the corresponding coherence time $\langle \tau_0 \rangle$ is given by

$$\langle \tau_0 \rangle = 1/\Delta \nu, \tag{12.11}$$

and the coherence length l_c is given by

$$l_c = v \langle \tau_0 \rangle. \tag{12.12}$$

12.2.3 *Spatial coherence*

We now consider a more general problem of spatial coherence associated with
the finite size of the electron source. Take a single quasi-monochromatic point
source S shown in Fig. 12.2a. The two receiving points P_1 and P_2 lie in the same
direction from the source. Accordingly the coherence between waves ψ_1 and ψ_2
represents the *longitudinal spatial coherence* of the wave field. On the other hand,
the receiving point P_3 is located at the same distance from S as P_1. In this case
the coherence between ψ_1 and ψ_3 represents the *transverse spatial coherence* of
the wave field.

The longitudinal coherence merely depends on how large r_{12} is in comparison
with the coherence length of the source. If $t_{12} = r_{12}/v \ll \tau_0$, then the degree of
coherence between ψ_1 and ψ_2 will be high, whereas if $t_{12} \gg \tau_0$ then there will
be little or no coherence. Regarding the transverse coherence, if S is a true point
source, then the time dependence of the two waves ψ_1 and ψ_3 will be precisely the
same, i.e. they will be completely mutually coherent. Partial coherence between
ψ_1 and ψ_3 occurs if the source has a finite spatial extent. To investigate the effect
of finite source size on coherence we consider two point sources S_a and S_b shown
in Fig. 12.2b and assume that they are identical, except for their phases varying
randomly and independently. In other words, they are mutually incoherent. We
have

$$\psi_1 = \psi_{1a} + \psi_{1b}, \quad \psi_2 = \psi_{2a} + \psi_{2b}.$$

The normalized correlation function for the two receiving points is given by

$$\nu_{12}(\tau) = \frac{1}{\sqrt{I_1 I_2}} \langle \psi_1(t)\psi_2^*(t+\tau)\rangle$$

$$= \frac{1}{\sqrt{I_1 I_2}} \{\langle \psi_{1a}(t)\psi_{2a}^*(t+\tau)\rangle + \langle \psi_{1b}(t)\psi_{2b}^*(t+\tau)\rangle\}.$$

In deriving the above result we have used the fact that S_a and S_b are mutually incoherent so that the cross-terms $\langle \psi_{1a}\psi_{2b}^*\rangle$ and $\langle \psi_{1b}\psi_{2a}^*\rangle$ both vanish.

If we assume that

$$\psi(t) = \psi_0 \exp(i\omega t)\exp[i\phi(t)],$$

we obtain

$$\nu_{12}(\tau) = \frac{1}{2}[\nu(\tau_a) + \nu(\tau_b)],$$

where

$$\nu(\tau) = \exp(i\omega\tau)(1 - \tau/\tau_0)$$

is the autocorrelation function of either source, and

$$\tau_a = \frac{1}{v}(r_{2a} - r_{1a}) + \tau, \quad \tau_b = \frac{1}{v}(r_{2b} - r_{1b}) + \tau.$$

Assuming that $\tau_a - \tau_b$ is small in comparison with τ_a and with τ_b, we find after some algebra that

$$|\nu_{12}(\tau)|^2 \approx \frac{1}{2}\{1 + \cos[\omega(\tau_a - \tau_b)]\}\left(1 - \frac{\tau_a}{\tau_0}\right)\left(1 - \frac{\tau_b}{\tau_0}\right).$$

The above analysis shows that the mutual coherence between a given fixed receiving point (say point P_1) and any other point (say P_2) illuminated by two mutually incoherent sources S_a and S_b exhibits a periodic spatial dependence that is somewhat similar to an interference pattern (note that the total intensity remains uniform). The mutual coherence is greatest at the centre where P_1 and P_2 coincide. The coherence drops to zero on either side of the central line at a distance l_t defined by the condition $\cos[\omega(\tau_b - \tau_a)] = -1$, i.e. $\omega(\tau_b - \tau_a) = \pi$. Assuming that s is the distance between the two sources, l is the distance between the two receiving points, r is the mean distance from the sources to the receiving points, and that $s \ll r$, we then have $\tau_b - \tau_a = (r_{2a} - r_{1b})/v$ and

$$\omega(\tau_b - \tau_a) = \pi = \frac{\omega(r_{2a} - r_{1b})}{v} = \omega \cdot \frac{sl}{2vr}. \tag{12.13}$$

Since $\omega = 2\pi v/\lambda$, we have

$$l = r\lambda/s. \tag{12.14}$$

In terms of the angular separation θ_s between the two sources, as seen from the receiving point P_1, we have $\theta_s \approx s/r$ and

$$l = \lambda/\theta_s. \tag{12.15}$$

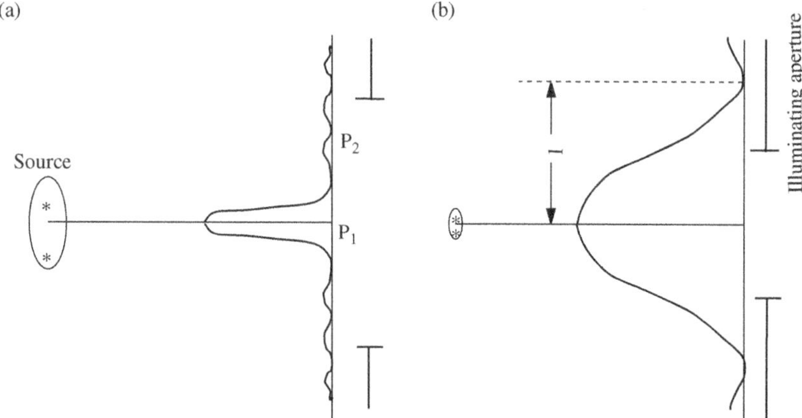

FIG. 12.3. Schematic diagram illustrating transverse spatial coherence for an electron source of (a) large and (b) small spatial extent.

This condition defines the width of the region of high mutual coherence between P_1 and P_2. We shall call it the transverse coherence width. For extended sources the *Van Cittert-Zernike theorem states that the complex degree of coherence between a fixed point P_1 and a variable point P_2 in a plane illuminated by an extended primary source is equal to the complex amplitude produced at P_2 by a spherical wave passing through an aperture of the same size and shape as the extended source and converging to P_1*. Figure 12.3 shows curves of $|\nu(r)|$. The mutual coherence is seen to be greatest at the centre where P_1 and P_2 coincide, i.e. where $r = 0$. The coherence drops to zero on either side of the central line at a distance l from the centre. In Fig. 12.3a the spatial extent of the electron source is large and the corresponding transverse coherence width is small. Each point within the illumination aperture may be regarded effectively as incoherent to each other. In Fig. 12.3b, on the other hand, the spatial extent of the electron source is small and every point within the illumination aperture may be treated as coherent.

12.3   Image formation in an electron microscope

In this section we will discuss image formation in an electron microscope. Assuming that in the image plane the electron wave function is given by $\psi(\mathbf{X})$, the image intensity distribution is

$$I(\mathbf{X}) = |\psi(\mathbf{X})|^2 = \psi(\mathbf{X})\psi^*(\mathbf{X}). \tag{12.16}$$

In a real electron microscope the electron wave function in the image plane is different from the electron wave function leaving the bottom surface of the specimen. The exit wave function $\psi_E(\mathbf{X})$ is affected by various aberrations of the electron lenses. Aberration effects can be best described in reciprocal space. Let $\phi(\mathbf{u})$ be the Fourier transform of $\psi(\mathbf{X})$, i.e.

$$\psi(\mathbf{X}) = \int \phi(\mathbf{u}) \exp\left(2\pi i \mathbf{u} \cdot \mathbf{X}\right) d\mathbf{u}, \qquad (12.17)$$

where $\mathbf{u}$ is a two-dimensional wave vector in the $\mathbf{X}$ plane. Using this definition we obtain for the Fourier transform $\mathcal{F}$ of the image intensity distribution $I(\mathbf{X})$

$$\begin{aligned}
J(\mathbf{u}) = \mathcal{F}\{I(\mathbf{X})\} &= \int \psi(\mathbf{X})\psi^*(\mathbf{X}) \exp(-2\pi i \mathbf{u} \cdot \mathbf{X}) d\mathbf{X} \\
&= \int \left[\int \phi(\mathbf{u}')e^{2\pi i \mathbf{u}' \cdot \mathbf{X}}d\mathbf{u}'\right]\left[\int \phi(\mathbf{u}'')e^{2\pi i \mathbf{u}'' \cdot \mathbf{X}}d\mathbf{u}''\right]^* e^{-2\pi i \mathbf{u} \cdot \mathbf{X}}d\mathbf{X} \\
&= \int\int \phi(\mathbf{u}')\phi^*(\mathbf{u}'')\left\{\int \exp[-2\pi i(\mathbf{u} - \mathbf{u}' + \mathbf{u}'') \cdot \mathbf{X}]d\mathbf{X}\right\}d\mathbf{u}'d\mathbf{u}'' \\
&= \int\int \phi(\mathbf{u}')\phi^*(\mathbf{u}'')\delta(\mathbf{u} - \mathbf{u}' + \mathbf{u}'')d\mathbf{u}'d\mathbf{u}'' \\
&= \int \phi(\mathbf{u}'' + \mathbf{u})\phi^*(\mathbf{u}'')d\mathbf{u}''. \qquad (12.18)
\end{aligned}$$

12.3.1 *Transmission cross-coefficient (TCC) for coherent illumination*

Since only small-angle scattering events and paraxial rays are involved in the formation of an image in an electron microscope, it is usually possible to ignore off-axis and higher order aberrations and consider only the axial aberrations. Here we consider only the most important axial aberrations including defocus, spherical aberration, and astigmatism.

For a point object on the lens axis close to the focal point, emitting electron rays at an angle α to the axis, the effect of spherical aberration is to reduce the effective focal length of the lens for such rays by an amount $C_s\alpha^2$, where C_s is the spherical aberration coefficient. As a result the rays are deflected more than in a perfect lens, and intersect the paraxial image plane at a distance $MC_s\alpha^3$ from the axis, where M is the lens magnification. This effect is most conveniently simulated by introducing an angular dependent negative phase change to the phase of the electron wave propagating at an angle to the lens axis, i.e. to multiply $\phi(\mathbf{u})$ by the following phase factor

$$\exp\left(-\frac{i\pi}{2}C_s\lambda^3\mathbf{u}^4\right). \qquad (12.19)$$

If the plane on which the microscope lens is focussed for paraxial rays is displaced by D from the object (D is positive for an over-focussed lens), the Fresnel diffraction effect introduces an additional phase change to the electron wave. This phase change is described by the Fourier transformation of the real-space Fresnel propagator

$$P(\mathbf{X}) = -\frac{i}{D\lambda} \exp\left(\frac{i\pi\mathbf{X}^2}{\lambda D}\right), \tag{12.20}$$

i.e. the effect of the defocus may be simulated by introducing the following phase factor multiplying $\phi(\mathbf{u})$:

$$\exp(-i\pi\lambda D\mathbf{u}^2). \tag{12.21}$$

In practice for a high-magnification objective lens, the object distance is almost equal to the focal length f, and therefore $D \approx \Delta f$, where Δf is the reduction in the focal length f due to over-focussing. The phase factor (12.21) therefore becomes $\exp(-i\pi\lambda\Delta f\mathbf{u}^2)$.

With our convention for plane waves propagating in the positive z direction, i.e. $\exp(i2\pi z/\lambda)$, the phase changes (12.19) and (12.21) are in fact negative (C_s is always positive and D is positive for an over-focussed lens). Thus the phases are phase *lags*, as is the total Scherzer phase shift given in eqn (12.28) below.

Astigmatism in an image is caused by the objective lens being of unequal strength for rays in different planes containing the lens axis. The defocus value Δf is then in general not a constant but depends on the azimuth about the axis of the plane of the rays. Assuming that the defocus values along the major and minor axes are given by Δf_1 and Δf_2 respectively, the defocus value along any direction making an angle θ to the x axis (perpendicular to the electron optical axis) is then given by (Henderson et al., 1986)

$$\Delta f = \Delta f(\theta) = \Delta f_1 \cos^2(\theta - \theta_0) + \Delta f_2 \sin^2(\theta - \theta_0), \tag{12.22}$$

in which θ_0 is the angle between the major axis and the x-axis. Let

$$C_A = \Delta f_1 - \Delta f_2, \quad Z = \frac{1}{2}(\Delta f_1 + \Delta f_2), \tag{12.23}$$

or

$$\Delta f_1 = \frac{1}{2}(2Z + C_A), \quad \Delta f_2 = \frac{1}{2}(2Z - C_A). \tag{12.24}$$

Substituting (12.24) and (12.22) into (12.21) and combining with (12.19), we then have for the Fourier transform of the electron wave function $\psi(\mathbf{r})$

$$\phi(\mathbf{u}) = \phi_E(\mathbf{u})t(\mathbf{u}), \tag{12.25}$$

in which ϕ_E is the Fourier transform of the electron wave function leaving the specimen exit surface (unaffected by the aberrations of the electron lens), and $t(\mathbf{u})$ describes the response of the electron microscope to a point source situated on the optical axis and usually called the point spread function (PSF)

$$t(\mathbf{u}, Z) = A(\mathbf{u}) \exp[-i\chi(\mathbf{u}, Z)], \tag{12.26}$$

where the effect of the condenser or illuminating aperture has been introduced and modelled by the aperture function

$$A(\mathbf{u}) = \begin{cases} 1 \text{ if } u \text{ is inside the limit defined by the objective aperture} \\ 0 \text{ otherwise} \end{cases} \qquad (12.27)$$

and $\chi(\mathbf{u})$ is the Scherzer phase shift (Scherzer, 1949) describing the combined effects of the defocus, spherical aberration, and astigmatism

$$\chi(\mathbf{u}, Z) = \pi \Delta f(\theta) \lambda \mathbf{u}^2 + \frac{1}{2} \pi C_s \lambda^3 \mathbf{u}^4$$

$$= \pi Z \lambda \mathbf{u}^2 + \frac{1}{2} \pi C_A \cos[2(\theta - \theta_0)] \lambda \mathbf{u}^2 + \frac{1}{2} \pi C_s \lambda^3 \mathbf{u}^4. \qquad (12.28)$$

Since in a modern electron microscope the axial astigmatism may usually be fully corrected, in what follows we shall therefore take $C_A = 0$ and consider only the effects resulting from the defocus and spherical aberration. Substituting expressions (12.25) and (12.26) for $\phi(\mathbf{u})$ and $t(\mathbf{u}; Z)$ into the general expression (12.18) for $J(\mathbf{u})$ we obtain

$$J(\mathbf{u}, Z) = \int d\mathbf{u}' \phi_E(\mathbf{u} + \mathbf{u}') \phi_E^*(\mathbf{u}') T(\mathbf{u} + \mathbf{u}', \mathbf{u}', Z), \qquad (12.29)$$

in which $T(\mathbf{u} + \mathbf{u}', \mathbf{u}', Z)$ is called the *transmission cross-coefficient* (TCC) and for perfect coherent illumination this coefficient is given by

$$T(\mathbf{u} + \mathbf{u}', \mathbf{u}', Z) = A(\mathbf{u} + \mathbf{u}') A(\mathbf{u}') \exp\left\{-i\left[\chi(\mathbf{u} + \mathbf{u}', Z) - \chi(\mathbf{u}', Z)\right]\right\}. \qquad (12.30)$$

12.3.2 *The TCC for incoherent illumination*

We now consider the effects associated with the finite size of the electron source, i.e. beam divergence effects (see Fig. 12.3a), and fluctuations of defocus associated with either the finite energy spread of the electron source or relative instabilities of the objective lens current. For the majority of conventional transmission electron microscopes (CTEMs), with the exception of those fitted with field-emission guns (FEGs), the spatial extent of the electron source is large, and images may be regarded as incoherent (see Fig. 12.3a) when formed with different defocus and with electrons emitted from different positions on the electron source. The resulting image is given by the average of the set of images formed for all angles of incidence and defocus values. Assuming that $B(\mathbf{s})$ and $f(Z)$ are the normalized distribution of the intensity of the electron source and the defocus spread, taking the appropriate average value gives

$$\langle J(\mathbf{u}) \rangle = \int \int d\mathbf{s} dZ B(\mathbf{s}) f(Z) J(\mathbf{u}, \mathbf{s}, Z) \qquad (12.31)$$

$$= \int \int d\mathbf{s} dZ B(\mathbf{s}) f(Z) \int d\mathbf{u}' \phi_E(\mathbf{u} + \mathbf{u}', \mathbf{s}) \phi_E^*(\mathbf{u}', \mathbf{s}) t(\mathbf{u} + \mathbf{u}', \mathbf{s}, Z) t^*(\mathbf{u}', \mathbf{s}, Z),$$

where $t(\mathbf{u}, \mathbf{s}, Z)$ denotes the response function of the electron microscope for a point source situated at point $\mathbf{s}$. Let $t(\mathbf{X})$ be the real-space representation of $t(\mathbf{u})$. We have

$$t(\mathbf{u}, \mathbf{s}, Z) = \mathcal{F}\left\{\exp(2\pi i \mathbf{s} \cdot \mathbf{X}) t(\mathbf{X})\right\} = \delta(\mathbf{u} + \mathbf{s}) * t(\mathbf{u}, Z) = t(\mathbf{u} + \mathbf{s}, Z), \qquad (12.32)$$

where $\mathcal{F}$ denotes the Fourier transform.

In general the electron wave function $\phi_E(\mathbf{u})$ at the exit surface of the sample depends on the angle of incidence, i.e. on $\mathbf{s}$. However, in the case of HREM imaging of thin specimens where the phase object approximation applies, this wave function $\phi_E(\mathbf{u})$ is approximately independent of the angle of incidence, i.e. $\phi_E(\mathbf{u}, \mathbf{s}) \approx \phi_E(\mathbf{u})$. We may therefore take the exit wave function as remaining effectively constant within the range of integration over $\mathbf{s}$. For small beam divergence values the aperture function $A(\mathbf{q})$ may also be treated as independent of $\mathbf{s}$. Using these approximations we obtain

$$\langle J(\mathbf{u}) \rangle = \int d\mathbf{u}' \phi_E(\mathbf{u} + \mathbf{u}') \phi_E^*(\mathbf{u}') T(\mathbf{u} + \mathbf{u}', \mathbf{u}', Z), \tag{12.33}$$

and the TCC now becomes

$$T(\mathbf{u}, \mathbf{u}', Z) = A(\mathbf{u}) A(\mathbf{u}') \int \int ds \, dZ \, B(\mathbf{s}) f(Z')$$
$$\times \exp\{-i[\chi(\mathbf{u} + \mathbf{s}, Z + Z') - \chi(\mathbf{u}' + \mathbf{s}, Z + Z')]\}. \tag{12.34}$$

Following Wade and Frank (Wade and Frank, 1977), we take both the effective source intensity distribution and the defocus spread distribution to be represented by normalized Gaussian functions

$$B(\mathbf{s}) = \frac{1}{\pi s_0^2} \exp\left(-\frac{s^2}{s_0^2}\right), \quad f(Z') = \frac{1}{\sqrt{\pi}\Delta} \exp\left[-\frac{(Z')^2}{\Delta^2}\right]. \tag{12.35}$$

To simplify the results even further we note that for most electron microscopes s_0 and Δ are very small, and the wave aberration $\chi(\mathbf{u} + \mathbf{s}, Z + Z')$ may be expanded as a Taylor series in the vicinity of $\mathbf{u}$ and Z

$$\chi(\mathbf{u} + \mathbf{s}; Z + Z') = \chi(\mathbf{u}, Z) + \mathbf{s} \cdot \nabla\chi(\mathbf{u}, Z) + Z' \frac{\partial \chi(\mathbf{u}, Z)}{\partial Z}, \tag{12.36}$$

where $\nabla\chi(\mathbf{u}, Z)$ denotes the gradient of the wave aberration function $\chi(\mathbf{u}, Z)$ with respect to $\mathbf{u}$.

Substitution of eqns (12.35) and (12.36) into eqn (12.34) gives

$$T(\mathbf{u}, \mathbf{u}', Z) = A(\mathbf{u}) A(\mathbf{u}') \exp\left\{-i[\chi(\mathbf{u}, Z) - \chi(\mathbf{u}', Z)]\right\} E_s(\mathbf{u}, \mathbf{u}', Z) E_f(\mathbf{u}, \mathbf{u}'; Z), \tag{12.37}$$

where

$$E_s(\mathbf{u}, \mathbf{u}', Z) = \int B(\mathbf{s}) \exp\{-i\mathbf{s} \cdot [\nabla\chi(\mathbf{u}, Z) - \nabla\chi(\mathbf{u}', Z)]\} ds$$
$$= \exp\left\{-\frac{1}{4} s_0^2 [\nabla\chi(\mathbf{u}, Z) - \nabla\chi(\mathbf{u}', Z)]^2\right\}, \tag{12.38}$$

and

$$E_f(\mathbf{u}, \mathbf{u}'; Z) = \int f(Z') \exp\left\{ -iZ'\left[\frac{\partial\chi(\mathbf{u}, Z)}{\partial Z} - \frac{\partial\chi(\mathbf{u}', Z)}{\partial Z}\right]\right\} dZ'$$

$$= \exp\left\{ -\frac{1}{4}\Delta^2\left[\frac{\partial\chi(\mathbf{u}, Z)}{\partial Z} - \frac{\partial\chi(\mathbf{u}', Z)}{\partial Z}\right]^2\right\}. \tag{12.39}$$

Since

$$\nabla\chi(\mathbf{u}; Z) = 2\pi(Z + C_s\lambda^2 u^2)\lambda\mathbf{u}, \qquad \frac{\partial}{\partial Z}\chi(\mathbf{u}; Z) = \pi\lambda u^2,$$

we obtain the following explicit expressions for E_s and E_f,

$$E_s(\mathbf{u}, \mathbf{u}'; Z) = \exp\left\{ -\pi^2 s_0^2\lambda^2\left[(Z + C_s\lambda^2 u^2)\mathbf{u} - (Z + C_s\lambda^2 (u')^2)\mathbf{u}'\right]^2\right\}, \tag{12.40}$$

and

$$E_f(\mathbf{u}, \mathbf{u}') = \exp\left\{ -\left(\frac{\pi\Delta\lambda}{2}\right)^2 [u^2 - (u')^2]^2\right\} \tag{12.41}$$

Effects associated with higher order terms in the wave aberration expansion (12.36) have been investigated by Wade and Frank (Wade and Frank, 1977). These authors showed that higher order terms modify $E_s(\mathbf{u}, \mathbf{u}'; Z)$ and $E_f(\mathbf{u}, \mathbf{u}')$ and introduce some additional multiplicative terms. Although these additional terms may be large, particularly in the limit of large $\mathbf{u}$, it was found that when these terms are large the two factors E_s and E_f are usually very small. In general these extra terms have little effect on the TCC and may therefore be neglected.

12.3.3 *The TCC for a partially coherent illumination*

In this section we discuss image formation in an FEG electron microscope. Since in the case of such an electron microscope the effective size of the source of electrons is very small, the wave field of electrons everywhere in the plane of the condenser or illuminating aperture may be considered to be entirely coherent (see Fig. 12.3b). Assuming that the amplitude function of the electron source is $\beta(\mathbf{s})$, we then have the following expression for the wave function:

$$\phi(\mathbf{u}, Z) = \int \beta(\mathbf{s})\phi_E(\mathbf{u}, \mathbf{s})t(\mathbf{u} + \mathbf{s}; Z)d\mathbf{s}. \tag{12.42}$$

Fourier transformation of the image intensity distribution $I(\mathbf{x})$ gives

$$J(\mathbf{u}, Z) = \int \psi(\mathbf{X})\psi^*(\mathbf{X})\exp(-2\pi i\mathbf{u}\cdot\mathbf{X})d\mathbf{X} = \int \phi(\mathbf{u}' + \mathbf{u}, Z)\phi^*(\mathbf{u}', Z)d\mathbf{u}'$$

$$= \int\int \beta(\mathbf{s})\beta^*(\mathbf{s}')\left\{\int \phi_E(\mathbf{u} + \mathbf{u}', \mathbf{s})\phi_E^*(\mathbf{u}', \mathbf{s}')T(\mathbf{u} + \mathbf{u}' + \mathbf{s}, \mathbf{u}' + \mathbf{s}'; Z)d\mathbf{u}'\right\} d\mathbf{s}d\mathbf{s}'.$$

The effect of fluctuations of defocus values on the image is treated in the same way as in the case of incoherent illumination discussed in the previous section. The average image intensity is therefore given by

$$\langle J(\mathbf{u}, Z)\rangle = \int f(Z')\langle J(\mathbf{u}, Z')\rangle dZ'$$

$$= \int dZ' f(Z') \int\int d\mathbf{s}d\mathbf{s}' \langle \beta(\mathbf{s})\beta^*(\mathbf{s}')\rangle \int d\mathbf{u}' \phi_E(\mathbf{u} + \mathbf{u}', \mathbf{s})\phi_E^*(\mathbf{u}', \mathbf{s}')$$

$$\times\, T(\mathbf{u} + \mathbf{u}' + \mathbf{s}, \mathbf{u}' + \mathbf{s}'; Z').$$

In the simplest case $\langle \beta(\mathbf{s})\beta^*(\mathbf{s}')\rangle = B(\mathbf{s})\delta(\mathbf{s} - \mathbf{s}')$, and both $B(\mathbf{s})$ and $f(Z)$ are given by (12.35). Making the same approximations as employed in the previous section we obtain the average image intensity

$$\langle J(\mathbf{u}; Z)\rangle = \int d\mathbf{u}' \phi_E(\mathbf{u} + \mathbf{u}')\phi_E^*(\mathbf{u}')T^{FEG}(\mathbf{u} + \mathbf{u}', \mathbf{u}'; Z), \qquad (12.43)$$

where T^{FEG} is the effective TCC for an FEG electron microscope:

$$T^{FEG}(\mathbf{u}, \mathbf{u}'; Z) = A(\mathbf{u})A(\mathbf{u}')\exp\left\{-i\left[\chi(\mathbf{u}, Z) - \chi(\mathbf{u}', Z)\right]\right\}$$

$$\times\, E_s^{FEG}(\mathbf{u}, \mathbf{u}'; Z)E_f(\mathbf{u}, \mathbf{u}'), \qquad (12.44)$$

E_s^{FEG} is the envelope function associated with the finite size of an FEG electron source

$$E_s^{FEG}(\mathbf{u}, \mathbf{u}'; Z) = \exp\left\{-\frac{s_0^2}{4}\left[|\nabla\chi(\mathbf{u}, Z)|^2 + |\nabla\chi(\mathbf{u}', Z)|^2\right]\right\} \qquad (12.45)$$

$$= \exp\left\{-\pi^2 s_0^2\lambda^2\left[(Z + C_s\lambda^2 u^2)\mathbf{u}^2 + (Z + C_s\lambda^2(u')^2)(\mathbf{u}')^2\right]\right\},$$

and E_f is the envelope function associated with the finite energy spread of the incoming electron beam. This function is the same as that defined in the case of incoherent electron illumination discussed previously (eqn (12.41)). It should be noted that the value of E_s defined in the case of a CTEM and the value of E_s^{FEG} defined for an FEG electron microscope are related through the following equations:

$$E_s^{FEG}(\mathbf{u}, 0; Z) = E_s(\mathbf{u}, 0; Z), \quad E_s^{FEG}(0, \mathbf{u}; Z) = E_s(0, \mathbf{u}; Z). \qquad (12.46)$$

A very important consequence of these relations is that as far as HREM imaging is concerned the TCCs remain the same in the case of both CTEMs and FEG electron microscopes. We shall discuss this point in more detail in the following section.

12.4   Exit electron wave function retrieval

In electron crystallography the ultimate goal is to retrieve the exit electron wave function $\psi_E(\mathbf{X})$, and then to determine the corresponding crystal structure. In this section we consider the first part of the problem, i.e. the retrieval of the exit electron wave function from digital electron micrographs using image processing. We start from the reciprocal space representation of the image intensity (12.29)

$$J(\mathbf{u} \neq 0, \Delta f) = \phi_E(\mathbf{u})\phi_E^*(0)T(\mathbf{u}, 0; \Delta f) + \phi_E(0)\phi_E^*(-\mathbf{u})T(0, -\mathbf{u}; \Delta f)$$

$$+ \int_{u' \neq 0, -u} \phi_E(\mathbf{u} + \mathbf{u}')\phi_E^*(\mathbf{u}')T(\mathbf{u} + \mathbf{u}', \mathbf{u}'; \Delta f)d\mathbf{u}', \quad (12.47)$$

where $\phi_E(\mathbf{u})$ is the Fourier transform of the real-space exit electron wave function $\psi_E(\mathbf{r})$ and $T(\mathbf{u}+\mathbf{u}', \mathbf{u}'; \Delta f)$ is the relevant TCC. In the case of perfectly coherent illumination this function is given by (12.30)

$$T(\mathbf{u}, \mathbf{u}'; \Delta f) = A(\mathbf{u})A(\mathbf{u}') \exp\{-i[\chi(\mathbf{u}, \Delta f) - \chi(\mathbf{u}', \Delta f)]\}. \quad (12.48)$$

Qualitatively speaking, specimens may be classified as weak or strong objects according to whether or not the scattered wave has an amplitude which is comparable with the amplitude of the incident wave. In the case of a weak object the scattered wave is much weaker than the transmitted wave and the corresponding HREM image is formed by the interference between the transmitted beam and the diffracted beams represented by the first two terms in eqn (12.47). Images obtained with this condition satisfied are called linear images. In the case of a strong object the scattered beams are strong, and their amplitudes are comparable with the amplitude of the transmitted beam. Interference between the transmitted and scattered beams and between scattered beams represented by the third term in eqn (12.47) now needs to be taken into account. Images of this type are called non-linear images.

12.4.1 *Linear image retrieval*

We first consider the problem of linear image retrieval. For a weak phase object illuminated by an incident plane electron wave the exit electron wave function may be approximated by (Cowley, 1990)

$$\psi_E(\mathbf{X}) = \exp\{-i\sigma V(\mathbf{X})t\} \approx 1 - i\sigma V(\mathbf{X})t, \quad (12.49)$$

where $\sigma \left(= m\lambda/(2\pi\hbar^2)\right)$ is the usual interaction constant, $V(\mathbf{X})$ is the potential field of the thin sample averaged in the direction of the incident electron beam (the z direction), and t is the thickness of the sample. The Fourier transformation of the exit electron wave function gives

$$\phi_E(\mathbf{u}) = \delta(\mathbf{u}) - i\sigma t F(\mathbf{u}), \quad (12.50)$$

where $F(\mathbf{u})$ is the Fourier transform of the projected potential field $V(\mathbf{X})$. For a thin sample we may neglect the effect of absorption on the electron wave function. Noticing that for a real crystal potential $V(\mathbf{X})$, $F(\mathbf{u}) = F^*(-\mathbf{u})$, we obtain from eqn (12.47)

$$J(\mathbf{u}, \Delta f) = \delta(\mathbf{u}) - i\sigma t F(\mathbf{u})\left[T(\mathbf{u}, 0; \Delta f) - T(0, -\mathbf{u}; \Delta f)\right]. \quad (12.51)$$

From our earlier discussion we know that for both the CTEM and the FEG electron microscopes,

$$T(\mathbf{u}, 0; \Delta f) = A(\mathbf{u}) \exp[-i\chi(\mathbf{u}; \Delta f)] E_s(\mathbf{u}, 0; \Delta f) E_f(\mathbf{u}, 0).$$

We therefore obtain for $u < u_0$

$$J(\mathbf{u}) = \delta(\mathbf{u}) - 2\sigma t F(\mathbf{u}) \sin[\chi(\mathbf{u}; \Delta f)] E_s(\mathbf{u}, 0; \Delta f) E_f(\mathbf{u}, 0). \qquad (12.52)$$

Since for both the CTEM and the FEG electron microscopes the envelope functions $E_s(\mathbf{u}, 0; Z)$ and $E_f(\mathbf{u}, 0)$ remain the same, in the weak phase object approximation the diffractogram $J(\mathbf{u})$ obtained using an FEG electron microscope is the same as that obtained using a CTEM. For diffracted beams with $\mathbf{u} \neq 0$, we obtain the crystal structure factors, or the Fourier transform of the projected crystal potential $V(\mathbf{X})$, as

$$F(\mathbf{u}) = -\frac{J(\mathbf{u})}{2\sigma t \sin[\chi(\mathbf{u}; \Delta f)] E_s(\mathbf{u}, 0; \Delta f) E_f(\mathbf{u}, 0)}. \qquad (12.53)$$

To a good approximation most quantities entering the above equation may be regarded as remaining constant during HREM experiments and may be measured separately. The most important variable is the defocus value Δf. A simple method of image deconvolution has been developed by Li and colleagues (Han et al., 1986) for the case of crystalline materials. In this method the spherical aberration coefficient C_s, the spatial extent of the electron source s_0, and the spread Δ of the wavelength of the incident electrons are assumed to be constant, and the defocus value Δf is determined via the method of trial and error. For a given set of $J(\mathbf{u})$, a set of defocus values Δf is assumed and a set of $\{F(\mathbf{u})\}$ is obtained for each value of Δf. In the case where the structure of the sample is crystalline and for the correct set of $\{F(\mathbf{u})\}$ it is assumed that the Sayre equation (Sayre, 1952) is satisfied

$$F(\mathbf{h}) = F_h = \frac{\theta_h}{\Omega_0} \sum_{h'} F_{h'} F_{h-h'}, \qquad (12.54)$$

in which Ω_0 is the volume of a unit cell of the crystalline sample, and θ_h relates the h-th Fourier coefficients of the projected potential $V(\mathbf{X})$ to that of the squared potential $V(\mathbf{X})^2$. Assuming that the latter is denoted by G_h we have $F_h = \theta_h G_h$. The Sayre equation is known to be valid for a structure containing equally resolved atoms. The meaning of this equation is simply that the distribution of the potential squared $[V(\mathbf{X})]^2$ has peaks at the same points where the original potential $V(\mathbf{x})$ (Woolfson, 1961) is a maximum. A figure of merit characterizing the degree of accuracy of the Sayre equation can now be defined and calculated for each set of $\{F(\mathbf{h})\}$. The correct set of $\{F(\mathbf{h})\}$ is taken to be the one corresponding to the minimum value of the figure of merit. For crystals composed of atoms of different types an alternative definition of the figure of merit

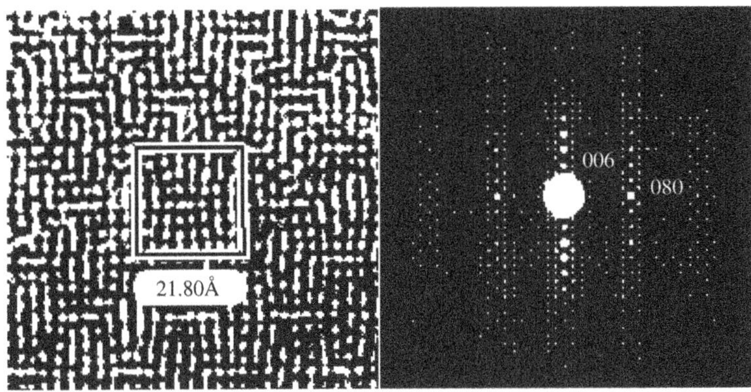

FIG. 12.4. (a) HREM image (left) and (b) the corresponding electron diffraction pattern (right) obtained from a $Bi_2(Sr_{0.9}La_{0.1})CoO_y$ crystal with the electron beam incident along the [100] zone axis. (Images courtesy of F. H. Li).

can be given that involves the principle of maximum entropy (Hu and Li, 1991).

Figure 12.4 shows an HREM image and the corresponding electron diffraction pattern obtained from a $Bi_2(Sr_{0.9}La_{0.1})CoO_y$ crystal with the electron beam incident along the [100] zone axis. The HREM image was taken using a JEM-2010 electron microscope operated at 200 keV. The spherical aberration coefficient of the objective lens of this electron microscope is 0.5 mm, and the standard deviation of the Gaussian distribution of defocus is approximately equal to 70 Å. Using these parameters the HREM image shown in Fig. 12.4 may be deconvoluted straightforwardly using eqn (12.53) to obtain Fig. 12.5a. While the basic structural features are clearly visible, some of the atoms (especially oxygen) are not clearly resolved. To a certain extent the resolution of the deconvoluted image may be enhanced by using the electron diffraction pattern shown in Fig. 12.4b. To achieve this a set of low-order structure factors is initially obtained from the HREM image (Fig. 12.4a) using eqn (12.4). Modules of higher order structure factors are then obtained from the corresponding electron diffraction pattern (Fig. 12.4b) and the phases are derived from the phases of lower order reflections using the direct method described in (Jiang et al., 1999). The resulting high-resolution image is called the projected potential map. An example of this kind of map is shown in Fig. 12.5b. In this [100] projected potential map weaker oxygen atoms are clearly seen between stronger scattering Co atoms as indicated in the figure by Co–O.

While it is good to be able to retrieve the exit wave function or its Fourier transform from a single micrograph, the method described above is not sufficiently accurate and stable in the regions where $\sin[\chi(u)] \approx 0$. The situation can be improved by making use of more than one micrograph, because each micrograph has different regions where $\sin[\chi(u)] \approx 0$. By combining micrographs recorded under different imaging conditions the exit wave function may be re-

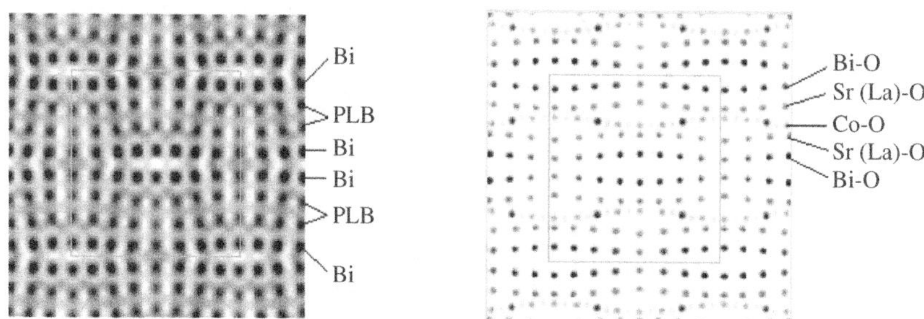

FIG. 12.5. (a) Deconvoluted image of 12.4 (left) and (b) [100] projected potential map (right) obtained using phase extension. (Images courtesy of F. H. Li).

trieved accurately over a wider frequency band. In what follows we will briefly discuss the defocus variation method (Van Dyck and Op de Beeck, 1990) where a series of micrographs $\{\psi_n(\mathbf{X})\}$ are recorded for different values of defocus Δf_n. Assuming that the amplitude of the scattered wave function is much weaker than the amplitude of the incident plane wave, we have

$$\psi(\mathbf{X}) = 1 + \psi_s(\mathbf{X}), \tag{12.55}$$

where the first term represents the incident plane wave and the second term $\psi_s(\mathbf{X})$ denotes the scattered wave. In particular, when the weak phase object approximation (WPOA) is valid $\psi_s(\mathbf{X}) \approx -i\sigma V(\mathbf{X})t$. The intensity of the image at a defocus value Δf is given by

$$I(\mathbf{X}, \Delta f) = |1 + \psi_s(\mathbf{X}, \Delta f)|^2 = 1 + \psi_s(\mathbf{X}, \Delta f) + \psi_s^*(\mathbf{X}, \Delta f) + h(\mathbf{X}, \Delta f), \tag{12.56}$$

where $h(\mathbf{X}) = |\psi_s(\mathbf{X})|^2$ is a term quadratic in ψ_s. Using a large objective aperture such that $A(\mathbf{u}) = 1$ for all beams and neglecting a phase factor $\exp(-i\pi C_s \lambda^3 u^4 / 2)$ common for all the micrographs, we obtain

$$\psi_s(\mathbf{u}, \Delta f) = \phi(\mathbf{u}) \exp(-\pi i \Delta f \lambda u^2), \tag{12.57}$$

with $\phi(\mathbf{u}) = \psi_s(\mathbf{u}, 0) = \phi_E(\mathbf{u})$. The two-dimensional Fourier transform of the image intensity distribution $I(\mathbf{X}, \Delta f)$ is given by

$$J(\mathbf{u}, \Delta f) = \delta(\mathbf{u}) + \phi(\mathbf{u}) \exp(-i\pi \Delta f \lambda u^2) + \phi^*(-\mathbf{u}) \exp(i\pi \Delta f \lambda u^2) + H(\mathbf{u}, \Delta f), \tag{12.58}$$

where $H(\mathbf{u}, \Delta f) = \mathcal{F}\{h(\mathbf{X}, \Delta f)\}$. For simplicity we assume that $\mathbf{u} \neq 0$. To extract $\phi(\mathbf{u})$ we multiply $J(\mathbf{u}, \Delta f_n)$ by a conjugate phase factor $\exp(i\pi \Delta f_n \lambda u^2)$ and sum the resulting expressions for a series of N images at equal steps δf of defocus ($\Delta f_n = n\delta f$)

$$\sum_n J(\mathbf{u}, \Delta f_n) \exp(\pi i \lambda \Delta f_n u^2) = N\phi(\mathbf{u}) + \phi^*(-\mathbf{u}) \sum_n \exp(2\pi i \lambda \Delta f_n u^2)$$

$$+ \sum_n H \exp(\pi i \Delta f_n \lambda u^2). \tag{12.59}$$

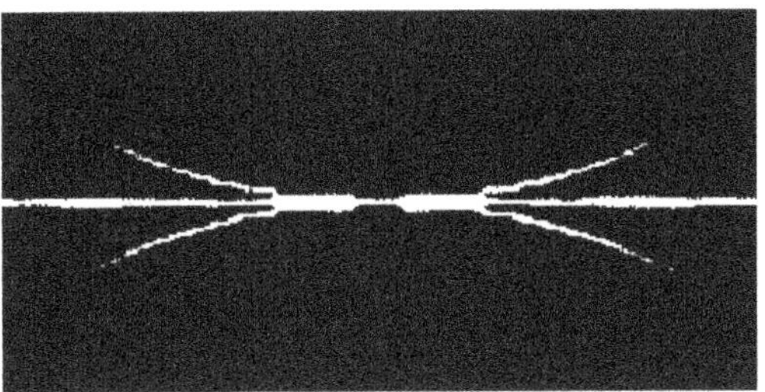

FIG. 12.6. Three-dimensional power spectrum $J(\mathbf{u}, k_z)$ constructed from a focal HRTEM series of a thin amorphous germanium specimen. $J(\mathbf{u}, k_z)$ is the Fourier transform of $J(\mathbf{u}, z)$. From (Op de Beeck et al., 1996).

The second and third terms scale like $O(\sqrt{N})$ as opposed to the linear order N scaling of the first term (Saxton, 1994). We therefore neglect the second and the third terms and obtain in the limit of large N

$$\phi(\mathbf{u}) \approx \frac{1}{N} \sum_{n=1}^{N} J(\mathbf{u}, n\delta f) \exp(i\pi n \delta f \lambda u^2). \tag{12.60}$$

This formula describes the paraboloid method developed by Op de Beeck et al. (Op de Beeck et al., 1996).

The name 'paraboloid' comes from the fact that the intensity in the diffractogram is mainly concentrated around two conjugate parabolas as shown in Fig. 12.6. This point may be clearly illustrated by Fourier transforming eqn (12.58) with respect to Δf or the z coordinate of $J(\mathbf{u}, z)$ (Van Dyck et al., 1993)

$$J(\mathbf{u}, k_z) = \int J(\mathbf{u}, z) \exp(ik_z z) dz = 2\pi \delta(\mathbf{u})\delta(k_z) + 2\pi \phi(\mathbf{u})\delta(k_z - \pi \lambda u^2)$$

$$+ 2\pi \phi^*(-\mathbf{u})\delta(k_z + \pi \lambda u^2) + \int H(\mathbf{u}, z) \exp(ik_z z) dz. \tag{12.61}$$

This expression shows that in reciprocal space $J(\mathbf{u}, k_z)$ is localized on two parabolas $k_z = \pm \pi \lambda u^2$ (see Fig. 12.6). The third term in the above expression denotes the contribution of non-linear terms that is in general spread out throughout reciprocal space. Utilizing the property of the δ-function, we obtain for $k_z = \pi \lambda u^2$ and $u \neq 0$

$$\int J(\mathbf{u}, z) \exp(ik_z z) dz = \int J(\mathbf{u}, z) \exp(i\pi \lambda u^2 z) dz \sim \phi(\mathbf{u}), \tag{12.62}$$

and this is equivalent to eqn (12.60). To prove the equivalence we notice that the integral may be approximated by the summation of a series of $J(\mathbf{u}, z)$ at different values of $z_n = n\delta f$.

In the most general form the linear methods may be described as (Saxton, 1994)

$$\phi(\mathbf{u}) = \sum_n J_n(\mathbf{u})r_n(\mathbf{u}), \tag{12.63}$$

where function $r_n(\mathbf{u})$ is called the restoring filter. By using more complex restoring filters the conjugate wave function (i.e. the second term in eqn (12.59)) can be eliminated completely rather than just being reduced by a factor of $\sqrt{N}$. One form of this type of filter was proposed by Schiske (see (Saxton, 1994)):

$$r_n(\mathbf{u}) = \exp\{-i\chi_n(\mathbf{u})\}\frac{N - \sum_m \exp\{-2i[\chi_m(\mathbf{u}) - \chi_n(\mathbf{u})]\}}{N - |\sum_m \exp\{-2i[\chi_m(\mathbf{u})]\}|^2}. \tag{12.64}$$

12.4.2 Non-linear image retrieval

We now consider a more general case of non-linear image retrieval. The first non-linear exit wave function retrieval method was developed in (Kirkland, 1984). The idea behind this and its more recent modifications (Huang and Ximen, 1991, Coene et al., 1992, Coene et al., 1993) consists in finding a wave function $\psi_E(\mathbf{X})$ at the specimen exit face that minimizes the error function measuring the difference between the experimental and the trial images. For a focal series of N images each characterized by a defocus value Δf_n, Coene et al. (Coene et al., 1993) used the error function S^2 defined in the following way:

$$S^2 = \sum_{n=1}^{N} \int |I_n^{exp}(\mathbf{u}) - I_n^{est}(\mathbf{u})|^2 d\mathbf{u} = \sum_{n=1}^{N} \int |\delta I_n(\mathbf{u})|^2 d\mathbf{u}, \tag{12.65}$$

where

$$\delta I_n(\mathbf{u}) = I_n^{exp}(\mathbf{u}) - I_n^{est}(\mathbf{u}) \tag{12.66}$$

and the superscripts *exp* and *est* denote the experimental and estimated images respectively, and the intensity distribution of the n-th estimated image is evaluated using the exit wave function $\phi_E(\mathbf{X})$ and the TCC $T(\mathbf{u}, \mathbf{u}'; \Delta f_n)$

$$I_n^{est}(\mathbf{u}) = \int d\mathbf{u}' \phi_E(\mathbf{u} + \mathbf{u}')\phi_E^*(\mathbf{u}')T(\mathbf{u} + \mathbf{u}', \mathbf{u}'; \Delta f_n). \tag{12.67}$$

The least-squares method is then applied to find the exit wave function $\phi_E(\mathbf{X})$ minimizing the error function S^2. This is achieved by setting the derivatives of S^2 with respect to $\phi_E^*(\mathbf{u})$ and $\phi_E(-\mathbf{u})$ to zero, leading to a set of maximum-likelihood (MAL) equations

$$\frac{\partial(S^2)}{\partial(\phi_E^*(\mathbf{u}))} = 0 = 2\sum_{n=1}^{N} \int d\mathbf{u}' \delta I_n(\mathbf{u}') \frac{\partial(\delta I_n(\mathbf{u}'))}{\partial(\phi_E^*(\mathbf{u}))}$$

$$\frac{\partial(S^2)}{\partial(\phi_E(-\mathbf{u}))} = 0 = 2\sum_{n=1}^{N} \int d\mathbf{u}' \delta I_n(\mathbf{u}') \frac{\partial(\delta I_n(\mathbf{u}'))}{\partial(\phi_E(-\mathbf{u}))}. \tag{12.68}$$

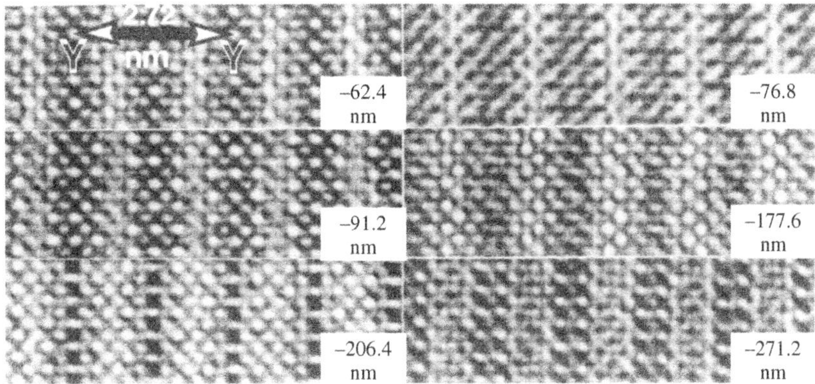

FIG. 12.7. Images of a focal series from $YBa_2Cu_4O_8$ oriented along [100]. The images were recorded on a CM20 FEG-Super TWIN, using 200 keV primary beam energy. From (Coene et al., 1993).

Using eqns (12.66) and (12.67) we have

$$\frac{\partial\big(\delta I_n(\mathbf{u}')\big)}{\partial\big(\phi_E^*(\mathbf{u})\big)} = -\frac{\partial}{\partial\big(\phi_E^*(\mathbf{u})\big)} \int d\mathbf{u}''\phi_E(\mathbf{u}' + \mathbf{u}'')\phi_E^*(\mathbf{u}'')T(\mathbf{u}' + \mathbf{u}'', \mathbf{u}''; \Delta f_n)$$

$$= -\int d\mathbf{u}''\phi_E(\mathbf{u}' + \mathbf{u}'')T(\mathbf{u}' + \mathbf{u}'', \mathbf{u}''; \Delta f_n)\delta(\mathbf{u}'' - \mathbf{u})$$

$$= -\phi_E(\mathbf{u}' + \mathbf{u})T(\mathbf{u}' + \mathbf{u}, \mathbf{u}; \Delta f_n),$$

and

$$\frac{\partial\big(\delta I_n(\mathbf{u}')\big)}{\partial\big(\phi_E(-\mathbf{u})\big)} = -\phi_E^*(-\mathbf{u} - \mathbf{u}')T(-\mathbf{u}, -\mathbf{u} - \mathbf{u}'; \Delta f_n).$$

The set of MAL equations (12.68) becomes

$$\sum_{n=1}^{N} \int d\mathbf{u}'T(\mathbf{u}', \mathbf{u}; \Delta f_n)\phi_E(\mathbf{u}')\delta I_n(\mathbf{u}' - \mathbf{u}) = 0$$

$$\sum_{n=1}^{N} \int d\mathbf{u}'T(-\mathbf{u}, -\mathbf{u}'; \Delta f_n)\phi_E^*(-\mathbf{u}')\delta I_n(\mathbf{u} - \mathbf{u}') = 0. \qquad (12.69)$$

Coene and colleagues (Coene et al., 1992, Coene et al., 1993) showed that this set of MAL equations may be solved recursively. Starting from an estimated wave function $\phi_E^0(\mathbf{u})$ (this starting wave function may be obtained using the linear method discussed in the previous section), this recursive scheme generates the next approximation $\phi_E^1(\mathbf{u})$. The j-th recursive step has the form

$$\phi_E^{j+1}(\mathbf{u}) = \phi_E^j(\mathbf{u}) + \frac{\gamma}{N} \sum_{n=1}^{N} \int d\mathbf{u}'T(\mathbf{u}', \mathbf{u}; \Delta f_n)\delta I_n^j(\mathbf{u} - \mathbf{u}'), \qquad (12.70)$$

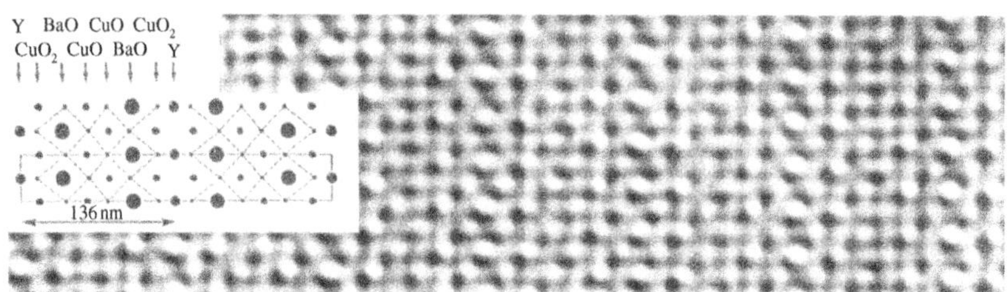

FIG. 12.8. Structural model and the reconstructed distribution of the phase of the exit wave function for [100]-oriented $YBa_2Cu_4O_8$. From (Coene et al., 1993).

where γ is a constant which controls the convergence of the scheme.

Figure 12.7 shows an experimental focal series of six images of a thin (the thickness is of the order of 2–3 nm) specimen of $YBa_2Cu_4O_8$ high-T_c super-conductor (Coene et al., 1993). This focal series was recorded on a CM20 FEG-Super TWIN microscope with a point resolution of 2.4 Å. In the images the configurations of the heavy atoms Ba, Y, and the Cu atoms in the CuO_2 planes are seen as prominent white dots for the defocus value of -91 nm. The faint crosses correspond to the CuO-planes, which are best seen for the large value of underfocus of -271 nm. Light oxygen atoms are not seen in the images.

Figure 12.8 shows a structural model and the phase distribution of the exit wave function reconstructed from Fig. 12.7. Since the specimen is very thin it satisfies the weak phase object approximation. The reconstructed phase distribution therefore provides a very good representation of the crystal structure. All atoms of the $YBa_2Cu_4O_8$ high-T_c superconductor can be identified directly in this single image. The O stacks in the BaO and CuO planes are clearly revealed, with the oxygen atoms represented by faint white dots seen between larger white dots corresponding to the heavier atoms (Cu and Ba). The yttrium atoms are seen as the largest white dots.

12.5 Summary

Depending on the type of filament fitted, the illumination in an electron microscope may be regarded as coherent, incoherent, or partially coherent. For different degrees of coherence the transmission cross-coefficient proves to be different. In some favourable cases high-resolution electron microscope images may be deconvoluted and effects associated with imperfections of the electron microscope optical system may be eliminated, allowing full retrieval of the electron wave function at the exit surface of the specimen.

13

THE ATOMIC SCATTERING FACTOR AND THE OPTICAL POTENTIAL

13.1 Introduction

Throughout this book we have frequently encountered the concept of the atomic scattering factor and the electron scattering potential $V(\mathbf{r})$ or the scaled potential $U(\mathbf{r})$. In this chapter we will give a comprehensive overview of these quantities.

In high-energy electron diffraction an incident electron interacts with the solid via the electrostatic Coulomb potential. To a good approximation the effects of inelastic scattering on the elastic scattering may be taken into account by representing the interaction between the incident electron and the solid by a complex potential (Yoshioka, 1957). This complex potential is usually called the *optical potential* (or the scattering potential as it was referred to earlier in this book), by analogy with the long-standing use of a complex refractive index often used in the description of optical properties of partially absorbing media. In an inelastic collision the incident electron loses energy while the solid is excited to a higher energy state. For high-energy electrons the probability that the inelastically scattered electron reappears in the elastic channel is very small (Rez, 1976). As far as elastic scattering is concerned, the inelastically scattered electron is 'absorbed', and therefore inelastic scattering is responsible for the appearance of an imaginary part of the optical potential.

For a solid consisting of N atoms a fair initial approximation is the so-called isolated atom superposition approximation, by which we assume that the distribution of electron density is not significantly affected by the formation of chemical bonds and that the optical potential may be approximated by the sum of contributions of individual atoms

$$V(\mathbf{r}) = \sum_{i=1}^{N} \varphi_i(\mathbf{r} - \mathbf{r}_i). \tag{13.1}$$

The contribution from the i-th atom $\varphi_i(\mathbf{r})$ is related to its Born atomic scattering factor $f_i^{(B)}(s)$ via the following Fourier transform equation,

$$f^{(B)}(s) = -\frac{m}{2\pi\hbar^2} \int \varphi(\mathbf{r}) \exp(-4\pi i \mathbf{s} \cdot \mathbf{r}) d\mathbf{r}, \tag{13.2}$$

where $s = \sin\theta/\lambda$, 2θ is the scattering angle. $f^{(B)}(s)$ is in turn related to the usual electron atomic scattering factor $f^{(e)}(s)$ tabulated in the International Tables for Crystallography (Wilson and Prince, 1999) via the relation

427

$$f^{(B)}(s) = \left(\frac{m}{m_0}\right) f^{(e)}(s), \tag{13.3}$$

where m_0 is the electron rest mass, m is its relativistically corrected mass and $\hbar \; (= h/2\pi)$ is the Planck constant. We then have for an atom by the inverse Fourier transform of (13.2)

$$\varphi(\mathbf{r}) = -\frac{16\pi\hbar^2}{m_0} \int f^{(e)}(\mathbf{s}) \exp(4\pi i \mathbf{s} \cdot \mathbf{r}) d\mathbf{s}. \tag{13.4}$$

It should be noted that the above expression is simply a definition of the electron atomic scattering factor $f^{(e)}$, i.e. it is an alternative way of expressing the electrostatic potential of a single atom. This definition does not depend on the accuracy of the Born approximation for the scattering of electrons by a single atom. In a real crystal the crystal potential does not have the exact form given by eqn (13.1) because of the presence of anisotropy of the valence electron charge distribution. In many materials this deviation is, however, very small and the error introduced by the approximation (13.1) is usually smaller than 1% for medium order reflections such as the 400 reflection of Si, and is negligible for higher order reflections. The formation of chemical bonds and the associated redistribution of charge density affect only the low-order reflections. Experimentally these effects may be measured by using quantitative electron diffraction (Smart and Humphreys, 1980, Spence and Zuo, 1992).

The electron atomic scattering factor $f^{(e)}(s)$ is a complex quantity. The real part of the scattering factor is called the *elastic scattering factor* and can be calculated using, for example, the relativistic Hartree–Fock atomic wave function (Coulthard, 1967, Doyle and Turner, 1968). The imaginary part of the complex scattering factor is called the *absorptive scattering factor*. It has been shown by several authors that for all $\mathbf{g} \neq 0$, the main contribution to the absorptive scattering factor comes from thermal diffuse scattering (TDS) (Whelan, 1965b, Yoshioka and Kainuma, 1962, Humphreys and Hirsch, 1968, Radi, 1970), and the TDS contribution can be calculated numerically using the published numerical routines (Bird and King, 1990, Weickenmeier and Kohl, 1991) or the parameters tabulated in (Peng et al., 1996e, Peng et al., 1996d) for isotropic thermal vibrations. Inner-shell ionization processes also contribute to the absorptive scattering factor, though the magnitude of these contributions is much smaller than that of TDS (Yoshioka, 1957, Radi, 1970, Oxley and Allen, 2000). Nevertheless, taking these contributions into account is important for accurate calculations of ionization cross sections for use in atom location by channelling enhanced microanalysis (ALCHEMI).

In this chapter we first define and derive an expression for the optical potential, and then discuss the numerical procedure for the calculation of electron atomic scattering factors and the optical potential for the general case of anisotropic thermal vibrations (Peng, 1997). For all neutral atoms and some commonly occurring ions the atomic scattering factors will be represented analytically by a sum of Gaussians. We will also derive convenient analytical expressions suitable for the calculation of the optical potential of crystalline materials.

13.2 The optical potential

In this section we give a formal derivation of the optical potential for high-energy electron diffraction by a crystal. Following Dederichs (Dederichs, 1972), we write the total Hamiltonian of the solid as

$$H = h_0 + H_0 + V, \qquad (13.5)$$

where $h_0 = -(\hbar^2/2m)\nabla^2$ is the free Hamiltonian of the incident electron, H_0 is the Hamiltonian of all the electrons and nuclei of the crystal, and V is the interaction between the incident electron and the crystal. In general V depends on all the coordinates of electrons and nuclei of the crystal, i.e. $V = V(\mathbf{r}, \mathbf{R}_1, ..., \mathbf{R}_n, ..., \mathbf{r}_j, ...)$, where $\mathbf{R}_n$ and $\mathbf{r}_j$ denote the coordinates of the n-th nucleus and the i-th electron respectively. Neglecting the interaction between the incident electron and the crystal we obtain the following system of eigenfunctions describing the state of the incident electron and the crystal:

$$\varphi_k = \exp(i\mathbf{k}\cdot\mathbf{r}), \quad h_0\varphi_k = \frac{\hbar^2 k^2}{2m}\varphi_k = E_k\varphi_k, \qquad (13.6)$$

and

$$\phi_\alpha = \phi_\alpha(\mathbf{R}_1, ..., \mathbf{R}_n, ..., \mathbf{r}_j, ...), \quad H_0\phi_\alpha = E_\alpha\phi_\alpha. \qquad (13.7)$$

The presence of interaction between the electron and the crystal makes the wave function of the electron inseparable from that of the crystal. Let $\Psi_{k,\alpha} = \Psi_{k,\alpha}(\mathbf{r}, \mathbf{R}_1, ..., \mathbf{R}_n, ..., \mathbf{r}_j, ...)$ be the total wave function of the system consisting of the high-energy electron and the crystal. This wave function satisfies the equation

$$(h_0 + H_0 + V)\Psi_{k,\alpha} = (E_k + E_\alpha)\Psi_{k,\alpha}. \qquad (13.8)$$

To solve eqn (13.8) formally, we write

$$\Psi_{k,\alpha} = \varphi_k\phi_\alpha + \Phi_{k,\alpha}. \qquad (13.9)$$

By substituting (13.9) into (13.8) we obtain the following differential equation

$$(E_k + E_\alpha - H_0 - h_0)\Phi_{k,\alpha} = V\Psi_{k,\alpha}. \qquad (13.10)$$

This equation may be solved using the free-particle Green's function G_0 defined by

$$G_0 = \frac{1}{E_k + E_\alpha + i\epsilon - H_0 - h_0}. \qquad (13.11)$$

This gives

$$\Phi_{k,\alpha} = G_0 V\Psi_{k,\alpha}, \qquad (13.12)$$

where in eqn (13.11) the limit $\epsilon \to 0^+$ is implied.

Substitution of (13.12) into (13.9) leads to the Lippmann-Schwinger equation (Schiff, 1968)

$$\Psi_{k,\alpha} = \varphi_k \phi_\alpha + G_0 V \Psi_{k,\alpha}. \tag{13.13}$$

The concept of the optical potential is introduced to take into account the effect of many-body excitations on the wave function of the high-energy electron. Elastic scattering processes are those where the initial and the final state of the crystal remain the same. It is possible to define the wave function of the high-energy electron by projecting the total wave function $\Psi_{k,\alpha}$ on one of the eigenstates of the crystal

$$\psi_{k,\alpha}(\mathbf{r}) = \langle \phi_\alpha | \Psi_{k,\alpha}(\mathbf{r}) \rangle$$

$$= \int \phi_\alpha^*(\mathbf{R}_1...\mathbf{R}_n...\mathbf{r}_j...)\Psi_{k,\alpha}(\mathbf{r},\mathbf{R}_1...\mathbf{R}_n...\mathbf{r}_j...)d\mathbf{R}_1...d\mathbf{R}_n...d\mathbf{r}_j... \tag{13.14}$$

By averaging (13.14) over the thermal equilibrium we find the wave function of the high-energy electron

$$\psi_k(\mathbf{r}) = \langle \psi_{k,\alpha} \rangle = \frac{1}{\mathcal{Z}} \sum_\alpha \exp\left(-\frac{E_\alpha}{k_B T}\right) \langle \phi_\alpha | \Psi_{k,\alpha}(\mathbf{r}) \rangle, \tag{13.15}$$

where the partition function $\mathcal{Z}$ is given by

$$\mathcal{Z} = \sum_\alpha \exp\left(-\frac{E_\alpha}{k_B T}\right). \tag{13.16}$$

Having defined the wave function (13.15), we now seek an equation that this wave function satisfies. The optical potential V^{op} is a quantity that enters this new equation. It is defined by a relationship that is similar to the Lippmann–Schwinger equation (13.13)

$$\psi_k = \varphi_k + G_0 V^{op} \psi_k, \tag{13.17}$$

which involves the wave function of the high-energy electron *only*. To derive (13.17) we write $V = V^{op} + (V - V^{op})$ and use eqn (13.13) to obtain the relation

$$\Psi_{k,\alpha} = \frac{1}{1 - G_0 V^{op}} \varphi_k \phi_\alpha + G_V (V - V^{op}) \Psi_{k,\alpha}, \tag{13.18}$$

where

$$G_V = \frac{1}{1 - G_0 V^{op}} G_0 = \frac{1}{E_k + E_\alpha + i\epsilon - H_0 - h_0 - V^{op}}. \tag{13.19}$$

We now multiply (13.17) from the right-hand side by ϕ_α and find

$$\psi_k \phi_\alpha = \frac{1}{1 - G_0 V^{op}} \varphi_k \phi_\alpha. \tag{13.20}$$

Substituting eqn (13.20) into eqn (13.18) we arrive at

$$\Psi_{k,\alpha} = \psi_k\phi_\alpha + G_V(V - V^{op})\Psi_{k,\alpha} = \frac{1}{1 - G_V(V - V^{op})}\psi_k\phi_\alpha$$

$$= \left\{1 + G_V(V - V^{op})\frac{1}{1 - G_V(V - V^{op})}\right\}\psi_k\phi_\alpha. \tag{13.21}$$

Multiplying (13.21) from the left-hand side by ϕ_α and averaging over thermal equilibrium we obtain the following equation:

$$\langle\psi_{k,\alpha}\rangle = \psi_k = \psi_k + \frac{1}{E_k + i\epsilon - h_0 - V^{op}}\Big\langle(V - V^{op})\frac{1}{1 - G_V(V - V^{op})}\Big\rangle\psi_k. \tag{13.22}$$

This is equivalent to

$$0 = \Big\langle(V - V^{op})\frac{1}{1 - G_V(V - V^{op})}\Big\rangle$$

$$= \Big\langle(V - V^{op})\big[1 + \frac{1}{1 - G_V(V - V^{op})}G_V(V - V^{op})\big]\Big\rangle, \tag{13.23}$$

from which we find a non-linear equation for the optical potential:

$$V^{op} = \langle V\rangle + \Big\langle(V - V^{op})\frac{1}{1 - G_V(V - V^{op})}G_V(V - V^{op})\Big\rangle$$

$$= \langle V\rangle + \Big\langle(V - V^{op})\frac{1}{E_k + E_\alpha + i\epsilon - h_0 - H_0 - V}(V - V^{op})\Big\rangle. \tag{13.24}$$

In the case of high-energy electron diffraction the interaction between the incident electron and the crystal is relatively weak and we can iterate eqn (13.24) up to the second order to obtain

$$V^{op} \approx V^{(0)} + V^{(1)}, \tag{13.25}$$

where

$$V^{(0)} = \langle V\rangle \tag{13.26}$$

is the averaged potential and

$$V^{(1)} = \Big\langle(V - \langle V\rangle)\frac{1}{E_k + E_\alpha + i\epsilon - h_0 - H_0}(V - \langle V\rangle)\Big\rangle \tag{13.27}$$

is the first-order correction due to diffuse scattering. Recent quantitative studies show that the accuracy of this approximation is sufficient for the majority of applications of high-energy electron diffraction (Spence and Zuo, 1992, Zuo et al., 1988, Zuo and Spence, 1991, Saunders et al., 1995, Ren et al., 1997).

To a good approximation the Hamiltonian of interaction between the incident high-energy electron and the crystal may be written as

$$V(\mathbf{r}, \mathbf{R}_1...\mathbf{R}_n...) = \sum_n \frac{1}{4\pi\epsilon_0}\left\{ -\frac{Z_n e^2}{|\mathbf{r} - \mathbf{R}_n|} + \int \frac{e^2}{|\mathbf{r} - \mathbf{R}_n - \mathbf{x}|} \rho_n^0(\mathbf{x})d\mathbf{x} \right\}$$

$$= \sum_n \int \varphi_n(\mathbf{r} - \mathbf{x})\delta(\mathbf{x} - \mathbf{R}_n)d\mathbf{x}, \tag{13.28}$$

where ϵ_0 is the permittivity of vacuum, Z_n is the atomic number, $\rho_n^0(\mathbf{x})$ is the electron density of the n-th atom in its ground state, and $\varphi_n(\mathbf{r})$ is given by

$$\varphi_n(\mathbf{r}) = -\frac{Z_n e^2}{4\pi\epsilon_0 r} + \int \frac{e^2}{4\pi\epsilon_0|\mathbf{r} - \mathbf{x}|} \rho_n^0(\mathbf{x})d\mathbf{x}. \tag{13.29}$$

13.3 The averaged potential

A more general treatment of the problem of interaction between an incident high-energy electron and the solid involving the density matrix formalism shows that the motion of nuclei and electrons in the crystal makes the many-body interaction potential a time-dependent quantity. Scattering involving this time-dependent interaction gives rise to energy losses of the incident high-energy electron. The optical potential is, however, a time-independent quantity and the first term on the right-hand side of the general expression (13.25) is just the time-averaged potential of the crystal. Scattering by this averaged potential does not give rise to energy losses and is usually called elastic scattering.

13.3.1 *Thermally averaged potential*

In the treatment of thermal vibrations we may assume that atomic electrons follow the motion of nuclei adiabatically and that all atomic electrons remain in their ground state (Ashcroft and Mermin, 1976). Let $\mathbf{R}_n = \mathbf{R}_n^{(0)} + \mathbf{u}_n$ where $\mathbf{R}_n^{(0)}$ denotes the equilibrium position of the n-th atom and $\mathbf{u}_n$ represents the thermal displacement of an atom from its equilibrium position. Equation (13.28) leads to the following expression for the averaged potential:

$$\langle V(\mathbf{r}) \rangle = \sum_n \int \varphi_n(\mathbf{r} - \mathbf{x})\langle \delta(\mathbf{x} - \mathbf{R}_n^{(0)} - \mathbf{u}_n)\rangle d\mathbf{x}. \tag{13.30}$$

The Fourier coefficients of the averaged potential are given by

$$V_g = \frac{1}{\Omega_{\text{cryst}}} \int \langle V(\mathbf{r}) \rangle e^{-i\mathbf{g}\cdot\mathbf{r}}d\mathbf{r} = -\frac{\hbar^2}{2m_0} \cdot \frac{4\pi}{\Omega_0} \sum_n f_n^{(e)}(s)\mathcal{T}_n(\mathbf{g})\exp(-i\mathbf{g}\cdot\mathbf{R}_n^{(0)}),$$

$$\tag{13.31}$$

where Ω_{cryst} and Ω_0 are the volumes of the entire crystal and of a unit cell respectively, $\mathbf{s} = \mathbf{g}/4\pi$, and $\mathcal{T}_n(\mathbf{g})$ is the temperature factor (Willis and Pryor, 1975) of the n-th atom. The summation over n is performed over a unit cell. In (13.31) the electron atomic scattering factor is related to $\varphi_n(\mathbf{r})$ via

$$f_n^{(e)}(\mathbf{s}) = -\frac{m_0}{2\pi\hbar^2} \int \varphi_n(\mathbf{r}) \exp(-i\mathbf{g} \cdot \mathbf{r}) d\mathbf{r}, \qquad (13.32)$$

and in the harmonic approximation for thermal vibrations the temperature factor is given by (Willis and Pryor, 1975)

$$T_n(\mathbf{g}) = \langle \exp(-\mathbf{g} \cdot \mathbf{u}_n) \rangle = \exp\left[-\frac{1}{2}\langle (\mathbf{g} \cdot \mathbf{u}_n)^2 \rangle \right]. \qquad (13.33)$$

In the following two sections we will consider how to evaluate the electron atomic scattering factor $f^{(e)}(s)$ and the temperature factor $T(\mathbf{g})$.

13.3.2 *Electron atomic scattering factor*

For an isolated spherically symmetric atom or an ion the atomic scattering factor for electrons (13.32) can be calculated from the atomic potential $\varphi(r)$. A detailed calculation of the atomic potential requires a knowledge of the total wave function of the atom. This wave function is known exactly only for the hydrogen atom. Scattering factors published for all other atoms were evaluated using various approximations. The published values of these factors are as reliable as the approximate representation of the electron wave functions of an atom. Various approximations for the total atomic wave function have been developed. The multi-configuration Dirac–Fock code (Grant et al., 1980) and the relativistic Hartree–Fock (RHF) equation provide the most accurate available atomic wave functions. The RHF equation given by Grant (Grant, 1961) was solved numerically and programmed by Coulthard (Coulthard, 1967) with the magnetic-interaction term and the off-diagonal Lagrange parameters omitted. Calculations performed by Doyle and Turner (Doyle and Turner, 1968) made it possible to find the total charge density $\rho(r)$ and the atomic potential $\phi(r)$, and through that, the atomic scattering factors for X-rays and electrons.

For all the neutral atoms and for many ions the numerical values of electron atomic scattering factors have been tabulated in the International Tables for Crystallography, Vol. C, table 4.3.1.1 (Cowley, 1992). In this table values of electron atomic scattering factors are given in units of Å and are tabulated for values of s up to 2 Å^{-1}. To find the values of scattering factors in the range 2 Å^{-1} < s < 6 Å^{-1}, the results of Dirac-Fock calculations of X-ray scattering factors $f^{(X)}(s)$ (Fox and Fisher, 1986, Rez et al., 1994) may be used to find values of $f^{(e)}(s)$ via the Mott formula

$$f^{(e)}(s) = \frac{m_0 e^2}{8\pi^2\hbar^2} \frac{1}{4\pi\epsilon_0} \frac{[Z - f^{(X)}(s)]}{s^2} = 0.023933754 \frac{[Z - f^{(X)}(s)]}{s^2}, \qquad (13.34)$$

where ϵ_0 is the permittivity of vacuum. It should be noted that for small values of s the use of the Mott formula should be avoided since the accuracy of this

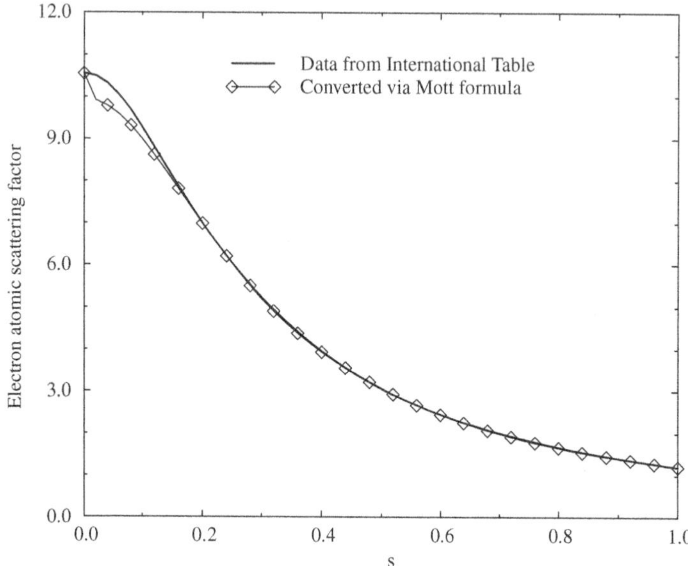

FIG. 13.1. Electron atomic scattering factor for Au as a function of s. The two curves in the figure are obtained from the data tabulated in the International Tables for Crystallography and from results of calculations using X-ray scattering factors of Rez et al. (Rez et al., 1994) and the Mott formula.

approximation becomes unacceptably low in the range of small angles of scattering (Peng and Cowley, 1988a). As an illustration we show in Fig. 13.1 two curves of $f^{(e)}(s)$. One curve uses the data given in the International Tables for Crystallography and the other uses the X-ray scattering factor data given by Rez et al. (Rez et al., 1994) and the Mott formula (13.34). For large s the two curves are practically indistinguishable, while for small s a noticeable difference between the two curves is evident.

At very large angles of scattering the scattering factor follows the screened Coulomb potential form

$$f^{(e)}(s) \to 0.023933754 \frac{Z}{s^2 + \alpha^2},$$
(13.35)

where α is related to the screening length of the Coulomb potential. Numerically α may be determined simply by matching the numerical value of $f^{(e)}(s)$ at a certain value of s with the form given by eqn (13.35).

13.3.3 Temperature factor

We now consider the temperature factor (13.33). Let $\mathbf{a}_1$, $\mathbf{a}_2$, $\mathbf{a}_3$ be the real space lattice vectors and $\mathbf{b}_1$, $\mathbf{b}_2$, $\mathbf{b}_3$ be the reciprocal space lattice vectors satisfying the relations $\mathbf{a}_i \cdot \mathbf{b}_j = 2\pi \delta_{ij}$. In terms of these vectors a real-space displacement vector $\mathbf{u}$ can be expressed as

$$\mathbf{u} = u_1\mathbf{a}_1 + u_2\mathbf{a}_2 + u_3\mathbf{a}_3, \tag{13.36}$$

and a reciprocal space vector $\mathbf{g}$ as

$$\mathbf{g} = h\mathbf{b}_1 + k\mathbf{b}_2 + l\mathbf{b}_3. \tag{13.37}$$

Using these equations, we find

$$\frac{\langle(\mathbf{g}\cdot\mathbf{u})^2\rangle}{4\pi^2} = \left\{ h^2\langle u_1^2\rangle + k^2\langle u_2^2\rangle + l^2\langle u_3^2\rangle + 2hk\langle u_1 u_2\rangle + 2hl\langle u_1 u_3\rangle + 2kl\langle u_2 u_3\rangle \right\}. \tag{13.38}$$

Using matrix notation, the above expression for $(\mathbf{g}\cdot\mathbf{u})^2$ can be written as

$$(\mathbf{g}\cdot\mathbf{u})^2 = 4\pi^2 (h,k,l) \begin{pmatrix} u_1 \\ u_2 \\ u_3 \end{pmatrix} (u_1, u_2, u_3) \begin{pmatrix} h \\ k \\ l \end{pmatrix} = 4\pi^2 \mathbf{G}^T (\mathbf{X}\mathbf{X}^T) \mathbf{G},$$

where $\mathbf{G}$ and $\mathbf{X}$ are 3×1 column vectors and their transposes are given by

$$\mathbf{G}^T = (h,k,l), \quad \mathbf{X}^T = (u_1, u_2, u_3).$$

The temperature factor (13.33) then becomes

$$T(\mathbf{g}) = \exp\{-2\pi^2 \mathbf{G}^T \boldsymbol{\beta}\, \mathbf{G}\}, \tag{13.39}$$

where the matrix $\boldsymbol{\beta} = \langle\mathbf{X}\mathbf{X}^T\rangle$ is symmetric and is given by

$$\boldsymbol{\beta} = \langle\mathbf{X}\mathbf{X}^T\rangle = \begin{pmatrix} \langle u_1^2\rangle & \langle u_1 u_2\rangle & \langle u_1 u_3\rangle \\ \langle u_2 u_1\rangle & \langle u_2^2\rangle & \langle u_2 u_3\rangle \\ \langle u_3 u_1\rangle & \langle u_3 u_2\rangle & \langle u_3^2\rangle \end{pmatrix}. \tag{13.40}$$

This matrix is usually referred to as the mean–square displacement matrix. In X-ray crystallography the general anisotropic vibration parameters are usually given as the elements of the $\mathbf{U}$ matrix which are related to that of the $\boldsymbol{\beta}$ matrix by the following equation:

$$\beta_{ij} = U_{ij}b_i b_j. \tag{13.41}$$

Explicitly the anisotropic temperature factor is given by

$$T(\mathbf{g}) = \exp\left\{ -2\pi^2 \left[U_{11}(hb_1)^2 + U_{22}(kb_2)^2 + U_{33}(lb_3)^2 \right.\right.$$

$$\left.\left. + 2U_{12}(hb_1)(kb_2) + 2U_{13}(hb_1)(lb_3) + 2U_{23}(kb_2)(lb_3) \right] \right\}. \tag{13.42}$$

The values of the matrix elements U_{ij} may be obtained by fitting the calculated X-ray beam intensities to the experimentally measured X-ray intensities using a general anisotropic temperature factor (Giacovazzo, 1992).

How does the general anisotropic temperature factor compare with the average isotropic temperature factor? The latter quantity can be obtained by defining the so-called $\mathbf{B}$ matrix with respect to Cartesian axes. We write

$$\mathbf{u} = u_x\mathbf{i} + u_y\mathbf{j} + u_z\mathbf{k},$$

and

$$\mathbf{g} = g_x\mathbf{i} + g_y\mathbf{j} + g_z\mathbf{k},$$

to obtain

$$\langle(\mathbf{u} \cdot \mathbf{g})^2\rangle = 4\pi^2\mathbf{H}^T\mathbf{B}\mathbf{H}, \tag{13.43}$$

in which $2\pi\mathbf{H}^T = (g_x, g_y, g_z)$ and

$$\mathbf{B} = \begin{pmatrix} \langle u_x^2\rangle & \langle u_x u_y\rangle & \langle u_x u_z\rangle \\ \langle u_y u_x\rangle & \langle u_y^2\rangle & \langle u_y u_z\rangle \\ \langle u_z u_x\rangle & \langle u_z u_y\rangle & \langle u_z^2\rangle \end{pmatrix}. \tag{13.44}$$

The general anisotropic temperature factor is then given by

$$\mathcal{T}(\mathbf{g}) = \exp\{-2\pi^2\mathbf{H}^T \cdot \mathbf{B} \cdot \mathbf{H}\}. \tag{13.45}$$

The relation between the $\mathbf{B}$ matrix and the $\boldsymbol{\beta}$ matrix may be obtained by comparing (13.39) and (13.45), giving

$$\mathbf{H}^T \cdot \mathbf{B} \cdot \mathbf{H} = \mathbf{G}^T \cdot \boldsymbol{\beta} \cdot \mathbf{G}. \tag{13.46}$$

Following Willis and Pryor (Willis and Pryor, 1975) we choose a special Cartesian system of coordinates to define the $\mathbf{B}$ matrix. Let the x axis lie along the $\mathbf{b}_1$ axis, the y axis lie in the plane containing $\mathbf{b}_1$ and $\mathbf{b}_2$, and the z axis complete a right-handed set. Vector $\mathbf{G}$ can be expressed in terms of vector $\mathbf{H}$ via a transformation matrix

$$\mathbf{G} = \mathbf{F} \cdot \mathbf{H}, \tag{13.47}$$

where

$$\mathbf{F} = 2\pi \begin{pmatrix} 1/b_1 & -\cot\alpha_3^*/b_1 & a_1\cos\alpha_2 \\ 0 & 1/(b_2\sin\alpha_3^*) & a_2\cos\alpha_1 \\ 0 & 0 & a_3 \end{pmatrix} \tag{13.48}$$

and α_i and α_i^* are the angles of the direct and reciprocal lattice unit cells. The substitution of (13.47) into (13.46) results in

$$\mathbf{B} = \mathbf{F}^T \cdot \boldsymbol{\beta} \cdot \mathbf{F}. \tag{13.49}$$

In terms of elements of the $\mathbf{B}$ matrix, the average mean–square displacement is given by

$$\langle u^2\rangle = \frac{1}{3}\text{Trace}(\mathbf{B}) = \frac{1}{3}(B_{11} + B_{22} + B_{33}) = \frac{1}{3}\left(\langle u_x^2\rangle + \langle u_y^2\rangle + \langle u_z^2\rangle\right), \tag{13.50}$$

and the usual Debye–Waller B–factor is defined as

$$B = 8\pi^2 \langle u^2 \rangle. \tag{13.51}$$

In terms of the Debye–Waller B-factor the average temperature factor is equal to

$$\mathcal{T}(s) = \exp\left\{ -\frac{1}{2}\langle (\mathbf{g} \cdot \mathbf{u})^2 \rangle \right\} = \exp(-Bs^2). \tag{13.52}$$

In general for isotropic vibrations $\langle u_x^2 \rangle = \langle u_y^2 \rangle = \langle u_z^2 \rangle$, and therefore $\langle u^2 \rangle = (1/3) \times 3\langle u_x^2 \rangle = \langle u_x^2 \rangle$. In particular we have for a cubic crystal $B_{ij} = \langle u^2 \rangle \delta_{ij}$. In this case the general anisotropic temperature factor (13.39) reduces to the isotropic form (13.52).

13.4 The absorptive potential

We now consider the first-order correction to the average potential. This correction is proportional to the square of the deviation of the many-body interaction V from the averaged potential $\langle V \rangle$, and is represented by the second term of eqn (13.25). In the real-space representation, the substitution of eqns (13.28) and (13.30) into (13.27) gives

$$V^{(1)}(\mathbf{r},\mathbf{r}') = \sum_{n,n'} \left\langle \int \int d\mathbf{x} d\mathbf{x}' \varphi_n(\mathbf{r}-\mathbf{x})[\delta(\mathbf{x}-\mathbf{R}_n) - \langle \delta(\mathbf{x}-\mathbf{R}_n)\rangle] \right. \tag{13.53}$$

$$\times \left. \langle \mathbf{r}| \frac{1}{E_k + E_\alpha + i\epsilon - h_0 - H_0}|\mathbf{r}'\rangle \varphi_{n'}(\mathbf{r}'-\mathbf{x}')[\delta(\mathbf{x}'-\mathbf{R}_{n'}) - \langle \delta(\mathbf{x}'-\mathbf{R}_{n'})\rangle] \right\rangle.$$

In the case of TDS the energies of phonons are very much smaller than the energy of the incident electrons, and we may neglect the difference $E_\alpha - H_0$ in (13.53) to find

$$V^{(1)}(\mathbf{r},\mathbf{r}') = \sum_{i,j} \int \int d\mathbf{x} d\mathbf{x}' \varphi_n(\mathbf{r}-\mathbf{x}) G(\mathbf{r},\mathbf{r}') \varphi_{n'}(\mathbf{r}'-\mathbf{x}')$$

$$\times \langle \delta(\mathbf{x}-\mathbf{R}_n)\delta(\mathbf{x}'-\mathbf{R}_{n'}) - \langle \delta(\mathbf{x}-\mathbf{R}_n)\rangle\langle \delta(\mathbf{x}'-\mathbf{R}_{n'})\rangle \rangle, \tag{13.54}$$

where

$$G(\mathbf{r},\mathbf{r}') = \left\langle \mathbf{r} \left| \frac{1}{E_k + i\epsilon - h_0} \right| \mathbf{r}' \right\rangle = -\frac{m}{2\pi\hbar^2} \cdot \frac{\exp(ik_0|\mathbf{r}-\mathbf{r}'|)}{|\mathbf{r}-\mathbf{r}'|}$$

$$= \frac{2m}{\hbar^2} \frac{1}{(2\pi)^3} \int \frac{\exp[i\mathbf{k}\cdot(\mathbf{r}-\mathbf{r}')]}{\mathbf{k}^2 - \mathbf{k}_0^2 + i0} d\mathbf{k} \tag{13.55}$$

is the free space Green's function. The absorptive potential is a non-local quantity, and the way it acts on the wave function of high-energy electrons is defined by the rule

$$V^{(1)}\psi_k = \int V^{(1)}(\mathbf{r},\mathbf{r}')\psi_k(\mathbf{r}')d\mathbf{r}'$$

$$= -\frac{m}{2\pi\hbar^2}\sum_{n,n'}\int d\mathbf{r}'\psi_k(\mathbf{r}')\frac{\exp(ik_0|\mathbf{r}-\mathbf{r}'|)}{|\mathbf{r}-\mathbf{r}'|}\int\int d\mathbf{x}d\mathbf{x}'\varphi_n(\mathbf{r}-\mathbf{x})\varphi_{n'}(\mathbf{r}'-\mathbf{x}')$$

$$\times\Big\langle\delta(\mathbf{x}-\mathbf{R}_n)\delta(\mathbf{x}'-\mathbf{R}_{n'})-\langle\delta(\mathbf{x}-\mathbf{R}_n)\rangle\langle\delta(\mathbf{x}'-\mathbf{R}_{n'})\rangle\Big\rangle. \qquad (13.56)$$

The result of integration depends on the phase difference between waves scattered at $\mathbf{r}$ and $\mathbf{r}'$ through a rapidly oscillating function $\exp(ik_0|\mathbf{r}-\mathbf{r}'|)$. Contributions from different atoms $n\neq n'$ largely cancel each other after the integration. We may therefore retain in the sum only the terms corresponding to $n=n'$. This approximation is equivalent to the Einstein model of TDS (Hall and Hirsch, 1965a). Equation (13.56) then becomes

$$V^{(1)}\psi_k = -\frac{m}{2\pi\hbar^2}\sum_{n}\int d\mathbf{r}'\psi_k(\mathbf{r}')\frac{\exp(ik_0|\mathbf{r}-\mathbf{r}'|)}{|\mathbf{r}-\mathbf{r}'|}\int\int d\mathbf{x}d\mathbf{x}'\varphi_n(\mathbf{r}-\mathbf{x})$$

$$\times\varphi_n(\mathbf{r}'-\mathbf{x}')\Big\langle\delta(\mathbf{x}-\mathbf{R}_n)\delta(\mathbf{x}'-\mathbf{R}_n)-\langle\delta(\mathbf{x}-\mathbf{R}_n)\rangle\langle\delta(\mathbf{x}'-\mathbf{R}_n)\rangle\Big\rangle. \qquad (13.57)$$

The Fourier transform of eqn (13.57) is given by

$$\mathcal{F}\{V^{(1)}\psi_k\} = \frac{1}{\Omega_{\mathrm{cryst}}}\int d\mathbf{r}\exp(-i\mathbf{g}\cdot\mathbf{r})\int V^{(1)}(\mathbf{r},\mathbf{r}')\psi_k(\mathbf{r}')d\mathbf{r}'$$

$$= \frac{\gamma\hbar^2}{\pi m_0\Omega_0}\int\frac{d\mathbf{k}}{k_0^2-k^2+i\epsilon}\sum_n\int d\mathbf{h}f_n^{(e)}(\mathbf{g}-\mathbf{k})f_n^{(e)}(\mathbf{k}-\mathbf{h})\psi(\mathbf{h})$$

$$\times\{\langle\exp[-i(\mathbf{g}-\mathbf{h})\cdot\mathbf{R}_n]\rangle-\langle\exp[-i(\mathbf{g}-\mathbf{k})\cdot\mathbf{R}_n]\rangle\langle\exp[-i(\mathbf{k}-\mathbf{h})\cdot\mathbf{R}_n]\rangle\}$$

$$= \int V^{TDS}(\mathbf{g},\mathbf{h})\psi(\mathbf{h})d\mathbf{h}, \qquad (13.58)$$

where $\gamma = m/m_0$ is the relativistic factor, and $\psi(\mathbf{h})$ is the Fourier transform of the wave function of the high-energy electron

$$\psi(\mathbf{h}) = \frac{1}{(2\pi)^3}\int\psi_k(\mathbf{r})\exp(-i\mathbf{h}\cdot\mathbf{r})d\mathbf{r}, \qquad (13.59)$$

and

$$V^{TDS}(\mathbf{g},\mathbf{h}) = \frac{\gamma\hbar^2}{\pi m_0\Omega_0}\int\frac{d\mathbf{k}}{k_0^2-k^2+i\epsilon}\sum_n f_n^{(e)}(\mathbf{g}-\mathbf{k})f_n^{(e)}(\mathbf{k}-\mathbf{h})e^{-i(\mathbf{g}-\mathbf{h})\cdot\mathbf{R}_n^{(0)}}$$

$$\times\left[\langle e^{-i(\mathbf{g}-\mathbf{h})\cdot\mathbf{u}_n}\rangle-\langle e^{-i(\mathbf{g}-\mathbf{k})\cdot\mathbf{u}_n}\rangle\langle e^{-i(\mathbf{k}-\mathbf{h})\cdot\mathbf{u}_n}\rangle\right],$$

$$(13.60)$$

where $\mathbf{R}_n = \mathbf{R}_n^{(0)} + \mathbf{u}_n$ and the summation over n is performed over a unit cell.

Using the formula

$$\frac{1}{k_0^2 - k^2 + i\epsilon} = \wp \frac{1}{k_0^2 - k^2} - i\frac{\pi}{2k_0}\delta(k_0 - k), \tag{13.61}$$

and substituting (13.61) into (13.60), we find that $V^{(1)}\psi_k$ is in general a complex quantity. The real part of this complex quantity results from *virtual diffuse scattering* and the imaginary part results from *real diffuse scattering*. The contribution from virtual diffuse scattering is almost an order of magnitude smaller than that associated with real diffuse scattering (Rez, 1976). In what follows we shall therefore consider only the imaginary part of the optical potential. Neglecting the real part in eqn (13.60) we arrive at the result

$$V^{TDS}(\mathbf{g},\mathbf{h}) = \frac{\gamma\hbar^2}{\pi m_0 \Omega_0}\left(\frac{-i\pi}{2k_0}\right)\sum_n \int d\mathbf{k}\delta(k - k_0)f_n^{(e)}(\mathbf{g} - \mathbf{k})f_n^{(e)}(\mathbf{k} - \mathbf{h})$$

$$\times e^{-i(\mathbf{g}-\mathbf{h})\cdot\mathbf{R}_n^{(0)}}\left\{\mathcal{T}_n(\mathbf{g} - \mathbf{h}) - \mathcal{T}_n(\mathbf{g} - \mathbf{k})\mathcal{T}_n(\mathbf{k} - \mathbf{h})\right\}. \tag{13.62}$$

By introducing two new parameters $\mathbf{s}$ and $\mathbf{s}'$

$$\mathbf{k} - \mathbf{h} = 4\pi\left(\frac{1}{2}\mathbf{s} + \mathbf{s}'\right), \quad \mathbf{g} - \mathbf{k} = 4\pi\left(\frac{1}{2}\mathbf{s} - \mathbf{s}'\right),$$

and neglecting the curvature of the Ewald sphere in (13.62) (i.e. using the high-energy approximation (Cowley, 1990)), we obtain from eqn (13.62)

$$V^{TDS}(\mathbf{g},\mathbf{h}) = -\frac{\hbar^2}{2m_0}\cdot\frac{4\pi}{\Omega_0}\sum_n if_n^{TDS}(\mathbf{s})\exp[-i(\mathbf{g} - \mathbf{h})\cdot\mathbf{R}_n^{(0)}] = V^{TDS}(\mathbf{g} - \mathbf{h}), \tag{13.63}$$

where

$$f_n^{TDS}(\mathbf{s}) = \frac{4\pi\hbar}{m_0 v}\int d\mathbf{s}' f_n^{(e)}(|\frac{\mathbf{s}}{2} + \mathbf{s}'|)f_n^{(e)}(|\frac{\mathbf{s}}{2} - \mathbf{s}'|)\left\{\mathcal{T}_n(\mathbf{s}) - \mathcal{T}_n(\frac{\mathbf{s}}{2} + \mathbf{s}')\mathcal{T}_n(\frac{\mathbf{s}}{2} - \mathbf{s}')\right\}, \tag{13.64}$$

and v is the velocity of the electron. In the case of isotropic thermal vibrations the temperature factor $\mathcal{T}_n(\mathbf{s})$ reduces to its isotropic form (13.52) and the above expression (13.64) becomes

$$f_n^{TDS}(\mathbf{s}) = \frac{4\pi\hbar}{m_0 v}\exp\{-B_n s^2\}\int d\mathbf{s}' f_n^{(e)}(|\frac{\mathbf{s}}{2} + \mathbf{s}'|)f_n^{(e)}(|\frac{\mathbf{s}}{2} - \mathbf{s}'|)$$

$$\times \left\{1 - \exp\{-2B_n[(s')^2 - s^2/4]\}\right\}. \tag{13.65}$$

The above expression was first derived by Hall and Hirsch (Hall and Hirsch, 1965a), where the result was obtained using a different argument based on physical diffraction principles. Substitution of (13.63) into (13.58) gives

$$\mathcal{F}\{V^{(1)}\psi_k\} = \int V^{TDS}(\mathbf{g} - \mathbf{h})\psi(\mathbf{h})d\mathbf{h} = V^{TDS}(\mathbf{g}) * \psi(\mathbf{g}). \tag{13.66}$$

Applying the inverse Fourier transform to (13.66), we find

$$V^{(1)}\psi_k = \mathcal{F}^{-1}\{V^{TDS}(\mathbf{g}) * \psi(\mathbf{g})\} = V^{(1)}(\mathbf{r})\psi_k(\mathbf{r}), \tag{13.67}$$

where

$$V^{(1)}(\mathbf{r}) = \int V^{TDS}(\mathbf{g})\exp(i\mathbf{g} \cdot \mathbf{r})d\mathbf{g}. \tag{13.68}$$

This expression shows that $V^{(1)}$ is a local potential, i.e. it depends only on the coordinate $\mathbf{r}$ of the electron, and the calculation of the matrix elements of this potential does not require integration over $\mathbf{r}'$ as in eqn (13.56).

13.5   Computation of the complex structure factor

In this section we consider the application of eqns (13.31) and (13.64) to the calculation of the general complex anisotropic structure factor. We shall illustrate the procedure by considering only one particular example. Combining (13.31) and (13.64), we arrive at a general expression for the complex anisotropic structure factor

$$F_g = \sum_n \left\{ f_n^{(e)}(s)\mathcal{T}_n(\mathbf{g}) + if_n^{TDS}(s) \right\} \exp(-i\mathbf{g} \cdot \mathbf{R}_n^{(0)}), \tag{13.69}$$

where $f_n^{(e)}(s)$ is the electron atomic scattering factor (13.32) of the n-th atom, $f_n^{TDS}(s)$ is the imaginary part of the atomic scattering factor due to thermal diffuse scattering (13.64), $\mathcal{T}_n(\mathbf{g})$ is the temperature factor (13.42), and the summation is performed over atoms in a unit cell. Consider the case where within the unit cell there is a total of t non-equivalent atoms and where all atoms are related by a total of m symmetry operations (m is the order of the relevant space group). In this case the structure factor may be represented in the form

$$F_g = \sum_{j=1}^{t} n_{n'} \sum_{s=1}^{m} \left\{ f_{n'}^{(e)}(s)\mathcal{T}_{js}(\mathbf{g}) + if_{js}^{TDS}(s) \right\} \exp(-i\mathbf{g} \cdot \mathbf{R}_{js}^{(0)}), \tag{13.70}$$

where $n_{n'}$ is the occupation number of the j-th atom defined as $n_{n'} = m_{n'}/m$, where $m_{n'}$ is the number of different atomic positions related through symmetry operations. For the n-th group of symmetry equivalent atoms, the s-th symmetry operator is given by $\mathcal{W}_s = (\mathcal{R}_s, \mathbf{T}_s)$, where $\mathcal{R}_s$ denotes a 3×3 rotation matrix and $\mathbf{T}_s$ a 3×1 translation vector. Using these operators we obtain for the s-th atom within the j-th group

$$\mathbf{R}_{js} = \mathcal{R}_s\mathbf{R}_{n'} + \mathbf{T}_s, \tag{13.71}$$

and

$$\mathcal{T}_{js}(\mathbf{g}) = \exp\{-\mathbf{G}^T \boldsymbol{\beta}_{js}\mathbf{G}\} = \exp\{-\mathbf{G}^T(\mathcal{R}_s\boldsymbol{\beta}_j\mathcal{R}_s^T)\mathbf{G}\}. \tag{13.72}$$

For an atom occupying a particular site there are n symmetry operations which leave this site invariant. The temperature matrices must remain invariant with respect to the symmetry operations. This requirement is expressed as

$$\boldsymbol{\beta}_s = \mathcal{R}_s\boldsymbol{\beta}\mathcal{R}_s^T = \boldsymbol{\beta}, \quad s = 1, ..., n. \tag{13.73}$$

These conditions can be used to derive constraints on $\boldsymbol{\beta}$ (Willis and Pryor, 1975).

13.5.1 *A worked example: strontium titanate*

Table 13.1 Structural and anisotropic vibration parameters (Å^{-2}) for strontium titanate (Abramov et al., 1995)

$a(\text{Å})$	$U(\text{Sr})$	$B(\text{Sr})$	$U(\text{Ti})$	$B(\text{Ti})$	$U_{11}(\text{O})$	$U_{22}(\text{O})$	$B(\text{O})$
3.901	0.00787	0.6214	0.00557	0.4398	0.004675	0.011575	0.7323

We now calculate the complex structure factors of strontium titanate $SrTiO_3$. This compound belongs to the important perovskite ABO_3 series, and it was recently studied by high-precision X-ray analysis (Abramov et al., 1995). At room temperature it has a cubic structure and its space group is Pm3m (number 221) (Hahn, 1983). Tables 13.1 summarises the relevant structural and anisotropic thermal vibration parameters. Figure 13.2 gives a graphic representation of thermal ellipsoids illustrating the character of thermal vibrations in this material.

In an $SrTiO_3$ crystal unit cell the Sr atom is situated at site $a = (0,0,0)$ and the Ti atom is located at site $b = (0.5, 0.5, 0.5)$. The restrictions on β_{ij} for both sites are given by case 17 of Willis and Pryor (Willis and Pryor, 1975), i.e. $\beta_{11} = \beta_{22} = \beta_{33}$ and $\beta_{12} = \beta_{13} = \beta_{23} = 0$ (Willis and Pryor, 1975). The $\mathbf{U}$ matrices (which are related to the $\boldsymbol{\beta}$ matrix via (13.41)) for the two atoms therefore have the isotropic form

$$\mathbf{U} = \begin{pmatrix} U & 0 & 0 \\ 0 & U & 0 \\ 0 & 0 & U \end{pmatrix}. \tag{13.74}$$

The three O atoms in $SrTiO_3$ occupy the c sites, and their coordinates are $(0, 0.5, 0.5)$, $(0.5, 0, 0.5)$, $(0.5, 0.5, 0)$. For the first oxygen atom situated at $(0, 0.5, 0.5)$ the conditions on β_{ij} are $\beta_{22} = \beta_{33}$ and $\beta_{12} = \beta_{13} = \beta_{23} = 0$. The $\mathbf{U}$ matrix for this oxygen atom has the form

$$\mathbf{U}(O_1) = \begin{pmatrix} U_{11} & 0 & 0 \\ 0 & U_{22} & 0 \\ 0 & 0 & U_{22} \end{pmatrix}. \tag{13.75}$$

The second oxygen atom situated at $(0.5, 0, 0.5)$ is related to the first oxygen atom by a rotation $3^+[111]$, i.e. rotation by $120°$ about $[111]$ anti-clockwise. The

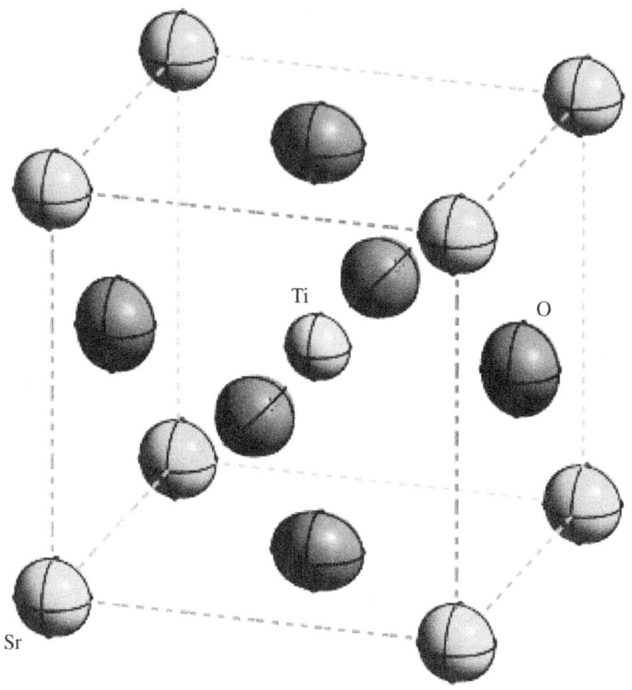

FIG. 13.2. Graphic representation of thermal ellipsoids for a SrTiO$_3$ single crystal at 296 K.

corresponding rotation matrix and the temperature matrix $\mathbf{U}(O_2)$ for this atom are given by

$$\mathcal{R}_2 = 3^+[111] = \begin{pmatrix} 0\ 0\ 1 \\ 1\ 0\ 0 \\ 0\ 1\ 0 \end{pmatrix},$$

$$\mathbf{U}(O_2) = \mathcal{R}_2 \mathbf{U}(O_1) \mathcal{R}_2^T = \begin{pmatrix} U_{22} & 0 & 0 \\ 0 & U_{11} & 0 \\ 0 & 0 & U_{22} \end{pmatrix}. \tag{13.76}$$

The third oxygen atom situated at $(0.5, 0.5, 0)$ is related to the first oxygen atom by the symmetry operation $3^-[111]$. The corresponding rotation matrix and the temperature matrix $\mathbf{U}(O_3)$ are given by

$$\mathcal{R}_3 = 3^-[111] = \begin{pmatrix} 0\ 1\ 0 \\ 0\ 0\ 1 \\ 1\ 0\ 0 \end{pmatrix},$$

$$\mathbf{U}(O_3) = \mathcal{R}_3 \mathbf{U}(O_1) \mathcal{R}_3^T = \begin{pmatrix} U_{22} & 0 & 0 \\ 0 & U_{22} & 0 \\ 0 & 0 & U_{11} \end{pmatrix}. \tag{13.77}$$

The mean–square displacement and the B-factor can be obtained using eqns (13.50) and (13.51). For the strontium atom we have $\langle u^2 \rangle = 0.00787$ Å^2 and $B = 0.6214$ Å^2. For the titanium atom $\langle u^2 \rangle = 0.00557$ Å^2 and $B = 0.4398$ Å^2. For the oxygen atom we have $\langle u^2 \rangle = 0.009275$ Å^2 and $B = 0.7323$ Å^2.

Table 13.2 Some [001] zone axis structure factors of SrTiO$_3$ (in Å^{-2} units) for isotropic and anisotropic thermal vibrations.

(h,	k,	l,)	$\Re(\mathrm{U}_{iso})$	$\Im(\mathrm{U}_{iso})$	$\Re(\mathrm{U}_{aniso})$	$\Im(\mathrm{U}_{aniso})$	Error(%)
2	0	0	.6269E-01	.2733E-02	.6269E-01	.2727E-02	.011
1	1	0	.4496E-01	.2335E-02	.4485E-01	.2377E-02	.264
4	0	0	.2380E-01	.2014E-02	.2381E-01	.2010E-02	.082
3	1	0	.1857E-01	.1908E-02	.1831E-01	.1931E-02	1.362
2	2	0	.4153E-01	.2451E-02	.4153E-01	.2445E-02	.016
6	0	0	.1039E-01	.1313E-02	.1043E-01	.1312E-02	.369
5	1	0	.8357E-02	.1361E-02	.8088E-02	.1364E-02	3.187
4	2	0	.1934E-01	.1839E-02	.1935E-01	.1835E-02	.068
3	3	0	.1158E-01	.1601E-02	.1132E-01	.1613E-02	2.236
8	0	0	.4921E-02	.7628E-03	.4973E-02	.7669E-03	1.061
7	1	0	.4249E-02	.8657E-03	.4020E-02	.8565E-03	5.305
6	2	0	.9176E-02	.1212E-02	.9206E-02	.1211E-02	.325
5	3	0	.6453E-02	.1165E-02	.6214E-02	.1165E-02	3.649
4	4	0	.1189E-01	.1424E-02	.1190E-01	.1420E-02	.081
8	2	0	.4498E-02	.7074E-03	.4542E-02	.7104E-03	.980
7	3	0	.3549E-02	.7488E-03	.3351E-02	.7406E-03	5.468
6	4	0	.6577E-02	.9587E-03	.6593E-02	.9562E-03	.252
5	5	0	.4249E-02	.8657E-03	.4053E-02	.8616E-03	4.532
8	4	0	.3488E-02	.5648E-03	.3516E-02	.5657E-03	.794
7	5	0	.2562E-02	.5622E-03	.2408E-02	.5553E-03	5.888
6	6	0	.4122E-02	.6562E-03	.4134E-02	.6539E-03	.298

Table 13.2 gives values of several calculated [001] zone axis structure factors for an SrTiO$_3$ single crystal at 296 K. These were calculated using an isotropic Debye–Waller factor and the general anisotropic vibration parameters listed in Table 13.1. The error introduced by using the isotropic Debye–Waller factor is estimated using the following expression:

$$\text{Error} = \frac{||U_{iso}| - |U_{aniso}||}{|U_{iso}|}. \tag{13.78}$$

The results listed in the table correspond to a 100 keV primary beam energy. They show that the error associated with the use of the isotropic vibration model varies from less than 1% to more than 5%, and in general an error of this order of magnitude cannot be neglected in quantitative electron diffraction (Peng, 1997).

13.6 Analytical representation of atomic scattering factors

While numerical values of scattering factors can be introduced into computer programs to carry out dynamical electron diffraction calculations, there are many cases where an analytic representation of the scattering factor is required. Examples include the case of reflection high-energy electron diffraction (RHEED) (Dudarev et al., 1995a) and the tensor method of determination of crystal structures by high-energy electrons (Peng and Dudarev, 1993a, Peng and Zuo, 1995). The most widely used analytical representation of $f^{(e)}(s)$ involves fitting the scattering factor by a sum of several Gaussian functions

$$f^{(e)}(s) = \sum_{i=1}^{n} a_i \exp(-b_i s^2),\tag{13.79}$$

where a_i and b_i are fitting parameters. This approximation was first used by Vand *et al.* (Vand et al., 1957) to represent the X-ray scattering factor. Later Smith and Burge (Smith and Burge, 1962), and subsequently Doyle and Turner (Doyle and Turner, 1968) used the Gaussian representation to fit the real part of the electron scattering factors of many elements. Dudarev, Peng and Whelan (Dudarev et al., 1995a) developed a similar approach to represent the absorptive part of electron scattering factors.

13.6.1   *The parameterization of the elastic atomic scattering factor for electrons*

13.6.1.1   *Scattering factors of neutral atoms*   Numerical values of elastic atomic scattering factors $f^{(e)}(s)$ are tabulated in the International Tables for Crystallography (Cowley, 1992) for all atoms of the periodic table. The majority of these values were obtained using RHF wave functions (Doyle and Turner, 1968, Coulthard, 1967). In several cases non-relativistic Hartree-Fock wave functions were used (Cowley, 1992). The tabulated values of elastic scattering factors have been parameterized using Gaussian representation and 10 parameters (i.e. $n=5$ in eqn (13.79)). The parameters are listed in Tables 13.3 and 13.4 at the end of the chapter. Errors associated with the Gaussian parameterization were evaluated by Peng *et al.* (Peng et al., 1996e) both for parameters given in Tables 13.3 and 13.4 and for those published previously (Doyle and Turner, 1968, Weickenmeier and Kohl, 1991). This analysis shows that the parameters given in Tables 13.3 and 13.4 are more accurate than those published previously. It should be noted that although the 10-parameter fitting is more accurate than that involving fewer parameters, many crystallographic computer routines developed by the X-ray diffraction community use eight fitting parameters for representing the X-ray atomic scattering factor $f(s)$. Parameterizations of elastic electron atomic scattering factors involving eight parameters are also available, and the interested reader can find the numerical values of these parameters in (Peng, 1999).

13.6.1.2   *Scattering factors of ions*   The scattering factor of an ion is different from that of a neutral atom. In the case of X-ray diffraction atomic scattering factors of both neutral atoms and ions satisfy the condition (James, 1948)

$$Z_0 = \lim_{s \to 0} f^{(X)}(s),\tag{13.80}$$

where Z_0 is the number of electrons associated with the atom, which can be either in a neutral or in a charged ionic state. For a neutral atom $Z_0 = Z$, where Z is the atomic number. For an ion $Z_0 \neq Z$ and the difference between the two quantities represents the excess or deficiency of the charge on the atom resulting from charge transfer associated with the formation of chemical bonds in the crystal (Cowley, 1992, Spence, 1993). The atomic scattering factor for electron diffraction is related to that for X-ray diffraction by the Mott formula (Mott and Massey, 1965)

$$f^{(e)}(s) = \frac{m_0 e^2}{8\pi^2 \hbar^2} \frac{1}{4\pi\epsilon_0} \frac{[Z - f^{(X)}(s)]}{s^2}.$$ (13.81)

For an ion where the number of electrons associated with the ion is not equal to the charge of the nucleus $Z \neq Z_0$, it follows from eqn (13.81) that as s approaches zero the scattering factor diverges asymptotically as $\sim (Z - Z_0)/s^2$.

It is well known (see for example (Doyle and Turner, 1968)) that the divergence of the electron scattering factor of an ion is associated with the contribution of scattering by the unscreened long-range Coulomb potential of the nucleus. This may be readily demonstrated using eqn (5.130) where $\Delta Z = Z - Z_0$ represents the ionic charge and the second term on the right-hand side represents the divergent contribution from the unscreened Coulomb potential of the ion. Electron scattering factors of ions have been calculated numerically and tabulated by several authors including Doyle and Turner (Doyle and Turner, 1968) and Rez et al. (Rez et al., 1994). In principle the ab-initio numerical values obtained by the above authors may be fitted using the same analytical expression (13.79) as for neutral atoms. However, since eqn (13.79) does not diverge in the limit of small angles of scattering as it should do for an ion, it is not suitable for accurate fitting of ionic scattering factors (for a discussion of the errors introduced by this procedure, see Peng et al. (Peng et al., 1998)). Instead, the electron scattering factor of an ion is represented as follows:

$$f^{(e)}(s) = \sum_j a_j \exp(-b_j s^2) + \frac{m_0 e^2}{8\pi^2 \hbar^2} \frac{1}{4\pi\epsilon_0} \frac{\Delta Z}{s^2},$$ (13.82)

where the amplitude of scattering of the electron by the screened atomic field $f_0^{(e)}(s)$ is represented by a sum of several Gaussian functions. In eqn (13.82) the pre-factor in the second term is equal to $(m_0 e^2)/(8\pi^2 \hbar^2) = 0.023934$ provided that s is given in units of Å^{-1} and $f^{(e)}(s)$ is given in Å units. In a general case where a crystal is characterized by an intermediate degree of ionicity α, the electron scattering factor may be represented as follows:

$$f^{(e)}(s) = (1 - \alpha)f^{(e)}_{neutral}(s) + \alpha f^{(e)}_{ion}(s),$$ (13.83)

where $f^{(e)}_{neutral}(s)$ and $f^{(e)}_{ion}(s)$ are scattering factors of neutral atoms and their ions, respectively. A recent study of RHEED from the (100) surface of NiO shows that this scheme works well in describing experimental data obtained for the case of systematic Bragg diffraction by atomic planes parallel to the surface (Peng et al., 1997).

Tables 13.5 and 13.6 at the end of the chapter give 106 sets of five fitting parameters a_i and b_i for 106 ions spanning the entire periodic table. In most cases fitting used *ab-initio* data for X-ray scattering factors of ions calculated by Rez *et al.* (Rez et al., 1994). For large angles of scattering these numerical data were converted into electron scattering factors using the Mott formula (13.81). For smaller angles of scattering where the X-ray data are less reliable (see for example (Peng and Cowley, 1988a)) we fitted $f_0^{(e)}(s)$ at $s = 0$ to the value given in Table 7 of Rez et al. (Rez et al., 1994). For ions where no reliable *ab-initio* data were available we used the comprehensive tabulation by Cowley (Cowley, 1992). Numerical electron scattering factors of 59 ions were taken from Table 4.3.1.2 of Cowley (Cowley, 1992). Values of $f_0^{(e)}(s)$ were first separated from the values given by Cowley (Cowley, 1992), and then they were fitted to a linear combination of five Gaussians by using eqn (13.82). A comprehensive analysis of errors was carried out by Peng *et al.* (Peng, 1998) where it was shown that for all the ions studied so far the quality of parameterization was satisfactory with the sole exception of Pd^{2+}. Further analysis showed that in this particular case the values of X-ray scattering factors given in Table 2 of Rez *et al.* (Rez et al., 1994) for small s were too large to be interpolated smoothly to the value of $f_0^{(e)}(0)$ given in Table 7 from the same source. We have therefore disregarded numerical data at $s = 0.05$ and fitted the remaining part by the expression (13.82). Although visually one cannot distinguish the *ab-initio* numerical data from the fitted data, our error analysis shows that the accuracy of fitting in this case is approximately one order of magnitude worse than for other ions. This suggests that for this particular ion the numerical data given by Rez *et al.* (Rez et al., 1994) were not sufficiently accurate.

13.6.2 *Parameterization of the absorptive atomic scattering factor*

A major difficulty encountered in the parameterization of atomic scattering factors by using algorithms based on the Marquardt–Levenberg procedure is the fact that the final values of parameters depend sensitively on their initially assigned approximate values. This problem is less important in the case of fitting of elastic scattering factors discussed earlier, where the procedure is less sensitive to the choice of starting values of a_j and b_j. But the fitting of absorptive scattering factors requires a more robust procedure. This is because for a given element the absorptive scattering factor depends both on the angle of scattering and on the Debye–Waller factor. It is difficult to make an exhaustive tabulation of all parameters for all elements and compounds. In Appendix D we present a computer routine which was used for parameterizing the absorptive atomic scattering factors. The procedure is able to fit a wide range of numerical scattering factors with high accuracy without any human interference, and it does not require guessing the initial set of values of fitting parameters as an input.

13.7 Analytical expressions for the optical potential of atoms and crystals

By definition the atomic potential is given by the inverse Fourier transform of the complex atomic scattering factors $f^{(e)}(s)$. In terms of the fitting parameters a_j and b_j defined by eqns (13.79) and (D.7) and the Debye–Waller factor we find

$$\varphi(\mathbf{r}) = -\frac{2\pi\hbar^2}{m_0} \cdot \frac{1}{(2\pi)^3} \int f^{(e)}(s) \exp(-Bs^2) \exp(i\mathbf{q} \cdot \mathbf{r}) d\mathbf{q}$$

$$= -\frac{2\pi\hbar^2}{m_0} \cdot \frac{1}{(2\pi)^3} \sum_{j=1}^{5} a_j \int \exp[-(b_j + B)q^2/(4\pi)^2] \exp(i\mathbf{q} \cdot \mathbf{r}) d\mathbf{q}$$

$$= -\frac{2\pi\hbar^2}{m_0} \sum_{j=1}^{5} a_j \left(\frac{4\pi}{b_j + B}\right)^{3/2} \exp\left(-\frac{4\pi^2 r^2}{b_j + B}\right)$$

where B is the relevant Debye–Waller factor. For the complex electron atomic scattering factor taken in the form (note that the thermal Debye–Waller factors have now been included in the definition $f^{(e)}(s)$)

$$f^{(e)}(s) = \sum_{j=1}^{5} \left\{ a_j^{(Re)} \exp\left[-(b_j^{(Re)} + B)s^2\right] + i a_j^{(Im)} \exp\left[-(b_j^{(Im)} + B/2)s^2\right] \right\},$$

we derive an analytical expression for the atomic potential

$$\varphi(r) = -\frac{2\pi\hbar^2}{m_0} \sum_{j=1}^{5} \left\{ a_j^{(Re)} \left(\frac{4\pi}{b_j^{(Re)} + B}\right)^{3/2} \exp\left(-\frac{4\pi^2 r^2}{b_j^{(Re)} + B}\right) \right. \tag{13.84}$$

$$\left. + i a_j^{(Im)} \left(\frac{4\pi}{b_j^{(Im)} + B/2}\right)^{3/2} \exp\left(-\frac{4\pi^2 r^2}{b_j^{(Im)} + B/2}\right) \right\},$$

where superscripts Re and Im refer to the real and imaginary fitting parameters.

For a solid consisting of N atoms, the optical potential may be expressed in the form of a sum of contributions of all the atoms. For a periodic crystal we have

$$V(\mathbf{r}) = \sum_{\ell} \sum_{k} \varphi_k(\mathbf{r} - \mathbf{R}_\ell - \mathbf{r}_k), \tag{13.85}$$

where the index ℓ refers to the ℓ-th unit cell of the crystal, and k refers to the k-th atom of the unit cell. In the case of a three-dimensionally periodic potential we may expand the potential as a Fourier series

$$V(\mathbf{r}) = \sum_{g} V_g \exp(i\mathbf{g} \cdot \mathbf{r}),$$

where the Fourier coefficients of the potential V_g are given by

$$V_g = \frac{1}{\Omega_{\text{cryst}}} \int V(\mathbf{r}) \exp(-i\mathbf{g} \cdot \mathbf{r}) d\mathbf{r},$$

and Ω_{cryst} is the volume of the crystal. Substituting eqn (13.85) into the above expression and noting that $\exp(-i\mathbf{g} \cdot \mathbf{R}_\ell) = 1$, we find

$$V_g = \frac{1}{\Omega_{\text{cryst}}} \sum_\ell \sum_k \int \varphi_k(\mathbf{r} - \mathbf{R}_\ell - \mathbf{r}_k) \exp(-i\mathbf{g} \cdot \mathbf{r}) d\mathbf{r} \tag{13.86}$$

$$= \frac{N}{\Omega_{\text{cryst}}} \sum_k \exp(-i\mathbf{g} \cdot \mathbf{r}_k) \int \varphi_k(\mathbf{r}') \exp(-i\mathbf{g} \cdot \mathbf{r}') d\mathbf{r}'$$

$$= -\frac{\hbar^2}{2m_0} \cdot \frac{4\pi}{\Omega_0} \sum_k f^{(e)}(s) \exp(-i\mathbf{g} \cdot \mathbf{r}_k),$$

where N is the total number of unit cells in the crystal, Ω_0 is the volume of a unit cell, and $s = g/(4\pi)$. If both the temperature and absorption effects are taken into account, we have

$$V_g = -\frac{\hbar^2}{2m_0} \cdot \frac{4\pi}{\Omega_0} \sum_k \exp(-i\mathbf{g} \cdot \mathbf{r}_k) \sum_{j=1}^{5} \left\{ a_{j,k}^{(Re)} \exp\left[-\left(b_{j,k}^{(Re)} + B_k \right) s^2 \right] \right.$$

$$\left. + i\, a_{j,k}^{(Im)} \exp\left[-\left(b_{j,k}^{(Im)} + B_k/2 \right) s^2 \right] \right\}. \tag{13.87}$$

In the case of dynamical RHEED calculations it is convenient to use a mixed real- and reciprocal space representation of the optical potential

$$V(\mathbf{x}, z) = \sum_G V_G(z) \exp(i\mathbf{G} \cdot \mathbf{x}), \tag{13.88}$$

where $\mathbf{G}$ denotes a two-dimensional reciprocal lattice vector parallel to the surface. The two-dimensional Fourier coefficients $V_G(z)$ are given by

$$V_G(z) = \frac{1}{A} \int V(\mathbf{x}, z) \exp(-i\mathbf{G} \cdot \mathbf{x}) d\mathbf{x} \tag{13.89}$$

$$= \frac{1}{A} \sum_{\ell,k} \exp(-i\mathbf{G} \cdot \mathbf{x}_k) \int \varphi_k(\mathbf{x}', z') \exp(-i\mathbf{G} \cdot \mathbf{x}') d\mathbf{x}',$$

where A is the surface area and $z' = z - z_k$. The two-dimensional integration over $\mathbf{x}'$ may be performed analytically

$$\int \{...\} d\mathbf{x}' = -\frac{\hbar^2}{m_0} \cdot \frac{1}{(2\pi)^2} \int \left\{ \int f^{(e)}(s) \exp(i\mathbf{q} \cdot \mathbf{r}') d\mathbf{q} \right\} \exp(-i\mathbf{G} \cdot \mathbf{x}') d\mathbf{x}'$$

$$= -\frac{\hbar^2}{m_0} \cdot \frac{1}{(2\pi)^2} \sum_{j=1}^{5} a_j \int \exp\left[-\frac{b_j}{(4\pi)^2} q^2 + iq_z z' \right] dq \int \exp[i(\mathbf{q}_t - \mathbf{G}) \cdot \mathbf{x}'] d\mathbf{x}'$$

$$= -\frac{4\pi\hbar^2}{m_0} \sum_{j=1}^{5} a_j \sqrt{\frac{\pi}{b_j}} \exp\left[-\frac{b_j}{(4\pi)^2} G^2 \right] \exp\left[-\frac{4\pi^2}{b_j} (z - z_k)^2 \right].$$

Denoting by A_0 the area of a surface unit cell we arrive at the following expression for the Fourier components of the potential:

$$V_G(z) = -\frac{4\pi}{A_0} \cdot \left(\frac{\hbar^2}{m_0}\right) \sum_{j,k} \exp(-i\mathbf{G} \cdot \mathbf{x}_k) \tag{13.90}$$

$$\left\{ a_{j,k}^{(Re)} \sqrt{\frac{\pi}{b_{j,k}^{(Re)} + B_k}} \exp\left[-\frac{\left(b_{j,k}^{(Re)} + B_k\right) G^2}{(4\pi)^2} - \frac{4\pi^2(z - z_k)^2}{b_{j,k}^{(Re)} + B_k} \right] \right.$$

$$\left. + i a_{j,k}^{(Im)} \sqrt{\frac{\pi}{b_{j,k}^{(Im)} + B_k/2}} \exp\left[-\frac{\left(b_{j,k}^{(Im)} + B_k/2\right) G^2}{(4\pi)^2} - \frac{4\pi^2(z - z_k)^2}{b_{j,k}^{(Im)} + B_k/2} \right] \right\}.$$

13.8 Summary

High-energy electrons may be elastically and inelastically scattered by a solid. To a good approximation the effect of inelastic scattering on the elastically scattered electrons may be taken into account using the concept of the complex optical potential. The optical potential may be calculated using electron atomic scattering factors, which in turn can be evaluated numerically using Hartree–Fock atomic wave functions and the Einstein model of atomic thermal vibrations. The numerical electron atomic scattering factors have been parameterized using five Gaussian functions for all the neutral atoms of the periodic table and for many important ions. A computer program given in Appendix D can be used to calculate the thermal diffuse absorptive atomic scattering factors and to fit the numerical results to a convenient analytical form.

Table 13.3 Elastic Atomic Scattering Factors of Neutral Atoms (a_i in Å, b_i in Å^2)

	Z	a_1	a_2	a_3	a_4	a_5	b_1	b_2	b_3	b_4	b_5
H	1	.0088	.0449	.1481	.2356	.0914	.1152	1.0867	4.9755	16.5591	43.2743
He	2	.0084	.0443	.1314	.1671	.0666	.0596	.5360	2.4274	7.7852	20.3126
Li	3	.0478	.2048	.5253	1.5225	.9853	.2258	2.1032	12.9349	50.7501	136.6280
Be	4	.0423	.1874	.6019	1.4311	.7891	.1445	1.4180	8.1165	27.9705	74.8684
B	5	.0436	.1898	.6788	1.3273	.5544	.1207	1.1595	6.2474	21.0460	59.3619
C	6	.0489	.2091	.7537	1.1420	.3555	.1140	1.0825	5.4281	17.8811	51.1341
N	7	.0267	.1328	.5301	1.1020	.4215	.0541	.5165	2.8207	10.6297	34.3764
O	8	.0365	.1729	.5805	.8814	.3121	.0652	.6184	2.9449	9.6298	28.2194
F	9	.0382	.1822	.5972	.7707	.2130	.0613	.5753	2.6858	8.8214	25.6668
Ne	10	.0380	.1785	.5494	.6942	.1918	.0554	.5087	2.2639	7.3316	21.6912
Na	11	.1260	.6442	.8893	1.8197	1.2988	.1684	1.7150	8.8386	50.8265	147.2073
Mg	12	.1130	.5575	.9046	2.1580	1.4735	.1356	1.3579	6.9255	32.3165	92.1138
Al	13	.1165	.5504	1.0179	2.6295	1.5711	.1295	1.2619	6.8242	28.4577	88.4750
Si	14	.0567	.3365	.8104	2.4960	2.1186	.0582	.6155	3.2522	16.7929	57.6767
P	15	.1005	.4615	1.0663	2.5854	1.2725	.0977	.9084	4.9654	18.5471	54.3648
S	16	.0915	.4312	1.0847	2.4671	1.0852	.0838	.7788	4.3462	15.5846	44.6365
Cl	17	.0799	.3891	1.0037	2.3332	1.0507	.0694	.6443	3.5351	12.5058	35.8633
Ar	18	.1044	.4551	1.4232	2.1533	.4459	.0853	.7701	4.4684	14.5864	41.2474
K	19	.2149	.8703	2.4999	2.3591	3.0318	.1660	1.6906	8.7447	46.7825	165.6923
Ca	20	.2355	.9916	2.3959	3.7252	2.5647	.1742	1.8329	8.8407	47.4583	134.9613
Sc	21	.4636	2.0802	2.9003	1.4193	2.4323	.3682	4.0312	22.6493	71.8200	103.3691
Ti	22	.2123	.8960	2.1765	3.0436	2.4439	.1399	1.4568	6.7534	33.1168	101.8238
V	23	.2369	1.0774	2.1894	3.0825	1.7190	.1505	1.6392	7.5691	36.8741	107.8517
Cr	24	.1970	.8228	2.0200	2.1717	1.7516	.1197	1.1985	5.4097	25.2361	94.4290
Mn	25	.1943	.8190	1.9296	2.4968	2.0625	.1135	1.1313	5.0341	24.1798	80.5598
Fe	26	.1929	.8239	1.8689	2.3694	1.9060	.1087	1.0806	4.7637	22.8500	76.7309
Co	27	.2186	.9861	1.8540	2.3258	1.4685	.1182	1.2300	5.4177	25.7602	80.8542
Ni	28	.2313	1.0657	1.8229	2.2609	1.1883	.1210	1.2691	5.6870	27.0917	83.0285
Cu	29	.3501	1.6558	1.9582	.2134	1.4109	.1867	1.9917	11.3396	53.2619	63.2520
Zn	30	.1780	.8096	1.6744	1.9499	1.4495	.0876	.8650	3.8612	18.8726	64.7016
Ga	31	.2135	.9768	1.6669	2.5662	1.6790	.1020	1.0219	4.6275	22.8742	80.1535
Ge	32	.2135	.9761	1.6555	2.8938	1.6356	.0989	.9845	4.5527	21.5563	70.3903
As	33	.2059	.9518	1.6372	3.0490	1.4756	.0926	.9182	4.3291	19.2996	58.9329
Se	34	.1574	.7614	1.4834	3.0016	1.7978	.0686	.6808	3.1163	14.3458	44.0455
Br	35	.1899	.8983	1.6358	3.1845	1.1518	.0810	.7957	3.9054	15.7701	45.6124
Kr	36	.1742	.8447	1.5944	3.1507	1.1338	.0723	.7123	3.5192	13.7724	39.1148
Rb	37	.3781	1.4904	3.5753	3.0031	3.3272	.1557	1.5347	9.9947	51.4251	185.9828
Sr	38	.3723	1.4598	3.5124	4.4612	3.3031	.1480	1.4643	9.2320	49.8807	148.0937
Y	39	.3234	1.2737	3.2115	4.0563	3.7962	.1244	1.1948	7.2756	34.1430	111.2079
Zr	40	.2997	1.1879	3.1075	3.9740	3.5769	.1121	1.0638	6.3891	28.7081	97.4289
Nb	41	.1680	.9370	2.7300	3.8150	3.0053	.0597	.6524	4.4317	19.5540	85.5011
Mo	42	.3069	1.1714	3.2293	3.4254	2.1224	.1101	1.0222	5.9613	25.1965	93.5831
Tc	43	.2928	1.1267	3.1675	3.6619	2.5942	.1020	.9481	5.4713	23.8153	82.8991
Ru	44	.2604	1.0442	3.0761	3.2175	1.9448	.0887	.8240	4.8278	19.8977	80.4566
Rh	45	.2713	1.0556	3.1416	3.0451	1.7179	.0907	.8324	4.7702	19.7862	80.2540
Pd	46	.2003	.8779	2.6135	2.8594	1.0258	.0659	.6111	3.5563	12.7638	44.4283
Ag	47	.2739	1.0503	3.1564	2.7543	1.4328	.0881	.8028	4.4451	18.7011	79.2633
Cd	48	.3072	1.1303	3.2046	2.9329	1.6560	.0966	.8856	4.6273	20.6789	73.4723
In	49	.3564	1.3011	3.2424	3.4839	2.0459	.1091	1.0452	5.0900	24.6578	88.0513
Sn	50	.2966	1.1157	3.0973	3.8156	2.5281	.0896	.8268	4.2242	20.6900	71.3399
Sb	51	.2725	1.0651	2.9940	4.0697	2.5682	.0809	.7488	3.8710	18.8800	60.6499
Te	52	.2422	.9692	2.8114	4.1509	2.8161	.0708	.6472	3.3609	16.0752	50.1724
I	53	.2617	1.0325	2.8097	4.4809	2.3190	.0749	.6914	3.4634	16.3603	48.2522
Xe	54	.2334	.9496	2.6381	4.4680	2.5020	.0655	.6050	3.0389	14.0809	41.0005
Cs	55	.5713	2.4866	4.9795	4.0198	4.4403	.1626	1.8213	11.1049	49.0568	202.9987

Table 13.4 Elastic Atomic Scattering Factors of Neutral Atoms — continued (a_i in Å, b_i in Å^2)

	Z	a_1	a_2	a_3	a_4	a_5	b_1	b_2	b_3	b_4	b_5
Ba	56	.5229	2.2874	4.7243	5.0807	5.6389	.1434	1.6019	9.4511	42.7685	148.4969
La	57	.5461	2.3856	5.0653	5.7601	4.0463	.1479	1.6552	10.0059	47.3245	145.8464
Ce	58	.2227	1.0760	2.9482	5.8496	7.1834	.0571	.5946	3.2022	16.4253	95.7030
Pr	59	.5237	2.2913	4.6161	4.7233	4.8173	.1360	1.5068	8.8213	41.9536	141.2424
Nd	60	.5368	2.3301	4.6058	4.6621	4.4622	.1378	1.5140	8.8719	43.5967	141.8065
Pm	61	.5232	2.2627	4.4552	4.4787	4.5073	.1317	1.4336	8.3087	40.6010	135.9196
Sm	62	.5162	2.2302	4.3449	4.3598	4.4292	.1279	1.3811	7.9629	39.1213	132.7846
Eu	63	.5272	2.2844	4.3361	4.3178	4.0908	.1285	1.3943	8.1081	40.9631	134.1233
Gd	64	.9664	3.4052	5.0803	1.4991	4.2528	.2641	2.6586	16.2213	80.2060	92.5359
Tb	65	.5110	2.1570	4.0308	3.9936	4.2466	.1210	1.2704	7.1368	35.0354	123.5062
Dy	66	.4974	2.1097	3.8906	3.8100	4.3084	.1157	1.2108	6.7377	32.4150	116.9225
Ho	67	.4679	1.9693	3.7191	3.9632	4.2432	.1069	1.0994	5.9769	27.1491	96.3119
Er	68	.5034	2.1088	3.8232	3.7299	3.8963	.1141	1.1769	6.6087	33.4332	116.4913
Tm	69	.4839	2.0262	3.6851	3.5874	4.0037	.1081	1.1012	6.1114	30.3728	110.5988
Yb	70	.5221	2.1695	3.7567	3.6685	3.4274	.1148	1.1860	6.7520	35.6807	118.0692
Lu	71	.4680	1.9466	3.5428	3.8490	3.6594	.1015	1.0195	5.6058	27.4899	95.2846
Hf	72	.4048	1.7370	3.3399	3.9448	3.7293	.0868	.8585	4.6378	21.6900	80.2408
Ta	73	.3835	1.6747	3.2986	4.0462	3.4303	.0810	.8020	4.3545	19.9644	73.6337
W	74	.3661	1.6191	3.2455	4.0856	3.2064	.0761	.7543	4.0952	18.2886	68.0967
Re	75	.3933	1.6973	3.4202	4.1274	2.6158	.0806	.7972	4.4237	19.5692	68.7477
Os	76	.3854	1.6555	3.4129	4.1111	2.4106	.0787	.7638	4.2441	18.3700	65.1071
Ir	77	.3510	1.5620	3.2946	4.0615	2.4382	.0706	.6904	3.8266	16.0812	58.7638
Pt	78	.3083	1.4158	2.9662	3.9349	2.1709	.0609	.5993	3.1921	12.5285	49.7675
Au	79	.3055	1.3945	2.9617	3.8990	2.0026	.0596	.5827	3.1035	11.9693	47.9106
Hg	80	.3593	1.5736	3.5237	3.8109	1.6953	.0694	.6758	3.8457	15.6203	56.6614
Tl	81	.3511	1.5489	3.5676	4.0900	2.5251	.0672	.6522	3.7420	15.9791	65.1354
Pb	82	.3540	1.5453	3.5975	4.3152	2.7743	.0668	.6465	3.6968	16.2056	61.4909
Bi	83	.3530	1.5258	3.5815	4.5532	3.0714	.0661	.6324	3.5906	15.9962	57.5760
Po	84	.3673	1.5772	3.7079	4.8582	2.8440	.0678	.6527	3.7396	17.0668	55.9789
At	85	.3547	1.5206	3.5621	5.0184	3.0075	.0649	.6188	3.4696	15.6090	49.4818
Rn	86	.4586	1.7781	3.9877	5.7273	1.5460	.0831	.7840	4.3599	20.0128	62.1535
Fr	87	.8282	2.9941	5.6597	4.9292	4.2889	.1515	1.6163	9.7752	42.8480	190.7366
Ra	88	1.4129	4.4269	7.0460	-1.0573	8.6430	.2921	3.1381	19.6767	102.0436	113.9798
Ac	89	.7169	2.5710	5.1791	6.3484	5.6474	.1263	1.2900	7.3686	32.4490	118.0558
Th	90	.6958	2.4936	5.1269	6.6988	5.0799	.1211	1.2247	6.9398	30.0991	105.1960
Pa	91	1.2502	4.2284	7.0489	1.1390	5.8222	.2415	2.6442	16.3313	73.5757	91.9401
U	92	.6410	2.2643	4.8713	5.9287	5.3935	.1097	1.0644	5.7907	25.0261	101.3899
Np	93	.6938	2.4652	5.1227	5.5965	4.8543	.1171	1.1757	6.4053	27.5217	103.0482
Pu	94	.6902	2.4509	5.1284	5.0339	4.8575	.1153	1.1545	6.2291	27.0741	111.3150
Am	95	.7577	2.7264	5.4184	4.8198	4.1013	.1257	1.3044	7.1035	32.4649	118.8647
Cm	96	.7567	2.7565	5.4364	5.1918	3.5643	.1239	1.2979	7.0798	32.7871	110.1512
Bk	97	.7492	2.7267	5.3521	5.0369	3.5321	.1217	1.2651	6.8101	31.6088	106.4853
Cf	98	.8100	3.0001	5.4635	4.1756	3.5066	.1310	1.4038	7.6057	34.0186	90.5226

Table 13.5 Electron Atomic Scattering Factors of Ions (a_i in Å, b_i in Å^2)

Ion	Z	a_1	a_2	a_3	a_4	a_5	b_1	b_2	b_3	b_4	b_5
H^{-1}	1	.140	.649	1.37	.337	.787	.984	8.67	38.9	111.	166.
Li^{+1}	3	.0046	.0165	.0435	.0649	.0270	.0358	.239	.879	2.64	7.09
Be^{2+}	4	.00340	.0103	.0233	.0325	.0120	.0267	.162	.531	1.48	3.88
O^{-1}	8	.205	.628	1.17	1.03	.290	.397	2.64	8.80	27.1	91.8
O^{-2}	8	.0421	.210	.852	1.82	1.17	.0609	.559	2.96	11.5	37.7
F^{-1}	9	.134	.391	.814	.928	.347	.228	1.47	4.68	13.2	36.0
Na^{1+}	11	.0256	.0919	.297	.514	.199	.0397	.287	1.18	3.75	10.8
Mg^{2+}	12	.0210	.0672	.198	.368	.174	.0331	.222	.838	2.48	6.75
Al^{3+}	13	.0192	.0579	.163	.284	.114	.0306	.198	.713	2.04	5.25
Si^{4+}	14	.192	.289	.100	-.0728	.00120	.359	1.96	9.34	11.1	13.4
Cl^{1-}	17	.265	.596	1.60	2.69	1.23	.252	1.56	6.21	17.8	47.8
K^{1+}	19	.199	.396	.928	1.45	.450	.192	1.10	3.91	9.75	23.4
Ca^{2+}	20	.164	.327	.743	1.16	.307	.157	.894	3.15	7.67	17.7
Sc^{3+}	21	.163	.307	.716	.880	.139	.157	.899	3.06	7.05	16.1
Ti^{2+}	22	.399	1.04	1.21	-.0797	.352	.376	2.74	8.10	14.2	23.2
Ti^{3+}	22	.364	.919	1.35	-.933	.589	.364	2.67	8.18	11.8	14.9
Ti^{4+}	22	.116	.256	.565	.772	.132	.108	.655	2.38	5.51	12.3
V^{2+}	23	.317	.939	1.49	-1.31	1.47	.269	2.09	7.22	15.2	17.6
V^{3+}	23	.341	.805	.942	.0783	.156	.321	2.23	5.99	13.4	16.9
V^{5+}	23	.0367	.124	.244	.723	.435	.0330	.222	.824	2.80	6.70
Cr^{2+}	24	.237	.634	1.23	.713	.0859	.177	1.35	4.30	12.2	39.0
Cr^{3+}	24	.393	1.05	1.62	-1.15	.407	.359	2.57	8.68	11.0	15.8
Cr^{4+}	24	.132	.292	.703	.692	9.59	.109	.695	2.39	5.65	14.7
Mn^{2+}	25	.0576	.210	.604	1.32	.659	.0398	.284	1.29	4.23	14.5
Mn^{3+}	25	.116	.523	.881	.589	.214	.0117	.876	3.06	6.44	14.3
Mn^{4+}	25	.381	1.83	-1.33	.995	.0618	.354	2.72	3.47	5.47	16.1
Fe^{2+}	26	.307	.838	1.11	.280	.277	.230	1.62	4.87	10.7	19.2
Fe^{3+}	26	.198	.387	.889	.709	.117	.154	.893	2.62	6.65	18.0
Co^{2+}	27	.213	.488	.998	.828	.230	.148	.939	2.78	7.31	20.7
Co^{3+}	27	.331	.487	.729	.608	.131	.267	1.41	2.89	6.45	15.8
Ni^{2+}	28	.338	.982	1.32	-3.56	3.62	.237	1.67	5.73	11.4	12.1
Ni^{3+}	28	.347	.877	.790	.0538	.192	.260	1.71	4.75	7.51	13.0
Cu^{1+}	29	.312	.812	1.11	.794	.257	.201	1.31	3.80	10.5	28.2
Cu^{2+}	29	.224	.544	.970	.727	.182	.145	.933	2.69	7.11	19.4
Zn^{2+}	30	.252	.600	.917	.663	.161	.161	1.01	2.76	7.08	19.0
Ga^{3+}	31	.391	.947	.690	.0709	.0653	.264	1.65	4.82	10.7	15.2
Ge^{4+}	32	.346	.830	.599	.0949	-.0217	.232	1.45	4.08	13.2	29.5
Br^{1-}	35	.125	.563	1.43	3.52	3.22	.0530	.469	2.15	11.1	38.9
Br^{1+}	37	.368	.884	1.14	2.26	.881	.187	1.12	3.98	10.9	26.6
Sr^{2+}	38	.346	.804	.988	1.89	.609	.176	1.04	3.59	9.32	21.4
Y^{3+}	39	.465	.923	2.41	-2.31	2.48	.240	1.43	6.45	9.97	12.2
Zr^{4+}	40	.234	.642	.747	1.47	.377	.113	.736	2.54	6.72	14.7
Nb^{3+}	41	.377	.749	1.29	1.61	.481	.184	1.02	3.80	9.44	25.7
Nb^{5+}	41	.0828	.271	.654	1.24	.829	.0369	.261	.957	3.94	9.44
Mo^{3+}	42	.401	.756	1.38	1.58	.497	.191	1.06	3.84	9.38	24.6
Mo^{5+}	42	.479	.846	15.6	-15.2	1.60	.241	1.46	6.79	7.13	10.4
Mo^{6+}	42	.203	.567	.646	1.16	.171	.0971	.647	2.28	5.61	12.4
Ru^{3+}	44	.428	.773	1.55	1.46	.486	.191	1.09	3.82	9.08	21.7
Ru^{4+}	44	.282	.653	1.14	1.53	.418	.125	.753	2.85	7.01	17.5
Rh^{3+}	45	.352	.723	1.50	1.63	.499	.151	.878	3.28	8.16	20.7
Rh^{4+}	45	.397	.725	1.51	1.19	.251	.177	1.01	3.62	8.56	18.9
Pd^{2+}	46	.935	3.11	24.6	-43.6	21.1	.393	4.06E	43.1	54.0	69.8
Pd^{4+}	46	.348	.640	1.22	1.45	.427	.151	.832E	2.85	6.59	15.6
Ag^{1+}	47	.503	.940	2.17	1.99	.726	.199	1.19	4.05	11.3	32.4
Ag^{2+}	47	.431	.756	1.72	1.78	.526	.175	.979	3.30	8.24	21.4

Table 13.6 Electron Atomic Scattering Factors of Ions — continued (a_i in Å, b_i in Å^2)

Ion	Z	a_1	a_2	a_3	a_4	a_5	b_1	b_2	b_3	b_4	b_5
Cd^{2+}	48	.425	.745	1.73	1.74	.487	.168	.944	3.14	7.84	20.4
In^{3+}	49	.417	.755	1.59	1.36	.451	.164	.960	3.08	7.03	16.1
Sn^{2+}	50	.797	2.13	2.15	-1.64	2.72	.317	2.51	9.04	24.2	26.4
Sn^{4+}	50	.261	.642	1.53	1.36	.177	.0957	.625	2.51	6.31	15.9
Sb^{3+}	51	.552	1.14	1.87	1.36	.414	.212	1.42	4.21	12.5	29.0
Sb^{5+}	51	.377	.588	1.22	1.18	.244	.151	.812	2.40	5.27	11.9
I^{1-}	53	.901	2.80	5.61	-8.69	12.6	.312	2.59	14.1	34.4	39.5
Cs^{1+}	55	.587	1.40	1.87	3.48	1.67	.200	1.38	4.12	13.0	31.8
Ba^{2+}	56	.733	2.05	23.0	-152.	134.	.258	1.96	11.8	14.4	14.9
La^{3+}	57	.493	1.10	1.50	2.70	1.08	.167	1.11	3.11	9.61	21.2
Ce^{3+}	58	.560	1.35	1.59	2.63	.706	.190	1.30	3.93	10.7	23.8
Ce^{4+}	58	.483	1.09	1.34	2.45	.797	.165	1.10	3.02	8.85	18.8
Pr^{3+}	59	.663	1.73	2.35	.351	1.59	.226	1.61	6.33	11.0	16.9
Pr^{4+}	59	.521	1.19	1.33	2.36	.690	.177	1.17	3.28	8.94	19.3
Nd^{3+}	60	.501	1.18	1.45	2.53	.920	.162	1.08	3.06	8.80	19.6
Pm^{3+}	61	.496	1.20	1.47	2.43	.943	.156	1.05	3.07	8.56	19.2
Sm^{3+}	62	.518	1.24	1.43	2.40	.781	.163	1.08	3.11	8.52	19.1
Eu^{2+}	63	.613	1.53	1.84	2.46	.714	.190	1.27	4.18	10.7	26.2
Eu^{3+}	63	.496	1.21	1.45	2.36	.774	.152	1.01	2.95	8.18	18.5
Gd^{3+}	64	.490	1.19	1.42	2.30	.795	.148	.974	2.81	7.78	17.7
Tb^{3+}	65	.503	1.22	1.42	2.24	.710	.150	.982	2.86	7.77	17.7
Dy^{3+}	66	.503	1.24	1.44	2.17	.643	.148	.970	2.88	7.73	17.6
Ho^{3+}	67	.456	1.17	1.43	2.15	.692	.129	.869	2.61	7.24	16.7
Er^{3+}	68	.522	1.28	1.46	2.05	.508	.150	.964	2.93	7.72	17.8
Tm^{3+}	69	.475	1.20	1.42	2.05	.584	.132	.864	2.60	7.09	16.6
Yb^{2+}	70	.508	1.37	1.76	2.23	.584	.136	.922	3.12	8.72	23.7
Yb^{3+}	70	.498	1.22	1.39	1.97	.559	.138	.881	2.63	6.99	16.3
Lu^{3+}	71	.483	1.21	1.41	1.94	.522	.131	.845	2.57	6.88	16.2
Hf^{4+}	72	.522	1.22	1.37	1.68	.312	.145	.896	2.74	6.91	16.1
Ta^{5+}	73	.569	1.26	.979	1.29	.551	.161	.972	2.76	5.40	10.9
W^{6+}	74	.181	.873	1.18	1.48	.562	.0118	.442	1.52	4.35	9.42
Os^{4+}	76	.586	1.31	1.63	1.71	.540	.155	.938	3.19	7.84	19.3
Ir^{3+}	77	.692	1.37	1.80	1.97	.804	.182	1.04	3.47	8.51	21.2
Ir^{4+}	77	.653	1.29	1.50	1.74	.683	.174	.992	3.14	7.22	17.2
Pt^{2+}	78	.872	1.68	2.63	1.93	.475	.223	1.35	4.99	13.6	33.0
Pt^{4+}	78	.550	1.21	1.62	1.95	.610	.142	.833	2.81	7.21	17.7
Au^{1+}	79	.811	1.57	2.63	2.68	.998	.201	1.18	4.25	12.1	34.4
Au^{3+}	79	.722	1.39	1.94	1.94	.699	.184	1.06	3.58	8.56	20.4
Hg^{1+}	80	.796	1.56	2.72	2.76	1.18	.194	1.14	4.21	12.4	36.2
Hg^{2+}	80	.773	1.49	2.45	2.23	.570	.191	1.12	4.00	10.8	27.6
Tl^{1+}	81	.820	1.57	2.78	2.82	1.31	.197	1.16	4.23	12.7	35.7
Tl^{3+}	81	.836	1.43	.394	2.51	1.50	.208	1.20	2.57	4.86	13.5
Pb^{2+}	82	.755	1.44	2.48	2.45	1.03	.181	1.05	3.75	10.6	27.9
Pb^{4+}	82	.583	1.14	1.60	2.06	.662	.144	.796	2.58	6.22	14.8
Bi^{3+}	83	.708	1.35	2.28	2.18	.797	.170	.981	3.44	9.41	23.7
Bi^{5+}	83	.654	1.18	1.25	1.66	.778	.162	.905	2.68	5.14	11.2
Ra^{2+}	88	.911	1.65	2.53	3.62	1.58	.204	1.26	4.03	12.6	30.0
Ac^{3+}	89	.915	1.64	2.26	3.18	1.25	.205	1.28	3.92	11.3	25.1
U^{3+}	92	1.14	2.48	3.61	1.13	.900	.250	1.84	7.39	18.0	22.7
U^{4+}	92	1.09	2.32	12.0	-9.11	2.15	.243	1.75	7.79	8.31	16.5
U^{6+}	92	.687	1.14	1.83	2.53	.957	.154	.861	2.58	7.70	15.9

14

TEMPERATURE-DEPENDENT DEBYE–WALLER FACTORS

14.1 Introduction and definitions

The *Debye–Waller factor* is a parameter widely used in both X-ray and electron crystallography to account for the effect of thermal vibrations of atoms in the lattice on the intensity of diffracted beams (Willis and Pryor, 1975). Thermal vibrations smear the periodic potential of the crystal. The instantaneous positions of atoms in the crystal are no longer periodic. However, an average periodic potential can be defined, which gives rise to discrete diffracted beams satisfying the same *Bragg law* that applies to the static crystal potential in the absence of thermal vibrations. The real-space distribution of the average potential is more smeared out in comparison with the distribution of potential in a static crystal. In real space within the harmonic approximation the effect may be described by a convolution of the potential corresponding to atoms frozen at their equilibrium positions with a Gaussian function describing the statistical distribution of thermal atomic displacements. Alternatively, in reciprocal space the effect of thermal vibrations on the average potential is described by the multiplication of the corresponding atomic electron scattering factor of the static crystal by the temperature factor $\mathcal{T}_k$ (James, 1948)

$$F_g = \sum_{k=1} f_k^{(e)} \mathcal{T}_k \exp(i\mathbf{g} \cdot \mathbf{r}_k), \tag{14.1}$$

where $f_k^{(e)}$ is the electron scattering factor of the k-th atom, the position of which with respect to the origin of the unit cell is given by the vector $\mathbf{r}_k$, and the temperature factor $\mathcal{T}_k$ is given by

$$\mathcal{T}_k = \exp\left\{-\frac{1}{2}\langle(\mathbf{g} \cdot \mathbf{u}_k)^2\rangle\right\}, \tag{14.2}$$

where $\langle...\rangle$ denotes the operation of averaging over thermal equilibrium, and $\mathbf{u}_k$ denotes the instantaneous displacement of the k-th atom from its equilibrium position. For isotropic thermal vibrations, $\langle(\mathbf{g} \cdot \mathbf{u}_k)^2\rangle = \frac{1}{3}g^2\langle\mathbf{u}_k^2\rangle$, and the temperature factor may be rewritten as

$$\mathcal{T}_k = \exp\left(-\frac{1}{2} \cdot \frac{1}{3}g^2\langle\mathbf{u}_k^2\rangle\right) = \exp\left(-B_k s^2\right),$$

with $s = g/(4\pi)$ and

$$B_k = \frac{8}{3}\pi^2\langle\mathbf{u}_k^2\rangle. \tag{14.3}$$

454

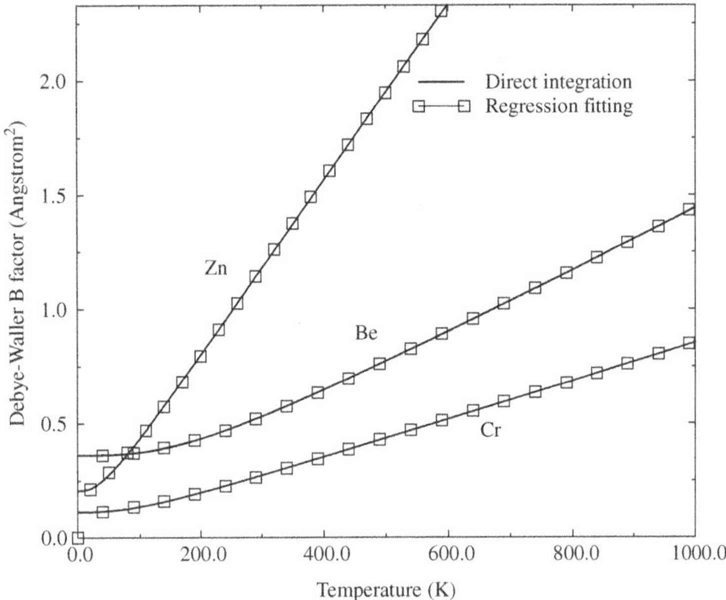

FIG. 14.1. Evaluated and fitted Debye–Waller B factors for Be, Cr and Zn.

The factor B is usually called the *Debye parameter* or the *Debye–Waller B factor*, and sometimes simply the Debye–Waller factor. For convenience we will call this parameter the Debye–Waller factor. Since the displacement vector $\mathbf{u}_k$ depends on temperature T, so does the Debye–Waller B factor, i.e. $B = B(T)$. Figure 14.1 shows the temperature dependence of the Debye–Waller $B(T)$ factors calculated for Be, Cr, and Zn crystals (see the following section for the method of calculation). The figure shows that, although for these crystals the dependence of the Debye–Waller factors on the temperature is strong, the shape of the $B(T)$ curves is simple. This rule is valid for all the elemental crystals and cubic compounds we have studied. These include 68 elemental crystals, 17 compounds with the zinc-blend structure, 19 compounds with the sodium chloride structure, and 5 compounds with the caesium chloride structure. This fact suggests that the temperature dependence of the Debye–Waller $B(T)$ factor may be fitted conveniently to a simple functional form by using polynomial regression fitting. For our purpose fourth degree polynomial regression fitting works well for all crystals (Gao and Peng, 1999):

$$B(T) = a_0 + a_1 T + a_2 T^2 + a_3 T^3 + a_4 T^4, \qquad (14.4)$$

where T is in K, $B(T)$ is in Å^2, and a_i are polynomial regression fitting parameters. The fitted curves shown in Fig. 14.1 are practically indistinguishable from those calculated numerically from first principles. To achieve the best accuracy we found that it was necessary to divide the entire temperature interval into the low-temperature region ($T < 80$ K) and the high-temperature region

$(T > 80 \text{ K})$ and fit $B(T)$ in these two regions separately. Detailed analysis of errors introduced by the fitting procedure of the temperature-dependent Debye–Waller factors was carried out for the entire set of data included in this chapter (Gao et al., 1999, Gao and Peng, 1999, Gao and Peng, 2000), with the general conclusion that the maximum relative error does not exceed 0.5%, which may be safely neglected.

14.2 Debye–Waller factors of elemental crystals

We will first discuss the simplest case of elemental crystals. An elemental crystal consists of identical atoms, and the displacement $\mathbf{u}_\ell$ of an atom at $\mathbf{r}_\ell$ due to a normal mode characterized by wave vector $\mathbf{q}$ and vibration direction $\mathbf{e}_{q,j}$ ($j = 1, 2, 3$) is given by

$$\mathbf{u}_\ell(\mathbf{q}, j) = a_{q,j} \mathbf{e}_{q,j} \cos\{\omega_{q,j} t - \mathbf{q} \cdot \mathbf{r}_\ell - \delta_{q,j}\}, \tag{14.5}$$

where $a_{q,j}$ is the amplitude of the wave, $\mathbf{e}_{q,j}$ is a unit vector defining the polarization direction of displacement, $\omega_{q,j}$ is the circular frequency, and $\delta_{q,j}$ is an arbitrary phase factor which allows for the fact that there are no phase relations among the modes of vibration. For a crystal consisting of a total of N unit cells the total number of wave vectors $\mathbf{q}$ of normal modes is equal to N. In the case where there is only one atom per unit cell for each wave vector $\mathbf{q}$ there are three acoustic phonon modes with mutually perpendicular polarization vectors $\mathbf{e}_{q,j}$, and the total number of modes is equal to $3N$. The resulting displacement of atom ℓ is equal to the sum of contributions of all the modes

$$\mathbf{u}_\ell = \sum_{q,j} \mathbf{u}_\ell(\mathbf{q}, j). \tag{14.6}$$

The total mean kinetic energy of the vibrating lattice is

$$\overline{E}_{kin} = \frac{1}{2} \sum_\ell m \left\langle \left(\frac{d\mathbf{u}_\ell}{dt}\right)^2 \right\rangle = \frac{1}{4} mN \sum_{q,j} a_{q,j}^2 \omega_{q,j}^2, \tag{14.7}$$

where m is the mass of the atom. The mean total energy $\overline{E}$ is twice the mean kinetic energy, i.e. $\overline{E} = 2\overline{E}_{kin}$. Suppose $\overline{E}_{q,j}$ is the average energy associated with the wave $(\mathbf{q}, j)$

$$\overline{E} = \sum_{q,j} \overline{E}_{q,j} = \frac{1}{2} mN \sum_{q,j} a_{q,j}^2 \omega_{q,j}^2. \tag{14.8}$$

Using this equation we obtain the mean square amplitude of the wave $(\mathbf{q}, j)$

$$a_{q,j}^2 = \frac{2\overline{E}_{q,j}}{mN\omega_{q,j}^2}. \tag{14.9}$$

On the other hand from the quantum theory of the harmonic oscillator we have

$$\overline{E}_{q,j} = \left(\overline{n}_{q,j} + \frac{1}{2}\right)\hbar\omega_{q,j}, \tag{14.10}$$

where $\overline{n}_{q,j}$ is the mean value of the quantum number of the harmonic oscillator at thermal equilibrium, i.e.

$$\overline{n}_{q,j} = \left[\exp\left(\frac{\hbar\omega_{q,j}}{k_BT}\right) - 1\right]^{-1}, \tag{14.11}$$

where T is the absolute temperature and k_B is the Boltzmann constant. $\overline{n}_{q,j}$ is the average number of phonons in the mode of vibration $(\mathbf{q}, j)$ at temperature T.

Substitution of (14.10) and (14.11) into (14.9) leads to

$$a_{q,j}^2 = \frac{\hbar}{mN\omega_{q,j}}\coth\left(\frac{\hbar\omega_{q,j}}{2k_BT}\right). \tag{14.12}$$

Using (14.5) we obtain the mean square displacement of the ℓ-th atom

$$\langle\mathbf{u}_\ell^2\rangle = \frac{1}{2}\sum_{q,j} a_{q,j}^2 = \frac{\hbar}{2mN}\sum_{q,j}\frac{1}{\omega_{q,j}}\coth\left(\frac{\hbar\omega_{q,j}}{2k_BT}\right). \tag{14.13}$$

In principle, if we know the frequencies $\omega_{q,j}$ of all the modes of vibration of the crystal, we can calculate the mean square thermal displacement of every atom and find the relevant Debye–Waller factors at a given temperature.

In the limit of a large crystal the discrete wave vectors $\mathbf{q}$ summed over in eqn (14.13) become densely distributed in the first Brillouin zone, and we can replace the sum over $\mathbf{q}$ by an integral. Neglecting the dependence of $\omega_{q,j}$ on the polarization vector $\mathbf{e}_{q,j}$ we obtain for the sum on the right of (14.13)

$$\sum_{q,j}\frac{1}{\omega_{q,j}}\coth\left(\frac{\hbar\omega_{q,j}}{2k_BT}\right) = 3\sum_{q}\frac{1}{\omega_q}\coth\left(\frac{\hbar\omega_q}{2k_BT}\right) = 3N\int_0^\infty\coth\left(\frac{\hbar\omega_q}{2k_BT}\right)\frac{g(\omega)}{\omega}d\omega,$$

where $g(\omega)$ is the normalized *phonon density of states* or frequency distribution of modes in the Brillouin zone. $Ng(\omega)d\omega$ gives the number of phonon modes with frequencies in the interval $\omega \rightarrow \omega + d\omega$.

For a crystal with only one atom per primitive unit cell only acoustic modes of vibration occur, and there exists an upper cut-off frequency ω_m for $g(\omega)$. The well-known Debye model relates this frequency to the *Debye temperature* Θ_D via the equation $\hbar\omega_m = k_B\Theta_D$. In this case we obtain

$$\langle\mathbf{u}_\ell^2\rangle = \frac{3\hbar}{2m}\int_0^{\omega_m}\coth\left(\frac{\hbar\omega}{2k_BT}\right)\left[\frac{g(\omega)}{\omega}\right]d\omega. \tag{14.14}$$

The normalized frequency distribution $g(\omega, j)$ depends not only on the frequency of normal modes but also on the polarization vector of the modes. In the case of a cubic crystal with one atom per primitive unit cell (e.g. in the fcc or bcc structures) $g(\omega)$ depends only on the phonon frequency. For crystals with other structures where the Debye–Waller factor is almost isotropic, such as for the diamond structure and for the hcp structure, the approximation $g(\omega, j) \approx g(\omega)$ also remains valid.

Experimentally the phonon density of states $g(\omega)$ can be determined by either measuring the cross-section of inelastic neutron scattering or evaluating it from the experimentally measured phonon dispersion curves (Lovesey, 1984). Using the available data the Debye–Waller factors have been calculated for 46 elemental crystals with the fcc, bcc, hcp, and the diamond structure, and represented analytically using fourth-degree polynomial regression fitting. In Tables 14.1 and 14.2 we have listed the resulting fitting parameters for these 46 elemental crystals for two temperature intervals. References describing the experimental inelastic neutron scattering data and results of error analysis can be found in (Gao and Peng, 1999). Where experimental phonon densities of states $g(\omega)$ were not available, the Debye–Waller factors of elemental crystals with a cubic lattice were estimated using the *Debye theory* that assumes that the phonon density of states is given by

$$g(\omega) = \begin{cases} 3\omega^2/\omega_m^3 & \text{for } 0 < \omega < \omega_m \\ 0 & \text{for } \omega > \omega_m \end{cases}. \tag{14.15}$$

For 22 elemental crystals the Debye–Waller factors have been estimated in this way (Peng et al., 1996d). The temperature dependence of these factors has been parameterized using polynomial regression fitting (Gao and Peng, 1999). The fitting parameters are given in Tables 14.3 and 14.4.

14.3 Debye–Waller factors of cubic compounds

A compound consists of more than one type of atom. We may write the equilibrium position of the k-th atom in the ℓ-th cell as

$$\mathbf{r}(k\ell) = \mathbf{R}_\ell + \boldsymbol{\tau}_k, \tag{14.16}$$

where $\mathbf{R}_\ell$ is the position of the ℓ-th lattice point with respect to the origin, $\boldsymbol{\tau}_k$ is the position of the k-th atom within the ℓ-th unit cell referred to the origin of the ℓ-th cell. The instantaneous displacement $\mathbf{u}(k\ell)$ of the k-th atom from its equilibrium position $\mathbf{r}(k\ell)$ is

$$\mathbf{u}(k\ell) = \sum_{q,j} a_{q,j}\mathbf{e}(k|qj)\exp\{i[\omega_{q,j}t - \mathbf{q}\cdot\mathbf{r}(k\ell)]\}, \tag{14.17}$$

where $\mathbf{e}(k|qj)$ is the polarization vector of the mode (q, j) and $a_{q,j}$ is given by (14.9) as

$$a_{q,j} = \sqrt{\frac{2}{Nm_k}}\left(\frac{\overline{E}_{q,j}}{\omega_{q,j}^2}\right)^{1/2},$$

where $\overline{E}_{\mathbf{q}j}$ is the mean energy in the mode (q,j), and m_k is the mass of the k-th atom. Suppose the number of atoms in a unit cell is equal to n, and the total number of unit cells in the crystal is equal to N. The total number of normal modes of thermal vibrations in this case is equal to $3nN$. The index $j = 1, 2, ..., 3n$ labels the branches of the dispersion curves, and the vector $\mathbf{q}$ samples N uniformly distributed points in the Brillouin zone. The temperature factor for the k-th atom is given by

$$\mathcal{T}_k = \exp(-\frac{1}{2}\langle(\mathbf{q}\cdot\mathbf{u}_k)^2\rangle)$$

$$= \exp\left\{ -\frac{1}{Nm_k}\left\langle \left| \sum_{q,j} \sqrt{\frac{\overline{E}_{q,j}}{\omega_{q,j}^2}}(\mathbf{q}\cdot\mathbf{e}(k|qj))\exp\{i\,[\omega_{q,j}t - \mathbf{q}\cdot\mathbf{r}(k\ell)]\} \right|^2 \right\rangle \right\}$$

$$= \exp\left\{ -\frac{1}{2Nm_k}\sum_{q,j}\frac{\overline{E}_{q,j}}{\omega_{q,j}^2}[\mathbf{q}\cdot\mathbf{e}(k|qj)]^2 \right\}.$$

For cubic compounds, $\langle(\mathbf{q}\cdot\mathbf{u}_x)^2\rangle = (1/3)q^2\langle\mathbf{u}^2\rangle$ and we obtain

$$\mathcal{T}_k = \exp\left\{ -\frac{1}{6Nm_k}\sum_{q,j}\frac{\overline{E}_{q,j}}{\omega_{q,j}^2}[\mathbf{e}(k|qj)]^2q^2 \right\} = \exp\{-B_ks^2\},$$

where $s = g/4\pi$, and the Debye–Waller B factor is given by

$$B_k = \frac{8\pi^2}{3m_kN}\sum_{q,j}\left(\frac{\overline{E}}{\omega^2}\right)_{q,j}|\mathbf{e}(k|\mathbf{q}j)|^2. \tag{14.18}$$

For cubic compounds both the frequency $\omega(\mathbf{q}j)$ and the displacement vector $\mathbf{e}(k|\mathbf{q}j)$ of the k-th atom and the mode $(\mathbf{q}j)$ may be obtained using one of the models of lattice dynamics (Born and Huang, 1954). Parameters of the model may be obtained by refining them against experimentally measured phonon dispersion curves (Gao et al., 1999, Gao and Peng, 2000) or against the experimentally measured Debye–Waller B factors (Reid, 1983). Given the frequencies $\omega_j(\mathbf{q})$ and the displacement vectors $\mathbf{e}(k|qj)$ for all modes (q,j), the Debye–Waller factors can then be calculated for the k-th atom and for any temperature following the approach outlined above.

Tables 14.5–14.10 give the polynomial regression fitting parameters for the Debye–Waller factors of 17 semiconducting compounds with the zinc-blende structure (Reid, 1983, Gao and Peng, 1999), 19 compounds with the sodium chloride structure (Gao et al., 1999), and 5 compounds with the caesium chloride structure (Gao and Peng, 2000). The calculated Debye–Waller factors have been compared with experimentally measured Debye–Waller B factors (Lawrence, 1973, Barron, 1977, Butt et al., 1988, Butt et al., 1993, Gopi and Sirdeshmukh, 1998, Krishna et al., 1998), and in most cases the agreement was found to be better than 10%.

14.4 Summary

Temperature-dependent Debye–Waller $B(T)$ factors are given in the form of fourth-order polynomials with regression fitting parameters determined for 66 elemental crystals, 17 cubic compounds with the zinc-blende structure, 19 compounds with the sodium chloride structure, and 5 compounds with the caesium chloride structure. The Debye–Waller $B(T)$ factors were obtained using either the experimentally measured frequency distributions or the dispersion curves of normal modes of lattice vibration. For 22 elemental crystals the Debye–Waller factors were estimated using the Debye theory.

Table 14.1 Parameterization of the Debye–Waller $B(T)$ factors of 46 elemental crystals (0-80 K)

Element	Z	Structure	a_0	a_1	a_2	a_3	a_4
He	2	HCP	16.03115	.1926E-01	.6846E-01	-.5670E-02	.9275E-03
He	2	FCC	4.55876	-.2830E-02	.9170E-02	-.1670E-02	.1334E-03
Li	3	BCC	1.20401	-.4142E-03	.1264E-03	-.3584E-06	-.5327E-09
Be	4	HCP	.36275	.6590E-05	-.7256E-06	.2249E-07	-.5366E-10
C	6	DIA	.11918	-.6360E-07	.1962E-06	.3167E-09	-.1858E-11
Ne	10	FCC	1.66115	.2430E-02	-.3280E-03	.1197E-03	-.2696E-05
Na	11	BCC	.81176	-.1380E-02	.3954E-03	-.3464E-05	.1240E-07
Mg	12	HCP	.42425	.1714E-03	.2935E-04	.1405E-06	-.1449E-08
Al	13	FCC	.27196	-.4254E-04	.7433E-05	.7042E-07	-.4391E-09
Si	14	DIA	.19284	.1670E-04	-.7475E-06	.1410E-06	-.8174E-09
Ar	18	FCC	.82048	-.2580E-02	.9105E-03	-.1047E-04	.4603E-07
K	19	BCC	.78847	-.2233E-03	.8963E-03	-.1055E-04	.4711E-07
Ca	20	FCC	.31376	-.1804E-03	.8118E-04	-.4502E-06	.8299E-09
Ca	20	BCC	.37894	-.4210E-03	.1313E-03	-.1006E-05	.3204E-08
Sc	21	HCP	.20201	-.9230E-04	.5810E-05	.1373E-06	-.9440E-09
Ti	22	HCP	.16677	.2669E-04	.2068E-05	.1022E-06	-.6030E-09
V	23	BCC	.15421	.6057E-04	.1253E-04	-.5130E-07	.1305E-09
Cr	24	BCC	.11316	-.4193E-05	.2064E-05	-.1451E-10	.3307E-10
Fe	26	BCC	.12132	-.1893E-04	.3513E-05	.1327E-07	-.8533E-10
Fe	26	FCC	.15517	-.1997E-04	.9983E-05	.8506E-08	-.2064E-09
Ni	28	FCC	.12573	-.1977E-04	.2775E-05	.3922E-07	-.2291E-09
Cu	29	FCC	.14616	-.1212E-04	.8220E-05	.4448E-07	-.4132E-09
Zn	30	HCP	.20745	-.2901E-03	.3690E-04	-.3618E-07	-.8717E-09
Ge	32	DIA	.13367	-.2609E-03	.2061E-04	-.4908E-07	-.2569E-09
Kr	36	FCC	.52667	.1250E-02	.8848E-03	-.1137E-04	.5367E-07
Rb	37	BCC	.57468	.7600E-02	.1160E-02	-.1563E-04	.7640E-07
Sr	38	BCC	.30725	-.1276E-03	.2858E-03	-.3173E-05	.1362E-07
Y	39	HCP	.15651	-.1590E-03	.2634E-04	-.3492E-07	-.4910E-09
Zr	40	HCP	.12491	-.6417E-04	.1215E-04	.3310E-07	-.5068E-09
Nb	41	BCC	.10862	-.4088E-04	.1003E-04	-.4529E-08	-.1664E-09
Mo	42	BCC	.07620	.3230E-04	.1210E-05	.2189E-07	-.1052E-09
Pd	46	FCC	.10288	-.1177E-03	.1154E-04	-.6362E-08	-.2215E-09
Ag	47	FCC	.12932	-.7537E-04	.2371E-04	-.5679E-07	-.2679E-09
Sn	50	DIA	.14057	-.8009E-03	.9321E-04	-.9629E-06	.3901E-08
Xe	54	FCC	.38452	.4820E-02	.7547E-03	-.1007E-04	.4870E-07
Cs	55	BCC	.50698	.2166E-01	.1350E-02	-.1933E-04	.9769E-07
Ba	56	BCC	.23013	-.3114E-03	.2572E-03	-.2981E-05	.1314E-07
La	57	FCC	.18446	-.6587E-03	.1378E-03	-.1364E-05	.5296E-08
Tb	65	HCP	.12655	-.2718E-03	.5559E-04	-.4050E-06	.1100E-08
Ho	67	HCP	.11529	-.2341E-03	.4000E-04	-.2202E-06	.2775E-09
Ta	73	BCC	.06738	.1172E-04	.5384E-05	.4135E-07	-.4006E-09
W	74	BCC	.04702	.3234E-04	.9579E-06	.2342E-07	-.1351E-09
Pt	78	FCC	.06770	-.4038E-04	.1223E-04	-.3998E-07	-.5083E-10
Au	79	FCC	.08767	-.2654E-03	.3587E-04	-.2836E-06	.9104E-09
Pb	82	FCC	.13625	.1480E-02	.1297E-03	-.1434E-05	.6111E-08
Th	90	FCC	.08981	-.2508E-03	.4298E-04	-.3296E-06	.9576E-09

Table 14.2 Parameterization of the Debye–Waller $B(T)$ factors of 42 elemental crystals ($T > 80$ K)

Element	Z	Structure	a_0	a_1	a_2	a_3	a_4
Li	3	BCC	.90169	.8780E-02	.3039E-04	-.5940E-07	.4448E-10
Be	4	HCP	.35851	-.1421E-03	.3225E-05	-.3136E-08	.1139E-11
C	6	DIA	.12034	-.2231E-04	.3348E-06	-.2108E-09	.5320E-13
Na	11	BCC	.38531	.1831E-01	.2176E-04	-.5389E-07	.5105E-10
Mg	12	HCP	.22668	.4830E-02	.3239E-05	-.3499E-08	.1362E-11
Al	13	FCC	.18496	.1630E-02	.2469E-05	-.2630E-08	.1015E-11
Si	14	DIA	.14236	.9261E-03	.1623E-05	-.1677E-08	.6351E-12
K	19	BCC	.24960	.3377E-01	.1624E-04	-.4263E-07	.4262E-10
Ca	20	FCC	.12238	.6070E-02	.1815E-05	-.1975E-08	.7726E-12
Ca	20	BCC	.14441	.8010E-02	.2131E-05	-.2317E-08	.9057E-12
Sc	21	HCP	.11892	.1650E-02	.1665E-05	-.1791E-08	.6952E-12
Ti	22	HCP	.10764	.1140E-02	.1470E-05	-.1573E-08	.6088E-12
V	23	BCC	.09905	.1370E-02	.1334E-05	-.1424E-08	.5500E-12
Cr	24	BCC	.08979	.3666E-03	.1137E-05	-.1198E-08	.4588E-12
Fe	26	BCC	.08725	.6233E-03	.1143E-05	-.1213E-08	.4670E-12
Fe	26	FCC	.09411	.1370E-02	.1302E-05	-.1398E-08	.5420E-12
Ni	28	FCC	.08579	.7271E-03	.1152E-05	-.1229E-08	.4745E-12
Cu	29	FCC	.08444	.1380E-02	.1185E-05	-.1276E-08	.4956E-12
Zn	30	HCP	.10549	.3030E-02	.3054E-05	-.5039E-08	.3072E-11
Ge	32	DIA	.06984	.1650E-02	.9448E-06	-.1010E-08	.3904E-12
Kr	36	FCC	.21138	.2794E-01	.5403E-04	-.3045E-06	.6835E-09
Rb	37	BCC	.11346	.4446E-01	.6973E-05	-.1760E-07	.1684E-10
Sr	38	BCC	.06924	.1227E-01	.1052E-05	-.1151E-08	.4515E-12
Y	39	HCP	.06510	.2520E-02	.9606E-06	-.1044E-08	.4082E-12
Zr	40	HCP	.06133	.1570E-02	.8855E-06	-.9584E-09	.3736E-12
Nb	41	BCC	.05889	.1190E-02	.8383E-06	-.9048E-09	.3521E-12
Mo	42	BCC	.05292	.4392E-03	.7142E-06	-.7625E-09	.2946E-12
Pd	46	FCC	.05200	.1220E-02	.7458E-06	-.8062E-09	.3140E-12
Ag	47	FCC	.05358	.2190E-02	.7901E-06	-.8588E-09	.3357E-12
Sn	50	DIA	.13541	.1900E-02	.1113E-04	-.2640E-07	.2193E-10
Xe	54	FCC	.10655	.2824E-01	.1613E-04	-.6878E-07	.1157E-09
Cs	55	BCC	.07343	.6136E-01	.4538E-05	-.1147E-07	.1098E-10
Ba	56	BCC	.04457	.1016E-01	.6803E-06	-.7445E-09	.2922E-12
La	57	FCC	.04370	.6040E-02	.6646E-06	-.7269E-09	.2852E-12
Tb	65	HCP	.03765	.3200E-02	.5682E-06	-.6205E-09	.2432E-12
Ho	67	HCP	.03625	.2630E-02	.5462E-06	-.5963E-09	.2337E-12
Ta	73	BCC	.03159	.9039E-03	.4625E-06	-.5021E-09	.1961E-12
W	74	BCC	.02908	.3748E-03	.4073E-06	-.4382E-09	.1701E-12
Pt	78	FCC	.02938	.1090E-02	.4310E-06	-.4681E-09	.1829E-12
Au	79	FCC	.02981	.1930E-02	.4444E-06	-.4842E-09	.1895E-12
Pb	82	FCC	.03639	.6980E-02	.1124E-05	-.1881E-08	.1157E-11
Th	90	FCC	.02586	.2330E-02	.3902E-06	-.4261E-09	.1670E-12

Table 14.3 Parameterization of the Debye–Waller $B(T)$ factors for 22 elemental crystals estimated using the Debye theory (0-80 K)

Element	Z	a_0	a_1	a_2	a_3	a_4
Mn	25	.12754	.2818E-05	.4722E-05	.8580E-08	-.8769E-10
Co	27	.10955	.1879E-05	.3467E-05	.5237E-08	-.5071E-10
Ga	31	.12877	.2241E-05	.7940E-05	.1539E-07	-.2170E-09
As	33	.13599	-.2725E-05	.1122E-04	.1326E-07	-.3003E-09
Se	34	.40458	-.5352E-03	.4310E-03	-.4870E-05	.2103E-07
Ru	44	.04738	.1585E-06	.8522E-06	.3805E-09	-.3294E-11
Rh	45	.05816	.7334E-06	.1596E-05	.1919E-08	-.1786E-10
Cd	48	.12239	-.4091E-04	.2150E-04	-.5408E-07	-.2069E-09
In	49	.23229	-.4124E-03	.1821E-03	-.1839E-05	.7271E-08
Sb	51	.11192	-.3555E-04	.1918E-04	-.4560E-07	-.1995E-09
Te	52	.14748	-.1611E-03	.5615E-04	-.3793E-06	.9330E-09
Gd	64	.09144	-.3805E-04	.1799E-04	-.5673E-07	-.1073E-09
Dy	66	.08425	-.2745E-04	.1462E-04	-.3576E-07	-.1464E-09
Hf	72	.06389	-.5951E-05	.6949E-05	.1648E-09	-.1566E-09
Ta	73	.06618	-.9267E-05	.8144E-05	-.4773E-08	-.1609E-09
Re	75	.03588	.7034E-06	.1212E-05	.2006E-08	-.1980E-10
Os	76	.03021	.3114E-06	.7675E-06	.7978E-09	-.7300E-11
Ir	77	.03559	.7481E-06	.1257E-05	.2188E-08	-.2193E-10
Hg	80	.19812	.2056E-03	.2895E-03	-.3592E-05	.1657E-07
Tl	81	.17865	-.3461E-04	.2324E-03	-.2793E-05	.1260E-07
Bi	83	.11585	-.1975E-03	.7561E-04	-.7050E-06	.2596E-08
U	92	.05836	-.2049E-04	.1051E-04	-.2789E-07	-.9295E-10

Table 14.4 Parameterization of the Debye–Waller $B(T)$ factors for 22 elemental crystals estimated using the Debye theory ($T > 80$ K)

Element	Z	a_0	a_1	a_2	a_3	a_4
Mn	25	.09004	.7491E-03	.1187E-05	-.1261E-08	.4858E-12
Co	27	.08122	.5452E-03	.1045E-05	-.1104E-08	.4238E-12
Ga	31	.11303	.3962E-03	.6605E-05	-.1772E-07	.1854E-10
As	33	.07665	.1450E-02	.1268E-05	-.1500E-08	.6424E-12
Se	34	.07777	.1750E-01	.1177E-05	-.1277E-08	.4974E-12
Ru	44	.04068	.1125E-03	.4621E-06	-.4754E-09	.1793E-12
Rh	45	.04497	.2454E-03	.5634E-06	-.5922E-09	.2265E-12
Cd	48	.06319	.1830E-02	.1861E-05	-.3082E-08	.1883E-11
In	49	.08212	.7660E-02	.4795E-05	-.1169E-07	.1066E-10
Sb	51	.04974	.1810E-02	.8538E-06	-.1017E-08	.4378E-12
Te	52	.05164	.3490E-02	.1075E-05	-.1424E-08	.6835E-12
Gd	64	.03722	.1610E-02	.5511E-06	-.5995E-09	.2344E-12
Dy	66	.03579	.1390E-02	.5278E-06	-.5736E-09	.2242E-12
Hf	72	.03167	.8288E-03	.4584E-06	-.4965E-09	.1936E-12
Ta	73	.03150	.9178E-03	.4586E-06	-.4972E-09	.1940E-12
Re	75	.02607	.1917E-03	.3390E-06	-.3591E-09	.1381E-12
Os	76	.02386	.1160E-03	.2944E-06	-.3084E-09	.1177E-12
Ir	77	.02550	.1993E-03	.3339E-06	-.3542E-09	.1363E-12
Hg	80	.05788	.1026E-01	.5610E-05	-.1853E-07	.2364E-10
Tl	81	.03756	.8810E-02	.1170E-05	-.1957E-08	.1204E-11
Bi	83	.03610	.3580E-02	.1111E-05	-.1855E-08	.1140E-11
U	92	.02449	.9819E-03	.3617E-06	-.3933E-09	.1538E-12

Table 14.5 Parameterization of the Debye–Waller $B(T)$ factors of 17 compounds with the zinc-blende structure (0-80 K)

Crystal	Atom	a_0	a_1	a_2	a_3	a_4
GaP	Ga	.12919	-.2611E-04	.4277E-05	.1318E-06	-.9790E-09
	P	.19714	-.4786E-04	.4958E-05	.1507E-06	-.1144E-08
GaSb	Ga	.16223	-.4262E-03	.4610E-04	-.3021E-06	.6715E-09
	Sb	.12072	-.3684E-03	.4148E-04	-.2710E-06	.5895E-09
GaAs	Ga	.13890	-.1719E-03	.1738E-04	.8267E-08	-.5764E-09
	As	.14029	-.2267E-03	.2059E-04	-.1258E-08	-.6303E-09
InP	In	.14043	-.3528E-03	.3512E-04	-.9239E-07	-.4990E-09
	P	.17317	-.1375E-03	.2216E-04	-.1676E-06	.5101E-09
InSb	In	.15730	-.8061E-03	.1070E-03	-.1097E-05	.4394E-08
	Sb	.13590	-.6231E-03	.8837E-04	-.9244E-06	.3805E-08
InAs	In	.13955	-.4468E-03	.4965E-04	-.3152E-06	.6372E-09
	As	.12433	-.1992E-03	.3181E-04	-.2601E-06	.8854E-09
ZnO	Zn	.14263	.1850E-04	.9258E-06	.1791E-06	-.1174E-08
	O	.21643	-.1364E-04	.4672E-05	.2922E-07	-.3009E-09
ZnS	Zn	.17311	-.1457E-03	.1957E-04	.7536E-07	-.1091E-08
	S	.20061	-.1104E-03	.1876E-04	-.3090E-07	-.3146E-09
ZnSe	Zn	.18397	-.3260E-03	.3395E-04	-.6329E-07	-.6199E-09
	Se	.13154	-.1820E-03	.2496E-04	-.1169E-06	.8580E-10
ZnTe	Zn	.20473	-.5841E-03	.7675E-04	-.6235E-06	.2023E-08
	Te	.12583	-.3925E-03	.6072E-04	-.5412E-06	.1937E-08
CdTe	Cd	.19189	-.1110E-02	.1689E-03	-.1886E-05	.8109E-08
	Te	.13937	-.5429E-03	.1052E-03	-.1182E-05	.5163E-08
HgSe	Hg	.21070	-.5317E-03	.4040E-03	-.5438E-05	.2651E-07
	Se	.15574	.1412E-03	.1214E-03	-.1608E-05	.7951E-08
HgTe	Hg	.22918	.2620E-02	.5125E-03	-.7157E-05	.3573E-07
	Te	.12889	.7283E-03	.1402E-03	-.1894E-05	.9419E-08
CuCl	Cu	.36726	-.2360E-02	.3149E-03	-.3370E-05	.1392E-07
	Cl	.31459	-.1090E-02	.1879E-03	-.2122E-05	.9284E-08
CuBr	Cu	.23794	-.6525E-03	.1545E-03	-.1725E-05	.7561E-08
	Br	.28486	-.1550E-02	.2473E-03	-.2714E-05	.1150E-07
CuI	Cu	.29083	-.1490E-02	.1944E-03	-.1993E-05	.8041E-08
	I	.15310	-.5973E-03	.1059E-03	-.1107E-05	.4561E-08
SiC	Si	.12859	.2243E-05	.6939E-06	.3985E-09	.2158E-10
	C	.16771	-.8428E-05	.9532E-06	-.1570E-08	.6547E-11

Table 14.6 Parameterization of the Debye–Waller factors of 17 compounds with the zinc-blende structure ($T > 80$ K)

Crystal	Atom	a_0	a_1	a_2	a_3	a_4
GaP	Ga	.07258	.1270E-02	.9754E-06	-.1042E-08	.4030E-12
	P	.13578	.1350E-02	.1554E-05	-.1607E-08	.6090E-12
GaSb	Ga	.07923	.2600E-02	.1271E-05	-.1496E-08	.6388E-12
	Sb	.04830	.2350E-02	.8070E-06	-.9583E-09	.4115E-12
GaAs	Ga	.07394	.1690E-02	.1003E-05	-.1074E-08	.4159E-12
	As	.06972	.1880E-02	.9534E-06	-.1021E-08	.3954E-12
InP	In	.05059	.2660E-02	.7411E-06	-.8060E-09	.3151E-12
	P	.13490	.1170E-02	.1539E-05	-.1584E-08	.5972E-12
InSb	In	.05476	.4180E-02	.1093E-05	-.1441E-08	.6892E-12
	Sb	.05115	.3450E-02	.1012E-05	-.1330E-08	.6348E-12
InAs	In	.05013	.2880E-02	.7275E-06	-.7896E-09	.3084E-12
	As	.06880	.1740E-02	.9253E-06	-.9870E-09	.3809E-12
ZnO	Zn	.08025	.1310E-02	.1107E-05	-.1189E-08	.4612E-12
	O	.19457	.5300E-03	.1662E-05	-.1595E-08	.5746E-12
ZnS	Zn	.08237	.2370E-02	.1152E-05	-.1242E-08	.4827E-12
	S	.13763	.1670E-02	.1656E-05	-.1728E-08	.6579E-12
ZnSe	Zn	.08353	.2810E-02	.1175E-05	-.1267E-08	.4926E-12
	Se	.06725	.1830E-02	.9242E-06	-.9912E-09	.3841E-12
ZnTe	Zn	.08544	.3970E-02	.1215E-05	-.1312E-08	.5105E-12
	Te	.04544	.2930E-02	.6712E-06	-.7409E-09	.2943E-12
CdTe	Cd	.05219	.6150E-02	.7666E-06	-.8345E-09	.3264E-12
	Te	.04582	.4060E-02	.6645E-06	-.7218E-09	.2821E-12
HgSe	Hg	.04160	.1254E-01	.1645E-05	-.3253E-08	.2390E-11
	Se	.09039	.4220E-02	.3197E-05	-.6104E-08	.4364E-11
HgTe	Hg	.04118	.1832E-01	.1588E-05	-.3081E-08	.2223E-11
	Te	.05916	.5360E-02	.2153E-05	-.4117E-08	.2942E-11
CuCl	Cu	.10769	.1143E-01	.2610E-05	-.3835E-08	.2058E-11
	Cl	.16536	.6700E-02	.3588E-05	-.5151E-08	.2723E-11
CuBr	Cu	.10301	.6030E-02	.2420E-05	-.3538E-08	.1893E-11
	Br	.08651	.9060E-02	.2106E-05	-.3101E-08	.1667E-11
CuI	Cu	.10593	.7530E-02	.2555E-05	-.3750E-08	.2011E-11
	I	.05425	.4260E-02	.1321E-05	-.1947E-08	.1047E-11
SiC	Si	.12047	.7436E-04	.1168E-05	-.1163E-08	.4301E-12
	C	.16536	.3385E-04	.7346E-06	-.5648E-09	.1713E-12

Table 14.7 Parameterization of the Debye–Waller $B(T)$ factors of compounds with the sodium chloride structure (0-80 K)

Crystal	Atom	a_0	a_1	a_2	a_3	a_4
KF	K	0.27569	-3.97001E-5	1.90082E-5	1.45132E-7	-1.36815E-9
	F	0.41449	-1.32321E-5	1.90294E-5	3.95364E-8	-4.43675E-10
KCl	K	0.34407	-2.71951E-4	5.20042E-5	-7.64989E-9	-1.40092E-9
	Cl	0.33970	-1.59544E-4	4.57814E-5	-5.53940E-8	-7.72741E-10
KBr	K	0.36082	-3.06273E-4	9.25366E-5	-5.74224E-7	1.42586E-9
	Br	0.25412	-6.33280E-4	1.18614E-4	-9.10515E-7	2.66978E-9
KI	K	0.38640	-2.04061E-4	1.26784E-4	-1.00081E-6	3.44973E-9
	I	0.22061	-8.82914E-4	1.92263E-4	-2.01719E-6	8.20798E-9
NaF	Na	0.31824	3.36863E-5	3.28677E-6	1.30257E-7	-6.58199E-10
	F	0.32146	5.17396E-6	6.33234E-6	3.29345E-8	-1.57649E-10
NaCl	Na	0.41303	-5.80361E-6	1.52557E-5	2.51515E-7	-1.81479E-9
	Cl	0.30318	-3.03046E-5	1.73283E-5	1.99679E-7	-1.66904E-9
NaBr	Na	0.46778	-5.23749E-5	4.03194E-5	5.83251E-8	-1.16063E-9
	Br	0.23260	-4.85717E-4	7.24684E-5	-3.24320E-7	-4.86362E-11
NaI	Na	0.51388	-9.52548E-5	7.31870E-5	-2.61936E-7	1.43879E-10
	I	0.19861	-7.92322E-4	1.32936E-4	-1.21423E-6	4.31378E-9
RbF	Rb	0.20574	-3.25100E-4	6.34232E-5	-3.20302E-7	2.71684E-10
	F	0.44368	3.77853E-6	4.20515E-5	-2.49335E-7	9.64524E-10
RbCl	Rb	0.24532	-6.61109E-4	1.22166E-4	-9.70945E-7	2.96674E-9
	Cl	0.36621	-1.50104E-4	8.05590E-5	-4.90654E-7	1.29254E-9
RbBr	Rb	0.25697	-6.16459E-4	1.52916E-4	-1.39827E-6	5.08502E-9
	Br	0.25998	-5.05979E-4	1.41578E-4	-1.26004E-6	4.50533E-9
RbI	Rb	0.27878	-5.25372E-4	2.07079E-4	-2.13499E-6	8.69368E-9
	I	0.22175	-5.39958E-4	2.10075E-4	-2.31512E-6	9.83668E-9
LiF	Li	0.55882	-4.51204E-6	4.05664E-6	-1.12501E-8	1.42384E-10
	F	0.28529	4.69857E-6	2.88026E-6	2.42906E-8	-4.19277E-11
FeO	Fe	0.14086	2.18745E-5	9.05213E-7	8.72665E-8	-4.85494E-10
	O	0.25160	-3.94824E-6	3.75333E-6	-1.88399E-9	-2.04531E-11
MnO	Mn	0.14816	2.53320E-5	3.21022E-7	1.11325E-7	-6.20955E-10
	O	0.25571	-2.31760E-6	3.87375E-6	-1.28513E-8	5.75966E-11
MgO	Mg	0.16642	-2.21183E-6	1.38712E-6	-5.48783E-9	6.78861E-11
	O	0.20771	-9.49815E-6	1.57687E-6	-5.95688E-9	3.92882E-11
CaO	Ca	0.15039	1.30475E-5	7.74278E-7	3.62603E-8	-1.30679E-10
	O	0.25202	-3.40839E-6	2.42189E-6	-1.14017E-8	1.02351E-10
SrO	Sr	0.11954	-1.11103E-5	4.27707E-6	1.11528E-7	-8.02252E-10
	O	0.29511	-1.53269E-5	6.92256E-6	-4.85607E-8	2.77940E-10
NiO	Ni	0.11079	1.07347E-5	8.56104E-7	2.94718E-8	-1.15121E-10
	O	0.21440	-2.63667E-6	1.91857E-6	2.36603E-9	-2.88934E-11

Table 14.8 Parameterization of the Debye–Waller $B(T)$ factors of compounds with the sodium chloride structure ($T > 80$ K)

Crystal	Atom	a_0	a_1	a_2	a_3	a_4
KF	K	0.14194	0.00318	2.02989E-6	-2.19253E-9	8.53360E-13
	F	0.27178	0.00297	3.69801E-6	-3.95570E-9	1.53053E-12
KCl	K	0.14761	0.00537	2.16668E-6	-2.35307E-9	9.19088E-13
	Cl	0.16026	0.00477	2.32750E-6	-2.52224E-9	9.83875E-13
KBr	K	0.14868	0.00659	2.19080E-6	-2.38050E-9	9.30057E-13
	Br	0.07524	0.00652	1.13348E-6	-1.23774E-9	4.85304E-13
KI	K	0.15024	0.00803	2.23466E-6	-2.43511E-9	9.53368E-13
	I	0.04829	0.00787	7.37853E-7	-8.08101E-10	3.17417E-13
NaF	Na	0.21945	0.00175	2.93528E-6	-3.12712E-9	1.20651E-12
	F	0.24767	0.00128	3.14520E-6	-3.31513E-9	1.27028E-12
NaCl	Na	0.23672	0.00378	3.34096E-6	-3.60005E-9	1.39937E-12
	Cl	0.15733	0.00338	2.25586E-6	-2.43726E-9	9.48641E-13
NaBr	Na	0.24414	0.00540	3.51803E-6	-3.80606E-9	1.48289E-12
	Br	0.07509	0.00507	1.13152E-6	-1.23607E-9	4.84749E-13
NaI	Na	0.24823	0.00712	3.61419E-6	-3.91746E-9	1.52803E-12
	I	0.04800	0.00609	7.30330E-7	-7.99154E-10	3.13594E-13
RbF	Rb	0.06966	0.00444	1.04785E-6	-1.14546E-9	4.49484E-13
	F	0.27795	0.00397	3.83827E-6	-4.11514E-9	1.59392E-12
RbCl	Rb	0.07044	0.00650	1.06161E-6	-1.15833E-9	4.53694E-13
	Cl	0.16240	0.00613	2.37878E-6	-2.58312E-9	1.00912E-12
RbBr	Rb	0.07064	0.00749	1.06457E-6	-1.16088E-9	4.54347E-13
	Br	0.07548	0.00725	1.13991E-6	-1.24437E-9	4.87347E-13
RbI	Rb	0.07118	0.00931	1.08335E-6	-1.18566E-9	4.65529E-13
	I	0.04818	0.00858	7.36715E-7	-8.07694E-10	3.17577E-13
LiF	Li	0.52670	2.69660E-4	5.38908E-6	-5.41633E-9	2.01237E-12
	F	0.23424	7.51248E-4	2.85120E-6	-2.97887E-9	1.13506E-12
FeO	Fe	0.09225	8.93403E-4	1.25697E-6	-1.34554E-9	5.20857E-13
	O	0.23132	3.27612E-4	2.37755E-6	-2.38946E-9	8.87476E-13
MnO	Mn	0.09459	9.91526E-4	1.29568E-6	-1.38792E-9	5.37390E-13
	O	0.23554	2.98103E-4	2.47013E-6	-2.49509E-9	9.29894E-13
MgO	Mg	0.15411	1.26015E-4	1.61232E-6	-1.62822E-9	6.06784E-13
	O	0.20298	1.11489E-5	1.81777E-6	-1.76510E-9	6.40742E-13
CaO	Ca	0.11567	5.40657E-4	1.46108E-6	-1.53965E-9	5.89934E-13
	O	0.23641	1.56284E-4	2.47508E-6	-2.49681E-9	9.29524E-13
SrO	Sr	0.06287	0.00126	8.94448E-7	-9.65280E-10	3.75579E-13
	O	0.26031	5.45236E-4	2.98862E-6	-3.08222E-9	1.16459E-12
NiO	Ni	0.08230	4.66033E-4	1.07087E-6	-1.13581E-9	4.37201E-13
	O	0.20747	8.48242E-5	1.89257E-6	-1.84447E-9	6.70888E-13

Table 14.9 Parameterization of the Debye–Waller $B(T)$ factors of five compounds with the caesium chloride structure (0-80 K)

Crystal	Atom	a_0	a_1	a_2	a_3	a_4
CsCl	Cs	0.19036	-9.10016E-4	1.22707E-4	-1.08235E-6	3.68061E-9
	Cl	0.35030	6.39211E-5	3.31992E-5	1.17168E-7	-1.50137E-9
CsBr	Cs	0.20417	-9.20228E-4	1.54877E-4	-1.50202E-6	5.66455E-9
	Br	0.24995	-4.19446E-4	9.76084E-5	-6.24174E-7	1.30696E-9
CsI	Cs	0.21434	-9.33589E-4	1.79157E-4	-1.82167E-6	7.18631E-9
	I	0.21745	-8.76931E-4	1.67403E-4	-1.63431E-6	6.20733E-9
TlCl	Tl	0.20095	-7.99987E-4	3.24559E-4	-4.15866E-6	1.96232E-8
	Cl	0.37477	1.09621E-4	2.56619E-5	4.13366E-7	-3.49634E-9
TlBr	Tl	0.19805	-0.00109	2.99804E-4	-3.75322E-6	1.74123E-8
	Br	0.25893	-4.81267E-4	1.06923E-4	-6.99310E-7	1.50929E-9

Table 14.10 Parameterization of the Debye–Waller $B(T)$ factors of five compounds with the caesium chloride structure ($T > 80$ K)

Crystal	Atom	a_0	a_1	a_2	a_3	a_4
CsCl	Cs	0.04587	0.00560	6.93610E-7	-7.57098E-10	2.96719E-13
	Cl	0.16295	0.00481	2.39007E-6	-2.59405E-9	1.01270E-12
CsBr	Cs	0.04346	0.00669	6.19356E-7	-6.64832E-10	2.57440E-13
	Br	0.07132	0.00617	1.00875E-6	-1.08079E-9	4.17783E-13
CsI	Cs	0.04624	0.00745	7.06922E-7	-7.74571E-10	3.04329E-13
	I	0.04840	0.00727	7.41666E-7	-8.12946E-10	3.19430E-13
TlCl	Tl	0.03036	0.01067	4.72673E-7	-5.22856E-10	2.07256E-13
	Cl	0.16407	0.00541	2.42180E-6	-2.63348E-9	1.02966E-12
TlBr	Tl	0.03032	0.00983	4.66100E-7	-5.11821E-10	2.01608E-13
	Br	0.07580	0.00654	1.14901E-6	-1.25640E-9	4.93055E-13

APPENDIX   A

SOME USEFUL MATHEMATICAL RELATIONS

A.1 Fourier transformation

A function, e.g. the electron wave function, may be described by its real-space representation $\psi(\mathbf{r})$, or in reciprocal space by its Fourier transform $\phi(\mathbf{q})$. The two forms $\psi(\mathbf{r})$ and $\phi(\mathbf{q})$ are two different *representations* of the same physical function, and we can go back and forth between the two representations by means of the integral transformations

$$\psi(\mathbf{r}) = (2\pi)^{-3} \int \phi(\mathbf{q}) \exp(i\mathbf{q} \cdot \mathbf{r})d\mathbf{q}$$

$$\phi(\mathbf{q}) = \int \psi(\mathbf{r}) \exp(-i\mathbf{q} \cdot \mathbf{r})d\mathbf{r}, \tag{A.1}$$

where here and below the integral signs imply integration over the whole spaces of the vectors $\mathbf{q}$ and $\mathbf{r}$.

Alternatively, by introducing $\mathbf{k} = \mathbf{q}/2\pi$, and defining

$$\Phi(\mathbf{k}) = \phi(2\pi\mathbf{k}),$$

we have

$$\psi(\mathbf{r}) = \int \Phi(\mathbf{k}) \exp(2\pi i\mathbf{k} \cdot \mathbf{r})d\mathbf{r},$$

$$\Phi(\mathbf{k}) = \int \psi(\mathbf{r}) \exp(-2\pi i\mathbf{k} \cdot \mathbf{r})d\mathbf{k}.$$

A.2 The Dirac delta function

The Dirac delta function is defined by the equations

$$\delta(\mathbf{r}) = 0, \quad \text{if } \mathbf{r} \neq 0,$$

and

$$\int \delta(\mathbf{r})d\mathbf{r} = 1.$$

Evidently the Dirac delta function is not an ordinary mathematical function. In fact this function has meaning only if it is used as an argument of an integral, e.g.

$$\int \delta(\mathbf{r} - \mathbf{a})f(\mathbf{r})d\mathbf{r} = f(\mathbf{a}).$$

It is often convenient to express the Dirac delta function in terms of its Fourier transform:

$$\delta(\mathbf{r}) = (2\pi)^{-3} \int \exp(i\mathbf{q} \cdot \mathbf{r})d\mathbf{q}. \tag{A.2}$$

Alternatively, we may define the Dirac delta function in reciprocal space

$$(2\pi)^3 \delta(\mathbf{k}) = \int \exp(-i\mathbf{k} \cdot \mathbf{r})d\mathbf{r}. \tag{A.3}$$

A.3 The Kronecker delta symbol

For a periodic function representable in the form of a Fourier series

$$\psi(\mathbf{r}) = \sum_{\mathbf{g}} \psi_{\mathbf{g}} \exp(i\mathbf{g} \cdot \mathbf{r}),$$

where $\mathbf{g}$ is a set of reciprocal lattice vectors, it is convenient to use the Kronecker delta symbol defined by

$$\delta_{\mathbf{g},\mathbf{g}'} = \begin{cases} 1 \text{ if } \mathbf{g} = \mathbf{g}' \\ 0 \text{ otherwise.} \end{cases}$$

For a three-dimensional crystal we have

$$\delta_{\mathbf{g},\mathbf{g}'} = \frac{1}{\Omega_0} \int_{\Omega_0} \exp[-i(\mathbf{g} - \mathbf{g}') \cdot \mathbf{r}]d\mathbf{r}, \tag{A.4}$$

where Ω_0 is the volume of a crystal unit cell, and the integration is performed over the unit cell. Ω_0 is related to the volume of the Brillouin zone (equal to the volume of the unit cell of the reciprocal lattice V_B) via the relation $\Omega_0 V_B = (2\pi)^3$.

For a two-dimensional net of reciprocal lattice vectors $\{\mathbf{G}\}$ we have

$$\int_{A_c} \exp[i(\mathbf{G} - \mathbf{G}') \cdot \mathbf{R}]d\mathbf{R} = A_c \delta_{\mathbf{G},\mathbf{G}'}, \tag{A.5}$$

where A_c is the area of a two-dimensional real-space unit cell, and the integration is performed over the unit cell. For a one-dimensional set of reciprocal lattice points $\{g\}$ we have

$$\int_{\ell} \exp[i(g - g')x]dx = \ell \delta_{g,g'}, \tag{A.6}$$

where ℓ is the repeat distance of points along the x axis in real-space, and the integration is performed over a single repeat distance.

We also have the following relations between the real and reciprocal lattice vectors $\mathbf{R}$ and $\mathbf{g}$:

$$\Omega_0 \sum_{\mathbf{R}} \delta(\mathbf{r} - \mathbf{R}) = \sum_{\mathbf{g}} \exp(i\mathbf{g} \cdot \mathbf{r}), \tag{A.7}$$

and

$$V_B \sum_{\mathbf{g}} \delta(\mathbf{q} - \mathbf{g}) = \sum_{R} \exp(-i\mathbf{q} \cdot \mathbf{R}). \tag{A.8}$$

A.4 Some useful integrals

$$\int_0^\infty \exp(-\alpha x^2)\cos(\beta x)dx = \frac{1}{2}\sqrt{\frac{\pi}{\alpha}}\exp\left(-\frac{\beta^2}{4\alpha}\right), \tag{A.9}$$

$$\int_0^\infty x\exp(-\alpha x^2)\sin(\beta x)dx = \frac{\beta\sqrt{\pi}}{4\alpha^{3/2}}\exp\left(-\frac{\beta^2}{4\alpha}\right), \tag{A.10}$$

$$\int_{-\infty}^{+\infty} \exp(-\alpha x^2 + i\beta x)dx = \sqrt{\frac{\pi}{\alpha}}\exp\left(-\frac{\beta^2}{4\alpha}\right), \tag{A.11}$$

$$\int_0^\infty \sin(x^2)dx = \int_0^\infty \cos(x^2)dx = \frac{1}{2}\sqrt{\frac{\pi}{2}}, \tag{A.12}$$

where $\alpha > 0$.

APPENDIX B

GREEN'S FUNCTIONS

The free space Green's function G is defined as the solution of the following Helmholtz equation:

$$\nabla^2 \mathsf{G}(\mathbf{r}, \mathbf{r}') + K^2 \mathsf{G}(\mathbf{r}, \mathbf{r}') = \delta(\mathbf{r} - \mathbf{r}'). \tag{B.1}$$

Taking the Fourier transform of both sides of this equation we find

$$(K^2 - k^2)\mathsf{G}(\mathbf{k}) = 1, \tag{B.2}$$

where

$$\mathsf{G}(\mathbf{k}) = \int \mathsf{G}(\mathbf{r}, \mathbf{r}') \exp[-i\mathbf{k} \cdot (\mathbf{r} - \mathbf{r}')] d(\mathbf{r} - \mathbf{r}') \tag{B.3}$$

is the Fourier transform of $\mathsf{G}(\mathbf{r}, \mathbf{r}') = \mathsf{G}(\mathbf{r} - \mathbf{r}')$, and where we have used the Fourier representation for the Dirac δ-function

$$\delta(\mathbf{r} - \mathbf{r}') = \frac{1}{(2\pi)^3} \int \exp[i\mathbf{k} \cdot (\mathbf{r} - \mathbf{r}')] d\mathbf{k}. \tag{B.4}$$

A general solution of eqn (B.2) has the form

$$\mathsf{G}(\mathbf{k}) = \frac{1}{K^2 - k^2} + C\delta(K^2 - k^2), \tag{B.5}$$

where C is an arbitrary constant. For any C we have

$$\mathsf{G}(\mathbf{r} - \mathbf{r}') = \int \mathsf{G}(\mathbf{k}) \exp[i\mathbf{k} \cdot (\mathbf{r} - \mathbf{r}')] \frac{d\mathbf{k}}{(2\pi)^3} = \frac{1}{(2\pi)^3} \int \frac{\exp[i\mathbf{k} \cdot (\mathbf{r} - \mathbf{r}')]}{K^2 - k^2} d\mathbf{k}. \tag{B.6}$$

By choosing the direction of the z axis in $\mathbf{k}$ space in the direction of the vector $\mathbf{r} - \mathbf{r}'$, we find

$$
\begin{aligned}
\mathsf{G}(\mathbf{r} - \mathbf{r}') &= \frac{1}{(2\pi)^3} \int_0^\infty 2\pi k^2 dk \int_{\theta=\pi}^{\theta=0} d(\cos\theta) \frac{\exp(ikR\cos\theta)}{K^2 - k^2} \\
&= \frac{1}{(2\pi)^2} \frac{1}{R} \frac{\partial}{\partial R} \int_{-\infty}^{\infty} dk \frac{\exp(ikR)}{k^2 - K^2},
\end{aligned}
$$

where $R = |\mathbf{r} - \mathbf{r}'|$.

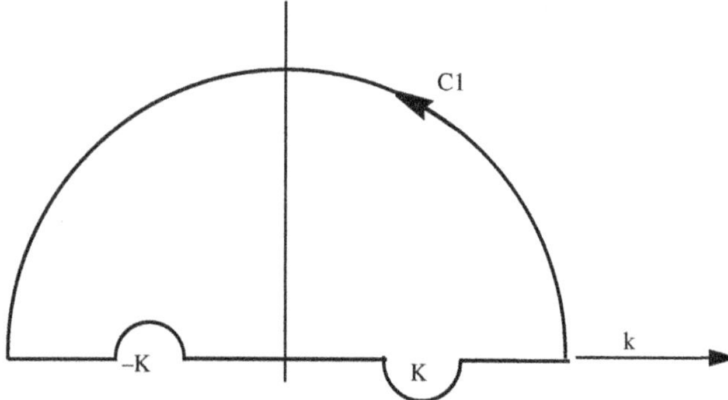

FIG. B.1. Schematic diagram showing the contour in the complex plane k used for evaluating the integral (B.7).

For the outgoing wave Green's function, we choose the contour C_1 as shown in Fig. B.1. Since $R > 0$, we can bend the contour into the upper half plane and find

$$\mathsf{G}(\mathbf{r}, \mathbf{r}') = \mathsf{G}(\mathbf{r} - \mathbf{r}') = \frac{1}{(2\pi)^2} \frac{1}{R} \frac{\partial}{\partial R} \int_{C_1} dk \frac{\exp(ikR)}{k^2 - K^2}$$

$$= \frac{1}{(2\pi)^2} \frac{1}{R} \frac{\partial}{\partial R} \left[2\pi i \frac{\exp(iKR)}{2K} \right] = -\frac{1}{4\pi} \frac{\exp(iKR)}{R}. \tag{B.7}$$

Alternatively, by using the Dirac formula, we can write the solution of eqn (B.2) in the form

$$\mathsf{G}(\mathbf{k}) = \frac{1}{K^2 - k^2 \pm i\epsilon} = \wp \frac{1}{K^2 - k^2} \mp i\pi \delta(K^2 - k^2), \tag{B.8}$$

where ϵ is a small positive quantity, $\wp$ denotes the Cauchy principal value of the integral, and the $\mp$ signs correspond to the outgoing and incoming wave Green's functions respectively.

Sometimes it is convenient to use a mixed real- and reciprocal space representation of the Green's function. Using the notation $\mathbf{r} = (\mathbf{X}, z)$, $\mathbf{k} = (\mathbf{Q}, k_z)$, we define the mixed space representation of the Green's function as

$$\mathsf{G}(\mathbf{Q}, \mathbf{X}'; z, z') = \int \mathsf{G}(\mathbf{X}, \mathbf{X}'; z, z') \exp(-i\mathbf{Q} \cdot \mathbf{X}) d\mathbf{X}. \tag{B.9}$$

Substitution of the reciprocal space form of Green's function (B.6)

$$\mathsf{G}(\mathbf{X}, z; \mathbf{X}', z') = \frac{1}{(2\pi)^3} \int d\mathbf{Q}' dk'_z \frac{\exp[i\mathbf{Q}' \cdot (\mathbf{X} - \mathbf{X}') + ik'_z(z - z')]}{K^2 + i\epsilon - (Q'^2 + k_z'^2)},$$

into (B.9) gives the following expression for the outgoing wave:

$$\begin{aligned}
\mathsf{G}(\mathbf{Q}, \mathbf{X}'; z, z') &= \frac{1}{(2\pi)^3} \int_{-\infty}^{\infty} dk_z' \int d\mathbf{Q}' \frac{\exp[ik_z'(z - z')]}{K^2 + i\epsilon - (Q'^2 + k_z'^2)} \\
&\quad \times \int d\mathbf{X} \exp[i(\mathbf{Q}' - \mathbf{Q}) \cdot \mathbf{X}] \exp(-i\mathbf{Q}' \cdot \mathbf{X}') \\
&= \frac{\exp(-i\mathbf{Q} \cdot \mathbf{X}')}{2\pi} \int_{-\infty}^{\infty} dk_z' \frac{\exp[ik_z'(z - z')]}{K^2 + i\epsilon - (Q^2 + k_z'^2)} \\
&= -\frac{i}{2|K_z|} \exp(-i\mathbf{Q} \cdot \mathbf{X}') \exp(iK_z|z - z'|),
\end{aligned}$$

(B.10)

where $K_z = \sqrt{K^2 - Q^2 + i\epsilon}$ and where we assume that $\Im K_z > 0$.

We now consider a general Green's function describing the motion of an electron in a potential field $U(\mathbf{r})$. The wave function of the fast electron in a potential field satisfies the Schrödinger equation

$$[\nabla^2 + K^2 + U(\mathbf{r})]\psi(\mathbf{r}) = S(\mathbf{r}), \tag{B.11}$$

where $S(\mathbf{r})$ is the source function. A general Green's function is defined as a function $\mathsf{G}_U(\mathbf{r}, \mathbf{r}')$ that satisfies the equation

$$[\nabla^2 + K^2 + U(\mathbf{r})]\mathsf{G}_U(\mathbf{r}, \mathbf{r}') = \delta(\mathbf{r} - \mathbf{r}'). \tag{B.12}$$

Provided that this general Green's function is known, the solution of eqn (B.11) can be found by integrating G_U with the source function as

$$\psi(\mathbf{r}) = \int \mathsf{G}_U(\mathbf{r}, \mathbf{r}')S(\mathbf{r}')d\mathbf{r}'. \tag{B.13}$$

At large distances from the region of space occupied by the potential $U(\mathbf{r})$ Green's function has the form

$$\mathsf{G}_U(\mathbf{r}, \mathbf{r}') = \mathsf{G}(\mathbf{r}, \mathbf{r}') + \mathsf{G}_s(\mathbf{r}, \mathbf{r}'), \tag{B.14}$$

where $\mathsf{G}(\mathbf{r}, \mathbf{r}')$ is given by (B.7) and where the term $\mathsf{G}_s(\mathbf{r}, \mathbf{r}')$ describes the part of Green's function associated with scattering by the potential $U(\mathbf{r})$. Although it is not possible to find an explicit expression for the Green's function $\mathsf{G}_U(\mathbf{r}, \mathbf{r}')$, which is valid for an arbitrarily chosen $U(\mathbf{r})$, we can still understand certain general properties of this function. Consider a two-dimensional Fourier transform of the Green's function

$$\mathsf{G}_U(\mathbf{Q}, \mathbf{X}'; z, z') = \int \mathsf{G}_U(\mathbf{X}, z; \mathbf{X}', z') \exp(-i\mathbf{Q} \cdot \mathbf{X})d\mathbf{X}. \tag{B.15}$$

Combining eqns (B.14) and (B.10) we see that

$$\begin{aligned}
\mathsf{G}_U(\mathbf{Q}, \mathbf{X}'; z, z') &= -\frac{i}{2|K_z|} \exp(-i\mathbf{Q} \cdot \mathbf{X}') \exp(iK_z|z - z'|) \\
&\quad + \int \mathsf{G}_s(\mathbf{X}, z; \mathbf{X}', z') \exp(-i\mathbf{Q} \cdot \mathbf{X})d\mathbf{X}.
\end{aligned}$$

(B.16)

In the case where $z > z'$ we find

$$\mathsf{G}_U(\mathbf{Q}, \mathbf{X}'; z, z') = -\frac{i}{2|K_z|} \exp(-i\mathbf{k} \cdot \mathbf{r}') \exp(iK_z z)$$
$$+ \int \mathsf{G}_s(\mathbf{X}, z; \mathbf{X}', z') \exp(-i\mathbf{Q} \cdot \mathbf{X}) d\mathbf{X}. \qquad (\text{B.17})$$

The first term in this equation, considered as a function of $\mathbf{r}'$, describes a freely propagating plane wave while the second term represents the wave field scattered by the potential U. Taking into account that scattering is a process that transforms a plane wave $\exp(i\mathbf{k} \cdot \mathbf{r}')$ into an exact solution $\psi_{\mathbf{k}}(\mathbf{r}')$ of the problem of scattering, we find that the Fourier transform of Green's function G_U is related to the solution of the Schrödinger equation describing scattering of a plane wave by the potential U via

$$\mathsf{G}_U(\mathbf{Q}, \mathbf{X}'; z, z') = -\frac{i}{2|K_z|} \exp(iK_z z)\psi_{-\mathbf{k}}(\mathbf{r}'), \qquad (\text{B.18})$$

where $\mathbf{k} = (\mathbf{Q}, K_z)$.

APPENDIX   C

FORTRAN LISTING OF RHEED ROUTINES

C.1   A FORTRAN routine for the calculation of $U_G(z)$

Below we give an example of the input file and a FORTRAN routine (2D-poten.f)
for calculating the potential $U_G(z)$ shown in Fig. 5.17. This routine 2D-poten.f
reads in the input file called 'potens.in', and generates three output data files:
rheed.dat, U0.dat, and UG.dat. The first output data file is an unformatted file
containing $U_G(z)$ and all the relevant structural data required for the subsequent
dynamical RHEED calculations. The other two data files are formatted, and the
first one (U0.dat) contains values of the average potential $U_0(z)$, and the second
one (UG.dat) contains values of several Fourier components of the potential
$U_G(z)$ corresponding to $G \neq 0$. Curves shown in Fig. 5.17 were calculated using
the following input file.

C.1.1   *The input file for the calculation of $U_G(z)$*

```
20.0     0.0045              ! Primary beam energy and mean absorption
2.892    2.892    16.36      ! Surface unit cell lattice parameters
1                            ! Types of atoms in a surface unit cell
AG                           ! Name of the 1st type of atoms
  0.6377        1.3790     2.8294       2.3631       1.4553  ! a_n for Re part
  0.2466        1.6974     5.7656      20.0943      76.7372  ! b_n for Re part
 -0.0087       -0.0677     0.3310       0.1765       0.0562  ! a_n for Im part
  0.0383        0.1552     0.5759       1.5350       4.9221  ! b_n for Im part
5   0.7341  1.0                 ! No of atoms in a surface unit cell,
0.0      0.0      0.375         ! the DW factor, occupation,
0.5      0.5      0.500         ! fractional coordinates x,y,z
0.0      0.0      0.625
0.5      0.5      0.750
0.0      0.0      0.875
0.0      7.1575  150            ! z1-z2 defines the surface selvage region, number of slices
7.1575 11.2475  100            ! z2-z3 defines the unit bulk slab, number of slices
7                              ! Number of rods to be included
0   0   1   0   -1   0   2   0   -2   0   3   0   -3   0   4   0   -4   0  ! Index
```

C.1.2   *FORTRAN routine for calculating $U_G(z)$*

```
C/NAME=2D-POTEN.FOR
C
C   ROUTINE FOR CALCULATING TWO-DIMENSIONAL FOURIER COMPONENTS OF
C   POTENTIAL U_G(Z). MAXIMUM NUMBER OF RODS IS DEFINED BY THE PARAMETER
C   NR, MAXIMUM NUMBER OF ATOMIC SPECIES IS DEFINED BY NATOM, AND
C   MAXIMUM NUMBER OF ATOMS OF EACH SPECIES IS BY ITYPE.
C
C   BOTH THE REAL AND IMAGINARY PARTS OF THE ATOMIC SCATTERING FACTORS
C   ARE REPRESENTED AS A SUM OF FIVE GAUSSIANS, AND THE RELEVANT FITTING
```

```fortran
C     PARAMETERS FOR ALL ATOMS OF THE PERIODIC TABLE ARE GIVEN IN:
C         ACTA CRYST. A52 (1996) pp257-276  AND 471-475
C     FOR IONS, THE FITTING PARAMETERS ARE GIVEN IN:
C         Acta Cryst. A64 (1998) pp481-485
C
C     THE INPUT PARAMETERS
C     A,B,C --- LATTICE CONSTANT
C     EV,ABS --- INCIDENT BEAM ENERGY (KEV) AND MEAN ABSORPTION
C     NTYPE --- NUMBER OF ATOMIC SPECIES
C     ENAME(I) --- NAME OF THE ITH TYPE OF ATOM SPECIES
C     NUM(I)   --- NUMBER OF THE ITH ATOM
C     TEMP(I)  --- DEBYE-WALLER FACTOR
C     OCCUP(I) --- OCCUPATION PROBABILITY (DEFAULT=1.0)
C     XX,YY,ZZ --- FRACTIONAL ATOMIC COORDINATES
C     HR,KR --- INDICES OF REFLECTIONS
C       NS --- NUMBER OF SLICES WITHIN THE SELVEDGE
C       ZS1 --- INITIAL Z COORDINATE OF THE SELVEDGE
C       ZS2 --- FINAL Z COORDINATES OF THE SELVEDGE
C       NB --- NUMBER OF SLICES WITHIN EACH REPEAT BULK UNIT SLAB
C       ZB1=ZS2 --- INITIAL Z COORDINATE OF THE REPEAT BULK SLAB
C       ZB2=ZB --- FINAL Z COORDINATE OF THE REPEAT BULK SLAB
C
      IMPLICIT REAL*8 (A-G,O-Z), INTEGER*4 (H-N)
      PARAMETER (PI=3.141592654,TWOPI=2.0*PI)
        PARAMETER (NATOM=20,ITYPE=10,NR=100)
C
      REAL*8 XX(ITYPE,NATOM),YY(ITYPE,NATOM),ZZ(ITYPE,NATOM)
      REAL*8 TEMP(ITYPE),OCCUP(ITYPE)
      REAL*8 AE(ITYPE,5),BE(ITYPE,5),AI(ITYPE,5),BI(ITYPE,5)
      COMPLEX*16 POTENZ
      INTEGER*4 HR(NR),KR(NR),NUM(ITYPE)
      CHARACTER ENAME(ITYPE)*2,NAM*20

      EXTERNAL POTENZ
C
      COMMON /PARA/A,B,C,RA,RB,RC
      COMMON /ELAS/AE,BE
      COMMON /ILAS/AI,BI
      COMMON /COOR/NTYPE,NUM,XX,YY,ZZ

      WRITE(6,10)
10    FORMAT(1X,' INPUT FILE NAME (potens.in)=?',$)
      READ(5,20)NAM
20    FORMAT(A20)
      IF(NAM.EQ.' ') NAM='potens.in'
      OPEN(2,FILE=NAM,FORM='FORMATTED',STATUS='OLD')
      REWIND 2

      READ(2,*)EV,ABS
      READ(2,*)A,B,C
      READ(2,*)NTYPE
      DO I=1,NTYPE
         READ(2,*)ENAME(I)
         READ(2,*)(AE(I,J),J=1,5)
         READ(2,*)(BE(I,J),J=1,5)
         READ(2,*)(AI(I,J),J=1,5)
```

```fortran
      READ(2,*)(BI(I,J),J=1,5)
      READ(2,*)NUM(I),TEMP(I),OCCUP(I)
      DO J=1,5
         AE(I,J)=AE(I,J)*OCCUP(I)        ! The occupation coefficient is
         AI(I,J)=AI(I,J)*OCCUP(I)        ! included in the definition of the a(i) factor
         BE(I,J)=BE(I,J)+TEMP(I)         ! The temperature factor is
         BI(I,J)=BI(I,J)+TEMP(I)/2.0     ! included in the definition of the b(i) factor
      ENDDO
      DO K=1,NUM(I)
         READ(2,*)XX(I,K),YY(I,K),ZZ(I,K)
         ZZ(I,K)=C*ZZ(I,K)               ! Converts ZZ into absolute units
      ENDDO
      ENDDO
C
C   DEFINE THE SURFACE SELVEDGE
      READ(2,*)ZS1,ZS2,NS
      DZS=(ZS2-ZS1)/FLOAT(NS)
C
C   DEFINE THE REPEAT BULK UNIT SLAB
      READ(2,*)ZB1,ZB2,NB
      DZB=(ZB2-ZB1)/FLOAT(NB)
C
      READ(2,*)NROD
      DO I=1,NROD
         READ(2,*)HR(I),KR(I)
      ENDDO
      CLOSE (2)

      WRITE(6,*)'NROD=',NROD
      WRITE(6,*)'H=',(HR(I),I=1,NROD)
      WRITE(6,*)'K=',(KR(I),I=1,NROD)
      WRITE(6,*)'NS,NB,DZS,DZB=',NS,NB,DZS,DZB
      WRITE(6,*)'NUM=',NUM(1)

C   LATTICE PARAMETERS IN RECIPROCAL SPACE
      RA=TWOPI/A
      RB=TWOPI/B

C   CALCULATE U_g(z) FOR ALL RODS AND STORE THE RESULTS IN FILE NAM
      WRITE(6,30)
30    FORMAT(1X,' OUTPUT FILE NAME (rheed.dat)=?',$)
      READ(5,20)NAM
      IF(NAM.EQ.' ') NAM='rheed.dat'
      OPEN(4,FILE=NAM,FORM='UNFORMATTED',STATUS='UNKNOWN')
      REWIND 4
C
      WRITE(4)EV,A,B
      WRITE(4)NROD
      WRITE(4)(HR(K),K=1,NROD)
      WRITE(4)(KR(K),K=1,NROD)
      WRITE(4)NS,NB,DZS,DZB
C
C   A  CONSTANT WHICH CONVERTS THE POTENTIAL INTO A^{-2} UNITS
      CONST=(1.0+1.9569341D-3*EV)*(8.0*PI)/(A*B)
         CONST1=CONST*47.878009*PI/(1.0+1.9569341D-3*EV)
```

```fortran
C
C     A CONSTANT THAT CONVERTS VALUES OF THE IMAGINARY PART OF THE
C     ATOMIC SCATTERING FACTORS GIVEN AT 100KeV INTO THOSE CORRESPONDING
C     TO THE ACTUAL ENERGY OF THE INCIDENT BEAM
      GAMMAEV=1.0+1.9569341D-3*EV
      GAMMA100=1.0+1.9569341D-3*100.0
      BETAEV=DSQRT(1-1.0/GAMMAEV**2)
      BETA100=DSQRT(1-1.0/GAMMA100**2)
      CONVERT=BETA100/BETAEV
C
C     OUTPUT AVERAGED ONE DIMENSIONAL POTENTIAL DISTRIBUTION UO and other UG
      OPEN(7,FILE='UO.dat',FORM='FORMATTED',STATUS='UNKNOWN')
      OPEN(10,FILE=UG.dat',FORM='FORMATTED',STATUS='UNKNOWN')
      REWIND 7
      REWIND 10

      Z=0.0
      DO I=1,NS+NB
        IF(I.LE.NS) THEN
           Z=Z+DZS
         ELSE
           Z=Z+DZB
        ENDIF
        DO H=1,NROD
          DO K=1,NROD
             HGH=HR(H)-HR(K)
             KGH=KR(H)-KR(K)
             WRITE(4)CONST*POTENZ(Z,CONVERT,HGH,KGH)
          ENDDO
        ENDDO
        WRITE(7,40)Z,DREAL(CONST*POTENZ(Z,CONVERT,0,0)),
     1                 DIMAG(CONST*POTENZ(Z,CONVERT,0,0))
40      FORMAT(3(1X,F12.6))

        WRITE(10,41)Z,(DREAL(CONST*POTENZ(Z,CONVERT,HR(H),0))
     1                                                  ,H=1,7,2)
41      FORMAT(1X,F8.3,9(1X,F8.4))
      ENDDO
C
      CLOSE (4)
      CLOSE (7)
      CLOSE (10)

      STOP
      END


      FUNCTION POTENZ(Z,CONVERT,HGH,KGH)
C
      IMPLICIT REAL*8 (A-G,O-Z), INTEGER*4 (H-N)
      PARAMETER (PI=3.141592654,FOURPI2=4.0*PI*PI)
        PARAMETER (NATOM=20,ITYPE=10)
C
      REAL*8 XX(ITYPE,NATOM),YY(ITYPE,NATOM),ZZ(ITYPE,NATOM)
      REAL*8 AE(ITYPE,5),BE(ITYPE,5),AI(ITYPE,5),BI(ITYPE,5)
      COMPLEX*16 POTEN,POTENZ,ONE
      INTEGER*4 NUM(ITYPE)
```

```fortran
C
      COMMON /PARA/A,B,C,RA,RB,RC
      COMMON /ELAS/AE,BE
      COMMON /ILAS/AI,BI
      COMMON /COOR/NTYPE,NUM,XX,YY,ZZ

      ONE=(0.0D0,1.0D0)
      G2=(FLOAT(HGH)*RA)**2+(FLOAT(KGH)*RB)**2

      POTENZ=(0.0D0,0.0D0)
      DO I=1,NTYPE
         DO 100 K=1,NUM(I)
            ZD=Z-ZZ(I,K)
C
C   CONTRIBUTION FROM ATOMS LYING BEYOND A CERTAIN DISTANCE IS NEGLECTED
            IF(ABS(ZD).GT.4.0) GOTO 100
            POTEN=(0.0D0,0.0D0)
C
C   CONTRIBUTION FROM THE REAL PART OF THE ATOMIC SCATTERING FACTOR
            DO L=1,5
               POTEN=POTEN+AE(I,L)*DSQRT(PI/BE(I,L))*
     1                     DEXP(-BE(I,L)*G2/(4.0*PI)**2)*
     2                     DEXP(-4.0*PI**2*ZD**2/BE(I,L))
            ENDDO
C
C   CONTRIBUTION FROM THE IMAGINARY PART OF THE SCATTERING FACTOR
            DO L=1,5
               POTEN=POTEN+ONE*CONVERT*AI(I,L)*DSQRT(PI/BI(I,L))*
     1                     DEXP(-BI(I,L)*G2/(4.0*PI)**2)*
     2                     DEXP(-4.0*PI**2*ZD**2/BI(I,L))
            ENDDO
            POTENZ=POTENZ+POTEN*CDEXP(-2.0*PI*ONE*
     1             (FLOAT(HGH)*XX(I,K)+FLOAT(KGH)*YY(I,K)))
100      CONTINUE
      ENDDO

      RETURN
      END
```

C.2   A FORTRAN routine for dynamical RHEED calculations

The FORTRAN routine RHEED98.f listed below reads the structural information and the two-dimensional Fourier coefficients $U_G(z)$ of the potential from an output data file created by the routine 2D-poten.f and solves a dynamical RHEED problem. The input file for the RHEED calculation is called rheed.in. This file contains information about the range of angles for which dynamical RHEED calculations will be performed. The direction of the incident electron beam can be scanned by varying either the glancing angle θ or the azimuthal angle ϕ. In a standard RHEED rocking curve calculation the beam azimuth is kept constant (i.e. by letting $\phi = 0$) and the glancing angle θ is varied starting from THETA0 over a total of NROK steps with the angular step defined by TSTEP. If necessary, the direction of the incident beam may be scanned by varying the azimuthal angle ϕ. In the example given below the input data file is the same as the output data file from routine 2D-poten.f and it is called rheed.dat. The beam azimuth is kept constant at $\phi = 0.0001$ (a small finite value is chosen in order to avoid numerical instabilities associated with degenerate eigenvalues appearing at points of high symmetry), and the glancing angle θ varies from $\theta = 0.1$ mrad through a sequence of 160 steps with the step size of 1.0 mrad. The surface selvedge layer is subdivided into 10 subslabs, where each subslab is composed of 15 slices, and the repeat bulk unit slab is subdivided into 10 subslabs, where each subslab consists of 10 slices. The last line of the input data file states that 5 beams must be treated fully dynamically while all other beams must be treated using the Bethe potential approximation.

C.2.1 *Example input data file for dynamical RHEED calculations*

```
rheed
160        0.1          1.0          !NROK, THETAO, TSTEP    (in mrad)
1          0.0001 0.3                !NROT, PHAO,   PSTEP    (in mrad)
10
15  15  15  15  15  15  15  15  15  15
10
10  10  10  10  10  10  10  10  10  10
5
```

C.2.2 *A FORTRAN routine for dynamical RHEED calculations*

Three subroutines ZGEDI.f, ZGEFA.f, and EISPACKCG.f are not listed in this appendix. The first two subroutines ZGEDI.f and ZGEFA.f are used in the subroutine PINVERSE.f for inverting a general complex matrix, which are LINPACK routines and may be downloaded from http://www.netlib.org. The subroutine EISPACKCG.f included in the main program RHEED98.f is indeed the EISPACK routine (Smith et al., 1976) cg.f which may also be downloaded from http://www.netlib.org and is used for solving a general complex eigensystem. Alternatively the commercial NAG (NAG93, 1993) routine F02AKF.f may be used for solving the general complex eigenvalue equations.

```
C/NAME=RHEED98
C
      include 'pinverse.f'
      include 'eispackcg.f'
C
      PROGRAM RHEED_98

      IMPLICIT REAL*8 (A-G,O-Z), INTEGER*4 (H-N)
      PARAMETER (IROD=21,IRODS=2*IROD,IR=21,IRS=2*IR,NIN=5,NOUT=6,
     1           NSLICES=400,NSLICEB=400,NTOT=IROD*IROD,NSL=20,
     2           PI=3.141592654)
      CHARACTER NAM*30
      INTEGER*4 HR(IROD),KR(IROD),NSUBS(NSL),NSUBB(NSL)
      REAL*8 POTENSR(NSLICES,NTOT),POTENSI(NSLICES,NTOT),
     1       POTENBR(NSLICEB,NTOT),POTENBI(NSLICEB,NTOT),
     2       KX,KY,KBX,KBY,KBZ,KZ(IROD),RINT(IROD)
       COMPLEX*16 MSS(NSL,IRS,IRS),MSB(NSL,IRS,IRS)
C
      COMMON /CONST/RA,RB,CHI,NROD,NRODS,NF,NFS,HR,KR
```

```fortran
C
C     THE INPUT PARAMETERS ARE STORED IN TAPE 3
      OPEN(UNIT=3,FILE='rheed.in',FORM='FORMATTED',STATUS='OLD')
      REWIND 3
C
C     INPUT PARAMETERS:
C        EV,A,B: ACCELERATION VOLTAGE, TWO-DIMENSIONAL LATTICE PARAMETERS
C        NROK,THETAO,TSTEP: NUMBER OF POINTS IN A ROCKING CURVE, THE
C                           STARTING ANGLE AND STEP INTERVAL (IN MRAD)
C        NROT,PHAO,PSTEP: NUMBER OF POINTS IN A ROTATION DIAGRAM, THE
C                         STARTING ANGLE AND STEP INTERVAL (IN MRAD)
C        ISUB: NUMBER OF SUBLAYERS
C        NSUM(I): SLICE NUMBER WITHIN THE Ith SUBLAYER
C        NF: NUMBER OF RODS TO BE TREATED FULLY
C
      READ(3,*)NAM
      READ(3,*)NROK,THETAO,TSTEP        !ANGLES IN MRAD
      READ(3,*)NROT,PHAO,PSTEP
C
      READ(3,*)ISUBS
      READ(3,*)(NSUBS(I),I=1,ISUBS)
      READ(3,*)ISUBB
      READ(3,*)(NSUBB(I),I=1,ISUBB)
      READ(3,*)NF
      CLOSE (3)
C
C     Open up the input potential data file
      NAM(11:14)='.dat'
      OPEN(UNIT=2,FILE=NAM(1:14),STATUS='OLD',FORM='UNFORMATTED')
      REWIND 2
C
      READ(2)EV,A,B
      READ(2)NROD
      READ(2)(HR(I),I=1,NROD)
      READ(2)(KR(I),I=1,NROD)
      READ(2)NSS,NSB,DZS,DZB
      WRITE(6,*)NSS,NSB,DZS,DZB
      RA=1.0D0/A
      RB=1.0D0/B
      NRODS=2*NROD
C
C     AMONG THE TOTAL NROD RODS, ONLY NF RODS WILL BE TREATED FULLY. THE
C     REMAINING NB RODS WILL BE TREATED USING BETHE'S PERTURBATION METHOD.
      NB=NROD-NF
      NFS=2*NF
      NBS=2*NB
C
      DO K=1,NSS
        DO H=1,NROD*NROD
           READ(2)POTENSR(K,H),POTENSI(K,H)
        ENDDO
      ENDDO
      DO K=1,NSB
        DO H=1,NROD*NROD
           READ(2)POTENBR(K,H),POTENBI(K,H)
        ENDDO
```

```fortran
      ENDDO
      CLOSE (2)
C
C    Relativistic electron wavelength
      WAVE=0.3878314/DSQRT(EV*(1.0+0.97846707D-3*EV))
      CHI=1.0/WAVE

C    THE INTENSITY FILE
      NAM(11:14)='.int'
      OPEN(UNIT=4,FILE=NAM(1:14),STATUS='UNKNOWN',FORM='FORMATTED')
      REWIND 4

      PHA=PHAO/1000.0D0
      DO I=1,NROT
         THETA=THETAO/1000.0D0
         DO J=1,NROK
            KBX=CHI*DCOS(THETA)*DSIN(PHA)
            KBY=CHI*DCOS(THETA)*DCOS(PHA)
            KBZ=CHI*DSIN(THETA)
            DO H=1,NROD
               KX=KBX+FLOAT(HR(H))*RA
               KY=KBY+FLOAT(KR(H))*RB
               KZ(H)=4.0*PI*PI*(CHI**2-KX**2-KY**2)
            ENDDO
            CALL SCATTM(KBZ,KZ,POTENSR,POTENSI,ISUBS,NSUBS,
     1                     DZS,NSLICES,MSS)
            CALL SCATTM(KBZ,KZ,POTENBR,POTENBI,ISUBB,NSUBB,
     1                     DZB,NSLICEB,MSB)
            CALL NSINT(NF,ISUBS,ISUBB,KZ,MSS,MSB,RINT)
            WRITE(4,12)THETA*1000.0,(RINT(NR),NR=1,NF,2)
12          FORMAT(1X,F7.3,8(1X,F8.6))
            THETA=THETA+TSTEP/1000.0D0
         ENDDO
         PHA=PHA+PSTEP/1000.0D0
      ENDDO
C
      WRITE(6,*)
      WRITE(6,*)' PROGRAM RHEED98 FINISHES HERE'
      WRITE(6,*)
C
      CLOSE (4)
C
      STOP
      END


C

      SUBROUTINE SCATTM(KBZ,KZ,POTENR,POTENI,ISUB,NSUB,DZ,NS,MS)
C
      IMPLICIT REAL*8 (A-G,O-Z), INTEGER*4 (H-N)
      PARAMETER (IROD=21,IRODS=2*IROD,PI=3.141592654,
     1               NTOT=IROD*IROD,IR=21,IRS=2*IR,NSL=20)
      REAL*8 AR(IROD,IROD),AI(IROD,IROD),ALFR(IR),ALFI(IR),
     1       ZR(IR,IR),ZI(IR,IR),KBZ,KZ(IROD),
     2       POTENR(NS,NTOT),POTENI(NS,NTOT),
     3       DENOR(IROD),DENOI(IROD),WK1(IROD),WK2(IROD),WK3(IROD)
      COMPLEX*16 CP(IR,IR),CI(IR,IR),M(IRS,IRS),EXP1,EXP2,CC(IR),CS(IR),
```

```fortran
     1                    MS(NSL,IRS,IRS),GAMMA(IR),SM(IRS,IRS)
      INTEGER*4 HR(IROD),KR(IROD),NSUB(ISUB)
C
      COMMON /CONST/ RA,RB,CHI,NROD,NRODS,NF,NFS,HR,KR
C
      NK=0
      DO IS=1,ISUB          !LOOP OVER SUBSLABS
         DO IH=1,NFS
            DO IG=1,NFS
               MS(IS,IH,IG)=(0.0D0,0.0D0)
            ENDDO
            MS(IS,IH,IH)=(1.0D0,0.0D0)
         ENDDO
         DO IK=1,NSUB(IS)
            NK=NK+1
C    Set up the NF X NF eigensystem for a thin layer
            IBEAM=0
            DO IG=1,NROD
               DO IH=1,NROD
                  IBEAM=IBEAM+1
                IF(IH.EQ.IG) THEN
                  AR(IH,IH)=POTENR(NK,IBEAM)+KZ(IH)
                     AI(IH,IH)=POTENI(NK,IBEAM)
                     DENOR(IH)=AR(IH,IH)-(2.0*PI*KBZ)**2
                     DENOI(IH)=AI(IH,IH)
                ELSE
                     AR(IG,IH)=POTENR(NK,IBEAM)
                     AI(IG,IH)=POTENI(NK,IBEAM)
                  ENDIF
               ENDDO
            ENDDO

C    NOW CALCULATE THE EFFECTIVE EXCITATION AND BETHE POTENTIAL FOR THE
C    STRONG BEAM USING BETHE'S METHOD
            DO IG=1,NF
               DO IH=1,NF
                  URCORR=0.0
                  UICORR=0.0
                  DO K=NF+1,NROD
                     TR=AR(IG,K)*AR(K,IH)-AI(IG,K)*AI(K,IH)
                     TI=AR(IG,K)*AI(K,IH)+AI(IG,K)*AR(K,IH)
                     CALL CDIV(TR,TI,DENOR(K),DENOI(K),URC,UIC)
                     URCORR=URCORR-URC
                     UICORR=UICORR-UIC
                  ENDDO
                  AR(IG,IH)=AR(IG,IH)+URCORR
                  AI(IG,IH)=AI(IG,IH)+UICORR
               ENDDO
            ENDDO
C
C    Call NAG routine F02AKF to solve the eigensystem and return the
C      eigenvalues and eigenvectors
C
C         IFAIL=0
C         CALL F02AKF(AR,IROD,AI,IROD,NF,ALFR,ALFI,ZR,IR,
```

```
C     1                         ZI,IR,INTGER,IFAIL)
C
C     Call eispack to solve the complex eigensystem
            MATZ=1
            CALL EISPACKCG(IROD,NF,AR,AI,ALFR,ALFI,MATZ,ZR,ZI,
     1                                  WK1,WK2,WK3,IFAIL)
C
C     Set up CP AND CI matrices
            DO H=1,NF
               DO K=1,NF
                  CP(H,K)=CMPLX(ZR(H,K),ZI(H,K))
                  CI(H,K)=CP(H,K)
               ENDDO
            ENDDO
            CALL PINVERSE(IR,NF,CI)
C
C     Constructing the scattering matrix for the ISth subslab
C
            DO I=1,NF
               GAMMA(I)=CDSQRT(DCMPLX(ALFR(I),ALFI(I)))
               EXP1=CDEXP((0.0D0,1.0D0)*GAMMA(I)*DZ)
               EXP2=1.0D0/EXP1
               CC(I)=(EXP1+EXP2)/2.0D0
               CS(I)=(EXP1-EXP2)/2.0D0
            ENDDO
            DO 1000 H=1,NF
               DO 1100 K=1,NF
                  M(H,K)=(0.0D0,0.0D0)
                  M(H,K+NF)=(0.0D0,0.0D0)
                  M(H+NF,K)=(0.0D0,0.0D0)
                  DO I=1,NF
                     M(H,K)=M(H,K)+CP(H,I)*CC(I)*CI(I,K)
                     M(H,K+NF)=M(H,K+NF)-CP(H,I)*CS(I)*CI(I,K)/GAMMA(I)
                     M(H+NF,K)=M(H+NF,K)-CP(H,I)*CS(I)*CI(I,K)*GAMMA(I)
               ENDDO
               M(H+NF,K+NF)=M(H,K)
 1100             CONTINUE
 1000          CONTINUE
C
            DO IG=1,NFS
               DO IH=1,NFS
                  SM(IG,IH)=(0.0D0,0.0D0)
                  DO I=1,NFS
                  SM(IG,IH)=SM(IG,IH)+MS(IS,IG,I)*M(I,IH)
               ENDDO
            ENDDO
            ENDDO
            DO IG=1,NFS
               DO IH=1,NFS
                  MS(IS,IG,IH)=SM(IG,IH)
               ENDDO
            ENDDO
         ENDDO
      ENDDO
C
```

```fortran
      RETURN
      END

      SUBROUTINE NSINT(NROD,NBS,NBB,KZ,MS,MB,RINT)
C
      IMPLICIT REAL*8 (A-G,O-Z), INTEGER*4 (H-N)
      PARAMETER (IR=21,IRS=2*IR,NIN=5,NOUT=6,NBULK=20,NSL=20)
      REAL*8 KZ(NROD),RINT(NROD)
      COMPLEX*16 MB(NSL,IRS,IRS),MS(NSL,IRS,IRS),
     1           TO(IR,IR),TOO,TO1,RAMP,GAMMA(IR),
     2           T1(IR,IR),SN(IR,IR)
C
      DO K=1,NROD
         GAMMA(K)=KZ(K)
         GAMMA(K)=CDSQRT(GAMMA(K))
      ENDDO

C     PROPAGATE THE R-MATRIX FROM THE EXIT FACE UPWARDS
C     1. ON THE EXIT FACE:
      DO H=1,NROD
         DO K=1,NROD
            TO(H,K)=MB(NBB,H,K)+MB(NBB,H,K+NROD)*GAMMA(K)
            T1(H,K)=MB(NBB,H+NROD,K)
     1              +MB(NBB,H+NROD,K+NROD)*GAMMA(K)
         ENDDO
      ENDDO
C
C     2. PROPAGATE THE R-MATRIX THROUGH THE PERIODIC SUBSTRATE
      KK=0
 10   KK=KK+1
C
      NS=NBB
      IF((KK.EQ.1).AND.(NBB.GT.1)) NS=NBB-1

      DO IS=NS,1,-1
         CALL PINVERSE(IR,NROD,TO)
         DO H=1,NROD
           DO K=1,NROD
           SN(H,K)=(0.0D0,0.0D0)
           DO L=1,NROD
              SN(H,K)=SN(H,K)+T1(H,L)*TO(L,K)
           ENDDO
            ENDDO
         ENDDO
C
         DO H=1,NROD
           DO K=1,NROD
           TO(H,K)=MB(IS,H,K)
           T1(H,K)=MB(IS,H+NROD,K)
           DO L=1,NROD
              TO(H,K)=TO(H,K)+MB(IS,H,L+NROD)*SN(L,K)
              T1(H,K)=T1(H,K)+MB(IS,H+NROD,L+NROD)*SN(L,K)
           ENDDO
            ENDDO
         ENDDO
      ENDDO
```

```
C
      IF(KK.LT.NBULK) GOTO 10
C
C  3. PROPAGATE THROUGH THE SELVEDGE
C
      DO IS=NBS,1,-1
         CALL PINVERSE(IR,NROD,TO)
         DO H=1,NROD
           DO K=1,NROD
           SN(H,K)=(0.0D0,0.0D0)
           DO L=1,NROD
              SN(H,K)=SN(H,K)+T1(H,L)*TO(L,K)
           ENDDO
            ENDDO
         ENDDO

         DO H=1,NROD
           DO K=1,NROD
           TO(H,K)=MS(IS,H,K)
           T1(H,K)=MS(IS,H+NROD,K)
           DO L=1,NROD
              TO(H,K)=TO(H,K)+MS(IS,H,L+NROD)*SN(L,K)
              T1(H,K)=T1(H,K)+MS(IS,H+NROD,L+NROD)*SN(L,K)
           ENDDO
            ENDDO
         ENDDO
      ENDDO
C
C  4. APPLY THE BOUNDARY CONDITION ON THE SURFACE (Z=0) TO GET THE SURFACE
C     REFLECTIVITIES
      CALL PINVERSE(IR,NROD,TO)
      CALL PINVERSE(IR,NROD,T1)
C
      DO H=1,NROD
         DO K=1,NROD
            TOO=TO(H,K)-T1(H,K)*GAMMA(K)
            TO1=TO(H,K)+T1(H,K)*GAMMA(K)
            TO(H,K)=TOO
            T1(H,K)=TO1
         ENDDO
      ENDDO
C
      CALL PINVERSE(IR,NROD,T1)
C
C  CALCULATE REFLECTED BEAM AMPLITUDES FOR ALL RODS
      DO H=1,NROD
         RAMP=(0.0D0,0.0D0)
         DO K=1,NROD
            RAMP=RAMP-T1(H,K)*TO(K,1)
         ENDDO
         RINT(H)=SQRT(RAMP*CONJG(RAMP))
C        RINT(H)=(REAL(GAMMA(H))/REAL(GAMMA(1)))*(RAMP*CONJG(RAMP))
      ENDDO
C
      RETURN
      END
```

```fortran
      SUBROUTINE CDIV(AR,AI,BR,BI,CR,CI)
      DOUBLE PRECISION AR,AI,BR,BI,CR,CI

      DOUBLE PRECISION S,ARS,AIS,BRS,BIS
      S = DABS(BR) + DABS(BI)
      ARS = AR/S
      AIS = AI/S
      BRS = BR/S
      BIS = BI/S
      S = BRS**2 + BIS**2
      CR = (ARS*BRS + AIS*BIS)/S
      CI = (AIS*BRS - ARS*BIS)/S
      RETURN
      END


      SUBROUTINE PINVERSE(IRODS,NRODS,A)
      PARAMETER (LWORK=6400)
      INTEGER*4 IRODS,NRODS,IPVT(LWORK),INFO
      COMPLEX*16 A(IRODS,IRODS),WORK(LWORK)

C     FACTORIZE A
      INFO=0
      CALL ZGEFA(A,IRODS,NRODS,IPVT,INFO)
C
      IF(INFO.EQ.0) THEN
C     COMPUTE INVERSE OF A
         JOB=1
         CALL ZGEDI(A,IRODS,NRODS,IPVT,DET,WORK,JOB)
       ELSE
         WRITE(6,*)' THE FACTOR U IS SINGULAR'
      ENDIF
C
      RETURN
      END
```

APPENDIX D

PARAMETERIZATION OF THE ELECTRON ATOMIC SCATTERING FACTOR

D.1 The parameterization algorithm

The algorithm given in this Appendix for the parameterization of the electron atomic scattering factors is based on a combination of the simulated annealing procedure and the least-squares method for solving a system of linear algebraic equations (Peng et al., 1996e). The method of simulated annealing is known to be suitable for large-scale optimization problems including the ones where a desired global minimum is hidden among many shallower local minima. The aim of our procedure is to find a set of parameters a_j and b_j that minimizes an objective function (the χ^2 function defined below) and gives the best fit of numerical values of scattering factors $\{f(s_i)\}$ in the form of a sum of n Gaussian functions (13.79).

The procedure starts by assigning random values to the n fitting parameters b_j $(j = 1, ..., n)$ in the range $0 < b_j < b_0$, where b_0 controls the area of the parameter space spanned by b_j searched by the fitting procedure. For a given set of m numerical scattering factors $f^{(e)}(s_i)$ $(i = 1, ...m)$, m linear algebraic equations can be written down for the set of n free parameters a_j $(j = 1, ..., n)$

$$f^{(e)}(s_i) = \sum_{j=1}^{n} a_j \exp(-b_j s_i^2).\tag{D.1}$$

The above equation can be represented in matrix form as

$$\mathbf{F} = \mathbf{BA},\tag{D.2}$$

where $\mathbf{F}$ is an m-dimensional vector with components $(F)_i = f^{(e)}(s_i)$, $\mathbf{A}$ is an n-dimensional parameter vector with components $(A)_j = a_j$, and $\mathbf{B}$ is an $m \times n$ matrix with elements $\{B\}_{ij} = \exp(-b_j s_i^2)$. Typically the number of data points m is much larger than the number n of unknown variables a_j $(j = 1, ..., n)$, and the problem (2.2) is therefore overdetermined. In general there exists no solution $\mathbf{A}$ that would satisfy eqns (D.2). However, we can find a compromise solution which comes very close to satisfying all the equations simultaneously. The degree of deviation from an ideal solution is defined on the basis of the least-squares approach by stating that the sum of squares of the difference between the left- and right-hand sides of eqn (D.2) is a minimum. The over-determined problem (D.2) is therefore reduced to the linear least-squares problem. The solution of this linear least-squares problem can be obtained by multiplying both sides of eqn (D.2) by $\mathbf{B}^T$, which is the transpose of matrix $\mathbf{B}$

$$(\mathbf{B}^T\mathbf{F}) = (\mathbf{B}^T\mathbf{B})\mathbf{A}. \tag{D.3}$$

The above equation is called the *normal equation* of the linear least-squares problem (D.2). Since the matrix $(\mathbf{B}^T\mathbf{B})$ is an $n \times n$ matrix, the solution can now be found as

$$\mathbf{A} = (\mathbf{B}^T\mathbf{B})^{-1}(\mathbf{B}^T\mathbf{F}). \tag{D.4}$$

In practice the $n \times n$ matrix $(\mathbf{B}^T\mathbf{B})$ can be singular giving rise to a divergent solution. This can be avoided by applying the singular value decomposition (SVD) method (see for example (Press et al., 1986)), and this is the method that we use in order to solve eqn (D.2).

In terms of solution $\mathbf{A}$ the χ^2 function can be defined as

$$\chi^2 = \sum_{i=1}^{m} \left[f^{(e)}(s_i) - \sum_{j=1}^{n} a_j \exp(-b_j s_i^2) \right]^2. \tag{D.5}$$

For a given value of χ^2 corresponding to a given set of b_j and a_j obtained from solving (D.4), values of parameters b_j are then moved randomly to the next step (here we denote it by b_j^*) following the rule

$$b_j^* = \begin{cases} b_j + (r - 0.5) \times b_t & \text{if } b_j^* > 0.0, \\ b_j + r \times b_t & \text{otherwise,} \end{cases} \tag{D.6}$$

where $b_t = b_0 \times T$ and T is the '*temperature*' of the system, and r is a random number belonging to the interval $0.0 < r < 1.0$. For this new set of parameters b_j^*, a set of linear algebraic equations is devised in the same way as eqn (D.1) and subsequently solved by the SVD method. A new value of χ^2 is then calculated and compared with the old value of χ^2. If the new value of χ^2 is lower than the previous value, values of parameters b_j are replaced with b_j^*. Otherwise, a new set of b_j^* is taken in accord with (D.6). This process is continued until the total number of random walks that fail to improve the value of χ^2 exceeds a certain threshold (say 200). The temperature is then decreased according to the law $T \to T \times \delta t$, where δt is in the range from 0.9 to 0.999, and the step b_t for the random walk of b_j is reduced. The process is terminated if the temperature falls below the threshold temperature (say 0.05).

It is interesting to note the similarity between this procedure and a phase transition, in particular with the way a liquid freezes and crystallizes. At a high temperature (high T), the molecules of a liquid move freely with respect to each other (large b_t). If the liquid is cooled slowly ($T \to T \times \delta t$), the thermal mobility of molecules gradually decreases ($b_t = b_0 \times T$ decreases with T). If cooling proceeds very slowly so that atoms have ample time to explore all the available configurations, then atoms arrange themselves in a regular lattice and form a crystal at low temperature. The structure formed in this way corresponds to the minimum of energy of the system (corresponding to the global minimum of χ^2).

A FORTRAN routine MCFIT listed below was developed for the purpose of fitting a tabulated scattering factor to a set of n Gaussian functions (see eqn (13.79)). In the routine the number of Gaussians is fixed at five since we found that a set of five Gaussians gives a satisfactory representation of both the elastic and absorptive scattering factors. The accuracy of the routine is controlled by three parameters, the initial temperature T_0 (T0), the initial step b_0 (DRTX0) for the random walk of b_j, and the parameter δt (DRT) that determines the rate of decrease of temperature. To accelerate the convergence the procedure is usually first applied to a subset of the complete data set (normally containing approximately one tenth of the data points). The resulting values of fitting parameters a_j and b_j are then used as the initial values for a more thorough search involving the complete data set.

The parameter b_0 determines the region of parameter space spanned by b_j ($j = 1, ..., 5$) where the routine MCFIT performs the search. In principle if the search is exhaustive (i.e. if b_0 is sufficiently large, the initial temperature T_0 is high enough, and the system is cooled down sufficiently slowly ($\delta t \rightarrow 1.0$)), the modified simulated annealing algorithm finds a unique set of parameters a_j, b_j corresponding to a global minimum of χ^2. For finite values of b_0, T_0, and δt, the algorithm is only approximate in the sense that only a subspace of the entire parameter space will be searched and the system may not be in the minimum energy state when the temperature reaches zero. This is analogous to the quenching process where a liquid does not freeze in the form of a single crystal but instead ends up in a polycrystalline or amorphous state characterized by a somewhat higher energy.

Fortunately for the parameterization of atomic scattering factors we do not have to find the true global minimum. The choice of controlling parameters is therefore dictated by a compromise between the accuracy and the computer time needed to reach the given level of accuracy. Our experience shows that satisfactory results may be obtained for $b_0 = 1.0$, $T_0 = 1.0$, and $\delta t = 0.9$. The amount of time spent by a HP735 workstation on performing a typical run is of the order of 1 minute. Better results may be obtained using $b_0 = 4.0$, $T_0 = 3.0$, and $\delta t = 0.99$. The amount of time required to accomplish the search in this case is of the order of 2 minutes. Very accurate results may be obtained using $T_0 = 30$, $b_0 = 1$, and $\delta t = 0.99$.

The routine MCFIT.f implements the combined least-squares and modified simulated annealing algorithm for fitting a given set of NP atomic scattering factor data FA given at a set of s values SS by five Gaussians, and returns values of fitting parameters a_i and b_i (A1 and B1). The input atomic scattering factors may be either elastic or absorptive atomic scattering factors, and the routine can be used for parameterization of either of the two factors. This routine uses subroutines RAN2.f, SVDCMP.f, and SVBKSB.f described in detail in sections

7.1 and 2.6 of Press et al. (Press et al., 1992).

```
C/NAME=MCFIT.FOR
C
C    THIS ROUTINE FINDS A SET OF DOYLE-TURNER PARAMETERS USING A COMBINED
C    LEAST-SQUARES AND A MODIFIED SIMULATED ANNEALING ALGORITHMS.
C
C                 LIAN-MAO PENG AND REN GANG, BEIJING, 1995
C
     SUBROUTINE MCFIT(A1,B1,N,SS,FA,NP,TO,DRTXO,DRT)
     IMPLICIT REAL*8 (A-G,O-Z), INTEGER*4 (H-N)
     PARAMETER (NLOC=5)
     REAL*8 A1(N),B1(N),A2(NLOC),B2(NLOC),T,SS(NP),FA(NP),
    1        DRT,DRTXO,DRTX,K1,K2,DRTK,RE,RAN
     REAL    RAN2
     INTEGER*4 NC,KS,IPC,IDUM,IR
C
     IDUM=-13     !SEED FOR GENERATING RANDOM NUMBER USING RAN2
C
C    ASSIGN RANDOM NUMBERS TO THE INITIAL VALUES OF THE DOYLE-TURNER PARAMETERS
     DO I=1,N
        B1(I)=RAN2(IDUM)*DRTXO
        A1(I)=0.0
     ENDDO
C
C    INITIALIZE THE CONTROL PARAMETERS
C    TO --- INITIAL TEMPERATURE
C    T  --- REAL TEMPERATURE
C    DRTXO --- INITIAL STEP OF RANDOM WALK
C    DRTX  --- REAL STEP
C
     T=TO
     DRTX=DRTXO
C
C    THE SUBROUTINE LSSVD SELECTS A SUBSET (1/10 IF IPC=0)
C    OF NUMERICAL DATA POINTS AND
C    SOLVES A1 FOR THE GIVEN B1 USING LEAST-SQUARES SVD METHOD
     IPC=0
     CALL LSSVD(A1,B1,N,SS,FA,NP,IPC)
C
C    FOR THE INITIAL SET OF A1,B1, CALCULATE THE \CHI^2 VALUE (K1)
     CALL CHISQ(A1,B1,N,K1,SS,FA,NP,IPC)
C
     K=0
     KS=0
     NC=0
     IR=0
 15  K=K+1
C
C    RANDOMLY MOVE THE PARAMETER SET B1 TO A NEW SET (THE STEP OR RADIUS
C    IS CONTROLLED BY PARAMETER DRTX)
     DO I=1,N
        RAN=RAN2(IDUM)
        B2(I)=B1(I)+(RAN-0.5)*DRTX
        IF(B2(I).LT.0.0) B2(I)=B1(I)+RAN*DRTX
     ENDDO
```

```fortran
C
C     FOR THE NEW SET B2, SOLVE A2 USING ONLY A SET OF 15 DATA POINTS (IF IPC=0)
C     OR THE COMPLETE SET, AND CALCULATE \CHI^2 FOR B2
      CALL LSSVD(A2,B2,N,SS,FA,NP,IPC)
      CALL CHISQ(A2,B2,N,K2,SS,FA,NP,IPC)
C
C     CALCULATE THE DIFFERENCE BETWEEN THE \CHI^2 VALUES FOR THE TWO SETS
C     OF B VALUES. IF \CHI^2 VALUE DECREASES AS A RESULT OF THE RANDOM MOVE,
C    THEN REPLACE   THE OLD A AND B WITH THE NEW VALUES,
C    OTHERWISE TRY ANOTHER MOVE
      DRTK=K2-K1
      RE=DRTK/K1
C
      IF(RE.LT.-1.E-3) THEN
         DO I=1,N
            A1(I)=A2(I)
            B1(I)=B2(I)
         ENDDO
         K1=K2
         IR=10
         NC=0
         KS=0      !RESET KS FOR ANOTHER 1000 SEARCH AT THIS TEMPERATURE
         GOTO 15
C
C     IF AT A CERTAIN TEMPERATURE, AFTER MAKING 200 RANDOM MOVES WE STILL CANNOT
C     REDUCE \CHI^2 VALUE BY MORE THAN THE THRESHOLD VALUE (-1.0E-3), THEN GO
C     TO A LOWER TEMPERATURE
       ELSE IF(NC.LT.200) THEN
             NC=NC+1
         GOTO 15
      ENDIF

      IF(KS.LT.10.AND.IR.EQ.10) THEN
         KS=KS+1
         NC=0
         GOTO 15
      ENDIF
      IR=0
      KS=0
      NC=0
C
C     NOW DECREASE THE TEMPERATURE AND REDUCE THE STEP OF THE RANDOM
C     WALK
C     HERE WE CREATED A DOUBLE LOOP
C     FIRST:    SELECT ONLY A SUBSET OF DATA POINTS (IPC=0), AND DECREASE THE
C               TEMPERATURE UNTIL T < 0.01
C     SECOND:   SELECT THE WHOLE SET OF NP DATA POINTS (IPC=9), AND DO THE ENTIRE
C          CALCULATION AGAIN
C     THE TEMPERATURE DECREASES FOLLOWING THE RULE: X(N+1) = X(N)*DRT
      T=T*DRT
      DRTX=DRTX0*T
      IF(T.GT.0.005) GOTO 15
      IF(IPC.EQ.0) THEN
         IPC=9
         T=T0
         DRTX=DRTX0
```

```fortran
      CALL CHISQ(A1,B1,N,K1,SS,FA,NP,IPC)
        GOTO 15
      ENDIF
D     WRITE(6,*)' K=',K,';    CHISQ=',K1

      RETURN
      END


C/NAME=CHISQ.FOR
C
C     THIS SUBROUTINE CALCULATES \CHI^2 FOR A SUBSET OF DATA POINTS
C     (THE SIZE OF THE SUBSET IS DETERMINED BY THE PARAMETER IPC)
C
      SUBROUTINE CHISQ(A,B,N,CHI2,S,FE,NP,IPC)
      REAL*8 A(N),B(N),CHI2,S(NP),FE(NP),FEC
      INTEGER*4 IPC
C
      CHI2=0.0
      DO I=1,NP,10-IPC
         FEC=0.0
         DO K=1,N
            FEC=FEC+A(K)*EXP(-B(K)*S(I)*S(I))
         ENDDO
         CHI2=CHI2+(FE(I)-FEC)*(FE(I)-FEC)
      ENDDO

      RETURN
      END


      SUBROUTINE LSSVD(A2,B2,N,S,FE,NP,IPC)
      IMPLICIT REAL*8 (A-G,O-Z), INTEGER*4 (H-N)
      PARAMETER (MP=151,NLOC=5)
      REAL*8 A2(N),B2(N),S(NP),FE(NP)
      REAL A(MP,NLOC),V(NLOC,NLOC),W(NLOC),B(MP),X(NLOC)
C
C     THIS ROUTINE FIRST SETS UP AN M BY N MATRIX, CALLS ROUTINE SVDCMP TO
C     COMPUTE ITS SINGULAR VALUE DECOMPOSITION, A=UWV^T. THE MATRIX U REPLACES
C     A ON OUTPUT. THE DIAGONAL MATRIX OF SINGULAR VALUES W IS OUTPUT AS A
C     VECTOR W(1:N). THE MATRIX V (NOT THE TRANSPOSE V^T) IS OUTPUT AS V(N,N).
C     THE ROUTINE SVBKSB THEN SOLVES AX=B FOR A VECTOR, WHERE A IS SPECIFIED
C     BY THE ARRAYS U,W,V AS RETURNED BY SVDCMP. X IS THE OUTPUT SOLUTION
C     VECTOR.
C
      TOL=1.0D-7
      IK=0
      DO I=1,NP,10-IPC
         IK=IK+1
         DO J=1,N
            A(IK,J)=EXP(-B2(J)*S(I)**2)
         ENDDO
         B(IK)=FE(I)
      ENDDO
      CALL SVDCMP(A,IK,N,MP,NLOC,W,V)
C
```

```
C    TO SET A LOW LIMIT TO ALL THE MATRIX ELEMENTS OF W, SO THAT THE MATRIX
C    W^-1 WOULD NOT DIVERGE
     WMAX=0.0
     DO J=1,N
        IF(W(J).GT.WMAX) WMAX=W(J)
     ENDDO
     WMIN=WMAX*TOL
     DO J=1,N
        IF(W(J).LT.WMIN) W(J)=0.0
     ENDDO

     CALL SVBKSB(A,W,V,IK,N,MP,NLOC,B,X)
     DO I=1,N
        A2(I)=X(I)
     ENDDO

     RETURN
     END
```

D.2   The absorptive atomic scattering factor

The absorptive scattering factor depends both on the angle of scattering and on the Debye–Waller factor. It is therefore difficult to make an exhaustive tabulation of this quantity for the same set of elements and compounds as the one for which we have tabulated the elastic atomic scattering factors. We choose instead to provide the text of the FORTRAN routine LMPTDS.f that evaluates the absorptive electron atomic scattering factor $f^{TDS}(s)$ for given s value SS, the atomic number Z0, the primary beam energy EV, and the Debye–Waller factor DW. These results can then be used by the main routine TDS98.f for the parameterization of the absorptive atomic scattering factor using the routine MCFIT.f described in the preceding section.

The numerical values of the absorptive structure factors are calculated using the isotropic formula (13.65). For small angles of scattering corresponding to $s < 2.0$ Å^{-1} the electron scattering factor $f^{(e)}(s)$ of Table 13.3 is used (and is referred to as 'neutral-atom.dat' in subroutine PARAMETER), while for $2 < s < 5$ Å^{-1} Table 5 of Rez et al. (Rez et al., 1994) is used (and is referred to as 'rez-table.dat' in subroutine PARAMETER). The numerical integration of (13.65) is performed using the NAG routine d01fcf.f (NAG93, 1993). It should be noted that in the limit of large s the absorptive atomic scattering factor $f^{TDS}(s)$ defined by formula (13.65) decreases exponentially with increasing s as $f^{TDS}(s) \sim -\exp(-Bs^2/2)$. To eliminate this trivial exponential behaviour, in the parameterization of the absorptive atomic scattering factors we consider the quantity $f^{TDS}(s)\exp(Bs^2/2)$, which we fit to a sum of Gaussian functions

$$f^{TDS}(s)\exp(Bs^2/2) = \sum_{i=1}^{5} a_i \exp(-b_i s^2). \tag{D.7}$$

Parameters describing the absorptive atomic electron scattering factors have been determined for a number of elements in the crystalline state and for several important materials with the zinc-blende structure. The interested reader can find the relevant parameters in (Peng et al., 1996e, Peng et al., 1996d). It should be noted that f^{TDS} depends on the primary beam energy, and the values given in (Peng et al., 1996e, Peng et al., 1996d) correspond to an incident beam energy of 100 keV. These parameters may, however, be converted easily from one voltage (e.g. 100 keV) to any other voltage using the formula

$$f^{TDS}(E) = \frac{\beta(100 \text{ keV})}{\beta(E)} f^{TDS}(100 \text{ keV}), \tag{D.8}$$

where $\beta = \sqrt{1 - \gamma^{-2}}$, $\gamma = (m/m_0) = 1.0 + 1.9569341 \times 10^{-3}E$ and the energy of the primary beam E is given in keV. We have also studied the absorptive atomic electron scattering factors of ions. It was found that while the absorptive scattering factors of ions deviate from those of neutron atoms, the deviations only become significant in the region of very small values of $s < 0.3$ Å^{-1} (Peng, 1998). Since for most materials the lattice spacing is smaller than 1.5 Å, we expect that in the majority of TED applications the effect of crystal ionicity on the absorptive scattering factors can be neglected. This conclusion is consistent with accurate measurements of crystal structure factors of crystalline MgO (Zuo et al., 1997).

```
C/NAME=TDS98.f
C
C   MAIN PROGRAM FOR THE CALCULATION OF THE DOYLE-TURNER
C   PARAMETERS FOR THE IMAGINARY PART
C   OF THE OPTICAL POTENTIAL FOR ANY OF THE 98 NEUTRAL ELEMENTS.
C
C   Reference: Acta Cryst. A52 (1996) pp257-276 by Peng et al.
C              Surface Science 330 (1995) pp86-100 by Dudarev et al.
C
C                    LIAN-MAO PENG, JUNE 95
C
    INCLUDE 'mcfit.f'
    INCLUDE 'd01fcf.f'
C
    IMPLICIT REAL*8 (A-G,O-Z), INTEGER*4 (H-N)
    PARAMETER (NOUT=6,NIN=5,NP=70,N=5,T0=10.0,X0=1,DRT=0.9,PI=3.141592654)
    REAL*8   EV,DW,FA(NP),SS(NP),A(N),B(N),CHI2
    INTEGER*4 IZ
C
    WRITE(NOUT,10)
10  FORMAT(//10X,' *********** T D S 9 8 ************'//
   1           1X,' Please input the atomic number (Z)  =',$)
    READ(NIN,*)IZ
    WRITE(NOUT,20)
20  FORMAT(1X,'      the acceleration voltage (KeV)  =',$)
    READ(NIN,*)EV
    WRITE(NOUT,30)
30  FORMAT(1X,'      and the Debye-Waller factor  =',$)
    READ(NIN,*)DW
C
C   CALCULATING TDS ABSORPTIVE SCATTERING FACTOR
    DO I=1,NP
```

```fortran
      SS(I)=2.0*FLOAT(I-1)/FLOAT(NP-1)
   ENDDO
   CALL LMPTDS(IZ,EV,DW,SS,FA,NP)
C
C  FIND THE INITIAL DOYLE-TURNER PARAMETERS USING THE
C  COMBINED LEAST-SQUARES
C  AND THE MODIFIED SIMULATED ANNEALING METHOD
        CALL MCFIT(A,B,N,SS,FA,NP,TO,XO,DRT)
   CALL SORT(A,B,N)
C
   WRITE(NOUT,70)(A(J),J=1,5)
   WRITE(NOUT,80)(B(J),J=1,5)
70 FORMAT('A=',5(1X,F10.4))
80 FORMAT('B=',5(1X,F10.4))
C
   STOP
   END


   SUBROUTINE SORT(A,B,N)
   REAL*8  A(N),B(N)
   INTEGER*4 N

   DO I=1,N
      DO J=I+1,N
         IF(B(J).LT.B(I)) THEN
            BTEMP=B(I)
            B(I)=B(J)
            B(J)=BTEMP
            ATEMP=A(I)
            A(I)=A(J)
            A(J)=ATEMP
         ENDIF
      ENDDO
   ENDDO

   RETURN
   END


   subroutine lmptds(z0,ev,dw,ss,f_tds,np)

   implicit real*8(a-g,o-z),integer*4(h-n)
   parameter (ndim=2,minpts=100,maxpts=1000000,lenwrk=1000000)
   parameter (absacc=1.0d-4)
   integer*4 z0,z1,np
   real*8 dw_1,ev,z,ae(5),be(5),ax(4),bx(4)
   real*8 xlow(2),xhigh(2),wrkstr(100000),ss(*),f_tds(*)
   common /para/z1,ae,be,ax,bx
   common /func/s1,dw_1,alpha2,z,xd
   external fxy

   dw_1=dw
   z1=z0
   z=float(z0)
   xd=100.0d0
   xlow(1)=-xd
```

```fortran
      xlow(2)=-xd
      xhigh(1)=xd
      xhigh(2)=xd

c     Get both a_j and b_j for electron and x-ray:
      call parameter(z1,ae,be,ax,bx)

c     Electron wavelength
      wave=0.3878314/dsqrt(ev*(1.0d0+0.97846707d-3*ev))
      write(6,*)' Wavelength=',wave
      const=2.0*wave*(1.0d0+1.9569341d-3*ev)

c     For large s (>5.0), fe(s)=0.023933754*Z/(s2+alpha^2)
c     Calculate alpha^2 using fe(s=5.0)
      alpha2=(0.023933754*z/sf_electron(5.0d0))-5.0**2

      do k=1,np
         s1=ss(k)
         ifail=-1
         call d01fcf(ndim,xlow,xhigh,minpts,maxpts,fxy,absacc,acc,
     1                  lenwrk,wrkstr,sum,ifail)
         f_tds(k)=const*sum*exp(-dw*s1**2/2)
      enddo

      return
      end

      function fxy(ndim,z)
      implicit real*8(a-g,o-z)
      integer*4 ndim
      real*8 s1,dw,z(ndim),fxy,sf_electron,fe1,fe2,z0,xd,alpha2
      external sf_electron
      common /func/s1,dw,alpha2,z0,xd

      if(sqrt(z(1)**2+z(2)**2).gt.xd) then
         fxy=0.0
         return
      endif

      arg1=dsqrt((z(1)-s1/2.0)**2+z(2)**2)
      arg2=dsqrt((z(1)+s1/2.0)**2+z(2)**2)
      temp=dexp(2.0d0*dw*(s1**2/4.0-z(1)**2-z(2)**2))
      if(arg1.lt.5.0) then
        fe1=sf_electron(arg1)
       else
         fe1=0.023933754*z0/(arg1**2+alpha2)
      endif
      if(arg2.lt.5.0) then
        fe2=sf_electron(arg2)
       else
         fe2=0.023933754*z0/(arg2**2+alpha2)
      endif
      fxy=fe1*fe2*(1.0d0-temp)
      return
      end
```

```fortran
      SUBROUTINE PARAMETER(Z,AE,BE,AX,BX)

      REAL*8       AE(5),BE(5),AX(4),BX(4)
      INTEGER*4    Z,Z0

      OPEN(2,FILE='neutral-atoms.dat',FORM='FORMATTED',STATUS='OLD')
      REWIND 2
      DO I=1,98
         READ(2,*)Z0
         READ(2,*)(AE(J),J=1,5),(BE(J),J=1,5)
         IF(Z0.EQ.Z) GOTO 100
      ENDDO
100   CLOSE (2)

      OPEN(3,FILE='rez-table.dat',FORM='FORMATTED',STATUS='OLD')
      REWIND 3
      DO I=2,92
         READ(3,*)Z0,(AX(J),BX(J),J=1,4)
         IF(Z0.EQ.Z) GOTO 200
      ENDDO
200   CLOSE (3)

      RETURN
      END


      FUNCTION SF_ELECTRON(S)

      IMPLICIT REAL*8(A-G,O-Z)
      REAL*8 SF_ELECTRON,AE(5),BE(5),AX(4),BX(4),S
      INTEGER*4 Z

      COMMON /PARA/Z,AE,BE,AX,BX

      SF=0.0D0
      IF(S.LT.2.0) THEN
         DO J=1,5
            SF=SF+AE(J)*DEXP(-BE(J)*S**2)
         ENDDO
       ELSE
         DO J=1,4
            SF=SF+AX(J)*DEXP(-BX(J)*S**2)
         ENDDO
         SF=0.02393367*(Z-SF)/S**2
      ENDIF
      SF_ELECTRON=SF

      RETURN
      END
```

REFERENCES

Abramov, Y. A., Tsirelson, V. G., Zavodnik, V. E., Ivanov, S. A., and Brown, I. D. (1995). The chemical bond and atomic displacements in $SrTiO_3$ from X-ray diffraction analysis. *Acta Crystallographica B*, 51:942–951.

Abrikosov, A. A., Gorkov, L. P., and Dzyaloshinski, I. E. (1975). *Methods of Quantum Field Theory in Statistical Physics*. Dover, New York, second edition.

Afanas'ev, A. M. and Kagan, Y. (1968). The role of lattice vibrations in dynamical theory of X-rays. *Acta Crystallographica A*, 24:163–170.

Afanas'ev, A. M., Kagan, Y., and Chukhovskii, F. N. (1968). Dynamical treatment of thermal diffuse scattering of X-rays. *physica status solidi*, 28:287–295.

Ahmed, J., Wilkinson, A. J., and Roberts, S. G. (1997). Characterising dislocation structures in bulk fatigued copper single crystals using electron channelling contrast imaging (ECCI). *Philosophical Magazine Letters*, 76:237 – 245.

Ajika, N., Hashimoto, H., Yamaguchi, K., and Endoh, H. (1985). Atomic structure images formed by plasma-loss electrons. *Japanese Journal of Applied Physics*, 24:L41 – L44.

Allen, L. J., Faulkner, H. M., and Leeb, H. (1999). Inversion of dynamical electron diffraction data including absorption. *Acta Crystallographica A*, 56:119–126.

Allen, L. J. and Josefsson, T. W. (1995). Inelastic scattering of fast electrons by crystals. *Physical Review B*, 52:3184–3198.

Allen, L. J., Josefsson, T. W., and Rossouw, C. J. (1994). Interaction delocalization in characteristic X-ray emission from light elements. *Ultramicroscopy*, 55:258 – 267.

Allen, L. J. and Rossouw, C. J. (1989). Effects of thermal diffuse scattering and surface tilt on diffraction and channelling of fast electrons in CdTe. *Physical Review B*, 39:8313–8321.

Allen, L. J. and Rossouw, C. J. (1993a). Crystal potential determination by inversion for centrosymmetric crystals. *Ultramicroscopy*, 48:341–346.

Allen, L. J. and Rossouw, C. J. (1993b). Delocalisation in electron-impact ionization in a crystalline environment. *Physical Review B*, 47:2446 – 2452.

Altmann, S. L. (1991). *Band Theory of Solids: An Introduction From the Point of View of Symmetry*. Oxford University Press, Oxford.

Andersen, S. K. and Howie, A. (1975). Diffraction effects in backscattering and Auger production near crystal surfaces. *Surface Science*, 50:197 – 214.

Anstis, G. R. and Gan, X. S. (1994). Calculation of RHEED intensities by a THEED algorithm. *Surface Science*, 314:L919–L924.

Artmann, K. (1948). Berechnung der Seitenversetzung des totalreflektierten Strahles. *Annalen der Physik*, 2:87–102.

Ashcroft, N. W. and Mermin, N. D. (1976). *Solid State Physics*. Saunders College, Philadephia.

Atwater, H. A., Wang, S. S., Ahn, C. C., Nikzad, S., and Frase, H. N. (1993). Analysis of monolayer thin films during molecular beam epitaxy by reflection electron energy loss spectroscopy. *Surface Science*, 298:273 – 283.

Authier, A., Lagomarsino, S., and Tanner, B. K., editors (1996). *X-ray Dynamical Diffraction: Theory and Applications*, volume 357 of *NATO Advanced Study Institute Series, Series B: Physics*, New York and London. Plenum Press.

Baba-Kishi, K. Z. (1998). Measurement of crystal parameters on backscatter Kikuchi diffraction patterns. *Scanning*, 20:117 – 127.

Baba-Kishi, K. Z. (2002). Electron backscattering Kikuchi diffraction in a scanning electron microscope for crystallographic analysis. *Journal of Materials Science*, 37:1715 – 1746.

Bakenfelder, A., Fromm, I., Reimer, L., and Rennekamp, R. (1990). Contrast of the electron spectroscopic imaging mode of a TEM. *Journal of Microscopy*, 159:161 – 177.

Barabanenkov, Y. N. (1968). Application of the smooth perturbation method to the solution of general equations of multiple wave scattering theory. *Soviet Physics JETP*, 27:954–959.

Barabanenkov, Y. N. (1969). On the spectral theory of radiation transport equations. *Soviet Physics JETP*, 29:679–684.

Barabanenkov, Y. N. and Finkelberg, V. M. (1968). Radiation transport equation for correlated scatterers. *Soviet Physics JETP*, 26:587–591.

Barron, T. H. (1977). Room-temperature Debye-Waller factors of magnesium oxide. *Acta Crystallographica A*, 33:602–604.

Bates, R. H. T. and McDonnell, M. J. (1986). *Image Restoration and Reconstruction*. Oxford University Press, Oxford.

Beenakker, C. W. J. (1997). Random-matrix theory of quantum transport. *Reviews of Modern Physics*, 69:731 – 808.

Berry, M. V. (1971). Diffraction in crystals at high energies. *Journal of Physics C: Solid State Physics*, 4:697–722.

Bethe, H. (1928). Theorie der Beugung von Elektronen an Kristallen. *Annalen der Physik*, 87:55–129.

Bilhorn, D. E., Foldy, L. L., Thaler, R. M., Tobocman, W., and Madsen, V. A. (1964). Remarks concerning reciprocity in quantum mechanics. *Journal of Mathematical Physics*, 5:435–441.

Bird, D. M. (1989). Theory of zone axis electron diffraction. *Journal of Electron Microscopy Technique*, 13:77–97.

Bird, D. M. and King, Q. A. (1990). Absorptive form factor for high-energy electron diffraction. *Acta Crystallographica A*, 46:202–208.

Bird, D. M. and Saunders, M. (1992). Inversion of convergent-beam electron diffraction patterns. *Acta Crystallographica A*, 48:555–562.

Blackman, M. (1939). On the intensities of electron diffraction rings. *Proceedings of the Royal Society of London A*, 173:68–82.

Bleloch, A. L., Howie, A., Milne, R. H., and Walls, M. G. (1989). Elastic and inelastic scattering effects in reflection electron microscopy. *Ultramicroscopy*, 29:175–182.

Blum, K. (1981). *Density Matrix: Theory and Applications*. Plenum, New York.

Booker, G. R., Shaw, A. M. B., Whelan, M. J., and Hirsch, P. B. (1967). Some comments on the interpretation of the Kikuchi-like reflection patterns observed by scanning electron microscopy. *Philosophical Magazine*, 16:1185 – 1191.

Born, M. and Huang, K. (1954). *Dynamical Theory of Crystal Lattice*. Oxford University Press, Oxford.

Born, M. and Wolf, E. (1980). *Principles of Optics*. Pergamon Press, Oxford, sixth edition.

Bragg, W. L., Phillips, D. C., and Lipson, H. (1975). *The Development of X-ray Analysis*. G. Bell and Sons, London.

Braun, W., Daweritz, L., and Ploog, K. H. (1998). Origin of electron diffraction oscillations during crystal growth. *Physical Review Letters*, 80:4935 – 4938.

Brigham, E. O. (1974). *The Fast Fourier Transform*. Prentice Hall, London.

Butt, N. M., Bashir, J., and Khan, M. N. (1993). Compilation of temperature factors of cubic compounds. *Acta Crystallographica A*, 49:171–174.

Butt, N. M., Bashir, J., Willis, B. T. M., and Heger, G. (1988). Compilation of temperature factors of cubic elements. *Acta Crystallographica A*, 44:396–398.

Buxton, B. F., Eades, J. A., Steeds, J. W., and Rackham, G. M. (1976). The symmetry of electron diffraction zone axis patterns. *Philosophical Transactions of the Royal Society of London*, 281:171–194.

Chambers, S. A. (1992). Elastic-scattering and interference of backscattered primary, Auger and X-ray photoelectrons at high kinetic energy - principles and applications. *Surface Science Reports*, 16:261–331.

Cherns, D., Howie, A., and Jacobs, M. H. (1973). Characteristic X-ray production in thin crystals. *Zeitschrift für Naturforschung*, a28:565 – 571.

Chitanvis, S. M. and Lax, M. (1984). Scattering of waves from rough surfaces. *Progress of Theoretical Physics*, 80(supplement):40 – 46.

Chukhovskii, F. N., Alexanjan, L. A., and Pinsker, Z. G. (1973). Dynamical treatment of Kikuchi patterns. *Acta Crystallographica A*, 29:38 – 45.

Clarke, D. R. (1971). Observation of crystal defects using the scanning electron microscope. *Philosophical Magazine*, 24:973 – 979.

Clarke, D. R. and Howie, A. (1971). Calculations of lattice defect images for scanning electron microscopy. *Philosophical Magazine*, 24:959 – 971.

Coates, D. G. (1967). Kikuchi-like reflection patterns obtained with the scanning electron microscope. *Philosophical Magazine*, 16:1179 – 1184.

Cochran, W. (1973). Theory of electron micrographs of amorphous materials. *Physical Review B*, 8:623 – 629.

Cochran, W. and Dyer, H. B. (1952). Some practical applications of generalized crystal-structure projections. *Acta Crystallographica*, 5:634–636.

Cockayne, D. J. H. and Mackenzie, D. R. (1988). Electron diffraction analysis of polycrystalline and amorphous thin films. *Acta Crystallographica A*, 44:870–878.

Coene, W., Janssen, G., Op de Beeck, M., and Van Dyck, D. (1992). Phase retrieval through focus variation for ultra-resolution in field-emission transmission electron microscopy. *Physical Review Letters*, 69:3743–3746.

Coene, W., Janssen, G., Op de Beeck, M., and Van Dyck, D. (1993). Ultra-resolution on a FEG-TEM by phase retrieval through focus variation. In Craven, A. J., editor, *Electron Microscopy and Analysis Group Conference EMAG93*, volume 138 of *Institute of Physics Conference Series*, pages 225–230, Bristol and Philadelphia. Institute of Physics.

Cohen, P. I., Petrich, G. S., Pukite, P. R., Whaley, G. J., and Arrott, A. S. (1989). Birth-death models of epitaxy. I. Diffraction oscillations from low index surfaces. *Surface Science*, 216:222 – 248.

Cohen, P. I., Pukite, P. R., and Batra, S. (1987). Diffraction studies of epitaxy: elastic, inelastic and dynamic contributions to RHEED. In Farrow, R. F. C., Parkin, S. S. P., Dobson, P. J., Neave, J. H., and Arrott, A. S., editors, *Thin Film Growth Techniques for Low-Dimensional Structures*, pages 69 – 94, New York and London. Plenum Press.

Colella, R. (1972). N-beam dynamical diffraction of high-energy electrons at glancing incidence. General theory and computational methods. *Acta Crystallographica A*, 28:11–15.

Coulthard, M. A. (1967). A relativistic Hartree-Fock atomic field calculation. *Proceedings of the Physical Society*, 91:44–49.

Cowley, J. M. (1969). Image contrast in a transmission scanning electron microscope. *Applied Physics Letters*, 15:58–59.

Cowley, J. M. (1982). Energy losses of fast electrons at crystal surfaces. *Physical Review B*, 25:1401–1404.

Cowley, J. M. (1988). Electron microscopy of crystals with time-dependent perturbations. *Acta Crystallographica A*, 44:847–853.

Cowley, J. M. (1990). *Diffraction Physics*. North-Holland, Amsterdam, second edition.

Cowley, J. M. (1992). Scattering factors for the diffraction of electrons by crystalline solids. In Wilson, A. J. C., editor, *International Tables for Crystallography, Volume C*. Kluwer Academic Publishers, Dordrecht, Boston and London.

Cowley, J. M., editor (1993a). *Electron Diffraction Techniques*, Oxford. Oxford University Press.

Cowley, J. M. (1993b). Twenty forms of electron holography. *Ultramicroscopy*, 41:335–348.

Cowley, J. M. (2002). Electron nanodiffraction methods for measuring medium-range order. *Ultramicroscopy*, 90:197 – 206.

Cowley, J. M. and Moodie, A. F. (1957). The scattering of electrons by atoms and crystals — I. A new theoretical approach. *Acta Crystallographica A*, 10:609–619.

Cowley, J. M. and Moodie, A. F. (1962). The scattering of electrons by thin crystals. *Journal of the Physical Society of Japan*, 17(Supplement B-II):86–91.

Crescenzi, M. D. (1985). Extended energy loss fine structure: a new structural probe for surfaces and interfaces. *Surface Science*, 162:838–846.

Crook, G. E., Eyink, K. G., Campbell, A. C., Hinson, D. R., and Streetman, B. G. (1989). Effects of Kikuchi scattering on reflection high energy electron diffraction intensities during molecular-beam epitaxy GaAs growth. *Journal of Vacuum Science and Technology*, A7:2549 – 2553.

Czernuszka, J. T., Long, N. J., Boyes, E. D., and Hirsch, P. B. (1990). Imaging of dislocations using backscattering electrons in a scanning electron microscope. *Philosophical Magazine Letters*, 62:227 – 232.

Daberkow, I., Hermann, K. H., Liu, L., and Rao, W. D. (1991). Performance of electron image converters with YAG single-crystal screen and CCD sensor. *Ultramicroscopy*, 38:215–223.

Daimon, H. and Ino, S. (1989). One-dimensional circular diffraction patterns. *Surface Science*, 222:274 – 282.

Darwin, C. G. (1914). The theory of X-ray reflection. *Philosophical Magazine and Journal of Science*, 27:315–333 and 675–690.

Davisson, L. and Germer, L. H. (1927). The scattering of electrons by a single crystal of nickel. *Nature*, 119:558–560.

de Broglie, L. (1923a). Ondes et quanta. *Comptes Rendus des Séances de l'Académie des Sciences Paris*, 177:507 – 510.

de Broglie, L. (1923b). Quanta de lumière, diffraction et interférences. *Comptes Rendus des Séances de l'Académie des Sciences Paris*, 177:548 – 550.

de Broglie, L. (1924). A tentative theory of light quanta. *Philosophical Magazine*, 47:446 – 458.

de Ruijter, W. J. and Weiss, J. K. (1992). Methods to measure properties of slow-scan CCD cameras for electron detection. *Review of Scientific Instruments*, 63:4314–4321.

Dederichs, P. H. (1972). Dynamical diffraction theory by optical potential methods. In Ehrenreich, H., Seitz, F., and Turnbull, D., editors, *Solid State Physics*, volume 27, pages 125–236, New York and London. Academic Press.

Deininger, C., Necker, G., and Mayer, J. (1994). Determination of structure factors, lattice strains and accelerating voltage by energy-filtered convergent beam electron diffraction. *Ultramicroscopy*, 54:15–30.

Derlet, P. and Smith, A. E. (1997). Perturbative approach to high-energy electron surface resonance scattering. *Physical Review B*, 55:7170–7180.

Derlet, P. and Smith, A. E. (1999). Real-space Green's-function approach within RHEED. *Acta Crystallographica A*, 55:133–142.

Dingley, D. J. and Field, D. P. (1997). Electron backscatter diffraction and orientation imaging microscopy. *Materials Science and Technology*, 13:69–78.

Dingley, D. J. and Randle, V. (1992). Microtexture determination by electron back-scatter diffraction. *Journal of Materials Science*, 27:4545–4566.

Dirac, P. A. M. (1981). *The Principles of Quantum Mechanics*. Clarendon Press, Oxford, UK, 4th edition.

Djurisic, A. B. (1998). Elite genetic algorithms with adaptive mutations for solving continuous optimization problems — application to modeling of the optical constants of solid. *Optics Communications*, 151:147–159.

Dmitrienko, V. E. and Kaganer, V. M. (1990). Bloch X-ray scattering in a slightly distorted crystal. *Physics of Metals*, 9:102–110.

Dobson, P. J., Joyce, B. A., Neave, J. H., and Zhang, J. (1988). Intensity oscillations in reflection high energy electron diffraction during epitaxial growth. In Howie, A. and Valdre, U., editors, *Surface and Inteface Characterization by Electron Optical Methods*, pages 185 – 193, New York and London. Plenum Press.

Doyle, P. A. and Turner, P. S. (1968). Relativistic Hartree-Fock X-ray and electron scattering factors. *Acta Crystallographica A*, 24:390–397.

Dudarev, S. L. (1988). Quantum kinetic equation for inelastic scattering of particles by crystals: beyond the Born approximation. *Soviet Physics JETP*, 94:2337 – 2347.

Dudarev, S. L. (1990). Kossel effect in potential and resonance scattering of neutrons by single crystals. *Soviet Physics Crystallography*, 35:348–350.

Dudarev, S. L. (1997). Incoherent scattering of electrons by a crystal surface. *Micron*, 28:139–158.

Dudarev, S. L. (1999). Diffraction imaging using backscattered electrons: fundamentals and applications. In Hirsch, P. B., editor, *Topics in electron diffraction and microscopy of materials*, pages 126 – 153, Bristol and Philadelphia. Institute of Physics.

Dudarev, S. L., Botton, G. A., Savrasov, S. Y., Humphreys, C. J., and Sutton, A. P. (1998). Electron energy-loss spectra and the structural stability of nickel oxide: an LSDA+U study. *Physical Review B*, 57:1505–1509.

Dudarev, S. L. and Peng, L.-M. (1991). The origin of electron backscattering circular patterns. *Surface Science*, 244:L133 – L136.

Dudarev, S. L. and Peng, L.-M. (1993a). Effects of bulk resonance diffraction on inelastical scattering of high energy electrons by crystals. *Proceedings of the Royal Society of London A*, 440:117–133.

Dudarev, S. L. and Peng, L.-M. (1993b). Theory of bulk resonance diffraction in THEED. *Proceedings of the Royal Society of London A*, 440:95–115.

Dudarev, S. L., Peng, L.-M., Savrasov, S. Y., and Zuo, J. M. (2000). Correlation effects in the ground-state charge density of Mott insulating NiO: a comparison of *ab-initio* calculations and high-energy electron diffraction measurements. *Physical Review B*, 61:2506–2512.

Dudarev, S. L., Peng, L.-M., and Whelan, M. J. (1992a). On the damping of coherence in the small-angle inelastic scattering of high-energy electrons by crystals. *Physics Letters A*, 170:111–115.

Dudarev, S. L., Peng, L.-M., and Whelan, M. J. (1992b). A treatment of RHEED from a rough surface of a crystal by an optical potential method. *Surface Science*, 279:380 – 394.

Dudarev, S. L., Peng, L.-M., and Whelan, M. J. (1993a). Correlations in space and time and dynamical diffraction of high energy electrons by crystals. *Physical Review B*, 48:13408–13429.

Dudarev, S. L., Peng, L.-M., and Whelan, M. J. (1993b). Distorted wave approach to diffuse scattering in THEED and RHEED. *Ultramicroscopy*, 52:393–399.

Dudarev, S. L., Peng, L.-M., and Whelan, M. J. (1993c). The effect of the surface on thermal diffuse intensities in RHEED. *Proceedings of the Royal Society of London A*, 440:567–588.

Dudarev, S. L., Peng, L.-M., and Whelan, M. J. (1993d). On the damping of coherence in the small-angle inelastic scattering of high-energy electrons by crystals: errata. *Physics Letters A*, 175:465.

Dudarev, S. L., Peng, L.-M., and Whelan, M. J. (1995a). On the Doyle-Turner representation of the optical potential for RHEED calculations. *Surface Science*, 330:86–100.

Dudarev, S. L., Rez, P., and Whelan, M. J. (1995b). Theory of electron backscattering from crystals. *Physical Review B*, 51:3397 – 3412.

Dudarev, S. L. and Ryazanov, M. I. (1985). Effect of variation of the cross-section of elastic scattering on multiple scattering of high-energy electrons by crystals. *Soviet Physics JETP*, 62:972–975.

Dudarev, S. L. and Ryazanov, M. I. (1988). Multiple scattering theory for fast electrons in single crystals and Kikuchi patterns. *Acta Crystallographica A*, 44:51–61.

Dudarev, S. L., Vvedensky, D. D., and Whelan, M. J. (1994). Statistical treatment of dynamical electron diffraction from a growing surface. *Physical Review B*, 50:14525 – 14538.

Dudarev, S. L. and Whelan, M. J. (1994a). Analytical treatment of surface resonance diffraction of high-energy electrons. *Surface Science*, 310:373–389.

Dudarev, S. L. and Whelan, M. J. (1994b). Surface resonance scattering of high-energy electrons. *Physical Review Letters*, 72:1032–1035.

Dudarev, S. L. and Whelan, M. J. (1994c). Temperature dependence of elastic backscattering of electrons from polycrystalline solids. *Surface Science*, 311:L687–L694.

Dudarev, S. L. and Whelan, M. J. (1995). Interference between resonance and potential scattering of high-energy electrons from crystal surfaces. *Surface Science*, 340:293–308.

Dudarev, S. L. and Whelan, M. J. (1996). Resonance scattering of high energy electrons by a crystal surface. *International Journal of Modern Physics B*, 10:133–168.

Dudarev, S. L. and Whelan, M. J. (1997). The origin of resonance scattering in RHEED: Beyond the tight-binding model. *Acta Crystallographica A*, 53:63–73.

Duncumb, P. (1962). Enhanced X-ray emission from extinction contours in a single-crystal gold film. *Philosophical Magazine*, 7:2101 – 2105.

Durham, P. J. (1988). Theory of XANES. In Koningsberger, D. C. and Prins, R., editors, *X-ray Absorption*, pages 53–84, New York. John Wiley and Sons.

Edwards, S. (1958). A new method for the evaluation of electric conductivity in metals. *Philosophical Magazine*, 3:1020–1031.

Egerton, R. F. (1986). *Electron Energy Loss Spectroscopy in the Electron Microscope*. Plenum Press, New York.

Egerton, R. F. (1996). Improvement of the hydrogenic model to give more accurate values of K-shell ionization cross sections. *Ultramicroscopy*, 63:11–13.

Emslie, A. G. (1934). Scattering of electrons by stibnite and galena. *Physical Review*, 45:43–46.

Ewald, P. P. (1916a). Zur Begründung der Kristalloptik, Part I. *Annalen der Physik*, 49:1–38.

Ewald, P. P. (1916b). Zur Begründung der Kristalloptik, Part II. *Annalen der Physik*, 49:117–143.

Ewald, P. P. (1916c). Zur Begründung der Kristalloptik, Part III. *Annalen der Physik*, 49:519–597.

Ewald, P. P. (1958). Group velocity and phase velocity in X-ray crystal optics. *Acta Crystallographica*, 11:888–891.

Fathers, D. and Rez, P. (1979). A transport equation theory of electron backscattering. In Johari, O., editor, *Scanning Electron Microscopy*, volume I, pages 55 – 66, AMF O Hare, IL. SEM Inc.

Fathers, D. and Rez, P. (1984). A transport theory of electron scattering in solids. In Kyzer, D. F., Neidrig, H., Newbury, D. E., and Shimizu, R., editors, *Electron Beam Interaction with Solids*, pages 193 – 208, AMF O Hare, IL. SEM Inc.

Feshbach, H. (1958). Unified theory of nuclear reactions. *Annals of Physics*, 5:357–390.

Fox, A. G. and Fisher, R. M. (1986). Accurate structure factor determination and electron charge distributions of binary cubic solid solutions. *Philosophical Magazine A*, 53:815–832.

Friedrich, W., Knipping, P., and von Laue, M. (1912). Interferenz-Erscheinungen bei Röntgenstrahlen. *Sitzung der Bayrischen Akademie der Wissenschaften*, 8:363–373.

Frisch, U. (1968). Wave propagation in random media. In Bharucha-Reid, A. T., editor, *Probabilistic Methods in Applied Mathematics*, page 75, New York. Academic Press.

Fuggle, J. C. and Inglesfield, J. E., editors (1992). *Unoccupied Electronic States*, Berlin and London. Springer-Verlag.

Fujimoto, F. (1959). Dynamical theory of electron diffraction in Laue-case — I. General theory. *Journal of the Physical Society of Japan*, 14:1558–1568.

Fujiwara, K. (1959). Application of higher order Born approximation to multiple elastic scattering of electrons by crystals. *Journal of the Physical Society of Japan*, 14:1513–1524.

Fujiwara, K. (1961). Relativistic dynamical theory of electron diffraction. *Journal of the Physical Society of Japan*, 16:2226–2238.

Fulde, P. (1995). *Electron Correlations in Molecules and Solids*. Springer, Berlin, third edition.

Gajdardziska-Josifovska, M. and Cowley, J. M. (1991). Brillouin zones and Kikuchi lines for crystals under electron channelling conditions. *Acta Crystallographica A*, 47:74–82.

Gao, H. X. and Peng, L.-M. (1999). Parameterization of the temperature dependence of the Debye-Waller factors. *Acta Crystallographica A*, 55:926–932.

Gao, H. X. and Peng, L.-M. (2000). Debye-Waller factors of compounds with the cesium chloride structure. *Acta Crystallographica A*, 56:519–524.

Gao, H. X., Peng, L.-M., and Zuo, J. M. (1999). Lattice dynamics and Debye-Waller factors of some compounds with the sodium chloride structure. *Acta Crystallographica A*, 55:1014–1025.

Giacovazzo, C. (1992). *Fundamentals of Crystallography*. Oxford University Press, Oxford.

Gibson, J. M. and Treacy, M. M. J. (1997). Diminished medium-range order observed in annealed amorphous germanium. *Physical Review Letters*, 78:1074 – 1077.

Gibson, J. M. and Treacy, M. M. J. (1998). Fluctuation microscopy: Atomic order revealed in seemingly random electron speckle patterns. In Kirkland, A. and Brown, P. D., editors, *The Electron*, pages 212 – 221, Cambridge. Cambridge University Press.

Gibson, J. M., Treacy, M. M. J., and Voyles, P. M. (2000). Atom pair persistence in disordered materials from fluctuation microscopy. *Ultramicroscopy*, 83:169 – 178.

Gjønnes, J. (1972). A note on the integral equation formulation of dynamical theory. *Zeitschrift für Naturforschung*, 27A:434–436.

Gjønnes, J. and Moodie, A. F. (1965). Extinction conditions in the dynamical theory of electron diffraction. *Acta Crystallographica*, 19:65–73.

Gjønnes, K., Gjønnes, J., Zuo, J., and Spence, J. C. H. (1988). Two-beam features in electron diffraction patterns - applications to refinement of low-order structure factors in GaAs. *Acta Crystallographica A*, 44:810–820.

Glauber, R. and Schomaker, V. (1953). The theory of electron diffraction. *Physical Review*, 89:667–671.

Goldberger, M. L. and Watson, K. M. (1964). *Collision Theory*. John Wiley & Sons, New York and London.

Gomoyunova, M. V., Dudarev, S. L., and Pronin, I. I. (1990). Incident beam diffraction effects in Auger electron emission from crystal surfaces. *Surface Science*, 235:156 – 168.

Gomoyunova, M. V., Zaslavskii, S. L., and Pronin, I. I. (1982). Anisotropy of interaction of intermediate energy electrons with single crystals of transition metals. *Soviet Physics - Solid State*, 24:221 – 224.

Goodman, J. W. (1968). *Introduction to Fourier Optics*. McGraw Hill, San Francisco.

Goodman, P. (1975). A practical method of three-dimensional space-group analysis using convergent-beam electron diffraction. *Acta Crystallographica A*, 31:126–133.

Goodman, P. and Moodie, A. F. (1974). Numerical evaluation of N-beam wave functions in electron scattering by the multi-slice method. *Acta Crystallographica A*, 30:280–290.

Gopi, N. and Sirdeshmukh, D. B. (1998). Compilation of temperature factors of hexagonal close packed elements. *Acta Crystallographica A*, 54:513–514.

Gorodnichev, E. E., Dudarev, S. L., and Rogozkin, D. B. (1990). Coherent backscattering of waves from a random medium: an exact solution of the albedo problem. *Physics Letters*, A144:48–54.

Gorodnichev, E. E., Dudarev, S. L., Rogozkin, D. B., and Ryazanov, M. I. (1987). Coherent effects in backscattering of waves from a randomly inhomogeneous medium. *Soviet Physics JETP*, 66:938–944.

Gorodnichev, E. E., Dudarev, S. L., Rogozkin, D. B., and Ryazanov, M. I. (1988). The origin of anomalous reflection of X-rays from a rough surface. *JETP Letters*, 48:147 – 150.

Gotsis, H. J. and Maksym, P. A. (1997). Investigation of multislice RHEED calculations. *Surface Science*, 385:15–23.

Grant, I. P. (1961). Relativistic self-consistent fields. *Proceedings of the Royal Society of London A*, 262:555–576.

Grant, I. P., McKenzie, B. J., Norrington, P. H., Mayers, D. F., and Pyper, N. C. (1980). An atomic multiconfigurational Dirac-Fock package. *Computational Physics Communications*, 21:207–231.

Gunning, J. and Goodman, P. (1992). Reciprocity in electron diffraction. *Acta Crystallographica A*, 48:591–595.

Hagemann, P. and Reimer, L. (1979). An experimental proof of the dependent Bloch wave model by large-angle scattering from thin crystals. *Philosophical Magazine*, A40:367 – 375.

Hahn, T., editor (1983). *International Tables for Crystallography, Volume A*, Dordrecht, the Netherlands. D. Reidel Publishing Company.

Hall, C. R. (1966). On the production of characteristic X-rays in thin metal crystals. *Proceedings of the Royal Society of London A*, 295:140 – 163.

Hall, C. R. (1970). On the thickness dependence of Kikuchi band contrast. *Philosophical Magazine*, 22:63 – 72.

Hall, C. R. and Hirsch, P. B. (1965a). Effect of thermal diffuse scattering on propagation of high energy electron through crystals. *Proceedings of the Royal Society of London A*, 286:158–177.

Hall, C. R. and Hirsch, P. B. (1965b). The effect of weak Bragg reflected beams on the absorption of electrons. *Philosophical Magazine*, 12:539–545.

Hall, C. R., Hirsch, P. B., and Booker, G. R. (1966). The effect of point defects on absorption of high-energy electrons passing through crystals. *Philosophical Magazine*, 14:979–989.

Hamada, T. and Fujita, F. E. (1986). Calculation of high-resolution electron microscopic images of crystalline embryos in amorphous metals. *Japanese Journal of Applied Physics*, 25:318 – 327.

Han, F. S., Fan, H. F., and Li, F. H. (1986). Image processing in high-resolution electron microscopy using the direct method – II. Image deconvolution. *Acta Crystallographica A*, 42:353–356.

Harris, J. J., Joyce, B. A., and Dobson, P. J. (1981). Oscillations in the surface structure of Sn-doped GaAs during growth by MBE. *Surface Science*, 103:L90–L96.

Hashimoto, H., Howie, A., and Whelan, M. J. (1962). Anomalous electron absorption effects in metal foils: theory and comparison with experiment. *Proceedings of the Royal Society of London A*, 269:80–103.

Hayakawa, K. and Miyake, S. (1974). Low-energy electron diffractometry from the cleavage face (001) of magnesium oxide. *Acta Crystallographica A*, 30:374–380.

Henderson, R., Baldwin, J. M., Downing, K. H., Lepault, J., and Zemlin, F. (1986). Structure of purple membrane from halobacterium-halobium - recording, measurement and evaluation of electron-micrographs at 3.5 Å resolution. *Ultramicroscopy*, 19:147–178.

Herrmann, K. H. and Krahl, D. (1984). Electronic image recording in conventional electron microscopy. *Advances in Optical and Electron Microscopy*, 9:1–64.

Hirsch, P. and Humphreys, C. J. (1970). The dynamical theory of scanning electron microscope channelling patterns. In Johari, O., editor, *Scanning Electron Microscopy*, pages 451 – 455, Chicago, IL. IIT Research Institute.

Hirsch, P. B., Howie, A., Nicholson, R. B., Pashley, D. W., and Whelan, M. J. (1977). *Electron Microscopy of Thin Crystals*. Krieger Publishing Company, Malabar and Florida, second edition.

Hirsch, P. B., Howie, A., and Whelan, M. J. (1960). A kinematical theory of diffraction contrast of electron transmission microscope images of dislocations and other defects. *Proceedings of the Royal Society of London A*, 252:499–529.

Hirsch, P. B., Howie, A., and Whelan, M. J. (1962). On the production of X-rays in thin metal foils. *Philosophical Magazine*, 7:2095 – 2100.

Hirsch, P. B. and Kellar, J. N. (1952). A study of cold-worked aluminium by an X-ray microbeam technique. I. Measurement of particle volume and misorientations. *Acta Crystallographica*, 5:162 – 167.

Hodgson, P. E. (1963). *The Optical Model of Elastic Scattering*. Oxford University Press, New York.

Hoerni, J. A. and Ibers, J. A. (1953). Complex amplitude for electron scattering by atoms. *Physical Review*, 91:1182–1185.

Høier, R. (1973). Multiple scattering and dynamical effects in diffuse electron scattering. *Acta Crystallographica A*, 29:663–672.

Høier, R., Bakken, L. N., Marthinsen, K., and Holmestad, R. (1993). Structure factor determination in non-centrosymmetric crystals by a two-dimensional CBED-based parameter refinement method. *Ultramicroscopy*, 49:159–170.

Holland, J. H. (1975). *Adaptation in Natural and Artificial Systems*. The University of Michigan Press, Ann Arbor.

Holmes, D. M., Sudijono, J. L., McConville, C. F., Jones, T. S., and Joyce, B. A. (1997). Direct evidence for the step density model in the initial stages of the layer-by-layer homoepitaxial growth of GaAs(111)A. *Surface Science*, 370: L173 – L178.

Horio, Y. (1996). Zero-loss reflection high-energy electron diffraction patterns and rocking curves of the Si(111)7×7 surface obtained by energy filtering. *Japanese Journal of Applied Physics*, 35:3559 – 3564.

Horio, Y. (1998). Fundamental relation between wave fields, rocking curves, and anomalous absorption for the RHEED of Si(111) crystals. *Physical Review B*, 57:4736–4746.

Horio, Y. and Hashimoto, Y. (1997). Observations of vicinal Si(111)7×7 surface by energy-filtered reflection high-energy electron diffraction. *Japanese Journal of Applied Physics*, 36:L808 – L810.

Horio, Y. and Ichimiya, A. (1983). Intensity anomalies of Auger electron signals observed by incident beam rocking method for Si(111) 7×7 and Si(111)$\sqrt{3}$ × $\sqrt{3}$Ag surfaces. *Physica B*, 117 & 118:792–794.

Horio, Y. and Ichimiya, A. (1993). Dynamical diffraction effect for RHEED intensity oscillations: phase shift of oscillations for glancing angles. *Surface Science*, 298:261 – 272.

Horio, Y. and Ichimiya, A. (1994a). Origin of phase shift phenomena in RHEED intensity oscillation curves. *Ultramicroscopy*, 55:321–328.

Horio, Y. and Ichimiya, A. (1994b). Phase shift and frequency doubling in intensity oscillations of RHEED: one-beam dynamical calculations for Ge on Ge(111) surface. *Japanese Journal of Applied Physics*, 33:L377 – L379.

Howie, A. (1962). Discussion of K. Fujiwara's paper by M.J. Whelan. *Journal of the Physical Society of Japan*, 17(Supplement BII):118.

Howie, A. (1963). Inelastic scattering of electrons by crystals I. The theory of small-angle inelastic scattering. *Proceedings of the Royal Society of London A*, 271:268–287.

Howie, A. (1974). Theory of diffraction contrast effects in the scanning electron microscope. In Holt, D. B., Muir, M. D., Grant, P. R., and Boswarva, I. M., editors, *Quantitative Scanning Electron Microscopy*, pages 183 – 211, London and New York. Academic Press.

Howie, A., Krivanek, O. L., and Rudee, M. L. (1973). Interpretation of electron micrographs and diffraction patterns of amorphous materials. *Philosophical Magazine*, 27:235 – 255.

Howie, A. and Whelan, M. J. (1961). Diffraction contrast of electron microscope images of crystal lattice defects — II. The development of a dynamical theory. *Proceedings of the Royal Society (London)*, 263:217–237.

Hsu, T. and Peng, L.-M. (1986). Experimental studies of atomic step contrast in reflection electron microscopy (REM). *Journal of Metals*, 38:30.

Hsu, T. and Peng, L.-M. (1987). Experimental studies of atomic step contrast in reflection electron microscopy (REM). *Ultramicroscopy*, 22:217–224.

Hu, G., Yu, Q. F., Lu, H. Q., and Peng, L.-M. (2000a). SYMCBED – a Windows program package for automatic identification of CBED pattern symmetry. In *Proceedings of the International Kunming Symposium on Microscopy*, pages 47–48. Chinese Electron Microscopy Society.

Hu, G. B., Peng, L.-M., Yu, Q. F., and Lu, H. Q. (2000b). Automated identification of symmetry in CBED patterns: a genetical approach. *Ultramicroscopy*, 84:47–56.

Hu, J. J. and Li, F. H. (1991). Maximum entropy image deconvolution in high resolution electron microscopy. *Ultramicroscopy*, 35:339–350.

Huang, K. (1947). X-ray reflections from dilute solid solutions. *Proceedings of the Royal Society of London A*, 190:102–117.

Huang, X. and Ximen, J. (1991). Discussions on the multiple-input maximum a-posteriori wave-function restoration method in high-resolution electron microscopy. *Journal of Electron Microscopy Technique*, 17:344–350.

Hüfner, S. (1994). Electronic structure of NiO and related 3d-transition metal compounds. *Advances in Physics*, 43:183–356.

Hull, R., Moore, M., Bahnck, D., Geva, M., Karlicek, R. F., Stevie, F. A., and Walker, J. F. (1995a). Observation of strong transmision electron microscope contrast from doped layers in InP-based structures. In Cullis, A. G. and Staton-Bevan, A. E., editors, *Microscopy of Semiconducting Materials*, volume 146 of *Institute of Physics Conference Series*, pages 613–616, Bristol and Philadelphia. Institute of Physics.

Hull, R., Stevie, F. A., and Bahnck, D. (1995b). Observation of strong contrast from doping variations in transmission electron microscopy of InP-based semiconductor-laser diodes. *Applied Physics Letters*, 66:341–343.

Humphreys, C. J. (1979). The scattering of fast electrons by crystals. *Reports on Progress in Physics*, 42:1825–1887.

Humphreys, C. J. and Hirsch, P. B. (1968). Absorption parameters in electron diffraction theory. *Philosophical Magazine*, 18:115–122.

Ibers, J. A. and Hoerni, J. A. (1954). Atomic scattering amplitudes for electron diffraction. *Acta Crystallographica*, 7:405–408.

Ichimaru, A. (1978). *Wave Propagation and Scattering in Random Media*. Academic Press, New York.

Ichimiya, A. (1983). Many-beam calculation of reflection high energy electron diffraction (RHEED) intensities by the multi-slice method. *Japanese Journal of Applied Physics*, 22:176–180.

Ichimiya, A. (1987). RHEED intensity analysis of Si(111) 7×7 at one-beam condition. *Surface Science*, 192:L893–L898.

Ichimiya, A., Kambe, K., and Lehmpfuhl, G. (1980). Observation of surface state resonance effect by the convergent beam RHEED technique. *Journal of the Physical Society of Japan*, 49:684–688.

Ichinokawa, T., Nishimura, M., and Wada, H. (1974). Contrast reversal of pseudo-Kikuchi bands and lines due to detector position in scanning electron microscopy. *Journal of the Physical Society of Japan*, 36:221 – 226.

Ignatovich, V. K. (1986). Étude on the one-dimensional periodic potential. *Soviet Physics Uspekhi*, 29:880–887.

Ishibashi, Y. (1991). Intensity oscillation of RHEED by a polynuclear growth model. *Journal of the Physical Society of Japan*, 60:3215 – 3217.

Ishizuka, K. (1982). Multislice formulation for inclined illumination. *Acta Crystallographica A*, 38:773–779.

Ishizuka, K. (1993). Analysis of electron image detection efficiency of slow-scan CCD cameras. *Ultramicroscopy*, 52:7–20.

Iwai, T., Voyles, P. M., Gibson, J. M., and Oono, Y. (1999). Method for detecting subtle spatial structures by fluctuation microscopy. *Physical Review B*, 60:191 – 200.

James, R., Bird, D. M., and Wright, A. G. (1989). Surface resonance curves in transmission and reflection diffraction. In Goodhew, P. J. and Elder, H. Y., editors, *EMAG-MICRO 89*, volume 98 of *Institute of Physics Conference Series*, pages 111–114, Bristol and New York. Institute of Physics.

James, R., Bird, D. M., and Wright, A. G. (1994). Smooth parabolas in transmission electron diffraction patterns. *Acta Crystallographica A*, 50:357–366.

James, R. W. (1948). *The Optical Principles of the Diffraction of X-rays*. Bell, London.

Jeffreys, H. and Jeffreys, B. S. (1950). *Methods of Mathematical Physics*. Cambridge University Press, Cambridge, England, second edition. p. 493.

Jenkins, M. L. and Kirk, M. (2001). *Characterization of Radiation Damage by Transmission Electron Microscopy*. Institute of Physics, Bristol and Philadelphia.

Jiang, H., Li, F. H., and Mao, Z. Q. (1999). Electron crystallographic study of $Bi_2(Sr_{0.9}La_{0.1})_2CoO_y$. *Micron*, 30:417–424.

Jong, K. A. D. (1975). *An Analysis of the Behaviour of a Class of Genetic Adaptive Systems*. PhD thesis, University of Michigan.

Josefsson, T. W., Allen, L. J., Miller, P. R., and Rossouw, C. J. (1994). K-shell ionization under zone-axis electron diffraction conditions. *Physical Review B*, 50:6673 – 6684.

Josefsson, T. W. E. (1993). *A Dielectric Matrix Calculation of the Electron Inelastic Scattering Potential for Silicon and Gallium Arsenide*. PhD thesis, University of Melbourne, Melbourne, Victoria, Australia.

Joy, D. C., Newbury, D. E., and Davidson, D. L. (1982). Electron channelling patterns in the scanning electron microscope. *Journal of Applied Physics*, 53:R81 – R122.

Joyce, B. A., Neave, J. H., Zhang, J., Dobson, P. J., Dawson, P., Moore, K. J., and Foxon, C. T. (1987). Dynamic RHEED techniques and interface quality in MBE-grown GaAs/(Al,Ga)As structures. In Farrow, R. F. C., Parkin, S. S. P., Dobson, P. J., Neave, J. H., and Arrott, A. S., editors, *Thin Film Growth Techniques for Low-Dimensional Structures*, pages 19 – 35, New York and London. Plenum Press.

Kagan, Y. and Kononets, Y. V. (1973). The theory of the channelling effect: the effect of inelastic collisions. *Soviet Physics JETP*, 37:530 – 540.

Kagan, Y. M. and Kononets, Y. V. (1970). The theory of the channelling effect. *Soviet Physics JETP*, 31:124–134.

Kainuma, Y. (1955). The theory of Kikuchi patterns. *Acta Crystallographica*, 8:247 – 257.

Kainuma, Y., Kashiwase, Y., and Kogiso, M. (1955). Thermal diffuse scattering in electron diffraction: I. General theory. *Journal of the Physical Society of Japan*, 40:1707 – 1712.

Kambe, K. (1967). Theory of electron diffraction by crystals. *Zeitschrift für Naturforschung*, 22a:422–431.

Kambe, K. (1988). Linearization of the basic equations of the dynamical theory of electron diffraction by crystals. *Acta Crystallographica A*, 44:885–890.

Kanamori, J. and Kotani, A., editors (1988). *Core-Level Spectroscopy in Condensed Systems*, volume 81 of *Springer Series in Solid State Sciences*, Berlin and London. Springer Verlag.

Kato, N. (1958). The flow of X-rays and material waves in ideally perfect single crystals. *Acta Crystallographica*, 11:885–887.

Kawamura, T., Ichimiya, A., and Maksym, P. A. (1988). Comparison of RHEED dynamical calculation methods. *Japanese Journal of Applied Physics*, 27:1098–1099.

Kellar, J. N., Hirsch, P. B., and Thorp, J. S. (1950). An X-ray microbeam examination of a plastically deformed metal. *Nature*, 165:554 – 556.

Kikuchi, S. and Nakagawa, S. (1933). Die anomale Reflexion der schnellen Elektronen an die Einkristalloberfläche. *Scientific Papers of the Institute of Physical and Chemical Research*, 21:256–265.

Kirkland, E. J. (1984). Improved high resolution image processing of bright field electron micrographs — I. Theory. *Ultramicroscopy*, 15:151–172.

Kohl, H. and Rose, H. (1985). Theory of image formation by inelastically scattered electrons in the electron microscope. In Hawkes, P., editor, *Advances in Electronics and Electron Physics*, volume 65, pages 173–227. Academic Press, New York.

Kohra, K., Moliére, K., Nakano, S., and Ariyama, M. (1962). Anomalous intensity of mirror reflection from the surface of a single crystal. *Journal of the Physical Society of Japan*, 17(Supplement B-II):82–85.

Korte, U. and Maksym, P. A. (1997). Role of the step density in reflection high-energy electron diffraction: questioning the step density model. *Physical Review Letters*, 78:2381–2384.

Korte, U., McCoy, J. M., Maksym, P. A., and Meyer-Ehmsen, G. (1996). Perturbation theory of diffuse RHEED applied to rough surfaces: comparison with supercell calculations. *Physical Review B*, 54:2121 – 2137.

Krishna, P. G., Subhadra, K. G., and Sireshmukh, D. B. (1998). X-ray determination of Debye-Waller factors of NaBr and NaI. *Acta Cryst. A*, 54:253–253.

Krivanek, O. L., Gaskell, P. H., and Howie, A. (1976). Seeing order in amorphous materials. *Nature*, 262:454 – 457.

Krivanek, O. L. and Howie, A. (1975). Kinematic theory of images from polycrystalline and random-network structures. *Journal of Applied Crystallography*, 8:213 – 219.

Krivanek, O. L. and Mooney, P. E. (1993). Applications of slow-scan CCD cameras in transmission electron microscopy. *Ultramicroscopy*, 49:95–108.

Krivanek, O. L., Mooney, P. E., Fan, G. Y., Leber, M. L., and Meyer, C. E. (1991). Slow-scan CCD cameras for transmission electron microscopy. In *Electron Microscopy and Analysis Group Conference EMAG91*, volume 119 of *Institute of Physics Conference Series*, pages 523–526, Bristol and London. Institute of Physics.

Lamla, V. E. (1938a). Zur Theorie der Electronenbeugung bei Berücksichtigung von mehr als 2 Strahlen und zur Erklärung der Kikuchi-Enveloppen. I. *Annalen der Physik*, 32:178–189.

Lamla, V. E. (1938b). Zur Theorie der Electronenbeugung bei Berücksichtigung von mehr als 2 Strahlen und zur Erklärung der Kikuchi-Enveloppen. II. *Annalen der Physik*, 32:225–241.

Landau, L. D. and Lifshitz, E. M. (1977). *Quantum Mechanics: Non-relativistic Theory*. Pergamon Press, Oxford, third edition.

Landau, L. D. and Lifshitz, E. M. (1993). *Statistical Physics, Part 1*. Pergamon Press, Oxford, third edition.

Landauer, R. and Martin, T. (1994). Barrier interaction time in tunneling. *Reviews of Modern Physics*, 66:217–228.

Larsen, P. K. and Meyer-Ehmsen, G. (1990). Influence of surface disorder on RHEED patterns from GaAs(001) 2×4 surfaces. *Surface Science*, 240:168 – 180.

Larsen, P. K., Meyer-Ehmsen, G., Bolger, B., and Hoeven, A.-J. (1987). Surface disorder induced Kikuchi features in reflection high-energy electron diffraction patterns of static and growing GaAs(001) films. *Journal of Vacuum Science and Technology A*, 5:611 – 614.

Lawrence, J. L. (1973). Debye-Waller factors for magnesium oxide. *Acta Crystallographica A*, 29:94–95.

Leapman, R. D. and Cosslett, V. E. (1976). Extended fine structure above the X-ray edge in electron energy loss spectra. *Journal of Physics D: Applied Physics*, 9:L29–L32.

Lehmpfuhl, G. and Dowell, W. C. T. (1986). Convergent-beam reflection high-energy electron diffraction (RHEED) observations from an Si(111) surface. *Acta Crystallographica A*, 42:569–577.

Lehmpfuhl, G., Ichimiya, A., and Nakahara, H. (1991). Interpretation of RHEED oscillations during MBE growth. *Surface Science*, 245:L159 – L162.

Lent, C. S. and Cohen, P. I. (1984). Diffraction from stepped surfaces. 1. Reversible surfaces. *Surface Science*, 139:121 – 154.

Lent, C. S. and Cohen, P. I. (1986). Quantitative analysis of streaks in reflection high-energy electron diffraction: GaAs and AlAs deposited on GaAs(001). *Physical Review B*, 33:8329 – 8335.

Lewis, A. L., Villagrana, R. F., and Metherall, A. J. F. (1978). A description of electron diffraction from higher order Laue zones. *Acta Crystallographica A*, 34:138–140.

Li, F. H. (1994). Two stages image processing in high resolution electron microscopy. In Jouffrey, B. and Colliex, C., editors, *Electron Microscopy 1994*, volume 1, pages 481–484. Electron Microscopy Society, France.

Lichte, H. (1992). Electron holography I – Can electron holography reach 0.1nm resolution? *Ultramicroscopy*, 47:223–230.

Lifshitz, E. M. and Pitaevskii, L. P. (1979). *Physical Kinetics*. Pergamon, Oxford.

Lovesey, S. W. (1984). *Theory of Neutron Scattering from Condensed Matter*, volume 1. Oxford University Press, Oxford.

Lu, P., Liu, J., and Cowley, J. M. (1991). Theoretical and experimental studies of electron resonance effects in RHEED. *Acta Crystallographica A*, 47:317–327.

Lynch, D. F. and Moodie, A. F. (1972). Numerical evaluation of low energy electron diffraction intensities -I. The perfect crystal with no upper layer lines and no absorption. *Surface Science*, 32:422–438.

Lynch, D. F. and Smith, A. E. (1983). A scattering matrix calculation of very low energy electron reflection intensities for barium and magnesium oxides. *physica status solidi b*, 119:355–360.

Ma, Y. (1991). A computational method for obtaining stationary solutions in RHEED and REM. *Acta Crystallographica A*, 47:137–139.

MacGillavry, C. H. (1940). Zur Prufung der dynamischen Theorie der Elektronenbeugung am Kristallgitter. *Physica*, 7:329–343.

Mahan, G. D. (1990). *Many-Particle Physics*. Plenum Press, New York and London.

Maksym, P. A. and Beeby, J. L. (1981). A theory of RHEED. *Surface Science*, 110:423–438.

Maksym, P. A., Korte, U., McCoy, J. M., and Gotsis, H. J. (1998). Calculation of RHEED intensities for imperfect surfaces. *Surface Review and Letters*, 5:873 – 880.

Marks, L. D. and Ma, Y. (1988). Current flow in reflection electron microscopy and RHEED. *Acta Crystallographica A*, 44:392–393.

Marshall, W. and Lovesey, S. W. (1971). *Theory of Thermal Neutron Scattering*. Clarendon Press, Oxford.

Marten, H. and Meyer-Ehmsen, G. (1985). Resonance effects in RHEED from Pt(111). *Surface Science*, 151:570–584.

Marten, H. and Meyer-Ehmsen, G. (1988). Effect of RHEED resonances on secondary and Auger emission of Pt(111) surfaces. *Acta Crystallographica A*, 44:853–857.

Marthinsen, K., Holmestad, R., and Høier, R. (1994). Analytical filtering of low-angle inelastic scattering contributions to CBED contrast. *Ultramicroscopy*, 55:268–275.

Maslen, V. W. (1983). Analytical angular integration of a product of hydrogenic bound-free first Born matrix elements. *Journal of Physics B*, 16:2065–2069.

Maslen, V. W. and Rossouw, C. J. (1983). The inelastic scattering matrix element and its application to electron energy loss spectroscopy. *Philosophical Magazine A*, 47:119–130.

Maslen, V. W. and Rossouw, C. J. (1984). Implications of (e,2e) scattering for inelastic electron diffraction in crystals. 1. theoretical. *Philosophical Magazine A*, 49:735–742.

Masuda, T., Yamamoto, K., and Yamada, H. (1993). Detection of partial symmetry using correlation with rotated-reflected images. *Pattern Recognition*, 26:1245–1253.

McRae, E. G. (1979). Electronic surface resonances of crystals. *Reviews of Modern Physics*, 51:541–568.

Messiah, A. (1972). *Quantum Mechanics.* North-Holland, Amsterdam.

Meyer-Ehmsen, G. (1988). Resonance effects in RHEED. In Larsen, P. K. and Dobson, P. J., editors, *Reflection High-Energy Electron Diffraction and Reflection Electron Imaging of Surfaces*, volume 188 of *NATO Advance Study Institute Series, series B: Physics*, pages 99–107, New York. Plenum Press.

Meyer-Ehmsen, G. (1998). Real-space dynamical calculation of diffuse RHEED intensities from disordered surfaces. *Surface Science*, 395:L189 – L195.

Miller, P. D. and Gibson, J. M. (1998). Connecting small-angle diffraction with real-space images by quantitative transmission electron microscopy of amorphous thin films. *Ultramicroscopy*, 74:221 – 235.

Mitsuishi, K., Hashimoto, I., Sakamoto, K., Sakamoto, T., and Watanabe, K. (1995). Mechanism of reflection high-energy electron diffraction intensity oscillations during molecular beam epitaxy on a Si(001) surface. *Physical Review B*, 52:10748 – 10751.

Mitura, Z., Daniluk, A., Strozak, M., Jalochowski, M., Smal, A., and Subotovwicz, M. (1991). Analysis of shapes of RHEED intensity oscillations observed for growing films. *Acta Physica Polonica A*, 80:365 – 368.

Mitura, Z., Dudarev, S. L., Peng, L.-M., Gladyszewski, G., and Whelan, M. J. (2002). The small terrace size approximation in the theory of RHEED oscillations. *Journal of Crystal Growth*, 235:79 – 88.

Mitura, Z., Dudarev, S. L., and Whelan, M. J. (1998). Phase of RHEED oscillations. *Physical Review B*, 57:6309 – 6312.

Mitura, Z., Dudarev, S. L., and Whelan, M. J. (1999). Interpretation of RHEED oscillation phase. *Journal of Crystal Growth*, 198/199:905 – 910.

Mitura, Z., Strozak, M., and Jalochowski, M. (1992). RHEED intensity oscillations with extra maxima. *Surface Science*, 276:L15–L18.

Miyake, S. and Hayakawa, K. (1970). Resonance effects in low and high energy electron diffraction by crystals. *Acta Crystallographica A*, 26:60–70.

Miyake, S., Kohra, K., and Takagi, M. (1954). The nature of the specular reflection of electrons from a crystal surface. *Acta Crystallographica*, 7:393–401.

Mooney, P. E., Fan, G. Y., Truong, K. V., Bui, D. B., and Krivanek, O. L. (1990). Slow-scan CCD camera for transmission electron microscopy. In *Proceedings of the XIIth International Congress on Electron Microscopy*, volume 1, pages 164–165, San Francisco. San Francisco Press.

Morin, P., Pitaval, M., Bernard, D., and Fontaine, G. (1979). Electron channelling imaging in scanning electron microscopy. *Philosophical Magazine A*, 40:511 – 524.

Mott, N. F. and Massey, H. S. W. (1965). *The Theory of Atomic Collisions*. Clarendon Press, Oxford.

Müller, K. A., Berlinger, W., and Waldner, F. (1968). Characteristic structural phase transition in perovskite-type compounds. *Physical Review Letters*, 21:814–817.

Mulvey, T. (1967). The history of the electron microscope. *Proceedings of the Royal Microscopical Society*, 2:201 – 227.

NAG93 (1993). *The NAG Fortran Library Manual, Mark 16*. The numerical algorithms group limited, Oxford, UK.

Neave, J. H., Dobson, P. J., Joyce, B. A., and Zhang, J. (1985). Reflection high-energy electron diffraction oscillations from vicinal surfaces - a new approach to surface diffusion measurements. *Applied Physics Letters*, 47:100 – 102.

Neave, J. H., Joyce, B. A., Dobson, P. J., and Norton, N. (1983). Dynamics of film growth of GaAs by MBE from RHEED observations. *Applied Physics A*, 31:1–8.

Newton, R. G. (1966). *Scattering Theory of Waves and Particles*. McGraw-Hill, New York.

Ochkur, V. I. (1964). The Born-Oppenheimer method in the theory of atomic collisions. *Soviet Physics JETP*, 18:503–508.

Ogura, N., Yoshida, K., Kojima, Y., and Saito, H. (1994). Development of the 25-micron pixel imaging plate system for TEM. In *Proceedings of the 13th International Congress on Electron Microscopy*, volume 1, pages 219–220. Societe Francaise de Microscopie Electronique.

Ohtsuki, Y. H. (1970). Theory of low energy electron diffraction. III. Inelastic scattering. *Journal of the Physical Society of Japan*, 29:398–403.

Op de Beeck, M., Van Dyck, D., and Coene, W. (1996). Wave function reconstruction in HRTEM: the parabola method. *Ultramicroscopy*, 64:167–183.

Oxley, M. P. and Allen, L. J. (2000). Atomic scattering factors for K-shell and L-shell ionization by fast electrons. *Acta Crystallographica A*, 56:470–490.

Oxley, M. P. and Allen, L. J. (2001). Atomic scattering factors for K-shell electron energy loss spectroscopy. *Acta Crystallographica A*, 57:713–728.

Parr, R. G. and Yang, W. (1989). *Density-Functional Theory of Atoms and Molecules*. Oxford University Press, Oxford.

Pauling, L. (1960). *The Nature of the Chemical Bond and the Structure of Molecules and Crystals*. Cornell University Press, Ithaca, third edition.

Pauling, L. and Brockway, L. O. (1935). Radial distribution method of interpretation of electron diffraction photographs of gas molecules. *Journal of the American Chemical Society*, 57:2684–2692.

Pendry, J. B. (1974). *Low Energy Electron Diffraction.* Academic Press, New York.

Pendry, J. B., Heinz, K., and Oed, W. (1988). Direct methods in surface crystallography. *Physical Review Letters*, 61:2953–2956.

Peng, L.-M. (1994). Bloch wave origin of surface resonance scattering in RHEED. *Surface Science*, 316:1049–1054.

Peng, L.-M. (1995). New developments of electron diffraction theory. In Hawkes, P. W., editor, *Advances in Imaging and Electron Physics*, volume 90, pages 205–351. Academic Press, London.

Peng, L.-M. (1996). Sampling theorem and digital microscopy. *Progress in Natural Sciences*, 7:110–114.

Peng, L.-M. (1997). Anisotropic thermal vibrations and dynamical electron diffraction by crystals. *Acta Crystallographica A*, 53:663–672.

Peng, L.-M. (1998). Parameterization of electron scattering factors of ions. *Acta Crystallographica A*, 54:481–485.

Peng, L.-M. (1999). Electron atomic scattering factors and scattering potential of crystals. *Micron*, 30:625–649.

Peng, L.-M. (2000). Kinetic equation, optical potential, tensor theory and structure factor refinement in high energy electron diffraction. In Mezey, P. G. and Robertson, B. E., editors, *Electron, Spin and Momentum Densities and Chemical Reactivity*, Dordrecht. Kluwer Academic Publishers.

Peng, L.-M. and Cowley, J. M. (1986). Dynamical diffraction calculations for RHEED and REM. *Acta Crystallographica A*, 42:545–552.

Peng, L.-M. and Cowley, J. M. (1988a). Errors arising from numerical use of the Mott formula in electron image simulation. *Acta Crystallographica A*, 44:1–5.

Peng, L.-M. and Cowley, J. M. (1988b). Experimental studies of surface resonance scattering processes in RHEED. *Surface Science*, 201:559–572.

Peng, L.-M. and Cowley, J. M. (1988c). A multislice approach to the RHEED and REM calculation. *Surface Science*, 199:609–622.

Peng, L.-M. and Cowley, J. M. (1988d). Surface resonance effects and beam convergence in REM. *Ultramicroscopy*, 26:161–168.

Peng, L.-M. and Cowley, J. M. (1989). Thermal diffuse scattering and REM image contrast preservation. *Ultramicroscopy*, 29:168–174.

Peng, L.-M., Cowley, J. M., and Yao, N. (1988). The observations of surface resonance effects in RHEED patterns. *Ultramicroscopy*, 26:189–194.

Peng, L.-M. and Dudarev, S. L. (1993a). Direct determination of crystal and surface structures in THEED. *Ultramicroscopy*, 52:312–317.

Peng, L.-M. and Dudarev, S. L. (1993b). Tensor theories of high energy electron diffraction and their use in surface crystallography. *Surface Science*, 298:316–330.

Peng, L.-M., Dudarev, S. L., and Whelan, M. J. (1993). Evidence for the damping of coherence in inelastic scattering of high-energy electrons by crystals. *Physics Letters A*, 175:461–464.

Peng, L.-M., Dudarev, S. L., and Whelan, M. J. (1996a). Approximate methods in dynamical RHEED calculations. *Acta Crystallographica A*, 52:909–922.

Peng, L.-M., Dudarev, S. L., and Whelan, M. J. (1996b). Bethe potentials in dynamical RHEED calculations. *Surface Science*, 351:L245–L252.

Peng, L.-M., Dudarev, S. L., and Whelan, M. J. (1996c). Dynamical RHEED calculations from the surface of a semi-infinite crystla. *Acta Crystallographica A*, 52:471–475.

Peng, L.-M., Dudarev, S. L., and Whelan, M. J. (1997). The ionicity of nickel monoxide: direct determination by reflection diffraction of high-energy electrons from the (100) surface. *Physical Review B*, 56:15314–15319.

Peng, L.-M., Dudarev, S. L., and Whelan, M. J. (1998). Electron scattering factors of ions and dynamical RHEED calculations from surfaces of ionic crystals. *Physical Review B*, 57:7259–7265.

Peng, L.-M. and Gjønnes, J. K. (1989). Bloch wave channeling and HOLZ effects in high energy electron diffraction. *Acta Crystallographica A*, 45:699–703.

Peng, L.-M., Gjønnes, K., and Gjønnes, J. (1992). Bloch wave treatment of symmetry and multiple beam cases in reflection high energy electron diffraction and reflection electron microscopy. *Microscopy Research and Technique*, 20:360–370.

Peng, L.-M. and Li, Y. J. (1992). Automated identification of symmetry elements in convergent-beam electron diffraction. In *Proceedings of the 5th Asia-Pacific Electron Microscopy Conference*, volume 1, pages 625–626. World Scientific.

Peng, L.-M., Ren, G., Dudarev, S. L., and Whelan, M. J. (1996d). Debye-Waller factors and absorptive scattering factors of elemental crystals. *Acta Crystallographica A*, 52:456–470.

Peng, L.-M., Ren, G., Dudarev, S. L., and Whelan, M. J. (1996e). Robust parameterization of elastic and absorptive electron atomic scattering factors. *Acta Crystallographica A*, 52:257–276.

Peng, L.-M. and Whelan, M. J. (1990a). Dynamical RHEED from MBE growing surfaces. *Surface Science*, 238:L446 – L452.

Peng, L.-M. and Whelan, M. J. (1990b). A general matrix representation of the dynamical theory of electron diffraction — I. General theory. *Proceedings of the Royal Society of London A*, 431:111–123.

Peng, L.-M. and Whelan, M. J. (1990c). A general matrix representation of the dynamical theory of electron diffraction — II. Applications to RHEED from relaxed and reconstructed surfaces. *Proceedings of the Royal Society of London A*, 431:125–142.

Peng, L.-M. and Whelan, M. J. (1991a). Dynamical calculations for RHEED from MBE growing surfaces — I. Growth on a low-index surface. *Proceedings of the Royal Society of London A*, 432:195–213.

Peng, L.-M. and Whelan, M. J. (1991b). Dynamical calculations for RHEED from MBE growing surfaces — II. Growth interruption and surface recovery. *Proceedings of the Royal Society of London A*, 435:257–267.

Peng, L.-M. and Whelan, M. J. (1991c). Dynamical calculations for RHEED from MBE growing surfaces — III. Heteroepitaxial growth and interface formation. *Proceedings of the Royal Society of London A*, 435:269–286.

Peng, L.-M. and Whelan, M. J. (1991d). Surface superlattice reflections and kinematic approximation in RHEED. *Acta Crystallographica A*, 47:95–101.

Peng, L.-M. and Zuo, J. M. (1995). Direct retrieval of crystal structure factors in THEED. *Ultramicroscopy*, 57:1–9.

Peng, L.-M. and Zuo, J. M. (1999). Anisotropic dispersion of the band structure and formation of ring patterns in CBED. *Acta Crystallographica A*, 55:1026–1033.

Pennycook, S. L. and Jesson, D. E. (1990). High-resolution incoherent imaging of crystals. *Physical Review Letters*, 64:938–941.

Perovic, D. D., Weatherly, G. C., Egerton, R. F., Houghton, D. C., and Jackman, T. E. (1991). On the electron microscope contrast of doped semiconductor layers. *Philosophical Magazine A*, 63:757–784.

Pettifor, D. G. (1995). *Bonding and Structure of Molecules and Solids*. Oxford University Press, Oxford.

Phillips, F. C. (1971). *An Introduction to Crystallography*. Longman Group Limited, Essex, fourth edition.

Pines, D. and Nozieres, P. (1966). *The Theory of Quantum Liquids*. W. A. Benjamin, Inc, New York and Amsterdam.

Platzman, P. M. and Wolf, P. A. (1973). Waves and interactions in solid state plasmas. In Ehrenreich, H., Seitz, F., and Turnbull, D., editors, *Solid State Physics*, volume Supplement 13, New York and London. Academic Press.

Pogany, A. P. and Turner, P. S. (1968). Reciprocity in electron diffraction and microscopy. *Acta Crystallographica A*, 24:103–109.

Press, W. H., Flannery, B. P., Teukolsky, S., and Vetterling, W. T. (1986). *Numerical Recipes*. Cambridge University Press, Cambridge.

Press, W. H., Teukolsky, S. A., Vetterling, W. T., and Flannery, B. P. (1992). *Numerical Recipes*. Cambridge University Press, Cambridge.

Pukite, P. R., Lent, C. S., and Cohen, P. I. (1985). Diffraction from stepped surfaces. 2. Arbitrary terrace distributions. *Surface Science*, 161:39 – 68.

Radi, G. (1970). Complex lattice potentials in electron diffraction calculated for a number of crystals. *Acta Crystallographica A*, 26:41–56.

Reginski, K., Lamin, M. A., Mashanov, V. I., Pchelyakov, O. P., and Sokolov, L. V. (1995). RHEED intensity oscillations from Si(111) surface in the presence of surface resonance. *Surface Science*, 327:93 – 99.

Reid, J. (1983). Debye-Waller factors of zinc-blende structure materials - a lattice dynamical comparison. *Acta Crystallographica A*, 39:1–13.

Reimer, L. (1989). *Transmission Electron Microscopy — Physics of Image Formation and Microanalysis*. Springer-Verlag, Berlin.

Reimer, L., Badde, H. G., Seidel, H., and Buhring, W. (1971). Orientierungsanisotropie des Ruckstreukoeffizienten und der Sekundarelektronenausbeute von 10-100 keV Elektronen. *Zeitschrift für angewandte Physik*, 31:145 – 151.

Ren, G., Zuo, J. M., and Peng, L.-M. (1997). Accurate measurements of crystal structure factors using a FEG electron microscope. *Micron*, 28:459–467.

Resh, J., Jamison, K. D., Strozier, J., Bensaoula, A., and Ignatiev, A. (1989). Phase of reflection high energy electron diffraction intensity oscillations during molecular-beam epitaxy growth of GaAs(100). *Physical Review B*, 40:11799 – 11803.

Rez, D., Rez, P., and Grant, I. (1994). Dirac-Fock calculations of X-ray scattering factors and contributions to the mean inner potential for electron scattering. *Acta Crystallographica A*, 50:481–497.

Rez, P. (1976). *The Theory of Inelastic Scattering in the Electron Microscopy of Crystals*. PhD thesis, St. Catherine's College, Oxford, Oxford.

Rez, P. (1978). Multiple inelastic scattering and dynamical diffraction. In Dobson, P. J., Pendry, J. B., and Humphreys, C. J., editors, *Electron Diffraction 1927-1977*, volume 41 of *Institute of Physics Conference Series*, pages 61–67, Bristol and London. Institute of Physics.

Rez, P. (1995). A matrix-operator approach to reflection high-energy electron diffraction theory. *Acta Crystallographica A*, 51:38–47.

Rez, P. (1999). Schemes to determine the crystal potential under dynamical conditions using voltage variation. *Acta Crystallographica A*, 55:160–167.

Rez, P., Bruley, J., Brohan, P., Payne, M., and Garvie, L. A. J. (1995). Review of methods for calculating near edge structure. *Ultramicroscopy*, 59:159–167.

Rez, P., Humphreys, C. J., and Whelan, M. J. (1977). The distribution of intensity in electron diffraction patterns due to phonon scattering. *Philosophical Magazine*, 35:81–96.

Rez, P., MacLaren, J. M., and Saldin, D. K. (1998). Application of the layer Korringa-Kohn-Rostoker method to the calculation of near-edge structure in X-ray-absorption and electron-energy-loss spectroscopy. *Physical Review B*, 57:2621–2627.

Rodriguez, E. (1989). Beyond the Kirchhoff approximation. *Radio Science*, 24:681 – 698.

Rogozkin, D. B. (1997). Angular intensity correlations in reflection of light from a random medium. *Physics Letters A*, 236:159 – 166.

Rossouw, C. J. and Bursill, L. A. (1985). Interpretation of dynamical diffuse scattering of fast electrons in rutile. *Acta Crystallographica A*, 41:320–327.

Rossouw, C. J., Dudarev, S. L., Josefsson, T. W., and Allen, L. J. (1998). Transmission resonance diffraction and low impact parameter inelastic scattering of high-energy electrons. *Ultramicroscopy*, 72:17–29.

Rossouw, C. J. and Maslen, V. W. (1984). Implications of (e,2e) scattering for inelastic electron diffraction in crystals. 2. application of the theory. *Philosophical Magazine A*, 49:743–757.

Rossouw, C. J., Spellward, P., Perovic, D. D., and Cherns, D. (1994). Dynamical zone-axis electron diffraction contrast of boron-doped silicon multilayers. *Philosophical Magazine A*, 69:255–265.

Rossouw, C. J. and Whelan, M. J. (1981). Diffraction contrast retained by plasmon and K-loss electrons. *Ultramicroscopy*, 6:53 – 66.

Rous, P. J. (1992). The tensor approximation and surface crystallography by low-energy electron diffraction. *Progress of Surface Science*, 39:3–63.

Rudee, M. L. and Howie, A. (1972). The structure of amorphous Si and Ge. *Philosophical Magazine*, 25:1001 – 1007.

Ruska, E. (1980). *The Early Development of Electron Lenses and Electron Microscopy*. (Translated by T. Mulvey). S. Hirzel Verlag, Stuttgart.

Saito, Y. (1984). The structural studies of amorphous Ge films prepared by vacuum deposition. *Journal of the Physical Society of Japan*, 53:4230 – 4240.

Sakamoto, T., Kawai, N. J., Nakagawa, T., Ohta, K., and Kojima, T. (1985). Intensity oscillations of reflection high-energy electron diffraction during silicon molecular beam epitaxial growth. *Applied Physics Letters*, 47:617 – 619.

Saldin, D. K. and Spence, J. C. H. (1994). On the mean inner potential in high and lower-energy electron diffraction. *Ultramicroscopy*, 55:397–405.

Saldin, D. K. and Ueda, Y. (1992). Dipole approximation in electron energy-loss spectroscopy - L-shell excitations. *Physical Review B*, 46:5100–5109.

Saldin, D. K. and Yao, J. M. (1990). Dipole approximation in electron energy-loss spectroscopy - K-shell excitations. *Physical Review B*, 41:52–61.

Sandstrom, R., Spencer, J. F., and Humphreys, C. J. (1974). A theoretical model for the energy dependence of electron channelling patterns in scanning electron microscopy. *Journal of Physics D: Applied Physics*, 7:1030 – 1046.

Sarikaya, M., Qian, M., and Stern, E. A. (1996). EXELFS revisited. *Micron*, 27:449–466.

Saunders, M., Bird, D. M., Zaluzec, N. J., Burgess, W. G., Preston, A. R., and Humphreys, C. J. (1995). Measurement of low-order structure factors for silicon from zone-axis CBED patterns. *Ultramicroscopy*, 60:311–323.

Saxton, W. O. (1994). What is the focus variation method? Is it new? Is it direct? *Ultramicroscopy*, 55:171–181.

Sayre, D. (1952). The squaring method: a new method for phase determination. *Acta Crystallographica*, 5:60–65.

Scherzer, O. (1949). The theoretical resolution limit of the electron microscope. *Journal of Applied Physics*, 20:20–29.

Schiff, L. I. (1968). *Quantum Mechanics*. McGraw-Hill, London.

Schmid, R., Gaukler, K. H., and Seiler, H. (1983). Measurement of elastically reflected electrons ($E < 2.5$ keV) for imaging of surfaces in a simple untra-high vacuum scanning electron microscope. In Johari, O., editor, *Scanning Electron Microscopy*, volume II, pages 501 – 509, Chicago, IL. IIT Research Institute.

Schomaker, V. and Glauber, R. (1952). The Born approximation in electron diffraction. *Nature*, 170:290–291.

Seah, M. P. (1986). Data compilations - their use to improve measurement certainty in surface analysis by AES and XPS. *Surface and Interface Analysis*, 9:85 – 98.

Self, P. G., O'Keefe, M. A., Buseck, P. R., and Spargo, A. E. C. (1983). Practical computation of amplitudes and phases in electron diffraction. *Ultramicroscopy*, 11:35–52.

Shitara, T., Vvedensky, D. D., Wilby, M. R., Zhang, J., Neave, J. H., and Joyce, B. A. (1992). Step-density variations and reflection high-energy electron diffraction intensity oscillations during epitaxial growth on vicinal GaAs(001). *Physical Review B*, 46:6815–6824.

Shmueli, U., editor (1993). *International Tables for Crystallography*, volume B, Dordrecht, the Netherlands. Kluwer.

Shubnikov, A. V. and Belov, N. V. (1964). *Coloured Symmetry*. Pergamon Press, Oxford.

Smart, D. J. and Humphreys, C. J. (1980). The application of electron diffraction to determining bonding charge densities in crystals. In Mulvey, T., editor, *Electron Microscopy and Analysis 1979*, volume 52 of *Institute of Physics Conference Series*, pages 211–214, Bristol and London. Institute of Physics.

Smith, A. and Lynch, D. F. (1988). Results of multislice matrix calculations for convergent-beam RHEED patterns. *Acta Crystallographica A*, 44:780–788.

Smith, A. E., Lehmpfuhl, G., and Uchida, Y. (1992). A comparison between experimental and calculated convergent beam RHEED patterns from the Pt(111) surface. *Ultramicroscopy*, 41:367–373.

Smith, B. T., Boyle, J. M., Dongarra, J. J., Garbow, B. S., Ikebe, Y., Klema, V. C., and Moler, C. B. (1976). *Matrix Eigensystem Routines-EISPACK Guide*. Springer-Verlag, Berlin.

Smith, G. H. and Burge, R. E. (1962). The analyticaly representation of atomic scattering amplitudes for electrons. *Acta Crystallographica*, 15:182–186.

Spargo, A. E. C., Beeching, M. J., and Allen, L. J. (1994). Inversion of electron scattering intensity for crystal structure analysis. *Ultramicroscopy*, 55:329–333.

Speer, S., Spence, J. C. H., and Ihrig, E. (1990). On differentiation of the scattering matrix in dynamical transmission electron diffraction. *Acta Crystallographica A*, 46:763–772.

Spence, J. C. H. (1988). *Experimental High Resolution Electron Microscopy*. Oxford University Press, Oxford.

Spence, J. C. H. (1993). On the accurate measurement of structure factor amplitude and phases by electron diffraction. *Acta Crystallographica A*, 49:231–260.

Spence, J. C. H. (1998). Direct inversion of dynamical electron diffraction patterns to structure factors. *Acta Crystallographica A*, 54:7–18.

Spence, J. C. H., Calef, B., and Zuo, J. M. (1999). Dynamical inversion by the method of generalized projections. *Acta Crystallographica A*, 55:112–118.

Spence, J. C. H. and Kim, Y. (1988). Atom site determination using channeling effects in RHEED on X-ray and Auger electron production. In Larsen, P. K. and Dobson, P. J., editors, *Reflection High-Energy Electron Diffraction and Reflection Electron Imaging of Surfaces*, volume 188 of *NATO Advance Study Institute Series, ser. B: Physics*, pages 117–129, New York. Plenum Press.

Spence, J. C. H. and Taftø, J. (1983). ALCHEMI - a new technique for locating atoms in small crystals. *Journal of Microscopy*, 130:147 – 154.

Spence, J. C. H. and Whelan, M. J. (1975). Multislice and the Darwin equations. *Acta Crystallographica A*, 31:S242.

Spence, J. C. H. and Zuo, J. M. (1988). Large dynamic range, parallel detection system for electron diffraction and imaging. *Review of Scientific Instruments*, 59:2102–2105.

Spence, J. C. H. and Zuo, J. M. (1992). *Electron Microdiffraction*. Plenum Press, New York and London.

Spencer, J. P. and Humphreys, C. J. (1980). A multiple scattering transport theory for electron channelling patterns. *Philosophical Magazine*, 42:433 – 451.

Spencer, J. P., Humphreys, C. J., and Hirsch, P. B. (1972). A dynamical theory for the contrast of perfect and imperfect crystals in the scanning electron microscope using backscattered electrons. *Philosophical Magazine*, 26:193 –213.

Staib, P., Tappe, W., and Contour, J. P. (1999). Imaging energy analyzer for RHEED: energy fitered diffraction patterns and in situ electron energy loss spectroscopy. *Journal of Crystal Growth*, 201/202:45 – 49.

Steeds, J. W. (1983). Electron crystallography. In Chapman, J. and Craven, A., editors, *Quantitative Electron Microscopy*. SUSSP publications, Edinburgh.

Stern, E. A. (1988). Theory of EXAFS. In Koningsberger, D. C. and Prins, R., editors, *X-ray absorption*, pages 3–84. Wiley, New York.

Sturkey, L. (1962). The calculation of electron diffraction intensities. *Proceedings of the Physical Society*, 80:321–354.

Sudijono, J., Johnson, M. D., Snyder, C. W., Elowitz, M. B., and Orr, B. G. (1992). Surface evolution during molecular beam epitaxy deposition of GaAs. *Physical Review Letters*, 69:2811 – 2814.

Taftø, J. and Spence, J. C. H. (1982). Crystal site location of iron and trace elements in a magnesium-iron olivine by a new crystallographic technique. *Science*, 218:49 – 51.

Tanaka, M. (1989). Symmetry analysis. *Journal of Electron Microscopy Technique*, 13:27–39.

Tanaka, M., Saito, R., and Sekii, H. (1983a). Point-group determination by convergent-beam electron diffraction. *Acta Crystallographica A*, 39:357–368.

Tanaka, M., Sekii, H., and Nagasawa, T. (1983b). Space-group determination by dynamical extinction in convergent-beam electron diffraction. *Acta Crystallographica A*, 39:825–837.

Tanaka, M. and Terauchi, M. (1985). *Convergent-Beam Electron Diffraction*. JEOL Ltd, Tokyo.

Tasker, P. W. (1979). The surface properties of uranium dioxide. *Surface Science*, 78:315–324.

Taylor, J. R. (1972). *Scattering Theory: The Quantum Theory of Nonrelativistic Collisions*. J. Wiley & Sons, New York and London.

Thomson, G. P. and Cochrane, W. (1939). *Theory and Practice of Electron Diffraction*. MacMillan and Co., London.

Thomson, G. P. and Reid, A. (1927). Diffraction of cathode rays by a thin film. *Nature*, 119:890.

Thomson, S. G. (1968). The early history of electron diffraction. *Contemporary Physics*, 9:1 – 15.

Tonomura, A. (1987). Applications of electron holography. *Reviews of Modern Physics*, 59:639–669.

Tournarie, M. (1962). Recent developments of the matrical and semi-reciprocal formulation in the field of dynamical theory. *Journal of the Physical Society of Japan*, 17(Supplement B-II):98–100.

Toyoshima, H., Shitara, T., Zhang, J., Neave, J. H., and Joyce, B. A. (1992). A systematic RHEED study of regular and random steps on GaAs(001) surfaces. *Surface Science*, 264:10 – 22.

Treacy, M. M. J. and Gibson, J. M. (1996). Variable coherence microscopy: a rich source of structural information from disordered materials. *Acta Crystallographica A*, 52:212 – 220.

Tsuda, K. and Tanaka, M. (1995). Refinement of crystal structure parameters using convergent-beam electron diffraction: the low-temperature phase of $SrTiO_3$. *Acta Crystallographica A*, 51:7–19.

Ueda, Y. and Saldin, D. K. (1992). Dipole approximation in electron energy-loss spectroscopy - M-shell excitations. *Physical Review B*, 46:13697–13701.

Unoki, H. and Sakudo, T. (1967). Electron spin resonance of Fe^{3+} in $SrTiO_3$ with special reference to the 110° phase transition. *Journal of the Physical Society of Japan*, 23:546–552.

Vainshtein, B. K. (1964). *Structure Analysis by Electron Diffraction*. Pergamon Press, Oxford.

Van Dyck, D. and Op de Beeck, M. (1990). New direct methods for phase and structure retrieval in HREM. In *Proceedings of the 12th International Congress on Electron Microscopy*, pages 64–65, San Francisco. San Francisco Press.

Van Dyck, D., Op de Beeck, M., and Coene, W. (1993). A new approach to object wavefunction reconstructruction in electron microscopy. *Optik*, 93:103–107.

Van Hove, J. M., Lent, C. S., Pukite, P. R., and Cohen, P. I. (1983). Damped oscillations in reflection high energy electron diffraction during GaAs MBE. *Journal of Vacuum Science and Technology*, B1:741–746.

Van Hove, L. (1954). Correlations in space and time and Born approximation scattering in systems of interacting particles. *Physical Review*, 95:249–262.

Vand, V., Eiland, T. F., and Pepinsky, R. (1957). Analytical representation of atomic scattering factors. *Acta Crystallographica*, 10:303–306.

Vincent, R., Bird, D. M., and Steeds, J. W. (1984). Structure of AuGeAs determined by convergent-beam electron diffraction — II. Refinement of structural parameters. *Philosophical Magazine A*, 50:765–786.

Vincent, R. and Walsh, T. D. (1997). Quantitative assessment of symmetry in CBED patterns. *Ultramicroscopy*, 70:83–94.

von Laue, M. (1912). Eine quantitative Prüfung der Theorie für die Interferenzerscheinungen bei Röntgenstrahlen. *Sitzung der Bayrischen Akademie der Wissenschaften*, 8:363–373.

von Laue, M. (1931). The diffraction of an electron-wave at a single layer of atoms. *Physical Review*, 37:53–59.

Voronovich, A. G. (1994). *Wave Scattering from Rough Surfaces.* Springer-Verlag, Berlin.

Voyles, P. M., Gerbi, J. E., Treacy, M. M. J., Gibson, J. M., and Abelson, J. R. (2001a). Increased medium-range order in amorphous silicon with increased substrate temperature. *Journal of Non-Crystalline Solids*, 293-295:45 – 52.

Voyles, P. M., Gibson, J. M., and Treacy, M. M. J. (2000). Fluctuation microscopy: a probe of atomic correlations in disordered materials. *Journal of Electron Microscopy*, 49:259 – 266.

Voyles, P. M. and Muller, D. A. (2002). Fluctuation microscopy in the STEM. *Ultramicroscopy*, 93:147 – 159.

Voyles, P. M., Zotov, N., Nakhmanson, S. M., Drabold, D. A., Gibson, J. M., Treacy, M. M. J., and Keblinski, P. (2001b). Structure and physical properties of paracrystalline atomistic models of amorphous silicon. *Journal of Applied Physics*, 90:4437 – 4451.

Vvedensky, D. D. (1992). Theory of X-ray absorption fine structure. In Fuggle, J. C. and Inglesfield, J. E., editors, *Unoccupied Electronic States*, Berlin and London. Springer-Verlag.

Vvedensky, D. D., Haider, N., Shitara, T., and Smilauer, P. (1993). Evolution of surface morphology during epitaxial growth. *Philosophical Transactions of the Royal Society of London A*, 344:493 – 505.

Wade, R. H. and Frank, J. (1977). Electron microscope transfer functions for partially coherent axial illumination and chromatic defocus spread. *Optik*, 49:81–92.

Wang, S. Q., Peng, L.-M., Duan, X. F., and Chu, Y. M. (1992a). Matrix description of dynamical HOLZ diffraction tested on the strained layer superlattice Si/GeSi. *Ultramicroscopy*, 45:405–409.

Wang, S. Q., Peng, L.-M., Xing, Y., Chu, Y. M., and Duan, X. F. (1992b). Many-beam simulations and observations of large-angle convergent-beam electron diffraction imaging of crystal defects. *Philosophical Magazine Letters*, 66:225–233.

Wang, Z. L. (1995). *Elastic and Inelastic Scattering in Electron Diffraction and Imaging.* Plenum Press, New York and London.

Wang, Z. L., Lu, P., and Cowley, J. M. (1987). Electron resonance channelling on crystal surfaces in RHEED geometry. *Ultramicroscopy*, 23:205–222.

Warren, B. E. (1990). *X-Ray Diffraction.* Dover Publications, New York.

Washburn, S. (1991). Resistance fluctuations in small samples: be careful when playing with Ohm's law. In Kramer, B., editor, *Quantum Coherence in Mesoscopic Systems*, pages 341 – 367, New York. Plenum Press.

Weickenmeier, A. and Kohl, H. (1991). Computation of absorptive form factors for high energy electron diffraction. *Acta Crystallographica A*, 47:590–597.

Weng, X. and Rez, P. (1989). Multiple-scattering approach to oxygen K near-edge structures in electron-energy-loss spectroscopy of alkaline earths. *Physical Review B*, 39:7405–7412.

Weng, X., Rez, P., and Ma, H. (1989a). Carbon K-shell near-edge structure: multiple scattering and band-theory calculations. *Physical Review B*, 40:4175–4178.

Weng, X., Rez, P., and Sankey, O. F. (1989b). Pseudo-atomic-orbital band theory applied to electron-energy-loss near-edge structure. *Physical Review B*, 40:5694–5704.

Whelan, M. J. (1965a). Inelastic scattering of fast electrons by crystals – I. Interband excitations. *Applied Physics*, 36:2099–2103.

Whelan, M. J. (1965b). Inelastic scattering of fast electrons by crystals–II. Phonon scattering. *Applied Physics*, 36:2103–2110.

Whelan, M. J. (1986). Reminiscences on the early observations of defects in crystals by electron microscopy. *Journal of Electron Microscopy Technique*, 3:109 – 129.

Whelan, M. J. (2002). The early observations of defects in metals by transmission electron microscopy. In Humphreys, C. J., editor, *Understanding Materials. A Festschrift for Sir Peter Hirsch*, pages 17 – 35, London. Institute of Materials.

Whelan, M. J. and Hirsch, P. B. (1957). Electron diffraction from crystals containing stacking faults: I. *Philosophical Magazine*, 2:1121–1142.

Wilkinson, A. J., Anstis, G. R., Czernuszka, J. T., Long, N. J., and Hirsch, P. B. (1993). Electron channelling contrast imaging of interfacial defects in strained silicon-germanium layers on silicon. *Philosophical Magazine A*, 68:59 – 80.

Wilkinson, A. J. and Hirsch, P. B. (1997). Electron diffraction based techniques in scanning electron microscopy of bulk materials. *Micron*, 28:279 – 308.

Williams, D. B. and Carter, C. B. (1996). *Transmission Electron Microscopy*. Plenum Press, New York and London.

Willis, B. T. M. and Pryor, A. W. (1975). *Thermal Vibrations in Crystallography*. Cambridge University Press, Cambridge.

Wilson, A. J. C. and Prince, E., editors (1999). *International Tables for Crystallography, Volume C: Mathematical, Physical and Chemical Tables*, Dordrecht, the Netherlands. Kluwer Academic Publishers.

Wolf, E. D., Coane, P. J., and Everhart, T. E. (1970). Coates-Kikuchi patterns. In Favard, P., editor, *Microscopie Electronique 1970*, volume II, pages 595 – 596, Grenoble and Paris. Societe Francaise de Microscopie Electronique.

Wollschläger, J., Schafer, F., and Schroder, K. M. (1998). Diffraction spot profile analysis for vicinal surfaces with long-range order. *Surface Science*, 396:94 – 106.

Wood, C. E. C. (1981). RED intensity oscillations during MBE from GaAs. *Surface Science*, 108:L441–L443.

Wood, E. A. (1964). Vocabulary of surface crystallography. *Journal of Applied Physics*, 35:1306–1312.

Woolfson, M. M. (1961). *Direct Methods in Crystallography*. Oxford University Press, Oxford.

Yao, N. and Cowley, J. M. (1989). The parabolas and circles in RHEED patterns. *Ultramicroscopy*, 31:149–157.

Yin, M. T. and Cohen, M. L. (1981). Theoretical determination of surface atomic geometry: Si(001)-(2 × 1). *Physical Review B*, 24:2303–2306.

Yoshioka, H. (1957). Effect of inelastic waves on electron diffraction. *Journal of the Physical Society of Japan*, 12:618–628.

Yoshioka, H. and Kainuma, Y. (1962). The effect of thermal vibrations on electron diffraction. *Journal of the Physical Society of Japan*, 17(Supplement BII):134–136.

Young, A. P. and Rez, P. (1975). Resonance errors and partial coherence in the inelastic scattering of fast electrons by crystal excitations. *Journal of Physics C: Solid State Physics*, 8:L1 – L7.

Zangwill, A. (1988). *Physics at Surfaces*. Cambridge University Press, Cambridge.

Zhang, J., Neave, J. H., Dobson, P. J., and Joyce, B. A. (1987). Effects of diffraction conditions and processes on RHEED intensity oscillations during the MBE growth of GaAs. *Applied Physics A*, 42:317–326.

Zhao, T. C., Poon, H. C., and Tong, S. Y. (1988). Invariant-embedding R-matrix scheme for reflection high-energy electron diffraction. *Physical Review B*, 38:1172–1182.

Zou, L. J. (1996). *Automated recognition of symmetry in CBED (MSc thesis)*. Institute of Physics, Chinese Academy of Sciences.

Zuo, J. M. (1991). Perturbation theory in high energy transmission electron diffraction. *Acta Crystallographica A*, 47:87–95.

Zuo, J. M. (1997). Electron detection characteristics of slow-scan CCD camera. *Ultramicroscopy*, 66:21–34.

Zuo, J. M. and Liu, J. (1992). Resonance effects in RHEED on GaAs(110) surface. *Surface Science*, 271:253–259.

Zuo, J. M., O'Keeffe, M., Rez, P., and Spence, J. C. H. (1997). Charge density of MgO: implications of precise new measurements for theory. *Physical Review Letters*, 78:4777–4780.

Zuo, J. M. and Spence, J. (1991). Automated structure factor refinement from convergent-beam patterns. *Ultramicroscopy*, 35:185–196.

Zuo, J. M., Spence, J. C. H., and O'Keeffe, M. (1988). Bonding in GaAs. *Physical Review Letters*, 61:353–356.

Subject Index

Printed and bound by CPI Group (UK) Ltd, Croydon, CR0 4YY